VIRAL HEMORRHAGIC FEVERS

CRC Press
Taylor & Francis Group
Boca Raton London New York

CRC Press is an imprint of the
Taylor & Francis Group, an **informa** business

VIRAL HEMORRHAGIC FEVERS

Edited by **Sunit K. Singh** • **Daniel Ruzek**

CRC Press
Taylor & Francis Group
Boca Raton London New York

CRC Press is an imprint of the
Taylor & Francis Group, an **informa** business

CRC Press
Taylor & Francis Group
6000 Broken Sound Parkway NW, Suite 300
Boca Raton, FL 33487-2742

First issued in paperback 2019

© 2014 by Taylor & Francis Group, LLC
CRC Press is an imprint of Taylor & Francis Group, an Informa business

No claim to original U.S. Government works

ISBN-13: 978-1-4398-8429-4 (hbk)
ISBN-13: 978-0-367-37979-7 (pbk)

This book contains information obtained from authentic and highly regarded sources. While all reasonable efforts have been made to publish reliable data and information, neither the author[s] nor the publisher can accept any legal responsibility or liability for any errors or omissions that may be made. The publishers wish to make clear that any views or opinions expressed in this book by individual editors, authors or contributors are personal to them and do not necessarily reflect the views/opinions of the publishers. The information or guidance contained in this book is intended for use by medical, scientific or health-care professionals and is provided strictly as a supplement to the medical or other professional's own judgement, their knowledge of the patient's medical history, relevant manufacturer's instructions and the appropriate best practice guidelines. Because of the rapid advances in medical science, any information or advice on dosages, procedures or diagnoses should be independently verified. The reader is strongly urged to consult the drug companies' printed instructions, and their websites, before administering any of the drugs recommended in this book. This book does not indicate whether a particular treatment is appropriate or suitable for a particular individual. Ultimately it is the sole responsibility of the medical professional to make his or her own professional judgements, so as to advise and treat patients appropriately. The authors and publishers have also attempted to trace the copyright holders of all material reproduced in this publication and apologize to copyright holders if permission to publish in this form has not been obtained. If any copyright material has not been acknowledged please write and let us know so we may rectify in any future reprint.

Library of Congress Cataloging-in-Publication Data

Viral hemorrhagic fevers / edited by Sunit K. Singh and Daniel Ruzek.
   p. ; cm.
Includes bibliographical references and index.
ISBN 978-1-4398-8429-4 (hardcover : alk. paper)
I. Singh, Sunit K. II. Ruzek, Daniel.
   [DNLM: 1. Hemorrhagic Fevers, Viral. 2. RNA Virus Infections. WC 534]

RA644.D4
614.5'8852--dc23   2013010253

**Visit the Taylor & Francis Web site at
http://www.taylorandfrancis.com**

**and the CRC Press Web site at
http://www.crcpress.com**

Contents

PART I   General Aspects

PART II   Specific Infections

Foreword

Viral hemorrhagic fevers comprise "a diverse group of virus diseases (as Korean hemorrhagic fever, Lassa fever, and Ebola) that are usually transmitted to humans by arthropods or rodents and are characterized by sudden onset, fever, aching, bleeding in the internal organs (as of the gastrointestinal tract), petechiae, and shock."* They include some of the first identified and studied viral diseases, such as yellow fever with its etiology determined over a century ago, as well as others discovered only within the last few years and whose pathogenesis, ecology, and epidemiology remain poorly understood. The etiologic agents are taxonomically diverse viruses, including members of the families *Arenaviridae*, *Bunyaviridae*, *Flaviviridae*, and *Filoviridae,* with varied natural histories, transmission cycles, and modes of human exposure. Interest in viral hemorrhagic fevers as naturally emerging zoonotic diseases peaked during the 1960s through the 1980s, with seminal discoveries of hemorrhagic arenaviruses and filoviruses. These viruses, especially the filoviruses, ultimately captured the imagination of the public and made their way into popular books and movies by virtue of their extreme virulence and mysterious origins. However, since 2001, concerns about the potential use of many hemorrhagic fever viruses as biological weapons have led to a resurgence in research to develop improved diagnostics, vaccines, and therapeutics, both for biodefense purposes and to treat naturally exposed persons. Surprisingly, this biodefense effort has not until now been accompanied by a major, recent, comprehensive publication to update the state of knowledge of viral hemorrhagic fevers.

In this book, Sunit Kumar Singh and Daniel Růžek have met the growing need for a single source of information, focusing on viral hemorrhagic fevers by editing 30 chapters that represent a major contribution to the virological literature and that are written by many experts worldwide. Their comprehensive treatment begins with a historical perspective, followed by general information on pathogenesis and immune responses, with appropriate emphasis on the endothelium that, when affected directly or indirectly by viral replication, leads to hemorrhagic manifestations, often with fatal outcomes. Chapters follow on animal models, critical to the development of vaccines and therapeutics, particularly for viruses with sporadic incidence that preclude human efficacy trials, as well as on natural reservoir hosts (mainly rodents and bats) and their role in viral maintenance and exposure of people. The critical roles of high-containment facilities and specially trained scientists in research on viral hemorrhagic fevers, as well as practical information on sample collection and transportation, rarely covered in the related literature, are also unique features of this book. Prevention and control, including diagnostics and vaccine development, round out Part I.

Part II begins with an overview chapter on the arenaviruses, followed by individual chapters on the major human pathogens, including the Old World Lassa and Lujo viruses; the New World Junin, Machupo, and Guanarito viruses; and a chapter on cellular receptors of the latter group. The bunyaviruses include Rift Valley Fever, Crimean–Congo hemorrhagic fever, and hantaviruses that cause hemorrhagic fever with renal syndrome follow, along with the infamous filoviruses Ebola and Marburg. Part II concludes with the flaviviruses, including dengue virus, which causes tens of thousands of hemorrhagic disease cases annually throughout the tropics and subtropics, usually

---

* *Merriam-Webster Dictionary*, http://www.merriam-webster.com//dictionary//hemorrhagic%20fever.

following secondary infections. Finally, yellow fever, Kyanasur Forest, Alkhurma, and Omsk hemorrhagic fever viruses, other flaviviruses with hemorrhagic potential during primary infections, complete the coverage of all major hemorrhagic viral pathogens.

Scott C. Weaver, PhD
John Sealy Distinguished University Chair in Human Infections and Immunity
Director, Institute for Human Infections and Immunity
Scientific Director, Galveston National Laboratory
University of Texas Medical Branch
Galveston, Texas

Preface

Viral hemorrhagic fevers (VHFs) represent a diverse group of human and animal illnesses that may be caused by four distinct families of RNA viruses: the families *Arenaviridae*, *Filoviridae*, *Bunyaviridae*, and *Flaviviridae*. In general, the term "viral hemorrhagic fever" is used to describe a severe multisystem syndrome. Characteristically, the overall vascular system is damaged, and the body's ability to regulate itself is impaired. All types of VHFs are characterized by fever and bleeding disorders, and all can progress to high fever, shock, and death in many cases. While some VHF viruses can cause relatively mild illnesses, many of these viruses cause severe life-threatening disease. Depending on the infecting virus, fatality rates can reach up to 90%. VHFs may occur as isolated case(s), such as imported case(s) from endemic areas, or may cause devastating lethal outbreaks. While significant progress has been made over the last few years in dissecting the molecular biology and pathogenesis of the hemorrhagic viruses, there are currently no licensed vaccines or drugs available for most VHFs.

This book is primarily aimed at health-care workers, virologists, clinicians, biomedical researchers, microbiologists, and students of medicine or biology wishing to gain a rapid overview of the nature of these widely varying agents linked only by their propensity of causing diseases with hemorrhagic manifestation. We hope that this book will serve as a useful resource for all others interested in the field of VHFs.

An inclusive and comprehensive book such as this is clearly beyond the capacity of an individual's efforts. Therefore, we are fortunate and honored to have a large panel of internationally renowned virologists as chapter contributors, whose detailed knowledge on VHFs has greatly enriched this book.

We are thankful to Dr. Jens H. Kuhn, lead virologist at Integrated Research Facility at Fort Detrick (IRF-Frederick), Maryland, for his enormous support in giving final shape to this book. The professional support provided by Taylor & Francis Group/CRC Press greatly contributed to the final presentation of the book. Our appreciations extend to our families and parents for their understanding and support during the compilation of this book.

Sunit Kumar Singh

Daniel Růžek

Editors

Dr. Sunit Kumar Singh completed his bachelor's program at the GB Pant University of Agriculture and Technology, Pantnagar, India. After earning his master's at the CIFE, Mumbai, India, he joined the Department of Paediatric Rheumatology, Immunology, and Infectious Diseases, Children's Hospital, University of Wuerzburg, Wuerzburg, Germany, as a biologist. He earned a PhD in the area of molecular infection biology from the University of Wuerzburg.

Dr. Singh completed his postdoctoral training in the Department of Internal Medicine, Yale University, School of Medicine, New Haven, Connecticut, and the Department of Neurology, University of California Davis Medical Center, Sacramento, California, in the areas of vector-borne infectious diseases and neuroinflammation, respectively. He has also worked as a visiting scientist in the Department of Pathology, Albert Einstein College of Medicine, New York, New York; Department of Microbiology, College of Veterinary Medicine, Chonbuk National University, Republic of Korea; Department of Arbovirology, Institute of Parasitology, Ceske Budejovice, Czech Republic; and Department of Genetics and Laboratory Medicine, University of Geneva, Switzerland. He now serves as a scientist and leads a research group in the area of neurovirology and inflammation biology at the prestigious Centre for Cellular and Molecular Biology, Hyderabad, India. His main areas of research interest are neurovirology and immunology. Dr. Singh has received several awards, including the Skinner Memorial Award, Travel Grant Award, NIH-Fogarty Fellowship, and Young Scientist Award. He is associated with several international journals as an associate editor and editorial board member.

Dr. Daniel Růžek is head of the Department of Virology and head of the Research Group on Emerging Viral Infections at the Veterinary Research Institute, Brno, Czech Republic, a research scientist at the Institute of Parasitology, Academy of Sciences of the Czech Republic, and an assistant professor in the Department of Medical Biology, Faculty of Science, University of South Bohemia, České Budějovice, Czech Republic. He earned a PhD in molecular and cellular biology and genetics at the Academy of Sciences of the Czech Republic and the University of South Bohemia. He received his postdoctoral training at the Department of Virology and Immunology, Texas Biomedical Research Institute (formerly Southwest Foundation for Biomedical Research), San Antonio, Texas. His primary interest is virology with a research emphasis on vector-borne viruses, especially tick-borne encephalitis virus, Omsk hemorrhagic fever virus, dengue virus, West Nile virus, and so on. In 2009, he was awarded the prestigious international Sinnecker–Kunz Award for young researchers.

Contributors

Thomas Arminio
Integrated Research Facility
Division of Occupational Health and Safety
Rocky Mountain Laboratories
National Institutes of Health
Hamilton, Montana

Tatjana Avšič-Županc
Faculty of Medicine
Institute of Microbiology and Immunology
University of Ljubljana
Ljubljana, Slovenia

Daniel G. Bausch
Department of Tropical Medicine
Tulane School of Public Health and Tropical
    Medicine
New Orleans, Louisiana

Sina Bavari
United States Army Medical Research Institute
    of Infectious Diseases
Fort Detrick, Maryland

Cecilia Bender
Facultad de Ciencias Exactas
Instituto de Biotecnología y Biología Molecular
Consejo Nacional de Investigaciones
    Científicas y Técnicas
Universidad Nacional de La Plata
La Plata, Argentina

Jason Botten
Division of Immunobiology
Department of Medicine
and
Department of Microbiology and Molecular
    Genetics
University of Vermont
Burlington, Vermont

Matthew L. Boulton
Division of Infectious Diseases
Department of Epidemiology and Preventive
    Medicine
School of Public Health
University Medical School
University of Michigan
Ann Arbor, Michigan

Gavin C. Bowick
Department of Microbiology and Immunology
University of Texas Medical Branch at
    Galveston
Galveston, Texas

Steven B. Bradfute
Department of Molecular Genetics and
    Microbiology
University of New Mexico
Albuquerque, New Mexico

Jennifer A. Cann
Integrated Research Facility at Fort Detrick
National Institute of Allergy and Infectious
    Diseases
National Institutes of Health
Frederick, Maryland

Ricardo Carrion Jr.
Department of Virology and Immunology
Texas Biomedical Research Institute
San Antonio, Texas

Hyeryun Choe
Department of Medicine
Boston Children's Hospital
Harvard Medical School
Boston, Massachusetts

Anna N. Clawson
Integrated Research Facility at Fort Detrick
National Institute of Allergy and Infectious
    Diseases
National Institutes of Health
Frederick, Maryland

Christel Cochez
Research Laboratory for Vector-Borne Diseases
and
Reference Laboratory for Hantavirus Infections
Queen Astrid Military Hospital
Brussels, Belgium

Nadine A. Dalrymple
Department of Molecular Genetics and
    Microbiology
Stony Brook University
Stony Brook, New York

Robert Davey
Department of Virology and Immunology
Texas Biomedical Research Institute
San Antonio, Texas

Fabian de Kok-Mercado
Integrated Research Facility at Fort Detrick
National Institute of Allergy and Infectious
    Diseases
National Institutes of Health
Frederick, Maryland

Gerhard Dobler
Department of Virology and Rickettsiology
Bundeswehr Institute of Microbiology
Munich, Germany

Mandy Elschner
Friedrich-Loeffler-Institut
Federal Research Institute for Animal Health
Institute of Bacterial Infections and Zoonoses
Jena, Germany

Onder Ergonul
Infectious Diseases Department
School of Medicine
Koç University
Istanbul, Turkey

Shamsudeen Fagbo
Zoonotic Diseases Unit
Preventive Medicine Department
Ministry of Health
Riyadh, Kingdom of Saudi Arabia

Miroslav Fajfr
Faculty of Military Health Sciences
University of Defense
and
Institute of Clinical Microbiology
University Hospital in Hradec Králové
Hradec Králové, Czech Republic
and
Center of Biological Defense Těchonín
Army of the Czech Republic
Těchonín, Czech Republic

Michael Farzan
Department of Microbiology and
    Immunobiology
Harvard Medical School
Boston, Massachusetts

M. Leticia Ferrelli
Facultad de Ciencias Exactas
Instituto de Biotecnología y Biología Molecular
Consejo Nacional de Investigaciones
    Científicas y Técnicas
Universidad Nacional de La Plata
La Plata, Argentina

Ramon Flick
BioProtection Systems Corporation
and
NewLink Genetics Corporation
Ames, Iowa

Alexander N. Freiberg
Department of Pathology
University of Texas Medical Branch at
    Galveston
Galveston, Texas

Irina Gavrilovskaya
Department of Molecular Genetics and
    Microbiology
Stony Brook University
Stony Brook, New York

Augustine Goba
Kenema Government Hospital
Ministry of Health and Sanitation
Kenema, Sierra Leone

Ricardo M. Gómez
Facultad de Ciencias Exactas
Instituto de Biotecnología y Biología Molecular
Consejo Nacional de Investigaciones
    Científicas y Técnicas
Universidad Nacional de La Plata
La Plata, Argentina

Elena Gorbunova
Department of Molecular Genetics and
    Microbiology
Stony Brook University
Stony Brook, New York

Donald S. Grant
Kenema Government Hospital
Ministry of Health and Sanitation
Kenema, Sierra Leone

Anthony Griffiths
Department of Virology and Immunology
Texas Biomedical Research Institute
San Antonio, Texas

Mary Kate Hart
DynPort Vaccine Company LLC
A CSC Company
Frederick, Maryland

Andrew Hayhurst
Department of Virology and Immunology
Texas Biomedical Research Institute
San Antonio, Texas

Paul Heyman
Research Laboratory for Vector-Borne Diseases
and
Reference Laboratory for Hantavirus Infections
Queen Astrid Military Hospital
Brussels, Belgium

Frank Hufert
Department of Virology
University Medical Center Göttingen
Göttingen, Germany

Louis M. Huzella
Integrated Research Facility at Fort Detrick
National Institute of Allergy and Infectious
    Diseases
National Institutes of Health
Frederick, Maryland

Peter B. Jahrling
Integrated Research Facility at Fort Detrick
National Institute of Allergy and Infectious
    Diseases
National Institutes of Health
Frederick, Maryland

Stephanie Jemielity
Department of Medicine
Boston Children's Hospital
Harvard Medical School
Boston, Massachusetts

Lyudmila S. Karan
Laboratory of Epidemology of Zoonoses
Central Research Institute of Epidemiology
Moscow, Russia

Sean G. Kaufman
Center for Public Health Preparedness
    and Response
Rollins School of Public Health
Emory University
Atlanta, Georgia

Humarr Khan
Kenema Government Hospital
Ministry of Health and Sanitation
Kenema, Sierra Leone

Benjamin King
Division of Immunobiology
Department of Medicine
University of Vermont
Burlington, Vermont

Chwan-Chuen King
Institute of Epidemiology and Preventive
    Medicine
College of Public Health
National Taiwan University
Taipei, Taiwan, Republic of China

Joseph Klaus
Division of Immunobiology
Department of Medicine
University of Vermont
Burlington, Vermont

William B. Klimstra
Department of Microbiology and Molecular
    Genetics
Center for Vaccine Research
University of Pittsburgh
Pittsburgh, Pennsylvania

Jens H. Kuhn
Integrated Research Facility at Fort Detrick
National Institute of Allergy and Infectious
    Diseases
National Institutes of Health
Frederick, Maryland

Matthew Lackemeyer
Integrated Research Facility at Fort Detrick
National Institute of Allergy and Infectious
    Diseases
National Institutes of Health
Frederick, Maryland

Jayashree Apurupa Lahoti
Laboratory of Neurovirology and Inflammation
    Biology
Centre for Cellular and Molecular Biology
Hyderabad, India

Ake Lundkvist
Department of Preparedness
Swedish Institute for Communicable Disease
    Control
and
Karolinska Institute
Solna, Sweden

Erich R. Mackow
Department of Molecular Genetics and
    Microbiology
Stony Brook University
Stony Brook, New York

Shannon S. Martin
DynPort Vaccine Company LLC
A CSC Company
Frederick, Maryland

Ziad A. Memish
WHO Collaborating Center for Mass Gathering
    Medicine
College of Medicine
Alfaisal University
Riyadh, Kingdom of Saudi Arabia

Ritu Mishra
Laboratory of Neurovirology and Inflammation
    Biology
Centre for Cellular and Molecular Biology
Hyderabad, India

Derek Morrow
United States Army Medical Research Institute
    of Infectious Diseases
Frederick, Maryland

Lina M. Moses
Department of Tropical Medicine
Tulane School of Public Health and Tropical
    Medicine
New Orleans, Louisiana

Jean L. Patterson
Department of Virology and Immunology
Texas Biomedical Research Institute
San Antonio, Texas

Janusz T. Paweska
Center for Emerging and Zoonotic Diseases
National Institute for Communicable Diseases
National Health Laboratory Service
and
Division Virology and Communicable Diseases
    Surveillance
School of Pathology
University of the Witwatersrand
Johannesburg, South Africa

Guey Chuen Perng
Department of Pathology and Laboratory
    Medicine
Emory Vaccine Center
School of Medicine
Emory University
Atlanta, Georgia

and

Department of Microbiology and Immunology
National Cheng Kung University Medical
    College
and
Center of Infectious Disease and Signaling
    Research
National Cheng Kung University
Tainan, Taiwan, Republic of China

Donna L. Perry
Integrated Research Facility at Fort Detrick
National Institute of Allergy and Infectious
    Diseases
National Institutes of Health
Frederick, Maryland

Matías L. Pidre
Facultad de Ciencias Exactas
Instituto de Biotecnología y Biología Molecular
Consejo Nacional de Investigaciones
    Científicas y Técnicas
Universidad Nacional de La Plata
La Plata, Argentina

Sheli R. Radoshitzky
United States Army Medical Research Institute
    of Infectious Diseases
Frederick, Maryland

George F. Risi
Rocky Mountain Laboratories
Hamilton, Montana

and

Infectious Disease Specialists, PC
Missoula, Montana

Víctor Romanowski
Facultad de Ciencias Exactas
Instituto de Biotecnología y Biología Molecular
Consejo Nacional de Investigaciones
    Científicas y Técnicas
Universidad Nacional de La Plata
La Plata, Argentina

Daniel Růžek
Institute of Parasitology
Biology Centre
Academy of Sciences of the Czech Republic
České Budějovice, Czech Republic

and

Department of Virology
Veterinary Research Institute
Brno, Czech Republic

Kate D. Ryman
Department of Microbiology and Molecular
    Genetics
Center for Vaccine Research
University of Pittsburgh
Pittsburgh, Pennsylvania

Ana Saksida
Faculty of Medicine
Institute of Microbiology and Immunology
University of Ljubljana
Ljubljana, Slovenia

Olena Shtanko
Department of Virology and Immunology
Texas Biomedical Research Institute
San Antonio, Texas

Amy C. Shurtleff
Integrated Toxicology Division
United States Army Medical Research Institute
    of Infectious Diseases
Fort Detrick, Maryland

Sunit Kumar Singh
Laboratory of Neurovirology and Inflammation
    Biology
Centre for Cellular and Molecular Biology
Hyderabad, India

Shannon L. Taylor
Virology Division
United States Army Medical Research Institute
    of Infectious Diseases
Fort Detrick, Maryland

Sergey E. Tkachev
Institute of Chemical Biology and Fundamental
    Medicine
Siberian Branch of the Russian Academy of
    Sciences
Novosibirsk, Russia

Petrus Jansen van Vuren
Center for Emerging and Zoonotic Diseases
National Institute for Communicable Diseases
National Health Laboratory Service
Johannesburg, South Africa

Victoria Wahl-Jensen
Integrated Research Facility at Fort Detrick
National Institute of Allergy and Infectious
    Diseases
National Institutes of Health
Frederick, Maryland

Travis K. Warren
United States Army Medical Research Institute
    of Infectious Diseases
Frederick, Maryland

Manfred Weidmann
Department of Virology
University Medical Center Göttingen
Göttingen, Germany

Eden V. Wells
Department of Epidemiology
School of Public Health
University of Michigan
Ann Arbor, Michigan

Jacqueline Weyer
Center for Emerging and Zoonotic Diseases
National Institute for Communicable Diseases
National Health Laboratory Service
Johannesburg, South Africa

Valeriy V. Yakimenko
Omsk Research Institute of Natural Foci
  Infections
Omsk, Russia

Ali Zaki
Virology Laboratory
Dr. Fakeeh Hospital
Jeddah, Kingdom of Saudi Arabia

Christopher Ziegler
Division of Immunobiology
Department of Medicine
University of Vermont
Burlington, Vermont

Part I

General Aspects

1 Viral Hemorrhagic Fevers
History and Definitions

Jens H. Kuhn, Anna N. Clawson, Sheli R. Radoshitzky,
Victoria Wahl-Jensen, Sina Bavari, and Peter B. Jahrling

CONTENTS

In the past, the term "viral hemorrhagic fever" (VHF) has been used by many researchers and clinicians for a variety of clinical syndromes that at least superficially bear resemblance to each other. However, despite more than 60 years of research, it remains unclear whether the various syndromes deemed to be VHFs truly share common denominators, such as shared effects on particular molecular pathways during pathogenesis or particular ecological factors that lead to their emergence. In a book on pathogens causing diseases currently referred to as VHFs, it therefore seems necessary to at least discuss the evolution of the term and to raise the question whether it is truly meaningful and thereby useful.

Although it is difficult to elucidate when the term "VHF" was introduced into the biomedical literature, it is with little doubt that it was famous Soviet virologist Čumakov Mihail Petrovič (1909–1993) who coined it. One of the first publications that uses the term was published by him in 1948 in Russian to address three diseases that on clinical grounds were challenging to differentiate (Čumakov, 1948a). At the time, these diseases, which today are known as hemorrhagic fever with renal syndrome (HFRS), Crimean–Congo hemorrhagic fever (CCHF), and Omsk hemorrhagic fever (OHF), had just been discovered in short succession in different geographic areas of the Soviet Union among humans (Čumakov, 1948b; Sokolov et al., 1945; Targanskaâ, 1935). Next to the similar clinical presentation (an influenza-like prodromal phase followed by more severe symptoms often ultimately resulting in shock due to multiorgan failure), HFRS, CCHF, and OHF were found to be zoonoses caused by distinct viruses, and the geographic distribution of the diseases was therefore found to be directly associated with the geographic distribution of their natural hosts and/or vectors. It was Pavlovskij Evgenij Nikanorovič (1884–1965) who developed the concept of "natural focus infections," today often referred to as "natural nidality," for the fact that the distribution of certain diseases can be described or predicted based on the knowledge of complex ecosystems (Pavlovskij, 1939, 1960). Many diseases today accepted as VHFs are indeed "natural focus infections."

Čumakov continued to develop the concept of VHFs, and in 1950, he published the first article on VHFs in a more widely distributed journal in Czech (Čumakov, 1950). HFRS, CCHF, and OHF were still the only major diseases comprising the VHF group at that time, but Čumakov recognized that there are other diseases which may have to be included (Čumakov, 1950). A first definition of VHF was then published in a Russian reference encyclopedia in 1951. Accordingly,

> [h]emorrhagic fevers are natural focus diseases of viral etiology developing with an acute febrile reaction, grave general symptoms, and generalized capillaropathy [translated from Russian] (Viskovskij, 1951).

In 1957, Čumakov finally published the VHF concept in English (Chumakov, 1957). In subsequent years, novel VHF-like diseases were discovered, and communication with researchers outside of the USSR increased and made the term "VHF" more widely known. This resulted in updated definitions published by at least four renowned scientists. Later Nobel laureate Daniel Carleton Gajdusek (1923–2008) initiated this process when he wrote in 1962 that VHFs are

> ... focal "place diseases" occurring in specific, well-localized geographic regions with peak incidence in a particular season of the year. Man appears to be involved in the cycle only accidentally when he upsets the ecologic balance and becomes the host for the vector arthropods which do not usually feed upon him (Gajdusek, 1962).

Based on ever increasing data regarding the ecology of the hosts and vectors carrying VHF agents, Gajdusek recognized three VHFs groups: those that were rodent-borne (HFRS and Argentinian hemorrhagic fever [AHF]), those that were tick-borne (CCHF, OHF), and those that were mosquito-borne (Dengue, yellow fever [YF]) (Gajdusek, 1962).

Čumakov followed Gajdusek in 1963. He added the newly discovered Indian Kyasanur Forest disease (KFD) to the group of tick-borne VHFs and defined VHFs as diseases with

> ... an acute course with fever, sometimes with symptoms of universal capillary toxicosis with disturbances or permeability of capillaries and small vessels; or with haemorrhagic syndrome of variable severity, as a rule, with changes in the peripheral nervous system, blood and viscera (Chumakov, 1963).

Furthermore, Čumakov realized that all definitions of VHFs available to that date were not necessarily ideal. While certain diseases, such as dengue and YF, more or less clearly belonged to the VHF complex based on his definition, others, such as Colorado tick fever, sandfly fever, and non-paralytic forms of tick-borne encephalitis, were not necessarily distinct enough to be excluded (Chumakov, 1963).

In the same year, Smorodincev Anatolij Aleksandrovič et al. (1901–1986) wrote that

> [viral h]emorrhagic fevers are natural focal diseases with a characteristic endemicity, a distinct geographical localization of the foci in rural areas and in sparsely populated regions, and where conditions capable of maintaining the circulation of the causative agent in certain species of carriers and warm-blooded animals prevail... man is included only sporadically and irregularly... (cited from the English translation of Smorodincev et al., 1963).

Finally, in 1970, Alexis Shelokov suggested that

> [t]he clinical picture [of VHF] is that of a febrile illness complicated by variable hemorrhagic phenomena (resulting from damage to the capillary and hematopoietic systems), shock, and sometimes death. Ecologically, each hemorrhagic disease occurs when man somehow disturbs the natural niche or nidus within which the virus circulates among local vertebrates, often with the aid of arthropod vectors (Shelokov, 1970b).

Even a cursory look at these definitions reveals their overall fuzziness. Gajdusek's, Smorodincev's, and Shelokov's definitions suggest that humans are only infected with VHF-causing pathogens by accident—yet dengue and yellow fever viruses (YFVs) are transmitted by female mosquitoes that routinely feed on humans (predominantly *Aedes (Stegomyia) aegypti*). Both the mosquito and the two viruses are endemic in tropical and subtropical areas worldwide, and it is well known that both viruses cause large urban outbreaks annually (Gardner and Ryman, 2010; Gubler, 2006). It is difficult to accept such a large geographic area as a disease "focus." The number of dengue and yellow fever virus infections does fluctuate with the number of mosquitoes in an area, which itself is indeed dependent on seasons. Yet, infections with both viruses occur throughout the year, and only their absolute number changes (Gardner and Ryman, 2010; Gubler, 2006). Does this suffice to classify dengue and YF as seasonal diseases? Čumakov probably realized this problem when he forwarded his definition of VHFs, which is mostly based on clinical rather than ecological/geographical characteristics. Indeed, clinically, patients with YF and severe dengue (previously known as dengue hemorrhagic fever [DHF]) resemble patients with HFRS, CCHF, or OHF. In all five diseases, the endothelium of blood vessels is affected directly or indirectly, resulting in increased permeability and therefore hemorrhages and/or edema. Hemorrhagic signs, if interpreted *sensu lato*, are typical for all five diseases as well. But Čumakov already added a classifier ("hemorrhagic syndrome of *variable* severity") and so did Shelokov ("*variable* hemorrhagic phenomena"), and until today it has never been clarified by clinicians at what extent certain manifestations should count as hemorrhagic signs worthy to fulfill a VHF inclusion criterion. For instance, are overt hemorrhages, such as bleeding from the gums, nose, rectum, or vagina, necessary inclusion criteria, and in which percentage of an infected population would they have to occur to make it a hallmark of a disease? Or do any disturbances in blood hemostasis, such as petechiae or subtle changes in the coagulative pathways, suffice? Given that "hemorrhage" is such a prominent word in the term "viral hemorrhagic fever," it is astounding that this issue has not yet been addressed. After all, "viral hemorrhagic fever" often evokes doomsday scenarios among the general public. These fears are supported by movies, books, and news reports describing patients who "bleed out"—despite the fact that hypovolemic shock due to blood loss is very rarely a cause of death among patients diagnosed with VHF. A stricter definition (overt hemorrhage necessary in majority of cases) would drastically reduce the number of diseases that should be classified as VHFs, whereas a laxer definition (any effect on blood hemostasis) may increase the number considerably. In the extreme, either direction may make the term "VHF" meaningless. Even the severity of VHFs is difficult to define. Lethality can be extremely high in one type of VHF (Ebola virus disease [EVD]) or very low (nephropathia epidemica subform of HRFS) and yet lethality does not seem to be correlated directly to the extent of hemorrhagic manifestations (which are much less pronounced in EVD than they are in the less lethal CCHF). These problems with the VHF definitions did not go unnoticed among the scientific and medical experts at the time, but their criticisms seem to have been largely forgotten. For instance, Jordi Casals wrote already in 1966 that the definitions

> … exposed many lacunae in clinical, virological, epidemiological, ecological, and zoological knowledge… [and that it] should therefore be considered a tenuous synthesis of certain demonstrated facts and speculations based on field observation (Casals et al., 1966),

followed shortly thereafter by Karl M. Johnson's view:

> Although a minority of the salient facts have been uncovered, current information leaves little doubt that beyond the gross common denominators of fever, presence of a virus infection and the occurrence of hemorrhage, hemoconcentration and thrombocytopenia, the use of the term hemorrhagic fever to characterize disease of South America [arenaviral VHFs] and Southeast Asia [DHF] is much more likely to confuse than to clarify thought and ideas regarding the respective syndromes (Johnson et al., 1967).

After Shelokov's VHF definition, further refinements were not made and no new definitions were brought forward. However, the list of viral diseases that ought to be classified as VHFs according to his and previous definitions increased of course. Shelokov himself was probably the first scientist to publish the logical conclusion that, with humans being animals, VHFs could also affect other animals (Shelokov, 1970a). In fact, neither Čumakov nor Gajdusek nor Smordincev ever explicitly stated that VHFs should be human diseases by definition—and indeed it would make little sense to do that. Shelokov therefore began to include animal diseases that have not been observed among humans into the list of VHFs, starting with what today is known as simian hemorrhagic fever (SHF)

(Shelokov, 1970a). Kirov et al. added epizootic hemorrhagic disease of deer (EHD) shortly thereafter (Kirov et al., 1972). Čumakov and others accepted these additions (Čumakov, 1979; Somov and Besednova, 1981) based on the fact that both diseases do not contradict the existing VHF definitions (Ševcova, 1967; Shope et al., 1955).

Today, most researchers seem to either have forgotten the evolution of VHF definitions and classifications or they are too young to ever been exposed to them. This then explains why review after review states that VHF-causing viruses are only found among four different RNA virus families (*Arenaviridae*, *Bunyaviridae*, *Flaviviridae*, and *Filoviridae*) when in fact SHF and EHD are caused by very different pathogens of the families *Arteriviridae* and *Reoviridae*, respectively (King et al., 2011). Several other agents clearly also cause VHFs and belong to yet other viral taxa. For instance, African swine fever (ASF) affects suids and is caused by a DNA virus of the "megaviral" family *Asfarviridae* (Gubler, 2006). Asfarvirologists have been referring to this disease as a VHF for quite some time as it fulfills all the VHF-defining criteria described previously (Sanchez-Cordon et al., 2008; Šubina et al., 1994)—yet this disease is rarely mentioned in general VHF reviews. The same is true for other important animal VHFs, such as rabbit hemorrhagic disease (Abrantes et al., 2012) or classical swine fever (CSF) (Lange et al., 2011).

To complicate the matters further, Gajdusek already had pointed out that "VHF" is a rather fluid syndrome that is almost always developing upon infection with some viruses (e.g., filoviruses, Crimean–Congo hemorrhagic fever virus [CCHFV], Machupo virus [MACV]), less commonly with others (e.g., Lassa virus [LASV], dengue viruses [DENVs]), and only as a rare complication with some (e.g., feline calicivirus, human herpesviruses 3 and 4, measles virus, rubella virus, and variola virus) (Gajdusek, 1962). Gajdusek was right when he wrote that

> [i]n many infections an increase in capillary permeability, associated often with small ruptures of capillaries, can produce petechiae and even more extensive cutaneous hemorrhages... or more severe extravasations of blood into the urinary or gastrointestinal tract, the endocrine and cardiorespiratory systems, the central nervous system, and other tissues of the body may occur. At times this syndrome may give rise to massive bleeding and produce symptoms of hemoptysis or hematemesis, hematuria or gastrointestinal bleeding, uterine hemorrhage, epistaxis, conjunctival hemorrhage, and ecchymoses (Gajdusek, 1962).

This, then, raises the question which VHF-like diseases should *not* be classified as VHFs. Walker et al., for instance, stated in 1982 with strong emphasis that "in fact, Lassa fever (LF) is pathologically not a hemorrhagic fever" based on postmortem examinations of 21 human cases of LF (Walker et al., 1982). Yet, LF is almost invariably listed as a *bona fide* VHF in almost any review published. This begs the question whether the scientific community did and does not agree with the assessment of Walker et al. (and for which reasons) or whether this study has simply been ignored. Severe dengue (DHF) seems to be due to sequential infection with two different dengue viruses—it may therefore be a fundamentally different disease than those that are associated with the infection of one particular agent. Other diseases are often still listed as VHFs although there is either no published evidence that they really fulfill the VHF inclusion criteria (e.g., Whitewater Arroyo virus infection) or because such evidence was published (e.g., chikungunya) but later studies demonstrated coinfection with viruses known to actually cause VHF (dengue viruses) (Shelokov, 1970b).

Table 1.1 lists the agents causing VHFs that fulfill most of the criteria for VHF inclusion described earlier, some of which are thought to be novel, and excludes those that are now considered dubious. Some agents cause significant mortality among humans (e.g., DENV) or are of significant concern to the livestock industries (e.g., ASF, rabbit hemorrhagic disease). Others are highly lethal for humans (e.g., filoviruses) or animals (e.g., simian hemorrhagic fever virus [SHFV]) but emerge only sporadically and are therefore considered exotic. Not listed are viruses that only cause VHFs under specific, most likely host-determined, conditions (e.g., Colorado tick fever virus, human herpesviruses 3 and 4, Ilesha virus, measles virus, rubella virus, and variola virus) or viruses that usually

TABLE 1.1

Pathogens Causing VHFs in Mammals

Virus (Abbreviation)	Taxonomic Classification	Vector	Reservoir	Mammal Affected/Name of Disease	Geographic Distribution
African swine fever virus (ASFV)	Family *Asfarviridae*, genus *Asfivirus*, species *African swine fever virus*	Argasid ticks (*Ornithodoros* sp.)	Bushpigs (*Potamochoerus larvatus*), common warthogs (*Phacochoerus africanus*)	Domestic wild boars (*Sus scrofa*)/ASF	Sub-Saharan Africa
Amur/Soochang virus (ASV)	Family *Bunyaviridae*, genus *Hantavirus*, species?	None	Korean field mice (*Apodemus peninsulae*)	Humans/HFRS	Far Eastern Russia
Bas-Congo virus (?)	Order *Mononegavirales*, family *Rhabdoviridae*, genus *Tibrovirus*? Species?	?	?	Humans?/?	Democratic Republic of the Congo
Bundibugyo virus (BDBV)	Order *Mononegavirales*, family *Filoviridae*, genus *Ebolavirus*, species *Bundibugyo ebolavirus*	None?	?	Humans/EVD	Uganda, Democratic Republic of Congo
Chapare virus (CHPV)	Family *Arenaviridae*, genus *Arenavirus*, species *Chapare virus*	None?	Cricetid rodents?	Humans/unnamed	Bolivia
Classical swine fever virus (CSFV)	Family *Flaviviridae*, genus *Pestivirus*, species *Classical swine fever virus*	None	Domestic wild boars (*Sus scrofa*)	Domestic wild boars (*Sus scrofa*)/ CSF, hog cholera	South and Central America, the Caribbean, Asia, and Eastern Europe
Crimean–Congo hemorrhagic fever virus (CCHFV)	Family *Bunyaviridae*, genus *Nairovirus*, species *Crimean–Congo hemorrhagic fever virus*	Predominantly ixodid ticks (predominantly *Hyalomma* sp.)	Cattle, dogs, goats, hares, hedgehogs, mice, ostriches, sheep	Humans/CCHF	Africa, Europe, Asia
Dengue viruses 1–4 (DENV-1–4)	Family *Flaviviridae*, genus *Flavivirus*, species *Dengue virus* 1–4	Mosquitoes (predominantly Aedes [*Stegomyia*] *aegypti*)	Humans, nonhuman primates	Humans/severe dengue (DHF)	>112 countries, throughout the Tropics
Dobrava–Belgrade virus (DOBV) genotype Dobrava	Family *Bunyaviridae*, genus *Hantavirus*, species *Dobrava–Belgrade virus*	None	Yellow-necked field mice (*Apodemus flavicollis*)	Humans/HFRS	Slovenia, Croatia, Serbia, Montenegro, Hungary, Slovakia, Bulgaria, Greece

(continued)

TABLE 1.1 (continued)
Pathogens Causing VHFs in Mammals

Virus (Abbreviation)	Taxonomic Classification	Vector	Reservoir	Mammal Affected/Name of Disease	Geographic Distribution
Dobrava–Belgrade virus (DOBV) genotype Kurkino	Family *Bunyaviridae*, genus *Hantavirus*, species *Dobrava–Belgrade virus*	None	Striped field mice (*Apodemus agrarius*)	Humans/HFRS	Germany, Slovakia, Russia, Hungary, Slovenia, Croatia, Denmark, mainland Estonia
Dobrava–Belgrade virus (DOBV) genotype Saaremaa	Family *Bunyaviridae*, genus *Hantavirus*, species *Dobrava–Belgrade virus*	None	Striped field mice (*Apodemus agrarius*)	Humans/HFRS	Estonia, Slovakia, Slovenia, Hungary, Denmark
Dobrava–Belgrade virus (DOBV) genotype Sochi	Family *Bunyaviridae*, genus *Hantavirus*, species *Dobrava–Belgrade virus*	None	Caucasus field mice (*Apodemus ponticus*)	Humans/HFRS	Russia
Ebola virus (EBOV)	Order *Mononegavirales*, family *Filoviridae*, genus *Ebolavirus*, species *Zaire ebolavirus*	None?	?	Western lowland gorillas (*Gorilla gorilla gorilla*), central chimpanzees (*Pan troglodytes troglodytes*), humans/EVD	Democratic Republic of the Congo, Gabon, Republic of the Congo
Epizootic hemorrhagic disease viruses 1–8 (EHDV-1–8)	Family *Reoviridae*, genus *Orbivirus*, species *Epizootic hemorrhagic disease virus 1–8*	Biting midges (*Culicoides* sp.)	?	Cattle? Mule deer (*Odocoileus hemionus*), pronghorns (*Antilocapra americana*), white-tailed deer (*Odocoileus virginianus*)/EHD	United States
European brown hare syndrome virus (EBHSV)	Family *Caliciviridae*, genus *Lagovirus*, species *European brown hare syndrome virus*	None	European hares (*Lepus* [*Eulagos*] *europaeus*), mountain hares (*Lepus* [*Lepus*] *timidus*)	European hares (*Lepus* [*Eulagos*] *europaeus*), mountain hares (*Lepus* [*Lepus*] *timidus*)/European brown hare syndrome (EBHS)	Europe
Guanarito virus (GTOV)	Family *Arenaviridae*, genus *Arenavirus*, species *Guanarito virus*	None	Short-tailed zygodonts (*Zygodontomys brevicauda*)	Humans/"Venezuelan hemorrhagic fever"	Venezuela

Virus	Classification	Vector	Reservoir host / Disease	Geographic location
Hantaan virus (HTNV)	Family *Bunyaviridae*, genus *Hantavirus*, species *Hantaan virus*	None	Striped field mice (*Apodemus agrarius coreae*) — Humans/HFRS	Asia, Eastern Russia
Junín virus (JUNV)	Family *Arenaviridae*, genus *Arenavirus*, species *Junín virus*	None	Drylands lauchas (*Calomys musculinus*) — Humans/Junín/AHF	Argentina
Kyasanur Forest disease virus (KFDV) including the Alkhurma/Alkhumra variant	Family *Flaviviridae*, genus *Flavivirus*, species *Kyasanur Forest disease virus*	Ixodid ticks (predominantly *Haemaphysalis spinigera*) and sand tampans (*Ornithodoros savignyi*)	Indomalayan vandeleurias (*Vandeleuria oleracea*), roof rats (*Rattus rattus*). Camels? Sheep? — Northern plains gray langur (*Semnopithecus entellus*), humans, bonnet macaque (*Macaca radiata*)/KFD, Alkhurma or Alkhumra hemorrhagic fever	India (KFDV), Saudi Arabia and Egypt (Alkhurma/Alkhumra variant)
Lassa virus (LASV)	Family *Arenaviridae*, genus *Arenavirus*, species *Lassa virus*	None	Natal mastomys (*Mastomys natalensis*) — Humans/LF	West Africa
Lujo virus (LUJV)	Family *Arenaviridae*, genus *Arenavirus*, species *Lujo virus*	None?	Murid rodent? — Humans/unnamed	Zambia
Machupo virus (MACV)	Family *Arenaviridae*, genus *Arenavirus*, species *Machupo virus*	None	Big lauchas (*Calomys callosus*) — Humans/Machupo/Bolivian hemorrhagic fever (BHF)	Bolivia
Marburg virus (MARV)	Order *Mononegavirales*, family *Filoviridae*, genus *Ebolavirus*, species *Marburg marburgvirus*	None?	Egyptian rousettes (*Rousettus* [*Rousettus*] *aegyptiacus*)? — Grivets (*Chlorocebus aethiops*)/Marburg virus disease (MVD)	Angola, Democratic Republic of the Congo, Kenya, Uganda, Zimbabwe
Ocozocoautla de Espinosa virus (OCEV)	Family *Arenaviridae*, genus *Arenavirus*, species?	None?	Mexican deermice (*Peromyscus mexicanus*) — Humans?/?	Mexico
Omsk hemorrhagic fever virus (OHFV)	Family *Flaviviridae*, genus *Flavivirus*, species *Omsk hemorrhagic fever virus*	Ixodid ticks (predominantly *Dermacentor* sp.)	Migratory birds, rodents — Common muskrats (*Ondatra zibethicus*), humans/OHF	Russia
Puumala virus (PUUV)	Family *Bunyaviridae*, genus *Hantavirus*, species *Puumala virus*	None	Bank voles (*Myodes glareolus*) — Humans/nephropathia epidemica subform of HFRS	Eurasia

(continued)

TABLE 1.1 (continued)
Pathogens Causing VHFs in Mammals

Virus (Abbreviation)	Taxonomic Classification	Vector	Reservoir	Mammal Affected/Name of Disease	Geographic Distribution
Rabbit hemorrhagic disease virus (RHDV)	Family *Caliciviridae*, genus *Lagovirus*, species *Rabbit hemorrhagic disease virus*	None	European rabbits (*Oryctolagus cuniculus*)	European rabbit (*Oryctolagus cuniculus*)/rabbit hemorrhagic disease (RHD)	Worldwide
Ravn virus (RAVV)	Order *Mononegavirales*, family *Filoviridae*, genus *Ebolavirus*, species *Marburg marburgvirus*	None?	Egyptian rousettes (*Rousettus [Rousettus] aegyptiacus*)?	Humans/Marburg virus disease (MVD)	Kenya, Democratic Republic of the Congo, Uganda
Reston virus (RESTV)	Order *Mononegavirales*, family *Filoviridae*, genus *Ebolavirus*, species *Reston ebolavirus*	None?	Domestic wild boars (*Sus scrofa*)?	Crab-eating macaques (*Macaca fascicularis*)/unnamed	Philippines, China?
Rift Valley fever virus (RVFV)	Family *Bunyaviridae*, genus *Phlebovirus*, species *Rift Valley fever virus*	Mosquitoes (predominantly *Aedes* sp.)	Cattle? Rodents?	Humans, ruminants (predominantly cattle, goats, and sheep)/Rift Valley fever (RVF)	Sub-Saharan Africa, Middle East
Sabiá virus (SABV)	Family *Arenaviridae*, genus *Arenavirus*, species *Sabiá virus*	None?	Cricetid rodents?	Humans/"Brazilian hemorrhagic fever"	Brazil
Seoul virus (SEOV)	Family *Bunyaviridae*, genus *Hantavirus*, species *Seoul virus*	None	Brown rats (*Rattus norvegicus*), roof rats (*Rattus rattus*)	Humans/HFRS	Worldwide
Severe fever and thrombocytopenia syndrome virus (SFTSV)/ Huaiyangshan virus (HYSV)/Henan fever virus (HNFV)	*Bunyaviridae*, genus *Phlebovirus*, species?	Ixodid ticks (*Haemaphysalis longicornis*, *Rhipicephalus microplus*)	?	Humans/"Severe fever and thrombocytopenia syndrome" (SFTS), "Henan fever," "Huaiyangshan hemorrhagic fever"	China, Japan

Virus	Taxonomy	Vector	Reservoir	Susceptible hosts/Disease	Distribution
Simian hemorrhagic fever virus (SHFV)	Order *Nidovirales*, family *Arteriviridae*, genus *Arterivirus*, species *Simian hemorrhagic fever virus*	None?	Patas monkeys (*Erythrocebus patas*)? Grivets (*Chlorocebus aethiops*)? Baboons?	Asian macaques/SHF	Africa?
Sudan virus (SUDV)	Order *Mononegavirales*, family *Filoviridae*, genus *Ebolavirus*, species *Sudan ebolavirus*	None?	?	Humans/EVD	Uganda, South Sudan
Taï Forest virus (TAFV)	Order *Mononegavirales*, family *Filoviridae*, genus *Ebolavirus*, species *Taï Forest ebolavirus*	None?	?	Western chimpanzees (*Pan troglodytes verus*), humans/EVD	Côte d'Ivoire
Tula virus (TULV)	Family *Bunyaviridae*, genus *Hantavirus*, species *Tula virus*	None	Common voles (*Microtus* [*Microtus*] *arvalis*), field voles (*Microtus* [*Microtus*] *agrestis*), East European voles (*Microtus* [*Microtus*] *levis*)	Humans/HFRS	Europe
Yellow fever virus (YFV)	Order *Flaviviridae*, genus *Flavivirus*, species *Yellow fever virus*	Mosquitoes (predominantly A. [*Stegomyia*] *aegypti*)	Nonhuman primates (*Alouetta* sp., *Ateles* sp., *Cebus* sp., *Cercopithecus* sp., *Colobus* sp.)	Humans/YF	Africa, South America

do not cause VHFs but can cause them if certain mutations arise (e.g., feline calicivirus [Pedersen et al., 2000], lymphocytic choriomeningitis virus [Smadel et al., 1942], Garissa isolate of Ngari virus [Bowen et al., 2001], and tick-borne encephalitis virus [Ternovoi et al., 2003]).

What becomes instantly clear from this table is that evolutionary distant agents can obviously cause disease with superficially similar clinical presentation in a variety of evolutionary distant mammals after either direct or vector-borne transmission in vastly different geographic areas. A further look at the clinical presentation and pathology/pathogenesis of the most notorious human VHF-causing pathogens (described in other chapters in this book) also emphasizes that a "typical" VHF presentation (i.e., hemorrhage) is not necessarily a frequent observation upon infection. The best example for this discrepancy is probably LF, of which hundreds of thousands of cases are recorded each year but of which only a few thousand develop a VHF-like phenotype (McCormick and Fisher-Hoch, 2002).

It will therefore be up to clinicians to decide whether the term "VHF" is useful. If a human or nonhuman patient presents with fever, disturbances in the hemostatic system, and overt hemorrhages (i.e., with a "VHF"), is there a common treatment regimen that can be applied in general to ameliorate or overcome the disease? If so, then should all diseases that present this way be classified as VHFs, possibly even including bacterial and parasitic diseases (rickettsioses, falciparum malaria)? Or is treatment based more on the agent causing the disease (e.g., ribavirin in the case of human arenavirus and maybe CCHFV infection and antibiotics in the case of bacterial infections)? If so, then what is the advantage of classifying a certain disease as "VHF?" It is time that these questions are addressed by the current virology and medical community in light of the most recent data—and that a clear definition of VHF is brought forward or that the term "VHF" is abolished.

DISCLAIMER

The content of this publication does not necessarily reflect the views or policies of the U.S. Department of the Army, the U.S. Department of Defense, or the U.S. Department of Health and Human Services, or of the institutions and companies affiliated with the authors. This work was funded by the Joint Science and Technology Office for Chem Bio Defense (proposal #TMTI0048_09_RD_T to SB). JHK and VWJ performed this work as employees of Tunnell Consulting, Inc., and ANC as the owner of Logos Consulting, both subcontractors to Battelle Memorial Institute, under its prime contract with NIAID, under contract no. HHSN272200200016I.

REFERENCES

Abrantes, J., van der Loo, W., Le Pendu, J., Esteves, P.J., 2012. Rabbit haemorrhagic disease (RHD) and rabbit haemorrhagic disease virus (RHDV): A review. *Vet. Res.* 43, 12.

Bowen, M.D., Trappier, S.G., Sanchez, A.J., Meyer, R.F., Goldsmith, C.S., Zaki, S.R., Dunster, L.M., Peters, C.J., Ksiazek, T.G., Nichol, S.T., and the RVF Task Force, 2001. A reassortant bunyavirus isolated from acute hemorrhagic fever cases in Kenya and Somalia. *Virology* 291, 185–190.

Casals, J., Hoogstraal, H., Johnson, K.M., Shelokov, A., Wiebenga, N.H., Work, T.H., 1966. A current appraisal of hemorrhagic fevers in the U.S.S.R. *Am. J. Trop. Med. Hyg.* 15, 751–764.

Chumakov, M.P., 1957. Etiology, epidemiology, and prophylaxis of hemorrhagic fevers. *Pub. Health Monogr.* 50(Supplement 1), 19–25.

Chumakov, M.P., 1963. Studies of virus haemorrhagic fevers. *J. Hyg. Epidemiol. Microbiol. Immunol.* VII, 125–135.

Čumakov, M.P., 1948a. Etiology and epidemiology of hemorrhagic fevers. *Ter. Arh.* 20, 85–86 [Russian].

Čumakov, M.P., 1948b. Results of an expedition of the Institute of Neurology regarding Omsk hemorrhagic fever (OF). *Vestn. Akad. Med. Nauk SSSR* (2), 19–26 [Russian].

Čumakov, M.P., 1950. Etiology, epidemiology, and prophylaxis of hemorrhagic fevers. *Čas. Lék. Česk.* 89, 1428–1430 [Czech].

Čumakov, M.P., 1979. Viral hemorrhagic fevers—Scientific overview. USSR Ministry of Health, All-Union Scientific-Research Institute of Medical and Medical-Technical Information, Moscow, RSFSR, USSR [Russian].

Gajdusek, D.C., 1962. Virus hemorrhagic fevers—Special reference to hemorrhagic fever with renal syndrome (epidemic hemorrhagic fever). *J. Pediatr.* 60, 841–857.

Gardner, C.L., Ryman, K.D., 2010. Yellow fever: A reemerging threat. *Clin. Lab. Med.* 30, 237–260.

Gubler, D.J., 2006. Dengue/dengue haemorrhagic fever: History and current status. *Novartis Found. Symp.* 277, 3–16.

Johnson, K.M., Halstead, S.B., Cohen, S.N., 1967. Hemorrhagic fevers of Southeast Asia and South America: A comparative appraisal, in: Melnick, J.L. (Ed.), *Progress in Medical Virology*, Vol. 9. S. Karger, Basel, Switzerland, pp. 105–158.

King, A.M.Q., Adams, M.J., Carstens, E.B., Lefkowitz, E.J., 2011. *Virus Taxonomy—Ninth Report of the International Committee on Taxonomy of Viruses.* Elsevier/Academic Press, London, U.K.

Kirov, I.D., Radev, M., Vasilenko, S., Varbanov, V., 1972. Current knowledge in viral hemorrhagic fever (VHF) research. *Vytresh. Bolest.* XI, 3–14 [Bulgarian].

Lange, A., Blome, S., Moennig, V., Greiser-Wilke, I., 2011. Pathogenesis of classical swine fever—Similarities to viral haemorrhagic fevers: A review. *Berl. Münch. Tierärztl. Wochenschr.* 124, 36–47 [German].

McCormick, J.B., Fisher-Hoch, S.P., 2002. Lassa fever, in: Oldstone, M.B.A. (Ed.), *Arenaviruses I—The Epidemiology, Molecular and Cell Biology of Arenaviruses. Current Topics in Microbiology and Immunology*, Vol. 262. Springer, Berlin, Germany, pp. 75–109.

Pavlovskij, E.N., 1939. On natural focus infections and parasitic diseases. *Vestn. Akad. Med. Nauk SSSR.* 10, 98–108 [Russian].

Pavlovskij, E.N., 1960. *Natural Focus Infections.* Medgiz, Moscow, RSFSR, USSR [Russian].

Pedersen, N.C., Elliott, J.B., Glasgow, A., Poland, A., Keel, K., 2000. An isolated epizootic of hemorrhagic-like fever in cats caused by a novel and highly virulent strain of feline calicivirus. *Vet. Microbiol.* 73, 281–300.

Sanchez-Cordon, P.J., Romero-Trevejo, J.L., Pedrera, M., Sanchez-Vizcaino, J.M., Bautista, M.J., Gomez-Villamandos, J.C., 2008. Role of hepatic macrophages during the viral haemorrhagic fever induced by African Swine Fever Virus. *Histol. Histopathol.* 23, 683–691.

Ševcova, Z.V., 1967. Studies on the etiology of hemorrhagic fever in monkeys. *Vopr. Virusol.* 12, 47–51 [Russian].

Shelokov, A., 1970a. New viruses associated with the hemorrhagic fever syndrome, in: Kundsin, R.B. (Ed.), *Unusual Isolates from Clinical Material. Annals of the New York Academy of Sciences*, Vol. 172(2). New York Academy of Sciences, New York, pp. 990–992.

Shelokov, A., 1970b. Viral hemorrhagic fevers. *J. Infect. Dis.* 122, 560–562.

Shope, R.E., MacNamara, L.G., Mangold, R., 1955. Report on the… Deer mortality—Epizootic hemorrhagic disease of deer. *New Jersey Outdoors.* 6, 17–21.

Smadel, J.E., Green, R.H., Paltauf, R.M., Gonzalez, T.A., 1942. Lymphocytic choriomeningitis: Two human fatalities following an unusual febrile illness. *Proc. Soc. Exp. Biol. Med.* 49, 683–686.

Smorodincev, A.A., Kazbincev, L.I., Čudakov, V.G., 1963. Viral hemorrhagic fevers. Gosudarstvennoe izdatel'stvo medicinskoj literatury—State publisher for medical literature, Leningrad, Leningrad Oblast, RSFSR, USSR [Russian].

Sokolov, A.E., Čumakov, M.P., Kolačev, A.A., 1945. Crimean hemorrhagic fever (acute infectious capillary toxicosis). Izdanie Otdel'noj Primorskoj Armii, Simferopol, Ukrainian SSR, USSR [Russian].

Somov, G.P., Besednova, N.N., 1981. Hemorrhagic fevers. Library for the practicing physician—Infectious and parasitic diseases. Medicina, Leningrad, RSFSR, USSR [Russian].

Šubina, N.G., Koloncov, A.A., Makarov, V.V., 1994. Delayed type hypersensitivity and pathogenesis of viral hemorrhagic fever (African swine fever). *Bûll. Èksp. Biol. Med.* 118, 168–170 [Russian].

Targanskaâ, V.A., 1935. The clinical presentation of acute nephritis, Trudy Dal'nevostočnogo Medinstituta, vol. II(1), Khabarovsk, Khabarovsk Krai, RFSFR, USSR [Russian].

Ternovoi, V.A., Kurzhukov, G.P., Sokolov, Y.V., Ivanov, G.Y., Ivanisenko, V.A., Loktev, A.V., Ryder, R.W., Netesov, S.V., Loktev, V.B., 2003. Tick-borne encephalitis with hemorrhagic syndrome, Novosibirsk region, Russia, 1999. *Emerg. Infect. Dis.* 9, 743–746.

Viskovskij, S.V., 1951. Hemorrhagic fevers, in: Zelenina, V.F., Kuršakova, N.A. (Eds.), *Therapy Reference in Two Volumes, Fifth Edition, Significantly Corrected and Expanded*, Vol. 1. Medgiz, Moscow, RSFSR, USSR, pp. 386–388 [Russian].

Walker, D.H., McCormick, J.B., Johnson, K.M., Webb, P.A., Komba-Kono, G., Elliott, L.H., Gardner, J.J., 1982. Pathologic and virologic study of fatal Lassa fever in man. *Am. J. Pathol.* 107, 349–356.

2 General Disease Pathology in Filoviral and Arenaviral Infections

Louis M. Huzella, Jennifer A. Cann, Matthew Lackemeyer, Victoria Wahl-Jensen, Peter B. Jahrling, Jens H. Kuhn, and Donna L. Perry

CONTENTS

2.1 INTRODUCTION

Viral hemorrhagic fever (VHF) is a virus-induced syndrome with a variable incubation period, followed by a prodromal period. The prodromal period is characterized by a nonspecific influenza-like illness. Some patients recover uneventfully, whereas others develop more severe symptoms that may include edema, hemorrhage, hypotension, circulatory collapse, and neurological signs. Humans are infected with VHF viruses through close contact with other infected people through mucous membranes, broken skin, aerosolization, and inhalation or through contact with reservoir hosts depending on the virus (Čumakov, 1950; Gajdusek, 1962). Several viruses assigned to the families *Filoviridae* (filoviruses) and *Arenaviridae* (arenaviruses) cause VHFs with extraordinarily high lethality in human populations. Absolute lethal case numbers usually range only in the lower hundreds during disease outbreaks, but the environmental and ecological factors that lead to outbreaks are poorly understood (Rollin et al., 2007). Consequently, there is concern that these viruses may spread and cause more substantial outbreaks in the future. At the same time, there is little to none that can be done for patients infected with filoviruses or arenaviruses other than supportive care (filoviruses) (Bausch et al., 2008) and application of convalescent immune sera or ribavirin (arenaviruses) (Charrel et al., 2011). The lack of therapy combined with the lack of generally approved or available vaccines may cause interest in these viruses among terrorist groups or hostile state actors, making the development of medical countermeasures against and general research on filoviruses and arenaviruses top priorities among biodefense professionals (Borio et al., 2002). Filovirus and arenavirus diseases occur sporadically in rural and/or underdeveloped areas in Africa or South America. This means that human clinical and pathological data are rarely collected with the most up-to-date technology. This, in turn, poses a problem for countermeasure development, as novel candidate antivirals or vaccines against filovirus and arenavirus diseases cannot be tested in controlled human clinical trials due to ethical concerns and therefore need to be developed based on animal models only. Under the U.S. Food and Drug Administrations' 2002 "Animal Efficacy Rule" (FDA 21 CFR 601.90), data collected from nonhuman primate (NHP) studies may be used for licensure of therapeutics, provided that the animal model faithfully recapitulates the pathogenesis of the human disease condition (Hensley et al., 2011a). The ultimate goal of these animal model studies is the development of effective medical countermeasures against these viruses. However, it is a daunting task to choose an appropriate animal model that adequately reflects the human condition in the absence of extensive human clinical data (Korch et al., 2011). Here we present an overview of what is known about human filovirus and arenavirus infections.

2.2 FILOVIRUSES

The family *Filoviridae* includes two genera, *Marburgvirus* (marburgviruses) and *Ebolavirus* (ebolaviruses). A third genus, *Cuevavirus*, was recently suggested to join the family (Table 2.1). Two marburgviruses, Marburg virus (MARV) and Ravn virus (RAVV), and four ebolaviruses, Bundibugyo virus (BDBV), Ebola virus (EBOV), Sudan virus (SUDV), and Taï Forest virus (TAFV), are known to cause VHF in humans (Adams and Carstens, 2012; Kuhn et al., 2010, 2011a). The best characterized human infections are those caused by MARV, EBOV, BDBV, and SUDV.

2.2.1 IN HUMANS

Only 10 MARV and 3 RAVV disease outbreaks have been recorded since the discovery of marburgviruses in 1967, and 2 of those were laboratory infections (Kuhn, 2008). The first MARV

TABLE 2.1
Filovirus Taxonomy

Current Taxonomy and Nomenclature (Ninth ICTV Report and Updates) (Adams and Carstens, 2012; Kuhn et al., 2010, 2011a)	Previous Taxonomy and Nomenclature (Eighth ICTV Report) (Feldmann et al., 2005)
Order *Mononegavirales*	Order *Mononegavirales*
Family *Filoviridae*	Family *Filoviridae*
Genus *Marburgvirus*	Genus *Marburgvirus*
Species *Marburg marburgvirus*	Species *Lake Victoria marburgvirus*
Virus 1: Marburg virus (MARV)	Virus: Lake Victoria marburgvirus (MARV)
Virus 2: Ravn virus (RAVV)	Genus *Ebolavirus*
Genus *Ebolavirus*	Species *Cote d'Ivoire ebolavirus* [sic]
Species *Taï Forest ebolavirus*	Virus: Cote d'Ivoire ebolavirus [sic] (CIEBOV)
Virus: Taï Forest virus (TAFV)	Species *Reston ebolavirus*
Species *Reston ebolavirus*	Virus: Reston ebolavirus (REBOV)
Virus: Reston virus (RESTV)	Species *Sudan ebolavirus*
Species *Sudan ebolavirus*	Virus: Sudan ebolavirus (SEBOV)
Virus: Sudan virus (SUDV)	Species *Zaire ebolavirus*
Species *Zaire ebolavirus*	Virus: Zaire ebolavirus (ZEBOV)
Virus: Ebola virus (EBOV)	
Species *Bundibugyo ebolavirus*	
Virus: Bundibugyo virus (BDBV)	
Genus *Cuevavirus* (proposed)	
Species *Lloviu cuevavirus* (proposed)	
Virus: Lloviu virus (LLOV)	

disease outbreaks occurred in 1967 simultaneously at three locations: in Marburg and Frankfurt in West Germany and in Belgrade, Yugoslavia. The outbreak is thought to have originated from *Chlorocebus aethiops* subsp. imported from Uganda, and it resulted in the death of seven people (Kissling et al., 1968; Siegert et al., 1967). This outbreak is noteworthy because it is thus far the only one that included several patients that could immediately be treated and studied in a First World clinical setting. Consequently, the most detailed clinical, macropathological, and histopathological data on marburgvirus disease still go back to this one outbreak. Other marburgvirus disease outbreaks that are noteworthy in regard to the accumulation of human clinical and pathology data are MARV infections that occurred in 1975 in Rhodesia/South Africa (three cases, one of them lethal) (Gear et al., 1975) and in 1980 in Kenya (three cases, one of them lethal) (Smith et al., 1982) and a single case of RAVV infection that occurred in Kenya in 1987 (Johnson et al., 1996).

There have been at least 23 reported disease outbreaks due to ebolavirus infections since their initial discovery in 1976. The majority of these, 16, were caused by EBOV in the Democratic Republic of the Congo/Zaire, Gabon, and Republic of the Congo and, in the form of laboratory accidents, in the United Kingdom and USSR/Russia. Five outbreaks were due to SUDV infection and occurred in Sudan/South Sudan and Uganda. One outbreak was due to BDBV infection in Uganda, and a single nonfatal TAFV infection occurred in Côte d'Ivoire (Kortepeter et al., 2011; Kuhn et al., 2010).

Disease severity resulting from filovirus infection seems to be dependent on the virulence of the virus, the host response, and the availability of supportive care. There does not seem to be a difference in virulence between MARV, RAVV, and EBOV, which cause disease outbreaks often characterized by lethalities of 89% in larger patient cohorts. On the other hand, SUDV and BDBV cause disease outbreaks with lethalities around 28%–50% (Kuhn et al., 2011b). Disease severity is also dependent on the route of virus exposure and dose, as well as largely unexplored endogenous characteristics of the infected host (human or other primate) (Hensley and Geisbert, 2005).

Filovirus disease outbreaks are currently unpredictable, and identification of the source of infection has been largely impossible. Although marburgviruses and ebolaviruses share less than 50% genetic similarity, the symptoms associated with early infection and the hemorrhagic diatheses that occur during the disease course are clinically indistinguishable (Kuhn et al., 2010, 2011b). These symptoms, however, are not pathognomonic for infection with filoviruses, as syndromes caused by enveloped, single-stranded RNA viruses of the families *Arenaviridae*, *Bunyaviridae*, and *Flaviviridae* may cause very similar VHFs (Bray, 2005; Maganga et al., 2011; Marty et al., 2006). The incubation period of filoviruses is 2–25 days during which replication in circulating monocytes, tissue macrophages, and dendritic cells occurs (Eichner et al., 2011; Hensley and Geisbert, 2005). Terminal cases have a mean survival time of 9 days; however, patients that survive the first 2 weeks have survival rates greater than 75% (Kortepeter et al., 2011), albeit with an extended period of convalescence. Expression of cell surface tissue factor and release of a variety of cytokines and chemokines including tumor necrosis factor-α (TNF-α), eotaxin, macrophage inflammatory protein (MIP)-1α, interleukin (IL)-1β, IL-6, IL-10, IL-15, IL-16, IL-1 receptor antagonist, and reactive oxygen species and nitrogen oxide, among others, are stimulated in infected cells resulting in influenza-like symptoms during the incubation period: fever (39°C–40°C), myalgia, fatigue, headache, sore throat, and, later in the disease course, anorexia, abdominal pain, emesis, and diarrhea, followed by conjunctivitis, jaundice, and unexplained hemorrhages. Epistaxis, hematemesis, and hematochezia are reported less commonly (<10%–20% of patients) (Adjemian et al., 2011; Heymann et al., 1980; Leroy et al., 2011; Ndambi et al., 1999; Roddy et al., 2010; Smith et al., 1982; Sureau, 1989). In 25%–50% of afflicted individuals, a rash follows the initial symptoms of fever, myalgia, fatigue, and headache, and it has been variably described as maculopapular, morbilliform, and scarletinoid. This rash typically develops focally, 5–8 days after the initial influenza-like symptoms, and becomes generalized as disease progresses (Ascenzi et al., 2008; Kissling et al., 1970; Kortepeter et al., 2011; Martini, 1973; Nikiforov et al., 1994; Roddy et al., 2010; Sureau, 1989; World Health Organization, 1978; Zampieri et al., 2007); its presence is seemingly coincident with systemic viral spread. Bradycardia is sometimes described (Kissling et al., 1970; Sureau, 1989). Peripheral lymphadenopathy involving the submandibular, axillary, cervical, and nuchal lymph nodes is observed in some patients (Martini, 1973; Murphy, 1978). Lymphocytes are not infected by filoviruses, but at the time of presentation, lymphopenia with granulocytosis has been described. This progresses to leukocytosis with a left shift during the course of illness, while thrombocytopenia is consistent throughout the period of illness (Geisbert and Hensley, 2004; Kortepeter et al., 2011; van Paassen et al., 2012; Richards et al., 2000; Sanchez et al., 2004). Hyposegmented neutrophils (pseudo-Pelger–Huët cells) and atypical lymphocytes and plasma cells have also been described as lymphoblasts and plasmacytoid lymphocytes (Kissling et al., 1970; Sanchez et al., 2004). The cause of lymphopenia is not known, but it is thought to impede the adaptive immune response, and the mechanism may be apoptotic or pyroptotic lympholysis rather than being secondary to virus-induced cytopathic effects (Mahanty and Bray, 2004; Sanchez et al., 2004). Release of tissue factor and cytokines from infected cells of the monocyte/macrophage system increases vascular permeability and promotes fibrin deposition from the disease outset that, in combination with thrombocytopenia and the production fibrin degradation products, leads to a hypercoagulable state that culminates in disseminated intravascular coagulation (DIC). Associated with this coagulopathy is widespread hemorrhage seen clinically as petechiation and ecchymoses, hyposphagma, epistaxis, hematemesis, melena, vaginal bleeding, and excessive hemorrhage from venipuncture sites. Consistent with these clinical observations, prolonged prothrombin and partial thromboplastin times have been measured in afflicted patients (Ascenzi et al., 2008; Mahanty and Bray, 2004; Richards et al., 2000; Zampieri et al., 2007). In fact, fibrin degradation products have been detected in nonhuman primates (NHPs) infected with EBOV within 24 h of the onset of clinical signs (Mahanty and Bray, 2004). The manifestation of bleeding diatheses secondary to infection is a characteristic, but not consistent, finding that is associated with a poorer prognosis (Martini, 1973; van Paassen et al., 2012; Roddy et al., 2010; Smith et al., 1982; Sureau, 1989).

Patients may also exhibit central and peripheral nervous system signs such as paresthesia, agitation, delirium, disinhibition, or convulsions, but it is often not known if these clinical signs are the result of intracerebral or intracranial hemorrhage or viral encephalitis (Adjemian et al., 2011; Martini, 1973). Cerebral edema followed by herniation was confirmed in one patient infected with MARV using transcranial Doppler testing (van Paassen et al., 2012), but it has also been reported without such conformation in others (Kissling et al., 1970). A female Gabonese patient infected with EBOV died following an intracranial hemorrhage (Richards et al., 2000). Excess protein (0.68 g/l) and WBCs (0–24) have been reported in the cerebral spinal fluid from patients infected with MARV, but histopathologic examination of the brain has varied from no significant histopathologic lesions to vascular congestion and hemorrhage to lesions consistent with a viral encephalitis, including lymphocytic perivascular infiltrates with gliosis and widespread microglial nodules. In most cases, histopathologic findings were not correlated with the patients' clinical neurologic exams. Psychosis without further description developed in one patient during convalescence, several weeks after the initial illness with MARV infection (Martini, 1973). It is not known whether filoviruses infect neurons or supporting glial cells (Jahrling, 2008; Kissling et al., 1970; Murphy, 1978). Filoviruses are able to infect a wide variety of parenchymal tissues including the spleen, liver, kidneys, adrenal cortical cells, and gonads, and the cytopathic effects associated with infection lead to widespread areas of necrosis (Kissling et al., 1970; Kortepeter et al., 2011; Zampieri et al., 2007). Women exhibit genital bleeding (Martini, 1973) and pregnant women abort (Solbrig and Naviaux, 1997). However, a pathologic assessment of the placenta or abortus has not been described, and it is therefore not known if there is a direct virus-associated placentitis or transplacental transmission.

Both marburgvirions and ebolavirions have been identified in hepatocytes, and hepatic infection results in terminal liver failure with reported mild to marked elevations in aspartate aminotransferase (AST), alanine aminotransferase (ALT), gamma glutamyltransferase, alkaline phosphatase, and lactate dehydrogenase concentrations (Adjemian et al., 2011; van Paassen et al., 2012; Rollin et al., 2007; Smith et al., 1982) and decreases in plasma protein concentrations (Kissling et al., 1970).

Hepatic failure is likely associated with the coagulopathies, secondary to loss of production of vitamin K–associated coagulation factors. Additional abnormalities secondary to systemic spread of these viruses and subsequent parenchymal organ infection include increases in pancreatic enzyme concentrations (serum amylase) and the muscle leakage enzyme, creatine kinase (Martini, 1973; van Paassen et al., 2012). Renal function has been reported to be preserved early in the course of infection. However, by the end of the first week of infection, azotemia and terminal oliguria commonly develop (Ndambi et al., 1999; Richards et al., 2000; Rollin et al., 2007; Smith et al., 1982). Hematuria and proteinuria have also been reported (Kortepeter et al., 2011). In the kidneys, MARV particles have been found in connective tissue macrophages and interstitial fibroblast-like supporting connective tissue cells, but not within the vascular glomerular endothelium or the tubular epithelium of the cortex and medulla (Geisbert and Jaax, 1998; Smith et al., 1982).

Filovirus disease outbreaks have yielded few autopsy reports and as the veterinary pathologist Frederick Murphy stated, "So few specimens of tissue from fatal cases of Ebola virus disease have been available for pathologic study that no description can be considered representative and any analysis of pathogenic mechanisms should be recognized as speculative" (Murphy, 1978; Zaki and Goldsmith, 1999). A total of eight autopsies of marburgvirus disease casualties have thus far been reported, five on patients that succumbed in 1967 and three on the lethal cases of 1975, 1980, and 1987. Twenty-one autopsies were performed on EBOV cases (3 in 1976 and 18 in 1995) and two on SUDV cases in 1976 (Korch et al., 2011; Kuhn, 2008). These few postmortem exams performed revealed, not surprisingly, widespread hemorrhage in the mucous membranes, skin, and parenchymal organs, including the stomach and intestines (Kissling et al., 1970; Kortepeter et al., 2011). Lymphadenopathy, splenomegaly, and brain swelling have also been described grossly (Murphy, 1978). The hepatic lesions are described as affecting all lobes and microscopically consist of random single cell to panlobular hepatocellular necrosis with prominent intracytoplasmic inclusion bodies (ICIBs) (Kissling et al., 1970; World Health Organization, 1978). Necrotic foci were described as growing by expansion and containing karyorrhectic debris. The degree of inflammation was considered disproportionately mild when compared to the extent of necrosis, although the character of the inflammatory cell infiltrate is not well described. In addition to the ICIBs, numerous councilman bodies were seen in hepatocytes in areas of hepatocellular coagulative necrosis, but differentiation from viral ICIBs has been achieved through transmission electron microscopy (TEM), confirming the presence of filamentous filovirions (Murphy, 1978; World Health Organization, 1978) in the space of Disse, and intracytoplasmically within hepatocytes and Kupffer cells (Geisbert and Jahrling, 1995). In an additional study of a single liver from a MARV disease patient, similar hepatocellular necrosis was seen with ICIBs, and these foci were immunohistochemically (IHC) positive for MARV antigen (Martini, 1973). In the same patient, diffuse, severe necrosis of the adrenal gland at the corticomedullary junction and multifocal necrosis within the remainder of the cortex and medulla was seen with eosinophilic ICIBs. These lesions also tested IHC positive for MARV antigen, and the degree of inflammation was also reported to be considered disproportionately mild when compared to the extent of necrosis, as has been reported in the liver. An almost identical lesion within the adrenal gland, described as corticomedullary necrosis, was reported in the adrenal gland of a 15-year-old male infected with RAVV in Kenya and confirmed via TEM. Filovirions were seen in interstitial fibroblast-like cells, adrenocortical cells, and within some subendothelial macrophages (Geisbert and Jaax, 1998). It is speculated that hypotension and shock associated with terminal cases of infection may be secondary to adrenocortical necrosis with loss of mineralocorticoids (aldosterone) and corticosteroids (cortisol) crucial for the maintenance of normal electrolyte concentrations and vascular tone leading to a terminal Addisonian-like crisis (Zampieri et al., 2007). Within the pancreas, identical lesions were also described in these two separate marburgvirus disease patients (Geisbert and Jaax, 1998; Martini, 1973). Both suffered from multifocal bacterial emboli/microabscessation. Interestingly, the neuroendocrine cells composing the islets of Langerhans tested IHC positive for RAVV antigen with the staining seen most intensely in beta cells, confirmed via TEM (Geisbert and Jaax, 1998). Epigastric pain in one patient with EBOV was thought to be secondary to pancreatitis induced by infection (Sureau, 1989).

Similar, randomly distributed, multifocal to coalescing necrosis characterizes the lesions seen in lymphoid tissues and the gonads in filovirus-infected individuals. Necrosis within the cortex and medulla of lymph nodes with "eosinophilic thrombic debris left in situ" has been described. Necrosis and congestion with fibrin deposition have also been described in both the red and the white pulp of the spleen (Geisbert and Jaax, 1998; Murphy, 1978). In the lungs, predominant lesions include edema, congestion, and hemorrhage with an increase in intra-alveolar macrophages. Immunohistochemistry has revealed few antigen-positive pneumocytes, interstitial monocytes, and intra-alveolar macrophages, but subsequent TEM to confirm the IHC results revealed no virus particles. Similarly, no viral antigen or virus particles were detected in the cardiac myocytes using IHC and TEM (Geisbert and Jaax, 1998).

2.2.2 In Nonhuman Primates

The sporadic nature of filovirus disease outbreaks and their high pathogenicity has made collection and analysis of samples from naturally occurring outbreaks difficult (Korch et al., 2011; Sanchez et al., 2004). This has necessitated the development of suitable animal models that recapitulate the human condition. An additional limitation is the high-containment facilities required to study them. NHPs, estimated to have a genetic similarity of 95%–99% to humans, are currently the preferred animal models to study filovirus pathogenesis, although that may change in the future (Korch et al., 2011).

Filovirus pathogenesis studies usually include NHPs assigned to Asian (*Macaca mulatta*, *Macaca fascicularis*), African (*C. aethiops* and *Papio hamadryas*), and South American species (*Callithrix jacchus*) (Carrion et al., 2011; Kuhn, 2008). The vast majority of NHP studies designed to elucidate different aspects of filovirus pathogenesis have been performed using EBOV and MARV. The clinical course varies somewhat with variant, dose, and route of infection, but it has been reported that a higher dose at the time of inoculation and an earlier and higher peak viremia postinoculation are associated with more severe clinical signs and a more rapid clinical course (Hensley and Geisbert, 2005; Kortepeter et al., 2011; Rowe et al., 1999), a finding that has been reported in a human outbreak of SUDV infection (Sanchez et al., 2004).

Cynomolgus macaques (*M. fascicularis*) and rhesus macaques (*M. mulatta*) infected with filoviruses typically develop a cutaneous rash similar to that seen in humans that starts focally and progresses to a generalized distribution approximately 5 days postinoculation depending on dose, animal species, and route of infection. To the contrary, a cutaneous rash is the not typically seen following filovirus inoculation of *Chlorocebus aethiops* subsp. (Hensley et al., 2011a; Lub et al., 1995; Ryabchikova et al., 1999). Additional clinical signs, also commonly observed in human cases of filovirus infection, include increased body temperature, anorexia, depression, and diarrhea (Hensley et al., 2011a; Lub et al., 1995). A peripheral lymphadenopathy that becomes evident 3–4 days postinoculation involving the axillary, inguinal, and submandibular lymph nodes is also consistent (Hensley et al., 2011a). In contrast to the bradycardia sometimes described during human infection, a sustained tachycardia has been reported in rhesus monkeys during the course of infection (Kortepeter et al., 2011). A neutrophilic leukocytosis has been described early in infection that progresses to a left shift with a concomitant lymphopenia and monocytopenia (Hensley et al., 2011a; Kortepeter et al., 2011; Lub et al., 1995). Lymphocytosis occurs in some survivors (Fisher-Hoch et al., 1992). Thrombocyte concentrations decrease through day 6 postinoculation (from $323 \times 10^3/mm^3$ to $194 \times 10^3/mm^3$) but may rebound later in the disease course, or the thrombocytopenia may persist (Hensley et al., 2011a; Kortepeter et al., 2011; Lub et al., 1995). Giant platelets have been reported during the course of infection (Fisher-Hoch et al., 1992). Coagulopathies are seen clinically as petechial and ecchymotic hemorrhages and persistent hemorrhage at venipuncture sites and clinicopathologically as prolonged prothrombin time (PT) and activated partial thromboplastin time (aPTT) and elevations in fibrin degradation product concentrations (D-dimers) (Hensley et al., 2011a; Lub et al., 1995). These changes in blood cell count and coagulation parameters are also described in human cases of filovirus infection. Frank hemorrhage is a feature most pronounced in hamadryad baboons (*P. hamadryas*), whereas fibrin deposition is the most prominent histopathologic feature seen in *Chlorocebus aethiops* subsp. (Hensley and Geisbert, 2005; Ryabchikova et al., 1999). In macaques, fibrin deposition is seen as early as day 2 postinoculation in the marginal zone of the spleen, and by day 5 fibrin deposits are so dense, they obscure the normal splenic architecture and can be found within the parenchyma and vessels of most tissues including the sinusoids of the liver (Geisbert et al., 2003b). Elevations in hepatic enzyme concentrations, including lactate dehydrogenase (Fisher-Hoch et al., 1992), serum AST, ALT, gamma-glutamyltransferase, and alkaline phosphatase, with widespread hepatic viral infection and hepatocellular necrosis, are also seen and are consistent with changes seen in human cases (Hensley et al., 2011a; Lub et al., 1995; Ryabchikova et al., 1999). Eosinophilic ICIBs are also reported (Hensley et al., 2011a). Replicating ebolaviruses have been demonstrated in the liver and adrenal glands of infected *Chlorocebus aethiops* subsp. 3–4 days postinoculation (Ryabchikova et al., 1999), and the highest titers of MARV have been reported in the liver, spleen, adrenal glands, and lymph nodes of macaques, with higher titers having a direct correlation with the degree of tissue destruction (Hensley et al., 2011a). Necrosis of the splenic white pulp and lymph nodes is consistent with that seen in human filovirus infection (Hensley et al., 2011a; Ryabchikova et al., 1999). A doubling of blood urea nitrogen (BUN) and creatinine concentrations three times above the normal range, consistent with terminal renal failure, has been described in macaques late in the disease course following EBOV inoculation, as has been described in humans. These increases in BUN and creatinine concentrations accompanied a terminal hypotension and lactic acidosis (metabolic acidosis) and fulfill the criteria of terminal shock (Hensley and Geisbert, 2005; Kortepeter et al., 2011). Histopathologically, renal lesions are reported as extensive thrombosis of glomerular capillaries, edema, and necrosis of glomeruli (Ryabchikova et al., 1999). As has been reported in human filovirus infections, a disproportionately minimal inflammatory response is seen in all NHPs tissues affected (Ryabchikova et al., 1999).

As in humans, cells of the mononuclear phagocytic system are well-recognized primary targets of virus infection and systemic spread (Ryabchikova et al., 1999; Stroher et al., 2001). In macaques, MARV has been detected in approximately 4% of circulating monocytes on day 2 postinoculation and 82% on day 8 postinoculation (Fritz et al., 2008). EBOV infection of monocytes and macrophages has been shown to increase the expression of tissue factor both *in vitro* and *in vivo* (Geisbert et al., 2003b). The role of endothelial cell infection in filovirus pathogenesis remains under debate. Infection of endothelial cells occurs readily *in vitro*, but gross and ultrastructural pathologic examination of human cases and additional insight gained from the use of NPH animal models in sequential sampling studies have failed to demonstrate evidence of vasculitis or changes in vascular endothelial cell ultrastructure, despite endothelial cell infection, which occurs only late in the disease course, after the onset of a clinically observable coagulopathy (Geisbert and Jaax, 1998; Geisbert et al., 2003c; Hensley and Geisbert, 2005; Schnittler et al., 1993). In fact, productive infection of endothelial cells *in vitro* does not result in significant cytopathology or induce apoptosis (Geisbert et al., 2003c). These findings indicate that destruction of the endothelium is not the underlying cause of the widespread hemorrhage and profound coagulopathies seen following filovirus infection. Widespread infection of circulating monocytes and macrophages results in systemic release of tissue factor, which in concert with the profoundly destructive infection of the liver, critical for the production of vitamin K–dependent coagulation factors, and of the adrenal glands, essential for the production of both corticosteroids and vasoactive amines (epinephrine and norepinephrine), appears to play a more central role in the development of the hypercoagulable state and hemorrhagic diatheses that are observed during the course of infection (Geisbert et al., 2003a,b; Rollin et al., 2007; Stroher et al., 2001). Interestingly, it has been proposed that filoviruses may enter endothelial cells from the basolateral surfaces, and fenestrated endothelial cells, such as those lining the sinusoids of the liver and adrenal gland, may provide easier access to the basolateral surface, making them easy, early targets of infection (Hensley and Geisbert, 2005). However, such fenestrated capillaries are not exclusive to the liver and adrenal glands; they are also present in the lymph nodes, spleen, and bone marrow.

Although endothelial cells do not appear to be primary targets of filovirus infection, endothelial cell dysfunction secondary to the release of a myriad of cytokines from parenchymal tissues, circulating leukocytes, and the endothelium itself during the course of infection is likely (Geisbert et al., 2003c). However, cytokine measures taken from sera during both NHP studies and human filovirus disease outbreaks are highly variable, and this makes interpretation of their precise role in the pathogenesis of filovirus infection difficult to determine. Increases in TNF-α concentration are most commonly reported to be modest but in combination with decreases in serum albumin secondary to extensive hepatic damage may work in concert, increasing vascular permeability and reducing plasma oncotic pressure. These derangements would be exacerbated by the vasoplegic effects of increased vessel-derived inducible nitric oxide (iNOS), which has also been shown to be induced by filovirus infection (Geisbert et al., 2003a,c; Hensley and Geisbert, 2005; van Paassen et al., 2012; Sanchez et al., 2004). Increases in IL-6 and IL-10 concentrations have also been reported, but the data are often conflicting with elevated, reduced, and undetectable concentrations reported in both survivors and those that succumb to infection (Baize et al., 2002; van Paassen et al., 2012; Sanchez et al., 2004). Similarly, type I interferons (INF-α) concentrations have been shown to increase in some cases but have been undetectable in others (Baize et al., 2002; Fritz et al., 2008; Villinger et al., 1999). Increases in monocyte chemoattractant molecule MIP-1α concentration occur in both humans and NHPs with filovirus infection but have been reported not to differ significantly in those who survive versus those who succumb to infection (Baize et al., 2002; Geisbert et al., 2003b). Further research is needed to decipher the role of each of these cytokines in the progression of filovirus infection and how their concentrations are associated with the progression of disease clinically. Cytokine measures are rarely, if ever, reported in the context of the circulating white blood cell numbers, which is problematic as both circulating and resident immune cells are where the majority of these detectable circulating cytokines originate.

2.3   ARENAVIRUSES

The family *Arenaviridae* includes single genus, *Arenavirus*, which currently includes 25 accepted species (Table 2.2) (Adams and Carstens, 2012; Salvato et al., 2011). Human VHFs are caused by two viruses of the Old World arenavirus serocomplex, Lassa virus (LASV) and Lujo virus (LUJV), and five viruses of the group B New World arenavirus serocomplex, Chapare virus (CHPV), Guanarito virus (GTOV), Junín virus (JUNV), Machupo virus (MACV), and Sabiá virus (SABV). Each of these viruses is carried in nature by a specific type of rodent, and it is the geographic distribution of the particular rodents that defines where in the world a particular arenavirus hemorrhagic fever may occur (Shurtleff et al., 2012).

The most significant arenavirus to cause VHF in humans is LASV, the causative agent of Lassa fever, which is endemic in certain regions of Africa. LASV annually infects from 100,000 to as many as 250,000 people in West Africa, killing up to 10,000 and leaving three times that many deaf (Jahrling, 2008). The reservoir of LASV is the Natal mastomys (*Mastomys natalensis*). JUNV is arguably the second most significant VHF-causing arenavirus. It causes Junín/Argentinian hemorrhagic fever (AHF) when transmitted to humans from its natural reservoir, the drylands laucha (*Calomys musculinus*). Other important South American hemorrhagic fever arenaviruses include MACV, the causative agent of Bolivian hemorrhagic fever (BHF); GTOV, the causative agent of "Venezuelan hemorrhagic fever"; and SABV, the causative agent of "Brazilian hemorrhagic fever" (Moraz and Kunz, 2011).

LASV is transmitted through contact with urine or feces of infected reservoirs. Human to human transmission can also occur through contact with contaminated blood and other body fluids. Clinical findings range from an asymptomatic infection to fatal hemorrhagic fever. After an incubation period of up to 18 days, fever, weakness, and general malaise develop. Other frequently

TABLE 2.2
Arenavirus Taxonomy

Taxa	Viruses (Abbreviations)
Family *Arenaviridae*	
Genus *Arenavirus*	
Old World Arenaviruses	
Species not yet proposed	Dandenong virus (DANV)
Species not yet proposed	Gbagroube virus (?)
Species *Ippy virus*	Ippy virus (IPPYV)
Species not yet proposed	Kodoko virus (KODV)
Species *Lassa virus*	Lassa virus (LASV)
Species *Lujo virus*	Lujo virus (LUJV)
Species *Luna virus*	Luna virus (LUNV)
Species *Lymphocytic choriomeningitis virus*	Lymphocytic choriomeningitis virus (LCMV)
Species not yet proposed	Menekre virus (?)
Species not yet proposed	Merino Walk virus (MWV)
Species *Mobala virus*	Mobala virus (MOBV)
Species *Mopeia virus*	Mopeia virus (MOPV)
	Morogoro virus (MORV)
New World Arenaviruses	
Group A	
Species *Allpahuayo virus*	Allpahuayo virus (ALLV)
Species *Flexal virus*	Flexal virus (FLEV)
Species *Paraná virus*	Paraná virus (PARV)
Species *Pichinde* [sic] *virus*	Pichindé virus (PICV)
Species *Pirital virus*	Pirital virus (PIRV)
Group B	
Species *Amapari* [sic] *virus*	Amaparí virus (AMAV)
Species *Chapare virus*	Chapare virus (CHPV)
Species *Cupixi virus*	Cupixi virus (CPXV)
Species *Guanarito virus*	Guanarito virus (GTOV)
Species *Junín virus*	Junín virus (JUNV)
Species *Machupo virus*	Machupo virus (MACV)
Species not yet proposed	Ocozocoautla de Espinosa virus (OCEV)
Species *Sabiá virus*	Sabiá virus (SABV)
Species *Tacaribe virus*	Tacaribe virus (TCRV)
Group C	
Species *Latino virus*	Latino virus (LATV)
Species *Oliveros virus*	Oliveros virus (OLVV)
Group Rec	
Species *Bear Canyon virus*	Bear Canyon virus (BCNV)
Species not yet proposed	Big Brushy Tank virus (BBTV?)
Species not yet proposed	Catarina virus (CTNV?)
Species not yet proposed	Skinner Tank virus (SKTV?)
Species *Tamiami virus*	Tamiami virus (TAMV)
Species not yet proposed	Tonto Creek virus (TTCV?)
Species *Whitewater Arroyo virus*	Whitewater Arroyo virus (WWAV)

Sources: Adams, M.J. and Carstens, E.B., *Arch. Virol.*, 157, 1411, 2012; Salvato, M.S. et al., Family *Arenaviridae*, in *Virus Taxonomy—Ninth Report of the International Committee on Taxonomy of Viruses*, King, A.M.Q., Adams, M.J., Carstens, E.B., and Lefkowitz, E.J., Eds., pp. 715–723, Elsevier/Academic Press, London, U.K., 2011.

occurring symptoms include headache, sore throat, cough, chest pain, myalgia, nausea, vomiting, and diarrhea. A poor prognosis is associated with the presence of facial edema and pleural effusions and is likely due to increased vascular permeability. Coma is another possible sequela and is typically preceded by signs of encephalopathy and seizures. Imminent death is reportedly associated with pulmonary edema, respiratory distress, bleeding from the mucus membranes, and shock. In those recovering, the viremia terminates approximately 2–3 weeks after the incubation period (Moraz and Kunz, 2011). Sensorineural hearing loss is reported to occur in both acute and convalescent stages of Lassa fever and is thought to be associated with the body's immune response to LASV (Okokhere et al., 2009). In experimental LASV infections of NHPs, the clinical signs of illness included increased body temperature, mild dehydration, and anorexia (Hensley et al., 2011b). The primary route of infection for JUNV is thought to be through skin abrasions; through inhalation of aerosolized urine, saliva, or blood of infected reservoir rodents; and through gastrointestinal mucosa. The incubation period is up to 2 weeks during which time the virus typically replicates in the lungs, enters the lymphoid system, and then spreads to the endothelium. Similar to other VHFs, initial symptoms are nonspecific, with fever, malaise, asthenia, myalgia, mild neurological symptoms, rashes of the skin and mucosa, and lymph node swelling. Within 10 days of disease onset, worsening cardiovascular, gastrointestinal, renal, and neurological symptoms are present, likely due to derangements of the endothelium and vasculature. Clinically there may be signs of severe hemorrhage, such as hematemesis, melena, hemoptysis, epistaxis, hematoma formation, uterine bleeding, and hematuria. In severe cases, the neurological symptoms may progress to include areflexia, muscular hypotonia, ataxia, and tremors, eventually followed by delirium, seizures, and coma (Gomez et al., 2011). MACV and GTOV infections are very similar to AHF both in clinical presentation and pathology, and there are a very limited number of clinical reports. In NHPs experimentally infected with MACV, clinical symptoms were typically more severe and evident by day 6 postinfection and consisted of conjunctivitis, depression, anorexia, dehydration, and elevated body temperature. Additional findings included constipation, followed sporadically by diarrhea; chronic spasms; nasal discharge, occasionally hemorrhagic; erythematous facial rash; and abdominal rash. A majority of surviving NHPs developed neurological signs including intention tremors, nystagmus, incoordination, paresis, and coma (Eddy et al., 1975).

Clinical pathology findings in Lassa fever include lymphopenia, thrombocytopenia, and elevated serum AST, serum ALT, and BUN concentrations when associated with hepatic and renal involvement. In addition, serum IL-8 and interferon (IFN)-inducible protein (IP) concentrations are low to undetectable in fatal LASV infections, whereas they are increased in patients with nonfatal LASV illness. Concentration increases in TNF-α are atypical in either fatal or nonfatal cases. IFN-γ, IL-6, IL-12, and RANTES concentrations are elevated in febrile patients (Mahanty et al., 2001). The most significant clinical pathology findings in AHF include thrombocytopenia, lymphopenia, and neutropenia, with a decrease in serum complement activity. Increased concentrations of IFNs, TNF-α (in contrast to LASV), factor V, von Willebrand factor (vWF), fibrinogen, prothrombin fragment $1+2$, thrombin–antithrombin complexes, plasminogen activator, and other inflammatory mediators have also been reported. In contrast, analysis of cytokine data in NHPs experimentally infected with LASV shows increases in monocyte chemoattractant protein (MCP)-1, eotaxin, IL-6, and IL-1β concentrations, but no increase in TNF-α concentration (similar to LASV findings) (Hensley et al., 2011b). Fibrin deposits are absent in circulation, i.e., there is no evidence of DIC. Hemorrhages noted in AHF can be attributed to thrombocytopenia and cytokine-induced endothelial damage (Kunz, 2009; Moraz and Kunz, 2011). Similarly, in experimental infections of NHPs infected with LASV, elevations of AST and ALT concentrations were reported in later stages of the disease as well as BUN concentration increases (Callis et al., 1982; Hensley et al., 2011b).

In postmortem examinations of LASV patients, the reported findings were nonspecific and consisted of soft tissue edema, congestion of viscera, and petechia of multiple tissues, especially the gastrointestinal tract. Other findings included pleural effusion, ascites, and small intestinal luminal hemorrhage. Similar findings were noted in human GTOV infection (Child et al., 1967;

McCormick et al., 1986; Winn and Walker, 1975). In experimental LASV infection in NHPs, postmortem findings typically included generalized lymphadenopathy and congestion of lymph nodes and visceral organs, including the liver, kidneys, and ileocecal junction (Hensley et al., 2011b). In experimental infections of NHPs with MACV, there were inconsistent hemorrhagic manifestations of disease including petechial hemorrhage and exudative rashes in the skin of the face, thorax, abdomen, and medial aspects of the thighs and forelimbs. There was also a generalized lymphadenopathy, and livers were often yellow, mottled, and friable (Terrell et al., 1973). Gross lesions often associated with fatal AHF include hemorrhages in the gastrointestinal mucosa, in subcutaneous tissues, and in multiple organs such as the liver, kidney, and lungs. Within the kidneys, there is often necrosis of the renal tubules and papillae. In the liver, there are often scattered necrotic hepatocytes. Occasionally there are CNS lesions, such as meningeal edema and perivascular hemorrhage, which reportedly coincide with neurological symptoms such as hyporeflexia and mental confusion (Gomez et al., 2011). Secondary bacterial infections resulting in pneumonia often occur as sequelae to AHF.

Histopathology is available for a limited number of organs from a few human cases of arenaviral hemorrhagic fever. Tissues, including the heart, skeletal muscle, lungs, kidneys, spleen, lymph nodes, liver, gastrointestinal tract, pancreas, brain, adrenal gland, uterus, ovary, placenta, skin, and breast, were examined. The most consistent lesion noted was multifocal hepatocellular necrosis, with councilman-like bodies, and a mononuclear cellular infiltrate (Child et al., 1967; McCormick et al., 1986; Winn and Walker, 1975). In NHPs experimentally infected with LASV, there were lymphoplasmacytic and neutrophilic infiltrates found in the liver, with some hepatic necrosis similar to what is described in human pathology. Other noteworthy lesions included interstitial pneumonia and neuronal necrosis, with little pathology in other organ systems (Callis et al., 1982; Hensley et al., 2011b). Experimental MACV infection in NHPs revealed hepatic necrosis, enteritis with necrosis, necrosis of the adrenal cortex, and necrosis of the epithelium (Terrell et al., 1973). Other histopathologic lesions reported in humans include myocarditis, interstitial pneumonia, and splenic necrosis of the marginal zone. The remaining histopathologic findings were nonspecific (Moraz and Kunz, 2011; Winn and Walker, 1975). Lesions consistently noted in fatal AHF included necrosis of the cortical and paracortical areas of lymph nodes, white pulp necrosis of the spleen, and generalized bone marrow depletion.

A key factor in of the pathogenesis of Lassa fever is suppression of the host's cellular and humoral immune system preventing an effective antiviral immune response. As with LASV, virally induced immunosuppression is also an important feature of fatal South American VHFs. Arenavirus nucleoproteins (NPs) act as a type I IFN antagonist and block the activation of beta IFN and IFN regulatory factor 3 (IRF-3)–dependent promoters, as well as the nuclear translocation of IRF-3. Other contributing factors are the potential for arenaviruses to impair dendritic cell function and for them to alter innate immune response (Martinez-Sobrido et al., 2007). In Lassa fever, the severity of viremia is a predictive factor for the outcome of infection (Kunz, 2009). In addition to the ability of arenavirus NP to block type I IFN induction, the matrix protein Z has the ability to disrupt the activation of retinoic acid–inducible gene I (RIG-I). RIG-I is an important cytoplasmic RNA helicase that induces the production of type I IFNs (IFN-α and IFN-β) when they recognize the genomic RNA of viruses (Fan et al., 2010). In experimental LASV of NHPs, dendritic cells were identified as targets of infection by the virus, as well as Kupffer cells, hepatocytes, adrenal cortical cells, and endothelial cells (Hensley et al., 2011b).

DISCLAIMER

The content of this publication does not necessarily reflect the views or policies of the institutions and companies affiliated with the authors. LH, JAC, and DLP performed this work as employees of Charles River Laboratories; JHK and VWJ as employees of Tunnell Consulting, Inc.; and ML as an employee of Lovelace Respiratory Research Institute; all were subcontractors to Battelle Memorial Institute under its prime contract HHSN2722007000161 with NIAID.

REFERENCES

Adams, M. J. and Carstens, E. B. (2012). Ratification vote on taxonomic proposals to the International Committee on Taxonomy of Viruses. *Arch Virol* **157**(7), 1411–1422.

Adjemian, J., Farnon, E. C., Tschioko, F., Wamala, J. F., Byaruhanga, E., Bwire, G. S., Kansiime, E. et al. (2011). Outbreak of Marburg hemorrhagic fever among miners in Kamwenge and Ibanda districts, Uganda, 2007. *J Infect Dis* **204**(Suppl 3), S796–S799.

Ascenzi, P., Bocedi, A., Heptonstall, J., Capobianchi, M. R., Di Caro, A., Mastrangelo, E., Bolognesi, M., and Ippolito, G. (2008). Ebolavirus and Marburgvirus: Insight the Filoviridae family. *Mol Aspects Med* **29**(3), 151–185.

Baize, S., Leroy, E. M., Georges, A. J., Georges-Courbot, M. C., Capron, M., Bedjabaga, I., Lansoud-Soukate, J., and Mavoungou, E. (2002). Inflammatory responses in Ebola virus-infected patients. *Clin Exp Immunol* **128**(1), 163–168.

Bausch, D. G., Sprecher, A. G., Jeffs, B., and Boumandouki, P. (2008). Treatment of Marburg and Ebola hemorrhagic fevers: a strategy for testing new drugs and vaccines under outbreak conditions. *Antiviral Res* **78**(1), 150–161.

Borio, L., Inglesby, T., Peters, C. J., Schmaljohn, A. L., Hughes, J. M., Jahrling, P. B., Ksiazek, T. et al. and Working Group on Civilian, B. (2002). Hemorrhagic fever viruses as biological weapons: medical and public health management. *JAMA* **287**(18), 2391–2405.

Bray, M. (2005). Pathogenesis of viral hemorrhagic fever. *Curr Opin Immunol* **17**(4), 399–403.

Callis, R. T., Jahrling, P. B., and DePaoli, A. (1982). Pathology of Lassa virus infection in the rhesus monkey. *Am J Trop Med Hyg* **31**(5), 1038–1045.

Carrion, R., Jr., Ro, Y., Hoosien, K., Ticer, A., Brasky, K., de la Garza, M., Mansfield, K., and Patterson, J. L. (2011). A small nonhuman primate model for filovirus-induced disease. *Virology* **420**(2), 117–124.

Charrel, R. N., Coutard, B., Baronti, C., Canard, B., Nougairede, A., Frangeul, A., Morin, B., Jamal, S., Schmidt, C. L., Hilgenfeld, R., Klempa, B., and de Lamballerie, X. (2011). Arenaviruses and hantaviruses: From epidemiology and genomics to antivirals. *Antiviral Res* **90**(2), 102–114.

Child, P. L., MacKenzie, R. B., Valverde, L. R., and Johnson, K. M. (1967). Bolivian hemorrhagic fever. A pathologic description. *Arch Pathol* **83**(5), 434–445.

Čumakov, M. P. (1950). Etiology, epidemiology and prophylaxis of hemorrhagic fever. *Čas Lék Česk* **89**(51), 1428–1430.

Eddy, G. A., Scott, S. K., Wagner, F. S., and Brand, O. M. (1975). Pathogenesis of Machupo virus infection in primates. *Bull World Health Organ* **52**(4–6), 517–521.

Eichner, M., Dowell, S. F., and Firese, N. (2011). Incubation period of Ebola hemorrhagic virus subtype Zaire. *Osong Public Health Res Perspect* **2**(1), 3–7.

Fan, L., Briese, T., and Lipkin, W. I. (2010). Z proteins of new world arenaviruses bind RIG-I and interfere with type I interferon induction. *J Virol* **84**(4), 1785–1791.

Feldmann, H., Geisbert, T. W., Jahrling, P. B., Klenk, H.-D., Netesov, S. V., Peters, C. J., Sanchez, A., Swanepoel, R., and Volchkov, V. E. (2005). Family *Filoviridae*. In *Virus Taxonomy–Eighth Report of the International Committee on Taxonomy of Viruses* (C. M. Fauquet, M. A. Mayo, J. Maniloff, U. Desselberger, and L. A. Ball, Eds.), pp. 645–653. Elsevier/Academic Press, San Diego, CA.

Fisher-Hoch, S. P., Brammer, T. L., Trappier, S. G., Hutwagner, L. C., Farrar, B. B., Ruo, S. L., Brown, B. G. et al. (1992). Pathogenic potential of filoviruses: role of geographic origin of primate host and virus strain. *J Infect Dis* **166**(4), 753–763.

Fritz, E. A., Geisbert, J. B., Geisbert, T. W., Hensley, L. E., and Reed, D. S. (2008). Cellular immune response to Marburg virus infection in cynomolgus macaques. *Viral Immunol* **21**(3), 355–363.

Gajdusek, D. C. (1962). Virus hemorrhagic fevers. Special reference to hemorrhagic fever with renal syndrome (epidemic hemorrhagic fever). *J Pediatr* **60**, 841–857.

Gear, J. S., Cassel, G. A., Gear, A. J., Trappler, B., Clausen, L., Meyers, A. M., Kew, M. C., Bothwell, T. H., Sher, R., Miller, G. B., Schneider, J., Koornhof, H. J., Gomperts, E. D., Isaacson, M., and Gear, J. H. (1975). Outbreak of Marburg virus disease in Johannesburg. *Br Med J* **4**(5995), 489–493.

Geisbert, T. W. and Hensley, L. E. (2004). Ebola virus: new insights into disease aetiopathology and possible therapeutic interventions. *Expert Rev Mol Med* **6**(20), 1–24.

Geisbert, T. W., Hensley, L. E., Larsen, T., Young, H. A., Reed, D. S., Geisbert, J. B., Scott, D. P., Kagan, E., Jahrling, P. B., and Davis, K. J. (2003a). Pathogenesis of Ebola hemorrhagic fever in cynomolgus macaques: Evidence that dendritic cells are early and sustained targets of infection. *Am J Pathol* **163**(6), 2347–2370.

Geisbert, T. W. and Jaax, N. K. (1998). Marburg hemorrhagic fever: Report of a case studied by immunohisto-chemistry and electron microscopy. *Ultrastruct Pathol* **22**(1), 3–17.

Geisbert, T. W. and Jahrling, P. B. (1995). Differentiation of filoviruses by electron microscopy. *Virus Res* **39**(2–3), 129–150.

Geisbert, T. W., Young, H. A., Jahrling, P. B., Davis, K. J., Kagan, E., and Hensley, L. E. (2003b). Mechanisms underlying coagulation abnormalities in Ebola hemorrhagic fever: overexpression of tissue factor in primate monocytes/macrophages is a key event. *J Infect Dis* **188**(11), 1618–1629.

Geisbert, T. W., Young, H. A., Jahrling, P. B., Davis, K. J., Larsen, T., Kagan, E., and Hensley, L. E. (2003c). Pathogenesis of Ebola hemorrhagic fever in primate models: Evidence that hemorrhage is not a direct effect of virus-induced cytolysis of endothelial cells. *Am J Pathol* **163**(6), 2371–2382.

Gomez, R. M., Jaquenod de Giusti, C., Sanchez Vallduvi, M. M., Frik, J., Ferrer, M. F., and Schattner, M. (2011). Junin virus. A XXI century update. *Microbes Infect* **13**(4), 303–311.

Hensley, L. E., Alves, D. A., Geisbert, J. B., Fritz, E. A., Reed, C., Larsen, T., and Geisbert, T. W. (2011a). Pathogenesis of Marburg hemorrhagic fever in cynomolgus macaques. *J Infect Dis* **204**(Suppl 3), S1021–S1031.

Hensley, L. E. and Geisbert, T. W. (2005). The contribution of the endothelium to the development of coagulation disorders that characterize Ebola hemorrhagic fever in primates. *Thromb Haemost* **94**(2), 254–261.

Hensley, L. E., Smith, M. A., Geisbert, J. B., Fritz, E. A., Daddario-DiCaprio, K. M., Larsen, T., and Geisbert, T. W. (2011b). Pathogenesis of Lassa fever in cynomolgus macaques. *Virol J* **8**, 205.

Heymann, D. L., Weisfeld, J. S., Webb, P. A., Johnson, K. M., Cairns, T., and Berquist, H. (1980). Ebola hemorrhagic fever: Tandala, Zaire, 1977–1978. *J Infect Dis* **142**(3), 372–376.

Jahrling, P. B., Marty, A. M., and Geisbert, T. W. (2008). Viral hemorrhagic fevers. In *Medical Aspects of Biological Warfare* (Z. F. Dembek, Ed.). Borden Institute (U.S. Army Walter Reed), Washington, DC.

Johnson, E. D., Johnson, B. K., Silverstein, D., Tukei, P., Geisbert, T. W., Sanchez, A. N., and Jahrling, P. B. (1996). Characterization of a new Marburg virus isolated from a 1987 fatal case in Kenya. *Arch Virol Suppl* **11**, 101–114.

Kissling, R. E., Murphy, F. A., and Henderson, B. E. (1970). Marburg virus. *Ann N Y Acad Sci* **174**(2), 932–945.

Kissling, R. E., Robinson, R. Q., Murphy, F. A., and Whitfield, S. G. (1968). Agent of disease contracted from green monkeys. *Science* **160**(3830), 888–890.

Korch, G. W., Jr., Niemi, S. M., Bergman, N. H., Carucci, D. J., Ehrlich, S. A., Gronvall, G. K., Hartung, T. et al. (2011). *Animal Models for Assessing Countermeasures to Bioterrorism Agents*. Institute for Laboratory Animal Research (ILAR), Division on Earth and Life Studies, The National Academies of Sciences (NAS), Washington, DC.

Kortepeter, M. G., Bausch, D. G., and Bray, M. (2011). Basic clinical and laboratory features of filoviral hemorrhagic fever. *J Infect Dis* **204**(Suppl 3), S810–S816.

Kortepeter, M. G., Lawler, J. V., Honko, A., Bray, M., Johnson, J. C., Purcell, B. K., Olinger, G. G., Rivard, R., Hepburn, M. J., and Hensley, L. E. (2011). Real-time monitoring of cardiovascular function in rhesus macaques infected with Zaire ebolavirus. *J Infect Dis* **204**(Suppl 3), S1000–S1010.

Kuhn, J. H. (2008). Filoviruses. A compendium of 40 years of epidemiological, clinical, and laboratory studies. *Arch Virol Suppl* **20**, 13–360.

Kuhn, J. H., Becker, S., Ebihara, H., Geisbert, T. W., Jahrling, P. B., Kawaoka, Y., Netesov, S. V. et al. (2011a). Family *Filoviridae*. In *Virus Taxonomy—Ninth Report of the International Committee on Taxonomy of Viruses* (A. M. Q. King, M. J. Adams, E. B. Carstens, and E. J. Lefkowitz, Eds.), pp. 665–671. Elsevier/Academic Press, London, U.K.

Kuhn, J. H., Becker, S., Ebihara, H., Geisbert, T. W., Johnson, K. M., Kawaoka, Y., Lipkin, W. I. et al. (2010). Proposal for a revised taxonomy of the family *Filoviridae*: Classification, names of taxa and viruses, and virus abbreviations. *Arch Virol* **155**(12), 2083–2103.

Kuhn, J. H., Dodd, L. E., Wahl-Jensen, V., Radoshitzky, S. R., Bavari, S., and Jahrling, P. B. (2011b). Evaluation of perceived threat differences posed by filovirus variants. *Biosecur Bioterror* **9**(4), 361–371.

Kunz, S. (2009). The role of the vascular endothelium in arenavirus haemorrhagic fevers. *Thromb Haemost* **102**(6), 1024–1029.

Leroy, E. M., Gonzalez, J. P., and Baize, S. (2011). Ebola and Marburg haemorrhagic fever viruses: Major scientific advances, but a relatively minor public health threat for Africa. *Clin Microbiol Infect* **17**(7), 964–976.

Lub, M., Sergeev, A. N., P'iankov, O. V., P'iankova, O. G., Petrishchenko, V. A., and Kotliarov, L. A. (1995). Certain pathogenetic characteristics of a disease in monkeys in infected with the Marburg virus by an airborne route. *Vopr Virusol* **40**(4), 158–161.

Maganga, G. D., Bourgarel, M., Ella, G. E., Drexler, J. F., Gonzalez, J. P., Drosten, C., and Leroy, E. M. (2011). Is Marburg virus enzootic in Gabon? *J Infect Dis* **204**(Suppl 3), S800–S803.

Mahanty, S., Bausch, D. G., Thomas, R. L., Goba, A., Bah, A., Peters, C. J., and Rollin, P. E. (2001). Low levels of interleukin-8 and interferon-inducible protein-10 in serum are associated with fatal infections in acute Lassa fever. *J Infect Dis* **183**(12), 1713–1721.

Mahanty, S. and Bray, M. (2004). Pathogenesis of filoviral haemorrhagic fevers. *Lancet Infect Dis* **4**(8), 487–498.

Martinez-Sobrido, L., Giannakas, P., Cubitt, B., Garcia-Sastre, A., and de la Torre, J. C. (2007). Differential inhibition of type I interferon induction by arenavirus nucleoproteins. *J Virol* **81**(22), 12696–12703.

Martini, G. A. (1973). Marburg virus disease. *Postgrad Med J* **49**(574), 542–546.

Marty, A. M., Jahrling, P. B., and Geisbert, T. W. (2006). Viral hemorrhagic fevers. *Clin Lab Med* **26**(2), 345–386, viii.

McCormick, J. B., Walker, D. H., King, I. J., Webb, P. A., Elliott, L. H., Whitfield, S. G., and Johnson, K. M. (1986). Lassa virus hepatitis: A study of fatal Lassa fever in humans. *Am J Trop Med Hyg* **35**(2), 401–407.

Moraz, M. L. and Kunz, S. (2011). Pathogenesis of arenavirus hemorrhagic fevers. *Expert Rev Anti Infect Ther* **9**(1), 49–59.

Murphy, F. A. (1978). Pathology of Ebola virus infection. In *Ebola Virus Haemorrhagic Fever: Colloquium Proceedings* (S. R. Pattyn, Ed.). Elsevier/North-Holland Biomedical Press, New York.

Ndambi, R., Akamituna, P., Bonnet, M. J., Tukadila, A. M., Muyembe-Tamfum, J. J., and Colebunders, R. (1999). Epidemiologic and clinical aspects of the Ebola virus epidemic in Mosango, Democratic Republic of the Congo, 1995. *J Infect Dis* **179**(Suppl 1), S8–S10.

Nikiforov, V. V., Turovskii, Iu. I., Kalinin, P. P., Akinfeeva, L. A., Katkova, L. R., Barmin, V. S., Riabchikova, E. I. et al. (1994). A case of a laboratory infection with Marburg fever. *Zh Mikrobiol Epidemiol Immunobiol* (3), 104–106.

Okokhere, P. O., Ibekwe, T. S., and Akpede, G. O. (2009). Sensorineural hearing loss in Lassa fever: Two case reports. *J Med Case Reports* **3**, 36.

van Paassen, J., Bauer, M. P., Arbous, M. S., Visser, L. G., Schmidt-Chanasit, J., Schilling, S., Olschlager, S. et al. (2012). Acute liver failure, multiorgan failure, cerebral oedema, and activation of proangiogenic and antiangiogenic factors in a case of Marburg haemorrhagic fever. *Lancet Infect Dis* **12**(8), 635–642.

Richards, G. A., Murphy, S., Jobson, R., Mer, M., Zinman, C., Taylor, R., Swanepoel, R., Duse, A., Sharp, G., De La Rey, I. C., and Kassianides, C. (2000). Unexpected Ebola virus in a tertiary setting: clinical and epidemiologic aspects. *Crit Care Med* **28**(1), 240–244.

Roddy, P., Thomas, S. L., Jeffs, B., Nascimento Folo, P., Pablo Palma, P., Moco Henrique, B., Villa, L. et al. (2010). Factors associated with Marburg hemorrhagic fever: Analysis of patient data from Uige, Angola. *J Infect Dis* **201**(12), 1909–1918.

Rollin, P. E., Bausch, D. G., and Sanchez, A. (2007). Blood chemistry measurements and D-Dimer levels associated with fatal and nonfatal outcomes in humans infected with Sudan Ebola virus. *J Infect Dis* **196**(Suppl 2), S364–S371.

Rollin, P. E., Nichol, S. T., Zaki, S., and Ksiazek, T. G. (2007). Arenaviruses and Filoviruses. In *Manual of Clinical Microbiology* (P. R. Murray, E. J. Baron, J. H. Jorgensen, M. L. Landry, and M. A. Pfaller, Eds.), 9th edn., Vol. 2, pp. 1510–1522. ASM Press, Washington, DC.

Rowe, A. K., Bertolli, J., Khan, A. S., Mukunu, R., Muyembe-Tamfum, J. J., Bressler, D., Williams, A. J. et al. (1999). Clinical, virologic, and immunologic follow-up of convalescent Ebola hemorrhagic fever patients and their household contacts, Kikwit, Democratic Republic of the Congo. Commission de Lutte contre les Epidemies a Kikwit. *J Infect Dis* **179**(Suppl 1), S28–S35.

Ryabchikova, E. I., Kolesnikova, L. V., and Luchko, S. V. (1999). An analysis of features of pathogenesis in two animal models of Ebola virus infection. *J Infect Dis* **179**(Suppl 1), S199–S202.

Salvato, M. S., Clegg, J. C. S., Buchmeier, M. J., Charrel, R. N., Gonzalez, J. P., Lukashevich, I. S., Peters, C. J., and Romanowski, V. (2011). Family Arenaviridae. In *Virus Taxonomy—Ninth Report of the International Committee on Taxonomy of Viruses* (A. M. Q. King, M. J. Adams, E. B. Carstens, and E. J. Lefkowitz, Eds.), pp. 715–723. Elsevier/Academic Press, London, U.K.

Sanchez, A., Lukwiya, M., Bausch, D., Mahanty, S., Sanchez, A. J., Wagoner, K. D., and Rollin, P. E. (2004). Analysis of human peripheral blood samples from fatal and nonfatal cases of Ebola (Sudan) hemorrhagic fever: Cellular responses, virus load, and nitric oxide levels. *J Virol* **78**(19), 10370–10377.

Schnittler, H. J., Mahner, F., Drenckhahn, D., Klenk, H. D., and Feldmann, H. (1993). Replication of Marburg virus in human endothelial cells. A possible mechanism for the development of viral hemorrhagic disease. *J Clin Invest* **91**(4), 1301–1309.

Shurtleff, A. C., Bradfute, S. B., Radoshitzky, S. R., Jahrling, P. B., Kuhn, J. H., and Bavari, S. (2012). Pathogens–Arenaviruses: Hemorrhagic fevers. *In BSL3 and BSL4 Agents—Epidemiology, Microbiology, and Practical Guidelines* (M. Elschner, S. Cutler, M. Weidman, and P. Butaye, Eds.), pp. 211–236. Wiley-Blackwell, Weinheim, Germany.

Siegert, R., Shu, H. L., Slenczka, W., Peters, D., and Muller, G. (1967). On the etiology of an unknown human infection originating from monkeys. *Dtsch Med Wochenschr* **92**(51), 2341–2343.

Smith, D. H., Johnson, B. K., Isaacson, M., Swanapoel, R., Johnson, K. M., Killey, M., Bagshawe, A., Siongok, T., and Keruga, W. K. (1982). Marburg-virus disease in Kenya. *Lancet* **1**(8276), 816–820.

Solbrig, M. D. and Naviaux, R. K. (1997). Review of the neurology and biology of Ebola and Marburg virus infections. *Neurolog Infect Epidemiol* **2**, 5–12.

Stroher, U., West, E., Bugany, H., Klenk, H. D., Schnittler, H. J., and Feldmann, H. (2001). Infection and activation of monocytes by Marburg and Ebola viruses. *J Virol* **75**(22), 11025–11033.

Sureau, P. H. (1989). Firsthand clinical observations of hemorrhagic manifestations in Ebola hemorrhagic fever in Zaire. *Rev Infect Dis* **11**(Suppl 4), S790–S793.

Terrell, T. G., Stookey, J. L., Eddy, G. A., and Kastello, M. D. (1973). Pathology of Bolivian hemorrhagic fever in the rhesus monkey. *Am J Pathol* **73**(2), 477–494.

Villinger, F., Rollin, P. E., Brar, S. S., Chikkala, N. F., Winter, J., Sundstrom, J. B., Zaki, S. R., Swanepoel, R., Ansari, A. A., and Peters, C. J. (1999). Markedly elevated levels of interferon (IFN)-gamma, IFN-alpha, interleukin (IL)-2, IL-10, and tumor necrosis factor-alpha associated with fatal Ebola virus infection. *J Infect Dis* **179**(Suppl 1), S188–S191.

Winn, W. C., Jr. and Walker, D. H. (1975). The pathology of human Lassa fever. *Bull World Health Organ* **52**(4–6), 535–545.

World Health Organization. (1978). Ebola haemorrhagic fever in Zaire, 1976. *Bull World Health Organ* **56**(2), 271–293.

Zaki, S. R. and Goldsmith, C. S. (1999). Pathologic features of filovirus infections in humans. *Curr Top Microbiol Immunol* **235**, 97–116.

Zampieri, C. A., Sullivan, N. J., and Nabel, G. J. (2007). Immunopathology of highly virulent pathogens: Insights from Ebola virus. *Nat Immunol* **8**(11), 1159–1164.

3 Interaction of the Host Immune System with Hemorrhagic Fever Viruses

Gavin C. Bowick

CONTENTS

3.1 INTRODUCTION

The immune system is a highly effective defense against infection. It's importance can be illustrated by the number of mechanisms in which viruses have evolved to evade it and in our increased susceptibility to infections when we are immunosuppressed, such as in HIV/AIDS, following measles infection, or during a number of medical interventions such as some cancer treatments or following transplantation. Despite huge advances in our knowledge of the basic structure, function, and regulation of the immune system, as well as how it responds to infection and how viruses interact with and try to disarm it, little of this knowledge has translated to the clinical setting. Indeed, the majority of vaccines in clinical use today are based on technologies developed in the first half of the last century.

The interaction of hemorrhagic fever viruses with the immune system is particularly interesting, given the role of immune mediators in contributing to the pathogenesis of hemorrhagic fever disease. Ebola virus infection, for example, can lead to a severe "cytokine storm," in which high levels of proinflammatory cytokine production is a key contributor to the disease. This chapter provides a summary of the immune responses to infection with the hemorrhagic fever-causing viruses with a particular emphasis on how these responses may differ between severe or fatal cases and mild or nonfatal cases. By studying these responses comparatively, it becomes possible to ascertain which immune responses are required to be induced for the development of a protective response. This chapter also discusses some of the mechanisms used by these viruses to inhibit these events to allow their replication in the face of our immune defenses.

3.2 MODELS OF INFECTION AND NATURAL INFECTION

Our understanding of the basis of hemorrhagic fever virus pathogenesis has been significantly advanced by the use of animal models to study aspects of their replication and disease progression. However, the development of animal models, small animal models in particular, can often be complicated by the fact that the hemorrhagic fevers are disease pathologies caused by infection of humans with a virus whose natural host is another species. For example, the natural hosts of the arenaviruses are rodents.

The development of small animal models for these diseases often follows three potential routes. The first is the adaptation of the virus to the animal model system. This has been performed for viruses such as Ebola virus, where a mouse-adapted virus has been developed (Bray et al., 1998, 2001). The second potential route is to use various recombinant or knockout mouse strains, such as the use of the STAT1 knockout mouse as a model system for Crimean–Congo hemorrhagic fever virus infection (Bente et al., 2010). The STAT1 knockout mouse model has also been used as an animal model for Machupo virus, the causative agent of Bolivian hemorrhagic fever (Bradfute et al., 2011). The third method is the use of a related virus that causes a similar disease in an animal model as a surrogate for the "authentic" virus. The use of lymphocytic choriomeningitis virus in macaques and Pichinde virus in guinea pigs as models for Lassa fever in humans has revealed a number of insights into virus host interactions and the immune mechanisms of protection.

High-throughput methods, mRNA expression microarrays in particular, are becoming more widely used to study changes in host gene expression, and use of animal models of infection has provided a wealth of information regarding the regulation of immune responses during infection. The use of lymphocytic choriomeningitis virus in macaques as a model for Lassa fever has allowed the comparison of responses to attenuated and virulent strains (Djavani et al., 2007, 2009). Functional genomics of a mouse model of Ebola infection identified metalloproteinase gene expression and further defined the regulation of the inflammatory responses in disease (Cilloniz et al., 2011).

While there are limitations associated with these approaches for animal model development, the importance of interaction between multiple cell types, cell location, and other variables in the development of immune responses requires the use of these systems as a complement to in vitro cell culture studies. Additionally, while the fact that these infections are zoonotic may complicate the development of small animal models. The study of the animals, which act as natural hosts or reservoirs of infection, can help in understanding the immune mechanisms responsible for the control of infection or prevention of virus-induced pathology.

3.3 PATTERN RECOGNITION RECEPTORS, SIGNALING, AND EARLY AND INNATE RESPONSES

Toll receptors are a class of proteins identified in the fruit fly *Drosophila melanogaster*, a workhorse system for genetics and developmental research. The 2011 Nobel Prize for physiology or medicine was awarded in part to Jules Hoffmann and Bruce Beutler for their work in characterizing the toll and toll-like receptors. This class of proteins has been shown to be critically important in mediating early nonspecific innate responses to pathogens. These receptors recognize particular components of pathogens that would not normally be seen in the absence of infection, such as lipopolysaccharide, flagellin, and CpG motifs present in bacteria or double-stranded RNA often present during viral replication. These receptors are coupled to signaling pathways central to the innate proinflammatory response including the NF-κB pathway.

Toll-like receptor (TLR4) is a recognition receptor for the lipopolysaccharide moiety of Gram-negative bacteria. Interestingly, the glycoprotein of Ebola virus has been shown to interact with this protein; this interaction leads to the downstream activation of NF-κB signaling and production of proinflammatory cytokines, as well as the suppressor of cytokine signaling SOCS1 (Okumura et al., 2010). This study also found that stimulation of TLR4 of infected cells led to an increase in virus budding.

The yellow fever vaccine, 17D, has proven to be a highly effective and successful vaccine since its development in the 1930s. Given its effectiveness in inducing strong and long-lasting protective immunity, it has also proved to be a useful workhorse in determining the immune mechanisms required for the production of this response (Querec and Pulendran, 2007). Use of 17D as a model system has indicated the importance of TLR activation and downstream signaling in the activation of dendritic cells and natural killer (NK) cells (Neves et al., 2009; Querec et al., 2006). Signaling through TLR3 and the RIG-I and MDA 5 proteins has also been shown to be important in the response to Dengue virus infection (Nasirudeen et al., 2009).

Monocytes and macrophages are central and early targets of hemorrhagic fever virus infection (Bray and Geisbert, 2005). Dendritic cells have been identified as important early cells infected by Ebola virus (Geisbert et al., 2003). The key roles of these cells in orchestrating the immune response following infection mean that modulation of the function of these cell types could lead to profound dysregulation in the establishment of downstream immune responses. Additionally, the capability of these cell types to secrete large quantities of proinflammatory cytokines means that incorrect regulation of these cells could be a major contributor to pathogenesis. Infection of these cell types is an early event in infection and, as such, could be a key determining factor in the eventual outcome of infection. Significant differences are seen following infection of these cell types with the pathogenic Lassa and Mopeia viruses. Infection of macrophages with Mopeia virus results in cell activation, whereas infection of dendritic cells does not (Pannetier et al., 2004). In contrast, infection of these cell types with Lassa virus does not induce activation, despite high levels of replication (Baize et al., 2004).

The type-I interferon (IFN-I) response is a critical early innate immune response in the control of viral infection. Activation of upstream signaling pathways following infection leads to the nuclear translocation and DNA binding of transcription factors such as IRF3. IFN secretion, following binding to its cellular receptor, activates the JAK/STAT signaling pathway leading to nuclear transloca-tion, of STAT1/STAT2 heterodimers in the case of IFN-I, and the expression of antiviral genes such as protein kinase R and $2',5'$-oligoadenylate synthetase. Given the ability of IFNs to induce so many antiviral genes, it is unsurprising that many viruses have mechanisms for inhibiting this pathway.

The arenaviruses have been a useful tool in determining differential immune responses. The large number of members of the family, which includes several viruses known to cause human disease as well as a number that do not, has allowed to perform many comparative studies. In the case of Lassa infection of nonhuman primates, IFN is produced in both fatal and nonfatal cases; however, it is expressed significantly earlier in nonfatal cases than in fatal cases of infection (Baize et al., 2009). The differential production of soluble mediators is a common theme among the are-naviruses, with the New World Tacaribe virus and Old World Mopeia virus often serving as the nonpathogenic representatives.

Studies in primary monocytes and macrophages have shown that high levels of IL-6, IL-10, and TNF-α are secreted following infection with Tacaribe virus, whereas, Junin do not induce secretion of these cytokines; use of UV-inactivated virus revealed that IL-6 and IL-10 production required viral replication, whereas TNF-α production did not (Groseth et al., 2011). This suggests that the pathways leading to TNF-α production are mediated by early events, likely downstream of pattern recognition receptors. Interestingly, this study showed that expression of these cytokines did not affect the growth of either Tacaribe virus or Junin virus (Groseth et al., 2011). This finding may illustrate the continuing importance of studying immune responses in mixed cell culture systems or animal models: the production of these cytokines may not affect growth in vitro but may be respon-sible for the induction of important downstream events.

A recent study has identified novel host method of inhibiting hemorrhagic fever virus release. BST-2/tetherin is an IFN-inducible cellular membrane protein that has been shown to inhibit the release of retroviruses. When the virus buds through the cellular membrane, it acquires one of the membrane domains of the protein; another domain remains "anchored" in the cellular membrane, preventing release of the newly budded virus. It has been shown that Lassa virus release is inhibited by this mechanism, but release of Marburg, Ebola, and Rift Valley fever viruses is not, although filovirus viruslike particles did show reduced budding (Radoshitzky et al., 2010).

Studies using the arenavirus Pichinde virus in guinea pigs as a model for Lassa fever have allowed the comparison of two virus variants, P2 and P18, which cause either a mild, self-limiting infection or a severe hemorrhagic disease, respectively. The comparative studies among variants in terms of cellular responses, provides the information about the signaling pathways activated/suppressed by the nonpathogenic virus and virulent virus. Studies have shown that numerous cellular pathways are activated following infection with these viruses (Bowick et al., 2006, 2007, 2009a,b). However, a striking feature of these studies is the significantly increased complexity of differentially activated signaling pathways following infection with the nonpathogenic virus compared to the pathogenic virus. This suggests that nonpathogenic virus initiates a coordinated, multipathway response that leads to the establishment of protective immunity and viral clearance. In contrast, infection with the pathogenic virus leads to either a failure to activate these pathways or an active suppression of their activation. Indeed, the P18 variant of Pichinde virus is able to effectively block NF-κB activation following stimulation of infected cells with lipopolysaccharide (Fennewald et al., 2002).

In addition to the NF-κB family, the AP1 family of transcription factors also plays a key role in controlling proinflammatory cytokine expression. As with NF-κB, the AP1 proteins are a large family of proteins, some of which are capable of transactivating gene expression and some of which are transcriptionally repressive. Differences in AP1 activation following Pichinde infection were observed by gel shift assay, and this suggested a strategy for a novel therapeutic approach. Double-stranded DNA sequences, modified with thiolated backbones, were synthesized to contain the AP1 binding site. These "thioaptamers" were then delivered to guinea pigs via liposomes. When used to treat guinea pigs infected with Pichinde virus, a significant increase in survival was observed (Fennewald et al., 2007). A suggested mechanism for this observation is that the virus induces activation of transcriptionally repressive forms of AP1, which bind to their promoter regions and inhibit transcription. The thioaptamers may compete for the active AP1, preventing them from repressing transcription of these genes and allowing transactivation by other active transcription factors. Thioaptamers designed specifically to bind to repressive forms of NF-κB increased the survival time, but did not influence overall survival (Bowick et al., 2009a).

3.4 HUMORAL AND CELL-MEDIATED RESPONSES

The humoral components of the adaptive immune system are those immune mediators that are components of fluids, such as antibody and the complement system. The humoral immune components comprise elements of both innate and adaptive immune responses. Following on from Section 3.3 on early and innate immune responses, it is becoming more clear that a strong early innate immune response, particularly one that activates multiple pathways, leads to stronger, more effective adaptive responses (Querec et al., 2006; Querec and Pulendran, 2007).

The complement system is a component of the innate immune system but can be further exploited as an additional effector system by the adaptive immune system. The complement system has been less extensively studied in the context of hemorrhagic fever virus infection, but a number of interesting effects have been observed. Activity of serum complement is significantly reduced in Argentine hemorrhagic fever, caused by Junin virus, with low C2, C3, and C5 levels but an increase in C4 (Debracco et al., 1978). It has also been shown that mannose-binding lectin is capable of binding to filovirus glycoproteins, leading to activation of the lectin-complement pathway and virus inactivation in a pseudotyped virus system (Ji et al., 2005).

The cell-mediated immune response refers to the direct killing of infected cells by cell types such as NK cells and cytotoxic CD8+ T lymphocytes. CD8+ T cells (CD8 cells) express the CD8 cell surface antigen, which recognizes MHC-I, expressed by virtually all cell types excluding neurons. The MHC-I pathway allows for the "scanning" of host cells to determine whether they are displaying "nonhost" peptide sequences. Proteins are degraded by the proteasome, and the resulting peptides are displayed on the cell surface via the MHC-I molecule.

In the majority of cases, the peptides scanned by the CD8 cell will be normal host-derived proteins and so not seen by the CD8 cell as cells which recognize these peptide sequences will have been removed by negative selection during T cell development. However, virally expressed proteins, or mutant cellular proteins, such as those that develop during cancer development, will also be degraded and expressed via MHC-I. CD8 T cells can recognize these cells and induce the death of these cells via Fas/FasL interactions and through the release of CD8 cell granule components including perforin and granzyme B. This mechanism of cell killing is an important immune mechanism for eliminating infected cells. NK cells operate via a similar mechanism of inducing cell death; however, they do not recognize peptide displayed via MHC-I as they do not express the T cell receptor; however, they are capable of detecting cells by mechanisms including the absence or downregulation of MHC-I, often triggered as an immune evasion mechanism by a number of viruses.

Comparing immune responses between fatal and nonfatal cases of Ebola virus infection reveals a significant deficiency in humoral immune mechanisms. In surviving patients, early production of IgG against the viral nucleoprotein and VP40 protein is observed, followed by CD8 T cell activation. In contrast, infections with a fatal outcome show a strikingly low level of IgM production and no detectable specific IgG (Baize et al., 1999). In the case of Lassa infection, there is a significant delay in the appearance of neutralizing antibodies, with low titers of specific neutralizing antibodies only appearing after recovery (Jahrling et al., 1985). However, the appearance of antibody does not correlate with survival, with cell-mediated immunity seemingly more important in controlling infection (ter Meulen et al., 2000, 2004). Serological surveys have shown that a high prevalence of both humoral and cell-mediated immunity exists in rural populations in Gabon (Becquart et al., 2010). This study identified existing IgG to Ebola virus in 15.3% of those individuals screened, suggesting that exposure may be more common than previously thought.

A further observation in hemorrhagic fevers is the apoptosis of significant numbers of uninfected bystander lymphocytes. Infection of nonhuman primates with Ebola and Marburg viruses results in the induction of lymphocyte apoptosis, particularly in response to infection with the highly pathogenic Zaire strain; apoptosis of mononuclear cells has also been observed (Geisbert et al., 2000; Gupta et al., 2007). A further study showed that lymphocyte apoptosis could be detected as early as 2 days post-infection, with an increase in Fas expression, suggesting that the Fas/FasL mechanism could be implicated in apoptotic induction and that, in addition to a 60%–70% reduction in CD4 and CD8 T cell populations, NK cells were also apoptotic (Reed et al., 2004). Lymphocyte apoptosis has also been reported in a mouse model of Ebola virus infection (Bradfute et al., 2007). The induction of bystander apoptosis does not appear to be due to the effect of secreted Ebola virus glycoprotein (Wolf et al., 2011).

A mouse model of Ebola virus infection has been used to define the protective capacity of cell-mediated mechanisms. Prior to the marked reduction in peripheral lymphocyte numbers by apoptosis, T cells show an increased expression of activation markers, and lymphoblasts were also observed in the spleen (Bradfute et al., 2008). Using adoptive transfer studies, it has been shown that splenocytes from symptomatic Ebola virus-infected mice are capable of protective naïve mice from challenge (Bradfute et al., 2008). This suggests that a functional cell-mediated response is induced during Ebola virus infection but that this response is insufficient, perhaps due to the massive levels of apoptosis, to clear infected cells and protect the host.

In human infections, numbers of peripheral CD4 and CD8 T cells are markedly reduced in fatal cases of Ebola virus infection, with a concomitant reduction in the levels of cytokines produced by these cell types including IL-2, IL-3, IL-4, IL-5, IL-9, and IL-13 (Wauquier et al., 2010). Given the significant effects on immune response generation due to these effects, not only in hemorrhagic fevers, but also in diseases such as anthrax and plague, inhibiting immune cell apoptosis has been postulated as a potential therapeutic strategy (Parrino et al., 2007). However, a recent study in which mice lacking the Fas-associated death domain or overexpressing the antiapoptotic protein Bcl-2 were infected with Ebola virus did not find an increase in survival, despite the reduction in lymphocyte apoptosis (Bradfute et al., 2010).

A recent study has also suggested that Ebola virus may display activity similar to bacterial superantigens: proteins that lead to amplification of specific T cell subsets followed by their rapid depletion and anergy. Analysis of levels of mRNA for various T cell receptor Vβ subsets during Ebola virus infection of humans revealed a reduction in message for three specific

Vβ transcripts, consistent with superantigen activity, a finding that was noted between fatal and nonfatal cases (Leroy et al., 2011).

Comparative studies between pathogenic and nonpathogenic or virulent and attenuated strains of viruses have shed light on the immune responses required for the development of a protective response, which may be dysregulated in disease. There are a number of examples of these comparative systems, including the Armstrong and WE strains of lymphocytic choriomeningitis virus; Lassa and Mopeia viruses; Junin, Machupo, Guanarito, and Sabia viruses with Tacaribe virus; and pathogenic Ebola virus strains with the Reston strain.

As a comparative model for protective responses to Old World arenaviruses, the nonpathogenic Mopeia virus has been an important tool. It has been shown that Mopeia virus induces significant CD4 and CD8 T cell responses early in infection, with memory and effector memory phenotypes observed, whereas the infection with the pathogenic Lassa virus led to only weak T cell responses and significantly decreased induction of memory responses (Pannetier et al., 2011).

Normal laboratory mice are resistant to disease caused by Lassa virus. However, mice humanized with human T cells are vulnerable to infection, with the striking observation that depletion of T cells in these mice prevented the onset of disease in the presence of high levels of virus replication (Flatz et al., 2010). The T-cell-depleted mice exhibited a lack of activation of monocytes and macrophages, suggesting an additional role for T cells in contributing to the dysregulated production of proinflammatory cytokines (Flatz et al., 2010).

In the case of Lassa virus infection of nonhuman primates, antibody responses develop earlier in survivors compared to nonsurvivors, along with increased T cell activation and higher levels of circulating activated monocytes (Baize et al., 2009). Interestingly, the IL-6 was only detectable in the plasma of fatal cases (Baize et al., 2009).

3.5   IMMUNOPATHOGENESIS

The immune system has developed numerous mediators and effectors to initiate an effective response to challenge and to eliminate or control infection. However, when these responses are dysregulated, these potent regulatory mechanisms can themselves lead to the symptomatic clinical disease. For example, the fever, lethargy, and myalgia that are characteristic of influenza infection are due to mediators of the immune response, rather than a direct result of viral cytopathic effect.

In the context of hemorrhagic fever pathogenesis, the roles of proinflammatory cytokines such as TNF-α and IL-1β cannot be understated. These cytokines are potent mediators in the inflammatory and fever responses. While these cytokines are critically important in establishing the early host responses to infection, aberrant and dysregulated expression can be detrimental to the host, leading to many of the symptoms seen during disease. During infection with Ebola virus, a number of cytokines are produced, including TNF-α, IL-1, IL-6, IL-8, TRAIL, IL-12, and IFN-α (Bray et al., 2001; Gupta et al., 2011; Hensley et al., 2002; Schnittler and Feldmann, 1999; Villinger et al., 1999). Clinical observations have shown high levels of IFN-α, IFN-γ, IL-2, IL-10, and TNF-α in fatal cases of Ebola infection, compared to survivors (Villinger et al., 1999). A more recent study has shown "cytokine storm" in fatal cases of Ebola infection, with strikingly high production of many cytokines including those listed earlier as well as many other cytokines, chemokines, and growth factors such as IL-15, IL-16, MIP-1 α and β, MCP-1, IP-10, GRO-α, and eotaxin (Wauquier et al., 2010).

While it is clear that proinflammatory cytokines can contribute significantly to the clinical symptoms and pathogenesis of hemorrhagic fever disease, it is likely that these effects are the result of "runaway" dysregulated responses and that inflammatory responses are critically important in early responses to these infections, which result in the establishment of effective adaptive humoral and cell-mediated immunity and clearance of the pathogen. A study of individuals who had close contact with Ebola virus-infected patients, but who never developed symptoms, revealed significant proinflammatory responses and production of specific IgM and IgG (Leroy et al., 2000).

3.6 IMMUNE EVASION BY HEMORRHAGIC FEVER VIRUSES

Given the effectiveness of the human immune system, it is no surprise that virtually every virus, from the small RNA picornaviruses to the large double-stranded DNA viruses, such as the poxviruses and herpesviruses, has evolved mechanisms to subvert, inhibit, and modulate these responses to allow their successful replication. Given their large coding capacity, it is unsurprising that the large DNA viruses exhibit perhaps the most diverse and overt mechanisms of immune evasion. During their evolution, these viruses have been able to acquire and mutate host genes, which now serve as important weapons in their arsenal against the immune system. For example, the poxviruses express a number of proteins, which are secreted from infected cells and bind cytokines and chemokines, preventing their binding to their cellular receptors and the activation of the downstream signaling cascades (Lalani et al., 2000; Nash et al., 1999; Smith et al., 1997; Smith and Alcami, 2000).

African swine fever virus (ASFV) is a particularly interesting example of a hemorrhagic fever virus. ASFV causes African swine fever in domestic pigs, whereas it is asymptomatic in its natural hosts, the warthog and the bushpig (Kleiboeker and Scoles, 2001; Tulman et al., 2009). In domestic swine, ASFV causes a fatal hemorrhagic disease, similar in clinical presentation to hemorrhagic fever in humans (Gomez-Villamandos et al., 2003). ASFV is the only known DNA hemorrhagic fever virus and is the only known DNA arbovirus, being transmitted by, and capable of infecting and persisting in, the soft ticks of the *Ornithodoros* genus (Kleiboeker and Scoles, 2001).

By virtue of the fact that ASFV is a large DNA virus of approximately 180 kb pairs, the virus encodes a number of proteins involved in modulating the host response and evading the immune system (Dixon et al., 2004). One such protein, A238L, has homology to the cellular inhibitor of the NF-κB transcription factor, IκB. NF-κB plays a central role in controlling immune and inflammatory responses, following induction by a number of stimuli. This transcription factor can be activated by many cytokines and chemokines, as well as activation of pattern recognition receptors. Once the pathway has been activated, IκB is degraded, allowing NF-κB to translocate to the nucleus and transactivate its target genes. However, the A238L is capable of acting as a functional homolog of IκB, inhibiting transcription of myriad genes important in controlling the response to infection (Powell et al., 1996; Silk et al., 2007; Tait et al., 2000).

However, in contrast to the example of ASFV, the viruses responsible for causing hemorrhagic fevers in humans are smaller RNA viruses, with more limited coding capacity. However, it does not appear that this reduced capacity to encode large numbers of genes has impaired the ability of the human hemorrhagic fever viruses to inhibit the host immune system. It seems that all of these viruses have mechanisms for inhibiting multiple immune responses. In particular, components of the IFN system are targeted by multiple viruses—concomitant with their central importance in controlling antiviral gene expression.

While it is known that Ebola virus is able to inhibit IFN production, the precise mechanisms for this inhibition had remained unclear. A study in 2009 revealed an elegant mechanism behind this observation. It was found that the viral VP35 protein was able to inhibit the transactivating activity of IRF7, which is required for IFN production (Chang et al., 2009). This inhibition of IRF7 is mediated by an elegant mechanism: VP35 is able to interact with the cellular PIAS1, a SUMO ligase, leading to an increase in IRF7 SUMOylation and an inhibition of transactivating ability (Chang et al., 2009).

The transcription factor IRF3 is a critical transcription factor, involved in the expression of the IFN-Is, as well as many IFN-stimulated genes. The nucleoproteins of the arenaviruses, from both the Old World and New World, have been shown to inhibit translocation of IRF-3 to the nucleus (Martinez-Sobrido et al., 2006, 2007). Interestingly, the nucleoprotein of the nonpathogenic Tacaribe virus did not show this ability (Martinez-Sobrido et al., 2007). Inhibition of the IFN-I system at this early stage could inhibit the production of IFN-α and IFN-β, preventing the downstream transcription of many antiviral genes. An additional mechanism of IFN induction has also been reported for the four pathogenic New World arenaviruses: Guanarito, Sabia, Junin, and Machupo. The Z proteins of these viruses are capable of binding the cellular sensor protein of viral infection, RIG-I, inhibiting its ability to activate upstream elements of the IFN system (Fan et al., 2010).

Despite the ability of many of these viruses to inhibit the production of IFN, inhibition of the pathways downstream of the IFN-I is also often observed. The VP24 protein of Ebola virus inhibits the nuclear accumulation of phosphorylated STAT1 by binding proteins of the karyopherin-α family of transporter proteins (Reid et al., 2006, 2007). In contrast, despite being in the same family, Marburg virus adopts a different strategy to inhibit STAT1-mediated antiviral gene expression. The VP40 of Marburg virus is able to inhibit phosphorylation of STAT1; as this antagonism of the pathway occurs upstream of Jak phosphorylation, phosphorylation of Jak1, STAT2, and STAT3 is also inhibited (Valmas et al., 2010). The NS4B protein of Dengue virus also inhibits the phosphorylation of STAT1 and leads to inhibition of antiviral gene expression (Munoz-Jordan et al., 2003). Global mRNA expression profiling following infection with Marburg virus and the Zaire and Reston strains of Ebola virus has shown that infection with the Reston strain, which appears not to be pathogenic in humans, leads to induction of a greater number of IFN-induced antiviral genes than infection with the pathogenic viruses (Kash et al., 2006), suggesting that the ability of these viruses to inhibit these immune mechanisms is an important determinant of pathogenesis.

As described previously, Pichinde virus is able to inhibit NF-κB activation. Interestingly, data also suggest that infection of cells with the pathogenic variant of the virus leads to an accumulation of the NF-κB family member p50 (Bowick et al., 2009). The p50 protein contains a DNA-binding domain, but not a transactivation domain, and so p50 homodimers are thought of as transcriptionally repressive. The selective expression of this protein may represent a novel mechanism to inhibit gene expression controlled by NF-κB.

The hemorrhagic fever viruses have also developed mechanisms to inhibit the complement system. The secreted NS1 protein of flaviviruses, which include the hemorrhagic fever-causing Yellow fever virus and Dengue virus, is capable of inhibiting both the classical and lectin-complement pathways (Avirutnan et al., 2010).

3.7   CONCLUSIONS

Hemorrhagic fevers have among the highest case fatality rates of any infectious disease; as has been discussed earlier, their ability to inhibit specific pathways of the immune response can lead to delayed, suppressed, dysregulated, and inappropriate immune responses, which not only prevent clearance of the virus by the host, but can also contribute to the pathogenesis of the disease.

The last few years have seen significant increases in the amount of research being performed on these pathogens, both with live, infectious viruses as new BSL-4 facilities are constructed and also with the availability of tools and reagents such as viruslike particles. Additionally, increasing numbers of transcriptomics-based studies are providing a wealth of information on host-cell responses to infection. In particular, evidence is continuing to suggest that early production of proinflammatory cytokines is critical in determining clinical outcome, with these early responses required to drive effective adaptive immune responses.

As our understanding of pathogenesis continues to grow, we may be in a position to utilize our knowledge regarding the molecular basis of protection in a number of ways to treat these diseases. Characterizing the correlation between cytokine production and disease severity may allow the development of prognostic biomarker panels that may be used to better triage patients and manage therapeutic options. This ability may be critical in managing the limited resources available during epidemics or in response to intentional release.

Given the continuing characterization of which molecular events are required for the development of a protective response and which may be dysregulated and lead to the development of clinical disease, we are in a position where we can begin to consider modulating host-cell signaling pathways as a potential therapeutic strategy. By targeting host-cell signaling, it may become possible to drive immune responses in a direction that is likely to lead to protection or inhibit pathways that are inappropriately activated.

Additionally, these types of therapeutics may be less vulnerable to the evolution of resistance mutants. However, these types of therapies may require the concurrent development of advanced biomarker panels in order to accurately determine the stage of infection. It is possible that the type of immune modulation required therapeutically may be very different at early stages of disease compared to late, post-symptomatic stages. The use of high-throughput "omics" techniques, combined with mathematical and computational "machine learning" approaches, is likely to become a strategy that sees increasing use. One study has used these types of "systems biology" methods to identify gene expression signatures, which are predictive of T cell and B cell responses following yellow fever vaccination (Querec et al., 2009).

In summary, our understanding of how hemorrhagic fever viruses evade the immune system to cause disease continues to grow. And, as basic immunological research identifies new regulators and effectors of the immune responses, it is likely that we will find ways in which these viruses have been manipulating them. Of particular interest in the cases of these zoonotic infections, however, is the lack of disease in the natural host. What can we learn from studying these interactions to enhance our knowledge of the underlying mechanisms of protection? With continuing investment in the additional high-containment laboratories, it is likely that we will see significant advances in our knowledge and that these advances will lead to additional treatments for these diseases.

REFERENCES

Avirutnan, P., A. Fuchs, R. E. Hauhart, P. Somnuke, S. Youn, M. S. Diamond, and J. P. Atkinson. 2010. Antagonism of the complement component C4 by flavivirus nonstructural protein NS1. *The Journal of Experimental Medicine* 207:793–806.

Baize, S., J. Kaplon, C. Faure, D. Pannetier, M. C. Georges-Courbot, and V. Deubel. 2004. Lassa virus infection of human dendritic cells and macrophages is productive but fails to activate cells. *Journal of Immunology* 172:2861–2869.

Baize, S., E. M. Leroy, M. C. Georges-Courbot, M. Capron, J. Lansoud-Soukate, P. Debre, S. P. Fisher-Hoch, J. B. McCormick, J. B. McCormick, and A. J. Georges. 1999. Defective humoral responses and extensive intravascular apoptosis are associated with fatal outcome in Ebola virus-infected patients. *Nature Medicine* 5:423–426.

Baize, S., P. Marianneau, P. Loth, S. Reynard, A. Journeaux, M. Chevallier, N. Tordo, V. Deubel, and H. Contamin. 2009. Early and strong immune responses are associated with control of viral replication and recovery in Lassa virus-infected cynomolgus monkeys. *Journal of Virology* 83:5890–5903.

Becquart, P., N. Wauquier, T. Mahlakoiv, D. Nkoghe, C. Padilla, M. Souris, B. Ollomo et al. 2010. High prevalence of both humoral and cellular immunity to zaire ebolavirus among rural populations in Gabon. *PLoS One* 5:e9126.

Bente, D. A., J. B. Alimonti, W. J. Shieh, G. Camus, U. Stroher, S. Zaki, and S. M. Jones. 2010. Pathogenesis and immune response of Crimean-Congo hemorrhagic fever virus in a STAT-1 knockout mouse model. *Journal of Virology* 84:11089–11100.

Bowick, G. C., S. M. Fennewald, B. L. Elsom, J. F. Aronson, B. A. Luxon, D. G. Gorenstein, and N. K. Herzog. 2006. Differential signaling networks induced by mild and lethal hemorrhagic fever virus infections. *Journal of Virology* 80:10248–10252.

Bowick, G. C., S. M. Fennewald, E. P. Scott, L. H. Zhang, B. L. Elsom, J. F. Aronson, H. M. Spratt, B. A. Luxon, D. G. Gorenstein, and N. K. Herzog. 2007. Identification of differentially activated cell-signaling networks associated with Pichinde virus pathogenesis by using systems kinomics. *Journal of Virology* 81:1923–1933.

Bowick, G. C., S. M. Fennewald, L. H. Zhang, X. B. Yang, J. F. Aronson, R. E. Shope, B. A. Luxon, D. G. Gorenstein, and N. K. Herzog. 2009. Attenuated and lethal variants of pichinde virus induce differential patterns of NF-kappa B activation suggesting a potential target for novel therapeutics. *Viral Immunology* 22:457–462.

Bowick, G. C., H. M. Spratt, A. E. Hogg, J. J. Endsley, J. E. Wiktorowicz, A. Kurosky, B. A. Luxon, D. G. Gorenstein, and N. K. Herzog. 2009. Analysis of the differential host cell nuclear proteome induced by attenuated and virulent hemorrhagic arenavirus infection. *Journal of Virology* 83:687–700.

Bradfute, S. B., D. R. Braun, J. D. Shamblin, J. B. Geisbert, J. Paragas, A. Garrison, L. E. Hensley, and T. W. Geisbert. 2007. Lymphocyte death in a mouse model of Ebola virus infection. *Journal of Infectious Diseases* 196:S296–S304.

Bradfute, S. B., K. S. Stuthman, A. C. Shurtleff, and S. Bavari. 2011. A STAT-1 knockout mouse model for Machupo virus pathogenesis. *Virology Journal* 8:300.

Bradfute, S. B., P. E. Swanson, M. A. Smith, E. Watanabe, J. E. McDunn, R. S. Hotchkiss, and S. Bavari. 2010. Mechanisms and consequences of Ebolavirus-induced lymphocyte apoptosis. *Journal of Immunology* 184:327–335.

Bradfute, S. B., K. L. Warfield, and S. Bavari. 2008. Functional CD8(+) T cell responses in lethal Ebola virus infection. *Journal of Immunology* 180:4058–4066.

Bray, M., K. Davis, T. Geisbert, C. Schmaljohn, and J. Huggins. 1998. A mouse model for evaluation of pro-phylaxis and therapy of Ebola hemorrhagic fever. *The Journal of Infectious Diseases* 178:651–661.

Bray, M. and T. W. Geisbert. 2005. Ebola virus: The role of macrophages and dendritic cells in the patho-genesis of Ebola hemorrhagic fever. *International Journal of Biochemistry and Cell Biology* 37: 1560–1566.

Bray, M., S. Hatfill, L. Hensley, and J. W. Huggins. 2001. Haematological, biochemical and coagulation changes in mice, guinea-pigs and monkeys infected with a mouse-adapted variant of Ebola Zaire virus. *Journal of Comparative Pathology* 125:243–253.

Chang, T. H., T. Kubota, M. Matsuoka, S. Jones, S. B. Bradfute, M. Bray, and K. Ozato. 2009. Ebola zaire virus blocks type I interferon production by exploiting the host SUMO modification machinery. *Plos Pathogens* 5:e1000493.

Cilloniz, C., H. Ebihara, C. Ni, G. Neumann, M. J. Korth, S. M. Kelly, Y. Kawaoka, H. Feldmann, and M. G. Katze. 2011. Functional genomics reveals the induction of inflammatory response and metallopro-teinase gene expression during lethal Ebola virus infection. *Journal of Virology* 85:9060–9068.

Debracco, M. M. E., M. T. Rimoldi, P. M. Cossio, A. Rabinovich, J. I. Maiztegui, G. Carballal, and R. M. Arana. 1978. Argentine hemorrhagic-fever—Alterations of complement-system and anti-Junin-virus humoral response. *New England Journal of Medicine* 299:216–221.

Dixon, L. K., C. C. Abrams, G. Bowick, L. C. Goatley, P. C. Kay-Jackson, D. Chapman, E. Liverani, R. Nix, R. Silk, and F. Q. Zhang. 2004. African swine fever virus proteins involved in evading host defence systems. *Veterinary Immunology and Immunopathology* 100:117–134.

Djavani, M., O. R. Crasta, Y. Zhang, J. C. Zapata, B. Sobral, M. G. Lechner, J. Bryant, H. Davis, and M. S. Salvato. 2009. Gene expression in primate liver during viral hemorrhagic fever. *Virology Journal* 6:20.

Djavani, M. M., O. R. Crasta, J. C. Zapata, Z. Fei, O. Folkerts, B. Sobral, M. Swindells et al. 2007. Early blood profiles of virus infection in a monkey model for Lassa fever. *Journal of Virology* 81:7960–7973.

Fan, L., T. Briese, and W. I. Lipkin. 2010. Z proteins of New World arenaviruses bind RIG-I and interfere with type I interferon induction. *Journal of Virology* 84:1785–1791.

Fennewald, S. M., J. F. Aronson, L. Zhang, and N. K. Herzog. 2002. Alterations in NF-kappaB and RBP-Jkappa by arenavirus infection of macrophages in vitro and in vivo. *Journal of Virology* 76:1154–1162.

Fennewald, S. M., E. P. Scott, L. Zhang, X. Yang, J. F. Aronson, D. G. Gorenstein, B. A. Luxon, R. E. Shope, D. W. Beasley, A. D. Barrett, and N. K. Herzog. 2007. Thioaptamer decoy targeting of AP-1 proteins influences cytokine expression and the outcome of arenavirus infections. *Journal of General Virology* 88:981–990.

Flatz, L., T. Rieger, D. Merkler, A. Bergthaler, T. Regen, M. Schedensack, L. Bestmann et al. 2010. T cell-depen-dence of Lassa Fever pathogenesis. *Plos Pathogens* 6:e1000836.

Geisbert, T. W., L. E. Hensley, T. R. Gibb, K. E. Steele, N. K. Jaax, and P. B. Jahrling. 2000. Apoptosis induced in vitro and in vivo during infection by Ebola and Marburg viruses. *Laboratory Investigation* 80:171–186.

Geisbert, T. W., L. E. Hensley, T. Larsen, H. A. Young, D. S. Reed, J. B. Geisbert, D. P. Scott, E. Kagan, P. B. Jahrling, and K. J. Davis. 2003. Pathogenesis of Ebola hemorrhagic fever in cynomolgus macaques: Evidence that dendritic cells are early and sustained targets of infection. *The American Journal of Pathology* 163:2347–2370.

Gomez-Villamandos, J. C., L. Carrasco, M. J. Bautista, M. A. Sierra, M. Quezada, J. Hervas, L. Chacon Mde et al. 2003. African swine fever and classical swine fever: a review of the pathogenesis. *Dtsch Tierarztl Wochenschr* 110:165–169.

Groseth, A., T. Hoenen, M. Weber, S. Wolff, A. Herwig, A. Kaufmann, and S. Becker. 2011. Tacaribe virus but not Junin virus infection induces cytokine release from primary human monocytes and macrophages. *Plos Neglected Tropical Diseases* 5:e1137.

Gupta, M., S. Mahanty, R. Ahmed, and P. E. Rollin. 2001. Monocyte-derived human macrophages and peripheral blood mononuclear cells infected with Ebola virus secrete MIP-1 alpha and TNF-alpha and inhibit poly-IC-induced IFN-alpha in vitro. *Virology* 284:20–25.

Gupta, M., C. Spiropoulou, and P. E. Rollin. 2007. Ebola virus infection of human PBMCs causes massive death of macrophages, CD4 and CD8 T cell sub-populations in vitro. *Virology* 364:45–54.

Hensley, L. E., H. A. Young, P. B. Jahrling, and T. W. Geisbert. 2002. Proinflammatory response during Ebola virus infection of primate models: possible involvement of the tumor necrosis factor receptor superfamily. *Immunology Letters* 80:169–179.

Jahrling, P. B., J. D. Frame, J. B. Rhoderick, and M. H. Monson. 1985. Endemic Lassa fever in Liberia.4. Selection of optimally effective plasma for treatment by passive-immunization. *Transactions of the Royal Society of Tropical Medicine and Hygiene* 79:380–384.

Ji, X., G. G. Olinger, S. Aris, Y. Chen, H. Gewurz, and G. T. Spear. 2005. Mannose-binding lectin binds to Ebola and Marburg envelope glycoproteins, resulting in blocking of virus interaction with DC-SIGN and complement-mediated virus neutralization. *Journal of General Virology* 86:2535–2542.

Kash, J. C., E. Muhlberger, V. Carter, M. Grosch, O. Perwitasari, S. C. Proll, M. J. Thomas, F. Weber, H. D. Klenk, and M. G. Katze. 2006. Global suppression of the host antiviral response by Ebola- and Marburgviruses: Increased antagonism of the type I interferon response is associated with enhanced virulence. *Journal of Virology* 80:3009–3020.

Kleiboeker, S. B. and G. A. Scoles. 2001. Pathogenesis of African swine fever virus in Ornithodoros ticks. *Animal Health Research Reviews* 2:121–128.

Lalani, A. S., J. W. Barrett, and G. McFadden. 2000. Modulating chemokines: More lessons from viruses. *Immunology Today* 21:100–106.

Leroy, E. M., S. Baize, V. E. Volchkov, S. P. Fisher-Hoch, M. C. Georges-Courbot, J. Lansoud-Soukate, M. Capron, P. Debre, J. B. McCormick, and A. J. Georges. 2000. Human asymptomatic Ebola infection and strong inflammatory response. *Lancet* 355:2210–2215.

Leroy, E. M., P. Becquart, N. Wauquier, and S. Baize. 2011. Evidence for Ebola virus superantigen activity. *Journal of Virology* 85:4041–4042.

Martinez-Sobrido, L., P. Giannakas, B. Cubitt, A. Garcia-Sastre, and J. C. de la Torre. 2007. Differential inhibition of type I interferon induction by arenavirus nucleoproteins. *Journal of Virology* 81: 12696–12703.

Martinez-Sobrido, L., E. I. Zuniga, D. Rosario, A. Garcia-Sastre, and J. C. de la Torre. 2006. Inhibition of the type I interferon response by the nucleoprotein of the prototypic arenavirus lymphocytic choriomeningitis virus. *Journal of Virology* 80:9192–9199.

Munoz-Jordan, J. L., G. G. Sanchez-Burgos, M. Laurent-Rolle, and A. Garcia-Sastre. 2003. Inhibition of interferon signaling by dengue virus. *Proceedings of the National Academy of Sciences of the United States of America* 100:14333–14338.

Nash, P., J. Barrett, J. X. Cao, S. Hota-Mitchell, A. S. Lalani, H. Everett, X. M. Xu et al. 1999. Immunomodulation by viruses: The myxoma virus story. *Immunological Reviews* 168:103–120.

Nasirudeen, A. M., H. H. Wong, P. Thien, S. Xu, K. P. Lam, and D. X. Liu. 2011. RIG-I, MDA5 and TLR3 synergistically play an important role in restriction of dengue virus infection. *Plos Neglected Tropical Diseases* 5:e926.

Neves, P. C., D. C. Matos, R. Marcovistz, and R. Galler. 2009. TLR expression and NK cell activation after human yellow fever vaccination. *Vaccine* 27:5543–5549.

Okumura, A., P. M. Pitha, A. Yoshimura, and R. N. Harty. 2010. Interaction between Ebola virus glycoprotein and host toll-like receptor 4 leads to induction of proinflammatory cytokines and SOCS1. *Journal of Virology* 84:27–33.

Pannetier, D., C. Faure, M. C. Georges-Courbot, V. Deubel, and S. Baize. 2004. Human macrophages, but not dendritic cells, are activated and produce alpha/beta interferons in response to Mopeia virus infection. *Journal of Virology* 78:10516–10524.

Pannetier, D., S. Reynard, M. Russier, A. Journeaux, N. Tordo, V. Deubel, and S. Baize. 2011. Human dendritic cells infected with the nonpathogenic Mopeia virus induce stronger T-cell responses than those infected with Lassa virus. *Journal of Virology* 85:8293–8306.

Parrino, J., R. S. Hotchkiss, and M. Bray. 2007. Prevention of immune cell apoptosis as potential therapeutic strategy for severe infections. *Emerging Infectious Diseases* 13:191–198.

Powell, P. P., L. K. Dixon, and R. M. Parkhouse. 1996. An IkappaB homolog encoded by African swine fever virus provides a novel mechanism for downregulation of proinflammatory cytokine responses in host macrophages. *Journal of Virology* 70:8527–8533.

Querec, T., S. Bennouna, S. Alkan, Y. Laouar, K. Gorden, R. Flavell, S. Akira, R. Ahmed, and B. Pulendran. 2006. Yellow fever vaccine YF-17D activates multiple dendritic cell subsets via TLR2, 7, 8, and 9 to stimulate polyvalent immunity. *The Journal of Experimental Medicine* 203:413–424.

Querec, T. D., R. S. Akondy, E. K. Lee, W. Cao, H. I. Nakaya, D. Teuwen, A. Pirani et al. 2009. Systems biology approach predicts immunogenicity of the yellow fever vaccine in humans. *Nature Immunology* 10:116–125.

Querec, T. D. and B. Pulendran. 2007. Understanding the role of innate immunity in the mechanism of action of the live attenuated Yellow Fever Vaccine 17D. *Advances in Experimental Medicine and Biology* 590:43–53.

Radoshitzky, S. R., L. A. Dong, X. O. Chi, J. C. Clester, C. Retterer, K. Spurgers, J. H. Kuhn, S. Sandwick, G. Ruthel, K. Kota, D. Boltz, T. Warren, P. J. Kranzusch, S. P. J. Whelan, and S. Bavari. 2010. Infectious Lassa Virus, but not Filoviruses, is restricted by BST-2/Tetherin. *Journal of Virology* 84:10569–10580.

Reed, D. S., L. E. Hensley, J. B. Geisbert, P. B. Jahrling, and T. W. Geisbert. 2004. Depletion of peripheral blood T lymphocytes and NK cells during the course of Ebola hemorrhagic fever in cynomolgus macaques. *Viral Immunology* 17:390–400.

Reid, S. P., L. W. Leung, A. L. Hartman, O. Martinez, M. L. Shaw, C. Carbonnelle, V. E. Volchkov, S. T. Nichol, and C. F. Basler. 2006. Ebola virus VP24 binds karyopherin alpha 1 and blocks STAT1 nuclear accumulation. *Journal of Virology* 80:5156–5167.

Reid, S. P., C. Valmas, O. Martinez, F. M. Sanchez, and C. F. Basler. 2007. Ebola virus VP24 proteins inhibit the interaction of NPI-1 subfamily karyopherin a proteins with activated STAT1. *Journal of Virology* 81:13469–13477.

Schnittler, H. J. and H. Feldmann. 1999. Molecular pathogenesis of filovirus infections: role of macrophages and endothelial cells. *Current Topics in Microbiology and Immunology* 235:175–204.

Silk, R. N., G. C. Bowick, C. C. Abrams, and L. K. Dixon. 2007. African swine fever virus A238L inhibitor of NF-kappaB and of calcineurin phosphatase is imported actively into the nucleus and exported by a CRM1-mediated pathway. *Journal of General Virology* 88:411–419.

Smith, G. L., J. A. Symons, A. Khanna, A. Vanderplasschen, and A. Alcami. 1997. Vaccinia virus immune evasion. *Immunological Reviews* 159:137–154.

Smith, V. P. and A. Alcami. 2000. Expression of secreted cytokine and chemokine inhibitors by ectromelia virus. *Journal of Virology* 74:8460–8471.

Tait, S. W., E. B. Reid, D. R. Greaves, T. E. Wileman, and P. P. Powell. 2000. Mechanism of inactivation of NF-kappa B by a viral homologue of I kappa b alpha. Signal-induced release of i kappa b alpha results in binding of the viral homologue to NF-kappa B. *The Journal of Biological Chemistry* 275:34656–34664.

ter Meulen, J., M. Badusche, K. Kuhnt, A. Doetze, J. Satoguina, T. Marti, C. Loeliger, K. Koulemou, L. Koivogui, H. Schmitz, B. Fleischer, and A. Hoerauf. 2000. Characterization of human CD4(+) T-cell clones recognizing conserved and variable epitopes of the Lassa virus nucleoprotein. *Journal of Virology* 74:2186–2192.

ter Meulen, J., M. Badusche, J. Satoguina, T. Strecker, O. Lenz, C. Loeliger, M. Sakho, K. Koulemou, L. Koivogui, and A. Hoerauf. 2004. Old and New World arenaviruses share a highly conserved epitope in the fusion domain of the glycoprotein 2, which is recognized by Lassa virus-specific human CD4+ T-cell clones. *Virology* 321:134–143.

Tulman, E. R., G. A. Delhon, B. K. Ku, and D. L. Rock. 2009. African swine fever virus. *Current Topics in Microbiology and Immunology* 328:43–87.

Valmas, C., M. N. Grosch, M. Schumann, J. Olejnik, O. Martinez, S. M. Best, V. Krahling, C. F. Basler, and E. Muhlberger. 2010. Marburg virus evades interferon responses by a mechanism distinct from Ebola virus. *Plos Pathogens* 6:e1000721.

Villinger, F., P. E. Rollin, S. S. Brar, N. F. Chikkala, J. Winter, J. B. Sundstrom, S. R. Zaki, R. Swanepoel, A. A. Ansari, and C. J. Peters. 1999. Markedly elevated levels of interferon (IFN)-gamma, IFN-alpha, interleukin (IL)-2, IL-10, and tumor necrosis factor-alpha associated with fatal Ebola virus infection. *Journal of Infectious Diseases* 179:S188–S191.

Wauquier, N., P. Becquart, C. Padilla, S. Baize, and E. M. Leroy. 2010. Human fatal Zaire Ebola virus infection is associated with an aberrant innate immunity and with massive lymphocyte apoptosis. *Plos Neglected Tropical Diseases* 4:e837.

Wolf, K., N. Beimforde, D. Falzarano, H. Feldmann, and H. J. Schnittler. 2011. The Ebola Virus Soluble Glycoprotein (sGP) does not affect lymphocyte apoptosis and adhesion to activated endothelium. *The Journal of Infectious Diseases* 204(Suppl 3):S947–S952.

4 Vascular Endothelium and Hemorrhagic Fever Viruses

*Nadine A. Dalrymple, Elena Gorbunova,
Irina Gavrilovskaya, and Erich R. Mackow*

CONTENTS

4.1 INTRODUCTION

Viremia disseminates virus and concomitantly exposes the vasculature to viral pathogens (Cines et al., 1998; Valbuena and Walker, 2006). Yet only a few viruses specifically target the endothelial cell (EC) lining of capillaries and cause vascular permeability deficits that result in acute edematous or hemorrhagic disease (Cines et al., 1998; Martina et al., 2009; Valbuena and Walker, 2006; Zaki et al., 1995). Hantaviruses, dengue viruses, and ebolaviruses clearly infect the endothelium and cause dramatic changes in capillary functions that result in vascular pathology (Basu and Chaturvedi, 2008; Cines et al., 1998; Diamond et al., 2000a; Gavrilovskaya et al., 2008; Halstead, 2008; Jessie et al., 2004; Koster and Mackow, 2012; Lahdevirta, 1982; Lee, 1982b; Lee et al., 1982; Zaki et al., 1995). The mechanisms by which these viruses alter the fluid barrier integrity of the endothelium are varied but commonly involve dysregulating EC functions and as a result emphasize the importance of vessel stability in preventing highly lethal viral diseases (Aird, 2004; Cines et al., 1998; Gorbunova et al., 2010, 2011; Halstead, 2008).

The vasculature is lined by a single layer of ECs that collectively form one of the largest tissues of the body (Aird, 2004; Valbuena and Walker, 2006). The endothelium forms a primary fluid barrier within vessels but serves as more than just a conduit for blood to reach and return from organs. The endothelium selectively restricts blood and plasma from entering tissues, regulates immune cell infiltration, and responds to damage by limiting leakage, repairing vessels and directing angiogenesis (Fishman, 1972; Pober et al., 2009). These ubiquitous functions require the endothelium to respond to a myriad of systemic and locally generated factors that alter inter-EC adherence and fluid barrier properties. ECs respond to platelets, immune cells, clotting cascades, chemokines, cytokines, growth factors, nitric oxide, and hypoxic conditions (Berger et al., 2005; Breen, 2007; Cines et al., 1998; Coller and Shattil, 2008; Dvorak et al., 1995). However, ECs also secrete chemokines, cytokines, complement, and growth factors that positively or negatively impact the adherence and activation of platelets and immune cells; regulate responses to hypoxia; and diminish or enhance extravasation of fluid into tissues (Bahram and Claesson-Welsh, 2010; Berger et al., 2005; Cines et al., 1998). Each of these responses is controlled by a diverse mesh of intertwined sensors and signals aimed at returning the endothelium to a resting state, countering permeabilizing effectors, repairing vessel damage, and restoring fluid and oxygenation levels within tissues (Aird, 2004; Matthay et al., 2002; Pober et al., 2009).

Arteries, veins, lymphatic vessels, and the blood–brain barrier each contain a unique endothelium that is central to functions of vast renal and pulmonary capillary beds as well as the selective emigration of blood constituents into the brain. Although edema results from increased endothelial permeability, the failure of lymphatic vessels to remove fluid from tissues is also a cause of edema and uniquely regulated by the function of lymphatic ECs (LECs) (Bahram and Claesson-Welsh, 2010; Breslin et al., 2007; Schraufnagel, 2010). The failure to regulate fluid accumulation in tissues has pathologic consequences resulting in edema, shock, hemorrhage, and, depending on the location of vascular dysfunction, encephalitis and acute renal or pulmonary insufficiency (Cines et al., 1998; Hanaoka et al., 2003; Pober et al., 2009; Zaki et al., 1995). Consequently, capillary barrier integrity is redundantly regulated by a constellation of EC-specific receptors, adherence junction proteins, phosphatases, and secreted factors that coordinately balance vascular fluid containment with tissue-specific needs (Broholm and Laursen, 2004; Coller and Shattil, 2008; Dvorak et al., 1995; Satchell et al., 2004).

Viral infections disrupt normal EC functions and have the potential to alter fluid barrier functions of the vasculature through lysis or by altering normal fluid barrier functions of capillary ECs (Basu and Chaturvedi, 2008; Gavrilovskaya et al., 2008; Valbuena and Walker, 2006; Zaki et al., 1995). It is clear that hemorrhagic disease resulting from *Ebola virus* is a consequence of lytic viral infection of ECs that physically disrupts the endothelium (Clark et al., 1998). Vascular permeability induced by nonlytic viruses is likely to result from a combination of virally altered EC responses, changes in activation of clotting factors, immune cells, and platelets, or a collaboration of several dysregulated systems that impact normal EC function. Here, we focus on the mechanisms by which nonlytic dengue virus and hantavirus infections of ECs induce vascular permeability, acute edema, and hemorrhagic disease.

4.2 VIRAL INFECTIONS OF THE ENDOTHELIUM

The ability of viruses to infect human ECs has been known for decades to affect the function of the endothelium (Friedman et al., 1981, 1986). Herpes viruses, coxsackie virus, simian immunodeficiency virus, human immunodeficiency virus, measles virus, Nipah virus, hantavirus, West Nile virus, dengue virus, arenaviruses, filoviruses, parainfluenza virus, poliovirus, adenovirus, and many others reportedly infect human ECs (Balsitis et al., 2009; Basu and Chaturvedi, 2008; Dalrymple and Mackow, 2011; Friedman et al., 1986; Jessie et al., 2004; Kunz, 2009; Schnittler et al., 1993; Steffan et al., 1992; Verma et al., 2009; Wong et al., 2002). Despite this, only a few viruses cause vascular permeability deficits resulting in hemorrhagic or edematous disease (Basu and Chaturvedi, 2008; Halstead, 2008; Lee, 1982a; Moraz and Kunz, 2010; Zaki et al., 1995). Studies of the infected endothelium are highly invasive, and as a result, the effects of viral infection on the endothelium remain largely unknown but central to understanding the mechanisms of virally induced vascular diseases.

4.3   REGULATION OF ENDOTHELIAL CELL PERMEABILITY

The endothelium lines a series of discrete vessel types that conduct fluids to and from tissues; direct the transfer of nutrients, wastes, and oxygen; and coordinate tissue responses to changing conditions and pathogens. Each of these vessels restricts the transfer of plasma and immune cells through uniquely regulated EC functions (Aird, 2004; Bongrazio et al., 2003; Dvorak et al., 1995; Schraufnagel, 2010; Valbuena and Walker, 2006). Vascular ECs serve mainly as a conduit for fluid in high-pressure arteries while permitting the selective permeability of low-pressure veins and capillaries that innervate every organ and tissue of the body (Aird, 2004; Schraufnagel, 2010). These diverse EC settings require discrete functions within kidney, liver, and pulmonary capillary beds where a high level of exchange is required. Some ECs are fenestrated to meet the needs of specific organ systems and EC functions within each organ that are regulated by unique conditions, requirements, and regulatory factors that locally distinguish EC responses and functions (Cines et al., 1998; Schraufnagel, 2010).

Lymphatic vessels have a primary role in draining fluid, proteins, and immune cells from tissues and returning these components to the venous circulation (Bahram and Claesson-Welsh, 2010; Breslin and Kurtz, 2009; Mutlu and Sznajder, 2005; Saharinen et al., 2004; Schraufnagel, 2010). The lymphatic endothelium has a discontinuous fenestrated basement membrane that is permeable to interstitial fluid and cells and that expands and contracts in response to excess fluid load. Depending on their location, lymphatic vessels serve discrete fluid barrier and regulatory functions, keeping pulmonary alveolar spaces dry and clearing fluid influx from the lungs (Breslin et al., 2007; Schraufnagel, 2010). Failure of lymphatic vessels to clear fluids results in lymphedema and has spawned a growing interest in lymphatic EC functions, regulation, and contributions to edematous disease. Although there is far less known about LECs, they express a unique constellation of cell-surface receptors, and their integrity is regulated by discrete growth factors (Bahram and Claesson-Welsh, 2010; Breslin et al., 2007; Schraufnagel, 2010).

Unless activated, the endothelium normally prevents immune cells and platelets from adhering to its surface (Aird, 2004; Nalbandian and Henry, 1978). Endothelial quiescence is maintained by several mediators that prevent activation of clotting factors, platelets, and ECs. Vascular injury activates clotting factors, platelets, and ECs resulting in the recruitment of platelets to the endothelium (Cines et al., 1998; Pober et al., 2009; Stoermer and Morrison, 2011). Immune cells recruited to the site of infection locally release chemokines and cytokines that can increase vascular permeability and contribute to disease (Cines et al., 1998; Dvorak, 2006; Hodivala-Dilke et al., 1999; Reynolds et al., 2002). Type I interferons (IFNs) also induce EC proliferation (Gomez and Reich, 2003), and thus, IFN induction may contribute to both pathogen protection and repair of the endothelium. EC migration and vessel remodeling also require changes in cell adherence within the endothelium, and a variety of carefully orchestrated receptor interactions and signaling responses are required to accomplish this without increasing vascular permeability (Aird, 2004; Coller and Shattil, 2008; Dvorak, 2010; Robinson et al., 2004; Weis and Cheresh, 2011).

Intracellular signaling pathways coordinately regulate the adherence of ECs to the extracellular matrix, anchor receptors to cytoskeletal elements, and induce growth factor–directed migration, proliferation, and permeability responses (Acevedo et al., 2008b; Borges et al., 2000; Galvagni et al., 2010; Lampugnani et al., 2006; Olsson et al., 2006; Soker et al., 2002; Weitzman et al., 2008). Signals and complexes of diverse receptors combine to direct or regulate pathways induced by additional factors and ligands (Garcia et al., 2001; Robinson et al., 2004; Soker et al., 2002; Stenmark et al., 2006). Permeabilizing responses are localized since circulating soluble receptors bind and inactivate released factors in order to prevent systemic vascular permeability (Dvorak et al., 1991). As a result, the regulation of vascular integrity is a cumulative response to signals that induce localized EC permeability, direct EC responses to immune cells, control angiogenesis, and regulate EC migration cues (Borges et al., 2000; Breen, 2007; Coller and Shattil, 2008;

Gavard, 2009; Robinson et al., 2004). This multifactorial coordination is the reason why there are so many factors and receptor responses that are capable of both stabilizing the endothelium and altering vascular permeability.

Lysis of ECs, apoptosis, or cytotoxic immune cell targeting are obvious means for viruses to disrupt the endothelium and cause hemorrhagic disease. However, there are far more subtle effects on EC functions that result in vascular permeability and contribute to pathogenesis. ECs contain a unique set of receptors that specifically regulate inter-EC adherence, binding to the extracellular matrix, EC–immune cell and EC–platelet interactions, and the migration and proliferation of ECs (Galvagni et al., 2010; Garcia et al., 2001; Robinson et al., 2004; Soker et al., 2002; Stenmark et al., 2006). Nearly all of these interactions positively or negatively impact vascular integrity by effecting changes in inter-EC adherens junctions (AJs). AJs provide the primary fluid barrier function of the endothelium and are composed of EC-specific vascular-endothelial cadherin (VE-cadherin). EC barrier functions are increased by the cell-surface expression of VE-cadherin and reduced by the dissociation and internalization of VE-cadherin (Dejana et al., 2008; Gavard and Gutkind, 2006; Wallez et al., 2006).

ECs uniquely respond to VE growth factor (VEGF) and angiopoietin growth factors that have opposing effects on vascular permeability. VEGF receptors (VEGFR1/2/3) on ECs respond to novel forms of VEGF (A–E) that control AJ disassembly (Breslin et al., 2007; Holmes et al., 2007; Neufeld et al., 1999). In response to VEGF-A, VEGFR2 activates a Src–Rac–Pak–VE-cadherin phosphorylation pathway resulting in increased vascular permeability (Dejana et al., 2008; Gavard and Gutkind, 2006). EC-specific phosphatases return VE-cadherin to an unphosphorylated resting state restoring AJ assembly and EC integrity (Broermann et al., 2011). Angiopoietin-1 (Ang-1) transdominantly inhibits VEGF-induced EC permeability through interactions with Tie-2 receptors (Baffert et al., 2006; Cascone et al., 2005; Gavard et al., 2008; Satchell et al., 2004; Thurston et al., 2000).

VEGF was originally identified as a vascular permeability factor for its ability to potently cause tissue edema ~50,000 times more effectively than histamine (Dvorak, 2006; Dvorak et al., 1995). VEGF is secreted by ECs and immune cells and acts within 0.5 mm of its release to induce localized EC dissociation required for vessel repair and angiogenesis (Dvorak et al., 1991). VEGFR2 forms ectodomain complexes with $\alpha_v\beta_3$ integrins that direct cellular migration and VEGFR2-$\alpha_v\beta_3$ complexes temper permeabilizing effects of VEGF. In fact, knocking out β_3 integrins or antagonizing $\alpha_v\beta_3$ results in enhanced VEGF-A that directed permeability responses within capillaries in vivo and in vitro (Borges et al., 2000; Robinson et al., 2004; Weis and Cheresh, 2011).

VEGF is also induced by hypoxia and causes high-altitude-induced pulmonary edema (HAPE) (Berger et al., 2005; Christou et al., 1998; Dehler et al., 2006; Stenmark et al., 2006). HAPE results from the activation of the hypoxia-induced VEGF transcription, and VEGF further directs the transcription of the HIF-1α amplifying permeability responses of the endothelium, causing edema (Berger et al., 2005; Stenmark et al., 2006).

Although VEGF is an extraordinarily potent vascular permeability factor, additional factors like TNFα, platelet-activating factor (PAF), histamine, and nitric oxide also contribute to localized vessel edema and vascular permeability (Coller and Shattil, 2008; Dvorak, 2006; Dvorak et al., 1995). Thus, altering platelets or EC receptor activation can have pathologic effects that are dependent on the location of vessel dysfunction. The extensive regulation of EC permeability demonstrates the central importance in preventing vascular dysfunction. Nevertheless, the intricacies of regulating vascular barrier functions highlight the potential for dysregulating EC responses to contribute to pathogenesis.

4.4 HANTAVIRUSES AND DENGUE VIRUSES INFECT HUMAN ENDOTHELIAL CELLS

Hemorrhagic diseases result from large inter-EC gaps or physical damage to the endothelium that permits both fluid and cellular effluxes into surrounding tissues (Halstead, 2008; Koster and

Mackow, 2012; Lee, 1982a; Martina et al., 2009). Virally induced hemorrhagic diseases may also result from acute thrombocytopenia, clotting factor deficits, or vasculitis (Cosgriff et al., 1991; Halstead, 2008; Lee, 1982a; Schmaljohn, 2001). The evidence of induced hemorrhagic disease is often present in mucosal membranes or limited to localized hemorrhagic petechia or purpura (Halstead, 2008; Lahdevirta, 1982). It is clear that *Ebola virus* causes hemorrhagic disease by physically disrupting the endothelium, although lysis may be preceded by edema (Feldmann et al., 1996; Kortepeter et al., 2011; Schnittler et al., 1993). In contrast, dengue virus and hantavirus infections nonlytically cause hemorrhagic and edematous diseases, yet dengue virus and hantavirus appear to cause vascular leakage through completely discrete nonlytic mechanisms.

Dengue virus is a mosquito-borne flavivirus that is not spread person to person (Halstead, 2008). Dengue virus infects a number of cell types, including ECs and immune and dendritic cells (Halstead, 2008; Jessie et al., 2004; Luplertlop et al., 2006). Dengue causes a mild febrile illness, dengue fever (DF), and two highly lethal vascular permeability–based diseases with 5%–30% mortality rates: dengue hemorrhagic fever (DHF) and dengue shock syndrome (DSS) (Halstead, 2008). Most DF patients are asymptomatic or display mild symptoms that include the rapid onset of fever, headache, pain, and rash (Halstead, 2008; Martina et al., 2009). However, DSS and DHF are edematous and hemorrhagic diseases, respectively, which occur in the absence of EC lysis (Halstead, 2008). The appearance of DHF and DSS is enhanced by the presence of antibodies to a second dengue virus serotype, termed antibody-dependent enhancement (ADE) (Halstead, 2008; Takada and Kawaoka, 2003). Patients with DHF and DSS display additional symptoms including increased edema, hemorrhage, thrombocytopenia, shock, and viremia. The hemorrhagic manifestations of dengue virus are evident as a rash and as more extended localized regions of disseminated intravascular coagulation (Gubler, 2006; Halstead, 2008; Martina et al., 2009).

Hantaviruses are the only members of the *Bunyaviridae* that are transmitted by the small mammal hosts (Schmaljohn, 2001). Hantaviruses nonlytically cause two highly lethal vascular leak–based diseases: hemorrhagic fever with renal syndrome (HFRS) (Lee, 1982a) and hantavirus pulmonary syndrome (HPS) (Schmaljohn, 2001). HFRS-causing hantaviruses are predominately Eurasian, while HPS-causing hantaviruses are present throughout the Americas. *Hantaan virus* is the namesake of hantavirus originally found as a cause of Korean hemorrhagic fever (KHF) (Lee, 1982a,b). There are several additional HFRS-associated hantaviruses including European Puumala virus (PUUV) and Dobrava–Belgrade viruses (DOBVs) that are present in discrete regions coincident with their primary small mammal hosts (Schmaljohn, 2001). HPS-causing viruses include North American *Sin Nombre virus* (SNV) and South American *Andes virus* (ANDV) (Enria et al., 1996; Lopez et al., 1996; Nichol et al., 1993). Except for ANDV, person-to-person spread is not observed during hantavirus infections (Enria et al., 1996; Padula et al., 1998).

Hantaviruses are one of only a few viruses that predominantly infect human ECs suggesting a direct effect of viral infection on EC functions that contribute to pathogenesis (Cosgriff and Lewis, 1991; Nolte et al., 1995; Pensiero et al., 1992; Yanagihara and Silverman, 1990; Zaki et al., 1995). Both diseases cause tissue edema in major organs along with hallmark acute thrombocytopenia. HPS results in acute pulmonary edema with fluid accumulation of up to a liter per hour that results in respiratory insufficiency (Bustamante et al., 1997; Duchin et al., 1994; Enria et al., 1996; Koster and Mackow, 2012; Lopez et al., 1996; Padula et al., 1998; Zaki et al., 1995). HFRS patients have prominent edema of organ capsules with pronounced effects on kidney function and localized mucosal capillary beds showing hemorrhagic sequelae. HFRS is also associated with prominent immune complex formation and deposition that may contribute to hemorrhagic disease and renal dysfunction (Gavrilovskaya et al., 1987; Lahdevirta, 1982; Lee, 1982a).

Although hantaviruses and dengue viruses nonlytically cause edematous and hemorrhagic diseases, they appear to permeabilize the endothelium through completely distinct mechanisms. Hantaviruses primarily infect ECs and alter the function of ECs as a primary means of permeabilizing capillaries (Gavrilovskaya et al., 2008; Gorbunova et al., 2010, 2011). In contrast, dengue infects both endothelial and immune cells and induces immune-enhanced permeability responses that act on the endothelium (Avirutnan et al., 2006; Dalrymple and Mackow, 2011, 2012; Halstead, 2008; Jessie et al., 2004). The mechanisms by which these nonlytic viruses impact EC functions and dysregulate vascular barrier functions are just beginning to be uncovered and underlie our ability to therapeutically address virally induced vascular diseases.

4.5 VIRAL-ENDOTHELIAL EDEMAGENIC MECHANISMS

The means by which viruses cause capillary leakage by altering the function of vascular ECs is complicated by layers of vascular barrier regulatory mechanisms, tissue-specific responses of ECs, and unique EC regulation within distinct vessel systems. Edema results from the accumulation of plasma fluids within tissues and is caused by vascular leakage, transcytosis, altered ion channel functions, or failure to clear fluid from tissues (Aird, 2004; Cines et al., 1998; Pober et al., 2009). Edema resulting from viral infections can lead to localized organ dysfunction or systemic fluid loss resulting in shock (Halstead, 2008; Koster and Mackow, 2012).

Severe dengue-induced disease results in DSS or DHF with edematous vascular leakage playing a central role in the pathogenesis of both diseases but with clear links to an immune-enhanced disease process (Gubler, 2006; Halstead, 2008; Martina et al., 2009). Edema is also a prominent component of HPS and HFRS diseases but appears to primarily be the result of virally induced changes in EC functions (Koster and Mackow, 2012). In fact, pulmonary and renal dysfunctions are components of both HPS and HFRS diseases and likely stem from hantavirus infection of ECs within vast alveolar and renal capillary beds (Koster and Mackow, 2012; Lahdevirta, 1982; Zaki et al., 1995). Although just beginning to be understood mechanistically, the causes of DHF, DSS, HPS, and HFRS are likely to be multifactorial in nature (Gavrilovskaya et al., 2010; Gorbunova et al., 2010; Halstead, 2008; Hammerbeck and Hooper, 2011; Kilpatrick et al., 2004; Koster and Mackow, 2012; Krakauer et al., 1995; Rothman, 2010). However, pathogenesis specifically revolves around the ability of viruses to infect ECs and alter EC responses that normally regulate capillary leakage (Dalrymple and Mackow, 2012; Gavrilovskaya et al., 2008, 2010; Shrivastava-Ranjan et al., 2010).

4.5.1 ENDOTHELIAL CELL RESPONSES TO HANTAVIRUS INFECTION

The absence of cytopathic effect following hantavirus infection suggests that nonlytic mechanisms direct edematous and hemorrhagic HFRS patient responses (Gavrilovskaya et al., 2008; Gorbunova et al., 2010; Koster and Mackow, 2012; Lahdevirta, 1982; Sundstrom et al., 2001). Clues to the mechanism of hantavirus-induced edema come from disparate findings on the role of hypoxia in acute pulmonary edema and the role of vascular permeability regulators that are uniquely altered during hantavirus infection of ECs (Gavrilovskaya et al., 2008; Koster and Mackow, 2012). Hantavirus infections result in a long prodromal period prior to respiratory or renal effects. HFRS patients are hospitalized with high fever and shock caused by vascular leakage from all organs (Lahdevirta, 1982). HPS patients seek medical attention when already in edematous or hypoxic states that rapidly progress. The volume and rate of fluid accumulation in HPS patient lungs are unprecedented, and recent findings of elevated VEGF levels within acute-stage HPS patient pulmonary edema fluid are consistent with a hypoxia-induced VEGF amplification loop during the early phase of HPS (Dehler et al., 2006; Gavrilovskaya et al., 2012; Koster and Mackow, 2012; Pham et al., 2002; Stenmark et al., 2006). Hypoxia is a prominent component of HPS patients and directs VEGF secretion from endothelial, epithelial, and immune cells (Dehler et al., 2006; Koster and Mackow, 2012; Zaki et al., 1995). Although this response is aimed at increasing gas exchange, in continued low-oxygen environments, these cellular responses instead cause pulmonary edema and respiratory distress (Berger et al., 2005; Hopkins et al., 2005; Stenmark et al., 2006). Although a demonstrated role for hypoxia in hantavirus-induced permeability has yet to be investigated, the ability of extracorporeal membrane oxygenation (ECMO) to reduce HPS patient mortality strongly suggests a role for hypoxia and VEGF in the acute pulmonary edema of HPS patients (Hanaoka et al., 2003; Koster and Mackow, 2012).

Curiously, monolayers of hantavirus-infected ECs are not permeabilized by infection alone (Gavrilovskaya et al., 2008; Sundstrom et al., 2001). This suggests that hantaviruses alter EC responses that control vascular permeability rather than directly inducing permeability. EC responses to infection by HFRS- and HPS-causing viruses demonstrate that both common and discrete EC transcriptional responses are elicited (Geimonen et al., 2002). Notably, pathogenic HTNV and NY-1V both regulate early IFN responses following infection of ECs compared to PHV that fails to replicate in human ECs (Alff et al., 2006, 2008; Geimonen et al., 2002). Thus, at one level, IFN regulation is required for hantaviruses to be pathogenic. Further changes demonstrated the specific induction of immune-enhancing chemokine responses within HTNV-infected ECs compared to HPS-causing NY-1V-infected ECs (Geimonen et al., 2002). Whether these responses contribute to differences in the hemorrhagic versus edemagenic permeability of the endothelium during infections has yet to be defined. However, these findings are consistent with the high-level production of immune complexes in HTNV infections that may contribute to both renal and hemorrhagic manifestations (Gavrilovskaya et al., 1987; Lahdevirta, 1982; Zaki et al., 1995).

Interestingly, thus far, pathogenic hantaviruses (HTNV, ANDV, SNV, NY-1V), but not nonpathogenic hantaviruses (PHV, TULV), all reportedly bind to and inactivate $\alpha_v\beta_3$ integrins (Gavrilovskaya et al., 1998, 1999, 2008; Raymond et al., 2005). Although this initially directs viral entry, hantavirus $\alpha_v\beta_3$ integrin interactions only inhibit $\alpha_v\beta_3$ integrin functions at late times post-infection (Gavrilovskaya et al., 2002, 2008). Days after infection, hantaviruses remain cell associated by binding to inactive, basal conformations of $\alpha_v\beta_3$ integrins (Gavrilovskaya et al., 2010; Goldsmith et al., 1995; Raymond et al., 2005). This blocks the migration of pathogenic hantavirus-infected ECs days after infection (Gavrilovskaya et al., 2002).

However, β_3 integrins normally regulate vascular permeability through the formation of ectodomain $\alpha_v\beta_3$–VEGFR2 complexes that restrict VEGF-directed permeabilizing responses (Borges et al., 2000). In fact, knocking out or antagonizing β_3 integrin functions results in the hyperpermeability of ECs in response to VEGF (Hodivala-Dilke et al., 1999; Reynolds et al., 2002; Robinson et al., 2004; Weis and Cheresh, 2011). Hantavirus inactivation of $\alpha_v\beta_3$ is consistent with several studies indicating that the addition of VEGF to pathogenic hantavirus-infected cells enhances VEGFR2 phosphorylation, VE-cadherin internalization, and EC permeability (Gavrilovskaya et al., 2008; Gorbunova et al., 2010, 2011). The ability of pathogenic hantaviruses to bind and inactivate $\alpha_v\beta_3$ days after infection (Gavrilovskaya et al., 2010; Raymond et al., 2005) may similarly drive VEGF-directed permeability in the infected endothelium, and hantavirus dysregulation of $\alpha_v\beta_3$ appears to be critical to increased capillary permeability.

4.5.2 POTENTIAL ROLE OF LECs IN HANTAVIRUS EDEMA

Unique VEGFR2 and VEGFR3 responses regulate pulmonary fluid clearance by lymphatic vessels (Bahram and Claesson-Welsh, 2010; Breslin et al., 2007; Schraufnagel, 2010). LECs contain $\alpha_v\beta_3$ and VEGFR2 and VEGFR3 receptors that act in concert to regulate LEC permeability and growth (Saharinen and Petrova, 2004; Schraufnagel, 2010). Interestingly, VEGFR3 receptors are uniquely present on LECs and respond to VEGF-C but not VEGF-A (Bahram and Claesson-Welsh, 2010; Breslin et al., 2007). VEGFR3 activation is associated with reduced tissue edema and inhibiting VEGFR3 signaling results in lymphedema (Bahram and Claesson-Welsh, 2010). However, VEGFR responses of LECs are further complicated by findings that VEGFR2 forms a heterodimeric complex with VEGFR3 that is uniquely regulated by VEGF-A and VEGF-C (Bahram and Claesson-Welsh, 2010; Nilsson et al., 2010; Olsson et al., 2006). It remains unclear whether inactivating $\alpha_v\beta_3$ integrins also impacts VEGFR2/3 responses of LECs. In addition, the role of lymphatic vessels and LEC responses in HPS is currently unknown, and it is unclear whether hantaviruses infect LECs or alter LEC barrier functions that regulate fluid clearance from the lung.

4.5.3 Role of Platelets in Viral-Induced Hemorrhagic Disease

There is a little understanding of how thrombocytopenia contributes to hemorrhagic manifestations of hantavirus infections since both HFRS and HPS patients are acutely thrombocytopenic (Schmaljohn, 2001). Both dengue and hantavirus infections also cause acute thrombocytopenia; however, it is unclear how these viruses alter platelet functions (Halstead, 2008; Zaki et al., 1995). Complement, chemokine, and cytokine responses may contribute to plasma leakage and as a result thrombocytopenia. Cosgriff has shown the absence of activated platelets in hantavirus patients (Cosgriff et al., 1991; Cosgriff and Lewis, 1991). Although it is reported that dengue virus–infected ECs recruit platelets to their surface, the evidence for this is less than convincing (Krishnamurti et al., 2002). Hantaviruses have been shown to directly bind platelets through interactions with platelet β_3 integrins that are highly expressed on the surface of platelets (Coller and Shattil, 2008; Gavrilovskaya et al., 2010). Furthermore, at late times post-infection cell–associated hantaviruses accumulate on the surface of infected ECs and reportedly recruit quiescent platelets to the EC surface (Gavrilovskaya et al., 2010). Thus, hantavirus-infected capillary beds may be covered by inactive platelets on the surface of the endothelium. This may effectively mask the presence of virus on ECs and deplete platelets from plasma. These reports demonstrate that hantaviruses on the surface of ECs functionally alter normal EC interactions with platelets and may modify the surface of the infected endothelium. Whether this is a means of inactivating platelets or contributing to thrombocytopenia has yet to be determined. Immune responses in HFRS and DHF have been reviewed extensively by others (Halstead, 2008, 2009; Kanerva et al., 1998; Kilpatrick et al., 2004; Rothman, 2010, 2011; Vapalahti et al., 2001) and in HFRS point to a role for the complement, thrombocytopenia, and immune complexes as collectively contributing to hemorrhagic manifestations (Gavrilovskaya et al., 1987; Lahdevirta, 1982).

4.5.4 Endothelial Cell Responses to Dengue Virus Infection

Unlike hantaviruses, the dengue virus–induced vascular permeability diseases, DSS and DHF, primarily occur through ADE, the result of preexisting maternal antibodies to dengue virus or a second dengue virus infection by a discrete dengue serotype (Fink et al., 2006; Halstead, 2008; Takada and Kawaoka, 2003). The role of the endothelium in this process is largely dismissed since antibodies facilitate infection of immune cells. However, this ignores the fact that ECs are themselves capable of enhancing immune responses and that ECs are the ultimate targets of immune responses that cause vascular leakage (Avirutnan et al., 1998; Azizan et al., 2006; Basu and Chaturvedi, 2008; Srikiatkhachorn, 2009). A severe dengue disease is found to occur in the presence of elevated levels of cytokines and chemokines in the blood of DHF and DSS patients including IP-10, ITAC, IL-1β, IL-2, IL-6, IL-8, IL-10, IL-12, and IL-13; TNFα, IFNα, and IFNγ; MIF; RANTES; histamine; and complement proteins properdin, C3, C3a, and C5a (Avirutnan et al., 2006; Fink et al., 2006, 2007; Gubler, 2006; Halstead, 2008; Stoermer and Morrison, 2011). Many of these cytokines, chemokines, and complement-activating factors can be elicited by and act on the endothelium, influencing EC regulation of fluid barrier function and vascular leakage.

When the dengue virus infection of ECs is examined kinetically in vitro, an important picture emerges for the role of the endothelium in a dengue virus disease. At an MOI of six, primary human ECs are ~80% infected by the dengue virus and rapidly produce virus by 12–24 h post-infection (hpi) (Dalrymple and Mackow, 2011). Dengue virus titers from ECs are nearly identical to viral titers observed in IFN-deficient VeroE6 cells suggesting that virus regulates early cellular IFN responses of ECs. Consistent with the rapid early replication of the dengue virus within ECs 12–24 hpi, dengue virus infection induces IFN responses 24–48 hpi (Dalrymple and Mackow, 2012). Prior reports found that dengue virus infected only 2%–10% of ECs and resulted in little or no cytokine induction (Avirutnan et al., 1998; Huang et al., 2000; Peyrefitte et al., 2006; Warke et al., 2003). However, these studies were performed on 90%–98% uninfected cells and complicated by the use of ECV304 cell lines that are known to be bladder carcinoma rather than human ECs (Avirutnan et al., 1998;

Dirks et al., 1999; Warke et al., 2003). Other dengue-induced EC responses were also analyzed in continuous cell lines HMEC-1, HPMEC-ST1.6R, or LSEC-1 that are not primary ECs (Avirutnan et al., 1998; Azizan et al., 2006, 2009; Lacroix, 2008; Talavera et al., 2004). Dengue infection of the HPMEC-ST1.6R cell line increased IL-6 and IL-8 as well as VEGF but only 6–8 days after infection when it is unclear if ECs are infected (Azizan et al., 2006, 2009). Others have found no induction of VEGF (Azizan et al., 2009; Dvorak, 2010) in primary human ECs or little or no IL-6 and IL-8 induction at early times post-infection of primary human ECs (Warke et al., 2003). Little is known about EC responses following natural dengue virus infection or how immune responses that target or alter EC functions contribute to dengue pathogenesis.

The ability of dengue virus to infect the endothelium intimates that additional mechanisms may contribute to vascular permeability deficits through both direct- and immune-enhanced disease processes (Balsitis et al., 2009; Dalrymple and Mackow, 2012; Halstead, 2008; Jessie et al., 2004). Preexisting antibodies may target viral proteins displayed by infected ECs, and dengue-infected ECs may elicit chemokine and cytokine responses that further activate or recruit immune cells to the endothelium (Fink et al., 2006; Halstead, 2003; Takada and Kawaoka, 2003). In particular, secreted dengue NS1 protein and cross-reactive NS1 antibodies are present in high quantities and have been shown to bind the surface of platelets and ECs (Libraty et al., 2002; Lin et al., 2005). Thus, EC expressed proteins may be targets of NS1 antibodies and contribute to immune-enhanced pathogenesis in the presence of cross-reactive preexisting NS1 antibodies (Lin et al., 2003). Additionally, a recent analysis of EC transcriptional responses indicates that dengue virus strongly induces ECs to secrete immune cell activating cytokines and chemokines that are likely to contribute to an immune-enhanced disease process (Dalrymple and Mackow, 2012). Other studies have also singled out RANTES, IL-6, and/or IL-8 as cytokines elicited by dengue-infected ECs as potential contributors to dengue pathogenesis (Avirutnan et al., 1998; Azizan et al., 2006; Bosch et al., 2002; Huang et al., 2000; Talavera et al., 2004; Warke et al., 2003). These findings foster the endothelium as a source of potent chemotactic cytokines in DSS/DHF patients and suggest that infected ECs contribute to an immune-enhanced disease process. Since permeability is ultimately the result of responses that act on endothelium, dengue-infected ECs are likely to be key elements in DSS and DHF diseases and need to be considered more fully within animal and in vitro models.

4.6 ENDOTHELIUM AND ANIMAL MODELS OF HANTAVIRUS AND DENGUE VIRUS PATHOGENESIS

Only ANDV infection of Syrian hamsters (*Mesocricetus auratus*) serves as a model of hantavirus pathogenesis that mimics human HPS in onset symptoms and lethal acute respiratory disease (Hammerbeck and Hooper, 2011; Hooper et al., 2001; Wahl-Jensen et al., 2007). The inoculation of Syrian hamsters with ANDV, but not SNV, HTNV, or other pathogenic hantaviruses, induces pathology approximating human disease (Hammerbeck and Hooper, 2011; Hooper et al., 2001; Wahl-Jensen et al., 2007). The disease is characterized by large pleural effusions, congested lungs, and interstitial pneumonitis in the absence of disrupted endothelium. The onset of pulmonary edema coincides with a rapid increase in viremia and large inclusion bodies and vacuoles in ultrastructural studies of infected pulmonary ECs (Hooper et al., 2001; Wahl-Jensen et al., 2007). Viral antigen was localized to capillary ECs, alveolar macrophages, and splenic follicular marginal zones populated by dendritic cells. Although it has been suggested that T cells contribute to hantavirus pathogenesis (Kilpatrick et al., 2004; Lindgren et al., 2011; Safronetz et al., 2011), recent findings demonstrate that depleting CD4 and CD8 T cells from Syrian hamsters had no effect on the onset, course, pulmonary symptoms, or pathogenic outcome of ANDV infection and indicates the absence of T-cell-mediated pathogenesis (Hammerbeck and Hooper, 2011). Consistent with the potential involvement of β_3 integrins and VEGF in this process, ANDV binds to conserved residues within PSI domains of both human and hamster β_3 integrins (Matthys et al., 2010; Raymond et al., 2005).

Thus, the mechanism of pathogenesis caused by ANDV remains unknown but is consistent with hypoxia–VEGF-directed acute pulmonary edema and appears to occur in the absence of T-cell-mediated pathology (Hammerbeck and Hooper, 2011). The mechanism of pathogenesis may be further elucidated by studies in Syrian hamsters and permits the evaluation of therapeutics that target barrier functions of the endothelium that regulate hemorrhagic disease.

The dengue virus infects ECs in vivo and in vitro, and this is prominently noted within pathology samples from dengue patients and dengue virus–infected animal models (Balsitis et al., 2009; Balsitis and Harris, 2010; Jessie et al., 2004; Shresta et al., 2006; Zellweger et al., 2010). However, the progress in understanding dengue pathogenesis has been hampered by the lack of suitable mouse models that fully replicate human cellular tropism and disease symptoms. In normal mice, dengue infection results in limb paralysis with little mortality (Balsitis and Harris, 2010). Recently mouse-adapted dengue strains mimic aspects of severe human disease in the IFN-receptor knockout AG129 mice and have been used as a dengue virus animal model (Balsitis and Harris, 2010; Balsitis et al., 2010; Shresta et al., 2006; Tan et al., 2010; Zellweger et al., 2010). Organ damage, hemorrhage, vascular leakage, viremia, and elevated cytokine levels analogous to that in humans are observed following dengue infection of AG129 mice (Balsitis et al., 2010; Tan et al., 2010). Nevertheless, IFN defective murine models further complicate our understanding of the dengue disease process since they lack IFN responses that would otherwise limit dengue spread and induce EC proliferation and repair (Gomez and Reich, 2003). Despite these limitations, the current models have provided new insights into dengue virus pathogenesis and allow for kinetic studies of dengue virus infection of ECs.

Vascular leakage occurs in AG129 mice infected with mouse-adapted dengue strains, and several studies have isolated murine-infected ECs within the spleen and liver (Balsitis et al., 2009; Zellweger et al., 2010). In support of a role for infected ECs in mediating severe dengue disease, Zellweger et al. reported that in the presence of sub-neutralizing levels of dengue antibody (ADE-mediated infection), a large percentage of murine-infected liver sinusoidal ECs were detected and correlated directly with disease severity (Zellweger et al., 2010). No evidence for ADE-mediated infection of ECs exists in vitro (Arevalo et al., 2009) although liver sinusoidal cells reportedly express Fcγ receptors that may contribute to immune-enhanced infection of liver ECs (Elvevold et al., 2008). The infection of mucosal macrophages was not enhanced by the presence of dengue antibodies and only occurred after detection of infected LSECs. This suggests that increased viral loads derived from the initial infection of ECs may enhance the infection of immune cells. In addition to increasing viremia, the ability of the dengue virus to infect ECs provides a means to directly alter capillary permeability, contribute to viremia, and induce EC responses that are immune-enhancing components of dengue pathogenesis.

Since IFN plays a significant role in the regulation of viral spread and the growth and repair of the endothelium (Diamond and Harris, 2001; Diamond et al., 2000b; Halstead, 2008; Munoz-Jordan et al., 2003), it is important to consider the consequences of dengue infections occurring in IFN unresponsive mouse models. Since IFN reportedly stimulates EC proliferation (Gomez and Reich, 2003), the IFN secretion by dengue-infected cells is also likely to contribute to vascular repair following dengue infection, and the absence of this response may explain the enhanced pathogenesis of dengue infections in IFN-receptor knockout (AG129) mice (Balsitis and Harris, 2010; Tan et al., 2010). Current work is ongoing to address the lack of IFN responses within mice and create knock-in mice that harbor functional human STAT2 (Ashour et al., 2010), but additional models of dengue pathogenesis await discovery.

4.7 TARGETED THERAPEUTIC APPROACHES FOR STABILIZING THE ENDOTHELIUM

There are currently no therapeutics for hantavirus or dengue infections. IFN is only effective prophylactically, and similarly, the nucleoside analog ribavirin is only effective at very early times post-infection when the virus begins to replicate and spread but well in advance of symptoms that take 1–2 weeks to present (Jonsson et al., 2008; Koster and Mackow, 2012). Since these viruses infect the endothelium and alter the fluid barrier functions, targeting EC responses that stabilize the vasculature has the potential to reduce the severity and mortality of disease. Targeting the endothelium has further advantages over antiviral approaches since they can be implemented at the onset of symptoms without the need to define viral etiology and may be broadly applicable to viral infections that cause vascular leakage.

Antibody to VEGFR2 reportedly suppresses VEGF-induced pulmonary edema and suggests the potential of therapeutically antagonizing VEGFR2–Src–VE-cadherin signaling pathways as a means of reducing acute pulmonary edema during HPS (Acevedo et al., 2008a; Gavard et al., 2008; Gorbunova et al., 2011; Peng et al., 2004; Porkka et al., 2010; Schmid et al., 2007; Teijaro et al., 2011; Xu et al., 2010). Several well-studied VEGFR2 and Src inhibitors are in human clinical trials or are used therapeutically to treat human cancers and have the potential to reduce the severity of viral permeability–based diseases. In vitro, Angiopoietin-1 (Ang-1), sphingosine-1-phosphate (S1P), pazopanib, and dasatinib inhibited EC permeability directed by pathogenic hantaviruses (Gavrilovskaya et al., 2008; Gorbunova et al., 2011). Ang-1 is an EC-specific growth factor that transdominantly blocks VEGFR2-directed permeability in vitro and in vivo by binding to Tie-2 receptors (Kim et al., 2002; Satchell et al., 2004; Thurston et al., 1999, 2000). S1P is a platelet-derived lipid mediator that enhances vascular barrier functions by binding to Edg-1 receptors on the endothelium (McVerry and Garcia, 2005; Schmid et al., 2007; Wang and Dudek, 2009), while pazopanib and dasatinib are drugs that inhibit VEGFR2–Src signaling (Podar et al., 2006; Porkka et al., 2010). Pazopanib, dasatinib, and the S1P analog FTY720 are already in clinical trials or used clinically for other purposes (McVerry et al., 2004). Targeting EC responses provides a potential means of stabilizing HPS patient vessels and reducing edema. The use of S1P-receptor agonists has also been shown to regulate the pathogenesis of influenza virus infection by acting on ECs and reducing immune cell recruitment and entry into the lung (Teijaro et al., 2011). These findings suggest the targeting of EC functions as a means of increasing capillary barrier functions and regulating immune responses that contribute to viral pathogenesis.

The regulation of additional EC receptors that stabilize inter-EC AJs and fluid barrier functions of the endothelium may be considered as therapeutic targets. The Robo4 receptor has been shown to inhibit VEGFR2 responses, stabilize vessels, and block vascular permeability (Acevedo et al., 2008b; Koch et al., 2011). This new potential target is highly expressed by lung microvascular ECs and is currently being evaluated as a therapeutic for a variety of vascular disorders (Alajez et al., 2011). However, Robo4-directed stability of inter-EC junctions may also be applicable to reducing HPS severity.

Several additional EC receptors that bind to VEGFR2 ectodomains positively or negatively regulate $\alpha_v\beta_3$–VEGFR2 functions and may provide additional therapeutic targets for regulating vascular permeability. For instance, neuropilin-1 forms an ectodomain complex with VEGFR2 that enhances VEGF binding and EC permeability that may be impacted by both syndecan1 (sdc1) and insulin-like growth factor1 receptor (IGF1R) that are recruited to $\alpha_v\beta_3$ ectodomain complexes (Acevedo et al., 2008a; Beauvais et al., 2009; Soker et al., 1998; Wang et al., 2003). Surfen is a heparan sulfate–containing protein that reportedly blocks EC permeability (Schuksz et al., 2008), and fibulin-5 is a matrix protein that reportedly promotes EC adherence by binding $\alpha_v\beta_3$ (Guadall et al., 2011). Furthermore, VEGF signaling is antagonized by fibulin-5 and the loss of fibulin-5 is associated with emphysema (Albig and Schiemann, 2004). Thus, targeting the VEGFR2 axis that regulates EC permeability may be a central mechanism for stabilizing the endothelium and reducing the severity of HPS.

These findings suggest a plethora of potential therapeutics that have the potential to regulate virally induced vascular permeability and are already clinically available and approved for other indications. Moreover, targeting these EC responses may be broadly applicable to reducing the severity of a wide range of viral infections that impact the endothelium and cause edematous diseases.

4.8 CONCLUSIONS AND FUTURE DIRECTIONS

These findings demonstrate the fundamental role of the endothelium in vascular leakage and the need for evaluating the ability of stabilizing the vasculature in reducing the severity and mortality of viral vascular diseases. This is especially important for viral infections that manifest disease 1–2 weeks after infection at time points when antiviral approaches are no longer viable. Although not directly addressed in this review, the role of EC function in virally induced neurologic sequelae also needs to be considered and analyzed as a target for reducing pathogenesis, especially in relation to viruses such as West Nile virus that can pass the blood–brain barrier to disrupt normal neurological functions. In addition, investigating the ability of edematous viruses to infect lymphatic ECs and prevent fluid clearance from tissues may provide a discretely regulated target for reducing viral pathology that has not been factored into disease processes. The ability of the endothelium to regulate platelet functions and complement activation and immune responses in virally induced hemorrhagic and edematous diseases warrants consideration as the central components of vascular leakage and as a target for reducing the severity of viral pathogenesis.

ACKNOWLEDGMENTS

We thank Valery Matthys for critical manuscript review and helpful discussions. This work is supported by NIH, NIAID grants R01AI47873, PO1AI055621, R21AI1080984, and U54AI57158 (NBC-Lipkin).

REFERENCES

Acevedo, L.M., Barillas, S., Weis, S.M., Gothert, J.R., Cheresh, D.A., 2008a. Semaphorin 3A suppresses VEGF-mediated angiogenesis yet acts as a vascular permeability factor. *Blood* 111, 2674–2680.
Acevedo, L.M., Weis, S.M., Cheresh, D.A., 2008b. Robo4 counteracts VEGF signaling. *Nat Med* 14, 372–373.
Aird, W.C., 2004. Endothelium as an organ system. *Crit Care Med* 32, S271–S279.
Alajez, N.M., Lenarduzzi, M., Ito, E., Hui, A.B., Shi, W., Bruce, J., Yue, S. et al., 2011. MiR-218 suppresses nasopharyngeal cancer progression through downregulation of survivin and the SLIT2-ROBO1 pathway. *Cancer Res* 71, 2381–2391.
Albig, A.R., Schiemann, W.P., 2004. Fibulin-5 antagonizes vascular endothelial growth factor (VEGF) signaling and angiogenic sprouting by endothelial cells. *DNA Cell Biol* 23, 367–379.
Alff, P.J., Gavrilovskaya, I.N., Gorbunova, E., Endriss, K., Chong, Y., Geimonen, E., Sen, N., Reich, N.C., Mackow, E.R., 2006. The pathogenic NY-1 hantavirus G1 cytoplasmic tail inhibits RIG-I- and TBK-1- directed interferon responses. *J Virol* 80, 9676–9686.
Alff, P.J., Sen, N., Gorbunova, E., Gavrilovskaya, I.N., Mackow, E.R., 2008. The NY-1 hantavirus Gn cytoplasmic tail coprecipitates TRAF3 and inhibits cellular interferon responses by disrupting TBK1-TRAF3 complex formation. *J Virol* 82, 9115–9122.
Arevalo, M.T., Simpson-Haidaris, P.J., Kou, Z., Schlesinger, J.J., Jin, X., 2009. Primary human endothelial cells support direct but not antibody-dependent enhancement of dengue viral infection. *J Med Virol* 81, 519–528.
Ashour, J., Morrison, J., Laurent-Rolle, M., Belicha-Villanueva, A., Plumlee, C.R., Bernal-Rubio, D., Williams, K.L. et al., 2010. Mouse STAT2 restricts early dengue virus replication. *Cell Host Microbe* 8, 410–421.
Avirutnan, P., Malasit, P., Seliger, B., Bhakdi, S., Husmann, M., 1998. Dengue virus infection of human endothelial cells leads to chemokine production, complement activation, and apoptosis. *J Immunol* 161, 6338–6346.
Avirutnan, P., Punyadee, N., Noisakran, S., Komoltri, C., Thiemmeca, S., Auethavornanan, K., Jairungsri, A. et al., 2006. Vascular leakage in severe dengue virus infections: A potential role for the nonstructural viral protein NS1 and complement. *J Infect Dis* 193, 1078–1088.
Azizan, A., Fitzpatrick, K., Signorovitz, A., Tanner, R., Hernandez, H., Stark, L., Sweat, M., 2009. Profile of time-dependent VEGF upregulation in human pulmonary endothelial cells, HPMEC-ST1.6R infected with DENV-1, -2, -3, and -4 viruses. *Virol J* 6, 49.

Azizan, A., Sweat, J., Espino, C., Gemmer, J., Stark, L., Kazanis, D., 2006. Differential proinflammatory and angiogenesis-specific cytokine production in human pulmonary endothelial cells, HPMEC-ST1.6R infected with dengue-2 and dengue-3 virus. *J Virol Methods* 138, 211–217.

Baffert, F., Le, T., Thurston, G., McDonald, D.M., 2006. Angiopoietin-1 decreases plasma leakage by reducing number and size of endothelial gaps in venules. *Am J Physiol Heart Circ Physiol* 290, H107–H118.

Bahram, F., Claesson-Welsh, L., 2010. VEGF-mediated signal transduction in lymphatic endothelial cells. *Pathophysiology* 17, 253–261.

Balsitis, S.J., Coloma, J., Castro, G., Alava, A., Flores, D., McKerrow, J.H., Beatty, P.R., Harris, E., 2009. Tropism of dengue virus in mice and humans defined by viral nonstructural protein 3-specific immunostaining. *Am J Trop Med Hyg* 80, 416–424.

Balsitis, S.J., Harris, E., 2010. Animal models of dengue virus infection and disease: Applications, insights and frontiers, in: Hanley, K.A., Weaver, S.C. (Eds.), *Frontiers in Dengue Virus Research*. Caister Academic Press, Norfolk, U.K., pp. 103–115.

Balsitis, S.J., Williams, K.L., Lachica, R., Flores, D., Kyle, J.L., Mehlhop, E., Johnson, S., Diamond, M.S., Beatty, P.R., Harris, E., 2010. Lethal antibody enhancement of dengue disease in mice is prevented by Fc modification. *PLoS Pathog* 6, e1000790.

Basu, A., Chaturvedi, U.C., 2008. Vascular endothelium: the battlefield of dengue viruses. *FEMS Immunol Med Microbiol* 53, 287–299.

Beauvais, D.M., Ell, B.J., McWhorter, A.R., Rapraeger, A.C., 2009. Syndecan-1 regulates alphavbeta3 and alphavbeta5 integrin activation during angiogenesis and is blocked by synstatin, a novel peptide inhibitor. *J Exp Med* 206, 691–705.

Berger, M.M., Hesse, C., Dehnert, C., Siedler, H., Kleinbongard, P., Bardenheuer, H.J., Kelm, M., Bartsch, P., Haefeli, W.E., 2005. Hypoxia impairs systemic endothelial function in individuals prone to high-altitude pulmonary edema. *Am J Respir Crit Care Med* 172, 763–767.

Bongrazio, M., Pries, A.R., Zakrzewicz, A., 2003. The endothelium as physiological source of properdin: Role of wall shear stress. *Mol Immunol* 39, 669–675.

Borges, E., Jan, Y., Ruoslahti, E., 2000. Platelet-derived growth factor receptor beta and vascular endothelial growth factor receptor 2 bind to the beta 3 integrin through its extracellular domain. *J Biol Chem* 275, 39867–39873.

Bosch, I., Xhaja, K., Estevez, L., Raines, G., Melichar, H., Warke, R.V., Fournier, M.V., Ennis, F.A., Rothman, A.L., 2002. Increased production of interleukin-8 in primary human monocytes and in human epithelial and endothelial cell lines after dengue virus challenge. *J Virol* 76, 5588–5597.

Breen, E.C., 2007. VEGF in biological control. *J Cell Biochem* 102, 1358–1367.

Breslin, J.W., Gaudreault, N., Watson, K.D., Reynoso, R., Yuan, S.Y., Wu, M.H., 2007. Vascular endothelial growth factor-C stimulates the lymphatic pump by a VEGF receptor-3-dependent mechanism. *Am J Physiol Heart Circ Physiol* 293, H709–H718.

Breslin, J.W., Kurtz, K.M., 2009. Lymphatic endothelial cells adapt their barrier function in response to changes in shear stress. *Lymphat Res Biol* 7, 229–237.

Broermann, A., Winderlich, M., Block, H., Frye, M., Rossaint, J., Zarbock, A., Cagna, G., Linnepe, R., Schulte, D., Nottebaum, A.F., Vestweber, D., 2011. Dissociation of VE-PTP from VE-cadherin is required for leukocyte extravasation and for VEGF-induced vascular permeability in vivo. *J Exp Med* 208, 2393–2401.

Broholm, H., Laursen, H., 2004. Vascular endothelial growth factor (VEGF) receptor neuropilin-1's distribution in astrocytic tumors. *APMIS* 112, 257–263.

Bustamante, E.A., Levy, H., Simpson, S.Q., 1997. Pleural fluid characteristics in hantavirus pulmonary syndrome. *Chest* 112, 1133–1136.

Cascone, I., Napione, L., Maniero, F., Serini, G., Bussolino, F., 2005. Stable interaction between alpha5beta1 integrin and Tie2 tyrosine kinase receptor regulates endothelial cell response to Ang-1. *J Cell Biol* 170, 993–1004.

Christou, H., Yoshida, A., Arthur, V., Morita, T., Kourembanas, S., 1998. Increased vascular endothelial growth factor production in the lungs of rats with hypoxia-induced pulmonary hypertension. *Am J Respir Cell Mol Biol* 18, 768–776.

Cines, D.B., Pollak, E.S., Buck, C.A., Loscalzo, J., Zimmerman, G.A., McEver, R.P., Pober, J.S. et al., 1998. Endothelial cells in physiology and in the pathophysiology of vascular disorders. *Blood* 91, 3527–3561.

Clark, I.A., Awburn, M.M., Cowden, W.B., 1998. Pathophysiology of Ebola haemorrhagic fever. *Trans R Soc Trop Med Hyg* 92, 469.

Coller, B.S., Shattil, S.J., 2008. The GPIIb/IIIa (integrin alphaIIbbeta3) odyssey: A technology-driven saga of a receptor with twists, turns, and even a bend. *Blood* 112, 3011–3025.

Cosgriff, T.M., Lee, H.W., See, A.F., Parrish, D.B., Moon, J.S., Kim, D.J., Lewis, R.M., 1991. Platelet dysfunction contributes to the haemostatic defect in haemorrhagic fever with renal syndrome. *Trans R Soc Trop Med Hyg* 85, 660–663.

Cosgriff, T.M., Lewis, R.M., 1991. Mechanisms of disease in hemorrhagic fever with renal syndrome. *Kidney Int Suppl* 35, S72–S79.

Dalrymple, N., Mackow, E.R., 2011. Productive dengue virus infection of human endothelial cells is directed By Heparan sulfate-containing proteoglycan receptors. *J Virol* 85, 9478–9485.

Dalrymple, N., Mackow, E.R., 2012. Endothelial cells elicit immune enhancing responses to dengue virus infection. *J. Virol* 86, 6408–6415.

Dehler, M., Zessin, E., Bartsch, P., Mairbaurl, H., 2006. Hypoxia causes permeability oedema in the constant-pressure perfused rat lung. *Eur Respir J* 27, 600–606.

Dejana, E., Orsenigo, F., Lampugnani, M.G., 2008. The role of adherens junctions and VE-cadherin in the control of vascular permeability. *J Cell Sci* 121, 2115–2122.

Diamond, M.S., Edgil, D., Roberts, T.G., Lu, B., Harris, E., 2000a. Infection of human cells by dengue virus is modulated by different cell types and viral strains. *J Virol* 74, 7814–7823.

Diamond, M.S., Harris, E., 2001. Interferon inhibits dengue virus infection by preventing translation of viral RNA through a PKR-independent mechanism. *Virology* 289, 297–311.

Diamond, M.S., Roberts, T.G., Edgil, D., Lu, B., Ernst, J., Harris, E., 2000b. Modulation of Dengue virus infection in human cells by alpha, beta, and gamma interferons. *J Virol* 74, 4957–4966.

Dirks, W.G., MacLeod, R.A., Drexler, H.G., 1999. ECV304 (endothelial) is really T24 (bladder carcinoma): Cell line cross- contamination at source. *In Vitro Cell Dev Biol Anim* 35, 558–559.

Duchin, J.S., Koster, F.T., Peters, C.J., Simpson, G.L., Tempest, B., Zaki, S.R., Ksiazek, T.G. et al., 1994. Hantavirus pulmonary syndrome: A clinical description of 17 patients with a newly recognized disease. The Hantavirus Study Group [see comments]. *N Engl J Med* 330, 949–955.

Dvorak, H.F., 2006. Discovery of vascular permeability factor (VPF). *Exp Cell Res* 312, 522–526.

Dvorak, H.F., 2010. Vascular permeability to plasma, plasma proteins, and cells: an update. *Curr Opin Hematol* 17, 225–229.

Dvorak, H.F., Brown, L.F., Detmar, M., Dvorak, A.M., 1995. Vascular permeability factor/vascular endothelial growth factor, microvascular hyperpermeability, and angiogenesis. *Am J Pathol* 146, 1029–1039.

Dvorak, H.F., Sioussat, T.M., Brown, L.F., Berse, B., Nagy, J.A., Sotrel, A., Manseau, E.J., Van de Water, L., Senger, D.R., 1991. Distribution of vascular permeability factor (vascular endothelial growth factor) in tumors: Concentration in tumor blood vessels. *J Exp Med* 174, 1275–1278.

Elvevold, K., Smedsrod, B., Martinez, I., 2008. The liver sinusoidal endothelial cell: A cell type of controversial and confusing identity. *Am J Physiol Gastrointest Liver Physiol* 294, G391–G400.

Enria, D., Padula, P., Segura, E.L., Pini, N., Edelstein, A., Posse, C.R., Weissenbacher, M.C., 1996. Hantavirus pulmonary syndrome in Argentina. Possibility of person to person transmission. *Medicina* 56, 709–711.

Feldmann, H., Bugany, H., Mahner, F., Klenk, H.D., Drenckhahn, D., Schnittler, H.J., 1996. Filovirus-induced endothelial leakage triggered by infected monocytes/macrophages. *J Virol* 70, 2208–2214.

Fink, J., Gu, F., Ling, L., Tolfvenstam, T., Olfat, F., Chin, K.C., Aw, P. et al., 2007. Host gene expression profiling of dengue virus infection in cell lines and patients. *PLoS Negl Trop Dis* 1, e86.

Fink, J., Gu, F., Vasudevan, S.G., 2006. Role of T cells, cytokines and antibody in dengue fever and dengue haemorrhagic fever. *Rev Med Virol* 16, 263–275.

Fishman, A.P., 1972. Pulmonary edema. The water-exchanging function of the lung. *Circulation* 46, 390–408.

Friedman, H.M., Macarak, E.J., MacGregor, R.R., Wolfe, J., Kefalides, N.A., 1981. Virus infection of endothelial cells. *J Infect Dis* 143, 266–273.

Friedman, H.M., Wolfe, J., Kefalides, N.A., Macarak, E.J., 1986. Susceptibility of endothelial cells derived from different blood vessels to common viruses. *In Vitro Cell Dev Biol* 22, 397–401.

Galvagni, F., Pennacchini, S., Salameh, A., Rocchigiani, M., Neri, F., Orlandini, M., Petraglia, F. et al., 2010. Endothelial cell adhesion to the extracellular matrix induces c-Src-dependent VEGFR-3 phosphorylation without the activation of the receptor intrinsic kinase activity. *Circ Res* 106, 1839–1848.

Garcia, J.G., Liu, F., Verin, A.D., Birukova, A., Dechert, M.A., Gerthoffer, W.T., Bamberg, J.R., English, D., 2001. Sphingosine 1-phosphate promotes endothelial cell barrier integrity by Edg-dependent cytoskeletal rearrangement. *J Clin Invest* 108, 689–701.

Gavard, J., 2009. Breaking the VE-cadherin bonds. *FEBS Lett* 583, 1–6.

Gavard, J., Gutkind, J.S., 2006. VEGF controls endothelial-cell permeability by promoting the beta-arrestin-dependent endocytosis of VE-cadherin. *Nat Cell Biol* 8, 1223–1234.

Gavard, J., Patel, V., Gutkind, J.S., 2008. Angiopoietin-1 prevents VEGF-induced endothelial permeability by sequestering Src through mDia. *Dev Cell* 14, 25–36.

Gavrilovskaya, I.N., Brown, E.J., Ginsberg, M.H., Mackow, E.R., 1999. Cellular entry of hantaviruses which cause hemorrhagic fever with renal syndrome is mediated by beta3 integrins. *J Virol* 73, 3951–3959.

Gavrilovskaya, I.N., Gorbunova, E.E., Koster, F., Mackow, E.R., 2012. Elevated VEGF levels in pulmonary edema fluid and PBMCs from patients with acute hantavirus pulmonary syndrome. *Adv Virol* 2012, Article ID 674360, 8 pages.

Gavrilovskaya, I., Gorbunova, E.E., Mackow, E.R., 2010. Pathogenic hantaviruses direct the adherence of quiescent platelets to infected endothelial cells. *J Virol* 84, 4832–4839.

Gavrilovskaya, I.N., Gorbunova, E.E., Mackow, N.A., Mackow, E.R., 2008. Hantaviruses direct endothelial cell permeability by sensitizing cells to the vascular permeability factor VEGF, while angiopoietin 1 and sphingosine 1-phosphate inhibit hantavirus-directed permeability. *J Virol* 82, 5797–5806.

Gavrilovskaya, I.N., Peresleni, T., Geimonen, E., Mackow, E.R., 2002. Pathogenic hantaviruses selectively inhibit beta3 integrin directed endothelial cell migration. *Arch Virol* 147, 1913–1931.

Gavrilovskaya, I., Podgorodnichenko, V., Apekina, N., Gordbachkova, E., Bogdanova, S., Kodkind, G., Linev, M., Chumakoc, M., 1987. Determination of specific immune complexes and dynamics of their circulation in patients suffering from hemorrhagic fever with a renal syndrome. *Microbiol J* 49, 71–76.

Gavrilovskaya, I.N., Shepley, M., Shaw, R., Ginsberg, M.H., Mackow, E.R., 1998. beta3 Integrins mediate the cellular entry of hantaviruses that cause respiratory failure. *Proc Natl Acad Sci USA* 95, 7074–7079.

Geimonen, E., Neff, S., Raymond, T., Kocer, S.S., Gavrilovskaya, I.N., Mackow, E.R., 2002. Pathogenic and nonpathogenic hantaviruses differentially regulate endothelial cell responses. *Proc Natl Acad Sci USA* 99, 13837–13842.

Goldsmith, C.S., Elliott, L.H., Peters, C.J., Zaki, S.R., 1995. Ultrastructural characteristics of Sin Nombre virus, causative agent of hantavirus pulmonary syndrome. *Arch Virol* 140, 2107–2122.

Gomez, D., Reich, N.C., 2003. Stimulation of primary human endothelial cell proliferation by IFN. *J Immunol* 170, 5373–5381.

Gorbunova, E., Gavrilovskaya, I.N., Mackow, E.R., 2010. Pathogenic hantaviruses Andes virus and Hantaan virus induce adherens junction disassembly by directing vascular endothelial cadherin internalization in human endothelial cells. *J Virol* 84, 7405–7411.

Gorbunova, E.E., Gavrilovskaya, I.N., Pepini, T., Mackow, E.R., 2011. VEGFR2 and Src kinase inhibitors suppress ANDV induced endothelial cell permeability. *J Virol* 85, 2296–2303.

Guadall, A., Orriols, M., Rodriguez-Calvo, R., Calvayrac, O., Crespo, J., Aledo, R., Martinez-Gonzalez, J., Rodriguez, C., 2011. Fibulin-5 is up-regulated by hypoxia in endothelial cells through a hypoxia-inducible factor-1 (HIF-1alpha)-dependent mechanism. *J Biol Chem* 286, 7093–7103.

Gubler, D.J., 2006. Dengue/dengue haemorrhagic fever: History and current status. *Novartis Found Symp* 277, 3–16; discussion 16–22, 71-13, 251–253.

Halstead, S.B., 2003. Neutralization and antibody-dependent enhancement of dengue viruses. *Adv Virus Res* 60, 421–467.

Halstead, S.B., 2008. Pathophysiology, in: Halstead, S.B. (Ed.), *Dengue*. Imperial College Press, London, U.K., pp. 265–326.

Halstead, S.B., 2009. Antibodies determine virulence in dengue. *Ann NY Acad Sci* 1171(Suppl 1), E48–E56.

Hammerbeck, C.D., Hooper, J.W., 2011. T cells are not required for pathogenesis in the Syrian hamster model of hantavirus pulmonary syndrome. *J Virol* 85, 9929–9944.

Hanaoka, M., Droma, Y., Naramoto, A., Honda, T., Kobayashi, T., Kubo, K., 2003. Vascular endothelial growth factor in patients with high-altitude pulmonary edema. *J Appl Physiol* 94, 1836–1840.

Hodivala-Dilke, K.M., McHugh, K.P., Tsakiris, D.A., Rayburn, H., Crowley, D., Ullman-Cullere, M., Ross, F.P., Coller, B.S., Teitelbaum, S., Hynes, R.O., 1999. Beta3-integrin-deficient mice are a model for Glanzmann thrombasthenia showing placental defects and reduced survival. *J Clin Invest* 103, 229–238.

Holmes, K., Roberts, O.L., Thomas, A.M., Cross, M.J., 2007. Vascular endothelial growth factor receptor-2: Structure, function, intracellular signalling and therapeutic inhibition. *Cell Signal* 19, 2003–2012.

Hooper, J.W., Larsen, T., Custer, D.M., Schmaljohn, C.S., 2001. A lethal disease model for hantavirus pulmonary syndrome. *Virology* 289, 6–14.

Hopkins, S.R., Garg, J., Bolar, D.S., Balouch, J., Levin, D.L., 2005. Pulmonary blood flow heterogeneity during hypoxia and high-altitude pulmonary edema. *Am J Respir Crit Care Med* 171, 83–87.

Huang, Y.H., Lei, H.Y., Liu, H.S., Lin, Y.S., Liu, C.C., Yeh, T.M., 2000. Dengue virus infects human endothelial cells and induces IL-6 and IL-8 production. *Am J Trop Med Hyg* 63, 71–75.

Jessie, K., Fong, M.Y., Devi, S., Lam, S.K., Wong, K.T., 2004. Localization of dengue virus in naturally infected human tissues, by immunohistochemistry and in situ hybridization. *J Infect Dis* 189, 1411–1418.

Jonsson, C.B., Hooper, J., Mertz, G., 2008. Treatment of hantavirus pulmonary syndrome. *Antiviral Res* 78, 162–169.

Kanerva, M., Mustonen, J., Vaheri, A., 1998. Pathogenesis of puumala and other hantavirus infections. *Rev Med Virol* 8, 67–86.

Kilpatrick, E.D., Terajima, M., Koster, F.T., Catalina, M.D., Cruz, J., Ennis, F.A., 2004. Role of specific CD8+ T cells in the severity of a fulminant zoonotic viral hemorrhagic fever, hantavirus pulmonary syndrome. *J Immunol* 172, 3297–3304.

Kim, I., Oh, J.L., Ryu, Y.S., So, J.N., Sessa, W.C., Walsh, K., Koh, G.Y., 2002. Angiopoietin-1 negatively regulates expression and activity of tissue factor in endothelial cells. *FASEB J* 16, 126–128.

Koch, A.W., Mathivet, T., Larrivee, B., Tong, R.K., Kowalski, J., Pibouin-Fragner, L., Bouvree, K. et al., 2011. Robo4 maintains vessel integrity and inhibits angiogenesis by interacting with UNC5B. *Dev Cell* 20, 33–46.

Kortepeter, M.G., Bausch, D.G., Bray, M., 2011. Basic clinical and laboratory features of filoviral hemorrhagic fever. *J Infect Dis* 204(Suppl 3), S810–S816.

Koster, F., Mackow, E.R., 2012. Pathogenesis of the hantavirus pulmonary syndrome. *Future Virol* 7, 41–51.

Krakauer, T., Leduc, J.W., Krakauer, H., 1995. Serum levels of tumor necrosis factor-alpha, interleukin-1, and interleukin-6 in hemorrhagic fever with renal syndrome. *Viral Immunol* 8, 75–79.

Krishnamurti, C., Peat, R.A., Cutting, M.A., Rothwell, S.W., 2002. Platelet adhesion to dengue-2 virus-infected endothelial cells. *Am J Trop Med Hyg* 66, 435–441.

Kunz, S., 2009. The role of the vascular endothelium in arenavirus haemorrhagic fevers. *Thromb Haemost* 102, 1024–1029.

Lacroix, M., 2008. Persistent use of "false" cell lines. *Int J Cancer* 122, 1–4.

Lahdevirta, J., 1982. Clinical features of HFRS in Scandinavia as compared with East Asia. *Scand J Infect Dis Suppl* 36, 93–95.

Lampugnani, M.G., Orsenigo, F., Gagliani, M.C., Tacchetti, C., Dejana, E., 2006. Vascular endothelial cadherin controls VEGFR-2 internalization and signaling from intracellular compartments. *J Cell Biol* 174, 593–604.

Lee, H.W., 1982a. Hemorrhagic fever with renal syndrome (HFRS). *Scand J Infect Dis Suppl* 36, 82–85.

Lee, H.W., 1982b. Korean hemorrhagic fever. *Prog Med Virol* 28, 96–113.

Lee, H.W., Baek, L.J., Johnson, K.M., 1982. Isolation of Hantaan virus, the etiologic agent of Korean hemorrhagic fever, from wild urban rats. *J Infect Dis* 146, 638–644.

Libraty, D.H., Young, P.R., Pickering, D., Endy, T.P., Kalayanarooj, S., Green, S., Vaughn, D.W., Nisalak, A., Ennis, F.A., Rothman, A.L., 2002. High circulating levels of the dengue virus nonstructural protein NS1 early in dengue illness correlate with the development of dengue hemorrhagic fever. *J Infect Dis* 186, 1165–1168.

Lin, C.F., Chiu, S.C., Hsiao, Y.L., Wan, S.W., Lei, H.Y., Shiau, A.L., Liu, H.S. et al., 2005. Expression of cytokine, chemokine, and adhesion molecules during endothelial cell activation induced by antibodies against dengue virus nonstructural protein 1. *J Immunol* 174, 395–403.

Lin, C.F., Lei, H.Y., Shiau, A.L., Liu, C.C., Liu, H.S., Yeh, T.M., Chen, S.H., Lin, Y.S., 2003. Antibodies from dengue patient sera cross-react with endothelial cells and induce damage. *J Med Virol* 69, 82–90.

Lindgren, T., Ahlm, C., Mohamed, N., Evander, M., Ljunggren, H.G., Bjorkstrom, N.K., 2011. Longitudinal analysis of the human T cell response during acute hantavirus infection. *J Virol* 85, 10252–10260.

Lopez, N., Padula, P., Rossi, C., Lazaro, M.E., Franze-Fernandez, M.T., 1996. Genetic identification of a new hantavirus causing severe pulmonary syndrome in Argentina. *Virology* 220, 223–226.

Luplertlop, N., Misse, D., Bray, D., Deleuze, V., Gonzalez, J.P., Leardkamolkarn, V., Yssel, H., Veas, F., 2006. Dengue-virus-infected dendritic cells trigger vascular leakage through metalloproteinase overproduction. *EMBO Rep* 7, 1176–1181.

Martina, B.E., Koraka, P., Osterhaus, A.D., 2009. Dengue virus pathogenesis: An integrated view. *Clin Microbiol Rev* 22, 564–581.

Matthay, M.A., Folkesson, H.G., Clerici, C., 2002. Lung epithelial fluid transport and the resolution of pulmonary edema. *Physiol Rev* 82, 569–600.

Matthys, V.S., Gorbunova, E.E., Gavrilovskaya, I.N., Mackow, E.R., 2010. Andes virus recognition of human and Syrian hamster beta3 integrins is determined by an L33P substitution in the PSI domain. *J Virol* 84, 352–360.

McVerry, B.J., Garcia, J.G., 2005. In vitro and in vivo modulation of vascular barrier integrity by sphingosine 1-phosphate: Mechanistic insights. *Cell Signal* 17, 131–139.

McVerry, B.J., Peng, X., Hassoun, P.M., Sammani, S., Simon, B.A., Garcia, J.G., 2004. Sphingosine 1-phosphate reduces vascular leak in murine and canine models of acute lung injury. *Am J Respir Crit Care Med* 170, 987–993.

Moraz, M.L., Kunz, S., 2010. Pathogenesis of arenavirus hemorrhagic fevers. *Expert Rev Anti Infect Ther* 9, 49–59.

Munoz-Jordan, J.L., Sanchez-Burgos, G.G., Laurent-Rolle, M., Garcia-Sastre, A., 2003. Inhibition of interferon signaling by dengue virus. *Proc Natl Acad Sci USA* 100, 14333–14338.

Mutlu, G.M., Sznajder, J.I., 2005. Mechanisms of pulmonary edema clearance. *Am J Physiol Lung Cell Mol Physiol* 289, L685–L695.

Nalbandian, R.M., Henry, R.L., 1978. Platelet-endothelial cell interactions. Metabolic maps of structures and actions of prostaglandins, prostacyclin, thromboxane and cyclic AMP. *Semin Thromb Hemost* 5, 87–111.

Neufeld, G., Cohen, T., Gengrinovitch, S., Poltorak, Z., 1999. Vascular endothelial growth factor (VEGF) and its receptors. *FASEB J* 13, 9–22.

Nichol, S.T., Spiropoulou, C.F., Morzunov, S., Rollin, P.E., Ksiazek, T.G., Feldmann, H., Sanchez, A., Childs, J., Zaki, S., Peters, C.J., 1993. Genetic identification of a hantavirus associated with an outbreak of acute respiratory illness [see comments]. *Science* 262, 914–917.

Nilsson, I., Bahram, F., Li, X., Gualandi, L., Koch, S., Jarvius, M., Soderberg, O. et al., 2010. VEGF receptor 2/-3 heterodimers detected in situ by proximity ligation on angiogenic sprouts. *EMBO J* 29, 1377–1388.

Nolte, K.B., Feddersen, R.M., Foucar, K., Zaki, S.R., Koster, F.T., Madar, D., Merlin, T.L., McFeeley, P.J., Umland, E.T., Zumwalt, R.E., 1995. Hantavirus pulmonary syndrome in the United States: A pathological description of a disease caused by a new agent. *Human Pathology* 26, 110–120.

Olsson, A.K., Dimberg, A., Kreuger, J., Claesson-Welsh, L., 2006. VEGF receptor signalling—in control of vascular function. *Nat Rev Mol Cell Biol* 7, 359–371.

Padula, P.J., Edelstein, A., Miguel, S.D., Lopez, N.M., Rossi, C.M., Rabinovich, R.D., 1998. Hantavirus pulmonary syndrome outbreak in Argentina: Molecular evidence for person-to-person transmission of Andes virus. *Virology* 241, 323–330.

Peng, X., Hassoun, P.M., Sammani, S., McVerry, B.J., Burne, M.J., Rabb, H., Pearse, D., Tuder, R.M., Garcia, J.G., 2004. Protective effects of sphingosine 1-phosphate in murine endotoxin-induced inflammatory lung injury. *Am J Respir Crit Care Med* 169, 1245–1251.

Pensiero, M.N., Sharefkin, J.B., Dieffenbach, C.W., Hay, J., 1992. Hantaan virus infection of human endothelial cells. *J Virol* 66, 5929–5936.

Peyrefitte, C.N., Pastorino, B., Grau, G.E., Lou, J., Tolou, H., Couissinier-Paris, P., 2006. Dengue virus infection of human microvascular endothelial cells from different vascular beds promotes both common and specific functional changes. *J Med Virol* 78, 229–242.

Pham, I., Uchida, T., Planes, C., Ware, L.B., Kaner, R., Matthay, M.A., Clerici, C., 2002. Hypoxia upregulates VEGF expression in alveolar epithelial cells in vitro and in vivo. *Am J Physiol Lung Cell Mol Physiol* 283, L1133–L1142.

Pober, J.S., Min, W., Bradley, J.R., 2009. Mechanisms of endothelial dysfunction, injury, and death. *Annu Rev Pathol* 4, 71–95.

Podar, K., Tonon, G., Sattler, M., Tai, Y.T., Legouill, S., Yasui, H., Ishitsuka, K. et al., 2006. The small-molecule VEGF receptor inhibitor pazopanib (GW786034B) targets both tumor and endothelial cells in multiple myeloma. *Proc Natl Acad Sci USA* 103, 19478–19483.

Porkka, K., Khoury, H.J., Paquette, R.L., Matloub, Y., Sinha, R., Cortes, J.E., 2010. Dasatinib 100 mg once daily minimizes the occurrence of pleural effusion in patients with chronic myeloid leukemia in chronic phase and efficacy is unaffected in patients who develop pleural effusion. *Cancer* 116, 377–386.

Raymond, T., Gorbunova, E., Gavrilovskaya, I.N., Mackow, E.R., 2005. Pathogenic hantaviruses bind plexin-semaphorin-integrin domains present at the apex of inactive, bent alphavbeta3 integrin conformers. *Proc Natl Acad Sci USA* 102, 1163–1168.

Reynolds, L.E., Wyder, L., Lively, J.C., Taverna, D., Robinson, S.D., Huang, X., Sheppard, D., Hynes, R.O., Hodivala-Dilke, K.M., 2002. Enhanced pathological angiogenesis in mice lacking beta3 integrin or beta3 and beta5 integrins. *Nat Med* 8, 27–34.

Robinson, S.D., Reynolds, L.E., Wyder, L., Hicklin, D.J., Hodivala-Dilke, K.M., 2004. Beta3-integrin regulates vascular endothelial growth factor-A-dependent permeability. *Arterioscler Thromb Vasc Biol* 24, 2108–2114.

Rothman, A.L., 2010. Cellular immunology of sequential dengue virus infection and its role in disease pathogenesis. *Curr Top Microbiol Immunol* 338, 83–98.

Rothman, A.L., 2011. Immunity to dengue virus: A tale of original antigenic sin and tropical cytokine storms. *Nat Rev Immunol* 11, 532–543.

Safronetz, D., Zivcec, M., Lacasse, R., Feldmann, F., Rosenke, R., Long, D., Haddock, E. et al., 2011. Pathogenesis and host response in Syrian Hamsters following intranasal infection with Andes virus. *PLoS Pathog* 7, e1002426.

Saharinen, P., Petrova, T.V., 2004. Molecular regulation of lymphangiogenesis. *Ann N Y Acad Sci* 1014, 76–87.

Saharinen, P., Tammela, T., Karkkainen, M.J., Alitalo, K., 2004. Lymphatic vasculature: Development, molecular regulation and role in tumor metastasis and inflammation. *Trends Immunol* 25, 387–395.

Satchell, S.C., Anderson, K.L., Mathieson, P.W., 2004. Angiopoietin 1 and vascular endothelial growth factor modulate human glomerular endothelial cell barrier properties. *J Am Soc Nephrol* 15, 566–574.

Schmaljohn, C., 2001. Bunyaviridae and their replication, in: Fields (Ed.), *Virology*. Lippincott-Raven, Philadelphia, PA, pp. 1581–1602.

Schmid, G., Guba, M., Ischenko, I., Papyan, A., Joka, M., Schrepfer, S., Bruns, C.J., Jauch, K.W., Heeschen, C., Graeb, C., 2007. The immunosuppressant FTY720 inhibits tumor angiogenesis via the sphingosine 1-phosphate receptor 1. *J Cell Biochem* 101, 259–270.

Schnittler, H.J., Mahner, F., Drenckhahn, D., Klenk, H.D., Feldmann, H., 1993. Replication of Marburg virus in human endothelial cells. A possible mechanism for the development of viral hemorrhagic disease. *J Clin Invest* 91, 1301–1309.

Schraufnagel, D.E., 2010. Lung lymphatic anatomy and correlates. *Pathophysiology* 17, 337–343.

Schuksz, M., Fuster, M.M., Brown, J.R., Crawford, B.E., Ditto, D.P., Lawrence, R., Glass, C.A., Wang, L., Tor, Y., Esko, J.D., 2008. Surfen, a small molecule antagonist of heparan sulfate. *Proc Natl Acad Sci USA* 105, 13075–13080.

Shresta, S., Sharar, K.L., Prigozhin, D.M., Beatty, P.R., Harris, E., 2006. Murine model for dengue virus-induced lethal disease with increased vascular permeability. *J Virol* 80, 10208–10217.

Shrivastava-Ranjan, P., Rollin, P.E., Spiropoulou, C.F., 2010. Andes virus disrupts the endothelial cell barrier by induction of vascular endothelial growth factor and downregulation of VE-cadherin. *J Virol* 84, 11227–11234.

Soker, S., Miao, H.Q., Nomi, M., Takashima, S., Klagsbrun, M., 2002. VEGF165 mediates formation of complexes containing VEGFR-2 and neuropilin-1 that enhance VEGF165-receptor binding. *J Cell Biochem* 85, 357–368.

Soker, S., Takashima, S., Miao, H.Q., Neufeld, G., Klagsbrun, M., 1998. Neuropilin-1 is expressed by endothelial and tumor cells as an isoform-specific receptor for vascular endothelial growth factor. *Cell* 92, 735–745.

Srikiatkhachorn, A., 2009. Plasma leakage in dengue haemorrhagic fever. *Thromb Haemost* 102, 1042–1049.

Steffan, A.M., Lafon, M.E., Gendrault, J.L., Schweitzer, C., Royer, C., Jaeck, D., Arnaud, J.P., Schmitt, M.P., Aubertin, A.M., Kirn, A., 1992. Primary cultures of endothelial cells from the human liver sinusoid are permissive for human immunodeficiency virus type 1. *Proc Natl Acad Sci USA* 89, 1582–1586.

Stenmark, K.R., Fagan, K.A., Frid, M.G., 2006. Hypoxia-induced pulmonary vascular remodeling: Cellular and molecular mechanisms. *Circ Res* 99, 675–691.

Stoermer, K.A., Morrison, T.E., 2011. Complement and viral pathogenesis. *Virology* 411, 362–373.

Sundstrom, J.B., McMullan, L.K., Spiropoulou, C.F., Hooper, W.C., Ansari, A.A., Peters, C.J., Rollin, P.E., 2001. Hantavirus infection induces the expression of RANTES and IP-10 without causing increased permeability in human lung microvascular endothelial cells. *J Virol* 75, 6070–6085.

Takada, A., Kawaoka, Y., 2003. Antibody-dependent enhancement of viral infection: Molecular mechanisms and in vivo implications. *Rev Med Virol* 13, 387–398.

Talavera, D., Castillo, A.M., Dominguez, M.C., Gutierrez, A.E., Meza, I., 2004. IL8 release, tight junction and cytoskeleton dynamic reorganization conducive to permeability increase are induced by dengue virus infection of microvascular endothelial monolayers. *J Gen Virol* 85, 1801–1813.

Tan, G.K., Ng, J.K., Trasti, S.L., Schul, W., Yip, G., Alonso, S., 2010. A non mouse-adapted dengue virus strain as a new model of severe dengue infection in AG129 mice. *PLoS Negl Trop Dis* 4, e672.

Teijaro, J.R., Walsh, K.B., Cahalan, S., Fremgen, D.M., Roberts, E., Scott, F., Martinborough, E., Peach, R., Oldstone, M.B., Rosen, H., 2011. Endothelial cells are central orchestrators of cytokine amplification during influenza virus infection. *Cell* 146, 980–991.

Thurston, G., Rudge, J.S., Ioffe, E., Zhou, H., Ross, L., Croll, S.D., Glazer, N., Holash, J., McDonald, D.M., Yancopoulos, G.D., 2000. Angiopoietin-1 protects the adult vasculature against plasma leakage. *Nat Med* 6, 460–463.

Thurston, G., Suri, C., Smith, K., McClain, J., Sato, T.N., Yancopoulos, G.D., McDonald, D.M., 1999. Leakage-resistant blood vessels in mice transgenically overexpressing angiopoietin-1. *Science* 286, 2511–2514.

Valbuena, G., Walker, D.H., 2006. The endothelium as a target for infections. *Annu Rev Pathol* 1, 171–198.

Vapalahti, O., Lundkvist, A., Vaheri, A., 2001. Human immune response, host genetics, and severity of disease. *Curr Top Microbiol Immunol* 256, 153–169.

Verma, S., Lo, Y., Chapagain, M., Lum, S., Kumar, M., Gurjav, U., Luo, H., Nakatsuka, A., Nerurkar, V.R., 2009. West Nile virus infection modulates human brain microvascular endothelial cells tight junction proteins and cell adhesion molecules: Transmigration across the in vitro blood-brain barrier. *Virology* 385, 425–433.

Wahl-Jensen, V., Chapman, J., Asher, L., Fisher, R., Zimmerman, M., Larsen, T., Hooper, J.W., 2007. Temporal analysis of Andes virus and Sin Nombre virus infections of Syrian hamsters. *J Virol* 81, 7449–7462.

Wallez, Y., Vilgrain, I., Huber, P., 2006. Angiogenesis: The VE-cadherin switch. *Trends Cardiovasc Med* 16, 55–59.

Wang, L., Dudek, S.M., 2009. Regulation of vascular permeability by sphingosine 1-phosphate. *Microvasc Res* 77, 39–45.

Wang, L., Zeng, H., Wang, P., Soker, S., Mukhopadhyay, D., 2003. Neuropilin-1-mediated vascular permeability factor/vascular endothelial growth factor-dependent endothelial cell migration. *J Biol Chem* 278, 48848–48860.

Warke, R.V., Xhaja, K., Martin, K.J., Fournier, M.F., Shaw, S.K., Brizuela, N., de Bosch, N. et al., 2003. Dengue virus induces novel changes in gene expression of human umbilical vein endothelial cells. *J Virol* 77, 11822–11832.

Weis, S.M., Cheresh, D.A., 2011. alphav Integrins in angiogenesis and cancer. *Cold Spring Harb Perspect Med* 1, a006478.

Weitzman, M., Bayley, E.B., Naik, U.P., 2008. Robo4: A guidance receptor that regulates angiogenesis. *Cell Adh Migr* 2, 220–222.

Wong, K.T., Shieh, W.J., Kumar, S., Norain, K., Abdullah, W., Guarner, J., Goldsmith, C.S. et al., 2002. Nipah virus infection: pathology and pathogenesis of an emerging paramyxoviral zoonosis. *Am J Pathol* 161, 2153–2167.

Xu, D., Fuster, M.M., Lawrence, R., Esko, J.D., 2010. Heparan sulfate regulates VEGF165- and VEGF121-mediated vascular hyperpermeability. *J Biol Chem* 286, 737–745.

Yanagihara, R., Silverman, D.J., 1990. Experimental infection of human vascular endothelial cells by pathogenic and nonpathogenic hantaviruses. *Arch Virol* 111, 281–286.

Zaki, S.R., Greer, P.W., Coffield, L.M., Goldsmith, C.S., Nolte, K.B., Foucar, K., Feddersen, R.M. et al., 1995. Hantavirus pulmonary syndrome. Pathogenesis of an emerging infectious disease. *Am J Pathol* 146, 552–579.

Zellweger, R.M., Prestwood, T.R., Shresta, S., 2010. Enhanced infection of liver sinusoidal endothelial cells in a mouse model of antibody-induced severe dengue disease. *Cell Host Microbe* 7, 128–139.

5 Vascular Endothelial Dysfunctions
Viral Attack and Immunological Defense

Jayashree Apurupa Lahoti, Ritu Mishra, and Sunit Kumar Singh

CONTENTS

5.1 INTRODUCTION

Viral hemorrhagic fever (VHF) is a term used to define acute febrile hemorrhagic diseases caused by RNA viruses mostly from four virus families, namely, *Arenaviridae*, *Bunyaviridae*, *Flaviviridae*, and *Filoviridae*. Except for *Filoviridae*, these viruses are transmitted by vectors such as insects and rodents. The vector responsible for the transmission of viruses belonging to *Filoviridae* is not known so far. VHFs are characterized by severe thrombocytopenia and hemoconcentration.

Endothelial cells (ECs) play important role in pathogenesis of VHFs. ECs form a thin layer in the interior surface of blood vessels and thus separate blood from underlying tissue by performing barrier function. Endothelial dysfunction can be defined as imbalance between endothelium-derived relaxing factors (EDRFs) and endothelium-derived constricting factors (EDCFs) and maintenance of barrier integrity (Yao et al., 2010). The barrier integrity is maintained by tight junctions (TJs) and adherens junctions (AJs) (Bazzoni and Dejana, 2004) (Figure 5.1). Major factors underlying pathology of VHFs are damage to vessels, endothelial dysfunction, and increased permeability leading to compromise in barrier function. Permeability of EC barrier can be determined in in vitro experiments by transendothelial electrical resistance (TEER) measurements. TEER is an estimate of electrical current, which can pass through EC layer and inverse to barrier resistance. Thus, when barrier is intact, it shows high resistance and decreased permeability, and when it is damaged, it shows low resistance and increased permeability. Vasculature damage can be direct mechanisms, which is due to replication of virus in ECs, or indirect mechanisms where

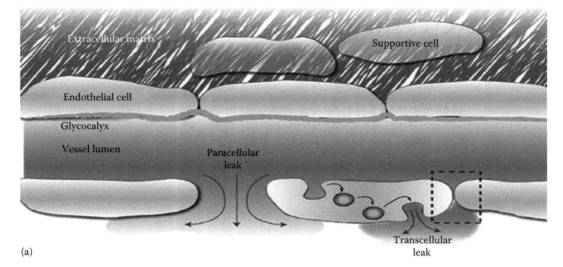

(a)

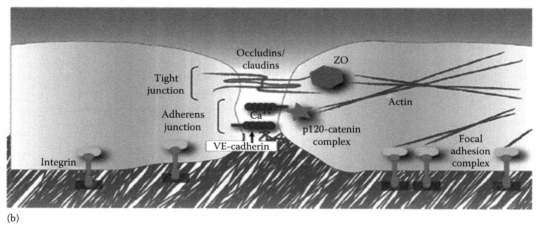

(b)

FIGURE 5.1 **(See color insert.)** Microvascular constituents and leak pathways. (a) The blood vessel lumen is lined by ECs, which comprise the primary component of the microvascular permeability barrier. Vessel integrity is further influenced by the interaction of ECs with the extracellular matrix (ECM), glycocalyx, and supporting cells. In certain physiological and pathological settings, microvascular leak pathways can be initiated via paracellular and transcellular mechanisms. (b) The principal interendothelial junctions are the TJ and AJ, which interface with the cell's cytoskeleton via specific adaptor proteins. The TJ is composed of claudin and occludin proteins, which interact with the zona occludens proteins (labeled ZO in the figure) along their cytoplasmic surface. The calcium-dependent homotypic interaction of endothelial VE-cadherin of the AJ is relayed to the actin cytoskeleton by multiple adaptor proteins. The p120-catenin complex is shown here. The importance of the different junctions in preventing leak varies across different vascular beds. The cytoskeleton also engages with integrin proteins that anchor ECs to the surrounding ECM. (Reproduced from Steinberg, B.E. et al., *Antiviral Res.*, 93, 2, 2012. With permission.)

host immune cells and complexes are involved. Both conditions might result into the leakage of blood to peripheral organs and this situation gets exacerbated by decreased hemostasis due to thrombocytopenia.

This chapter describes various mechanisms by which viruses of *Arenaviridae*, *Bunyaviridae*, *Flaviviridae*, and *Filoviridae* families cause endothelial dysfunction and platelet depletion resulting in hemorrhage and edema.

5.2   VIRUSES OF FAMILY *ARENAVIRIDAE* AND ENDOTHELIAL DYSFUNCTIONS

Arenaviruses can be divided into two subgroups: Old World and New World arenaviruses (Clegg, 2002; Kunz, 2009). Arenaviruses are known to infect capillary ECs. Old World arenaviruses include Lassa virus (LASV), Lujo virus, and lymphocytic choriomeningitis virus (LCMV). Old World arenaviruses bind to a host cell receptor, α-dystroglycan (α-DG), with very high affinity in order to gain entry into the cells. α-DG, a cellular receptor for extracellular matrix (ECM) proteins, is ubiquitously expressed in cells (Cao et al., 1998). Upon binding to the receptor, arenaviruses are internalized by endocytosis to endosome, where viral particles are delivered to a low-pH environment. Clathrin-mediated endocytosis has been reported as a major mode of entry into cell by New World arenaviruses. This low pH triggers the fusion of virus to the endosomal membrane and that finally results into the release of ribonucleoprotein (RNP) complex containing viral genome into the cytoplasm (Rojek and Kunz, 2008). Histological analysis of infected sites in Lassa fever patients showed no signs of infiltration of inflammatory cells (Walker et al., 1982). A hemorrhagic shock syndrome is noticed in fatal cases of Lassa fever, where no evident signs of necrosis and vascular damage are visible. This implicates the impairment of host cellular function rather than the direct effect on morphological changes in the ECs in case of Lassa fever infection (Moraz and Kunz, 2011). ECs are reported to be infected only at the late stages of LASV infection (Hensley et al., 2011). No specific cytopathic effect has been reported in ECs infected with LASV because LASV has a nonlytic cell cycle (Lukashevich et al., 1999).

New World arenaviruses, South American HF viruses, Junín virus (JUNV), Machupo virus (MACV), Guanarito virus (GTOV), and Sabia virus (SABV) mostly use transferrin receptor 1 (TfR1) to infect the ECs and activated immune cells (Geisbert and Jahrling, 2004; Radoshitzky et al., 2007). In vitro infection of ECs by JUNV leads to the activation of ECs, which results into increased expression of vascular cell adhesion molecule-1 (VCAM-1) and intercellular adhesion molecule-1 (ICAM-1) on ECs. Increased levels of vasoactive nitric oxide (NO), prostaglandin PGI_2, and endothelial NO synthase have been reported in the culture supernatants of in vitro experiments (Gomez et al., 2003).

In case of Argentine hemorrhagic fever, higher levels of TNF-α and IFN have been reported and this might result to additional damage to ECs leading to permeability defects (Marta et al., 1999). These proinflammatory mediators may aid in compromising the vascular permeability of ECs in hemorrhagic fevers caused by arenaviruses.

5.3   VIRUSES OF FAMILY *BUNYAVIRIDAE* AND ENDOTHELIAL DYSFUNCTIONS

Bunyaviridae family of viruses is divided into five genera: Bunyavirus, Hantavirus, Nairovirus, Phlebovirus, and Tospovirus. Hantavirus, Crimean–Congo hemorrhagic fever virus (CCHFV), and Rift Valley fever virus are the major hemorrhagic fever-causing viruses in this family. Hantavirus causes two vascular permeability-based diseases: hemorrhagic fever with renal syndrome (HFRS) and hantavirus cardiopulmonary syndrome (HCPS).

Rodents are the natural hosts for hantaviruses that cause HFRS and HCPS, and humans are usually infected by aerosolized virus-contaminated rodent excreta (Schonrich et al., 2008; Vapalahti et al., 2003). Except for Andes virus (ANDV), human-to-human transmission of hantaviruses does not seem to occur (Montgomery et al., 2007). Hantavirus primarily infects and replicates in ECs, forming the lining of blood capillaries (Niikura et al., 2004; Nolte et al., 1995; Pensiero et al., 1992; Yanagihara et al., 1987; Zaki et al., 1995). Internalization of vascular endothelial cadherin (VE-cadherin) is reported as a prominent factor, responsible for increased permeability of ECs (Shrivastava-Ranjan et al., 2010) (Figure 5.2). Entry of pathogenic hantaviruses, Hantaan virus (HTNV), Seoul virus (SEOV), Dobrava virus (DOBV), and Puumala virus (PUUV) into ECs, is mediated by the presence of $\alpha_v\beta_3$ integrins expressed on the EC membrane. Hantavirus binds with high affinity to β_3 integrins (Gavrilovskaya et al., 1998, 1999). β_3 integrins

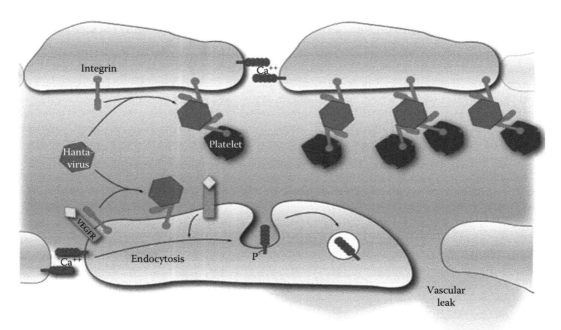

FIGURE 5.2   (See color insert.) Binding and entry of Hantavirus onto the integrins of ECs. Hantavirus binds to the β_3 integrin of the $\alpha_v\beta_3$ integrins on the EC membrane to gain entry by means of endocytosis. The binding of hantavirus to the $\alpha_v\beta_3$ integrins impairs its function of reducing the VEGFR2-mediated internalization of VE-cadherin. The β_3 integrin, in its inactive conformation, cannot mediate the VEGF signaling through its receptor, which leads to phosphorylation and internalization of VE-cadherin by endocytosis. VE-cadherin maintains the AJ integrity. Upon VEGF stimulation, VE-cadherin is phosphorylated and internalized reducing the adherence between ECs and aiding in angiogenesis and leukocyte extravasation and ultimately causes paracellular leakage. Also, ECs covered by hantavirus also bind to quiescent platelets by binding to the integrin on them. This can directly lead to thrombocytopenia and coagulation defects. (Reproduced from Steinberg, B.E. et al., *Antiviral Res.*, 93, 2, 2012. With permission.)

primarily help in regulating vascular integrity, EC permeability, and platelet functions (Byzova et al., 2000; Hodivala-Dilke et al., 1999; Hynes, 2003; Reynolds et al., 2002; Robinson et al., 2004). Hantavirus covers the entire surface of the ECs within few days of infection (Goldsmith et al., 1995). (The $\alpha_v\beta_3$ integrins play a key role in the regulation of vascular endothelial growth factor receptor 2 (VEGFR2)-based internalization of VEGF (Gavrilovskaya et al., 2008).) VEGF is a vascular permeability factor involved in directing localized tissue edema (Dvorak et al., 1995). Once hantavirus binds to $\alpha_v\beta_3$ integrins, it locks the $\alpha_v\beta_3$ integrin in an inactive bent conformation. The functions related to regulation of the vascular permeability through the interaction of $\alpha_v\beta_3$ integrin to VEGFR2 are dysregulated in this inactive bent conformation (Raymond et al., 2005; Takagi et al., 2002) (Figure 5.2). The migration through ECs requires VEGF-mediated loosening of adherent ECs and integrins to provide a vector for movement (Ridley et al., 2003). Binding of hantavirus to the $\alpha_v\beta_3$ integrins blocks the EC migration on β_3 integrin ligands (Gavrilovskaya et al., 2002). Generally, quiescent platelets do not adhere to ECs, but ECs, occupied by hantavirus, adhere to quiescent platelets (Mackow and Gavrilovskaya, 2009). This could directly lead to thrombocytopenia. This clearly indicates that infected ECs have a much altered adherence property compared to uninfected ECs.

Intense CD8+ T cell response has been reported against hantavirus in HCPS patients (Kilpatrick et al., 2004). This intense immunological response is having an impact on the exacerbation of the capillary leakage, observed during HCPS and HFRS (Terajima et al., 2007). TNF-α, IL-6, and

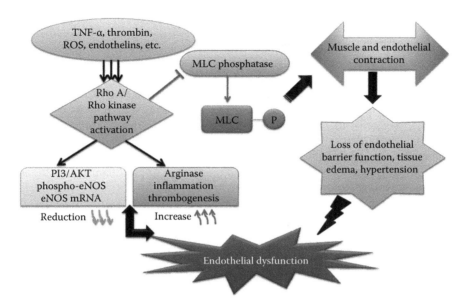

FIGURE 5.3 Schematic of the events that lead to vascular leakage and dysfunction in ECs. Various factors such as TNF-α, thrombin, endothelin, and mechanical shear stress can trigger off the activation of Rho A pathway. This might lead to the activation of myosin light chain (MLC) by phosphorylation, which causes actin–myosin interaction. Increase in inflammation and thrombus formation is also evident in activated Rho A pathway. EDRFs such as NO decrease due to (i) the increase in expression/activity of arginase, which reduces the availability of L-arginine for endothelial nitric oxide synthase (eNOS) activity. (ii) The decrease in activation of PI3/AKT pathway leads to decrease in phosphorylation of eNOS and reduction of eNOS mRNA stability. This cytoskeletal rearrangement ensues an imbalance, which leads to the smooth muscle and EC contraction. In turn, this might result into the loss of endothelial barrier function and vascular leakage.

IL-10 are elevated among HFRS patients during acute phases of infection (Linderholm et al., 1996). The amount of TNF-α released during hantavirus infection determines the severity of HFRS (Maes et al., 2006). TNF-α might induce paracellular permeability in a Rho pathway-dependent manner as shown in Figure 5.3 (Kakiashvili et al., 2009). Human lung microvascular ECs infected with hantavirus produce RANTES and IP-10 leads to the activation of IRFs and NF-kB contributing to induction of TH1 immune response. Hantavirus infection induces the expression of RANTES and IP-10 without causing increased permeability in human lung microvascular ECs (Sundstrom et al., 2001). ANDV-infected monocyte-derived dendritic cells are reported to secrete matrix metalloproteinases 9 (MMP9). MMPs are endopeptidases involved in remodeling of ECM (Marsac et al., 2011). Thus, the vascular leakage associated with pathogenesis of HFRS and HPS is due to host immune mechanisms rather than direct damage to ECs.

CCHFV has been reported to infect human ECs in vitro resulting in the production of infectious viral particles. Endothelial damage has been reported among CCHF patients (Bodur et al., 2010). Increased levels of expression of E-selectin, VCAM-1, and ICAM-1 in ECs during CCHFV infection indicate the activation of ECs. CCHFV-infected ECs express higher levels of TNF-α, IL-6, and IL-8 (Connolly-Andersen et al., 2011). IL-8 is a potential chemoattractant and increases transendothelial migration of immune cells by increasing the permeability of ECs (Biffl et al., 1995; Talavera et al., 2004). ECs are known to produce C-type natriuretic peptide (CNP) (Garbers et al., 2006). CNP is a member of natriuretic peptides, which are known for its vasodilatory properties and maintaining the vascular tone (Heublein et al., 1992; Rubattu et al., 2008). The presence of CNP in CCHF patients varies with the severity of the disease. The more severe the CCHF condition, the more the amount of CNP has been reported. Level of CNP is used as a prognostic risk assessment marker for CCHF patients (Turkdogan et al., 2012).

5.4 VIRUSES OF FAMILY *FILOVIRIDAE* AND ENDOTHELIAL DYSFUNCTIONS

Ebola and Marburg viruses are the major filoviruses responsible for hemorrhagic fevers. Cellular attachment factors such as the C-type lectins DC-SIGN and DC-SIGNR can augment viral infection and might promote viral dissemination in hosts. The other lectin LSECtin is encoded in the same chromosomal locus as DC-SIGN/R and is co-expressed with DC-SIGNR on sinusoidal ECs in liver and lymph nodes (Liu et al., 2004). Entry of filoviruses into ECs is mediated by binding of the filovirus glycoproteins (GPs), which contain higher amount of mannose oligosaccharides to the LSECtin (Powlesland et al., 2008). Primary infection of the filovirus occurs in monocytes/macrophages and dendritic cells (Geisbert et al., 2003b; Hensley et al., 2002). This results into their activation and production of proinflammatory cytokines and chemokines such as TNF-α, which leads to the disorganization and increase in permeability of ECs (Feldmann et al., 1996; Stroher et al., 2001). Direct infection of ECs by filoviruses has been reported at the advanced stages of infections (Geisbert et al., 2003b). The expression of cellular adhesion molecules, such as plakoglobin, cadherin, and catenins, gets dysregulated, in ECs. This leads to disseminated intravascular coagulation (DIC). Geisbert et al. have reported that the ECs derived from human lung and umbilical vein were highly permissive for Ebola virus (EBOV) replication (Geisbert and Hensley, 2004), but EBOV infection did not result into any major damage to ECs. This suggests that the severe pathogenesis of EBOV hemorrhagic fever might be due to immune-mediated mechanisms. Similar kind of interaction has been reported in primates, where the endothelium remained undamaged even until the advanced stages of disease (Geisbert et al., 2003b). This suggests that ECs do not get damaged by EBOV infections in-vivo.

The full-length type I transmembrane $GP_{1,2}$ of EBOV has been shown to lead to cytopathic effects on the ECs. Serine–threonine-rich, mucin-like domain of this $GP_{1,2}$ is involved into severe permeability defect of ECs and hemorrhage (Yang et al., 2000). Transcriptional RNA editing of $GP_{1,2}$ protein transcript is not required for efficient EBOV replication, but it seems to be an important means to decrease the cytotoxicity of this protein. In turn, this helps the virus to increase the viral load in the host and promote further viral infection (Volchkov et al., 2001).

The RNA transcript codes for a smaller secreted glycoprotein (sGP) without any RNA editing. The event of RNA editing is rare; hence, the amount of sGP is more compared to the full-length GP (Volchkov et al., 1995). sGP helps to restore the barrier function of the ECs, which get affected adversely by the effect of various cytokines (Wahl-Jensen et al., 2005). The anti-inflammatory role of sGP further helps the virus to modulate its pathogenesis and survive for the longer duration in the host. Therefore, severe pathogenesis in case of filoviral infection seems to be a concerted effect of proinflammatory cytokines (hypercytokinemia) and up-to some extent the direct endothelial damage caused by viral replication (Villinger et al., 1999; Yang et al., 2000).

5.5 VIRUSES OF FAMILY *FLAVIVIRIDAE* AND ENDOTHELIAL DYSFUNCTIONS

Dengue virus (DENV), yellow fever virus (YFV), and Kyasanur Forest disease virus (KFDV) are capable of causing hemorrhagic fevers. Endothelial dysfunctions are much pronounced in dengue hemorrhagic fever (DHF). DENV causes two severe manifestations such as DHF and dengue shock syndrome (DSS). These manifestations are related to vascular dysfunctions due to loss of barrier function of the ECs. The morphology of ECs remains unaltered, but profuse vascular leakage is evident in DHF and DSS. Although DENV infection of ECs is not the main reason for pathogenesis of DHF or DSS, the EC dysfunction seems to be due to various immune-mediated mechanisms (Basu and Chaturvedi, 2008).

DENV-mediated activation of ECs is through a cell surface protein on DENV called protein disulfide isomerase (PDI). After DENV infection, PDI colocalizes with β1 and β3 integrins, implicating the activation of integrins (Wan et al., 2012). Levels of soluble cell adhesion molecules such as sICAM-1 and sVCAM-1 are significantly increased in circulation in DENV infection due to EC damage and activation (Cardier et al., 2006). Dewi et al. (2004) reported that DENV infection of ECs at a multiplicity of infection (m.o.i) 5 decreased the TEER value of ECs, indicating increase in EC permeability. They also reported the increase in EC permeability even at a low viral titer, due to cytosolic mediators such as TNF-α (Carr et al., 2003). In another study, addition of peripheral blood mononuclear cells (PBMCs) decreased the TEER value and increased the trans-albumin permeability in human umbilical vein endothelial cells (HUVECs) infected with DENV-2 (Dewi et al., 2008).

VEGF plays a pivotal role in maintenance of vascular permeability and angiogenesis (Ferrara and Davis-Smyth, 1997). VEGF regulates the localization of VE-cadherin, which maintains the integrity of AJs (Figure 5.3). Upon VEGF stimulation, VE-cadherin gets phosphorylated and internalized, which results into reduced adherence between ECs and aiding in leukocyte extravasations (Carr et al., 2003). VEGF binds to the VEGFR2 and initiates the Src-Vav2-Rac1-PAK pathway, which leads to the tyrosine phosphorylation of VE-cadherin and eventually leads to internalization of VE-cadherin via clathrin-coated pits (Figure 5.4). This results into dismantling of the AJs, which leads to increase in EC permeability (Carmeliet and Collen, 2000; Dejana et al., 2008; Gavard and Gutkind, 2006; Monaghan-Benson and Burridge, 2009; Mukherjee et al., 2006; Wallez et al., 2006). VEGF is capable of inducing endothelial migration and angiogenesis by Rho A/Rho kinase signaling, which helps in cytoskeletal rearrangements required for cell migration (van Nieuw Amerongen et al., 2003).

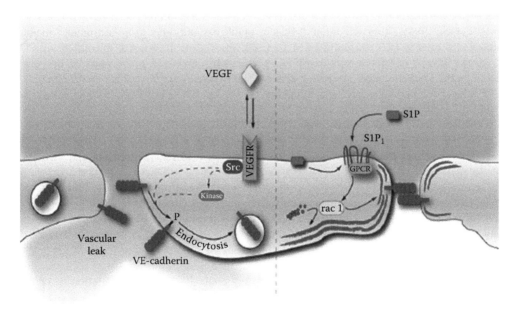

FIGURE 5.4 **(See color insert.)** Barrier disruptive and protective signaling pathways at the AJ. VEGF signaling through its receptor leads to the activation of Src family kinases and downstream phosphorylation of the VE-cadherin, which is subsequently endocytosed. Removal of VE-cadherin from the interendothelial junction results in paracellular leak (Carr et al., 2003). Sphingosine-1-phosphate (S1P) is produced within ECs as well as in the circulation. Binding to the S1P1 receptor (S1P1) leads to G-protein-coupled receptor activation and downstream stabilization of the cortical actin network and maintenance of intercellular junctional proteins. (Reproduced from Steinberg, B.E. et al., *Antiviral Res.*, 93, 2, 2012. With permission.)

Plasma VEGF levels are significantly higher in patients with DHF. D-dimer level, an activation marker of fibrinolysis, reported to be significantly higher and varies with the changes in VEGF levels. TNF-α has been reported to be associated with the levels of D-dimer, a fibrin degradation product implicated in coagulopathies leading to DIC, in patient samples suffering from DSS (Suharti et al., 2002).

The activation of the fibrinolytic system might play a role in VEGF production among DHF patients (Tseng et al., 2005). TJs are also disrupted during DHF and DSS along with AJs. Dengue-infected monocytes and lymphocytes secrete chemokine called monocyte chemoattractant protein-1 (MCP-1). Exposure of ECs with recombinant MCP-1 and culture supernatant from dengue-infected monocytes has been reported to disrupt tight junction protein (TJP) such as ZO-1 (Lee et al., 2006).

In another study, DENV-infected human dermal microvascular EC line (HMEC-1) showed increase in actin-containing structures and endothelial permeability, without having any noticeable cytopathic effect. The tight junction protein (TJP) occludin was displaced to the cytoplasm. This change in transendothelial permeability in infected cells was attributed to higher levels of IL-8. Very high levels of IL-8 have been reported in plasma and pleural fluids of patients (Avirutnan et al., 1998). This effect on cytoskeleton and TJs was partly inhibited by antibodies against IL-8; implicating a role of IL-8 in impairing vascular permeability in DHF (Bosch et al., 2002; Talavera et al., 2004). Many other factors also play role in vascular damage during DHF. Infection of ECs by DENV leads to the activation of ECs and expression of cytokines such as BAFF, IL-6/8, CXCL9/10/11, RANTES, and IL-7. Alternative complement pathway activator, properdin factor B, also increases significantly in DENV infection. This suggests that the proinflammatory factors released from activated endothelium aid in the exacerbation of the severity of DHF/DSS (Avirutnan et al., 1998; Brown et al., 2006; Dalrymple and Mackow, 2012). TNF-α released by mast cells activates ECs leading to increased expression of ICAM-1 and VCAM-1. VCAM-1 helps in recruitment of PBMCs to ECs (Henseleit et al., 1995). Upon activation of ECs by cytokines and other proinflammatory mediators, the leukocytes bind to ECs through ICAM-1/LFA-1 interactions to migrate into the tissues (Yang et al., 2005). TNF-α is known to increase EC permeability and aids in the vascular dysfunctions mediated by the viral infection (Anderson et al., 1997; Brown et al., 2011; Carr et al., 2003).

TNF-α has been known to have proangiogenic properties, whereas VEGF is known as a vascular permeability regulator. The production of VEGF is induced by inflammatory cytokines, such as TNF-α (Hangai et al., 2006; Nabors et al., 2003; Robinson et al., 2011; Ryuto et al., 1996; Shweiki et al., 1992; Wojciak-Stothard et al., 1998). Therefore, VEGF is probably one of the major mediators, which potentiates the TNF-α-induced cellular actomyosin rearrangements and formation of intercellular gaps in ECs. Higher levels of IFN-γ are reported in DENV infections (Bozza et al., 2008; Chaturvedi et al., 2000). IFN-γ decreases the expression of $\alpha_V\beta_3$ integrins. This leads to vascular defects such as decrease in adhesion and survival of ECs (Ruegg et al., 1998).

TNF-α is known to induce the expression of platelet tissue factor (TF)/thrombokinase in ECs leading to increased local concentration of thrombin, which results into higher fibrin deposition. TF-dependent pathway is the dominant route of thrombin generation that leads to fibrin deposition (Taylor et al., 1991) (Geisbert et al., 2003a). This mediates a transition from antihemostatic to procoagulant state (Hirokawa et al., 2000; Polunovsky et al., 1994). TF upregulation is reported to play a role in DIC (Drake et al., 1993; Semeraro et al., 1993), which is commonly observed in VHFs. This may be one of the ways by which the DENV facilitates the hemorrhage. DIC, also known as disseminated intravascular coagulopathy or consumptive coagulopathy, is a pathological activation of coagulation (blood clotting) mechanisms that happens in response to a variety of factors. DIC is a syndrome characterized by activation of the coagulation system, either in a localized region or throughout the entire vascular system, leading to histologically visible microthrombi generated in the microvasculature (Levi, 2001; Levi et al., 1999). These microthrombi may hamper adequate blood supply, thereby contributing to the multiple organ dysfunction. As the small clots consume coagulation proteins and platelets, normal coagulation is disrupted and abnormal bleeding occurs from the vital parts.

Antibodies against NS1 also cross-react with ECs and platelets to cause damage (Lin et al., 2006). These cross-reactive autoantibodies induce EC apoptosis in a caspase-dependent manner. Epitope of the autoantibody is targeted to the LYRIC protein on human ECs (Lin et al., 2003; Liu et al., 2011). Upon induction by the cross-reactive anti-DENV NS1 antibodies, the ECs get activated through the NF-kB pathway and release a plethora of cytokines and chemokines such as IL-6, IL-8, and MCP-1, which could trigger inflammatory pathways leading to vascular damage (Lin et al., 2005). DENV nonstructural protein-1 (NS1) binds to prothrombin and inhibits its activation into thrombin (Lin et al., 2012). Anti-DENV NS1 antibodies cross-react with human platelets, leading to damage of these cells, which might result into immune-mediated thrombocytopenia (Lin et al., 2006, 2011).

Decreased levels of circulating activated protein C (aPC) are reported in DENV infection (Cabello-Gutierrez et al., 2009). aPC plays a vital role during sepsis, by reducing the inflammation and maintaining vascular coagulopathy through stabilizing the endothelial barrier and by acting as an antiapoptotic factor. aPC mainly downregulates the production of thrombin, which is proinflammatory, procoagulant, and antifibrinolytic in nature. aPC also reduces the production of IL-1, IL-6, and TNF-α by monocytes and inactivates PAI-1, thereby acting as a profibrinolytic agent (Griffin et al., 2007). The antithrombotic and cytoprotective roles of aPC are altered during dengue infections. This gives an insight as to how DENV causes such severe vasculopathy (Cabello-Gutierrez et al., 2009).

Increased levels of TNF-α have been reported in various VHFs. Higher levels of TNF-α have been shown to promote apoptosis in vivo and in vitro in ECs (Ceconi et al., 2007) (Valgimigli et al., 2003). TNF-α induces the expression of Notch-2 receptor and suppresses the Notch-4 receptor. Notch signaling regulates activities such as cell proliferation, differentiation, and apoptosis (Basuroy et al., 2006; Miele et al., 2006; Quillard et al., 2009). Functionally, this switch to Notch-2 receptor leads to the priming of ECs and decrease in the expression of antiapoptotic regulator for caspase-3 (Quillard et al., 2010).

Binding of TNF-α to its receptor (p60 or TNFR1) has been reported to activate sphingomyelinase to generate ceramide. Ceramide in turn engages specific protein kinases and phosphatases, which activates to the downstream signaling pathways (Hirokawa et al., 2000). A decrease in F-actin levels and increase in G-actin level have been reported in ECs exposed to TNF-α, resulting into F-actin depolymerization increasing EC barrier dysfunction (Goldblum et al., 1993).

5.6 PLATELET DYSFUNCTION AND DEPLETION IN VIRAL HEMORRHAGIC FEVERS

Endothelial dysfunctions leading to hemorrhages in VHFs are mostly accompanied by thrombocytopenia or severe platelet dysfunction. Role of platelets has been well established in thrombosis and hemostasis. Overstimulation of platelets during infection leads to DIC. Thrombocytopenia is commonly seen in VHFs such as DHF and HFRS. The key hemostatic reaction of fibrin deposition and platelet adhesion takes place at the sites of vessel wall damage (Kuijper et al., 1997). Once the vascular integrity is disrupted, ECM components such as collagens get exposed and start interacting with platelet receptors. Activated endothelium interacts with platelets in promoting inflammation. Platelet receptors interact with platelet cell adhesion molecule (PCAM-1), P-selectin, and VCAM. Furthermore, P-selectin binds to von Willebrand factor (vWf) and potentiates platelet interactions via collagen receptor GPIbα (Cruz et al., 2005; Morrell et al., 2008; Padilla et al., 2004).

Platelets do not normally adhere to resting ECs, but an increased binding of platelets to stimulated ECs is reported in DHF (Butthep et al., 1993; Krishnamurti et al., 2002). Additionally, activated ECs have increased expression of ICAM-1 and PCAM-1 in DHF patients (Cardier et al., 2006). This increase in CAMs would aid in increased platelet adhesion to ECs, which could lead to severe thrombocytopenia in DHF patients.

TABLE 5.1

Summary of the Various Steps Involved in Viral Entry and Pathogenesis of the Disease Causing Endothelial Activation and Damage

Virus	EC Infection and Factors Contributing to Endothelial Dysfunction	References
Dengue virus	Mode of entry into endothelium.	
	Cell surface protein called PDI is known to aid in viral entry and activation of ECs.	Wan et al. (2012)
	DEN-4 entry has been shown to be mediated by heparan sulfate-containing cell surface receptors on ECs.	Dalrymple and Mackow (2011)
	Activation of ECs: infection of ECs by DENV causes EC activation.	Brown et al. (2006)
	Cytokines such as TNF-α increase activation of ECs.	Wan et al. (2012)
	Endothelial dysfunction: soluble adhesion molecules such as sICAM-1 and sVCAM-1 are increased.	Cardier et al. (2006)
	VE-cadherin is internalized leading to the dismantling of the AJs.	Dewi et al. (2008)
	Viral factors: NS-1 binds to prothrombin aiding in the coagulation defect.	Lin et al. (2012)
	The cleavage of NS4A polyprotein during viral replication significantly increases levels of immunomediators such as TNF-α and IL-8, which can lead to vascular defect.	Kelley et al. (2012)
Filovirus (Ebola and Marburg viruses)	Mode of entry into endothelium: entry of the enveloped viruses such as filoviruses into ECs is mediated by binding of the surface GP, which contains high amount of mannose oligosaccharides to the LSECtin.	Powlesland et al. (2008)
	Endothelial dysfunction.	
	TNF-α released from activated immune cells due to viral infection increases permeability of ECs.	Feldmann et al. (1996); Stroher et al. (2001).
	Direct infection of ECs results in damage in vitro and may lead to endothelial dysfunction.	Yang et al. (2000)
	Viral factors: type I transmembrane $GP_{1,2}$ of EBOV has been indicated to cause cytopathic effect in the ECs.	Yang et al. (2000)
	sGP has an anti-inflammatory role slightly curbing the vascular defect caused by TNF-α.	Wahl-Jensen et al. (2005)
Hantavirus	Mode of entry into endothelium.	
	Entry of pathogenic Hantavirus (HTNV, SEOV, PUUV, DOBV) into the ECs is mediated by the presence of $α_vβ_3$ integrins on the EC membrane.	Gavrilovskaya et al. (1999)
	Endothelial dysfunction: binding of Hantavirus to $α_vβ_3$ integrins of ECs leads to dysregulation of EC function.	Raymond et al. (2005); Takagi et al. (2002)
	Loss of barrier function has been shown due to the internalization of VE-cadherin.	Shrivastava-Ranjan et al. (2010)
	Loss of infected EC defense against CD8+ T cells leads to capillary leakage.	Terajima et al. (2007)
Crimean–Congo hemorrhagic fever virus (CCHFV)	Mode of entry: CCHFV is taken up by clathrin-dependent endocytosis, which is cholesterol and pH dependent.	Simon et al. (2009)
	Human cell surface nucleolin has been suggested to be a putative CCHFV entry factor.	Xiao et al. (2011)

(continued)

TABLE 5.1 (continued)
Summary of the Various Steps Involved in Viral Entry and Pathogenesis of the Disease Causing Endothelial Activation and Damage

Virus	EC Infection and Factors Contributing to Endothelial Dysfunction	References
	Activation of ECs: EC activation is evident in clinical patients of CCHV and in in vitro models.	Bodur et al. (2010); Connolly-Andersen et al. (2011)
	Endothelial dysfunction: activated ECs release IL-8, which increases vascular permeability.	Connolly-Andersen et al. (2011)
	CNP levels in CCHF patients are high indicating dysregulation of vascular functions.	Turkdogan et al. (2012)
Old World arenaviruses (LASV)	Mode of entry: Old World arenaviruses bind to a host cell receptor, α-DG, with very high affinity in order to gain entry into the cells.	Cao et al. (1998)
	Pathogenesis: in Lassa fever, no cytokine storm ensues.	Lukashevich et al. (1999)
	Infected cells bind to DG with high affinity. This interferes β1 integrin signaling affecting the ECM and in turn the vascular system.	Rojek et al. (2012)
	ECs are only infected at the late stage of infection.	Hensley et al. (2011)
New World arenaviruses	Mode of entry: these viruses use TfR1 to bind and infect endothelial and activated immune cells.	Geisbert and Jahrling (2004); Radoshitzky et al. (2007)
	Endothelial activation and dysfunction.	
	VCAM-1 and ICAM-1 expression was increased upon EC infection of JUNV.	Gomez et al. (2003)
	Expression of coagulation factors was decreased and an increase in levels of vasoactive NO, prostaglandin PGI_2, and endothelial NO synthase was observed.	
	Cytokines such as TNF-α and IFN are increased during JUNV infection.	Marta et al. (1999)

Platelet dysfunction results into defective release of adenosine diphosphate in DHF patients (Mitrakul et al., 1977). Adenosine diphosphate released by maturing megakaryocytes plays a vital role in proplatelet formation (Balduini et al., 2012). Platelet dysfunction is well documented in HFRS, which results into a defect in aggregation and release reactions of platelets (Cosgriff et al., 1991; Guang et al., 1989; Lee et al., 1989). Hantavirus infection in ECs results into severe thrombocytopenia in HFRS. Hantavirus aids in the attachment of quiescent platelets to infected ECs due to its inherent property to bind to β3 integrins. β3 integrins are the integral adhesive receptors on platelets and ECs regulating adhesion, platelet activation, and vascular permeability (Figure 5.2) (Gavrilovskaya et al., 2010; Mackow and Gavrilovskaya, 2001). Cytokine-mediated increase in activation of ECs might lead to apoptosis, which may also be a critical factor involved in depleting the circulating platelets. The important events taking place during different VHFs have been summarized in Table 5.1.

5.7 CONCLUSION

Pathogenesis of hemorrhagic fever viruses is governed by three responses: the immune response, hemostatic response, and virus-mediated response in the form of secreted proteins. Local immune response to viral infection leads to the release of various proinflammatory cytokines and complement activation that leads to tissue destruction by disseminated intravascular inflammation (DII). Sometimes, virus-encoded proteins play a major role in pathogenesis of VHF by activating ECs or by interfering directly with the coagulation machinery of the body. Similarly, hemostatic response to viruses leads to accelerated coagulation, which eventually leads to DIC. This ultimately results into tissue ischemia and multiple organ failures. Participation of ECs in pathogenesis of hemorrhagic fever viruses is one of the most important and complex factors governing the clinical outcome of the disease. Current molecular biology techniques have greatly aided in understanding the exact mechanism of how the viruses target different cell types and how their replication impedes normal cellular functions. Understanding the unique characteristics of different VHFs will be quite helpful in developing therapeutic tools against these infections.

REFERENCES

Anderson, R., Wang, S., Osiowy, C., and Issekutz, A. C. (1997). Activation of endothelial cells via antibody-enhanced dengue virus infection of peripheral blood monocytes. *J Virol* **71**(6), 4226–4232.

Avirutnan, P., Malasit, P., Seliger, B., Bhakdi, S., and Husmann, M. (1998). Dengue virus infection of human endothelial cells leads to chemokine production, complement activation, and apoptosis. *J Immunol* **161**(11), 6338–6346.

Balduini, A., di Buduo, C. A., Malara, A., Lecchi, A., Rebuzzini, P., Currao, M., Pallotta, I., Jakubowski, J., and Cattaneo, M. (2012). Constitutively released adenosine diphosphate regulates proplatelet formation by human megakaryocytes. *Haematologica* 97(11), 1657–1665.

Basu, A. and Chaturvedi, U. C. (2008). Vascular endothelium: The battlefield of dengue viruses. *FEMS Immunol Med Microbiol* **53**(3), 287–299.

Basuroy, S., Bhattacharya, S., Tcheranova, D., Qu, Y., Regan, R. F., Leffler, C. W., and Parfenova, H. (2006). HO-2 provides endogenous protection against oxidative stress and apoptosis caused by TNF-alpha in cerebral vascular endothelial cells. *Am J Physiol Cell Physiol* **291**(5), C897–C908.

Bazzoni, G. and Dejana, E. (2004). Endothelial cell-to-cell junctions: Molecular organization and role in vascular homeostasis. *Physiol Rev* **84**(3), 869–901.

Biffl, W. L., Moore, E. E., Moore, F. A., Carl, V. S., Franciose, R. J., and Banerjee, A. (1995). Interleukin-8 increases endothelial permeability independent of neutrophils. *J Trauma* **39**(1), 98–102; discussion 102–103.

Bodur, H., Akinci, E., Onguru, P., Uyar, Y., Basturk, B., Gozel, M. G., and Kayaaslan, B. U. (2010). Evidence of vascular endothelial damage in Crimean-Congo hemorrhagic fever. *Int J Infect Dis* **14**(8), e704–e707.

Bosch, I., Xhaja, K., Estevez, L., Raines, G., Melichar, H., Warke, R. V., Fournier, M. V., Ennis, F. A., and Rothman, A. L. (2002). Increased production of interleukin-8 in primary human monocytes and in human epithelial and endothelial cell lines after dengue virus challenge. *J Virol* **76**(11), 5588–5597.

Bozza, F. A., Cruz, O. G., Zagne, S. M., Azeredo, E. L., Nogueira, R. M., Assis, E. F., Bozza, P. T., and Kubelka, C. F. (2008). Multiplex cytokine profile from dengue patients: MIP-1beta and IFN-gamma as predictive factors for severity. *BMC Infect Dis* **8**, 86.

Brown, M. G., Hermann, L. L., Issekutz, A. C., Marshall, J. S., Rowter, D., Al-Afif, A., and Anderson, R. (2011). Dengue virus infection of mast cells triggers endothelial cell activation. *J Virol* **85**(2), 1145–1150.

Brown, M. G., King, C. A., Sherren, C., Marshall, J. S., and Anderson, R. (2006). A dominant role for FcgammaRII in antibody-enhanced dengue virus infection of human mast cells and associated CCL5 release. *J Leukoc Biol* **80**(6), 1242–1250.

Butthep, P., Bunyaratvej, A., and Bhamarapravati, N. (1993). Dengue virus and endothelial cell: a related phenomenon to thrombocytopenia and granulocytopenia in dengue hemorrhagic fever. *Southeast Asian J Trop Med Public Health* **24**(Suppl 1), 246–249.

Byzova, T. V., Goldman, C. K., Pampori, N., Thomas, K. A., Bett, A., Shattil, S. J., and Plow, E. F. (2000). A mechanism for modulation of cellular responses to VEGF: Activation of the integrins. *Mol Cell* **6**(4), 851–860.

Cabello-Gutierrez, C., Manjarrez-Zavala, M. E., Huerta-Zepeda, A., Cime-Castillo, J., Monroy-Martinez, V., Correa, B. B., and Ruiz-Ordaz, B. H. (2009). Modification of the cytoprotective protein C pathway during Dengue virus infection of human endothelial vascular cells. *Thromb Haemost* **101**(5), 916–928.

Cao, W., Henry, M. D., Borrow, P., Yamada, H., Elder, J. H., Ravkov, E. V., Nichol, S. T., Compans, R. W., Campbell, K. P., and Oldstone, M. B. (1998). Identification of alpha-dystroglycan as a receptor for lymphocytic choriomeningitis virus and Lassa fever virus. *Science* **282**(5396), 2079–2081.

Cardier, J. E., Rivas, B., Romano, E., Rothman, A. L., Perez-Perez, C., Ochoa, M., Caceres, A. M., Cardier, M., Guevara, N., and Giovannetti, R. (2006). Evidence of vascular damage in dengue disease: Demonstration of high levels of soluble cell adhesion molecules and circulating endothelial cells. *Endothelium* **13**(5), 335–340.

Carmeliet, P. and Collen, D. (2000). Molecular basis of angiogenesis. Role of VEGF and VE-cadherin. *Ann N Y Acad Sci* **902**, 249–262; discussion 262–264.

Carr, J. M., Hocking, H., Bunting, K., Wright, P. J., Davidson, A., Gamble, J., Burrell, C. J., and Li, P. (2003). Supernatants from dengue virus type-2 infected macrophages induce permeability changes in endothelial cell monolayers. *J Med Virol* **69**(4), 521–528.

Ceconi, C., Fox, K. M., Remme, W. J., Simoons, M. L., Bertrand, M., Parrinello, G., Kluft, C., Blann, A., Cokkinos, D., Ferrari, R., Investigators, E., Investigators, P., and the Statistical, Committee. (2007). ACE inhibition with perindopril and endothelial function. Results of a substudy of the EUROPA study: PERTINENT. *Cardiovasc Res* **73**(1), 237–246.

Chaturvedi, U. C., Agarwal, R., Elbishbishi, E. A., and Mustafa, A. S. (2000). Cytokine cascade in dengue hemorrhagic fever: Implications for pathogenesis. *FEMS Immunol Med Microbiol* **28**(3), 183–188.

Clegg, J. C. (2002). Molecular phylogeny of the arenaviruses. *Curr Top Microbiol Immunol* **262**, 1–24.

Connolly-Andersen, A. M., Moll, G., Andersson, C., Akerstrom, S., Karlberg, H., Douagi, I., and Mirazimi, A. (2011). Crimean-Congo hemorrhagic fever virus activates endothelial cells. *J Virol* **85**(15), 7766–7774.

Cosgriff, T. M., Lee, H. W., See, A. F., Parrish, D. B., Moon, J. S., Kim, D. J., and Lewis, R. M. (1991). Platelet dysfunction contributes to the haemostatic defect in haemorrhagic fever with renal syndrome. *Trans R Soc Trop Med Hyg* **85**(5), 660–663.

Cruz, M. A., Chen, J., Whitelock, J. L., Morales, L. D., and Lopez, J. A. (2005). The platelet glycoprotein Ib-von Willebrand factor interaction activates the collagen receptor alpha2beta1 to bind collagen: Activation-dependent conformational change of the alpha2-I domain. *Blood* **105**(5), 1986–1991.

Dalrymple, N. and Mackow, E. R. (2011). Productive dengue virus infection of human endothelial cells is directed by heparan sulfate-containing proteoglycan receptors. *J Virol* **85**(18), 9478–9485.

Dalrymple, N. A. and Mackow, E. R. (2012). Endothelial cells elicit immune-enhancing responses to dengue virus infection. *J Virol* **86**(12), 6408–6415.

Dejana, E., Orsenigo, F., and Lampugnani, M. G. (2008). The role of adherens junctions and VE-cadherin in the control of vascular permeability. *J Cell Sci* **121**(Pt 13), 2115–2122.

Dewi, B. E., Takasaki, T., and Kurane, I. (2004). In vitro assessment of human endothelial cell permeability: Effects of inflammatory cytokines and dengue virus infection. *J Virol Methods* **121**(2), 171–180.

Dewi, B. E., Takasaki, T., and Kurane, I. (2008). Peripheral blood mononuclear cells increase the permeability of dengue virus-infected endothelial cells in association with downregulation of vascular endothelial cadherin. *J Gen Virol* **89**(Pt 3), 642–652.

Drake, T. A., Cheng, J., Chang, A., and Taylor, F. B., Jr. (1993). Expression of tissue factor, thrombomodulin, and E-selectin in baboons with lethal Escherichia coli sepsis. *Am J Pathol* **142**(5), 1458–1470.

Dvorak, H. F., Brown, L. F., Detmar, M., and Dvorak, A. M. (1995). Vascular permeability factor/vascular endothelial growth factor, microvascular hyperpermeability, and angiogenesis. *Am J Pathol* **146**(5), 1029–1039.

Feldmann, H., Bugany, H., Mahner, F., Klenk, H. D., Drenckhahn, D., and Schnittler, H. J. (1996). Filovirus-induced endothelial leakage triggered by infected monocytes/macrophages. *J Virol* **70**(4), 2208–2214.

Ferrara, N. and Davis-Smyth, T. (1997). The biology of vascular endothelial growth factor. *Endocr Rev* **18**(1), 4–25.

Garbers, D. L., Chrisman, T. D., Wiegn, P., Katafuchi, T., Albanesi, J. P., Bielinski, V., Barylko, B., Redfield, M. M., and Burnett, J. C., Jr. (2006). Membrane guanylyl cyclase receptors: An update. *Trends Endocrinol Metab* **17**(6), 251–258.

Gavard, J. and Gutkind, J. S. (2006). VEGF controls endothelial-cell permeability by promoting the beta-arrestin-dependent endocytosis of VE-cadherin. *Nat Cell Biol* **8**(11), 1223–1234.

Gavrilovskaya, I. N., Brown, E. J., Ginsberg, M. H., and Mackow, E. R. (1999). Cellular entry of hantaviruses which cause hemorrhagic fever with renal syndrome is mediated by beta3 integrins. *J Virol* **73**(5), 3951–3959.

Gavrilovskaya, I. N., Gorbunova, E. E., and Mackow, E. R. (2010). Pathogenic hantaviruses direct the adherence of quiescent platelets to infected endothelial cells. *J Virol* **84**(9), 4832–4839.

Gavrilovskaya, I. N., Gorbunova, E. E., Mackow, N. A., and Mackow, E. R. (2008). Hantaviruses direct endothelial cell permeability by sensitizing cells to the vascular permeability factor VEGF, while angiopoietin 1 and sphingosine 1-phosphate inhibit hantavirus-directed permeability. *J Virol* **82**(12), 5797–5806.

Gavrilovskaya, I. N., Peresleni, T., Geimonen, E., and Mackow, E. R. (2002). Pathogenic hantaviruses selectively inhibit beta3 integrin directed endothelial cell migration. *Arch Virol* **147**(10), 1913–1931.

Gavrilovskaya, I. N., Shepley, M., Shaw, R., Ginsberg, M. H., and Mackow, E. R. (1998). beta3 Integrins mediate the cellular entry of hantaviruses that cause respiratory failure. *Proc Natl Acad Sci USA* **95**(12), 7074–7079.

Geisbert, T. W. and Hensley, L. E. (2004). Ebola virus: New insights into disease aetiopathology and possible therapeutic interventions. *Expert Rev Mol Med* **6**(20), 1–24.

Geisbert, T. W. and Jahrling, P. B. (2004). Exotic emerging viral diseases: Progress and challenges. *Nat Med* **10**(12 Suppl), S110–S121.

Geisbert, T. W., Young, H. A., Jahrling, P. B., Davis, K. J., Kagan, E., and Hensley, L. E. (2003a). Mechanisms underlying coagulation abnormalities in Ebola hemorrhagic fever: Overexpression of tissue factor in primate monocytes/macrophages is a key event. *J Infect Dis* **188**(11), 1618–1629.

Geisbert, T. W., Young, H. A., Jahrling, P. B., Davis, K. J., Larsen, T., Kagan, E., and Hensley, L. E. (2003b). Pathogenesis of Ebola hemorrhagic fever in primate models: Evidence that hemorrhage is not a direct effect of virus-induced cytolysis of endothelial cells. *Am J Pathol* **163**(6), 2371–2382.

Goldblum, S. E., Ding, X., and Campbell-Washington, J. (1993). TNF-alpha induces endothelial cell F-actin depolymerization, new actin synthesis, and barrier dysfunction. *Am J Physiol* **264**(4 Pt 1), C894–C905.

Goldsmith, C. S., Elliott, L. H., Peters, C. J., and Zaki, S. R. (1995). Ultrastructural characteristics of Sin Nombre virus, causative agent of hantavirus pulmonary syndrome. *Arch Virol* **140**(12), 2107–2122.

Gomez, R. M., Pozner, R. G., Lazzari, M. A., D'Atri, L. P., Negrotto, S., Chudzinski-Tavassi, A. M., Berria, M. I., and Schattner, M. (2003). Endothelial cell function alteration after Junin virus infection. *Thromb Haemost* **90**(2), 326–333.

Griffin, J. H., Fernandez, J. A., Gale, A. J., and Mosnier, L. O. (2007). Activated protein C. *J Thromb Haemost* **5**(Suppl 1), 73–80.

Guang, M. Y., Liu, G. Z., and Cosgriff, T. M. (1989). Hemorrhage in hemorrhagic fever with renal syndrome in China. *Rev Infect Dis* **11**(Suppl 4), S884–S890.

Hangai, M., He, S., Hoffmann, S., Lim, J. I., Ryan, S. J., and Hinton, D. R. (2006). Sequential induction of angiogenic growth factors by TNF-alpha in choroidal endothelial cells. *J Neuroimmunol* **171**(1–2), 45–56.

Henseleit, U., Steinbrink, K., Sunderkotter, C., Goebeler, M., Roth, J., and Sorg, C. (1995). Expression of murine VCAM-1 in vitro and in different models of inflammation in vivo: Correlation with immigration of monocytes. *Exp Dermatol* **4**(5), 249–256.

Hensley, L. E., Smith, M. A., Geisbert, J. B., Fritz, E. A., Daddario-DiCaprio, K. M., Larsen, T., and Geisbert, T. W. (2011). Pathogenesis of Lassa fever in cynomolgus macaques. *Virol J* **8**, 205.

Hensley, L. E., Young, H. A., Jahrling, P. B., and Geisbert, T. W. (2002). Proinflammatory response during Ebola virus infection of primate models: Possible involvement of the tumor necrosis factor receptor superfamily. *Immunol Lett* **80**(3), 169–179.

Heublein, D. M., Clavell, A. L., Stingo, A. J., Lerman, A., Wold, L., and Burnett, J. C., Jr. (1992). C-type natriuretic peptide immunoreactivity in human breast vascular endothelial cells. *Peptides* **13**(5), 1017–1019.

Hirokawa, M., Kitabayashi, A., Kuroki, J., and Miura, A. B. (2000). Induction of tissue factor production but not the upregulation of adhesion molecule expression by ceramide in human vascular endothelial cells. *Tohoku J Exp Med* **191**(3), 167–176.

Hodivala-Dilke, K. M., McHugh, K. P., Tsakiris, D. A., Rayburn, H., Crowley, D., Ullman-Cullere, M., Ross, F. P., Coller, B. S., Teitelbaum, S., and Hynes, R. O. (1999). Beta3-integrin-deficient mice are a model for Glanzmann thrombasthenia showing placental defects and reduced survival. *J Clin Invest* **103**(2), 229–238.

Hynes, R. O. (2003). Structural biology. Changing partners. *Science* **300**(5620), 755–756.

Kakiashvili, E., Speight, P., Waheed, F., Seth, R., Lodyga, M., Tanimura, S., Kohno, M., Rotstein, O. D., Kapus, A., and Szaszi, K. (2009). GEF-H1 mediates tumor necrosis factor-alpha-induced Rho activation and myosin phosphorylation: Role in the regulation of tubular paracellular permeability. *J Biol Chem* **284**(17), 11454–11466.

Kelley, J. F., Kaufusi, P. H., and Nerurkar, V. R. (2012). Dengue hemorrhagic fever-associated immunomediators induced via maturation of dengue virus nonstructural 4B protein in monocytes modulate endothelial cell adhesion molecules and human microvascular endothelial cells permeability. *Virology* **422**(2), 326–337.

Kilpatrick, E. D., Terajima, M., Koster, F. T., Catalina, M. D., Cruz, J., and Ennis, F. A. (2004). Role of specific CD8+ T cells in the severity of a fulminant zoonotic viral hemorrhagic fever, hantavirus pulmonary syndrome. *J Immunol* **172**(5), 3297–3304.

Krishnamurti, C., Peat, R. A., Cutting, M. A., and Rothwell, S. W. (2002). Platelet adhesion to dengue-2 virus-infected endothelial cells. *Am J Trop Med Hyg* **66**(4), 435–441.

Kuijper, P. H., Gallardo Torres, H. I., Lammers, J. W., Sixma, J. J., Koenderman, L., and Zwaginga, J. J. (1997). Platelet and fibrin deposition at the damaged vessel wall: cooperative substrates for neutrophil adhesion under flow conditions. *Blood* **89**(1), 166–175.

Kunz, S. (2009). The role of the vascular endothelium in arenavirus haemorrhagic fevers. *Thromb Haemost* **102**(6), 1024–1029.

Lee, M., Kim, B. K., Kim, S., Park, S., Han, J. S., Kim, S. T., and Lee, J. S. (1989). Coagulopathy in hemorrhagic fever with renal syndrome (Korean hemorrhagic fever). *Rev Infect Dis* **11**(Suppl 4), S877–S883.

Lee, Y. R., Liu, M. T., Lei, H. Y., Liu, C. C., Wu, J. M., Tung, Y. C., Lin, Y. S., Yeh, T. M., Chen, S. H., and Liu, H. S. (2006). MCP-1, a highly expressed chemokine in dengue haemorrhagic fever/dengue shock syndrome patients, may cause permeability change, possibly through reduced tight junctions of vascular endothelium cells. *J Gen Virol* **87**(Pt 12), 3623–3630.

Levi, M. (2001). Pathogenesis and treatment of disseminated intravascular coagulation in the septic patient. *J Crit Care* **16**(4), 167–177.

Levi, M., de Jonge, E., van der Poll, T., and ten Cate, H. (1999). Disseminated intravascular coagulation. *Thromb Haemost* **82**(2), 695–705.

Lin, C. F., Chiu, S. C., Hsiao, Y. L., Wan, S. W., Lei, H. Y., Shiau, A. L., Liu, H. S., Yeh, T. M., Chen, S. H., Liu, C. C., and Lin, Y. S. (2005). Expression of cytokine, chemokine, and adhesion molecules during endothelial cell activation induced by antibodies against dengue virus nonstructural protein 1. *J Immunol* **174**(1), 395–403.

Lin, C. F., Lei, H. Y., Shiau, A. L., Liu, C. C., Liu, H. S., Yeh, T. M., Chen, S. H., and Lin, Y. S. (2003). Antibodies from dengue patient sera cross-react with endothelial cells and induce damage. *J Med Virol* **69**(1), 82–90.

Lin, C. F., Wan, S. W., Cheng, H. J., Lei, H. Y., and Lin, Y. S. (2006). Autoimmune pathogenesis in dengue virus infection. *Viral Immunol* **19**(2), 127–132.

Lin, S. W., Chuang, Y. C., Lin, Y. S., Lei, H. Y., Liu, H. S., and Yeh, T. M. (2012). Dengue virus nonstructural protein NS1 binds to prothrombin/thrombin and inhibits prothrombin activation. *J Infect* **64**(3), 325–334.

Lin, Y. S., Yeh, T. M., Lin, C. F., Wan, S. W., Chuang, Y. C., Hsu, T. K., Liu, H. S., Liu, C. C., Anderson, R., and Lei, H. Y. (2011). Molecular mimicry between virus and host and its implications for dengue disease pathogenesis. *Exp Biol Med (Maywood)* **236**(5), 515–523.

Linderholm, M., Ahlm, C., Settergren, B., Waage, A., and Tarnvik, A. (1996). Elevated plasma levels of tumor necrosis factor (TNF)-alpha, soluble TNF receptors, interleukin (IL)-6, and IL-10 in patients with hemorrhagic fever with renal syndrome. *J Infect Dis* **173**(1), 38–43.

Liu, I. J., Chiu, C. Y., Chen, Y. C., and Wu, H. C. (2011). Molecular mimicry of human endothelial cell antigen by autoantibodies to nonstructural protein 1 of dengue virus. *J Biol Chem* **286**(11), 9726–9736.

Liu, W., Tang, L., Zhang, G., Wei, H., Cui, Y., Guo, L., Gou, Z., Chen, X., Jiang, D., Zhu, Y., Kang, G., and He, F. (2004). Characterization of a novel C-type lectin-like gene, LSECtin: Demonstration of carbohydrate binding and expression in sinusoidal endothelial cells of liver and lymph node. *J Biol Chem* **279**(18), 18748–18758.

Lukashevich, I. S., Maryankova, R., Vladyko, A. S., Nashkevich, N., Koleda, S., Djavani, M., Horejsh, D., Voitenok, N. N., and Salvato, M. S. (1999). Lassa and Mopeia virus replication in human monocytes/macrophages and in endothelial cells: Different effects on IL-8 and TNF-alpha gene expression. *J Med Virol* **59**(4), 552–560.

Mackow, E. R. and Gavrilovskaya, I. N. (2001). Cellular receptors and hantavirus pathogenesis. *Curr Top Microbiol Immunol* **256**, 91–115.

Mackow, E. R. and Gavrilovskaya, I. N. (2009). Hantavirus regulation of endothelial cell functions. *Thromb Haemost* **102**(6), 1030–1041.

Maes, P., Clement, J., Groeneveld, P. H., Colson, P., Huizinga, T. W., and Van Ranst, M. (2006). Tumor necrosis factor-alpha genetic predisposing factors can influence clinical severity in nephropathia epidemica. *Viral Immunol* **19**(3), 558–564.

Marsac, D., Garcia, S., Fournet, A., Aguirre, A., Pino, K., Ferres, M., Kalergis, A. M., Lopez-Lastra, M., and Veas, F. (2011). Infection of human monocyte-derived dendritic cells by ANDES Hantavirus enhances pro-inflammatory state, the secretion of active MMP-9 and indirectly enhances endothelial permeability. *Virol J* **8**, 223.

Marta, R. F., Montero, V. S., Hack, C. E., Sturk, A., Maiztegui, J. I., and Molinas, F. C. (1999). Proinflammatory cytokines and elastase-alpha-1-antitrypsin in Argentine hemorrhagic fever. *Am J Trop Med Hyg* **60**(1), 85–89.

Miele, L., Golde, T., and Osborne, B. (2006). Notch signaling in cancer. *Curr Mol Med* **6**(8), 905–918.

Mitrakul, C., Poshyachinda, M., Futrakul, P., Sangkawibha, N., and Ahandrik, S. (1977). Hemostatic and platelet kinetic studies in dengue hemorrhagic fever. *Am J Trop Med Hyg* **26**(5 Pt 1), 975–984.

Monaghan-Benson, E. and Burridge, K. (2009). The regulation of vascular endothelial growth factor-induced microvascular permeability requires Rac and reactive oxygen species. *J Biol Chem* **284**(38), 25602–25611.

Montgomery, J. M., Ksiazek, T. G., and Khan, A. S. (2007). Hantavirus pulmonary syndrome: The sound of a mouse roaring. *J Infect Dis* **195**(11), 1553–1555.

Moraz, M. L. and Kunz, S. (2011). Pathogenesis of arenavirus hemorrhagic fevers. *Expert Rev Anti Infect Ther* **9**(1), 49–59.

Morrell, C. N., Murata, K., Swaim, A. M., Mason, E., Martin, T. V., Thompson, L. E., Ballard, M., Fox-Talbot, K., Wasowska, B., and Baldwin, W. M., 3rd. (2008). In vivo platelet-endothelial cell interactions in response to major histocompatibility complex alloantibody. *Circ Res* **102**(7), 777–785.

Mukherjee, S., Tessema, M., and Wandinger-Ness, A. (2006). Vesicular trafficking of tyrosine kinase receptors and associated proteins in the regulation of signaling and vascular function. *Circ Res* **98**(6), 743–756.

Nabors, L. B., Suswam, E., Huang, Y., Yang, X., Johnson, M. J., and King, P. H. (2003). Tumor necrosis factor alpha induces angiogenic factor up-regulation in malignant glioma cells: A role for RNA stabilization and HuR. *Cancer Res* **63**(14), 4181–4187.

van Nieuw Amerongen, G. P., Koolwijk, P., Versteilen, A., and van Hinsbergh, V. W. (2003). Involvement of RhoA/Rho kinase signaling in VEGF-induced endothelial cell migration and angiogenesis in vitro. *Arterioscler Thromb Vasc Biol* **23**(2), 211–217.

Niikura, M., Maeda, A., Ikegami, T., Saijo, M., Kurane, I., and Morikawa, S. (2004). Modification of endothelial cell functions by Hantaan virus infection: Prolonged hyper-permeability induced by TNF-alpha of hantaan virus-infected endothelial cell monolayers. *Arch Virol* **149**(7), 1279–1292.

Nolte, K. B., Feddersen, R. M., Foucar, K., Zaki, S. R., Koster, F. T., Madar, D., Merlin, T. L., McFeeley, P. J., Umland, E. T., and Zumwalt, R. E. (1995). Hantavirus pulmonary syndrome in the United States: A pathological description of a disease caused by a new agent. *Hum Pathol* **26**(1), 110–120.

Padilla, A., Moake, J. L., Bernardo, A., Ball, C., Wang, Y., Arya, M., Nolasco, L., Turner, N., Berndt, M. C., Anvari, B., Lopez, J. A., and Dong, J. F. (2004). P-selectin anchors newly released ultralarge von Willebrand factor multimers to the endothelial cell surface. *Blood* **103**(6), 2150–2156.

Pensiero, M. N., Sharefkin, J. B., Dieffenbach, C. W., and Hay, J. (1992). Hantaan virus infection of human endothelial cells. *J Virol* **66**(10), 5929–5936.

Polunovsky, V. A., Wendt, C. H., Ingbar, D. H., Peterson, M. S., and Bitterman, P. B. (1994). Induction of endothelial cell apoptosis by TNF alpha: modulation by inhibitors of protein synthesis. *Exp Cell Res* **214**(2), 584–594.

Powlesland, A. S., Fisch, T., Taylor, M. E., Smith, D. F., Tissot, B., Dell, A., Pohlmann, S., and Drickamer, K. (2008). A novel mechanism for LSECtin binding to Ebola virus surface glycoprotein through truncated glycans. *J Biol Chem* **283**(1), 593–602.

Quillard, T., Devalliere, J., Chatelais, M., Coulon, F., Seveno, C., Romagnoli, M., Barille Nion, S., and Charreau, B. (2009). Notch2 signaling sensitizes endothelial cells to apoptosis by negatively regulating the key protective molecule survivin. *PLoS One* **4**(12), e8244.

Quillard, T., Devalliere, J., Coupel, S., and Charreau, B. (2010). Inflammation dysregulates Notch signaling in endothelial cells: implication of Notch2 and Notch4 to endothelial dysfunction. *Biochem Pharmacol* **80**(12), 2032–2041.

Radoshitzky, S. R., Abraham, J., Spiropoulou, C. F., Kuhn, J. H., Nguyen, D., Li, W., Nagel, J., Schmidt, P. J., Nunberg, J. H., Andrews, N. C., Farzan, M., and Choe, H. (2007). Transferrin receptor 1 is a cellular receptor for New World haemorrhagic fever arenaviruses. *Nature* **446**(7131), 92–96.

Raymond, T., Gorbunova, E., Gavrilovskaya, I. N., and Mackow, E. R. (2005). Pathogenic hantaviruses bind plexin-semaphorin-integrin domains present at the apex of inactive, bent alphavbeta3 integrin conformers. *Proc Natl Acad Sci USA* **102**(4), 1163–1168.

Reynolds, L. E., Wyder, L., Lively, J. C., Taverna, D., Robinson, S. D., Huang, X., Sheppard, D., Hynes, R. O., and Hodivala-Dilke, K. M. (2002). Enhanced pathological angiogenesis in mice lacking beta3 integrin or beta3 and beta5 integrins. *Nat Med* **8**(1), 27–34.

Ridley, A. J., Schwartz, M. A., Burridge, K., Firtel, R. A., Ginsberg, M. H., Borisy, G., Parsons, J. T., and Horwitz, A. R. (2003). Cell migration: integrating signals from front to back. *Science* **302**(5651), 1704–1709.

Robinson, R., Ho, C. E., Tan, Q. S., Luu, C. D., Moe, K. T., Cheung, C. Y., Wong, T. Y., and Barathi, V. A. (2011). Fluvastatin downregulates VEGF-A expression in TNF-alpha-induced retinal vessel tortuosity. *Invest Ophthalmol Vis Sci* **52**(10), 7423–7431.

Robinson, S. D., Reynolds, L. E., Wyder, L., Hicklin, D. J., and Hodivala-Dilke, K. M. (2004). Beta3-integrin regulates vascular endothelial growth factor-A-dependent permeability. *Arterioscler Thromb Vasc Biol* **24**(11), 2108–2114.

Rojek, J. M. and Kunz, S. (2008). Cell entry by human pathogenic arenaviruses. *Cell Microbiol* **10**(4), 828–835.

Rojek, J. M., Moraz, M. L., Pythoud, C., Rothenberger, S., Van der Goot, F. G., Campbell, K. P., and Kunz, S. (2012). Binding of Lassa virus perturbs extracellular matrix-induced signal transduction via dystroglycan. *Cell Microbiol* **14**(7), 1122–1134.

Rubattu, S., Sciarretta, S., Valenti, V., Stanzione, R., and Volpe, M. (2008). Natriuretic peptides: An update on bioactivity, potential therapeutic use, and implication in cardiovascular diseases. *Am J Hypertens* **21**(7), 733–741.

Ruegg, C., Yilmaz, A., Bieler, G., Bamat, J., Chaubert, P., and Lejeune, F. J. (1998). Evidence for the involvement of endothelial cell integrin alphaVbeta3 in the disruption of the tumor vasculature induced by TNF and IFN-gamma. *Nat Med* **4**(4), 408–414.

Ryuto, M., Ono, M., Izumi, H., Yoshida, S., Weich, H. A., Kohno, K., and Kuwano, M. (1996). Induction of vascular endothelial growth factor by tumor necrosis factor alpha in human glioma cells. Possible roles of SP-1. *J Biol Chem* **271**(45), 28220–28228.

Schonrich, G., Rang, A., Lutteke, N., Raftery, M. J., Charbonnel, N., and Ulrich, R. G. (2008). Hantavirus-induced immunity in rodent reservoirs and humans. *Immunol Rev* **225**, 163–189.

Semeraro, N., Triggiani, R., Montemurro, P., Cavallo, L. G., and Colucci, M. (1993). Enhanced endothelial tissue factor but normal thrombomodulin in endotoxin-treated rabbits. *Thromb Res* **71**(6), 479–486.

Shrivastava-Ranjan, P., Rollin, P. E., and Spiropoulou, C. F. (2010). Andes virus disrupts the endothelial cell barrier by induction of vascular endothelial growth factor and downregulation of VE-cadherin. *J Virol* **84**(21), 11227–11234.

Shweiki, D., Itin, A., Soffer, D., and Keshet, E. (1992). Vascular endothelial growth factor induced by hypoxia may mediate hypoxia-initiated angiogenesis. *Nature* **359**(6398), 843–845.

Simon, M., Johansson, C., and Mirazimi, A. (2009). Crimean-Congo hemorrhagic fever virus entry and replication is clathrin-, pH- and cholesterol-dependent. *J Gen Virol* **90**(Pt 1), 210–215.

Steinberg, B. E. et al. (2012). Do viral infections mimic bacterial sepsis? The role of microvascular permeability: A review of mechanisms and methods. *Antiviral Res* **93**, 2–15.

Stroher, U., West, E., Bugany, H., Klenk, H. D., Schnittler, H. J., and Feldmann, H. (2001). Infection and activation of monocytes by Marburg and Ebola viruses. *J Virol* **75**(22), 11025–11033.

Suharti, C., van Gorp, E. C., Setiati, T. E., Dolmans, W. M., Djokomoeljanto, R. J., Hack, C. E., ten Cate, H., and van der Meer, J. W. (2002). The role of cytokines in activation of coagulation and fibrinolysis in dengue shock syndrome. *Thromb Haemost* **87**(1), 42–46.

Sundstrom, J. B., McMullan, L. K., Spiropoulou, C. F., Hooper, W. C., Ansari, A. A., Peters, C. J., and Rollin, P. E. (2001). Hantavirus infection induces the expression of RANTES and IP-10 without causing increased permeability in human lung microvascular endothelial cells. *J Virol* **75**(13), 6070–6085.

Takagi, J., Petre, B. M., Walz, T., and Springer, T. A. (2002). Global conformational rearrangements in integrin extracellular domains in outside-in and inside-out signaling. *Cell* **110**(5), 599–511.

Talavera, D., Castillo, A. M., Dominguez, M. C., Gutierrez, A. E., and Meza, I. (2004). IL8 release, tight junction and cytoskeleton dynamic reorganization conducive to permeability increase are induced by dengue virus infection of microvascular endothelial monolayers. *J Gen Virol* **85**(Pt 7), 1801–1813.

Taylor, F. B., Jr., Chang, A., Ruf, W., Morrissey, J. H., Hinshaw, L., Catlett, R., Blick, K., and Edgington, T. S. (1991). Lethal *E. coli* septic shock is prevented by blocking tissue factor with monoclonal antibody. *Circ Shock* **33**(3), 127–134.

Terajima, M., Hayasaka, D., Maeda, K., and Ennis, F. A. (2007). Immunopathogenesis of hantavirus pulmonary syndrome and hemorrhagic fever with renal syndrome: Do CD8+ T cells trigger capillary leakage in viral hemorrhagic fevers? *Immunol Lett* **113**(2), 117–120.

Tseng, C. S., Lo, H. W., Teng, H. C., Lo, W. C., and Ker, C. G. (2005). Elevated levels of plasma VEGF in patients with dengue hemorrhagic fever. *FEMS Immunol Med Microbiol* **43**(1), 99–102.

Turkdogan, K. A., Zorlu, A., Engin, A., Guven, F. M., Polat, M. M., Turgut, O. O., and Yilmaz, M. B. (2012). C-type natriuretic peptide is associated with the severity of Crimean-Congo hemorrhagic fever. *Int J Infect Dis* **16**(8), e616–e620.

Valgimigli, M. et al. (2003). Serum from patients with acute coronary syndromes displays a proapoptotic effect on human endothelial cells: A possible link to pan-coronary syndromes. *Circulation* **107**(2), 264–270.

Vapalahti, O., Mustonen, J., Lundkvist, A., Henttonen, H., Plyusnin, A., and Vaheri, A. (2003). Hantavirus infections in Europe. *Lancet Infect Dis* **3**(10), 653–661.

Villinger, F., Rollin, P. E., Brar, S. S., Chikkala, N. F., Winter, J., Sundstrom, J. B., Zaki, S. R., Swanepoel, R., Ansari, A. A., and Peters, C. J. (1999). Markedly elevated levels of interferon (IFN)-gamma, IFN-alpha, interleukin (IL)-2, IL-10, and tumor necrosis factor-alpha associated with fatal Ebola virus infection. *J Infect Dis* **179**(Suppl 1), S188–S191.

Volchkov, V. E., Becker, S., Volchkova, V. A., Ternovoj, V. A., Kotov, A. N., Netesov, S. V., and Klenk, H. D. (1995). GP mRNA of Ebola virus is edited by the Ebola virus polymerase and by T7 and vaccinia virus polymerases. *Virology* **214**(2), 421–430.

Volchkov, V. E., Volchkova, V. A., Muhlberger, E., Kolesnikova, L. V., Weik, M., Dolnik, O., and Klenk, H. D. (2001). Recovery of infectious Ebola virus from complementary DNA: RNA editing of the GP gene and viral cytotoxicity. *Science* **291**(5510), 1965–1969.

Wahl-Jensen, V. M., Afanasieva, T. A., Seebach, J., Stroher, U., Feldmann, H., and Schnittler, H. J. (2005). Effects of Ebola virus glycoproteins on endothelial cell activation and barrier function. *J Virol* **79**(16), 10442–10450.

Walker, D. H., McCormick, J. B., Johnson, K. M., Webb, P. A., Komba-Kono, G., Elliott, L. H., and Gardner, J. J. (1982). Pathologic and virologic study of fatal Lassa fever in man. *Am J Pathol* **107**(3), 349–356.

Wallez, Y., Vilgrain, I., and Huber, P. (2006). Angiogenesis: The VE-cadherin switch. *Trends Cardiovasc Med* **16**(2), 55–59.

Wan, S. W., Lin, C. F., Lu, Y. T., Lei, H. Y., Anderson, R., and Lin, Y. S. (2012). Endothelial cell surface expression of protein disulfide isomerase activates beta1 and beta3 integrins and facilitates dengue virus infection. *J Cell Biochem* **113**(5), 1681–1691.

Wojciak-Stothard, B., Entwistle, A., Garg, R., and Ridley, A. J. (1998). Regulation of TNF-alpha-induced reorganization of the actin cytoskeleton and cell-cell junctions by Rho, Rac, and Cdc42 in human endothelial cells. *J Cell Physiol* **176**(1), 150–165.

Xiao, X., Feng, Y., Zhu, Z., and Dimitrov, D. S. (2011). Identification of a putative Crimean-Congo hemorrhagic fever virus entry factor. *Biochem Biophys Res Commun* **411**(2), 253–258.

Yanagihara, R., Daum, C. A., Lee, P. W., Baek, L. J., Amyx, H. L., Gajdusek, D. C., and Gibbs, C. J., Jr. (1987). Serological survey of Prospect Hill virus infection in indigenous wild rodents in the USA. *Trans R Soc Trop Med Hyg* **81**(1), 42–45.

Yang, L., Froio, R. M., Sciuto, T. E., Dvorak, A. M., Alon, R., and Luscinskas, F. W. (2005). ICAM-1 regulates neutrophil adhesion and transcellular migration of TNF-alpha-activated vascular endothelium under flow. *Blood* **106**(2), 584–592.

Yang, Z. Y., Duckers, H. J., Sullivan, N. J., Sanchez, A., Nabel, E. G., and Nabel, G. J. (2000). Identification of the Ebola virus glycoprotein as the main viral determinant of vascular cell cytotoxicity and injury. *Nat Med* **6**(8), 886–889.

Yao, L., Romero, M. J., Toque, H. A., Yang, G., Caldwell, R. B., and Caldwell, R. W. (2010). The role of RhoA/Rho kinase pathway in endothelial dysfunction. *J Cardiovasc Dis Res* **1**(4), 165–170.

Zaki, S. R. et al. (1995). Hantavirus pulmonary syndrome. Pathogenesis of an emerging infectious disease. *Am J Pathol* **146**(3), 552–579.

6 Animal Models of Viral Hemorrhagic Fevers

Amy C. Shurtleff, Travis K. Warren, Derek Morrow,
Sheli R. Radoshitzky, Jens H. Kuhn, and Sina Bavari

CONTENTS

6.1 INTRODUCTION: ANIMAL MODELS AS TOOLS

Animal models of viral diseases are an integral part of the process to develop drugs and vaccines against highly pathogenic microorganisms. In addition to testing and developing therapeutic approaches, animal models can also provide important scientific insights on the mechanisms of disease progression, the transmission and spread of infectious agents, the key correlates of immunity, and the identification of future drug targets. Animals ranging in size from small mice and other rodents to larger nonhuman primates (NHPs) provide a variety of platforms in which to investigate disease pathogenesis, immune responses, and the therapeutic potential of small molecules, immunotherapeutics, or vaccines.

The U.S. Food and Drug Administration (FDA) Center for Drug Evaluation and Research (CDER) along with the Center for Biologics Evaluation and Research (CBER) are responsible for ensuring that new small molecules, vaccines, and other therapeutics are safe and efficacious before granting approval for widespread clinical use. Whereas clinical trials are typically how new therapies are evaluated, such an approach is not feasible for many viral hemorrhagic fevers as infrequent and sporadic outbreaks often only occur in remote areas. An alternative approach—intentionally exposing human volunteers to these highly pathogenic viruses for the sake of evaluating a novel therapy—is clearly unethical. In response to these concerns, the FDA issued a rule in 2002, codified 21 CFR 314.600, also known as "the Animal Rule," which allows data from efficacy studies in animal models to be used in place of human clinical data when such trials would be unfeasible or unethical.

Several important considerations must be addressed for animal efficacy studies to receive approval under the Animal Rule: (1) the pathophysiology of disease and the mechanism by which the drug prevents or reduces the disease must be "reasonably well understood"; (2) efficacy must be demonstrated in animals of more than one species, unless the animal model is sufficiently well characterized for predicting the response in humans; (3) the animal study endpoint must be clearly related to the desired human efficacy endpoint, such as enhancement of survival; and (4) pharmacokinetic and pharmacodynamic data must be generated in the animal studies to allow selection of an effective dose in humans (FDA U.S. Department of Health and Human Services, 2009).

It is important to emphasize just how difficult it may be to receive approval through the Animal Rule. Although in place for 10 years, the Animal Rule has been used successfully only twice to license products (pyridostigmine bromide for pretreatment of exposure to soman and Cyanokit for the treatment of cyanide poisoning). Researchers are just now starting to learn about the strict requirements that the FDA will place upon studies. In response to this, intragovernmental working groups and funding agencies in the United States are forming committees and partnerships to investigate animal models for infections involving filoviruses and arenaviruses in sufficient detail to satisfy queries from the FDA. An additional challenge to complying with the Animal Rule is that the pivotal animal efficacy studies must be conducted in compliance with good laboratory practices (GLP; 21 CFR 58) that can be particularly challenging in biosafety level 4 containment laboratories in which most hemorrhagic fever viruses must be handled. Finally, although traditional human clinical efficacy trials require evidence that the therapy is effective, the Animal Rule imposes an additional burden on investigators to establish the mode of action for the drug candidate in at least one animal model that reproduces accurate human disease pathology.

6.2 FILOVIRUSES

Filovirus disease outbreaks in human populations are rare and often self-limiting as person-to-person transmission only occurs after close skin-to-skin contact or contact with blood or other bodily fluids (Feldmann et al., 2005). Researchers though have found that aerosol preparations of filoviruses are highly infectious in the laboratory, leading to concerns that these agents could be used for nefarious purposes (Bazhutin et al., 1992; Leffel and Reed, 2004). As there are currently no licensed antiviral therapies or vaccines for the treatment or prevention of Ebola virus disease (EVD) or Marburg virus disease (MVD), only supportive care can be offered to suspected or confirmed cases. Several types of NHPs are susceptible to infection with filoviruses, although the majority of research is performed using rhesus monkeys (*Macaca mulatta*) and crab-eating macaques (*Macaca fascicularis*) as these animal models at least appear to reproduce human disease based on the little human clinical data that are available. Mice, guinea pigs, and hamsters do not succumb to infection with wild-type filoviruses unless the viruses have been extensively passaged and adapted to virulence in these animals (Bray et al., 1999; Connolly et al., 1999; Warfield et al., 2007a,b, 2009; Webb et al., 1975). Although the resulting rodent infection models are valuable screening tools, particularly when evaluating virus-specific therapeutic agents, the NHP models are considered the most reliable predictor of therapeutic efficacy and are currently the most suitable model to satisfy the FDA's Animal Rule.

The family *Filoviridae* includes two genera. The genus *Ebolavirus* currently consists of five species with one member virus each: *Zaire ebolavirus* (Ebola virus, EBOV), *Sudan ebolavirus* (Sudan virus, SUDV), *Reston ebolavirus* (Reston virus, RESTV), *Taï Forest ebolavirus* (Taï Forest virus, TAFV), and *Bundibugyo virus* (Bundibugyo virus, BDBV) (Kuhn et al., 2009, 2010). Ebolaviruses were first discovered during simultaneous outbreaks in Zaire (today the Democratic Republic of the Congo) and Sudan (today Republic of South Sudan) in 1976. Since then, these viruses have caused sporadic outbreaks of EVD with case fatality rates of approximately 50%–90% in both countries and also in Gabon, Republic of Congo, and Côte d'Ivoire (Kuhn, 2008). Because of the historical prevalence of EBOV and SUDV during outbreaks and their comparatively high case fatality rates, these two ebolaviruses have been the focus of most of the animal studies to date. The genus *Marburgvirus* includes only a single species, *Marburg marburgvirus*, which has two members, Marburg virus (MARV) and Ravn virus (RAVV) (Kuhn et al., 2009, 2010). MARV was the first filovirus discovered when in 1967 German and Yugoslavian laboratory workers manipulating grivets (*Chlorocebus aethiops*) imported from Uganda became infected and initiated a 31-person outbreak of viral hemorrhagic fever in Marburg, West Germany (Siegert et al., 2007; Todorović et al., 1969).

Since that time, outbreaks of MVD have occurred in several African countries such as Kenya, the Democratic Republic of the Congo, Zimbabwe, and most recently in Angola in 2005 and Uganda in 2007–2008 (Kuhn, 2008; Leroy et al., 2011; Towner et al., 2006). The case fatality rate of MVD ranges from 23% to 90%, and several variants of MARV, such as Ci67, Angola, and Musoke, have become the subject of intensive laboratory investigations.

Immunocompetent adult mice are resistant to wild-type filoviruses likely because of their inherent strong innate immune response (Bray et al., 1999; Gonchar et al., 1991). In contrast, neonates and mouse strains with an impaired type I interferon (IFN) response will succumb to filovirus infection delivered by subcutaneous inoculation (Bray, 2001; Raymond et al., 2011; van der Groen et al., 1979). Adult mice with severe combined immunodeficiency (SCID) will also succumb to wild-type filovirus infection over a period of 1–2 months (Bray, 2001; Warfield et al., 2007). However, the course of disease and pathology in these wild-type infected mice do not mirror that of humans or NHPs and therefore do not provide useful animal models.

Certain filoviruses have been adapted to cause lethal infections in adult immunocompetent mice. EBOV was adapted to mice by passaging the virus first in young, and then in progressively older suckling mice. The resulting virus was plaque purified twice and is now used to infect adult immunocompetent mice resulting in uniform lethality (Bray, 2001; Bray et al., 1999). Three to four days after intraperitoneal inoculation, the mice appear ill and will succumb to the infection 4–6 days later with high viral burdens in the liver, spleen, and other tissues. A similar process resulted in mouse-adapted lethal marburgviruses. RAVV was serially passaged 10 times through liver homogenates from SCID mice followed by another 14 passages in adult BALB/c mice resulting in a virus lethal to mice 7–10 days after intraperitoneal inoculation with 1000 plaque-forming units (pfu) (Warfield et al., 2009). Both the EVD and MVD mouse models demonstrate rapid viremia, high viral burden in the spleen, liver and multiple organ tissues, lymphopenia, thrombocytopenia, and liver damage resulting in high serum levels of aspartate aminotransferase (AST) and alanine aminotransferase (ALT) (Warfield et al., 2009). In both mouse models, the early inflammatory cytokine response is similar to observations in guinea pig and NHP infections. It is important to note that these mice do not demonstrate disseminated intravascular coagulation (DIC), fibrin deposits, or skin rashes such as petechiae (which are typical findings in NHPs and humans) although some coagulation abnormalities are found (Warfield et al., 2009).

During the adaptation process, several key yet different mutations occurred for each virus. The mouse-adapted EBOV had changes in the nucleoprotein (NP), secondary matrix protein VP24, and the RNA-dependent RNA polymerase (L) gene allowing for increased virulence. Specifically, changes in NP and VP24 more effectively dampen murine type I IFN responses, thereby enabling viral replication and disease progression (Ebihara et al., 2006). Mouse-adapted RAVV was mutated in the matrix protein VP40 and the polymerase cofactor VP35 and to a lesser extent in NP and the transcriptional activator VP30 (Lofts et al., 2011; Warfield et al., 2009). Importantly, changes in VP40 have been reported for both mouse-adapted RAVV and MARV, and for guinea pig–adapted RAVV, suggesting that VP40 plays a key role promoting viral fitness in these rodent models (Lofts et al., 2007; Swenson et al., 2004) likely through its known ability to antagonize the host type I IFN response (Lofts et al., 2011; Valmas et al., 2010).

Guinea pigs inoculated with wild-type filoviruses do not succumb to infection, but rather demonstrate a febrile, self-limiting disease and eventually recover (Gonchar et al., 1991). Passaging EBOV four to eight times in spleen tissue led to a virus with uniform lethality within 8–11 days after inoculation (Connolly et al., 1999; Ryabchikova et al., 1996, 1999b). Similarly, guinea pigs infected with MARV that had been passaged four to eight times succumbed to infection 7–17 days after inoculation (Robin et al., 1971; Simpson et al., 1968). Five days after infection with either of the guinea pig–passaged filoviruses, infected animals appeared ill showing signs of dehydration but with no apparent hemorrhage. Viral replication in the spleen and liver was first detected 2 days after inoculation and was subsequently seen in the pancreas, lungs, adrenal glands, and kidneys. Although skin rashes did not appear, infected guinea pigs did show coagulation abnormalities, such as prolongation of prothrombin and activated partial thromboplastin times. In these models, filoviruses initially replicate in monocytes and macrophages, and as the disease progresses histopathologic lesions are observed in the liver, spleen, adrenal gland, and kidney tissues. There are reports of a granulomatous cellular response in lesions in the guinea pig compared to the NHP (Connolly et al., 1999; Ryabchikova et al., 1996, 1999b). This response indicates that mononuclear cells remain functional in this model, whereas the activity of this cell population appears suppressed in NHPs.

In contrast to the rodent models, filovirus infections of NHPs lead to outcomes very similar to what is seen in human infections. Disease parameters such as clinical presentation, hematology, serum clinical chemistry, anatomic pathology findings, and some aspects of humoral and cellular immune responses are similar. Rhesus monkeys and crab-eating macaques are typically utilized in laboratory studies due to their size, availability, and the reproducibility of the results (Carrion, 2010; Stroher and Feldmann, 2006). Hamadryas baboons have been used extensively in Russia, but not in the United States (Gonchar et al., 1991; Mikhailov et al., 1994).

Published studies and data from historical control animals infected at the U.S. Army Medical Research Institute of Infectious Diseases (USAMRIID; Fort Detrick, Maryland, United States) indicate that the average time to death following intramuscular (i.m.) challenge with a target dose of 1000 pfu of wild-type EBOV is approximately 8 days for rhesus monkeys and 6 days for crab-eating macaques (Geisbert et al., 2003c, 2007; Hensley et al., 2002, 2011). For wild-type MARV, the average time to death is 7 days for rhesus monkeys and 8 days for crab-eating macaques (Alves et al., 2010; Geisbert et al., 2007; Hensley et al., 2011). A challenge dose of 1000 pfu is commonly used for therapeutic evaluations and vaccine development evaluations involving NHPs. This dose is usually lethal and probably near the LD_{99} although this has not actually been calculated due to small experimental group sizes. The kinetics in the these NHP models matches closely what is seen in human infections during which the incubation periods for EBOV and MARV infection in humans is 4–7 days with a reported range of 2–25 days and death often occurs within 6 days of symptom onset (Bente et al., 2009; Eichner et al., 2011).

There are differences though between the typical outcome of a natural human infection compared to a laboratory infection of NHPs. Symptoms of filovirus disease are quite pronounced for NHPs and survival of untreated or negative-control animals is rare (Shurtleff et al., 2011). In contrast, human disease may be mild or even subclinical in a substantial number of cases (Baize et al., 1999; Leroy et al., 2000, 2001, 2002; Sanchez et al., 2004). Humans receive an unknown amount of virus upon exposure, and it is unknown whether the exposure dose can result in different disease outcomes (Alves et al., 2010). In controlled laboratory settings, animals are administered an intentionally lethal dose, seeking to generate highly reproducible results. However, this experimental approach does not seek to address why humans have a broad range of infection outcomes.

In humans, symptoms of filovirus disease, such as fever, chills, headache, malaise, and myalgia, appear on average within 4 days after infection (1978; Bente et al., 2009; Bruce and Brysiewicz, 2002; Bwaka et al., 1999; McLeod et al., 1978). Nausea, vomiting, abdominal pain, diarrhea, anorexia, a dry nonproductive cough, chest pain, and shortness of breath have all been reported as the predominant gastrointestinal and respiratory symptoms (2001; 2005; Formenty et al., 1999). Vascular abnormalities include conjunctival injection, postural hypotension, and edema (2001). Hemorrhagic signs include development of petechiae on the torso and arms, ecchymoses, bleeding from venipuncture sites and mucous membranes, gingival bleeding, epistaxis, and visceral hemorrhaging (2001; Bente et al., 2009).

In experimental infections of NHPs, fever is detected 3–4 days after infection and is coincident with onset of viremia (Alves et al., 2010; Geisbert et al., 2003a,c, 2007; Hensley et al., 2002). A maculopapular rash generally appears around the same time. Dehydration, anorexia, and diarrhea are often noted in animals, beginning approximately 4 days after infection (Alves et al., 2010; Baskerville et al., 1978; Bowen et al., 1978). Five or six days after infection, nearly all infected animals appear ill exhibiting moderate to severe depression in activity, dehydration, hunched posture, and rashes (Alves et al., 2010; Fisher-Hoch et al., 1983; Geisbert et al., 2003a, 2007; Hensley et al., 2011; Ryabchikova et al., 1999).

White blood cell counts and blood chemistry parameters are drastically altered during human and NHP infection with EBOV or MARV. Lymphocytopenia occurs within 3–4 days after infection (Geisbert et al., 2003a, 2007; Hensley et al., 2011; Johnson et al., 1995; Reed et al., 2004). Neutrophil counts increase becoming the majority of the circulating leukocyte population as monocytes, macrophages, and natural killer (NK) populations decline. Thrombocytopenia may not be as prominent for MARV infection of NHPs as it is for EBOV infection (Geisbert et al., 2003a; Hensley et al., 2011). Serum concentrations of AST, ALT, alkaline phosphatase (ALP), and gamma glutamyltransferase (GGT) are typically highly elevated 5 or 6 days after infection, which is indicative of liver damage (Geisbert et al., 2003a, 2007; Johnson et al., 1995). Blood urea nitrogen (BUN) and creatinine concentrations, signs of decreased glomerular filtration, whether by dehydration or renal dysfunction, progressively increase 5 or 6 days after infection (Geisbert et al., 2003a; Johnson et al., 1995). Laboratory findings include proteinuria related to inflammatory vascular permeability of the urinary system, along with hyperproteinemia likely related to combined dehydration and increased positive acute phase proteins (e.g., fibrinogen, haptoglobin, C-reactive protein). Decreases in serum albumin, commonly associated with inflammation, are detectable in the NHP model, indicating a loss of circulating smaller molecular weight proteins due to general inflammatory vascular permeability. A summary of clinical pathology changes and their clinical significance is presented in Table 6.1.

Histopathologic changes in humans and NHPs include extensive necrosis of parenchymal cells in multiple target organs, especially liver, lymph nodes, and spleen (Baskerville et al., 1978; Bowen et al., 1978; Geisbert et al., 2007; Jaax et al., 1996; Zaki and Goldsmith, 1999) with remarkably little influx of inflammatory cells to these sites of necrosis (Alves et al., 2010; Davis et al., 1997; Geisbert et al., 2003a,c, 2007; Hensley et al., 2011; Jaax et al., 1996). Filoviruses replicate in many cell types: monocytes, macrophages, and dendritic cells appear to be early targets (Bosio et al., 2003; Bray and Geisbert, 2005; Fritz et al., 2008; Fuller et al., 2007; Geisbert et al., 2003a; Gibb et al., 2002; Gupta et al., 2001) followed later by endothelial cells, fibroblasts, hepatocytes, adrenal cortical cells, and epithelial cells (Baskerville et al., 1985; Bray et al., 1999; Geisbert et al., 2007; Hensley et al., 2011; Ryabchikova et al., 1999; Zaki et al., 1999). The extent of hepatocellular necrosis and the high liver viral burden suggest the liver is a major target organ and that liver damage may result from direct viral cytopathic effect, especially for MARV (Baskerville et al., 1978, 1985; Geisbert et al., 2003a; Rollin et al., 1999; Zaki and Goldsmith, 1999). Spleen and adrenal glands are also important target organs with high viral loads (Bowen et al., 1978; Geisbert et al., 2003a). Lymphoid depletion and necrosis are typically noted in the spleen, thymus, and lymph nodes of fatal human cases and NHPs (Fisher-Hoch et al., 1983, 1985; Geisbert et al., 2000, 2007; Hensley et al., 2011; Ryabchikova et al., 1999; Zaki and Goldsmith, 1999).

The vascular and coagulation systems are interrelated, and both are severely affected by filovirus infection. The hemorrhagic manifestations include alterations in blood coagulation, fibrinolysis, thrombocytopenia, consumption of coagulation factors, deficiency of anticoagulation, and increased levels of fibrin degradation products (Alves et al., 2010; Fisher-Hoch et al., 1983; Geisbert et al., 2003a, 2007; Hensley et al., 2011; Ryabchikova et al., 1999). DIC is a phenomenon resulting from the activation of the procoagulant and fibrinolytic systems, leading to hemorrhage, diffuse thrombosis, and end-stage organ damage and shock. Endothelial cells lining the vasculature were initially thought to be infected as early viral targets, succumbing to cytopathic effects leading to the leakiness, DIC, and hemorrhagic signs prevalent during late disease (Alves et al., 2010; Geisbert et al., 2003a,c, 2007; Hensley et al., 2011). While the virus is capable of infecting and replicating in these cells, more recent immunohistochemical studies in crab-eating macaques suggest endothelial cells do not exhibit viral antigen expression until approximately 4–5 days after infection (Geisbert et al., 2003c). In addition, there are very few reports of vascular lesions in human autopsies or in NHP studies, and DIC is not apparent in NHPs until approximately 1 day after EBOV is first detected in endothelial cells, further suggesting that endothelial cytopathicity is not the cause of DIC (Geisbert et al., 2003c). D-dimers, an indicator

TABLE 6.1

Observations of Changes in Clinical Parameters during Hemorrhagic Fever Virus Infection of Nonhuman Primates

Laboratory Parameter	Characteristic Change(s) for Hemorrhagic Fever	Interpretation(s)	Example(s) for Specific Pathogen	References
White blood cell count	Initial variability, then rebound increases of PMNs at later stages	Some early tissue demand, some lymphocyte apoptosis; regeneration/proliferation of marrow storage pool.	EBOV and MARV	Fisher-Hoch et al. (1985) and Hensley et al. (2011)
Lymphocyte count	Decrease	1. Acute systemic infection: generalized distribution of antigen with resulting lymphocyte trapping in nodes; inflammatory disruption of nodes. 2. Filovirus-induced apoptosis.	EBOV and MARV	Fisher-Hoch et al. (1985), Geisbert et al. (2000), and Hensley et al. (2011)
Platelet count	Decrease	Activation and consumption; disrupted anticoagulation. Possible relationship to impaired antithrombin production if hepatic functional mass is decreased (e.g., MARV).[a]	Occurs around days 4–6 for filoviruses; around day X for LASV, MACV	Geisbert et al. (2003)
Activated partial thromboplastin time (APTT)	Prolonged	DIC; possible relationship to impaired factor production if decreased hepatic functional mass	10–20 s prolongation for MARV	Hensley et al. (2011)
Prothrombin time (PT)	Prolonged in parallel with APTT, usually much subtler in magnitude	DIC; possible relationship to impaired factor production if decreased hepatic functional mass.	10–20 s prolongation for MARV	Hensley et al. (2011)
Fibrinogen	Decreases	In parallel with APTT/PT prolongations, consistent with DIC; a difficult analyte to interpret in infected states as it increases with inflammation		
Sodium (Na), chloride (Cl) ions	Decreases	Likely related to combinations of decreased feed consumption, thoracic/abdominal effusion, and renal insult if present.		
Alkaline phosphatase (ALP)	Variable increases: inconsequential to substantial	While ALP is found in several tissues, large increases in infected animals usually accompany noteworthy/adverse TBIL increases, consistent with cholestasis.	MARV, EBOV	Geisbert et al. (2003) and Johnson et al. (1995)
Alanine aminotransferase (ALT)	Massive increases in parallel with AST	Hepatocellular injury; likely decreased hepatic functional mass when accompanied by adverse TBIL increase.	Late stage increase for MARV, EBOV, and LASV	Carrion et al. (2007), Geisbert et al. (2003), and Hensley et al. (2011)
Aspartate aminotransferase (AST)	Massive increases	Nonspecific tissue injury; measurement can be much higher than ALT.	Late stage increase for MARV, EBOV, and LASV	Geisbert et al. (2003), Gunther and Lenz (2004), and Hensley et al. (2011)

TABLE 6.1 (continued)

Observations of Changes in Clinical Parameters during Hemorrhagic Fever Virus Infection of Nonhuman Primates

Laboratory Parameter	Characteristic Change(s) for Hemorrhagic Fever	Interpretation(s)	Example(s) for Specific Pathogen	References
Total bilirubin (TBIL)	Increases variably depending on pathogen	Noteworthy increases likely result from both cholestasis (conjugated) and decreased uptake by injured hepatocytes (unconjugated)	More adverse in MARV compared to EBOV	Shurtleff et al. (2011)
Albumin (ALB), calcium (Ca)	Decreases	ALB loss is likely a combination of its role as a negative acute-phase reactant and component of transudative effusion. ALB serves as a carrier of approximately 40% of serum Ca.	EBOV, LASV	Carrion et al. (2011), Geisbert et al. (2003), and Shurtleff et al. (2011)
Blood urea nitrogen (BUN)	Increases in moribund animals	Most likely pre-renal (dehydration) azotemia predominating with likely, variable renal contributions varying across animals.	MARV, EBOV	Hensley et al. (2011) and Johnson et al. (1995)
Creatinine	Increases in most moribund animals	Increases, when present, occur in tandem with BUN corroborating pre-renal/renal azotemia, likely combined with muscle wasting due to decreased or nonconsumption of feed.	Late stage disease increase for MARV	Hensley et al. (2011)

Note: DIC: disseminated intravascular coagulation.

[a] Decrease in hepatic functional mass is preceded by massive hepatocellular damage; however, the liver has a substantial functional reserve and evidence of hepatocellular injury is not in itself evidence of impaired hepatic function.

of coagulation abnormalities, are a product of plasmin digestion of fibrin and are detected early in infection in humans and around 3–5 days after infection in NHPs (Geisbert et al., 2003a, 2007; Hensley et al., 2011; Rollin et al., 2007). Tissue factor, or factor III, expression and release from filovirus-infected monocytes and macrophages has been shown in the NHP model and is linked to the formation of DIC (Geisbert et al., 2003b, 2007). Vascular permeability and leakiness seen in late infection may be due less to viral cytopathic effects and more to the elaboration of cytokines and inflammatory mediators.

6.3 ARENAVIRUSES

The Old World arenaviruses, the Lassa–lymphocytic choriomeningitis serocomplex, include Lassa virus (LASV), which causes a viral hemorrhagic fever, Lassa fever, in Africa and the ubiquitous lymphocytic choriomeningitis virus (LCMV). Recently, a novel pathogenic arenavirus, Lujo virus (LUJV), has been isolated from patients suffering from viral hemorrhagic fever (Briese et al., 2009). The New World arenaviruses, the Tacaribe serocomplex, include viruses indigenous to the Americas, such as the viral hemorrhagic fever–causing Junín (JUNV), Chapare (CHPV), Guanarito (GTOV), Machupo (MACV), and Sabiá (SABV) viruses (Bowen et al., 1996; Cajimat and Fulhorst, 2004; Clegg 2002; Rowe et al., 1970).

6.3.1 OLD WORLD ARENAVIRAL HEMORRHAGIC FEVERS

The signs and symptoms of human Lassa fever vary, but the disease commonly presents with fever, malaise, abdominal pain, pharyngitis, sore throat, cough, vomiting, diarrhea, high AST to ALT concentration ratios, and proteinuria (McCormick et al., 1987). Clinical complications include pleural and pericardial effusions, neurological manifestations, bleeding from mucosal surfaces, and shock, yet frank hemorrhage is not generally a hallmark of Old World arenavirus infection. Viremia has been noted by day 3 of disease, and a high risk of death is closely associated with high viremia; survival is lowest in patients with both high viremia and high AST concentrations. Patients infected with LUJV presented with many similar symptoms including no overt hemorrhage besides gingival bleeding, petechial rash, and some oozing from injection sites (Briese et al., 2009; Paweska et al., 2009).

The histopathology of fatal human Lassa fever reveals minimal necrosis of hepatocytes, necrosis of the splenic red pulp and adrenal glands, focal tubular necrosis of the kidney, and interstitial pneumonia (Walker et al., 1975, 1982; Winn and Walker, 1975). Macroscopic changes observed during Lassa fever in humans are pulmonary edema, pleural effusion, ascites, and signs of gastrointestinal hemorrhage (Gunther and Lenz, 2004). Microscopically, hepatocellular necrosis with phagocytic macrophage reaction, but with minimal lymphocyte infiltration, is typical with splenic red pulp necrosis also common. Renal tubular injury, interstitial nephritis, interstitial pneumonitis, and limited myocarditis are additional hallmarks of the disease, but central nervous system lesions are generally absent. Autopsies of patients who succumbed to LUJV infection revealed thrombocytopenia, granulocytosis, increased serum ALT and AST concentrations, hepatocyte necrosis without prominent inflammatory cell infiltrates, and skin vasculitis (Briese et al., 2009; Paweska et al., 2009).

Animal models of Lassa fever include infection of guinea pigs, common marmosets (*Callithrix jacchus*), and rhesus monkeys, but as of yet there are no models for LUJV infection.

LASV will replicate to modest levels in the blood and tissues of several strains of wild-type or knockout mice, but the mice will survive the infection (Peters et al., 1987). These models then only serve as in vivo replication systems for investigating virus growth in certain experimental conditions and are not models of human disease. Although outbred Hartley guinea pigs are not uniformly susceptible to lethal infection with LASV, strain 13 inbred guinea pigs are susceptible to a lethal infection with a low dose (2 pfu) of unpassaged, wild-type LASV, strain Josiah (Jahrling et al., 1982). In the strain 13 model, LASV infects a wide variety of cells outside the nervous system just as seen in human LASV infection. In the spleen and lymph nodes, the virus replicates to high titers (10^7 and 10^8 pfu/g tissue) within 8–9 days after infection. Virus is also present in the salivary glands, lungs, liver, adrenal gland, kidneys, pancreas, and heart. Low viral titers are commonly found in plasma and the brain (10^4 pfu/g tissue). The most severe lesion in the LASV-infected guinea pig is interstitial pneumonia, which is not common in cases of infected humans (Jahrling et al., 1982).

LASV infection in common marmosets mimics the course of disease seen in fatally infected humans (Carrion et al., 2007) including fever, weight loss, and high viral burden. Histological lesions are mostly limited to the organs of the reticuloendothelial system (liver, spleen, and lymph nodes). Multifocal hepatic necrosis, mild to moderate interstitial pneumonitis, and edematous changes in lung vasculature are typical with lymphoid depletion occurring in the spleen and lymph nodes (Avila et al., 1985). These animals provide a smaller, more cost-effective alternative to other NHP models that are expensive, harder to house due to size, and available in only limited quantities.

The rhesus monkey model of LASV infection is the most similar to human disease both in terms of clinical course and histopathologic findings (Peters et al., 1989). Onset of viremia typically occurs 4 days after infections and viremia remains detectable for the next 2 weeks (Jahrling et al., 1980). As in human infections, high viremia in the rhesus model directly correlates with fatality. Cellular immune responses fail to prevent viral replication resulting in high viral titers and dissemination throughout all tissues. The elevated levels of IFN seen 4–6 days after infection do not appear to influence viremia or disease outcome (Johnson et al., 1987; Peters et al., 1989). Despite the wide range of tissues infected, histological lesions are minor or largely absent and do not usually correlate with fatality (Kunz, 2009).

Because LASV must be handled at maximum biocontainment in BSL-4 laboratories, animal model research is dangerous and limited. Pichindé virus (PICV) is a New World arenavirus that can be handled in a BSL-2 laboratory and causes a disease similar to human Lassa fever in guinea pigs, thereby providing a safer and more widely available animal model for LASV infection research (Jahrling et al., 1981). After being serially passaged 8 times in strain 13 guinea pigs, the resulting virus was uniformly lethal upon infection with less than 10 pfu. Guinea pigs died 13–19 days after inoculation with serum, and viremia reached greater than 100,000 pfu/ml. Infected animals exhibited severe lymphopenia, and virus was found throughout the body except in the brain and nervous system. Along with moderate hepatocellular necrosis, there was also necrosis of splenic red pulp and adrenal gland tissues. The lung tissue exhibited interstitial pneumonia. Cardiac and pulmonary functions have also been thoroughly investigated after infection of catheterized guinea pigs (Qian et al., 1994). Severe hypotension, circulatory shock, and pulmonary edema could all be attributed to cardiovascular disturbances. Cardiac output, heart rate, cardiac work, cardiac power, and stroke volume were all significantly decreased, whereas pulmonary functions were not particularly altered (Qian et al., 1994). In addition, there was marked wasting of greater than 25% of the animal's body weight over the course of the infection. Along with pulmonary distress syndrome, there is also a circulatory response to the fluid load that is probably due to the capillary leakiness syndrome. Despite the cardiac involvement, no viral antigen was found in cardiac tissues or cells 14 days after infection (Jahrling et al., 1981; Qian et al., 1994). The similarity in pathogenesis of PICV-infected guinea pigs to Lassa fever in humans and macaques makes this an acceptable surrogate.

The WE strain of LCMV, which can be handled in a BSL-2 or BSL-3 containment laboratory, also provides a more accessible model for the study of LASV infection in some strains of hamsters, guinea pigs, and NHPs. Outbred Hartley guinea pigs succumb to a low-dose LCMV-WE infection within 12–14 days with high levels of virus replication seen throughout the body (Riviere et al., 1985). LCMV-WE was also fatal for rhesus monkeys and crab-eating macaques by a variety of routes of administration (Lukashevich et al., 2002, 2003; Peters et al., 1987). Infected monkeys exhibited fever, anorexia, and high viremias, which persisted from 3 days after infection through to death, which generally occurred 13–14 days after exposure. Infected animals had lymphopenia, thrombocytopenia, petechial hemorrhage, and elevated AST and ALT concentrations; the virus was largely hepatotropic. Increases in IL-6, sIL-6R, sTNFR1 and sTNFR2, and IFN-γ have been documented in LCMV-WE-infected rhesus monkeys (Lukashevich et al., 2003). The Armstrong strain of LCMV is comparatively avirulent in these animal models. The LCMV-WE models are similar to human Lassa fever and reminiscent of the rare cases of human LCMV hemorrhagic fever that have been previously reported (Smadel et al., 1942).

6.3.2 New World Arenaviral Hemorrhagic Fevers

Argentinian, Bolivian, "Venezuelan," and "Brazilian" hemorrhagic fevers are clinically similar to each other (Arribalzaga, 1955; de Manzione et al., 1998; Harrison et al., 1999; Molteni et al., 1961; Rugiero et al., 1964; Stinebaugh et al., 1996) and distinct from Old World arenavirus infections. These diseases are caused by JUNV, MACV, GTOV, and SABV, respectively. After an incubation period of 1–2 weeks, there is onset of fever and malaise, headache, myalgia, epigastric pain, and anorexia. Three to four days later, the symptoms become increasingly severe with multisystem involvement: prostration, abdominal pain, nausea and vomiting, and constipation or mild diarrhea. Dizziness, photophobia, retro-orbital pain, and disorientation may also appear in some cases, as well as the earliest signs of vascular damage, such as conjunctival injection, flushing over the head and upper torso, skin petechiae, and mild (postural) hypotension. The more severe cases, one-third of patients, begin to exhibit hemorrhagic or neurologic manifestations and secondary bacterial infections during the second week of illness. Neurological signs may be prominent and include tremor of the hands and tongue, coma, and convulsions. Hemorrhages from mucous membranes (gums, nose, vagina/uterus, and gastrointestinal tract) and ecchymoses at needle puncture sites develop and shock follows. Capillary leakiness is a hallmark of the disease, with minor blood loss overall, and elevated hematocrit during the peak of capillary leak syndrome (Charrel and de Lamballerie, 2003). Death usually occurs 7–12 days after disease onset in a subset of critically ill patients. However, petechiae, erythema, facial edema, and hyperesthesia are less common signs of the disease.

Presenting more hemorrhagic manifestations than Old World arenaviral infections, severe cases of New World arenaviral hemorrhagic fevers have widespread hemorrhage, particularly in the skin and mucous membranes (gastrointestinal tract), as well as intracranial hemorrhages (Virchow–Robin space) and hemorrhages in the kidney, pericardium, spleen, adrenal glands, and lung (the last four common in JUNV infections). Microscopic lesions include acidophilic bodies and focal necrosis in the liver, acute tubular and papillary necrosis in the kidney, reticular hyperplasia of the spleen and lymph nodes, and secondary bacterial lung infections in the case of Argentinian hemorrhagic fever (AHF) (acute bronchitis and bronchopneumonia, myocardial and lung abscesses) and interstitial pneumonia in the case of Bolivian hemorrhagic fever (Child et al., 1967; Elsner et al., 1973). In AHF, the sites of cellular necrosis (hepatocytes, renal tubular epithelium, macrophages, dendritic reticular cells of the spleen and lymph nodes) have been shown to correspond to sites of viral antigen accumulation, and both JUNV and MACV could be isolated from blood, spleen, and lymph nodes of patients (Cossio et al., 1975; Gonzalez et al., 1980; Johnson, 1965; Maiztegui, 1975). However, the overall histopathologic findings in human and animal model infections are relatively subtle with comparatively little necrosis and do not reflect the severity of disease (Child et al., 1967; Eichner et al., 2011; McLeod et al., 1976, 1978; Salas et al., 1991; Terrell et al., 1973).

6.3.2.1 Junín Virus

Guinea pigs, NHPs, and marmosets are all susceptible to JUNV infection and provide reasonable similarities to human disease. Guinea pigs infected intramuscularly with the prototype JUNV, strain XJ, die 13–18 days after infection with hemorrhagic manifestations, thrombocytopenia, leukopenia, and a heavy virus burden in spleen, lymph nodes, and bone marrow (Carballal et al., 1977, 1981; Molinas et al., 1978). Patterns of human clinical AHF illness are JUNV strain-specific and can be "hemorrhagic" (Espindola strain), "neurologic" (Ledesma strain), "mixed" (P-3551 strain), or "common" (Romero strain) (Maiztegui, 1975). These disease forms are replicated faithfully in experimental animal infections. Guinea pigs and NHPs infected with Espindola strain ("hemorrhagic") exhibit a pronounced bleeding diathesis, with disseminated cutaneous and mucous membrane hemorrhage. In contrast, Ledesma-infected animals ("neurologic") show little or no hemorrhagic manifestations but develop overt and generally progressive signs of neurologic dysfunction: limb paresis, ataxia, tremors, and hyperactive startle reflexes. In guinea pigs, the Espindola strain replicates predominantly in spleen, lymph nodes, and bone marrow, the major sites of necrosis, while lower levels of virus are present in blood and brain. The Ledesma strain, however, is found predominantly in the brain, and only low amounts of virus are recovered from the spleen and lymph nodes. The route of exposure has very little effect on the disease form, which instead appears to be driven by strain-specific factor. For instance, strains that produce hemorrhagic disease still do so even when inoculated intracranially retaining their tropism for visceral tissues such as the spleen and lymph nodes (Peters, 1987). In Espindola-infected rhesus monkeys, the appearance of high viral titers 7–8 days after infection correlates with the appearance of vascular disturbances such as facial flushing, rash, conjunctivitis, progressing to hemorrhagic diathesis, petechiae, and bleeding from the mucous membranes. The disease is 100% fatal in these animals with a mean time to death of 33 days. These animals also shed more virus from the oropharynx, and respond more slowly with the production of antibodies than do those infected with the Ledesma strain (Kenyon et al., 1988; McKee et al., 1985, 1987).

6.3.2.2 Guanarito Virus

Humans suffering from "Venezuelan hemorrhagic fever," caused by GTOV infection, present with severe disease and hemorrhagic manifestations that are associated with marked thrombocytopenia (Salas et al., 1981). In an effort to establish a mouse model, incomplete mortality at 10 days post infection was observed when neonatal outbred Swiss mice were infected by intracranial injection. Intraperitoneal infection of adult mice resulted in no illness or death (Tesh et al., 1994). In the guinea pig model, hemorrhage is minimal, but the histopathology is similar in both Hartley and strain 13 guinea pigs (Hall et al., 1996). Strain 13 guinea pigs die between 12 and 23 days after infection, depending on the GTOV isolate, when exposed by the subcutaneous route (Tesh et al., 1994). Several notable lesions occur due to viral infection such as epithelial necrosis in the gastrointestinal tract, lung interstitial pneumonia and hemorrhage, and necrosis or depletion of lymphoid tissues of spleen, lymph nodes, intestines, and lungs. There are also signs of adrenal cortical necrosis and congestion along with splenic congestion. Viral antigen is present in all lymphoid cells of all organs. In the spleen, viral antigen is abundant in large macrophages and lymphoblastic cells of the marginal zone. Virus levels are higher in homogenates of organ tissues than in guinea pig serum, suggesting viral replication in these organ tissues. The highest viral titers are seen in the spleen (Hall et al., 1996). There were no lesions in the brain, although some antigen was present. In general, these findings indicate that GTOV infection of guinea pigs pathologically resembles infection with other arenaviruses, leading to lymphoid necrosis, bone marrow depletion interstitial pneumonia, and platelet thrombi. Comparatively, PICVs and LASVs viruses seem to cause liver necrosis to a higher extent than GTOV infection in this guinea pig model. Infection of guinea pigs with GTOV closely resembles infection with viscerotropic strains of Junín virus, where there are similar virus titers and distribution of virus in the tissues (Kenyon et al., 1988).

A small study of three adult rhesus monkeys given 25,000 pfu of GTOV subcutaneously exhibited clinical illness and lethargy, decreased appetite, fever, and over 100 pfu/ml of virus in blood samples taken at days 7 and 11 postinfection (Tesh et al., 1994). Although all three monkeys survived, the case fatality rate in human cases caused by an outbreak of this strain was only about 3%–4%; therefore, this small monkey study may have been consistent with human disease.

6.3.2.3 Machupo Virus

MACV, the causative agent of Bolivian hemorrhagic fever, is spread by inhalation of aerosols generated from excretions of its carrier rodent, the big laucha (*Calomys callosus*), although human-to-human transmission can also occur (1994; Charrel and de Lamballerie, 2003; Peters et al., 1974). Case fatality rates are approximately 20% in humans. The disease course in humans is similar to other New World arenavirus infections, with fever and malaise, headache, dizziness, back pain, petechia, erythema, and myalgia (Charrel and de Lamballerie, 2003).

Early studies showed that MACV was also lethal in guinea pigs, hamsters, and suckling mice (Webb et al., 1975), but these models are not well characterized. Guinea pigs can be lethally infected with MACV, but data characterizing viral replication, pathology, and host responses are lacking (Webb et al., 1975). Subcutaneous inoculation of NHPs results in a fairly similar course of disease to that seen in humans, with death ranging from 8 to 25 days after virus exposure (Eddy et al., 1975b; Kastello et al., 1976; McLeod et al., 1976, 1978; Terrell et al., 1973; Webb et al., 1975). NHPs that survive past this time point develop neurological symptoms, with some eventually succumbing to disease and others surviving. Histopathologic comparison of tissue from NHPs or humans that succumb to MACV infection revealed hemorrhaging and necrosis in various organs, but none of these manifestations were thought to be severe enough to cause death (Child et al., 1967; Terrell et al., 1973). In a set of companion studies describing one of the only MACV NHP studies, limited viremia data are available for small cohorts of experimental NHPs belonging to various species showing viremia over 5–21 days after infection (Eddy et al., 1975a; Kastello et al., 1976). Animals in these cohorts also show signs of hepatic necrosis, lymphoid depletion, and adrenal cortical necrosis. Other characteristic symptoms, such as thrombocytopenia and decreased hematocrit, hematological changes, clinical chemistry, and immune responses to infection have not been reported.

MACV is not lethal in adult immunocompetent mice (Webb et al., 1975), but recent work described the use of STAT-1 knockout mice as a model for MACV pathogenesis and immunity (Bradfute et al., 2011). Mice lacking type I IFN responses have been used as models for infections with other hemorrhagic fever viruses, including filoviruses, arenaviruses, and bunyaviruses (Bente et al., 2010; Bereczky et al., 2010; Bray, 2001; Kolokoltsova et al., 2010). Signal transducer and activator of transcription 1 (STAT-1) knockout mice are defective in not just type I but type II and type III IFN signaling as well. Intraperitoneal infection of STAT-1 knockout mice with ≈1000 pfu of MACV resulted in 100% lethality with mean time to death around 7 days. Subcutaneous and intranasal routes of infection were not uniformly lethal. Using this model, it was possible to characterize MACV growth kinetics, hematologic and metabolic alterations, cytokine responses, and histopathologic changes. High viral loads were detected in the plasma, spleen, lungs, liver, and kidneys 5–7 days after infection. ALT, AST, and BUN concentrations rose dramatically 7 days after infection, indicative of possible liver and kidney damage. Hematology revealed a substantial increase in white blood cells and a decrease in platelets during the course of infection. Concentrations of certain cytokines and chemokines were elevated during the course of infection: IFN-gamma, IL-5, KC (mouse orthologue of IL-8), IL-6, IL-10, MIP-1α, MIP-1β, TNF-α, G-CSF, and RANTES. Mild to moderate hepatocellular degeneration and necrosis were found 7 days after infection. Dead lymphocytes were present in the thymus, spleen, and lymph nodes as early as 3 days after infection and increased throughout the course of disease. Although it was not observed in the STAT-1 knockout mouse model, NHPs tend to have a biphasic course of disease after MACV infection, with most deaths occurring in the first "acute" phase. Surviving NHPs tend to undergo a second, prolonged disease state, characterized by neurological disease (Eddy et al., 1975a). This second phase is not found in human MACV infection or in the STAT-1 knockout model. Overall, due to a lack of data, it is not feasible to conclusively compare the faithfulness of MACV infections in different species to each other or the human disease course.

DISCLAIMER

The content of this publication does not necessarily reflect the views or policies of the U.S. Department of Defense, the U.S. Department of the Army, or the U.S. Department of Health and Human Services or of the institutions and companies affiliated with the authors. JHK performed this work as an employee of Tunnell Consulting, Inc., a subcontractor to Battelle Memorial Institute under its prime contract with NIAID, under Contract No. HHSN272200200016I.

REFERENCES

1994. Bolivian hemorrhagic fever—El Beni Department, Bolivia, 1994. *MMWR Morb Mortal Wkly Rep* 43:943–946.

1978. Ebola haemorrhagic fever in Sudan, 1976. Report of a WHO/International Study Team. *Bull World Health Organ* 56:247–270.

2001. Outbreak of Ebola hemorrhagic fever Uganda, August 2000–January 2001. *MMWR Morb Mortal Wkly Rep* 50:73–77.

2005. Outbreak of Marburg virus hemorrhagic fever—Angola, October 1, 2004–March 29, 2005. *MMWR Morb Mortal Wkly Rep* 54:308–309.

Alves, D. A., A. R. Glynn, K. E. Steele, M. G. Lackemeyer, N. L. Garza, J. G. Buck, C. Mech, and D. S. Reed. 2010. Aerosol exposure to the Angola strain of Marburg virus causes lethal viral hemorrhagic fever in cynomolgus macaques. *Vet Pathol* 47:831–851.

Arribalzaga, R. A. 1955. New epidemic disease due to unidentified germ: Nephrotoxic, leukopenic and enanthematous hyperthermia. *Dia Med* 27:1204–1210.

Avila, M. M., M. J. Frigerio, E. L. Weber, S. Rondinone, S. R. Samoilovich, R. P. Laguens, L. B. de Guerrero, and M. C. Weissenbacher. 1985. Attenuated Junin virus infection in *Callithrix jacchus*. *J Med Virol* 15:93–100.

Baize, S., E. M. Leroy, M. C. Georges-Courbot, M. Capron, J. Lansoud-Soukate, P. Debre, S. P. Fisher-Hoch, J. B. McCormick, and A. J. Georges. 1999. Defective humoral responses and extensive intravascular apoptosis are associated with fatal outcome in Ebola virus-infected patients. *Nat Med* 5:423–426.

Baskerville, A., E. T. Bowen, G. S. Platt, L. B. McArdell, and D. I. Simpson. 1978. The pathology of experimental Ebola virus infection in monkeys. *J Pathol* 125:131–138.

Baskerville, A., S. P. Fisher-Hoch, G. H. Neild, and A. B. Dowsett. 1985. Ultrastructural pathology of experimental Ebola haemorrhagic fever virus infection. *J Pathol* 147:199–209.

Bazhutin, N. B., E. F. Belanov, V. A. Spiridonov, A. V. Voitenko, N. A. Krivenchuk, S. A. Krotov, N. I. Omel'chenko, A. Tereshchenko, and V. V. Khomichev. 1992. The effect of the methods for producing an experimental Marburg virus infection on the characteristics of the course of the disease in green monkeys. *Vopr Virusol* 37:153–156.

Bente, D., J. Gren, J. E. Strong, and H. Feldmann. 2009. Disease modeling for Ebola and Marburg viruses. *Dis Model Mech* 2:12–17.

Bente, D. A., J. B. Alimonti, W. J. Shieh, G. Camus, U. Stroher, S. Zaki, and S. M. Jones. 2010. Pathogenesis and immune response of Crimean-Congo hemorrhagic fever virus in a STAT-1 knockout mouse model. *J Virol* 84:11089–11100.

Bereczky, S., G. Lindegren, H. Karlberg, S. Akerstrom, J. Klingstrom, and A. Mirazimi. 2010. Crimean-Congo hemorrhagic fever virus infection is lethal for adult type I interferon receptor-knockout mice. *J Gen Virol* 91:1473–1477.

Bosio, C. M., M. J. Aman, C. Grogan, R. Hogan, G. Ruthel, D. Negley, M. Mohamadzadeh, S. Bavari, and A. Schmaljohn. 2003. Ebola and Marburg viruses replicate in monocyte-derived dendritic cells without inducing the production of cytokines and full maturation. *J Infect Dis* 188:1630–1638.

Bowen, E. T., G. S. Platt, D. I. Simpson, L. B. McArdell, and R. T. Raymond. 1978. Ebola haemorrhagic fever: Experimental infection of monkeys. *Trans R Soc Trop Med Hyg* 72:188–191.

Bowen, M. D., C. J. Peters, and S. T. Nichol. 1996. The phylogeny of New World (Tacaribe complex) arenaviruses. *Virology* 219:285–290.

Bradfute, S. B., K. S. Stuthman, A. C. Shurtleff, and S. Bavari. 2011. A STAT-1 knockout mouse model for Machupo virus pathogenesis. *Virol J* 8:300.

Bray, M. 2001. The role of the Type I interferon response in the resistance of mice to filovirus infection. *J Gen Virol* 82:1365–1373.

Bray, M., K. Davis, T. Geisbert, C. Schmaljohn, and J. Huggins. 1999. A mouse model for evaluation of prophylaxis and therapy of Ebola hemorrhagic fever. *J Infect Dis* 178:651–661; 179(Suppl 1):S248–S258.

Bray, M. and T. W. Geisbert. 2005. Ebola virus: The role of macrophages and dendritic cells in the pathogenesis of Ebola hemorrhagic fever. *Int J Biochem Cell Biol* 37:1560–1566.

Briese, T., J. T. Paweska, L. K. McMullan, S. K. Hutchison, C. Street, G. Palacios, M. L. Khristova, J. Weyer, R. Swanepoel, M. Egholm, S. T. Nichol, and W. I. Lipkin. 2009. Genetic detection and characterization of Lujo virus, a new hemorrhagic fever-associated arenavirus from southern Africa. *PLoS Pathog* 5:e1000455.

Bruce, J. and P. Brysiewicz. 2002. Ebola fever: The African emergency. *Int J Trauma Nurs* 8:36–41.

Bwaka, M. A., M. J. Bonnet, P. Calain, R. Colebunders, A. De Roo, Y. Guimard, K. R. Katwiki et al. 1999. Ebola hemorrhagic fever in Kikwit, Democratic Republic of the Congo: Clinical observations in 103 patients. *J Infect Dis* 179(Suppl 1):S1–S7.

Cajimat, M. N. and C. F. Fulhorst. 2004. Phylogeny of the Venezuelan arenaviruses. *Virus Res* 102:199–206.

Carballal, G., P. M. Cossio, R. P. Laguens, C. Ponzinibbio, J. R. Oubina, P. C. Meckert, A. Rabinovich, and R. M. Arana. 1981. Junin virus infection of guinea pigs: Immunohistochemical and ultrastructural studies of hemopoietic tissue. *J Infect Dis* 143:7–14.

Carballal, G., M. Rodriguez, M. J. Frigerio, and C. Vasquez. 1977. Junin virus infection of guinea pigs: Electron microscopic studies of peripheral blood and bone marrow. *J Infect Dis* 135:367–373.

Carrion, R. 2010. The common marmoset as a model for filovirus induced disease. Presented at the *XIVth International Conference on Negative Strand Viruses*, Brugge, Belgium.

Carrion, R., Jr., K. Brasky, K. Mansfield, C. Johnson, M. Gonzales, A. Ticer, I. Lukashevich, S. Tardif, and J. Patterson. 2007. Lassa virus infection in experimentally infected marmosets: Liver pathology and immunophenotypic alterations in target tissues. *J Virol* 81:6482–6490.

Charrel, R. N. and X. de Lamballerie. 2003. Arenaviruses other than Lassa virus. *Antiviral Res* 57:89–100.

Child, P. L., R. B. MacKenzie, L. R. Valverde, and K. M. Johnson. 1967. Bolivian hemorrhagic fever. A patho-logic description. *Arch Pathol* 83:434–445.

Clegg, J. C. 2002. Molecular phylogeny of the arenaviruses. *Curr Top Microbiol Immunol* 262:1–24.

Connolly, B. M., K. E. Steele, K. J. Davis, T. W. Geisbert, W. M. Kell, N. K. Jaax, and P. B. Jahrling. 1999. Pathogenesis of experimental Ebola virus infection in guinea pigs. *J Infect Dis* 179(Suppl 1):S203–S217.

Cossio, P., R. Laguens, R. Arana, A. Segal, and J. Maiztegui. 1975. Ultrastructural and immunohistochemical study of the human kidney in Argentine haemorrhagic fever. *Virchows Arch A Pathol Anat Histol* 368:1–9.

Davis, K. J., A. O. Anderson, T. W. Geisbert, K. E. Steele, J. B. Geisbert, P. Vogel, B. M. Connolly, J. W. Huggins, P. B. Jahrling, and N. K. Jaax. 1997. Pathology of experimental Ebola virus infection in African green monkeys. Involvement of fibroblastic reticular cells. *Arch Pathol Lab Med* 121:805–819.

de Manzione, N., R. A. Salas, O. Godoy, L. Rojas, F. Araoz, C. F. Fulhorst, T. G. Ksiazek, J. N. Mills, B. A. Ellis, C. J. Peters, and R. B. Tesh. 1998. Venezuelan hemorrhagic fever: Clinical and epide-miological studies of 165 cases. *Clin Infect Dis* 26:308–313.

Ebihara, H., A. Takada, D. Kobasa, S. Jones, G. Neumann, S. Theriault, M. Bray, H. Feldmann, and Y. Kawaoka. 2006. Molecular determinants of Ebola virus virulence in mice. *PLoS Pathog* 2:e73.

Eddy, G. A., S. K. Scott, F. S. Wagner, and O. M. Brand. 1975. Pathogenesis of Machupo virus infection in primates. *Bull World Health Organ* 52:517–521.

Eddy, G. A., F. S. Wagner, S. K. Scott, and B. J. Mahlandt. 1975. Protection of monkeys against Machupo virus by the passive administration of Bolivian haemorrhagic fever immunoglobulin (human origin). *Bull World Health Organ* 52:723–727.

Eichner, M., S. F. Dowell, and N. Firese. 2011. Incubation period of Ebola hemorrhagic virus subtype Zaire. *Osong Public Health Res Perspect* 2:3–7.

Elsner, B., E. Schwarz, O. G. Mando, J. Maiztegui, and A. Vilches. 1973. Pathology of 12 fatal cases of Argentine hemorrhagic fever. *Am J Trop Med Hyg* 22:229–236.

FDA U.S. Department of Health and Human Services (ed.). 2009. *Draft Guidance for Industry—Essential Elements for Address Efficacy under the Animal Rule*, Vol. 74. Federal Register, Silver Spring, MD, pp. 3610–3611.

Feldmann, H., S. M. Jones, H. J. Schnittler, and T. Geisbert. 2005. Therapy and prophylaxis of Ebola virus infections. *Curr Opin Investig Drugs* 6:823–830.

Fisher-Hoch, S. P., G. S. Platt, G. Lloyd, D. I. Simpson, G. H. Neild, and A. J. Barrett. 1983. Haematological and biochemical monitoring of Ebola infection in rhesus monkeys: Implications for patient management. *Lancet* 2:1055–1058.

Fisher-Hoch, S. P., G. S. Platt, G. H. Neild, T. Southee, A. Baskerville, R. T. Raymond, G. Lloyd, and D. I. Simpson. 1985. Pathophysiology of shock and hemorrhage in a fulminating viral infection (Ebola). *J Infect Dis* 152:887–894.

Formenty, P., C. Hatz, B. Le Guenno, A. Stoll, P. Rogenmoser, and A. Widmer. 1999. Human infection due to Ebola virus, subtype Cote d'Ivoire: Clinical and biologic presentation. *J Infect Dis* 179(Suppl 1): S48–S53.

Fritz, E. A., J. B. Geisbert, T. W. Geisbert, L. E. Hensley, and D. S. Reed. 2008. Cellular immune response to Marburg virus infection in cynomolgus macaques. *Viral Immunol* 21:355–363.

Fuller, C. L., G. Ruthel, K. L. Warfield, D. L. Swenson, C. M. Bosio, M. J. Aman, and S. Bavari. 2007. NKp30-dependent cytolysis of filovirus-infected human dendritic cells. *Cell Microbiol* 9:962–976.

Geisbert, T. W., K. M. Daddario-DiCaprio, J. B. Geisbert, H. A. Young, P. Formenty, E. A. Fritz, T. Larsen, and L. E. Hensley. 2007. Marburg virus Angola infection of rhesus macaques: Pathogenesis and treatment with recombinant nematode anticoagulant protein c2. *J Infect Dis* 196(Suppl 2):S372–S381.

Geisbert, T. W., L. E. Hensley, T. R. Gibb, K. E. Steele, N. K. Jaax, and P. B. Jahrling. 2000. Apoptosis induced in vitro and in vivo during infection by Ebola and Marburg viruses. *Lab Invest* 80:171–186.

Geisbert, T. W., L. E. Hensley, T. Larsen, H. A. Young, D. S. Reed, J. B. Geisbert, D. P. Scott, E. Kagan, P. B. Jahrling, and K. J. Davis. 2003. Pathogenesis of Ebola hemorrhagic fever in cynomolgus macaques: Evidence that dendritic cells are early and sustained targets of infection. *Am J Pathol* 163:2347–2370.

Geisbert, T. W., H. A. Young, P. B. Jahrling, K. J. Davis, E. Kagan, and L. E. Hensley. 2003. Mechanisms under-lying coagulation abnormalities in Ebola hemorrhagic fever: Overexpression of tissue factor in primate monocytes/macrophages is a key event. *J Infect Dis* 188:1618–1629.

Geisbert, T. W., H. A. Young, P. B. Jahrling, K. J. Davis, T. Larsen, E. Kagan, and L. E. Hensley. 2003. Pathogenesis of Ebola hemorrhagic fever in primate models: Evidence that hemorrhage is not a direct effect of virus-induced cytolysis of endothelial cells. *Am J Pathol* 163:2371–2382.

Gibb, T. R., D. A. Norwood, Jr., N. Woollen, and E. A. Henchal. 2002. Viral replication and host gene expres-sion in alveolar macrophages infected with Ebola virus (Zaire strain). *Clin Diagn Lab Immunol* 9:19–27.

Gonchar, N. I., V. A. Pshenichnov, V. A. Pokhodiaev, K. L. Lopatov, and I. V. Firsova. 1991. The sensitivity of different experimental animals to the Marburg virus. *Vopr Virusol* 36:435–437.

Gonzalez, P. H., P. M. Cossio, R. Arana, J. I. Maiztegui, and R. P. Laguens. 1980. Lymphatic tissue in Argentine hemorrhagic fever. Pathologic features. *Arch Pathol Lab Med* 104:250–254.

Gunther, S. and O. Lenz. 2004. Lassa virus. *Crit Rev Clin Lab Sci* 41:339–390.

Gupta, M., S. Mahanty, R. Ahmed, and P. E. Rollin. 2001. Monocyte-derived human macrophages and peripheral blood mononuclear cells infected with Ebola virus secrete MIP-1alpha and TNF-alpha and inhibit poly-IC-induced IFN-alpha in vitro. *Virology* 284:20–25.

Hall, W. C., T. W. Geisbert, J. W. Huggins, and P. B. Jahrling. 1996. Experimental infection of guinea pigs with Venezuelan hemorrhagic fever virus (Guanarito): A model of human disease. *Am J Trop Med Hyg* 55:81–88.

Harrison, L. H., N. A. Halsey, K. T. McKee, Jr., C. J. Peters, J. G. Barrera Oro, A. M. Briggiler, M. R. Feuillade, and J. I. Maiztegui. 1999. Clinical case definitions for Argentine hemorrhagic fever. *Clin Infect Dis* 28:1091–1094.

Hensley, L. E., D. A. Alves, J. B. Geisbert, E. A. Fritz, C. Reed, T. Larsen, and T. W. Geisbert. 2011. Pathogenesis of Marburg hemorrhagic fever in cynomolgus macaques. *J Infect Dis* 204(Suppl 3):S1021–S1031.

Hensley, L. E., H. A. Young, P. B. Jahrling, and T. W. Geisbert. 2002. Proinflammatory response during Ebola virus infection of primate models: Possible involvement of the tumor necrosis factor receptor superfamily. *Immunol Lett* 80:169–179.

Jaax, N. K., K. J. Davis, T. J. Geisbert, P. Vogel, G. P. Jaax, M. Topper, and P. B. Jahrling. 1996. Lethal experimental infection of rhesus monkeys with Ebola-Zaire (Mayinga) virus by the oral and conjunctival route of exposure. *Arch Pathol Lab Med* 120:140–155.

Jahrling, P. B., R. A. Hesse, G. A. Eddy, K. M. Johnson, R. T. Callis, and E. L. Stephen. 1980. Lassa virus infection of rhesus monkeys: Pathogenesis and treatment with ribavirin. *J Infect Dis* 141:580–589.

Jahrling, P. B., R. A. Hesse, J. B. Rhoderick, M. A. Elwell, and J. B. Moe. 1981. Pathogenesis of a pichinde virus strain adapted to produce lethal infections in guinea pigs. *Infect Immun* 32:872–880.

Jahrling, P. B., S. Smith, R. A. Hesse, and J. B. Rhoderick. 1982. Pathogenesis of Lassa virus infection in guinea pigs. *Infect Immun* 37:771–778.

Johnson, E., N. Jaax, J. White, and P. Jahrling. 1995. Lethal experimental infections of rhesus monkeys by aerosolized Ebola virus. *Int J Exp Pathol* 76:227–236.

Johnson, K. M. 1965. Epidemiology of Machupo virus infection. 3. Significance of virological observations in man and animals. *Am J Trop Med Hyg* 14:816–818.

Johnson, K. M., J. B. McCormick, P. A. Webb, E. S. Smith, L. H. Elliott, and I. J. King. 1987. Clinical virology of Lassa fever in hospitalized patients. *J Infect Dis* 155:456–464.

Kastello, M. D., G. A. Eddy, and R. W. Kuehne. 1976. A rhesus monkey model for the study of Bolivian hemorrhagic fever. *J Infect Dis* 133:57–62.

Kenyon, R. H., D. E. Green, J. I. Maiztegui, and C. J. Peters. 1988. Viral strain dependent differences in experimental Argentine hemorrhagic fever (Junin virus) infection of guinea pigs. *Intervirology* 29:133–143.

Kolokoltsova, O. A., N. E. Yun, A. L. Poussard, J. K. Smith, J. N. Smith, M. Salazar, A. Walker, C. T. Tseng, J. F. Aronson, and S. Paessler. 2010. Mice lacking interferon {alpha}/{beta} and {gamma} receptors are susceptible to Junin virus infection. *J Virol* 84(24):13063–13067.

Kuhn, J. H. 2008. *Filoviruses: A Compendium of 40 years of Epidemiological, Clinical, and Laboratory Studies*. SpringerWienNewYork, Vienna, Austria.

Kuhn, J. H., S. Becker, H. Ebihara, T. W. Geisbert, P. B. Jahrling, Y. Kawaoka, S. V. Netesov et al. 2011. Family *Filoviridae*, In A. M. Q. King, M. J. Adams, E. B. Carstens, and E. J. Lefkowitz (ed.), *Virus Taxonomy—Ninth Report of the International Committee on Taxonomy of Viruses*. Elsevier/Academic Press, London, U.K., pp. 665–671.

Kuhn, J. H., S. Becker, H. Ebihara, T. W. Geisbert, K. M. Johnson, Y. Kawaoka, W. I. Lipkin et al. Proposal for a revised taxonomy of the family *Filoviridae*: Classification, names of taxa and viruses, and virus abbreviations. *Arch Virol* 155:2083–2103.

Kunz, S. 2009. The role of the vascular endothelium in arenavirus haemorrhagic fevers. *Thromb Haemost* 102:1024–1029.

Leffel, E. K. and D. S. Reed. 2004. Marburg and Ebola viruses as aerosol threats. *Biosecur Bioterror* 2:186–191.

Leroy, E. M., S. Baize, P. Debre, J. Lansoud-Soukate, and E. Mavoungou. 2001. Early immune responses accompanying human asymptomatic Ebola infections. *Clin Exp Immunol* 124:453–460.

Leroy, E. M., S. Baize, E. Mavoungou, and C. Apetrei. 2002. Sequence analysis of the GP, NP, VP40 and VP24 genes of Ebola virus isolated from deceased, surviving and asymptomatically infected individuals during the 1996 outbreak in Gabon: Comparative studies and phylogenetic characterization. *J Gen Virol* 83:67–73.

Leroy, E. M., S. Baize, V. E. Volchkov, S. P. Fisher-Hoch, M. C. Georges-Courbot, J. Lansoud-Soukate, M. Capron, P. Debre, J. B. McCormick, and A. J. Georges. 2000. Human asymptomatic Ebola infection and strong inflammatory response. *Lancet* 355:2210–2215.

Leroy, E. M., J. P. Gonzalez, and S. Baize. 2011. Ebola and Marburg haemorrhagic fever viruses: Major scientific advances, but a relatively minor public health threat for Africa. *Clin Microbiol Infect* 17:964–976.

Lofts, L. L., M. S. Ibrahim, D. L. Negley, M. C. Hevey, and A. L. Schmaljohn. 2007. Genomic differences between guinea pig lethal and nonlethal Marburg virus variants. *J Infect Dis* 196(Suppl 2):S305–S312.

Lofts, L. L., J. B. Wells, S. Bavari, and K. L. Warfield. 2011. Key genomic changes necessary for an in vivo lethal mouse marburgvirus variant selection process. *J Virol* 85:3905–3917.

Lukashevich, I. S., M. Djavani, J. D. Rodas, J. C. Zapata, A. Usborne, C. Emerson, J. Mitchen, P. B. Jahrling, and M. S. Salvato. 2002. Hemorrhagic fever occurs after intravenous, but not after intragastric, inoculation of rhesus macaques with lymphocytic choriomeningitis virus. *J Med Virol* 67:171–186.

Lukashevich, I. S., I. Tikhonov, J. D. Rodas, J. C. Zapata, Y. Yang, M. Djavani, and M. S. Salvato. 2003. Arenavirus-mediated liver pathology: Acute lymphocytic choriomeningitis virus infection of rhesus macaques is characterized by high-level interleukin-6 expression and hepatocyte proliferation. *J Virol* 77:1727–1737.

Maiztegui, J. I. 1975. Clinical and epidemiological patterns of Argentine haemorrhagic fever. *Bull World Health Organ* 52:567–575.

McCormick, J. B., I. J. King, P. A. Webb, K. M. Johnson, R. O'Sullivan, E. S. Smith, S. Trippel, and T. C. Tong. 1987. A case-control study of the clinical diagnosis and course of Lassa fever. *J Infect Dis* 155:445–455.

McKee, K. T., Jr., B. G. Mahlandt, J. I. Maiztegui, G. A. Eddy, and C. J. Peters. 1985. Experimental Argentine hemorrhagic fever in rhesus macaques: Viral strain-dependent clinical response. *J Infect Dis* 152:218–221.

McKee, K. T., Jr., B. G. Mahlandt, J. I. Maiztegui, D. E. Green, and C. J. Peters. 1987. Virus-specific factors in experimental Argentine hemorrhagic fever in rhesus macaques. *J Med Virol* 22:99–111.

McLeod, C. G., J. L. Stookey, G. A. Eddy, and K. Scott. 1976. Pathology of chronic Bolivian hemorrhagic fever in the rhesus monkey. *Am J Pathol* 84:211–224.

McLeod, C. G., J. L. Stookey, J. D. White, G. A. Eddy, and G. A. Fry. 1978. Pathology of Bolivian hemorrhagic fever in the African green monkey. *Am J Trop Med Hyg* 27:822–826.

Mikhailov, V. V., I. V. Borisevich, N. K. Chernikova, N. V. Potryvayeva, and V. P. Krasnyanskii. 1994. An evaluation of the possibility of Ebola fever specific prophylaxis in baboons (*Papio hamadryas*). *Vopr Virusol* 39:82–84.

Molinas, F. C., R. A. Paz, M. T. Rimoldi, and M. M. de Bracco. 1978. Studies of blood coagulation and pathology in experimental infection of guinea pigs with Junin virus. *J Infect Dis* 137:740–746.

Molteni, H. D., H. C. Guarinos, C. O. Petrillo, and F. Jaschek. 1961. Clinico-statistical study of 338 patients with epidemic hemorrhagic fever in the northwest of the province of Buenos Aires. *Sem Med* 118:839–855.

Paweska, J. T., N. H. Sewlall, T. G. Ksiazek, L. H. Blumberg, M. J. Hale, W. I. Lipkin, J. Weyer et al. 2009. Nosocomial outbreak of novel arenavirus infection, southern Africa. *Emerg Infect Dis* 15:1598–1602.

Peters, C. J., P. B. Jahrling, C. T. Liu, R. H. Kenyon, K. T. McKee, Jr., and J. G. Barrera Oro. 1987. Experimental studies of arenaviral hemorrhagic fevers. *Curr Top Microbiol Immunol* 134:5–68.

Peters, C. J., R. W. Kuehne, R. R. Mercado, R. H. Le Bow, R. O. Spertzel, and P. A. Webb. 1974. Hemorrhagic fever in Cochabamba, Bolivia, 1971. *Am J Epidemiol* 99:425–433.

Peters, C. J., C. T. Liu, G. W. Anderson, Jr., J. C. Morrill, and P. B. Jahrling. 1989. Pathogenesis of viral hemorrhagic fevers: Rift Valley fever and Lassa fever contrasted. *Rev Infect Dis* 11(Suppl 4):S743–S749.

Qian, C., P. B. Jahrling, C. J. Peters, and C. T. Liu. 1994. Cardiovascular and pulmonary responses to Pichinde virus infection in strain 13 guinea pigs. *Lab Anim Sci* 44:600–607.

Raymond, J., S. Bradfute, and M. Bray. 2011. Filovirus infection of STAT-1 knockout mice. *J Infect Dis* 204(Suppl 3):S986–S990.

Reed, D. S., L. E. Hensley, J. B. Geisbert, P. B. Jahrling, and T. W. Geisbert. 2004. Depletion of peripheral blood T lymphocytes and NK cells during the course of Ebola hemorrhagic fever in cynomolgus macaques. *Viral Immunol* 17:390–400.

Riviere, Y., R. Ahmed, P. J. Southern, M. J. Buchmeier, and M. B. Oldstone. 1985. Genetic mapping of lymphocytic choriomeningitis virus pathogenicity: Virulence in guinea pigs is associated with the L RNA segment. *J Virol* 55:704–709.

Robin, Y., P. Bres, and R. Camain. 1971. Passage of Marburg virus in guinea pigs, In G. A. Martini and R. Siegert (ed.), *Marburg Virus*. Springer-Verlag, New York, pp. 117–122.

Rollin, P. E., D. G. Bausch, and A. Sanchez. 2007. Blood chemistry measurements and D-Dimer levels associated with fatal and nonfatal outcomes in humans infected with Sudan Ebola virus. *J Infect Dis* 196(Suppl 2):S364–S371.

Rollin, P. E., R. J. Williams, D. S. Bressler, S. Pearson, M. Cottingham, G. Pucak, A. Sanchez et al. 1999. Ebola (subtype Reston) virus among quarantined nonhuman primates recently imported from the Philippines to the United States. *J Infect Dis* 179(Suppl 1):S108–S114.

Rowe, W. P., W. E. Pugh, P. A. Webb, and C. J. Peters. 1970. Serological relationship of the Tacaribe complex of viruses to lymphocytic choriomeningitis virus. *J Virol* 5:289–292.

Rugiero, H. R., H. Ruggiero, C. Gonzalezcambaceres, F. A. Cintora, F. Maglio, C. Magnoni, L. Astarloa, G. Squassi, A. Giacosa, and D. Fernandez. 1964. Argentine hemorrhagic fever. Ii. Descriptive clinical study. *Rev Assoc Med Argent* 78:281–294.

Ryabchikova, E., L. Kolesnikova, M. Smolina, V. Tkachev, L. Pereboeva, S. Baranova, A. Grazhdantseva, and Y. Rassadkin. 1996. Ebola virus infection in guinea pigs: Presumable role of granulomatous inflammation in pathogenesis. *Arch Virol* 141:909–921.

Ryabchikova, E. I., L. V. Kolesnikova, and S. V. Luchko. 1999. An analysis of features of pathogenesis in two animal models of Ebola virus infection. *J Infect Dis* 179(Suppl 1):S199–S202.

Ryabchikova, E. I., L. V. Kolesnikova, and S. V. Netesov. 1999. Animal pathology of filoviral infections. *Curr Top Microbiol Immunol* 235:145–173.

Salas, R., N. de Manzione, R. B. Tesh, R. Rico-Hesse, R. E. Shope, A. Betancourt, O. Godoy, R. Bruzual, M. E. Pacheco, B. Ramos et al. 1991. Venezuelan haemorrhagic fever. *Lancet* 338:1033–1036.

Sanchez, A., M. Lukwiya, D. Bausch, S. Mahanty, A. J. Sanchez, K. D. Wagoner, and P. E. Rollin. 2004. Analysis of human peripheral blood samples from fatal and nonfatal cases of Ebola (Sudan) hemorrhagic fever: Cellular responses, virus load, and nitric oxide levels. *J Virol* 78:10370–10377.

Shurtleff, A. C., T. Warren, and S. Bavari. 2011. Nonhuman primates as models for the discovery and development of ebolavirus therapeutics. *Expert Opin Drug Discov* 6:1–18.

Siegert, R., H. Shu, W. Slenczka, D. Peters, and G. Müller. 1967. The aetiology of an unknown human infection transmitted by monkeys. *Dtsch Med Wochenschr* 92:2341–2343.

Simpson, D. I., I. Zlotnik, and D. A. Rutter. 1968. Vervet monkey disease. Experiment infection of guinea pigs and monkeys with the causative agent. *Br J Exp Pathol* 49:458–464.

Smadel, J. E., R. H. Green, R. M. Paltauf, and T. A. Gonzalez. 1942. Lymphocytic choriomeningitis: Two human fatalities following an unusual febrile illness. *Proc Soc Exp Biol Med* 49:683–686.

Stinebaugh, B. J., F. X. Schloeder, K. M. Johnson, R. B. Mackenzie, G. Entwisle, and E. De Alba. 1966. Bolivian hemorrhagic fever. A report of four cases. *Am J Med* 40:217–230.

Stroher, U. and H. Feldmann. 2006. Progress towards the treatment of Ebola haemorrhagic fever. *Expert Opin Investig Drugs* 15:1523–1535.

Swenson, D. L., K. L. Warfield, K. Kuehl, T. Larsen, M. C. Hevey, A. Schmaljohn, S. Bavari, and M. J. Aman. 2004. Generation of Marburg virus-like particles by co-expression of glycoprotein and matrix protein. *FEMS Immunol Med Microbiol* 40:27–31.

Terrell, T. G., J. L. Stookey, G. A. Eddy, and M. D. Kastello. 1973. Pathology of Bolivian hemorrhagic fever in the rhesus monkey. *Am J Pathol* 73:477–494.

Tesh, R. B., P. B. Jahrling, R. Salas, and R. E. Shope. 1994. Description of Guanarito virus (Arenaviridae: Arenavirus), the etiologic agent of Venezuelan hemorrhagic fever. *Am J Trop Med Hyg* 50:452–459.

Todorović, K., M. Mocić, R. Klašnja, L. Stojković, M. Bordjoški, A. Gligić, and Ž. Stefanović. 1969. An unknown virus disease transmitted from infected green–monkeys to men. *Glas Srpske Akademije Nauka i Umetnosti, Odeljenje Meditsinskikh Nauka* 275:91–101.

Towner, J. S., M. L. Khristova, T. K. Sealy, M. J. Vincent, B. R. Erickson, D. A. Bawiec, A. L. Hartman et al. 2006. Marburgvirus genomics and association with a large hemorrhagic fever outbreak in Angola. *J Virol* 80:6497–6516.

Valmas, C., M. N. Grosch, M. Schumann, J. Olejnik, O. Martinez, S. M. Best, V. Krahling, C. F. Basler, and E. Muhlberger. 2010. Marburg virus evades interferon responses by a mechanism distinct from Ebola virus. *PLoS Pathog* 6:e1000721.

van der Groen, G., W. Jacob, and S. R. Pattyn. 1979. Ebola virus virulence for newborn mice. *J Med Virol* 4:239–240.

Walker, D. H., K. M. Johnson, J. V. Lange, J. J. Gardner, M. P. Kiley, and J. B. McCormick. 1982. Experimental infection of rhesus monkeys with Lassa virus and a closely related arenavirus, Mozambique virus. *J Infect Dis* 146:360–368.

Walker, D. H., H. Wulff, and F. A. Murphy. 1975. Experimental Lassa virus infection in the squirrel monkey. *Am J Pathol* 80:261–278.

Warfield, K. L., D. A. Alves, S. B. Bradfute, D. K. Reed, S. VanTongeren, W. V. Kalina, G. G. Olinger, and S. Bavari. 2007. Development of a model for marburgvirus based on severe-combined immunodeficiency mice. *Virol J* 4:108.

Warfield, K. L., S. B. Bradfute, J. Wells, L. Lofts, M. T. Cooper, D. A. Alves, D. K. Reed, S. A. VanTongeren, C. A. Mech, and S. Bavari. 2009. Development and characterization of a mouse model for Marburg hemorrhagic fever. *J Virol* 83:6404–6415.

Warfield, K. L., D. L. Swenson, G. G. Olinger, W. V. Kalina, M. J. Aman, and S. Bavari. 2007. Ebola virus-like particle-based vaccine protects nonhuman primates against lethal Ebola virus challenge. *J Infect Dis* 196 Suppl 2:S430–S437.

Webb, P. A., G. Justines, and K. M. Johnson. 1975. Infection of wild and laboratory animals with Machupo and Latino viruses. *Bull World Health Organ* 52:493–499.

Winn, W. C. Jr. and D. H. Walker. 1975. The pathology of human Lassa fever. *Bull World Health Organ* 52:535–545.

Zaki, S. R. and C. S. Goldsmith. 1999. Pathologic features of filovirus infections in humans. *Curr Top Microbiol Immunol* 235:97–116.

Zaki, S. R., W. J. Shieh, P. W. Greer, C. S. Goldsmith, T. Ferebee, J. Katshitshi, F. K. Tshioko et al. 1999. A novel immunohistochemical assay for the detection of Ebola virus in skin: implications for diagnosis, spread, and surveillance of Ebola hemorrhagic fever. Commission de Lutte contre les Epidemies a Kikwit. *J Infect Dis* 179(Suppl 1):S36–S47.

7 Role of Rodents and Bats in Human Viral Hemorrhagic Fevers

Victoria Wahl-Jensen, Sheli R. Radoshitzky,
Fabian de Kok-Mercado, Shannon L. Taylor,
Sina Bavari, Peter B. Jahrling, and Jens H. Kuhn

CONTENTS

7.1 RODENTS

The reservoir hosts of those hemorrhagic fever viruses that are not transmitted by arthropods or bats are assigned to the mammalian order Rodentia (rodents), which includes some 2000 different species. Its superfamily Muroidea (hamsters, gerbils true mice, and rats) includes over 1300 species (Wilson and Reeder, 2005). Of the hemorrhagic fever viruses, only those of the family *Arenaviridae* (arenaviruses) and the bunyaviral genus *Hantavirus* (hantaviruses) infect rodents, and all those rodents are muroid.

7.1.1 ARENAVIRUSES

The family *Arenaviridae* includes a single genus, *Arenavirus*, currently comprising 25 recognized species (Charrel et al., 2003; Clegg, 2002; Erickson et al., 2006; Jay et al., 2005; Lecompte et al., 2007;

Oldstone, 2002; Salvato et al., 2011). Based on antigenic properties and sequence phylogeny, arenaviruses have been divided into two distinct groups: The Old World arenaviruses include viruses indigenous to Africa, and the New World arenaviruses include viruses indigenous to the Americas (Bowen et al., 1996; Clegg, 2002; Rowe et al., 1970).

Most arenaviruses are not known to cause human disease, but several arenaviruses have been identified as the etiological agents of viral hemorrhagic fevers (VHFs) with case-fatality rates as high as 30%. Lassa virus (LASV) is an Old World arenavirus that causes Lassa fever (LF) in West Africa. More than 300,000 LASV infections (most of which do not manifest as overt VHFs) are reported in endemic areas per year with several thousand deaths (McCormick and Fisher-Hoch, 2002). Another Old World arenavirus, Lujo virus (LUJV), has been recently isolated from severe cases of undiagnosed viral hemorrhagic fevers in southern Africa (Zambia) (Briese et al., 2009). Machupo (MACV), Guanarito (GTOV), Junín (JUNV), and Sabiá (SABV) viruses are New World arenaviruses that cause Bolivian (BHF), "Venezuelan" ("VHF"), Argentinian (AHFV), and "Brazilian" hemorrhagic fevers, respectively (Buchmeier et al., 2006). Another virus, Chapare virus (CHPV), also causes VHF, but this disease has not yet received a name (Delgado et al., 2008). Among these arenaviruses, JUNV is the most important pathogen causing annual outbreaks in a progressively expanding region in north central Argentina, with almost five million individuals at risk of infection (Enria et al., 2008). There is a remarkable rodent specificity seen among arenaviruses in nature. Field studies strongly support the concept of only a single major reservoir rodent host for each virus (Salazar-Bravo et al., 2002b). Non-reservoir rodents might at times develop chronic infection and viruria, such as has been observed following experimental MACV infection of golden hamsters (*Mesocricetus auratus*) (Johnson et al., 1965). Rodents of the superfamily Muroidea are the natural hosts of arenaviruses (with the possible exception of Tacaribe virus, which might be transmitted by bats, and CHPV, LUJV, and SABV, for which no reservoirs have yet been identified). Old World arenaviruses are found in rodents of the muroid family Muridae, subfamily Murinae (Old World rats and mice), in sub-Saharan Africa, whereas New World arenaviruses are found in rodents of the muroid family Cricetidae, subfamily Sigmodontinae (New World rats and mice), in specialized ecological niches in South and North America. The geographic distribution of each arenavirus is determined by the range of its corresponding rodent host (Clegg, 2002; Salazar-Bravo et al., 2002b). Current evidence suggests a long-term "diffuse coevolution" between the arenaviruses and their rodent hosts. According to this model, a parallel phylogeny between the viruses and their corresponding rodent host(s) allows for host switches between rodents of closely related taxa (Hugot et al., 2001; Salazar-Bravo et al., 2002b). Arenaviruses establish chronic infections in their respective reservoirs accompanied by chronic viremia or viruria without clinical signs of disease (Fulhorst et al., 1999; Johnson et al., 1965; Sabattini et al., 1977; Walker et al., 1975). The chronic carrier state in rodents usually results from exposure to infectious virus early in ontogeny or later in life through aggressive or venereal behavior (Coetzee, 1975; Mills et al., 1992; Webb et al., 1975). Humans become infected through contact with infected rodent reservoirs or inhalation of aerosolized virus from contaminated rodent excreta or secreta and from rodents caught in mechanical harvesters and probably via consumption of rodent meat (Charrel and de Lamballerie, 2003; Maiztegui, 1975; TerMeulen et al., 1996). In fact, AHF, BHF, and "VHF" are typically seasonal diseases and outbreak frequency peaks during the harvest season with the majority of infected cases being male agricultural workers. BHF outbreak frequency peaks during April–July, AHF during the corn-harvesting season (March–June), and "VHF" between November and January. Direct human-to-human transmission, though possible, is probably not the principal mode of disease dissemination.

7.1.1.1 Lassa Virus

The Natal mastomys (*Mastomys natalensis*) is the natural reservoir host of LASV (Lecompte et al., 2006; Monath et al., 1974). The Natal mastomys is widely distributed in sub-Saharan Africa, breeds to high numbers, and is semi-commensal, i.e., lives both in the wild and in and around human dwellings and houses, which the rodents seek out especially during the rainy season (Coetzee, 1975;

Isaacson, 1975; Keenlyside et al., 1983; McCormick et al., 1987; Monath et al., 1974). Rainfall and to lesser extent temperature variability influence the prevalence of LASV (Fichet-Calvet and Rogers, 2009). For instance, the prevalence of LASV is higher during the rainy season than during the dry season (Fichet-Calvet et al., 2007). The reproductive activity of Natal mastomys is also highest during the rainy season (Fichet-Calvet et al., 2008). LASV can be found in Natal mastomys of all age groups, but LASV prevalence increases with age (Fichet-Calvet et al., 2007). Rainfalls could therefore lead to increased breeding of Natal mastomys and thereby increase the likelihood of LF outbreaks (Fichet-Calvet et al., 2008; Sluydts et al., 2007). The absence of rain, on the other hand, could decrease Natal mastomys populations because of lack of food, which could then force the rodents to enter human homes, especially into grain storage areas and kitchens. This would then again increase the risk of LASV transmission to humans. The Natal mastomys is the most common rodent found in the sub-Saharan region. Curiously, LF cases seem to be restricted to focal areas, however (Demby et al., 2001). A survey to examine the distribution of LASV, conducted in households and bush sited across the savannah, forest, urban, and coastal regions of Guinea, found a LASV prevalence in Natal mastomys ranging from 0% to 9% per region examined (Demby et al., 2001). The distribution and prevalence of LASV-infected rodents did not appear to localize to anyone particular region but rather resembled focal spots by clustering in houses (Demby et al., 2001). A recent evolutionary sequence analysis of LASV suggests that the virus has appeared between 750 and 900 years ago in Nigeria but only recently spread to western Africa (150–250 years). The study identified a close relationship between civil war–related mass movements of refugees into new areas that were subsequently environmentally changed (deforestation), a decline in Natal mastomys populations in these areas due to these environmental disruptions, and an increase in LF cases due to increased human contact with rodents (including consumption) (Lalis et al., 2012).

7.1.1.2 Machupo Virus

The big laucha (*Calomys callosus*) is the principal host for MACV. The virus was recovered repeatedly from this small pastoral and peridomestic mouse (Johnson et al., 1966). Little information has been published about the ecology or natural history of this rodent. Although its exact geographic range has not been determined, available evidence indicates that the big laucha is distributed through the grasslands and along the forest edges from San Joaquín (lowlands and open biomes of eastern Bolivia) south to northern Argentina, in the northern portion of Paraguay, as well as in the continuous western fringe of Mato Grosso State in Brazil (Olds, 1988). It is preadapted for peridomestic living; it readily invades houses and gardens, where it lives in much the same manner as the house mouse (*Mus musculus*), and reaches population densities under these circumstances that are never observed in the absence of man (Mercado, 1975). During the 1960s, when the first known BHF outbreaks occurred, human settlements were almost invariably located in either of two types of ecological settings in the BHF epidemic region. The first included port villages, located on elevated riverbanks in gallery forest, where domestic rodents, if present, included usually roof rats (*Rattus rattus*) and house mice, with small numbers of rice rats (*Oryzomys bicolor*). Lauchas were not found in this ecological setting, presumably for lack of an avenue of suitable habitat between neighboring grasslands and the houses surrounded by forest. No human cases of BHF have been reported from port villages. The second ecological setting involved elevated sites between river systems, on which the richest agricultural developments were located in clearings in the climax forests. Such elevated areas were known locally as "Alturas," and their elevation was sufficient to escape inundation during all except the most severe flood conditions. Farmhouses were usually located at the edge of the forest overlooking the grass-covered marshlands or "savannas." Lauchas usually infested such farm villages. This was the ecological setting characteristic of all villages and isolated settlements from which human cases of BHF have been reported (Kuns, 1965). MACV induces a viremic immunotolerant infection in suckling lauchas and a split response in animals more than 9 days of age (Justines and Johnson, 1969; Webb et al., 1975). The "immunocompetent" response of 50% of the mice is characterized by clearance of viremia, minimal or absent viruria, and presence of circulating neutralizing antibodies. The other "immunotolerant" mice develop persistent viremia, viruria, little or no neutralizing antibodies, anemia, and splenomegaly. MACV antigen can be detected in most tissues of these animals, including the reproductive organs (Justines and Johnson, 1969; Webb et al., 1975), and virus can be isolated from blood, spleen, and kidneys (Johnson et al., 1965; Kuns, 1965). The long-term effects of tolerant infection include mild runting, reduced survival rate, and almost total sterility among females, largely caused by virus infection of embryos. Selective breeding experiments in lauchas demonstrated that a complex polygenic inheritance accounts for the split response following MACV infection, suggesting a host genetic component as a determinant (Justines and Johnson, 1969; Webb et al., 1975). In these experiments, the infection of newly born lauchas could occur neonatally through the milk, and adult mice were infected through sexual transmission of MACV, suggesting that horizontal transmission through venereal encounters might be an important natural mechanism for virus maintenance (Webb et al., 1975). These studies also predict a model for MACV maintenance in its reservoir: virus infection would be more common in larger wild colonies of lauchas where increasing venereal transmission occurs, and infected colonies would eventually pass through a phase of reduced population with near complete, tolerant infection as young infected females are rendered sterile.

7.1.1.3 Junín Virus

The drylands laucha (*Calomys musculinus*), a wild rodent that inhabits crop (corn, wheat, and soybeans) fields, pastures, and stable linear habitats (adjacent roadsides and fence lines), is the reservoir of JUNV (Sabattini and Gonzalez, 1967; Sabattini et al., 1977). It is rarely captured in or around houses. The populations of these rodents reach maximum densities in autumn in Argentina, coinciding with the harvest of the principal summer crops. Furthermore, the patchy spatial distribution of these rodents has been suggested to account for the focal distribution of AHF. The transmission to humans is believed to occur predominantly by inhaling aerosolized viral particles from contaminated soil and plant litter, which are disturbed during the mechanized harvesting process (Carballal et al., 1988; Maiztegui, 1975). Both horizontal and vertical transmissions have been reported as possible maintenance mechanisms of JUNV (Sabattini et al., 1977). Drylands lauchas infected at birth with JUNV exhibit decreased survival, body growth, and fertility, whereas animals that are inoculated with the virus as adults are usually asymptomatic and do not show altered body weight, reproduction, and survival (Vitullo et al., 1987; Vitullo and Marani, 1987; Vitullo and Merani, 1990). Furthermore, 50% of drylands lauchas infected as adults (90–120 days) develop persistent infections with JUNV isolated from urine, saliva, blood, or brain (and infection observed in brain, kidney, and spleen). The others develop serum antibodies and appear to clear the virus within 21 days after infection. The virus cannot transgress the placenta and reach the embryos (Vitullo and Merani, 1990). This suggests that vertical transmission might contribute, to some extent, to the maintenance of infection. In terms of a natural population of drylands lauchas, it may be assumed that animals vertically infected (during lactation), if unable to transfer the infection satisfactorily to the next generation, contribute toward intra-generation infection by horizontal transmission. Horizontally infected adults may secure the intergeneration transmission by both vertical and horizontal means. Under these circumstances, JUNV maintenance may arise from an equilibrium between both modes of transmission, with the horizontal route representing the main route resulting in viral persistence in nature and vertical transmission being an added option for intergeneration transfer that may support the infection when population numbers are reduced and horizontal transmission is precluded (Sabattini et al., 1977; Vitullo and Merani, 1988; Vitullo et al., 1987). A 30 month field study in the epidemic area of AHF estimated the total prevalence of JUNV infection to be 10.9% in drylands lauchas. Serum antibody and viral antigen were detected in blood and saliva of these rodents. JUNV-infected animals were predominantly males in the older age and heavier body mass classes. Seropositive males were twice as likely to have body scars as the overall population. JUNV-infected animals were also strongly associated with the relatively rare roadside and fence-line habitats (Mills et al., 1991, 1992, 1994). These observations implicate horizontal transmission as the primary mode of infection in drylands laucha populations and suggest that aggressive encounters among adult, male lauchas in relatively densely populated roadside and fence-line habitats are an important mechanism of transmission of JUNV within its reservoir population. The high rate of virus production in salivary glands of drylands lauchas, as well as virus isolation from saliva of infected reservoirs (Peralta et al., 1979; Sabattini et al., 1977), makes JUNV transmission following a bite highly likely. Furthermore, laboratory studies with drylands lauchas showed that the transmission of JUNV was generally horizontal, taking place between rodents in close contact with each other (Sabattini et al., 1977). There are some differences in the maintenance mechanism of JUNV and MACV. First, horizontal venereal transmission does not seem to be predominant in JUNV transmission, as it would not account for the greater prevalence of infection in male drylands lauchas. Second, while viral infection is hypothesized to be an important driving force in reservoir population dynamics in the MACV big laucha model, JUNV infection in drylands lauchas should have a much less severe effect. This is because in contrast to MACV-infected female mice, which abort (Webb et al., 1975), chronically JUNV-infected rodent females, when infected as adults, have a normal number of pups (Vitullo and Merani, 1990). Finally, all pups born to MACV viremic female mice in the laboratory are infected neonatally through the milk (Webb et al., 1975), whereas JUNV-infected female rodents transmit the virus to only half of their pups (Vitullo and Merani, 1990; Vitullo et al., 1987).

7.1.1.4 Guanarito Virus

Experimental work identified the nocturnal short-tailed zygodont (*Zygodontomys brevicauda*) as the reservoir of GTOV. Alston's cotton rats (*Sigmodon alstoni*) were indicated as a secondary host, and fulvous colilargos, Guaira spiny rats, and roof rats (*Oligoryzomys fulvescens, Proechimys guairae* and *R. rattus*, respectively) were found to be seropositive (Tesh et al., 1993). GTOV could be isolated from throat swabs, urine, spleens, lungs, or kidneys of infected animals, and antibodies were found in the sera (Milazzo et al., 2011). Short-tailed zygodonts are native to the plains of western Venezuela and can reach high densities in tall grassy (weedy) areas found in pastoral and agricultural areas along roadsides and fence lines and in the naturally occurring savannah that dominates the landscape of the "VHF" endemic region (Fulhorst et al., 1997; Fulhorst et al., 1999; Salazar-Bravo et al., 2002b; Tesh et al., 1993). The presence of GTOV-infected short-tailed zygodonts in Apure, Barinas, Cojedes, and Guárico indicates that GTOV was enzootic in Venezuela's Portuguesa State long before GTOV was discovered in 1989. As such, the emergence of "VHF" was likely a consequence of demographic and/or ecological changes in rural areas of Portuguesa State that eventually resulted in a significant increase in the frequency of contact between humans and GTOV-infected rodents (Fulhorst et al., 2008). However, during four years of rodent trapping in the region of "VHF," neither short-tailed zygodonts nor Alston's cotton rats were ever found within houses or farm buildings (de Manzione et al., 1998). Presumably, human infection therefore occurs outdoors. Thus, one might expect persons having frequent contact with rodent-infested grassland habitats to be at higher risk of contracting "VHF." The laboratory infection of short-tailed zygodonts with GTOV produces chronic viremia characterized by persistent shedding of infectious virus in oral and respiratory secretions and urine without clinical signs of disease through day 208 postinoculation (Fulhorst et al., 1999). The analyses of field and laboratory data suggest that horizontal transmission is the dominant mode of GTOV transmission in short-tailed zygodonts, as most GTOV infections in these mice are acquired in an age-dependent manner. Therefore, the chronic carrier state in short-tailed zygodonts most likely results from exposure to infectious virus later in life through aggressive or venereal behavior such as allogrooming, mating, intraspecies aggression, and other activities that entail close physical contact. Evidence also suggests that male and female mice contribute equally to GTOV transmission (Milazzo et al., 2011).

7.1.1.5 Conclusion

LF, BHF, AHF, and "VHF" are all examples of natural nidality; zoonoses within the host reservoir occur focally and have an incomplete pattern of overlap with the host species range. Natal mastomys, big lauchas, drylands lauchas, and short-tailed zygodonts, the hosts of LASV, MACV, JUNV, and GTOV, respectively, are found in larger distribution areas than the endemic areas of BHF, AHF, and "VHF" (Demby et al., 2001; Fulhorst et al., 1997; Mills et al., 1992; Weaver et al., 2000). At least in the case of BHF, it has been demonstrated that the population of big lauchas responsible for the maintenance and transmission of MACV represent an independent monophyletic lineage, different from that in other areas of South America (Salazar-Bravo et al., 2002a), which could explain the phenomenon of natural nidality for BHF. Interestingly, although drylands lauchas are found in most of central and northwestern Argentina, a gradient of infection in these rodents has been described in JUNV surveys across the boundaries of AHF endemic–epidemic regions. The prevalence of JUNV infection in drylands lauchas is highest in endemic regions and is reported to be nonexistent or low outside the endemic zone. Nonetheless, JUNV has been isolated from rodents in areas where human cases have not been reported (deVillafañe et al., 1977; Garcia et al., 1996; Mills et al., 1991). Similarly, some GTOV variants were isolated from locations outside of the endemic–epidemic regions (outlying locations in Cojedes, Barinas, and Apure States of Venezuela) and yet were found to belong to genotypes that included variants isolated from human cases of "VHF" from areas surrounding Guanarito. This suggests that pathogenic GTOV variants occur in these outlying areas, but do not frequently infect people and/or cause inapparent disease there. Furthermore, "VHF" does not appear to be associated with a specific GTOV genotype that is restricted to a particular rodent host (Weaver et al., 2000). Most of the rodents associated with arenaviruses, such as those assigned to the genera Mus, Mastomys, and Calomys, are found in grassland/brush habitats and frequently come in contact with human dwellings and therefore humans. Contact opportunities are increased when rodents infest field crops or invade storage areas for grains. The association of rodents with agricultural practices often results in cyclic reproductive patterns linked to the harvesting of crops, and the invasion of barns and other storage areas increases when rodent food opportunities decrease.

7.1.2 Hantaviruses

Hantaviruses represent a diverse group of viruses within a separate genus (*Hantavirus*) of the family *Bunyaviridae*, each carried by a specific rodent, bat or eulipotyphlan host (Heyman et al., 2011; Klempa et al., 2003; Okumura et al., 2007; Pensiero et al., 1992; Vapalahti et al., 1996). Rodent-borne hantaviruses are associated with two disease syndromes with varying degrees of severity: hemorrhagic fever with renal syndrome (HFRS) or hantavirus pulmonary syndrome (HPS). Of the two syndromes, only HFRS is considered a viral hemorrhagic fever and will be discussed here. Human infections with HFRS-causing hantaviruses are exclusively zoonotic, with transmission occurring through contact with or inhalation of excreted or secreted virus from rodents. Vascular leakage, acute thrombocytopenia, and kidney dysfunction are associated with HFRS (Lee et al., 1999; Zaki and Nolte, 1999). Those viruses that cause HFRS in humans are listed in Table 7.1. Four hantaviruses cause the majority of cases of HFRS: Hantaan virus (HTNV), Seoul virus (SEOV), Dobrava–Belgrade virus (DOBV), and Puumala virus (PUUV). The most prevalent and lethal HFRS-associated hantavirus is HTNV (>100,000 cases/year, mostly in Asia) with a case-fatality rate of 10%–15%. The host reservoirs for HFRS-causing viruses are assigned to the muroid family Muridae, subfamily Murinae, and the muroid family Cricetidae, subfamily Arvicolinae (lemmings, voles, and muskrats). Assigned to the subfamily Murinae are the reservoir hosts for DOBV, HTNV, and SEOV. PUUV is the only HFRS-causing virus harbored by an arvicoline rodent. There are no vaccines or specific antiviral drugs licensed by the U.S. Food and Drug Administration to treat or prevent HFRS, but vaccines of varying quality and efficacy are available in countries other than the United States. For instance, inactivated virus vaccines against HFRS are licensed for use in China and South Korea, which may account for the reduced incidence of HFRS in these countries in the past 10 years.

TABLE 7.1

Hantaviruses Causing HFRS

Virus (Abbreviation)	Case-Fatality Rate (%)	Rodent (Species)	Geographic Location
Puumala virus (PUUV)	0.1–0.4	Bank vole (*Myodes glareolus*)	Eurasia
Dobrava–Belgrade virus (DOBV) genotype Sochi	>6	Caucasus field mouse (*Apodemus ponticus*)	Russia
Dobrava–Belgrade virus (DOBV) genotype Dobrava	10–12	Yellow-necked field mouse (*Apodemus flavicollis*)	Slovenia, Croatia, Serbia, Montenegro, Hungary, Slovakia, Bulgaria, Greece
Dobrava–Belgrade virus (DOBV) genotype Saaremaa	?	Striped field mouse (*Apodemus agrarius*)	Estonia, Slovakia, Slovenia, Hungary, Denmark
Dobrava–Belgrade virus (DOBV) genotype Kurkino	0.3–0.9	Striped field mouse (*A. agrarius*)	Germany, Slovakia, Russia, Hungary, Slovenia, Croatia, Denmark, mainland Estonia
Tula virus (TULV)	Not known	Common vole (*Microtus arvalis*), field vole (*Microtus agrestis*), East European vole (*Microtus levis*)	Europe
Amur/Soochang virus (ASV)	?	Korean field mouse (*Apodemus peninsulae*)	Far Eastern Russia
Hantaan virus (HTNV)	≥10	Striped field mouse (*A. agrarius*)	Asia, Eastern Russia
Seoul virus (SEOV)	<1	Brown rat (*Rattus norvegicus*), roof rat (*R. rattus*)	Worldwide

7.1.2.1 Brief History of Hemorrhagic Fever with Renal Syndrome

Although described in the Soviet and Japanese literature since the late 1930s, the western world became aware of HFRS for the first time during an outbreak of what was then referred to as Korean hemorrhagic fever (KHF) that began in 1951 during the United States–Korean War. KHF had affected nearly 3000 United Nations troops by 1954 and exhibited a case-fatality rate of 7% (Johnson, 2001). Despite extensive investigations and reports on this outbreak, the etiologic agent was not identified at that time. Indeed, the mystery agent for KHF remained elusive until HTNV, named for the river that runs near the border of today's North and South Korea, was isolated in 1978 by Lee Ho-wang and colleagues (Lee et al., 1978). Shortly following the discovery of HTNV, other hantaviruses were identified in Eurasia (Brummer-Korvenkontio et al., 1980; Brummer-Korvenkontio et al., 1982; Gresikova et al., 1984).

7.1.2.2 Epidemiology

In general, the infection of the natural hantavirus reservoir hosts results in a chronic carrier state without pronounced pathology or signs of disease. However, in-depth histological studies identified lesions within the lungs of North American deer mice (*Peromyscus maniculatus*) infected with HPS-causing Sin Nombre virus (SNV) and white-footed deer mice (*Peromyscus leucopus*) infected with HPS-causing New York virus (NYV) (Lyubsky et al., 1996; Netski et al., 1999). There is also a report that PUUV-infected animals are less likely to survive winter, suggesting infection has a negative effect on host fitness (Kallio et al., 2007). Disease in humans occurs when persons are exposed to contaminated rodent feces, urine, or saliva. The most common mode of transmission is thought to be the inhalation of aerosolized rodent droppings; however, contact with open wounds, rodent bites, and ingestion of contaminated material are also possible modes of transmission. The ingestion of virus as a mode of infection is not well documented; however, laboratory hamsters can be readily infected through the gut (gavage needle) with the HPS-associated Andes virus (ANDV), supporting the possibility that hantaviruses could be transmitted by the ingestion of contaminated food or the consumption of rodents (Hooper et al., 2008). In humans, hantaviruses cause disease in the young and old, male and female. In most studies, HFRS occurred predominantly in working-age males. The preponderance of the disease in this population is likely related to occupational exposure. The epidemiologic studies of HFRS report increased incidence of hantavirus disease in persons working or sleeping in environments inhabited by rodents, which include agricultural workers, forest workers, and soldiers (Abu Sin et al., 2007; Mulic and Ropac, 2002; Sinclair et al., 2007; Vapalahti et al., 2003). In many regions, hantavirus disease has a seasonal peak. For example, most of the cases in the 2005 outbreaks in Europe occurred in June and July (Heyman et al., 2007). The geographic range of HFRS is shown in Figure 7.1 and compared to the geographic range of the rodent hosts for HFRS-causing hantaviruses. Although disease has not been detected in all of these regions, the presence of the rodent reservoirs indicates that the potential for hantavirus disease exists. HFRS has been documented in China, the Korean peninsula, Russia, Northern Europe/Scandinavia, and Southern Europe/Balkans. Approximately 40 countries have reported hantavirus disease, the presence of virus, or the serological evidence of infection, whereas several other countries have reported rare and sporadic HFRS in port cities that can probably be attributed to SEOV infections spread by rats carried port to port on ships.

7.1.2.3 China, Korea, and Far Eastern Russia

HFRS is a significant public health concern in China and has been a notifiable disease there since 1950. During the period of 1950–2007, a total of 1,557,622 cases of HFRS in humans and 46,427 deaths (3%) were reported in China (Fang et al., 2006; Kariwa et al., 2007; Yan et al., 2007; Zhang et al., 2010). Most of the severe cases that occur in rural areas are caused by HTNV, whereas SEOV is the major cause of a less severe form of HFRS carried by anthropophilic urban species of rodents. Although HFRS cases have been found in 29 Chinese provinces, the disease remains most prevalent in Shāndōng, Hēilóngjiāng, Jílín, Liáoníng, Héběi, Jiāngsū, Zhèjiāng, Ānhuī, Hénán, Jiāngxī, Húběi, Húnán, Shǎnxī, Sìchuān, and Guìzhōu provinces (Zhang et al., 2010).

South Korea reported 3039 cases of HFRS between 1997 and 2006 (DisWeb, 2003). During that time, there has been a trend of increasing numbers of cases with more than 400 cases per year for the last 3 years. Based on its geographic location, it is very likely that North Korea has a significant number of HFRS cases; however, as is the case for many countries, the number of HFRS cases is not readily available. One review from 1996 reported 316 HFRS cases in North Korea since 1961 (Lee, 1996).

In Far Eastern Russia, there were 4442 cases of HFRS between 1978 and 1997 (Tkachenko et al., 1998). ASV, HTNV, and SEOV are causative agents of HFRS in Far Eastern Russia. Several other countries in the area, including Australia, Fiji, Hong Kong, India, Indonesia, Japan, Malaysia, Mongolia, Myanmar, Singapore, Sri Lanka, Taiwan, Thailand, and Vietnam, have reported rare or sporadic cases of HFRS or sero-epidemiological evidence that hantaviruses exist and can cause infections (reviewed in Kariwa et al., 2007). The major viruses that cause HFRS in the Far East are HTNV and SEOV.

7.1.2.4 Western Russia and Eastern Europe

HFRS has been a reportable disease in Russia since 1978. In a review of HFRS in Russia, E. Tkachenko reports that between 1978 and 1997, there were 109,082 cases in western Russia (Tkachenko et al., 1998). Specific regions of Russia have reported relatively large outbreaks. The Bashkiria region in particular has regularly reported high numbers of cases, including epidemics in 1993 and 1997, where the numbers of cases approached 150 and 287 cases/1,000,000 population, respectively (Niklasson et al., 1993; Tkachenko et al., 1998). In 2007, there were outbreaks of greater than 3000 cases in the vicinity of Voronezh and Lipetsk (Dybas, 2007). As in China and Korea, most of the HFRS cases in Russia occur in rural areas; however, there can be epidemics in urban areas, as was the situation in the 1997 Bashkiria outbreak (Tkachenko et al., 1998). In western Russia, most of the HFRS is caused by PUUV, carried by bank voles (*Myodes glareolus*). Case-fatality rates indicate

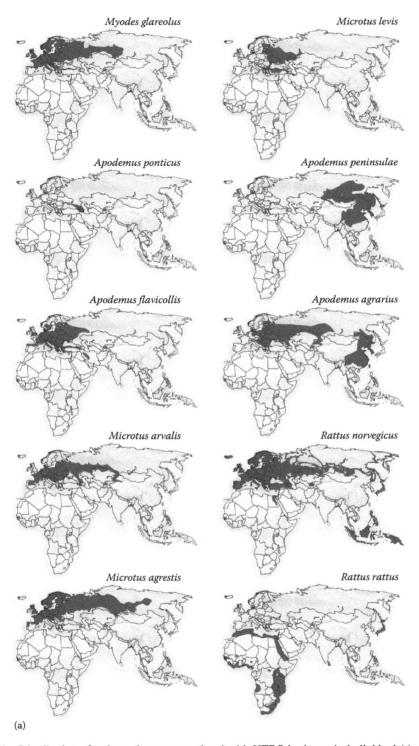

(a)

FIGURE 7.1 Distribution of rodents that are associated with HFRS is shown in individual (a) as well as a single composite map (b). All data used to generate maps were obtained from IUCN 2012. (From IUCN, The IUCN red list of threatened species, version 2012.1, 2012, http://www.iucnredlist.org.)

(continued)

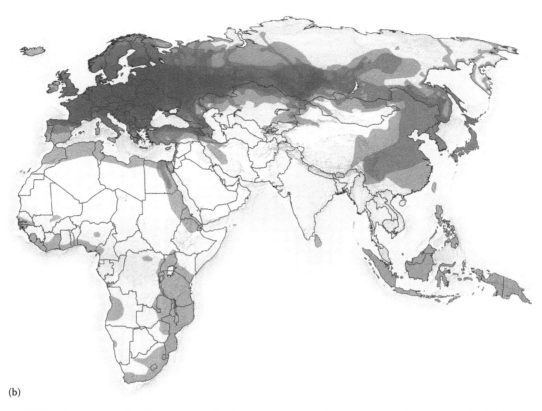

(b)

FIGURE 7.1 (continued) Distribution of rodents that are associated with HFRS is shown in individual (a) as well as a single composite map (b). All data used to generate maps were obtained from IUCN 2012. (From IUCN, The IUCN red list of threatened species, version 2012.1, 2012, http://www.iucnredlist.org.)

that strains of PUUV in Russia, such as strain K27 isolated from a fatal human case, cause a more severe form of disease (0.4% case fatality) than the strains of PUUV found in Finland and Sweden (0.1% case fatality) (Tkachenko et al., 1998). Several other countries in Eastern Europe, including Belorus, Estonia, Georgia, Latvia, Lithuania, Poland, Romania, and Ukraine, have reported rare or sporadic cases of HFRS or serologic evidence that pathogenic hantaviruses are endemic (reviewed in Avsic-Zupanc, 1998; Tkachenko et al., 1999; Vapalahti et al., 2003).

7.1.2.5 Northern and Central Europe/Scandinavia

A review published in 2003 indicated that 12 countries in Europe reported five or more cases of HFRS per year (Vapalahti et al., 2003). Other countries in Europe with rare or sporadic cases of HFRS or seroepidemiologic evidence of infection include Austria, Czech Republic, Great Britain, Portugal, and Switzerland (Lee, 1996; Vapalahti et al., 2003). It is likely that HFRS in Europe is underdiagnosed because the disease is relatively mild as compared to HFRS in Asia. A study in Belgium reported that the seroprevalence was at 3.8%, well above the number of diagnosed cases (van der Groen et al., 1983). Finland and Sweden have the most cases (1000 or more/year, respectively) followed by Germany, France, and Belgium each with ≈100 cases/year. These national numbers mask the health problem HFRS causes in specific geographic regions. For example, the number of cases in northern Sweden is much higher than in the south. Most of the cases are in Vasterbotten County where the incidence is normally ≈23.5 cases/100,000 (Settergren et al., 1988). In Finland, the number of cases in the province of Mikkeli was 70 cases/100,000, mostly in farmers (Vapalahti et al., 1999). PUUV and DOBV are believed to cause most of the cases of HFRS in Northern Europe and Scandinavia.

Recently, HFRS has become more of a health concern in Europe. In 2005, there was a relatively large series of outbreaks in Belgium, France, the Netherlands, Luxembourg, and Germany (1114 cases), caused by PUUV (Heyman et al., 2007). In Germany, many of the cases occurred in relatively large cities, including Osnaburg and Cologne, where the annual incidence was 8.5 and 4.2 cases/100,000, respectively (AbuSin et al., 2007). In 2007, Sweden experienced an almost 10-fold increase in the annual number of HFRS cases (Pettersson et al., 2008). Vasterbotten County accounted for 800 of the 2200 cases, with two fatalities. In Europe and elsewhere, there is a suspected link between climate change, food-source production, reservoir population, and incidence of human disease. Climate conditions that result in increased production of food source (e.g., mast years resulting in high yields of oak and beechnuts) can, in turn, result in an increase in reservoir population and an increase in the incidence of human disease (Heyman et al., 2007; Vapalahti et al., 2003). Although HFRS in Europe has not resulted in high numbers of fatalities, this disease does have a substantial impact on the health-care system because many patients are hospitalized and some require hemodialysis.

7.1.2.6 Southern Europe/Balkans

There are approximately 100 HFRS cases per year reported in the Balkans (Avsic-Zupanc, 1998), including Albania, Bosnia, Bulgaria, Croatia, Greece, Macedonia, Montenegro, Serbia, and Slovenia. The military conflicts in the area of former Yugoslavia resulted in increased numbers of HFRS cases, as would be expected based on the increased numbers of military personnel and displaced civilians serving, working, and sleeping outdoors (Avsic-Zupanc, 1998). Severe cases of HFRS in the Balkans are usually caused by DOBV (Avsic-Zupanc et al., 1992). PUUV is also circulating in this region and has been associated with disease (Lundkvist et al., 1997).

7.1.2.7 Prospects for the Future

As climatic changes affect the environment, we can expect to see changes in rodent populations and distributions. It is likely that these changes will alter the ways in which humans come in contact with hantaviruses and these changes could result in changes in the rates and distribution of HFRS cases. Moreover, changes in interactions between rodents of different species could, theoretically, increase the possibility that different hantaviruses might coinfect the same rodent host and produce reassortant progeny viruses. There is an evidence that hantaviruses have reassorted in nature (Henderson et al., 1995; Li et al., 1995). Reassortant viruses could exhibit altered biological properties including virulence, as is the case for influenza viruses (genetic shift). Thus, the threat posed by naturally occurring hantavirus disease is dynamic and must be carefully monitored.

7.2 BATS

The mammalian order Chiroptera includes over 1000 established bat species (Nowak and Walker, 1994; Wilson and Reeder, 2005). One arenavirus (Downs et al., 1963) and several hantaviruses (Jung and Kim, 1995; Sumibcay et al., 2012; Weiss et al., 2012) are associated with bats in nature, but none of these agents have been identified as human pathogens. Filoviruses (order *Mononegavirales*, family *Filoviridae*) are the only VHF-causing pathogens for which a bat association has been plausibly documented.

7.2.1 FILOVIRUSES

Filoviruses are the etiological agents of geographically isolated severe VHFs among human and other ape populations (Kuhn, 2008). The most current filovirus classification distinguishes three separate genera based on molecular and genomic properties. Two viruses, Marburg virus (MARV) and Ravn virus (RAVV), are assigned to the genus *Marburg virus*; five viruses, Bundibugyo virus (BDBV), Ebola virus (EBOV), Reston virus (RESTV), Sudan virus (SUDV), and Taï Forest virus (TAFV), belong to the genus *Ebolavirus*. Finally, Lloviu virus (LLOV) is the sole member of the genus *Cuevavirus* (Adams and Carstens, 2012; Kuhn et al., 2010, 2011). With the exception of RESTV and LLOV, all filoviruses have been identified as the causes of human VHFs.

Natural disease outbreaks among humans and animals indicate that filoviruses are endemic in equatorial Africa (MARV, RAVV, BDBV, EBOV, SUDV, TAFV), the Philippines (RESTV), and southern Europe (LLOV) (Kuhn, 2008; Negredo et al., 2011). Numerous surveys of collected human and animal sera suggest that filoviruses are also endemic in regions from where disease outbreaks have not yet been reported, but the results of many of these studies are considered controversial (Kuhn, 2008). However, ecological niche studies support the idea that the natural filovirus distribution is broader than the areas of recorded disease outbreaks (Peterson et al., 2004, 2006). Interestingly, these studies also imply that marburgviruses and ebolaviruses circulate in different ecological zones. Marburgviruses have thus far only been detected in arid woodlands, whereas ebolaviruses seem to be endemic in humid rain forests. Furthermore, marburgvirus disease outbreaks were often associated with people visiting or working in caves, whereas such a connection has not been made in the case of ebolavirus disease outbreaks (Peterson et al., 2004, 2006).

The limited number and temporal separation of filovirus disease outbreaks and the rapidly fatal disease these viruses cause in primates (Kuhn, 2008) indicate that they are maintained in nature in organisms other than primates over long periods of time. Since all other viruses known to cause VHFs in humans are known to be arthropod-borne (arboviruses) or rodent-borne, it was quickly hypothesized that this would be the case for filoviruses as well. However, the natural filovirus reservoirs proved elusive despite numerous animal sampling studies (Kuhn, 2008).

In 1996, Swanepoel et al. reported a study during which several types of plants and animals were inoculated with EBOV to evaluate whether they could support virus replication and therefore aid in the identification of the filovirus reservoir hosts. Surprisingly, sustained EBOV replication was detected in the absence of clinical signs in individual wild-caught little free-tailed bats (*Chaerephon pumilus*), Angolan free-tailed bats (*Mops condylurus*), and Wahlberg's epauletted fruit bats (*Epomophorus wahlbergi*) after subcutaneous infection of $10^{4.6}$ ffu EBOV. EBOV antigen could be detected in lung endothelial cells of one insectivorous bat. More importantly, EBOV could be recovered from the feces of a Wahlberg's epauletted fruit bat on day 21 post-infection, and viral titers of $10^{4.6}$–$10^{7.0}$ ffu/ml and $10^{2.0}$–$10^{6.5}$ ffu/ml were detected in the sera and pooled viscera of several of these bats, respectively. EBOV isolation was successful up to day 12 (Angolan free-tailed bats), day 14 (little free-tailed bats), and day 21 post-infection (Wahlberg's epauletted fruit bats), and the study was terminated soon thereafter (Swanepoel et al., 1996). Bats had been collected before these experiments in areas affected by ebolavirus disease outbreaks, but filoviruses were never isolated or otherwise detected (Table 7.2). However, as this was the first study that experimentally proved that sustained replication of filoviruses is possible in animals in the absence of clinical signs, bats became the prime suspects for harboring filoviruses in nature.

Approximately 20% of all mammalian species have bats as their members (Teeling et al., 2005). Traditionally, bats (order Chiroptera) have been divided into two major clades, the megabats (suborder Megachiroptera) and the microbats (suborder Microchiroptera). Megabats, often also referred to as fruits bats or flying foxes, are typically frugivorous or nectarivorous, often quite large, and do not use echolocation, whereas microbats are typically insectivorous, generally smaller, and use echolocation for orientation (Simmons, 2005). This division and the monophyly of individual suprageneric taxa are, however, currently hotly debated (Agnarsson et al., 2011; Simmons, 2005). Bats have long been known to carry viruses, including important human pathogens, such as rabies virus (Calisher et al., 2006, 2008; Laminger and Prinz, 2010; Messenger et al., 2003; Wang et al., 2011; Wong et al., 2007). However, they had not been associated with the transmission of VHFs.

A study performed around Kikwit in Zaire (today Democratic Republic of the Congo), which in 1995 was the epicenter of one of the largest human ebolavirus disease outbreaks thus far recorded (317 cases, 245 deaths), revealed the presence of bats belonging to at least 18 different species in that area alone (van Cakenberghe et al., 1999). This exemplifies the vast diversity of bats in general and implies that it will not be easy to pinpoint particular bats as filovirus hosts.

TABLE 7.2

Bats Screened for Filoviruses with Negative Results

Bat Species (Vernacular Name)	Type of Bat	Sampling Location (Year)	Number Screened	Assay	Reference
Casinycteris argynnis (golden short-palated fruit bat)	Frugivorous (megachiropteran pteropodid)	~Kikwit, Zaire (1995)	2	Filovirus isolation, ELISA (EBOV)	Leirs et al. (1999)
C. argynnis (golden short-palated fruit bat)	Frugivorous (megachiropteran pteropodid)	Gabon and COG (2005–2006)	2	qRT-PCR (MARV), nested PCR (MARV), ELISA (MARV)	Towner et al. (2007)
Chaerephon ansorgei (Ansorge's free-tailed bat)	Insectivorous (microchiropteran molossid)	~ Kikwit, Zaire (1995)	121	Filovirus isolation, ELISA (EBOV)	Leirs et al. (1999)
Chaerephon major (lapped-eared free-tailed bat)	Insectivorous (microchiropteran molossid)	~ Tandala, Zaire (1976)	26	Filovirus isolation, IFA (EBOV)	Breman et al. (1999)
Chaerephon pumilus (little free-tailed bat)	Insectivorous (microchiropteran molossid)	~ Kikwit, Zaire (1995)	213	Filovirus isolation, ELISA (EBOV)	Leirs et al. (1999)
Chalinolobus sp. (wattled bat)	Insectivorous (microchiropteran vespertilionid)	~ Tandala, Zaire (1976)	15	Filovirus isolation, IFA (EBOV)	Breman et al. (1999)
Coleura afra (African sheath-tailed bat)	Insectivorous (microchiropteran emballonurid)	Gabon (2009–2010)	31	qRT-PCR (MARV), nested PCR (MARV)	Maganga et al. (2011)
Eidolon helvum (African straw-colored fruit bat)	Frugivorous (megachiropteran pteropodid)	~ Tandala, Zaire (1976)	6	Filovirus isolation, IFA (EBOV)	Breman et al. (1999)
E. helvum (African straw-colored fruit bat)	Frugivorous (megachiropteran pteropodid)	Gabon and COG (2005–2006)	35	qRT-PCR (MARV), nested PCR (MARV), ELISA (MARV)	Towner et al. (2007)
E. franqueti (Franquet's epauletted fruit bat)	Frugivorous (megachiropteran pteropodid)	~ Tandala, Zaire (1976)	21	Filovirus isolation, IFA (EBOV)	Breman et al. (1999)
E. franqueti (Franquet's epauletted fruit bat)	Frugivorous (megachiropteran pteropodid)	~ Kikwit, Zaire (1995)	2	Filovirus isolation, ELISA (EBOV)	Leirs et al. (1999)
E. franqueti (Franquet's epauletted fruit bat)	Frugivorous (megachiropteran pteropodid)	Gabon and COG (2005–2006)	296	qRT-PCR (MARV), nested PCR (MARV), ELISA (MARV)	Towner et al. (2007)
Eptesicus sp. (house bat)	Insectivorous (microchiropteran vespertilionid)	~ Tandala, Zaire (1976)	22	Filovirus isolation, IFA (EBOV)	Breman et al. (1999)
Hipposideros caffer (Sundevall's leaf-nosed bat)	Insectivorous (microchiropteran hipposiderid)	Gabon (2009–2010)	521	qRT-PCR (MARV), nested PCR (MARV)	Maganga et al. (2011)

(continued)

TABLE 7.2 (continued)
Bats Screened for Filoviruses with Negative Results

Bat Species (Vernacular Name)	Type of Bat	Sampling Location (Year)	Number Screened	Assay	Reference
Hipposideros cyclops (cyclops leaf-nosed bat)	Insectivorous (microchiropteran hipposiderid)	~ Tandala, Zaire (1976)	52	Filovirus isolation, IFA (EBOV)	Breman et al. (1999)
H. gigas (giant leaf-nosed bat)	Frugivorous (megachiropteran pteropodid)	Gabon and COG (2005–2006)	1	qRT-PCR (MARV), nested PCR (MARV), ELISA (MARV)	Towner et al. (2007)
H. gigas (giant leaf-nosed bat)	Frugivorous (megachiropteran pteropodid)	Gabon (2009–2010)	233	qRT-PCR (MARV), nested PCR (MARV)	Maganga et al. (2011)
H. ruber (Noack's leaf-nosed bat)	Insectivorous (microchiropteran hipposiderid)	~ Tandala, Zaire (1976)	17	Filovirus isolation, IFA (EBOV)	Breman et al. (1999)
Hypsignathus (hammer-headed fruit bat)	Frugivorous (megachiropteran pteropodid)	~ Yambuku, Zaire (1976)	1	Filovirus isolation	Germain (1978)
H. monstrosus (hammer-headed fruit bat)	Frugivorous (megachiropteran pteropodid)	Gabon and COG (2005–2006)	56	qRT-PCR (MARV), nested PCR (MARV), ELISA (MARV)	Towner et al. (2007)
Megaloglossus woermanni (Woermann's long-tongued fruit bat)	Frugivorous (megachiropteran pteropodid)	~ Kikwit, Zaire (1995)	43	Filovirus isolation, ELISA (EBOV)	Leirs et al. (1999)
M. woermanni (Woermann's long-tongued fruit bat)	Frugivorous (megachiropteran pteropodid)	Gabon and COG (2005–2006)	37	qRT-PCR (MARV), nested PCR (MARV), ELISA (MARV)	Towner et al. (2007)
Micropteropus pusillus (Peters's lesser epauletted fruit bat)	Frugivorous (megachiropteran pteropodid)	~ Kikwit, Zaire (1995)	78	Filovirus isolation, ELISA (EBOV)	Leirs et al. (1999)
M. pusillus (Peters's lesser epauletted fruit bat)	Frugivorous (megachiropteran pteropodid)	Gabon and COG (2005–2006)	149	qRT-PCR (MARV), nested PCR (MARV), ELISA (MARV)	Towner et al. (2007)
Miniopterus inflatus (greater long-fingered bat)	Insectivorous (microchiropteran vespertilionid)	Gabon (2009–2010)	186	qRT-PCR (MARV), nested PCR (MARV)	Maganga et al. (2011)
Miniopterus minor (least long-fingered bat)	Insectivorous (microchiropteran vespertilionid)	~ Kikwit, Zaire (1995)	2	Filovirus isolation, ELISA (EBOV)	Leirs et al. (1999)
M. condylurus (Angolan free-tailed bat)	Insectivorous (microchiropteran molossid)	~ Tandala, Zaire (1976)	54	Filovirus isolation, IFA (EBOV)	Breman et al. (1999)
M. condylurus (Angolan free-tailed bat)	Insectivorous (microchiropteran molossid)	~ Kikwit, Zaire (1995)	10	Filovirus isolation, ELISA (EBOV)	Leirs et al. (1999)

Species	Diet (type)	Location (date)	No.	Assay (virus)	Reference
Mops congicus (Congo free-tailed bat)	Insectivorous (microchiropteran molossid)	~ Tandala, Zaire (1976)	20	Filovirus isolation, IFA (EBOV)	Breman et al. (1999)
Mops niveiventer (white-bellied free-tailed bat)	Insectivorous (microchiropteran molossid)	~ Kikwit, Zaire (1995)	3	Filovirus isolation, ELISA (EBOV)	Leirs et al. (1999)
Mops (Xiphonycteris) nanulus (dwarf free-tailed bat)	Insectivorous (microchiropteran molossid)	~ Tandala, Zaire (1976)	15	Filovirus isolation, IFA (EBOV)	Breman et al. (1999)
M. (Xiphonycteris) nanulus (dwarf free-tailed bat)	Insectivorous (microchiropteran molossid)	~ Kikwit, Zaire (1995)	14	Filovirus isolation, ELISA (EBOV)	Leirs et al. (1999)
Mops (Xiphonycteris) thersites (railer free-tailed bat)	Insectivorous (microchiropteran molossid)	~ Tandala, Zaire (1976)	69	Filovirus isolation, IFA (EBOV)	Breman et al. (1999)
Mops (Xiphonycteris) thersites (railer free-tailed bat)	Insectivorous (microchiropteran molossid)	~ Kikwit, Zaire (1995)	1	Filovirus isolation, ELISA (EBOV)	Leirs et al. (1999)
M. torquata (little collared fruit bat)	Frugivorous (megachiropteran pteropodid)	Gabon and COG (2005–2006)	264	qRT-PCR (MARV), nested PCR (MARV), ELISA (MARV)	Towner et al. (2007)
Myopterus whitleyi (bini winged-mouse bat)	Insectivorous (microchiropteran molossid)	~ Kikwit, Zaire (1995)	2	Filovirus isolation, ELISA (EBOV)	Leirs et al. (1999)
Myotis bocagii (rufous myotis)	Insectivorous (microchiropteran vespertilionid)	~ Tandala, Zaire (1976)	17	Filovirus isolation, IFA (EBOV)	Breman et al. (1999)
M. bocagii (rufous myotis)	Insectivorous (microchiropteran vespertilionid)	~ Kikwit, Zaire (1995)	22	Filovirus isolation, ELISA (EBOV)	Leirs et al. (1999)
Neoromicia nanus (banana pipistrelle)	Insectivorous (microchiropteran vespertilionid)	~ Tandala, Zaire (1976)	73	Filovirus isolation, IFA (EBOV)	Breman et al. (1999)
N. nanus (banana pipistrelle)	Insectivorous (microchiropteran vespertilionid)	~ Kikwit, Zaire (1995)	2	Filovirus isolation, ELISA (EBOV)	Leirs et al. (1999)
Neoromicia somalicus (Somali serotine)	Insectivorous (microchiropteran vespertilionid)	~ Kikwit, Zaire (1995)	1	Filovirus isolation, ELISA (EBOV)	Leirs et al. (1999)
Neoromicia tenuipinnis (white-winged serotine)	Insectivorous (microchiropteran vespertilionid)	~ Kikwit, Zaire (1995)	1	Filovirus isolation, ELISA (EBOV)	Leirs et al. (1999)
Nycteris hispida (hairy slit-faced bat)	Insectivorous (microchiropteran nycterid)	~ Kikwit, Zaire (1995)	2	Filovirus isolation, ELISA (EBOV)	Leirs et al. (1999)
Nycteris sp. (slit-faced bat)	Insectivorous (microchiropteran nycterid)	~ Tandala, Zaire (1976)	14	Filovirus isolation, IFA (EBOV)	Breman et al. (1999)
Rhinolophus sp. (horseshoe bat)	Insectivorous (microchiropteran rhinolophid)	Gabon (2009–2010)	17	qRT-PCR (MARV), nested PCR (MARV)	Maganga et al. (2011)

(continued)

TABLE 7.2 (continued)
Bats Screened for Filoviruses with Negative Results

Bat Species (Vernacular Name)	Type of Bat	Sampling Location (Year)	Number Screened	Assay	Reference
Saccolaimus peli (Pel's pouched bat)	Insectivorous (microchiropteran emballonurid)	~ Tandala, Zaire (1976)	9	Filovirus isolation, IFA (EBOV)	Breman et al. (1999)
Scotophilus dinganii (yellow-bellied house bat)	Insectivorous (microchiropteran vespertilionid)	~ Kikwit, Zaire (1995)	20	Filovirus isolation, ELISA (EBOV)	Leirs et al. (1999)
Scotophilus sp. (yellow bat)	Insectivorous (microchiropteran vespertilionid)	~ Tandala, Zaire (1976)	10	Filovirus isolation, IFA (EBOV)	Breman et al. (1999)
Chaerephon chapini (pale free-tailed bat)	Insectivorous (microchiropteran molossid)	~ Tandala, Zaire (1976)	Total of 23	Filovirus isolation, IFA (EBOV)	Breman et al. (1999)
Hipposideros commersoni (Commerson's leaf-nosed bat)	Insectivorous (microchiropteran hipposiderid)				
Hipposideros sp. (leaf-nosed bat)	Insectivorous (microchiropteran hipposiderid)				
H. monstrosus (hammer-headed fruit bat)	Frugivorous (megachiropteran pteropodid)				
Kerivoula lanosa (lesser woolly bat)	Insectivorous (microchiropteran vespertilionid)				
M. woermanni (Woermann's long-tongued fruit bat)	Frugivorous (megachiropteran pteropodid)				
Mops sp. (free-tailed bat)	Insectivorous (microchiropteran molossid)				
M. torquata (little collared fruit bat)	Frugivorous (megachiropteran pteropodid)				
Taphozous (Taphozous) mauritianus (Mauritian tomb bat)	Insectivorous (microchiropteran emballonurid)				
Unspecified microbats	Insectivorous	Gabon and COG (2005–2006)	15	qRT-PCR (MARV), nested PCR (MARV), ELISA (MARV)	Towner et al. (2007)

COD, Republic of the Congo; bats for which a connection with filoviruses could be established in other studies are printed bold.

More and more scientific evidence accumulated in recent years supports the hypothesis that bats are indeed at least involved in the sustenance of filoviruses, if not being their reservoir hosts. The first set of data that revealed a direct link between bats and filoviruses was published in 2005 (Leroy et al., 2005). Specifically, the data indicated that at least three types of bats were in contact with EBOV during the 2002–2003 ebolavirus disease outbreaks in Gabon and Republic of the Congo, as anti-EBOV IgG antibodies could be detected by ELISA in sera of 8 of 117 collected Franquet's epauletted fruit bats (*Epomops franqueti*), 4 of 17 collected hammer-headed fruit bats (*Hypsignathus monstrosus*), and 4 of 58 collected little collared fruit bats (*Myonycteris torquata*). Intriguingly, Leroy et al. were also able to detect short fragments of what appeared to be filovirus polymerase (L) genes using nested RT-PCT in liver and spleen tissues from 5 of 117 Franquet's epauletted fruit bats, 4 of 21 hammer-headed fruit bats, and 4 of 141 little collared fruit bats. The team did, however, not succeed in isolating filoviruses from these animals (Leroy et al., 2005) despite the fact that filoviruses grow rapidly in most mammalian cell cultures (Kuhn, 2008). A follow-up study in the same area confirmed the serological results and established an overall 5% anti-EBOV IgG seroprevalence in bats belonging to all three species when they were collected during ebolavirus disease outbreaks among humans (2003–2005) (Pourrut et al., 2007). In 2007, Towner et al. published data that revealed the presence of anti-MARV IgG and MARV VP40, VP35, and NP gene-specific RNA sequences in 4 out of 283 Egyptian rousettes (*Rousettus aegyptiacus*) that were caught in Gabon and Republic of the Congo, but not in numerous bats of other species, including those suggested previously with EBOV endemicity (Table 7.2; Towner et al., 2007). Others detected MARV-specific genomic (VP35 and VP40 gene) fragments in 9 out of 1257 bats of the same species collected between 2009 and 2010 in Gabon (Maganga et al., 2011). These studies not only implicated a fourth type of bat in filovirus transmission but also suggested that MARV, which had not previously been known to cause disease outbreaks in these geographic locations, may be present in Gabon and Republic of Congo. A large serological survey arrived at the same results and extended the possible host spectrum of filoviruses even further. In 2009, Pourrut et al. reported anti-MARV IgG in Egyptian rousettes and hammer-headed fruit bats collected in these two countries and anti-EBOV IgG not only in Franquet's epauletted fruit bats, hammer-headed fruit bats, and little collared fruit bats but also in Egyptian rousettes, Angolan free-tailed bats, and Peters's lesser epauletted fruit bats (*Micropteropus pusillus*). Filovirus genomic fragments were not detected, and filoviruses were not isolated from samples (Pourrut et al., 2009). However, in 2009, Towner et al. reported the isolation of infectious MARV (two times) and RAVV (three times) from Egyptian rousettes caught in Kitaka Cave, Uganda, the place of a limited cluster of human MARV/RAVV infections in 2007 (Towner et al., 2009). In 2010, Kuzmin et al. reported the detection of MARV NP gene fragments in Egyptian rousettes collected around Kitum Cave in Kenya, where MARV and RAVV infections were reported in 1980 and 1987. Also in 2010, Hayman et al. reported anti-EBOV IgG in an African straw-colored fruit bat (*Eidolon helvum*) collected in Ghana (where filovirus disease outbreaks have not been reported) (Hayman et al., 2010). The same team later also reported the presence of anti-EBOV antibodies in three Franquet's epauletted fruit bats, four Gambian epauletted fruit bats (*Epomophorus gambianus*), and two hammer-headed fruit bats and anti-RESTV antibodies in two Gambian epauletted fruit bats and one hammer-headed fruit bat. A single Gambian epauletted fruit bat tested positive for antibodies against both EBOV and RESTV (Hayman et al., 2012). Other studies revealed the presence of anti-RESTV IgG in Geoffrey's rousettes (*Rousettus amplexicaudatus*) in the Philippines, where RESTV is known to be endemic (Taniguchi et al., 2011), and the detection of a previously entirely unknown filovirus, LLOV, in deceased Schreiber's long-fingered bats (*Miniopterus schreibersii*) in caves of Southern Europe (Negredo et al., 2011).

Together, the findings of all these studies raise many questions. MARV and RAVV disease outbreaks have almost always been associated with people visiting or working in caves or mines (Adjemian et al., 2011; Bausch et al., 2006; Centers for Disease Control and Prevention, 2009; Johnson et al., 1996; Peterson et al., 2006; Smith et al., 1982; Timen et al., 2009; Towner et al., 2009).

Egyptian rousettes are strictly cavernicolous, and MARV and RAVV isolation succeeded from individual bats captured in a cave that was implicated in human infections (Towner et al., 2009). Therefore, the hypothesis of Egyptian rousettes being marburgvirus hosts stands on relatively solid ground. On the other hand, ebolavirus disease outbreaks have never been associated with caves and usually occur in areas of tropical rainforest, rather than arid woodlands (Peterson et al., 2004, 2006). It is difficult to interpret the finding of anti-EBOV IgG, anti-MARV IgG, and genomic fragments in Egyptian rousettes in Gabon and Republic of the Congo, in particular, because marburgvirus disease outbreaks have never been reported from these countries. This could mean that marburgviruses are endemic in Gabon and Republic of the Congo, as some speculate (Maganga et al., 2011). But if this is the case, then it is surprising that human cases of MARV or RAVV infection have not been detected in these countries. It seems unlikely that they have been overlooked given that ebolavirus disease outbreaks are recorded in these countries on a regular basis—clinically, the diseases caused by marburgviruses and ebolaviruses cannot be differentiated upon presentation. It is of course possible that people in these countries do not get in contact with Egyptian rousettes.

Vice versa, if Egyptian rousettes are ebolavirus reservoirs based on the described IgG findings, one cannot help but wonder why there have not been any ebolavirus disease outbreaks in Uganda or Kenya, where marburgviruses are endemic. Egyptian rousettes are frugivorous animals that are widely distributed over Africa, the Middle East, and even Southwest Asia. Are EBOV, MARV, and RAVV endemic in these countries and just have never been reported (or never caused human infections)? One explanation could be that not all Egyptian rousettes carry filoviruses. For instance, Egyptian rousettes are currently assigned to six different *R. aegyptiacus* subspecies. Bats of the subspecies *aegyptiacus* are found in Egypt only; those of the subspecies *arabicus* are found exclusively in Iran, Pakistan, and southern Arabia, and those of the subspecies *tomensis* and *princes* live in São Tomé and Príncipe. Only animals assigned to the subspecies *leachii* and *unicolor* are widely distributed across Africa (Benda et al., 2008). If filoviruses would be able to only infect animals of the latter two subspecies, then this would explain the absence of human filovirus infections outside of Africa. At the same time, if marburgviruses and ebolaviruses would preferentially infect animals of the one or the other subspecies, then this could possibly explain the different human case distributions. Subspecies specificity could also play a major role in the possible association of other filoviruses with bats. For instance, Geoffrey's rousettes, the presumed hosts of RESTV, are assigned to five different subspecies (Csorba et al., 2008). Maybe a species-specific association restricts RESTV to the Philippines. Or RESTV is endemic all over Southeast Asia—the range of Geoffrey's rousettes—and has just not been detected yet outside of the Philippines.

One must exert caution when speculating about an association of ebolaviruses and bats. It needs to be stressed that in contrast to marburgviruses, ebolaviruses have thus far not been isolated from any bat in the wild, and in three cases (BDBV, SUDV, TAFV), not even epidemiological links to bats have been uncovered. Instead, the hypothesis of bats being hosts of ebolaviruses is based on antibody detection in the case of two ebolaviruses (EBOV, RESTV), the detection of genomic RNA fragments in the case of one EBOV, and a speculated epidemiological link between an ebolavirus disease outbreak and the consumption of fruit bats in the Democratic Republic of the Congo (Grard et al., 2011; Leroy et al., 2009). Interestingly, anti-EBOV IgG was discovered in bats of various species, not only those of the species *R. aegyptiacus* (Table 7.3). This could mean that in contrast to what is known about the single host/single virus relationship found in the case of arenaviruses and hantaviruses, individual filoviruses may be able to colonize several hosts belonging to different genera at the same time. Likewise, bats belonging to one particular species could be reservoirs of filoviruses belonging to different species.

In the case of Franquet's epauletted fruit bats, hammer-headed fruit bats, and little collared fruit bats, only anti-EBOV IgG or EBOV genomic RNA fragments could be detected in individual animals, but never both together in the same individual animal (Leroy et al., 2005). The performers of the study hypothesized that IgG-positive animals were once infected but had cleared the infection, whereas RNA-positive animals had been infected recently, but had not then mounted an immune

TABLE 7.3

Evidence for an Association of Filoviruses with Bats

Filovirus	Suspected Bat Host (Species)	Supporting Data for Bat Association	Bat Geographic Distribution	Known Filovirus Endemicity (Based on Disease Outbreaks and Virus Isolation/Full Genome Detection)
BDBV	N/A	N/A	N/A	Uganda
EBOV	African straw-colored fruit bat	Anti-EBOV IgG (Hayman et al., 2010)	Sub-Saharan Africa	Democratic Republic of the Congo, Gabon, Republic of the Congo
	Angolan free-tailed bat Egyptian rousette	Anti-EBOV IgG (Pourrut et al., 2009) Anti-EBOV IgG (Pourrut et al., 2009)	Sub-Saharan Africa Africa Middle East Southwest Asia	
	Franquet's epauletted fruit bat	Anti-EBOV IgG (Hayman et al., 2012; Leroy et al., 2005; Pourrut et al., 2007; Pourrut et al., 2009) RT-PCR positive for EBOV L gene fragment (Leroy et al., 2005)	Equatorial Africa	
	Gambian epauletted fruit bat	Anti-EBOV IgG (Hayman et al., 2012)	Angola, Benin, Cameroon, Côte d'Ivoire, Democratic Republic of the Congo, Gabon, Ghana, Guinea, Nigeria, Republic of the Congo	
	Hammer-headed fruit bat	Anti-EBOV IgG (Hayman et al., 2012; Leroy et al., 2005; Pourrut et al., 2007; Pourrut et al., 2009) RT-PCR positive for EBOV L gene fragment (Leroy et al., 2005; Pourrut et al., 2009)		
	Little collared fruit bat	Anti-EBOV IgG (Leroy et al., 2005; Pourrut et al., 2007, 2009) RT-PCR positive for EBOV L gene fragment (Leroy et al., 2005)	Central Africa, West Africa	
	Peters's lesser epauletted fruit bat	Anti-EBOV IgG (Pourrut et al., 2009)	Equatorial Africa	
LLOV	Schreiber's long-fingered bat	Complete genome detection (Negredo et al., 2011)	Caucasus, North and West Africa, Southwestern Europe	France, Portugal, Spain

(continued)

TABLE 7.3 (continued)
Evidence for an Association of Filoviruses with Bats

Filovirus	Suspected Bat Host (Species)	Supporting Data for Bat Association	Bat Geographic Distribution	Known Filovirus Endemicity (Based on Disease Outbreaks and Virus Isolation/Full Genome Detection)
MARV	Egyptian rousette	Anti-MARV IgG (Pourrut et al., 2009; Towner et al., 2007)	Africa	Angola, Democratic Republic of the Congo, Kenya, Uganda, Zimbabwe
		RT-PCR positive for VP35, VP40, and NP gene fragment (Kuzmin et al., 2011; Maganga et al., 2011; Towner et al., 2007)	Middle East	
		Virus isolation (Towner et al., 2009)	Southwest Asia	
	Hammer-headed fruit bat	Anti-MARV IgG (Pourrut et al., 2009)	Angola, Benin, Cameroon, Côte d'Ivoire, Democratic Republic of the Congo, Gabon, Ghana, Guinea, Nigeria, Republic of the Congo	
RAVV	Egyptian rousette	Virus isolation (Towner et al., 2009)	Africa	Democratic Republic of the Congo, Kenya, Uganda
			Middle East	
			Southwest Asia	
RESTV	Gambian epauletted fruit bat	Anti-RESTV IgG (Hayman et al., 2012)	Equatorial Africa	Philippines
	Geoffroy's rousette	Anti-RESTV IgG (Taniguchi et al., 2011)	Cambodia, East Timor, Indonesia, Malaysia, Myanmar, Papua New Guinea, Philippines, Thailand, Vietnam	
	Hammer-headed fruit bat	Anti-RESTV IgG (Hayman et al., 2012)	Côte d'Ivoire, Guinea, Ghana, Benin, Nigeria, Cameroon, Angola, Republic of the Congo, Democratic Republic of the Congo, Gabon	
SUDV	N/A	N/A	N/A	South Sudan, Uganda
TAFV	N/A	N/A	N/A	Côte d'Ivoire

N/A, not applicable; bats for which a connection with filoviruses could not be established in other studies (Table 7.1) are printed bold.

response (Leroy et al., 2005). It is possible that filoviruses infect a bat only for a short period of time and are then transmitted to a co-roosting bat before the immune system of the first bat eliminates it. Especially in large cohorts of bats (hundreds of thousands of individuals), developing sterilizing immunity in individuals may not cause a bottleneck for efficient filovirus transmission. But in such a case, one would expect IgG antibodies in a large percentage of bats of a colony, which thus far has not been demonstrated (Kuzmin et al., 2011). On the other hand, if the natural maintenance of arenaviruses and hantaviruses is any indication, one would expect a more time-stable filovirus–bat relationship in the form of a subclinical persistent infection of a host. Also, one should not forget that all the studies published thus far on bats and filoviruses do not specifically address the order of events during a human outbreak, i.e., it is also possible that bats become infected with filoviruses because of an ongoing epizootic/epidemic, rather than being the factor that started it. A study by Biek et al., for instance, revealed that the EBOV-specific L gene fragments from the Franquet's epauletted fruit bats, hammer-headed fruit bats, and collected little collared fruit bats "appear to be direct descendents of viruses seen during previous outbreaks," i.e., this study revealed a direct connection between the few detected sequences and the EBOV known from human outbreaks (Biek et al., 2006). If these bats were the natural reservoirs of EBOV, then one could expect EBOV diversity to be by broader, with some sequences having only indirect connections to known viruses. Biek et al. offered a hypothesis why this does not necessarily have to be the case: a unknown event could have led to an extremely small population of infected bats, thereby selecting only a particular genetic EBOV lineage. But the authors and others also came to the conclusion that it is possible that a circulating disease outbreak–causing virus was introduced into the bat population (Biek et al., 2006; Kuzmin et al., 2011). Such a scenario is also supported by the finding of Pourrut et al. that 5% of sampled Franquet's epauletted fruit bats, hammer-headed fruit bats, and collected little collared fruit bats contained anti-EBOV IgG when they were collected during human ebolavirus disease outbreaks in either epidemic or nonepidemic regions but considerably lower (0.9%) in bats collected between ebolavirus disease outbreaks (Pourrut et al., 2007).

An alternative to the hypothesis that bats are the natural reservoir hosts of filoviruses is that bats may be merely in close contact with that host. For instance, bat ectoparasites such as winged or wingless bat flies (families Streblidae and Nycteribiidae) or other arthropods could be primary filovirus hosts (Monath, 1999). If that were the case, genomic RNA fragments could be detected in bats that had been bitten by infected arthropods, and IgG antibodies could be detected in bats that had been bitten in the past. Virus isolation could be possible if a bat was caught right around the time of the bite. Bat ectoparasites could also migrate from bats of one species to those of another, especially among co-roosting populations, and thereby explain why bats of different species tested positive for anti-EBOV IgG. If a filovirus-infected arthropod is not bat specific, but rather feeds on bats only occasionally, then this could explain the rarity of filovirus disease outbreaks. Arthropod-transmitted filoviruses could also better explain the discovery of LLOV in Spain (Negredo et al., 2011). LLOV was discovered after massive die-offs of Schreiber's long-fingered bats in Spain, France, and Portugal. Koch's postulates could not yet be fulfilled, which leaves the door open for three scenarios. One, these bats are subclinically and persistently infected with LLOV (natural reservoir hosts), and some of them died due to something unrelated to LLOV infection. Two, these bats became sick because of unknown reasons and this sickness allowed LLOV to infect them. And, finally, three, LLOV infected the bats and killed them. The second and third scenario would suggest that at least these bats are not reservoirs of LLOV.

The last intriguing question is how filoviruses are transmitted to humans from bats, if indeed bats are filovirus reservoir hosts or at least amplifying hosts. In the case of marburgviruses, humans probably simply come into direct contact with bats or their secreta or excreta in colonized caves or mines. In the case of ebolaviruses, and the absence of the cave connection, transmission is much less clear. Hunting of bats and subsequent slaughtering and food preparation and/or consumption could further transmission of filoviruses to humans, but convincing data supporting this hypothesis are lacking despite the fact that certain types of bats are part of the general human diet in many

African countries. Alternatively, nonhuman primates, which are often epidemiologically linked to ebolavirus disease outbreaks (Kuhn, 2008), could get in close contact with bats when feasting in fruit trees, thereby becoming intermediary hosts until they get killed by humans for food. Pourrut et al. pointed out that ebolavirus disease outbreaks occur most often at the end of the dry season, coinciding with the birthing season of many frugivorous bats. As fruit are scarce around that time in the forest, bats and nonhuman primates may forage in the same trees and even on the same fruit, thereby furthering transmission (Pourrut et al., 2006). This then, would leave the question open how insectivorous bats, which also have been associated with filovirus disease outbreaks (Angolan free-tailed bats, Schreiber's long-fingered bats) fit into filovirus epidemiology/epizootiology. Most likely, many other factors need to be evaluated to understand why filovirus disease outbreaks are rare despite an abundance of hosts. If bats truly are filovirus hosts, bat roost population size (hundreds of thousands of animals vs. few individuals), migration patterns (short vs. long range), roosting location (caves vs. forest or savannah), and even age and sex may be important factors. Laboratory studies will be necessary to better understand why certain bats can maintain filovirus replication without developing disease (Omatsu et al., 2007), how filoviruses could establish persistent infections and under which circumstances they could emerge from their hosts (Strong et al., 2008), and how filoviruses interact with bat cells (Jordan et al., 2009, 2012; Krähling et al., 2010; Kuhl et al., 2011).

DISCLAIMER

The content of this publication does not necessarily reflect the views or policies of the U.S. Department of the Army, the U.S. Department of Defense, or the U.S. Department of Health and Human Services or of the institutions and companies affiliated with the authors. This work was funded by the Joint Science and Technology Office for Chem Bio Defense (proposal #TMTI0048_09_RD_T) to SB and SRR. JHK and VWJ performed this work as employees of Tunnell Consulting, Inc., a subcontractor to Battelle Memorial Institute, and FKM as an employee of Battelle Memorial Institute, under its prime contract with NIAID, under Contract No. HHSN272200200016I.

REFERENCES

Abu Sin, M., Stark, K., van Treeck, U., Dieckmann, H., Uphoff, H., Hautmann, W., Bornhofen, B., Jensen, E., Pfaff, G., Koch, J., 2007. Risk factors for hantavirus infection in Germany, 2005. *Emerg Infect Dis* **13**, 1364–1366.

Adams, M.J., Carstens, E.B., 2012. Ratification vote on taxonomic proposals to the International Committee on Taxonomy of Viruses (2012). *Arch Virol* **157**, 1411–1422.

Adjemian, J., Farnon, E.C., Tschioko, F., Wamala, J.F., Byaruhanga, E., Bwire, G.S., Kansiime, E. et al., 2011. Outbreak of Marburg hemorrhagic fever among miners in Kamwenge and Ibanda districts, Uganda, 2007. *J Infect Dis* **204**(suppl. 3), S796–S799.

Agnarsson, I., Zambrana-Torrelio, C.M., Flores-Saldana, N.P., May-Collado, L.J., 2011. A time-calibrated species-level phylogeny of bats (Chiroptera, Mammalia). *PLoS Curr* **3**, RRN1212.

Avsic-Zupanc, 1998. Hantaviruses and hemorrhagic fever with renal syndrome in the Balkans, in: J.F. Saluzzo, B.D. (Ed.), *Emergence and Control of Rodent-borne Viral Diseases (Hantaviral and Arenal Diseases)*. Elsevier, Veyrier-du-lac, Annecy, France, pp. 93–98.

Avsic-Zupanc, T., Xiao, S.Y., Stojanovic, R., Gligic, A., van der Groen, G., LeDuc, J.W., 1992. Characterization of Dobrava virus: A Hantavirus from Slovenia, Yugoslavia. *J Med Virol* **38**, 132–137.

Bausch, D.G., Nichol, S.T., Muyembe-Tamfum, J.J., Borchert, M., Rollin, P.E., Sleurs, H., Campbell, P. et al., for the International Scientific and Technical Committee for Marburg Hemorrhagic Fever Control in the Democratic Republic of the Congo, 2006. Marburg hemorrhagic fever associated with multiple genetic lineages of virus. *New Engl J Med* **355**, 909–919.

Benda, P., Aulagnier, S., Hutson, A.M., Amr, Z.S., Kock, D., Sharifi, M., Karataş, A., Mickleburgh, S., Bergmans, W., Howell, K., 2008. *Rousettus aegyptiacus*, IUCN 2012. IUCN Red List of Threatened Species. Version 2012.1. www.iucnredlist.org. Downloaded on July 16, 2012.

Biek, R., Walsh, P.D., Leroy, E.M., Real, L.A., 2006. Recent common ancestry of Ebola Zaire virus found in a bat reservoir. *PLoS Pathog* **2**, e90.

Bowen, M.D., Peters, C.J., Nichol, S.T., 1996. The phylogeny of New World (Tacaribe complex) arenaviruses. *Virology* **219**, 285–290.

Breman, J.G., Johnson, K.M., van der Groen, G., Robbins, C.B., Szczeniowski, M.V., Ruti, K., Webb, P.A. et al., 1999. A search for Ebola virus in animals in the democratic Republic of the Congo and Cameroon: Ecologic, Virologic, and Serologic Surveys, 1979–1980. *J Infect Dis* **179**(suppl. 1), S139–S147.

Briese, T., Paweska, J.T., McMullan, L.K., Hutchison, S.K., Street, C., Palacios, G., Khristova, M.L. et al., 2009. Genetic detection and characterization of Lujo virus, a new hemorrhagic fever-associated arenavirus from southern Africa. *PLoS Pathog* **5**, e1000455.

Brummer-Korvenkontio, M., Henttonen, H., Vaheri, A., 1982. Hemorrhagic fever with renal syndrome in Finland: Ecology and virology of nephropathia epidemica. *Scand J Infect Dis Suppl* **36**, 88–91.

Brummer-Korvenkontio, M., Vaheri, A., Hovi, T., von Bonsdorff, C.H., Vuorimies, J., Manni, T., Penttinen, K., Oker-Blom, N., Lahdevirta, J., 1980. Nephropathia epidemica: Detection of antigen in bank voles and serologic diagnosis of human infection. *J Infect Dis* **141**, 131–134.

Buchmeier, M.J., de La Torre, J.C., Peters, C.J., 2006. Arenaviridae:The viruses and their replication. In: Knipe, D.M., Howley, P.M. (Eds.), *Fields Virology*, 5th edn. Lippincott Williams & Wilkins, Philadelphia, PA, pp. 1791–1827.

van Cakenberghe, V., de Vree, F., Leirs, H., 1999. On a collection of bats (Chiroptera) from Kikwit, Democratic Republic of the Congo. With French abstract. *Mammalia* **63**, 291–322.

Calisher, C.H., Childs, J.E., Field, H.E., Holmes, K.V., Schountz, T., 2006. Bats: Important reservoir hosts of emerging viruses. *Clin Microbiol Rev* **19**, 531–545.

Calisher, C.H., Holmes, K.V., Dominguez, S.R., Schountz, T., Cryan, P., 2008. Bats prove to be rich reservoirs for emerging viruses. *Microbe* (Washington, DC) **3**, 521–528.

Carballal, G., Videla, C.M., Merani, M.S., 1988. Epidemiology of Argentine hemorrhagic fever. *Eur J Epidemiol* **4**, 259–274.

Centers for Disease Control and Prevention, 2009. Imported case of Marburg hemorrhagic fever— Colorado, 2008. *MMWR Morb Mortal Wkly Rep* **58**, 1377–1381.

Charrel, R.N., de Lamballerie, X., 2003. Arenaviruses other than Lassa virus. *Antiviral Res* **57**, 89–100.

Charrel, R.N., Lemasson, J.J., Garbutt, M., Khelifa, R., De Micco, P., Feldmann, H., de Lamballerie, X., 2003. New insights into the evolutionary relationships between arenaviruses provided by comparative analysis of small and large segment sequences. *Virology* **317**, 191–196.

Clegg, J.C., 2002. Molecular phylogeny of the arenaviruses. *Curr Top Microbiol Immunol* **262**, 1–24.

Coetzee, C.G., 1975. The biology, behaviour, and ecology of *Mastomys natalensis* in southern Africa. *Bull WHO* **52**, 637–644.

Csorba, G., Rosell-Ambal, G., Ingle, N., 2008. *Rousettus amplexicaudatus*, IUCN 2012. IUCN Red List of Threatened Species. Version 2012.1. www.iucnredlist.org. Downloaded on July 16, 2012.

Delgado, S., Erickson, B.R., Agudo, R., Blair, P.J., Vallejo, E., Albarino, C.G., Vargas, J. et al., 2008. Chapare virus, a newly discovered arenavirus isolated from a fatal hemorrhagic fever case in Bolivia. *PLoS Pathog* **4**, e1000047.

Demby, A.H., Inapogui, A., Kargbo, K., Koninga, J., Kourouma, K., Kanu, J., Coulibaly, M. et al., 2001. Lassa fever in Guinea: II. Distribution and prevalence of Lassa virus infection in small mammals. *Vector Borne Zoonotic Dis* **1**, 283–297.

DisWeb, C.D.I.S., 2003. Communicable diseases statistics, Major statistics, reported cases by year National Institutes of Health.

Downs, W.G., Anderson, C.R., Spence, L., Aitken, T.H., Greenhall, A.H., 1963. Tacaribe virus, a new agent isolated from Artibeus bats and mosquitoes in Trinidad, West Indies. *Am J Trop Med Hyg* **12**, 640–646.

Dybas, C.L., 2007. Russia Sees Ill Effects of 'General Winter's' Retreat, Washington Post.

Enria, D.A., Briggiler, A.M., Sanchez, Z., 2008. Treatment of Argentine hemorrhagic fever. *Antiviral Res* **78**, 132–139.

Erickson, B.R., Delgado, S., Aguda, R., Vallejo, E., Vergas, J., Blair, P.J., Albarino, C. et al., 2006. A newly discovered arenavirus associated with a fatal hemorrhagic fever case in Bolivia, Abstracts of the *XIIIth International Conference on Negative Strand Viruses*, Salamanca, Spain, p. 165.

Fang, L., Yan, L., Liang, S., de Vlas, S.J., Feng, D., Han, X., Zhao, W. et al., 2006. Spatial analysis of hemorrhagic fever with renal syndrome in China. *BMC Infect Dis* **6**, 77.

Fichet-Calvet, E., Lecompte, E., Koivogui, L., Daffis, S., Ter Meulen, J., 2008. Reproductive characteristics of Mastomys natalensis and Lassa virus prevalence in Guinea, West Africa. *Vector Borne Zoonotic Dis* **8**, 41–48.

Fichet-Calvet, E., Lecompte, E., Koivogui, L., Soropogui, B., Dore, A., Kourouma, F., Sylla, O., Daffis, S., Koulemou, K., Ter Meulen, J., 2007. Fluctuation of abundance and Lassa virus prevalence in Mastomys natalensis in Guinea, West Africa. *Vector Borne Zoonotic Dis* **7**, 119–128.

Fichet-Calvet, E., Rogers, D.J., 2009. Risk maps of Lassa fever in west Africa. *PLoS Negl Trop Dis* **3**, e388.

Fulhorst, C.E., Bowen, M.D., Salas, R.A., de Manzione, N.M., Duno, G., Utrera, A., Ksiazek, T.G. et al., 1997. Isolation and characterization of pirital virus, a newly discovered South American arenavirus. *Am J Trop Med Hyg* **56**, 548–553.

Fulhorst, C.F., Cajimat, M.N.B., Milazzo, M.L., Paredes, H., de Manzione, N.M.C., Salas, R.A., Rollin, P.E., Ksiazek, T.G., 2008. Genetic diversity between and within the arenavirus species indigenous to western Venezuela. *Virology* **378**, 205–213.

Fulhorst, C.F., Ksiazek, T.G., Peters, C.J., Tesh, R.B., 1999. Experimental infection of the cane mouse *Zygodontomys brevicauda* (family Muridae) with Guanarito virus (*Arenaviridae*), the etiologic agent of Venezuelan hemorrhagic fever. *J Infect Dis* **180**, 966–969.

Garcia, J., Calderon, G., Sabattini, M., Enria, D., 1996. Infeccion por virus Junin (JV) de Calomys musculinus (Cm) en áreas con diferente situación epidemiológica para la Fiebre Hemorrágica Argentina (FHA). *Medicina* (B Aires) **56**, 624.

Germain, M., 1978. Collection of mammals and arthropods during the epidemic of haemorrhagic fever in Zaire. In: Pattyn, S.R. (Ed.), *Ebola Virus Haemorrhagic Fever—Proceedings of an International Colloquium on Ebola Virus Infection and Other Haemorrhagic Fevers held in Antwerp, Belgium*, December 6–8, 1977. Elsevier, Amsterdam, the Netherlands, pp. 185–189.

Grard, G., Biek, R., Muyembe Tamfum, J.J., Fair, J., Wolfe, N., Formenty, P., Paweska, J., Leroy, E., 2011. Emergence of divergent Zaire Ebola virus strains in Democratic Republic of the Congo in 2007 and 2008. *J Infect Dis* **204**(suppl. 3), S776–S784.

Gresikova, M., Rajcani, J., Sekeyova, M., Brummer-Korvenkontio, M., Kozuch, O., Labuda, M., Turek, R., Weismann, P., Nosek, J., Lysy, J., 1984. Haemorrhagic fever virus with renal syndrome in small rodents in Czechoslovakia. *Acta Virol* **28**, 416–421.

van der Groen, G., Piot, P., Desmyter, J., Colaert, J., Muylle, L., Tkachenko, E.A., Ivanov, A.P., Verhagen, R., van Ypersele de Strihou, C., 1983. Seroepidemiology of Hantaan-related virus infections in Belgian populations. *Lancet* **2**, 1493–1494.

Hayman, D.T., Emmerich, P., Yu, M., Wang, L.F., Suu-Ire, R., Fooks, A.R., Cunningham, A.A., Wood, J.L., 2010. Long-term survival of an urban fruit bat seropositive for Ebola and Lagos bat viruses. *PLoS One* **5**, e11978.

Hayman, D.T., Yu, M., Crameri, G., Wang, L.F., Suu-Ire, R., Wood, J.L., Cunningham, A.A., 2012. Ebola virus antibodies in fruit bats, Ghana, West Africa. *Emerg Infect Dis* **18**, 1207–1209.

Henderson, W.W., Monroe, M.C., St Jeor, S.C., Thayer, W.P., Rowe, J.E., Peters, C.J., Nichol, S.T., 1995. Naturally occurring Sin Nombre virus genetic reassortants. *Virology* **214**, 602–610.

Heyman, P., Ceianu, C.S., Christova, I., Tordo, N., Beersma, M., Joao Alves, M., Lundkvist, A. et al., 2011. A five-year perspective on the situation of haemorrhagic fever with renal syndrome and status of the hantavirus reservoirs in Europe, 2005–2010. *Euro Surveill.* **16**(36).

Heyman, P., Cochez, C., Ducoffre, G., Mailles, A., Zeller, H., Abu Sin, M., Koch, J. et al., 2007. Haemorrhagic Fever with Renal Syndrome: An analysis of the outbreaks in Belgium, France, Germany, the Netherlands and Luxembourg in 2005. *Euro surveillance: Bulletin europeen sur les maladies transmissibles = Eur. Commun. Dis. Bull.* **12**, E15–E16.

Hooper, J.W., Ferro, A.M., Wahl-Jensen, V., 2008. Immune serum produced by DNA vaccination protects hamsters against lethal respiratory challenge with Andes virus. *J Virol* **82**, 1332–1338.

Hugot, J.P., Gonzalez, J.P., Denys, C., 2001. Evolution of the Old World Arenaviridae and their rodent hosts: Generalized host-transfer or association by descent? *Infect Genet Evol* **1**, 13–20.

Isaacson, M., 1975. The ecology of *Praomys* (*Mastomys*) *natalensis* in southern Africa. *Bull World Health Organ* **52**, 629–636.

Jay, M.T., Glaser, C., Fulhorst, C.F., 2005. The arenaviruses. *J Am Vet Med Assoc* **227**, 904–915.

Johnson, K.M., 2001. Hantaviruses: History and overview. *Curr Top Microbiol Immunol* **256**, 1–14.

Johnson, E.D., Johnson, B.K., Silverstein, D., Tukei, P., Geisbert, T.W., Sanchez, A., Jahrling, P.B., 1996. Characterization of a new Marburg virus isolate from a 1987 fatal case in Kenya. In: Schwarz, T.F., Siegl, G. (Eds.), *Imported Virus Infections*. Springer-Verlag, Vienna, Austria, pp. 101–114.

Johnson, K.M., Kuns, M.L., Mackenzie, R.B., Webb, P.A., Yunker, C.E., 1966. Isolation of Machupo virus from wild rodent *Calomys callosus*. *Am J Trop Med Hyg* **15**, 103–106.

Johnson, K.M., Mackenzie, R.B., Webb, P.A., Kuns, M.L., 1965. Chronic infection of rodents by Machupo virus. *Science* (New York) **150**, 1618–1619.

Jordan, I., Horn, D., Oehmke, S., Leendertz, F.H., Sandig, V., 2009. Cell lines from the Egyptian fruit bat are permissive for modified vaccinia Ankara. *Virus Res* **145**, 54–62.

Jordan, I., Munster, V.J., Sandig, V., 2012. Authentication of the R06E fruit bat cell line. *Viruses* **4**, 889–900.

Jung, Y.T., Kim, G.R., 1995. Genomic characterization of M and S RNA segments of hantaviruses isolated from bats. *Acta Virol* **39**, 231–233.

Justines, G., Johnson, K.M., 1969. Immune tolerance in *Calomys callosus* infected with Machupo virus. *Nature* **222**, 1090–1091.

Kallio, E.R., Voutilainen, L., Vapalahti, O., Vaheri, A., Henttonen, H., Koskela, E., Mappes, T., 2007. Endemic hantavirus infection impairs the winter survival of its rodent host. *Ecology* **88**, 1911–1916.

Kariwa, H., Yoshimatsu, K., Arikawa, J., 2007. Hantavirus infection in East Asia. *Comp Immunol Microbiol Infect Dis* **30**, 341–356.

Keenlyside, R.A., McCormick, J.B., Webb, P.A., Smith, E., Elliott, L., Johnson, K.M., 1983. Case-control study of Mastomys natalensis and humans in Lassa virus-infected households in Sierra Leone. *Am J Trop Med Hyg* **32**, 829–837.

Klempa, B., Meisel, H., Rath, S., Bartel, J., Ulrich, R., Kruger, D.H., 2003. Occurrence of renal and pulmonary syndrome in a region of northeast Germany where Tula hantavirus circulates. *J Clin Microbiol* **41**, 4894–4897.

Krähling, V., Dolnik, O., Kolesnikova, L., Schmidt-Chanasit, J., Jordan, I., Sandig, V., Gunther, S., Becker, S., 2010. Establishment of fruit bat cells (*Rousettus aegyptiacus*) as a model system for the investigation of filoviral infection. *PLoS Negl Trop Dis* **4**, e802.

Kuhl, A., Hoffmann, M., Muller, M.A., Munster, V.J., Gnirss, K., Kiene, M., Tsegaye, T.S. et al., 2011. Comparative analysis of Ebola virus glycoprotein interactions with human and bat cells. *J Infect Dis* **204**(suppl. 3), S840–S849.

Kuhn, J.H., 2008. Filoviruses. A compendium of 40 years of epidemiological, clinical, and laboratory studies. *Arch Virol Suppl* 20. Springer Wien New York, Vienna, Austria.

Kuhn, J.H., Becker, S., Ebihara, H., Geisbert, T.W., Jahrling, P.B., Kawaoka, Y., Netesov, S.V. et al., 2011. Family *Filoviridae*. In: King, A.M.Q., Adams, M.J., Carstens, E.B., Lefkowitz, E.J. (Eds.), *Virus Taxonomy—Ninth Report of the International Committee on Taxonomy of Viruses*. Elsevier/Academic Press, London, U.K., pp. 665–671.

Kuhn, J.H., Becker, S., Ebihara, H., Geisbert, T.W., Johnson, K.M., Kawaoka, Y., Lipkin, W.I. et al., 2010. Proposal for a revised taxonomy of the family *Filoviridae*: Classification, names of taxa and viruses, and virus abbreviations. *Arch Virol* **155**, 2083–2103.

Kuns, M.L., 1965. Epidemiology of Machupo virus infection. II. Ecological and control studies of hemorrhagic fever. *Am J Trop Med Hyg* **14**, 813–816.

Kuzmin, I.V., Bozick, B., Guagliardo, S.A., Kunkel, R., Shak, J.R., Tong, S., Rupprecht, C.E., 2011. Bats, emerging infectious diseases, and the rabies paradigm revisited. *Emerg Health Threats J* **4**, 7159.

Lalis, A., Leblois, R., Lecompte, E., Denys, C., Ter Meulen, J., Wirth, T., 2012. The impact of human conflict on the genetics of Mastomys natalensis and Lassa virus in West Africa. *PLoS One* **7**, e37068.

Laminger, F., Prinz, A., 2010. Bats and other reservoir hosts of *Filoviridae*. Danger of epidemic on the African continent? A deductive literature analysis. *Wien Klin Wochenschr* **122**(suppl. 3), 19–30.

Lecompte, E., Fichet-Calvet, E., Daffis, S., Koulemou, K., Sylla, O., Kourouma, F., Dore, A. et al., 2006. Mastomys natalensis and Lassa fever, West Africa. *Emerg Infect Dis* **12**, 1971–1974.

Lecompte, E., Ter Meulen, J., Emonet, S., Daffis, S., Charrel, R.N., 2007. Genetic identification of Kodoko virus, a novel arenavirus of the African pigmy mouse (Mus Nannomys minutoides) in West Africa. *Virology* **364**, 178–183.

Lee, H.W., 1996. Hantaviruses: An emerging disease. *Phil J Microbiol Infect Dis* **25**, S19–S24.

Lee, H.W., Lee, P.W., Johnson, K.M., 1978. Isolation of the etiologic agent of Korean hemorrhagic fever. *J Infect Dis* **137**, 298–308.

Lee, J.S., Lahdevirta, J., Koster, F., Levy, H., 1999. Clinical manifestations and treatment of HFRS and HPS. In: Lee, H.W., Calisher, C., Schmaljohn, C. (Eds.), *Manual of Hemorrhagic Fever with Renal Syndrome and Hantavirus Pulmonary Syndrome*. ASAN Institute for Life Sciences, Seoul, South Korea, pp. 17–38.

Leirs, H., Mills, J.N., Krebs, J.W., Childs, J.E., Akaibe, D., Woollen, N., Ludwig, G. et al., 1999. Search for the Ebola virus reservoir in Kikwit, democratic Republic of the Congo: Reflections on a vertebrate collection. *J Infect Dis* **179**(suppl. 1), S155–S163.

Leroy, E.M., Epelboin, A., Mondonge, V., Pourrut, X., Gonzalez, J.P., Muyembe-Tamfum, J.J., Formenty, P., 2009. Human Ebola outbreak resulting from direct exposure to fruit bats in Luebo, Democratic Republic of Congo, 2007. *Vector Borne Zoonotic Dis* **9**, 723–728.

Leroy, E.M., Kumulungui, B., Pourrut, X., Rouquet, P., Hassanin, A., Yaba, P., Délicat, A., Paweska, J.T., Gonzalez, J.-P., Swanepoel, R., 2005. Fruit bats as reservoirs of Ebola virus. *Nature* **438**, 575–576.

Li, D., Schmaljohn, A.L., Anderson, K., Schmaljohn, C.S., 1995. Complete nucleotide sequences of the M and S segments of two hantavirus isolates from California: Evidence for reassortment in nature among viruses related to hantavirus pulmonary syndrome. *Virology* **206**, 973–983.

Lundkvist, A., Hukic, M., Horling, J., Gilljam, M., Nichol, S., Niklasson, B., 1997. Puumala and Dobrava viruses cause hemorrhagic fever with renal syndrome in Bosnia-Herzegovina: Evidence of highly cross-neutralizing antibody responses in early patient sera. *J Med Virol* **53**, 51–59.

Lyubsky, S., Gavrilovskaya, I., Luft, B., Mackow, E., 1996. Histopathology of *Peromyscus leucopus* naturally infected with pathogenic NY-1 hantaviruses: Pathologic markers of HPS viral infection in mice. *Lab Invest J Tech Methods Pathol* **74**, 627–633.

Maganga, G.D., Bourgarel, M., Ebang Ella, G., Drexler, J.F., Gonzalez, J.P., Drosten, C., Leroy, E.M., 2011. Is Marburg virus enzootic in Gabon? *J Infect Dis* **204**(suppl. 3), S800–S803.

Maiztegui, J.I., 1975. Clinical and epidemiological patterns of Argentine haemorrhagic fever. *Bull World Health Organ* **52**, 567–575.

de Manzione, N., Salas, R.A., Paredes, H., Godoy, O., Rojas, L., Araoz, F., Fulhorst, C.F. et al., 1998. Venezuelan hemorrhagic fever: Clinical and epidemiological studies of 165 cases. *Clin Infect Dis* **26**, 308–313.

McCormick, J.B., Fisher-Hoch, S.P., 2002. Lassa fever. *Curr Top Microbiol Immunol* **262**, 75–109.

McCormick, J.B., Webb, P.A., Krebs, J.W., Johnson, K.M., Smith, E.S., 1987. A prospective study of the epidemiology and ecology of Lassa fever. *J Infect Dis* **155**, 437–444.

Mercado, R., 1975. Rodent control programmes in areas affected by Bolivian haemorrhagic fever. *Bull World Health Organ* **52**, 691–696.

Messenger, S.L., Rupprecht, C.E., Smith, J.S., 2003. Bats, emerging virus infections, and the rabies paradigm. In: Kunz, T.H., Fenton, M.B. (Eds.), *Bat Ecology*. The University of Chicago Press, Chicago, IL, pp. 622–679.

Milazzo, M.L., Cajimat, M.N.B., Duno, G., Duno, F., Utrera, A., Fulhorst, C.F., 2011. Transmission of Guanarito and Pirital viruses among wild rodents, Venezuela. *Emerg Infect Dis* **17**, 2209–2215.

Mills, J.N., Ellis, B.A., Childs, J.E., McKee, K.T., Jr., Maiztegui, J.I., Peters, C.J., Ksiazek, T.G., Jahrling, P.B., 1994. Prevalence of infection with Junin virus in rodent populations in the epidemic area of Argentine hemorrhagic fever. *Am J Trop Med Hyg* **51**, 554–562.

Mills, J.N., Ellis, B.A., McKee, K.T., Jr., Calderon, G.E., Maiztegui, J.I., Nelson, G.O., Ksiazek, T.G., Peters, C.J., Childs, J.E., 1992. A longitudinal study of Junin virus activity in the rodent reservoir of Argentine hemorrhagic fever. *Am J Trop Med Hyg* **47**, 749–763.

Mills, J.N., Ellis, B.A., McKee, K.T., Jr., Ksiazek, T.G., Oro, J.G., Maiztegui, J.I., Calderon, G.E., Peters, C.J., Childs, J.E., 1991. Junin virus activity in rodents from endemic and nonendemic loci in central Argentina. *Am J Trop Med Hyg* **44**, 589–597.

Monath, T.P., 1999. Ecology of Marburg and Ebola Viruses: Speculations and Directions for Future Research. *J Infect Dis* **179**(suppl. 1), S127–S138.

Monath, T.P., Newhouse, V.F., Kemp, G.E., Setzer, H.W., Cacciapuoti, A., 1974. Lassa virus isolation from *Mastomys natalensis* rodents during an epidemic in Sierra Leone. *Science* (Washington, DC) **185**, 263–265. PMID: 4833828.

Mulic, R., Ropac, D., 2002. Epidemiologic characteristics and military implications of hemorrhagic fever with renal syndrome in Croatia. *Croat Med J* **43**, 581–586.

Negredo, A., Palacios, G., Vazquez-Moron, S., Gonzalez, F., Dopazo, H., Molero, F., Juste, J. et al., 2011. Discovery of an ebolavirus-like filovirus in Europe. *PLoS Pathog* **7**, e1002304.

Netski, D., Thran, B.H., St Jeor, S.C., 1999. Sin Nombre virus pathogenesis in *Peromyscus maniculatus*. *J Virol* **73**, 585–591.

Niklasson, B., Hornfeldt, B., Mullaart, M., Settergren, B., Tkachenko, E., Myasnikov Yu, A., Ryltceva, E.V., Leschinskaya, E., Malkin, A., Dzagurova, T., 1993. An epidemiologic study of hemorrhagic fever with renal syndrome in Bashkirtostan (Russia) and Sweden. *Am J Trop Med Hyg* **48**, 670–675.

Nowak, R.M., Walker, E.P., 1994. *Walker's Bats of the World*. Johns Hopkins University Press, Baltimore, MD.

Okumura, M., Yoshimatsu, K., Kumperasart, S., Nakamura, I., Ogino, M., Taruishi, M., Sungdee, A. et al., 2007. Development of serological assays for Thottapalayam virus, an insectivore-borne Hantavirus. *Clin Vaccine Immunol* **14**, 173–181.

Olds, N., 1988. A revision of the genus *Calomys* (*Rodentia: Muridae*). PhD dissertation. University of New York, New York.

Oldstone, M.B., 2002. Arenaviruses. I. The epidemiology molecular and cell biology of arenaviruses. Introduction. *Curr Top Microbiol Immunol* **262**, V-XII.

Omatsu, T., Watanabe, S., Akashi, H., Yoshikawa, Y., 2007. Biological characters of bats in relation to natural reservoir of emerging viruses. *Comp Immunol Microbiol Infect Dis* **30**, 357–374.

Pensiero, M.N., Sharefkin, J.B., Dieffenbach, C.W., Hay, J., 1992. Hantaan virus infection of human endothelial cells. *J Virol* **66**, 5929–5936.

Peralta, L.A., Laguens, R.P., Cossio, P.M., Sabattini, M.S., Maiztegui, J.I., Arana, R.M., 1979. Presence of viral particles in the salivary gland of Calomys musculinus infected with Junin virus by a natural route. *Intervirology* **11**, 111–116.

Peterson, A.T., Bauer, J.T., Mills, J.N., 2004. Ecologic and geographic distribution of Filovirus disease. *Emerg Infect Dis* **10**, 40–47.

Peterson, A.T., Ryan Lash, R., Carroll, D.S., Johnson, K.M., 2006. Geographical potential for outbreaks of Marburg hemorrhagic fever. *Am J Trop Med Hyg* **75**, 9–15.

Pettersson L., K.J., Hardestam J, Lundkvist Å, Ahlm C, Evander M., 2008. Hantavirus RNA in saliva from patients with hemorrhagic fever with renal syndrome. *Emerg Infect Dis* **14**(3), 406–411.

Pourrut, X., Delicat, A., Rollin, P.E., Ksiazek, T.G., Gonzalez, J.P., Leroy, E.M., 2007. Spatial and temporal patterns of Zaire ebolavirus antibody prevalence in the possible reservoir bat species. *J Infect Dis* **196**(suppl. 2), S176–S183.

Pourrut, X., Leroy, E., Gonzalez, J.P., 2006. Ebola: From bats to Gorillas. *Gorilla J* 19.

Pourrut, X., Souris, M., Towner, J.S., Rollin, P.E., Nichol, S.T., Gonzalez, J.P., Leroy, E., 2009. Large serological survey showing cocirculation of Ebola and Marburg viruses in Gabonese bat populations, and a high seroprevalence of both viruses in *Rousettus aegyptiacus*. *BMC Infect Dis* **9**, 159.

Rowe, W.P., Pugh, W.E., Webb, P.A., Peters, C.J., 1970. Serological relationship of the Tacaribe complex of viruses to lymphocytic choriomeningitis virus. *J Virol* **5**, 289–292.

Sabattini, M.S., Gonzalez, L.E., 1967. Direct identification of Junin's virus in infected wild rodents in nature. *Rev Soc Argent Biol* **43**, 252–260.

Sabattini, M.S., Gonzles del Rio, L., Diaz, G., Vega, V.R., 1977. Infeccion natural y experimental de roedores con virus Junin. *Medicina* (B Aires) **37** (suppl. 3), 149–161.

Salazar-Bravo, J., Dragoo, J.W., Bowen, M.D., Peters, C.J., Ksiazek, T.G., Yates, T.L., 2002a. Natural nidality in Bolivian hemorrhagic fever and the systematics of the reservoir species. *Infect Genet Evol* **1**, 191–199.

Salazar-Bravo, J., Ruedas, L.A., Yates, T.L., 2002b. Mammalian reservoirs of arenaviruses. *Curr Top Microbiol Immunol* **262**, 25–63.

Salvato, M.S., Clegg, J.C.S., Buchmeier, M.J., Charrel, R.N., Gonzalez, J.P., Lukashevich, I.S., Peters, C.J., Romanowski, V., 2011. Family *Arenaviridae*. In: King, A.M.Q., Adams, M.J., Carstens, E.B., Lefkowitz, E.J. (Eds.), *Virus Taxonomy—Ninth Report of the International Committee on Taxonomy of Viruses*. Elsevier, London, U.K., pp. 715–723.

Settergren, B., Juto, P., Wadell, G., Trollfors, B., Norrby, S.R., 1988. Incidence and geographic distribution of serologically verified cases of nephropathia epidemica in Sweden. *Am J Epidemiol* **127**, 801–807.

Simmons, N.B., 2005. Order Chiroptera. In: Wilson, D.E., Reeder, D.M. (Eds.), *Mammal Species of the World: A Taxonomic and Geographic Reference*, 3rd edn. Johns Hopkins University Press, Baltimore, MD.

Sinclair, J.R., Carroll, D.S., Montgomery, J.M., Pavlin, B., McCombs, K., Mills, J.N., Comer, J.A. et al., 2007. Two cases of hantavirus pulmonary syndrome in Randolph County, West Virginia: A coincidence of time and place? *Am J Trop Med Hyg* **76**, 438–442.

Sluydts, V., Crespin, L., Davis, S., Lima, M., Leirs, H., 2007. Survival and maturation rates of the African rodent, *Mastomys natalensis*: Density-dependence and rainfall. *Integr Zool* **2**, 220–232.

Smith, D.H., Johnson, B.K., Isaacson, M., Swanepoel, R., Johnson, K.M., Killey, M., Bagshawe, A., Siongok, T., Koinange Keruga, W., 1982. Marburg-virus disease in Kenya. *Lancet* **319**, 816–820.

Strong, J.E., Wong, G., Jones, S.E., Grolla, A., Theriault, S., Kobinger, G.P., Feldmann, H., 2008. Stimulation of Ebola virus production from persistent infection through activation of the Ras/MAPK pathway. *Proc Natl Acad Sci U S A* **105**, 17982–17987.

Sumibcay, L., Kadjo, B., Gu, S.H., Kang, H.J., Lim, B.K., Cook, J.A., Song, J.W., Yanagihara, R., 2012. Divergent lineage of a novel hantavirus in the banana pipistrelle (*Neoromicia nanus*) in Cote d'Ivoire. *Virol J* **9**, 34.

Swanepoel, R., Leman, P.A., Burt, F.J., Zachariades, N.A., Braack, L.E.O., Ksiazek, T.G., Rollin, P.E., Zaki, S.R., Peters, C.J., 1996. Experimental inoculation of plants and animals with Ebola virus. *Emerg Infect Dis* **2**, 321–325.

Taniguchi, S., Watanabe, S., Masangkay, J.S., Omatsu, T., Ikegami, T., Alviola, P., Ueda, N. et al., 2011. Reston ebolavirus antibodies in bats, the Philippines. *Emerg Infect Dis* **17**, 1559–1560.

Teeling, E.C., Springer, M.S., Madsen, O., Bates, P., O'Brien S, J., Murphy, W.J., 2005. A molecular phylogeny for bats illuminates biogeography and the fossil record. *Science* (New York) **307**, 580–584.

Ter Meulen, J., Lukashevich, I., Sidibe, K., Inapogui, A., Marx, M., Dorlemann, A., Yansane, M.L., Koulemou, K., Chang-Claude, J., Schmitz, H., 1996. Hunting of peridomestic rodents and consumption of their meat as possible risk factors for rodent-to-human transmission of Lassa virus in the Republic of Guinea. *Am J Trop Med Hyg* **55**, 661–666.

Tesh, R.B., Wilson, M.L., Salas, R., De Manzione, N.M., Tovar, D., Ksiazek, T.G., Peters, C.J., 1993. Field studies on the epidemiology of Venezuelan hemorrhagic fever: Implication of the cotton rat Sigmodon alstoni as the probable rodent reservoir. *Am J Trop Med Hyg* **49**, 227–235.

Timen, A., Koopmans, M.P.G., Vossen, A.C.T.M., van Doornum, G.J.J., Gunther, S., van den Berkmortel, F., Verduin, K.M. et al., 2009. Response to imported case of Marburg hemorrhagic fever, the Netherlands. *Emerg Infect Dis* **15**, 1171–1175.

Tkachenko, E., Dekonenko, A., Ivanov, A., Dzaguova, T., Ivanov, L., Slonova, R., Nurgaleeva, R., Stepanenko, A., Ivanidze, E., Zagidullin, I., 1998. Hemorrhagic fever with renal syndrome and hantaviruses in Russia. In: J.F. Saluzzo, B.D. (Ed.), *Emergence and Control of Rodent-Borne Viral Diseases (Hantaviral and Arenal Diseases)*. Elsevier, Veyrier-du-lac, Annecy, France, pp. 63–72.

Tkachenko, E., Ivanidze, E., Zagidullin, I., Dekonenko, A., 1999. HFRS in Eurasia: Russia and the Republics of the former USSR. In: H. W. Lee, C.C., C. Schmaljohn (Ed.), *Manual of Hemorrhagic Fever with Renal Syndrome and Hantavirus Pulmonary Syndrome*. WHO Collaborating Center for Virus Reference and Research (Hantaviruses), Asian Institute for Life Sciences, Seoul, pp. 49–57.

Towner, J.S., Amman, B.R., Sealy, T.K., Carroll, S.A., Comer, J.A., Kemp, A., Swanepoel, R. et al., 2009. Isolation of genetically diverse Marburg viruses from Egyptian fruit bats. *PLoS Pathog* **5**, e1000536.

Towner, J.S., Pourrut, X., Albariño, C.G., Nkogue, C.N., Bird, B.H., Grard, G., Ksiazek, T.G., Gonzalez, J.-P., Nichol, S.T., Leroy, E.M., 2007. Marburg virus infection detected in a common African bat. *PLoS One* **2**, e764.

Vapalahti, O., Lundkvist, A., Kukkonen, S.K., Cheng, Y., Gilljam, M., Kanerva, M., Manni, T. et al., 1996. Isolation and characterization of Tula virus, a distinct serotype in the genus Hantavirus, family *Bunyaviridae*. *J Gen Virol* **77** (Pt 12), 3063–3067.

Vapalahti, O., Mustonen, J., Lundkvist, A., Henttonen, H., Plyusnin, A., Vaheri, A., 2003. Hantavirus infections in Europe. *Lancet Infect Dis* **3**, 653–661.

Vapalahti, K., Paunio, M., Brummer-Korvenkontio, M., Vaheri, A., Vapalahti, O., 1999. Puumala virus infections in Finland: Increased occupational risk for farmers. *Am J Epidemiol* **149**, 1142–1151.

de Villafañe, G., Kravetz, F.O., Donadio, O., Persich, R., Knecher, L., Torrs, M.P., Fernandez, N., 1977. Dinámica de las comunidades de roedores en agroecosistemas pampásicos. *Medicina* (B Aires) **37**, 128–140.

Vitullo, A.D., Hodara, V.L., Merani, M.S., 1987. Effect of persistent infection with Junin virus on growth and reproduction of its natural reservoir, *Calomys musculinus*. *Am J Trop Med Hyg* **37**, 663–669.

Vitullo, A.D., Marani, M.S., 1987. Mecanismos de transmisión del virus Junín en su resrvorio natural, Calomys musculinus. *Medicina* (B Aires) **47**, 440.

Vitullo, A.D., Merani, M.S., 1988. Is vertical transmission sufficient to maintain Junin virus in nature? *J Gen Virol* **69**, 1437–1440.

Vitullo, A.D., Merani, M.S., 1990. Vertical transmission of Junin virus in experimentally infected adult Calomys musculinus. *Intervirology* **31**, 339–344.

Walker, D.H., Wulff, H., Lange, J.V., Murphy, F.A., 1975. Comparative pathology of Lassa virus infection in monkeys, guinea-pigs, and *Mastomys natalensis*. *Bull World Health Organ* **52**, 523–534.

Wang, L.F., Walker, P.J., Poon, L.L., 2011. Mass extinctions, biodiversity and mitochondrial function: Are bats 'special' as reservoirs for emerging viruses? *Curr Opin Virol* **1**, 649–657.

Weaver, S.C., Salas, R.A., de Manzione, N., Fulhorst, C.F., Duno, G., Utrera, A., Mills, J.N., Ksiazek, T.G., Tovar, D., Tesh, R.B., 2000. Guanarito virus (*Arenaviridae*) isolates from endemic and outlying localities in Venezuela: Sequence comparisons among and within strains isolated from Venezuelan hemorrhagic fever patients and rodents. *Virology* **266**, 189–195.

Webb, P.A., Justines, G., Johnson, K.M., 1975. Infection of wild and laboratory animals with Machupo and Latino viruses. *Bull World Health Organ* **52**, 493–499.

Weiss, S., Witkowski, P.T., Auste, B., Nowak, K., Weber, N., Fahr, J., Mombouli, J.V. et al., 2012. Hantavirus in bat, Sierra Leone. *Emerg Infect Dis* **18**, 159–161.

Wilson, D.E., Reeder, D.M., 2005. *Mammal Species of the World: A Taxonomic and Geographic Reference*. Johns Hopkins University Press, Baltimore, MD.

Wong, S., Lau, S., Woo, P., Yuen, K.-Y., 2007. Bats as a continuing source of emerging infections in humans. *Rev Med Virol* **17**, 67–91.

Yan, L., Fang, L.Q., Huang, H.G., Zhang, L.Q., Feng, D., Zhao, W.J., Zhang, W.Y., Li, X.W., Cao, W.C., 2007. Landscape elements and Hantaan virus-related hemorrhagic fever with renal syndrome, People's Republic of China. *Emerg Infect Dis* **13**, 1301–1306.

Zaki, S.R., Nolte, K.B., 1999. Pathology, Immunohistochemistry and in situ hybridization. In: Lee, H.W., Calisher, C.E., Schmaljohn, C. (Eds.), *Manual of Hemorrhagic Fever with Renal Syndrome and Hantavirus Pulmonary Syndrome*. ASAN Institute for Life Sciences, Seoul, South Korea, pp. 143–154.

Zhang, Y.Z., Zou, Y., Fu, Z.F., Plyusnin, A., 2010. Hantavirus infections in humans and animals, China. *Emerg Infect Dis* **16**, 1195–1203.

8 Biosafety Issues and Clinical Management of Hemorrhagic Fever Virus Infections

George F. Risi and Thomas Arminio

CONTENTS

8.1 INTRODUCTION

The term "viral hemorrhagic fever" (VHF) is used to describe the spectrum of disease caused by members of four different families of enveloped RNA viruses (Bray, 2005). Currently, there are over 20 members in this group (Table 8.1). However, novel agents continue to be described, and additional viruses capable of causing VHF will likely be discovered. The illness is an acute systemic disease characterized by fever, malaise and prostration, and, depending on the agent and the severity of disease, generalized signs of increased vascular permeability and coagulation abnormalities (Jahrling et al., 2007).

Essential factors in pathogenesis include rapid infection of dendritic cells, monocytes, and macrophages leading to impaired innate and adaptive immune mechanisms, massive apoptotic death of lymphocytes, and death of cells in various target tissues resulting in the release of several mediators that alter vascular function and trigger coagulation disorders (Jahrling et al., 2007; Sanchez et al., 2007).

The incidence of disease caused by agents of VHF is increasing for several reasons including destruction of or encroachment on habitats that are reservoirs for the agents, human changes to the environment such as dam building and deforestation, global climate change, failures in public health policy, and increased international travel (Gubler, 2005; Morse, 1995; Peters, 1997; Woolhouse, 2006). For example, dengue and dengue hemorrhagic fever have been increasing

TABLE 8.1

Agents of VHF in Humans

Family	Virus	Disease	Geographic Distribution
Arenaviridae			
	Lassa	Lassa fever	West Africa
	Lujo	Lujo	Central Africa
	Junin	Argentine hemorrhagic fever	Argentina
	Machupo	Bolivian hemorrhagic fever	Bolivia
	Sabia	Brazilian hemorrhagic fever	Brazil
	Guanarito	Venezuelan hemorrhagic fever	Venezuela
	Whitewater arroyo	Hemorrhagic fever	Southwestern United States
Bunyaviridae			
	CCHF	CCHF	Africa, Central Asia, Eastern Europe, Middle East
	Rift Valley fever	Rift Valley fever	Africa, Saudi Arabia, Yemen
	Hantaan, Seoul, Puumala, others	Hemorrhagic fever with renal syndrome	Asia, Balkans, Europe
	Huaiyangshan virus (proposed)	Severe fever and thrombocytopenia syndrome	Eastern China
Filoviridae			
	Ebola (five species)	Ebola hemorrhagic fever	Africa
	Marburg	Marburg hemorrhagic fever	Africa
Flaviviridae			
	Dengue	Dengue hemorrhagic fever	Asia, Africa, Central and South America
	Yellow fever	Yellow fever	Africa, Central and South America
	Omsk	Omsk hemorrhagic fever	Central Asia
	Kyasanur forest disease	Kyasanur forest disease	India
	Alkhurma virus	Hemorrhagic fever	Saudi Arabia

steadily for more than 50 years, with now 50–100 million cases annually, 500,000 hospitalizations, and 22,000 deaths, mostly in children (Morens and Fauci, 2008; WHO, 2011b). Rift Valley fever is now a major threat throughout Africa (CDC, 2007) and has also spread to the Middle East, with large outbreaks in Saudi Arabia and Yemen (CDC, 2000). Due to improved surveillance, increased geographic range, and reestablishment of the mosquito vector in areas where it had been eradicated, yellow fever cases are increasing both in South America and Africa (CDC, 2010b). The World Health Organization estimates that 200,000 cases with 30,000 deaths occur annually worldwide (WHO, 1992). There have been at least 30 outbreaks of filovirus (Ebola virus and Marburg virus [MARV]) infection involving nearly 2800 laboratory-confirmed, suspect, or putative cases since the viruses were discovered in 1967 (Kortepeter et al., 2011a; Roddy et al., 2011). There are annual outbreaks of 300–1000 cases of Argentine hemorrhagic fever (Enria et al., 2008). And there has been expansion of the geographic range of Crimean–Congo hemorrhagic fever (CCHF). Since 2001, outbreaks have been described in Kosovo, Albania, Iran, Pakistan, and South Africa (WHO, 2011a). Over 1000 cases occur annually in Turkey alone (Maltezou et al., 2010).

These expansions, coupled with the ease of international travel allowing an infected individual to reach virtually any location in the world within the incubation period, make it increasingly likely that Western clinicians will encounter a patient suffering from a VHF. This happened recently in Germany (Haas et al., 2003), the Netherlands (Timen et al., 2009), New Jersey (CDC, 2004a), and Colorado (Paddock, 2009).

Because VHF agents are studied in several laboratories and clinical centers around the world, considerable experience has evolved regarding containment and manipulation. Clinical experience has allowed development of practices for rendering care to infected individuals with minimal provider risk. Additionally, there have been advances in therapy that have improved the prospects for survival from infection by many of these agents.

This chapter reviews key aspects of these topics from the perspectives of a biosafety officer responsible for safe management of a high-containment laboratory and the clinical consultant responsible for management of infected individuals and the training of a health-care work force to safely care for VHF patients.

8.2   BIOSAFETY AND BIOSECURITY ISSUES WITH VHF AGENTS

8.2.1   LABORATORY SAFETY ISSUES

Biosafety refers to the level of training required for safe manipulation of an agent, the proper functioning of life safety systems, plans for decontamination, and emergency response. Good microbiologic practices are the foundation upon which all levels of biosafety are layered.

With few exceptions, most agents of VHF are studied at biosafety level (BSL)-4. BSL-4 designates a complex containment design, elaborate personal protective equipment (PPE), and increased security requirements.

Biocontainment and biosecurity are other key aspects of these laboratories. Biocontainment refers to the design aspects that function to prevent the release of a biologic agent either into the laboratory or the outside environment. This includes biological safety cabinets, heating, ventilation, and air conditioning (HVAC) systems with directional airflow, high-efficiency particulate air filtration, and effluent decontamination. Biosecurity refers to the regulatory aspects of a program, proper vetting of new employees in terms of personal reliability and medical qualifications, and assurance that a new employee has met the training requirements and that proper safeguards are in place to ensure the safe storage, transport, and use of the agents. For institutions that fall under the U.S. Select Agent Program, biosecurity programs assure that all regulatory aspects of this program are met.

A safe program also includes a well-defined emergency medical response and the ability to provide high-quality medical care to an exposed individual in a designated hospital with modern isolation capability and trained health-care workers (HCWs).

There are many comprehensive references covering biosafety, biocontainment, and biosecurity (Chosewood and Wilson, 2009; Fleming and Hunt, 2006; Salerno and Gaudioso, 2007). Additionally, the American Biological Safety Association has excellent information and links to texts, regulations, guidelines, training, and certification. Thus, for the remainder of this review, we will focus on issues that have arisen in our experience of managing the biosafety program in a maximum-containment laboratory and developing the medical response and medical management in a designated medical facility.

To be most effective, the safety program should operate on an equal footing with the scientific program in terms of design, operation, and daily decision making in the facility.

There are two categories of BSL-4 laboratory: the suit lab where individuals wear air-supplied positive pressure suits and the cabinet lab where work is performed in a class III biological safety cabinet. In the United States, the suit lab predominates; however, cabinet labs require less infrastructure and also are better suited for aerosol studies. Suit labs are housed within sophisticated buildings with computerized building automation systems. These systems monitor and ensure proper operation of all systems in the laboratory. Many are considered vital to "life safety" and need to be checked daily before personnel enter the lab. All personnel who enter the lab should possess the knowledge to evaluate the systems for proper functioning. A mechanism for reporting systems that are out of range should be established, and facilities should determine what systems need to be cleared before individuals enter the lab. At a minimum, the following systems need to be checked daily: HVAC, liquid effluent decontamination, autoclaves, breathing air, and chemical shower disinfectant.

Training is vital to any laboratory program. Whether working in a suit or cabinet lab, the workers must be familiar and comfortable with the equipment they are using. The BSL-4 laboratory can be a challenging and stressful environment (Kaufman and Berkelman, 2007); a "mock laboratory" enables workers to explore this environment without the associated risk and should be a part of any BSL-4 laboratory design. Training should be designed to cover all of the unique facility and laboratory issues, be individualized so that previous experience is taken into consideration, and utilize mentorship where the individual is guided by an experienced worker and competency is documented before independent access to the laboratory is given (Delany et al., 2011).

Maintenance personnel are often overlooked in biosafety training programs, even though many of the life systems they oversee can become contaminated. The biosafety department needs to ensure adequate training for these individuals. Maintenance staff need to understand the nature of the work being done and how decontamination is performed and validated so that systems are safe to work on.

Since the anthrax attacks in the United States in 2001, there has been increased scrutiny of laboratories working with agents having the potential to be used as bioweapons, which includes all of the VHF agents. The Personal Reliability Program ensures that all workers who have access to the laboratory or critical infrastructure are appropriately vetted. For programs that fall under the U.S. Select Agent Program, this is clearly defined (www.selectagents.gov). Other aspects of personal reliability that should be considered are medical evaluations for individuals who will be wearing positive pressure suits. Some individuals may not be suited to work in this environment because of medical conditions, claustrophobia, etc. Behavioral health screenings may reveal problems that an individual may have performing in this environment. A qualified psychologist should administer and evaluate these screenings using validated tools, with results included in the individual's medical record (Skvorc and Wilson, 2011).

Laboratories should decide how personnel will be allowed to work in the BSL-4 laboratory and consider what has been called the "buddy system." Because of the unique environment of the BSL-4 laboratory, a worker can be isolated and in an emergency may not be able to help him or herself. There are several interpretations of the buddy system. With a two-person rule there are always at least two workers in the lab. Modifications allow a risk assessment to be performed and certain activities allowed with one person in the lab working with an individual on the outside. While these are policies that should be made by individual institutions, it is important to realize that the timeliness of an emergency response may be impacted (Ippolito et al., 2009; LeDuc et al., 2009).

A comprehensive emergency response plan needs to be developed that anticipates the broad range of events that might occur during the daily operation of the lab. Risk assessments can help determine the likelihood of occurrences. Periodic drills to test the plan against situations such as spills, fires, system failures, natural disasters, and medical emergencies are necessary. In the case of medical emergencies, BSL-4 workers must act as first responders. Egress from the laboratory is through multiple sealed doors, and in some emergencies, egress may not be through normal exit routes. A mock laboratory will allow for ongoing emergency response training, but when the actual lab is inactive and decontaminated, it should be used for training. This allows for the workers and emergency medical system (EMS) personnel to become familiar with the actual space.

When planning for medical response, two types of situations can occur in the BSL-4 laboratory. Situations such as a heart attack, a traumatic injury, or a serious animal bite will require the worker to be urgently transported to the emergency department for stabilization with possible subsequent transfer to a designated care facility for the potential exposure. In situations such as sharps injury or a suit breach that are not immediately life threatening, risk assessment can be performed, the degree of exposure determined, and the necessary response taken in a more reflective fashion.

Occupational medical personnel are important to the operation of the BSL-4 laboratory. They play a key role in the initial medical screening of the workers, perform any ongoing medical surveillance, administer vaccines, and should be involved in any medical emergency that occurs. These personnel thus need to be trained and involved in emergency response and care. A dedicated area should be established near the lab where triage can be performed, care can be delivered, and individuals can remain until a decision is made on the level of care needed.

Ongoing training and communication with the HCWs who will be responsible for caring for a potentially exposed BSL-4 worker is vital. This includes the EMS team responding and the hospital team receiving the worker. Questions will arise regarding the agent the individual has been working with and whether they have been adequately decontaminated. A list of the pathogens being studied should be available as well as information regarding modes of transmission, the appropriate PPE to be worn when transporting the individual, and discussion of the potential exposure risk.

Because all pathogens have a latent period of at least a few days between exposure and becoming contagious, in the situation of a recent laboratory exposure, once proper decontamination has been performed, that worker can be safely transported to the hospital with standard precautions only.

The circumstance may arise of the need to transport an individual who is displaying symptoms compatible with VHF. An example might be a worker who is recently returned from fieldwork in an endemic area who becomes ill several days after returning home. In deciding to transport the individual, such issues as whether specialized equipment will be utilized, the type of PPE required, and willingness of the local EMS providers to transport the individual (Mackler et al., 2007) should be considered in advance.

8.3   BIOSAFETY IN CLINICAL MANAGEMENT OF INFECTIONS

The receiving facility for a patient with suspected or confirmed infection with an agent of VHF must be appropriately designed and have trained staff to treat such individuals. Several models for providing care have been developed. The U.S. Army Medical Research Institute for Infectious Diseases duplicated the BSL-4 laboratory environment to care for the exposed worker. This approach was utilized at the Ft Detrick facility in 2007 (Kortepeter et al., 2008). A similar approach was employed in Germany in 2009 (Gunther et al., 2011). Drawbacks to this approach include the substantial time required to don and doff special PPE as well as to enter and exit the unit and the fact that many otherwise standard interventions such as venipuncture while wearing this unaccustomed and inconvenient gear may pose an increased exposure risk to the HCWs. Furthermore, clinical experience from several situations has documented that nosocomial transmission can be prevented by implementing standard, contact, and airborne isolation procedures such as those outlined in the U.S. Centers for Disease Control and Prevention's (CDC's) isolation guidelines (Siegel et al., 2007). The CDC has published guidelines for management of patients infected with VHFs in the conventional hospital setting (CDC, 1988, 1995). Furthermore, care has been safely rendered on multiple occasions using conventional barrier precautions alone to persons infected with Ebola (Emond et al., 1977; Richards et al., 2000), Marburg (Paddock, 2009; Timen et al., 2009), Lassa (Helmick et al., 1986), Machupo (CDC, 1994), Sabia (Armstrong et al., 1999), and CCHF viruses (Ergonul et al., 2007).

Despite this, even well-trained HCWs may make mistakes due to anxiety, fatigue, or other stressors, so additional facility enhancements that augment safety are desirable when dealing with these agents. Thus, when the National Institute of Allergy and Infectious Diseases constructed an integrated research facility with BSL-4 research space at its Rocky Mountain Laboratories in Hamilton, Montana, they contracted with a local hospital to support the BSL-4 research program. This includes a comprehensive educational program as well as facility enhancements. Details of this program are described elsewhere (Risi, 2010).

In addition to the increase in operational BSL-4 laboratories globally (Risi, 2010), virtually everywhere in the world is accessible within the incubation period of a VHF agent. Thus, all hospitals should be prepared for an individual presenting to their facility harboring an exotic disease.

Despite the fact that modern hospitals have all of the resources required to protect the HCW from exposure, the sensational misrepresentation of these agents by commercial media as well as the high lethality of the agents may result in reluctance among HCWs to care for these individuals. Studies have demonstrated that during a pandemic or bioterrorism attack, first responders and other HCWs may choose not to respond for a variety of reasons (Masterson et al., 2009). In developing a relationship with a health-care facility for the care and treatment of potentially exposed BSL-4 workers, the need for transparency with the HCWs is of the utmost importance. Typically, intensive care units (ICUs) are the areas that would be utilized. The ICU physicians, nurses, and all ancillary staff need to "buy in" and be able to clearly state their concerns before being asked to care for these patients. The current configuration of the rooms may be adequate or remodeling may be needed (Beam et al., 2010; Risi, 2010).

The designated hospital should have procedures in place to verify that respiratory isolation systems are working and that negative air pressure is being maintained. Procedures need to be established for decontaminating a room after occupancy. Access to the areas where an exposed worker is being cared for must be controlled. Public relations protocols should be developed and coordinated with the laboratory so that conflicting messages are not released to the public.

Training programs need to be developed that have both didactic and hands-on components. Since admitting a potentially exposed BSL-4 worker will be a rare occurrence, refresher courses need to be offered periodically so that staff can retain their knowledge and skill. Programs need to focus on the basics of hospital infection control and the effectiveness of primary barrier precautions using current guidelines. Pre- and posttests are useful to gauge capture and retention of knowledge. With VHF patients, the PPE used may be more elaborate, and donning and doffing procedures must be established and practiced.

A final issue is development of community relationships and trust. Failure to involve the community in planning or failure to be transparent in reporting exposures has caused delays in the implementation of programs (Lawler, 2005). The community must be informed and have the ability to comment on programs that might be construed as being dangerous or potentially causing community risk. Ongoing meetings where ideas and concerns can be shared foster good relations. Transparency is a requirement of any successful risk communication strategy (Dickman et al., 2009; Lawler, 2005).

8.4   CLINICAL MANAGEMENT OF VHFS

VHFs are found on every continent with the exception of Antarctica, but the distribution of a specific virus is restricted by the distribution of its natural reservoir or arthropod vector. Humans become infected in different ways depending on the virus, either through exposure to blood/excreta of an animal or by the bite of an arthropod. Thus, for a patient presenting with clinical symptoms of VHF, the most likely pathogen(s) can often be ascertained with a travel and potential exposure history. For patients infected through lab exposure, the agent will be known, unless dealing with clinical specimens from an as yet unidentified disease.

In the early stages of disease, the clinical presentation is nonspecific, and there is also a considerable overlap with mundane agents. Thus, it is important to maintain a broad differential diagnosis, which may be narrowed as the disease progresses or as laboratory information becomes available (Table 8.2). And until the diagnosis is confirmed, it is imperative to treat the mundane possible diseases as well. Failure to do this may result in fatalities from such diseases as malaria, typhoid fever, and dysentery due to premature narrowing of the differential.

There is frequently a paucity of the information clinicians use in making treatment decisions. Basic vital signs and their evolution over the course of the disease, urine output, and basic blood tests are frequently unavailable (Kortepeter et al., 2011a). This is often due to the occurrence of disease in resource-limited settings or in outbreak settings, where control of the outbreak has taken precedence over the provision of aggressive medical care (Bausch et al., 2007). And due to

TABLE 8.2

Differential Diagnosis of VHF Infection

Malaria

Influenza, other acute mundane viral diseases

Dysentery (*Shigella*, *Campylobacter*, enterohemorrhagic *Escherichia coli*)

Sepsis from bacterial causes (*partial listing only*)

    Typhoid fever (*Salmonella enterica* subspecies Typhi, Paratyphi, and others)

    Plague (*Yersinia pestis*)

    Tularemia (*Francisella tularensis*)

    Brucellosis (*Brucella* sp.)

    Q fever (*Coxiella burnetii*)

    Severe leptospirosis (Weil's disease, *Leptospira* sp.)

    Rickettsial diseases (Rocky Mountain spotted fever, typhus, and others)

    Staphylococcus aureus

    Streptococcus pyogenes

Viral hepatitis

Noninfectious causes of disseminated intravascular coagulopathy

Acute leukemia

concerns of potential nosocomial transmission, invasive procedures such as central hemodynamic monitoring that might be done to obtain such information are often not performed (Kortepeter et al., 2011a). However, all of these agents induce microvascular damage and capillary leak syndrome (Peters et al., 1991), and thus tachypnea, tachycardia, and mild hypotension, as well as fever and myalgia, are common to all of these diseases (McCormick et al., 1987; Peters and Zaki, 2006). In many circumstances, illness progresses to involve the respiratory system as well as coagulopathy, renal failure, and shock (Bossi et al., 2004).

Animal models have until recently provided data primarily on hematologic parameters, but due to the constraints of operating in high containment, very little has been described regarding hemodynamic monitoring, urine output, etc. (Basler and Amarasinghe, 2009; Kortepeter et al., 2011b). In a recent pilot study, rhesus macaques were implanted with multi-sensor devices and internal jugular vein catheters prior to being infected with the Zaire strain of Ebola virus (ZEBOV). Three of the animals survived exposure to 50 plaque-forming units (pfu) of virus given by intramuscular injection, and the remaining six succumbed to disease. After the onset of fever, lethal illness was characterized by a decline in mean arterial blood pressure, an increase in pulse and respiratory rate, lactic acidosis, and renal failure. Survivors showed less-pronounced change in those parameters. Interestingly, four animals were randomized to receive infusions of normal saline when they became hypotensive, and this treatment did not appear to affect the overall outcome. However, the authors described significant logistical challenges in delivering intravenous fluids (IVFs) in the high-containment laboratory, leading in some cases to delay (Kortepeter et al., 2011b). Further laboratory evaluations should elucidate similar hemodynamic parameters for other virus infections, and the effects of various interventions can be monitored more precisely. Unfortunately, good animal models do not yet exist for some diseases, such as CCHF (Whitehouse, 2004).

8.5 SUPPORT MEASURES AND GOAL-DIRECTED THERAPY

The extensive organ involvement, severe immunosuppressive lymphoid tissue lesions, and lack of evidence of a humoral immune response in fatal cases taken together suggest that limitation of virus replication must occur before improvements in survival can be expected (Sanchez et al., 2007). However, even in the absence of specific antiviral therapy, there are several examples where support measures such as IVF and management of electrolyte imbalance and coagulopathy have had a favorable effect on mortality, buying time for the patient to develop an adaptive immune response to infection. This has been most clearly demonstrated in dengue shock syndrome, where volume replacement has been the mainstay of therapy since 1975 (WHO, 1975) and several trials have documented improved mortality rates using a variety of fluids (Dung et al., 1999; Wills et al., 2005). Additionally, in the 1967 outbreak of Marburg hemorrhagic fever in Germany, patients were managed in an ICU setting including IVF and replacement of clotting factors. The overall mortality was 23% (Slenczka, 1999) compared to the overall mortality in Marburg outbreaks of 82.5% (CDC, 2010a). In the outbreak of ZEBOV that occurred in Kikwit, Democratic Republic of the Congo in 1995, the initial case fatality rate was approximately 93%. However, the authors noted higher frequencies of coughing, and hemoptysis among survivors that they postulated might have resulted from pulmonary edema due to the "more aggressive parenteral treatment given later in the outbreak, when patient care had improved…" Concomitant with this, the mortality rate dropped to 69% (Bwaka et al., 1999). In the large outbreak of Marburg hemorrhagic fever that occurred in Uige, Angola, in 2005, the treating physicians had the impression that the introduction of IVF provided a considerable benefit for some patients (Jeffs et al., 2007). In a case of Marburg hemorrhagic fever managed in a Colorado community hospital managed with multiple interventions, the patient survived (CDC, 2009). And aggressive interventional therapy is commonly employed in Turkey for the treatment of CCHFV infection (Ergonul, 2008).

Aggressive interventions have not always proved successful however. In 1996, a 46-year-old anesthetics assistant with ZEBOV infection received intensive support including mechanical ventilation, Swan–Ganz catheterization, and platelet transfusions. Despite this, she expired after a 23 day illness (Richards et al., 2000). In 2004, a 38-year-old businessman returned from Liberia to the United States with Lassa fever. Despite aggressive medical care, he expired after a 5 day hospitalization (CDC, 2004b). In 2009, a U.S. soldier died from CCHF despite aggressive care in a military hospital (ProMED, 2009). Several other instances have been recorded of failure to survive despite modern medical treatment of Lassa (Atkin et al., 2009), Lujo (Paweska, 2009), Marburg (Gear, 1989), and others. In addition to the lack of effective antiviral therapy, additional determinants of the outcome of disease may include the source of the exposure (needlestick poses a much higher risk of death vs. mucous membrane exposure) (WHO, 1978), nosocomial exposure associated with a higher risk vs. tick bite (Whitehouse, 2004), inoculum size (Whitehouse, 2004), time between exposure and receiving treatment, and viral load (Cevik et al., 2007). Pregnancy, especially in the third trimester, poses a high risk of fatality to both the fetus and the mother (Mupapa et al., 1999; Price et al., 1988).

Because of the examples of survival described earlier and because the clinical features of hemorrhagic fever virus infection are clinically indistinguishable from septic shock caused by bacterial infection (Bray and Mahanty, 2003; Monath, 2008), a consensus has emerged that feasible patients with suspected or proven VHF infection should be managed in a fashion similar to patients septic from bacterial infection (Bausch, 2008; Bossi et al., 2004; Roddy et al., 2011; Solomon and Thomson, 2009). This includes supportive measures such as IVF, inotropic agents, use of blood products and management of disseminated intravascular coagulation, oxygenation and ventilation, stress ulcer prophylaxis, and attention to secondary bacterial infections. The decision to place a central venous catheter for direct measurements of filling pressures and cardiac output and for ease of blood draws must be weighed against the risk of bleeding at the venipuncture site and should be done on a case-by-case basis. In settings where a high-risk exposure has occurred through needlestick or other sharps exposure but while the subject is still in the incubation period of disease, central line placement before onset of viremia, clinical illness, and the associated coagulopathy develops would reduce the likelihood of nosocomial transmission and might lessen the risk of bleeding.

International guidelines for management of severe bacterial sepsis and septic shock were published in 2004 and updated in 2008 (Dellinger et al., 2008). Sepsis and severe sepsis were defined utilizing a combination of general variables, hemodynamic, organ dysfunction, and tissue perfusion variables and blood tests for markers of inflammation. The goals of resuscitation in the first 6 h include all of the following: achieving a central venous pressure of 8–12 mmHg, mean arterial pressure of $\geq$65 mmHg, urine output of $\geq$0.5 mL/kg/h, and either central venous O_2 saturation of $\geq$70% or mixed venous O_2 saturation of $\geq$65%. The use of early goal-directed resuscitation has been shown in a randomized, controlled, single-center study to improve 28 day mortality for emergency department patients presenting with septic shock (Rivers et al., 2001).

The major difference between the treatment of bacteria-induced septic shock and that induced by the agents of VHF is whether specific antimicrobial therapy is available. In the absence of appropriate antibacterial therapy, the outcome in bacterial septic shock is also poor and is actually no better than the outcome of patients with VHF (Ibrahim, 2000). Specific antiviral therapy is currently available for some agents and is an ongoing area of research for others.

8.5.1 RIBAVIRIN

Ribavirin is a synthetic purine nucleoside analog first synthesized in 1972 (Sidwell et al., 1972). Ribavirin exerts its antiviral effect by several mechanisms, including inhibition of inosine monophosphate dehydrogenase, a critical enzyme in the synthesis of guanosine monophosphate (GMP). Depletion of intracellular pools of GMP and its di- and triphosphorylated derivatives results in inhibition of both DNA and RNA synthesis. Ribavirin also interferes with RNA capping, which is necessary for RNA stability and translation, inhibits viral polymerase, and induces lethal mutagenesis. Finally, ribavirin serves as an immunomodulator, inducing a switch from a T-helper (Th) 2 to a Th1 phenotype that results in more effective T cell–mediated killing. Because of its several mechanisms of action, ribavirin has a broad spectrum of antiviral activity and is effective against several DNA and RNA viruses (Graci and Cameron, 2006). A variety of in vitro and animal models have demonstrated significant inhibitory effects of ribavirin against Lassa, Junin, Machupo, CCHF, Rift Valley fever, and Hantaan viruses (Canonico et al., 1984; Huggins, 1989).

On the basis of in vitro and animal data as well as clinical data that are stronger for some infections than others, ribavirin is indicated for the treatment of Lassa fever, New World arenavirus infections, CCHF, and perhaps Hantaan and Rift Valley fever (Peters and Zaki, 2006). It is not efficacious and should not be used for filovirus infections. Flaviviruses, in general, are less susceptible than are arenaviruses, bunyaviruses, and hantaviruses (Monath, 2008). The concentration of ribavirin that is effective in in vitro system against arenaviruses and bunyaviruses is approximately 25 µg/mL, whereas 10-fold higher concentrations are required to suppress yellow fever virus (Huggins et al., 1984). There are no convincing data that ribavirin can improve survival in flavivirus infections.

Ribavirin was demonstrated to be effective in monkeys lethally infected with Lassa and other arenaviruses (Peters et al., 1987). A human trial of treatment of Lassa fever was performed (McCormick et al., 1986) in Sierra Leone, West Africa. The IV administration of ribavirin to patients with aspartate transaminase elevations of at least 150 IU reduced mortality from 55% (33 of 60 patients) to 5% (1 of 20 patients) if treatment was begun before day 7 of the disease. Patients whose treatment began 7 or more days after the onset of fever had a mortality rate of 26% (11 of 43 patients). Viremia with levels of $\geq$10$^{3.6}$ tissue culture infectious doses (TCID) per mL on admission had a mortality rate of 76% (35 of 46 patients). Patients with this risk factor who were treated with IV ribavirin within the first 6 days after onset of fever had a mortality rate of 9% (1 of 11 patients), whereas the rate for those treated after 7 days was 47% (9 of 19 patients). Oral ribavirin was more effective than no therapy but less effective than IV treatment in all groups. Thus, ribavirin is indicated for treatment of disease caused by Lassa at all stages of illness. Oral therapy should be used if IV therapy is not available.

Ribavirin has some activity against New World arenaviruses, and limited data favor use of this drug for these infections. It was studied in two small trials of Junin infection. In the first open study, six patients received the drug and the mortality rate was 50%. The second double-blind placebo-controlled trial involved 18 patients of whom 8 received ribavirin and 10 placebos.

Seven of the 8 ribavirin recipients survived. Overall for the 14 ribavirin recipients in whom the diagnosis of infection was confirmed, 10 (71.4%) survived (Enria et al., 2008). An even smaller experience with Machupo is also favorable. Two patients were treated in 1994 with IV ribavirin using the same dosing protocol as used by McCormick for Lassa fever. One had laboratory-confirmed infection and one had clinically suspected infection. The first patient began receiving the drug 12 days after the appearance of the first symptoms, and the second not until 15 days after the appearance of the first symptoms. Both recovered; however, since most patients who die of this disease succumb between the 7th and 12th days of illness, the contribution of ribavirin to their recovery may have been small (Kilgore et al., 1997). In 1994, a 37-year-old laboratory technician suffered an exposure as the result of aerosol exposure to blood from a broken test tube in a centrifuge. She became ill and had initiation of ribavirin the following day. She had no hemorrhagic manifestations and recovered (CDC, 1994). In that same year, an agricultural worker from Bolivia developed a febrile hemorrhagic illness subsequently confirmed as Machupo. He was treated with IV ribavirin and recovered (CDC, 1994). The Working Group for Civilian Biodefense has recommended the use of ribavirin for all New World Arenaviruses in the setting of a contained casualty situation in which a modest number of patients require therapy (Borio et al., 2002).

For CCHFV infection, a number of observational studies have indicated that treatment with ribavirin is beneficial, but no placebo-controlled trials have been performed. Many consider the data from these observational studies to be strong enough that it would be unethical to conduct a placebo-controlled trial (Ergonul, 2008).

During an outbreak of CCHF at a teaching hospital near Cape Town, South Africa, six of nine HCWs who had penetrating injuries from contaminated needles were given ribavirin prophylactically (Van de Wal et al., 1985). One of these workers had a mild clinical course; the other five remained asymptomatic and did not develop measurable antibodies to the virus. Two of the three HCWs who were not treated with ribavirin developed severe disease. The fact that one worker and an additional 42 individuals who had contact with contaminated blood and who received no treatment remained free of illness makes it impossible to draw firm conclusions about the prophylactic value of ribavirin in this outbreak situation (Huggins, 1989). A report from South Africa in 1985 described the use of IV ribavirin in one patient in a nosocomial outbreak. The patient survived, but seven of the remaining eight patients in the outbreak who did not receive ribavirin also survived (VanEeden et al., 1985). In Pakistan in 1995, three HCWs who were severely ill with CCHF were given oral ribavirin, as the IV form was unavailable. The authors state that the expected mortality rate of these patients would have been 90% based on published literature. All three survived (Fisher-Hoch et al., 1995). In a series of 35 patients with confirmed CCHF at a tertiary care referral hospital in Ankara, Turkey, 27 had severe disease. All patients received supportive therapy. Eight of these patients were given oral ribavirin therapy and all survived. Of the remainder there was one fatality. The authors suggested that ribavirin should be administered to patients with severe illness (Ergonul et al., 2004).

For Rift Valley fever, ribavirin has been shown to be efficacious in both rodent and nonhuman primate (NHP) models both for postexposure prophylaxis (Peters et al., 1986) and treatment of infection (Huggins, 1989). Additionally, oral ribavirin has been successful in prophylaxis of the related Sicilian sandfly fever virus in humans (Huggins, 1989). The Working Group for Civilian Biodefense has recommended IV ribavirin be administered in the case of a contained casualty situation with Rift Valley fever infections, and in the case of mass casualties, an oral regimen is recommended (Borio et al., 2002).

The recommended dose for ribavirin therapy in patients with VHF is a loading dose of 30 mg/kg IV (maximum 2 g) followed by 16 mg/kg (maximum 1 g per dose) every 6 h for 4 days, followed by 8 mg/kg (maximum 500 mg) every 8 h for 6 days (Borio et al., 2002). The drug should be diluted in 150 mL of 0.9% saline and slowly infused (Bausch, 2008). Recommended dose for oral therapy is 2 g once followed by 600 mg twice daily for 10 days. For persons ≤75 kg, the dose is 400 mg in AM and 600 mg in PM (Borio et al., 2002). The main side effects of ribavirin are hemolytic anemia, which infrequently requires transfusion, as well as rigors if the IV drug is infused too rapidly. Ribavirin exhibits significant bioaccumulation and can still be found in the system for up to 6 months after the last dose. It is teratogenic and women receiving the drug should not conceive for at least 6 months after cessation of treatment. If treatment of pregnant women with VHF infection is contemplated, the risks and benefits of the drug must be thoroughly considered and discussed. Further information on use of ribavirin can be obtained from the package inserts from any one of the several manufacturers of the drug.

8.5.2 Modulation of the Procoagulant Response

Mention has been made of the similarities between the clinical presentation of VHF infection and sepsis induced by bacteria. Severe sepsis results from a generalized inflammatory and procoagulant response to infection (Bone et al., 1997), effecting an imbalance between pro- and anticoagulant forces leading to microvascular thrombosis, depletion of clotting factors, disseminated intravascular coagulation, and end-organ hypoxia. A number of anticoagulant therapies have been investigated for septic shock that have not resulted in benefit for a variety of reasons (Angus and Crowther, 2003). In November 2001, the U.S. Food and Drug Administration approved recombinant human activated protein C (rhAPC) for the treatment of patients with severe sepsis or septic shock. This approval was based primarily on the results of the rhAPC Worldwide Evaluation in Severe Sepsis (PROWESS) study that demonstrated improved survival in patients with sepsis regardless of the microbial source (Bernard et al., 2001). Subsequently, Hensley et al. investigated this drug in primates infected with ZEBOV. Fourteen macaques were challenged with a uniformly fatal dose of virus. Eleven of them were then treated with rhAPC beginning 30–60 min after challenge and continued for 7 days. Three were given saline. Two of the 11 treated monkeys survived. All three control monkeys died on day 8, whereas the mean survival for the treated non-survivors was 12.6 days. There was a significant difference in D-dimer levels between controls (highest), non-surviving treated animals (intermediate), and survivors (lowest). Lower levels of interleukin (IL)-6 and IL-10 were also found in survivors (Hensley et al., 2007).

After approval of rhAPC in 2001, early surveys and post marketing studies suggested that the risk of bleeding during clinical use might be greater than was noted in the original PROWESS study (Eichacker and Natanson, 2007). Additionally, two additional randomized controlled sepsis trials conducted post approval have failed to show any significant benefit from the drug (Abraham et al., 2005; Eichacker and Natanson, 2007). The major problem in comparing sepsis trials in humans with the potential use in VHF is that patients with bacterial sepsis are older, frequently have comorbidities (Martin et al., 2003), and present in advanced stages of illness, all of which put them at higher risk for adverse outcomes. The efficacy of rhAPC in VHF in humans has not been evaluated. Earlier use of the drug in select circumstances, particularly in the setting of high-risk occupational exposure to disease warrants further investigation.

A second approach modulating the coagulation cascade has been with the use of an inhibitor of the factor VIIa/tissue factor complex. Recombinant nematode anticoagulant protein c2 (rNAPc2) directly inhibits the fVIIa/TF complex by binding to factor X (Vlasuk and Rote, 2002). The antithrombotic potential of this drug had been shown in phase II trials for prevention of venous thromboembolism in orthopedic surgery (Lee et al., 2001) and in cardiovascular surgery (Moons et al., 2003). Geisbert and colleagues used a rhesus model of Ebola hemorrhagic fever that produces near 100% mortality. RNAPc2 was given either 10 min or 24 h after high-dose lethal injection of virus. Both treatment regimens prolonged survival time, with 33% survival in each group. The mean survival time for treated monkeys that ultimately died was 11.7 days compared to 8.3 days for untreated controls. Additionally, coagulation factors such as D-dimer levels were reduced in all treated animals (Geisbert et al., 2003). And for the Angola strain of MARV (considered to be the most lethal strain), a trial in rhesus monkeys resulted in mean survival of approximately 1.5 days in five of six challenged animals and survival in one, compared to 100% mortality in six control monkeys (Geisbert et al., 2007). The upregulation of tissue factor in monkeys infected with Marburg was found not to be as striking as for Ebola, providing a partial explanation for these less impressive results.

When a laboratory worker in Hamburg received a needlestick injury resulting in the potential transmission of ZEBOV infection, an expert international panel recommended that RNAPc2 be made available in the event of clinical illness. ARCA Biopharma released a batch of the drug to these clinicians for potential use. Fortunately for the patient, clinical illness did not develop (Gunther et al., 2011) and thus the drug was not utilized.

8.5.3 ANTISENSE THERAPY

There have been some very promising studies in the use of virus gene–specific oligonucleotides to interfere with translation of viral messenger RNA (mRNA). These studies have taken place using filoviruses but have broad potential applicability. RNA interference, also known as RNA silencing or posttranscriptional gene silencing, is an evolutionarily conserved mechanism found to operate in fungi, plants, and animals. In nature, double-stranded RNA (dsRNA) molecules are processed by a dsRNA-specific RNAse-named dicer into small fragments of 20–25 base pairs known as short-interfering RNAs or siRNAs. These siRNAs selectively associate with a multiprotein complex called RISC (RNA-induced silencing complex) that unwinds the siRNA duplex to produce a single-stranded RNA, which guides RISC to the target mRNA. The target mRNA is then cleaved by a ribonuclease, silencing the expression of the gene (Cullen, 2004). Geisbert and colleagues identified an siRNA, EK-1, that targeted the polymerase L protein of ZEBOV. When formulating this in stable nucleic acid lipid particles (SNALPs), guinea pigs were completely protected after a lethal challenge with ZEBOV (Geisbert et al., 2006). Subsequently, a combination of siRNAs targeting the polymerase L, VP 24, and VP 35 was tested in an NHP model. When given 30 min and 1, 3, and 5 days after lethal challenge, two of three rhesus macaques were protected from death. When treatment was given 30 min and 1, 2, 3, 4, 5, and 6 days after challenge, four of four macaques were protected. Animals became viremic but viremia never exceeded 2.4 $\log_{10}$ pfu/mL, whereas infected and untreated animals developed viremia of >4.5 $\log_{10}$ pfu before succumbing to disease (Geisbert et al., 2010). A similar study was performed using modified oligonucleotides (phosphorodiamidate morpholino oligomers or PMOs). These are (+) charged, ssDNA analogs that bind to complementary sequences of mRNA. In this study, 60% protection was conferred against lethal ZEBOV challenge and 100% protection against lethal Lake Victoria Marburgvirus infection in NHPs (Warren et al., 2010). PMOs have advanced into human trials. In a phase one trial, a combination of two PMOs with positive charges (PMOplus™, AVI BioPharma) that specifically target mRNA encoding MARV proteins has been given in ascending doses up to 6 mg/kg to healthy human volunteers. The compound was well tolerated with no clinically significant abnormalities detected (Axtelle et al., 2011). A parallel dose-escalation trial using two PMOs directed against Ebola virus, with similar results, is ongoing (Tim Axtell, AVI BioPharma, personal communication October 25, 2011).

Antisense therapy is currently considered the most promising avenue of research into treatment of filovirus infection either as postexposure or for established disease. Additionally, it may have broad applicability in treatment of a variety of infectious diseases (Castanotto and Rossi, 2009), including other agents of VHF. In the German experience described earlier, the international expert panel also advised that siRNA treatment be made available, and the drug was provided by Tekmira Pharmaceuticals (Gunther et al., 2011).

8.5.4 THERAPEUTIC VACCINES

Use of a vaccine after exposure represents another promising avenue in the early treatment of VHF infection. The majority of the work has been done for the filoviruses. A variety of methodologies have been employed (Bausch and Geisbert, 2007; Reed and Mohamadzadeh, 2007). Briefly, all attempts using inactivated virus were unsuccessful. Several strategies using recombinant viruses and/or DNA vaccination strategies were successful in rodents, but not in NHPs. The first vaccine of proven efficacy against NHPs was a combined DNA prime and adenovirus boost approach; however, this protocol required >6 months to provide protective immunity and is, therefore, of limited utility. Subsequent studies using only a single dose of NP and glycoprotein (GP) expressing recombinant adenovirus protected NHPs just 28 days after immunization (Sullivan et al., 2003). A modification of the adenovirus serotype 5 Ebola vaccine was developed to enhance the expression of the envelope antigen. This vaccine provided full protection to mice when given 30 min after to 1000 X LD50 of mouse-adapted ZEBOV (Richardson et al., 2009). Use of this vaccine concomitant with an adenovirus-based vaccine expressing interferon-α (IFN-α) gave full protection to guinea pigs 30 min and 24 h after lethal exposure. In NHPs, intramuscular vaccination with the combination therapy protected six of nine animals 30 min postexposure and one of three when administered 24 h postexposure. Coadministration of both vaccine and IFN-α coding adenovirus at the same site versus different sites potentiated postexposure survival, suggesting that IFN-α may be an adjuvant to adenovirus-based vaccines (Richardson et al., 2011).

A larger body of literature has been developed using recombinant vesicular stomatitis virus (rVSV) as the viral vector. Three trials, using the same platform but expressing GP from three different filoviruses, have been reported. Daddario-DiCaprio et al. achieved complete protection in five rhesus macaques challenged with 1000 pfu of the Musoke strain of MARV (MARV-Musoke) by administering vaccine (1×10^7 pfu of rVSV expressing the GP of Marburg) 20–30 min after challenge (Daddario-DiCaprio et al., 2006). Three control animals immunized with the same dose of rVSV expressing the GP of ZEBOV as a control succumbed to infection, demonstrating that protection requires expression of the appropriate surface glycoprotein.

For Ebola virus, Feldmann et al. gave eight macaques 2×10^7 pfu of vaccine 20–30 min after lethal challenge with 1000 pfu of ZEBOV. Survival was 50%, but one of the non-surviving animals actually succumbed at day 18 from septicemia and peritonitis caused by *Streptococcus pneumoniae* 18 days post challenge. This animal had cleared the ZEBOV infection by day 10. The three animals that succumbed to infection lived one to two days longer than the controls who had received Marburg or Lassa GPC prior to ZEBOV challenge (Feldmann et al., 2007). A third study reported complete protection from Sudan Ebola virus (SEBOV) of four rhesus macaques immunized with SEBOVGP 20–30 min after lethal challenge of 1000 pfu of SEBOV (Geisbert et al., 2008).

The reason for increased survival rates in the MARV and SEBOV vaccine trials compared to ZEBOV trial may be due to the length of the disease course in this animal model. In the study with SEBOV, the control animal did not succumb to illness until day 17. Similarly, the mean time to demise after challenge with MARV is 11.6 days. This is in comparison to a mean of 8.3 days survival after challenge with ZEBOV (Geisbert et al., 2008).

The exact mechanism of postexposure protection conferred by the VSVΔG/filovirus GP vector remains unclear. Interference between the rVSVs and wild-type filoviruses is a possibility. Several observations are pertinent (Garbutt et al., 2004): (1) Replication-competent virus is necessary. (2) Expression of specific GP is essential. (3) All of the rVSV vectors expressing filovirus GPs replicate much faster than their counterpart wild-type filovirus. As these rVSV vectors contain a full-length filoviral GP, they presumably target and infect the same preferred host cells as wild-type filoviruses. However, it is clear that a species-specific adaptive immune response is required, as MARV-infected macaques treated with ZEBOV GP died in parallel with untreated controls, showing that even though MARV and ZEBOV in general infect the same populations of host cells, an adaptive response is required to clear the virus and protect the host (Geisbert et al., 2008).

The rVSV expressing the ZEBOV GP was given to a researcher who received a needlestick injury from a syringe containing ZEBOV. A dose of 5×10^7 pfu was administered by IM injection 48 h after the exposure. The vaccinee developed fever 12 h later, and rVSV viremia was detectable for 2 days. Otherwise the person remained healthy, and ZEBOV RNA, except for the GP gene expressed in the vaccine, was never detected in serum and peripheral mononuclear cells during the 3 week observation period (Gunther et al., 2011).

The rVSV platform was utilized to make a vaccine against Lassa. It was given to four NHPs prior to lethal challenge. All four vaccinated animals survived vs. death in both controls (Geisbert et al., 2005). There are no published data on its use postexposure.

8.5.5 ADDITIONAL MODALITIES

A large number of products are in preclinical investigation for the treatment of VHF. The interested reader is referred to several reviews (deWit et al., 2011; Kunz and de la Torre, 2005; Monath, 2008; Sidwell and Smee, 2003; Yen et al., 2011).

Specific antiviral therapy has not been developed for the flaviviruses Omsk hemorrhagic fever virus and Kyasanur forest disease virus. However, it was recently reported that a currently licensed vaccine for tick-borne encephalitis virus (TBEV) (FSME-Immun, Baxter) may provide protection against Omsk. Serum from 41 healthy adults aged 16–65 who received three doses of vaccine had serum obtained 21 days after their third injection. High titers of neutralizing antibody against four variants of TBEV as well as against Omsk were seen (Orlinger et al., 2011). It is unlikely that this formalin-inactivated whole-virus vaccine would provide postexposure protection. Significant protection is not generated until after two doses of vaccine. Neutralization against TBEV was obtained in 89.3% of 56 volunteers 3 days after a second dose of vaccine, given 14 days after the first (Loew-Baselli et al., 2006).

Several platforms have been under development for vaccines against Rift Valley fever (Ikegami and Makino, 2009). None of these are yet licensed for humans.

8.6 SUMMARY AND CONCLUSIONS

From the earlier discussion, it can be seen that huge progress has been made in the study of the agents responsible for VHFs. Research and experience have led to an understanding of modes of transmission and methods for prevention of transmission both in the laboratory and in the health-care setting. While these agents must always be treated with respect, diligent application of tried and proven techniques will allow these agents to continue to be studied and infections treated. And while treatment modalities are still less than ideal, the situation is not nearly as dismal as one is led to believe from the popular press accounts of disease. One can be optimistic that further developments in both vaccines and therapies will allow even greater control of the agents of hemorrhagic fever.

REFERENCES

Abraham, E., Laterre, F., and Garg, R., 2005. Drotrecogin alfa (activated) for adults with severe sepsis and a low risk of death. *The New England Journal of Medicine* 353, 1332–1341.

Angus, D.C. and Crowther, M.A., 2003. Unraveling severe sepsis. Why did OPTIMIST fail and what's next? *Journal of the American Medical Association* 290, 256–258.

Armstrong, L., Dembry, L., Rainey, P., Russi, M., Khan, A., and Fischer, S., 1999. Management of a Sabia virus infected patient in a US hospital. *Infection Control and Hospital Epidemiology* 20, 176–182.

Atkin, S., Nanakari, S., Gothard, P., Walsh, A., Brown, D., 2009. The first case of Lassa fever imported from Mali to the United Kingdom, February, 2009. *Eurosurveillance* 14, 1–3.

Axtelle, T., Vutikullird, A., Heald, A., and Shrewsbury, S., 2011. A phase 1, single ascending dose study of AVI-6003, a combination of two PMOplus™ compounds with activity against Marburg virus. Abstracts of the *Infectious Disease Society of America's 49th Annual Meeting*, Boston, MA, abstract #1183.

Basler, C.F. and Amarasinghe, G.K., 2009. Evasion of interferon responses by Ebola and Marburg viruses. *Journal of Interferon and Cytokine Research: The Official Journal of the International Society for Interferon and Cytokine Research* 29, 511–520.

Bausch, D.G., 2008. Viral Hemorrhagic Fevers, in: Schlossberg, D. (Ed.), *Clinical Infectious Disease*. Cambridge University Press, New York, pp. 1319–1332.

Bausch, D.G., Feldmann, H., Geisbert, T.W., Bray, M., and Sprecher, A., 2007. Outbreaks of filovirus hemorrhagic fever: Time to refocus on the patient. *Journal of Infectious Disease* 196, S136–S141.

Bausch, D.G., Geisbert, T.W., 2007. Development of vaccines for Marburg hemorrhagic fever. *Expert Review of Vaccines* 6, 57–74.

Beam, E.L., Boulter, K.C., Freihaut, F., Schwedhelm, S., and Smith, P.W., 2010. The Nebraska experience in biocontainment patient care. *Public Health Nursing* 27, 140–147.

Bernard, G.R., Vincent, J.-L., Laterre, P.-F., LaRosa, S.P., and Dhainaut, J.-F., 2001. Efficacy and safety of recombinant human activated protein c for severe sepsis. *The New England Journal of Medicine* 344, 699–709.

Bone, R.C., Grodzin, C., and Balk, R., 1997. Sepsis: A new hypothesis for pathogenesis of the disease process. *Chest* 112, 235–243.

Borio, L., Inglesby, T., Peters, C.J., Schmaljohn, A.L., Hughes, J.M., and Jahrling, P.B., 2002. Hemorrhagic fever viruses as biological weapons. Medical and public health management. *Journal of the American Medical Association* 287, 2391–2405.

Bossi, P., Tegnell, A., Baka, A., Van Loock, F., and Hendriks, J., 2004. Bichat guidelines for the clinical management of haemorrhagic fever viruses and bioterrorism-related haemorrhagic fever viruses. *Eurosurveillance* 9, 1–8.

Bray, M., 2005. Pathogenesis of viral hemorrhagic fever. *Current Opinion in Immunology* 17, 399–403.

Bray, M. and Mahanty, S., 2003. Ebola hemorrhagic fever and septic shock. *The Journal of Infectious Diseases* 188, 1613–1617.

Bwaka, M.A., Bonnet, M.-J., Calain, P., Colebunders, R., DeRoo, A., and Guimard, Y., 1999. Ebola hemorrhagic fever in Kikwit, Democratic Republic of the Congo: Clinical observations in 103 patients. *Journal of Infectious Disease* 179, S1–S7.

Canonico, P.G., Kende, M., Luscri, B., and Huggins, J.W., 1984. In-vivo activity of antivirals against exotic RNA viral infections. *Journal of Antimicrobial Chemotherapy* 14, 27–41.

Castanotto, D. and Rossi, J.J., 2009. The promises and pitfalls of RNA-interference-based therapeutics. *Nature* 457, 426–433.

CDC, 1988. Management of patients with suspected viral hemorrhagic fever. *Morbidity and Mortality Weekly Report* 37, 1–15.

CDC, 1994. International notes: Bolivian hemorrhagic fever- El-Beni Dept, Bolivia, 1994. *Morbidity and Mortality Weekly Report* 43, 943–946.

CDC, 1995. Update: Management of patients with suspected viral hemorrhagic fever- United States. *Morbidity and Mortality Weekly Report* 44, 475–479.

CDC, 2000. Update: Outbreak of Rift Valley fever—Saudi Arabia, August-November, 2000. *Morbidity and Mortality Weekly Report* 49, 982–985.

CDC, 2004a. Imported Lassa fever—New Jersey, 2004. *Morbidity and Mortality Weekly Report* 53, 894–897.

CDC, 2004b. Imported Lassa fever—New Jersey, 2004. *Morbidity and Mortality Weekly Report* 53, 894–897.

CDC, 2007. Update: Outbreak of Rift Valley fever—Kenya. *Morbidity and Mortality Weekly Report* 56, 73–77.

CDC, 2009. Imported case of Marburg hemorrhagic fever—Colorado, 2008. *Morbidity and Mortality Weekly Report* 58, 1377–1381.

CDC, 2010a. Known cases of Marburg hemorrhagic fever, p. www.cdc.gov/ncidod/dvrd/spb/mnpages/dlspages/marburg/marburgtable.htm.

CDC, 2010b. Yellow fever vaccine: Recommendations of the Advisory Committee on Immunization Practices. *Morbidity and Mortality Weekly Report* 59, 1–22.

Cevik, M.A., Erbay, A., Bodur, H., Eran, S.S., and Akinci, A., 2007. Viral load as a predictor of outcome in Crimean-Congo hemorrhagic fever. *Clinical Infectious Diseases* 45, e96–e100.

Chosewood, L.C. and Wilson, D.E., 2009. *Biosafety in Microbiological and Biomedical Laboratories*, 5th edn. U.S. Department of Health and Human Services, Washington, DC.

Cullen, B., 2004. Derivation and function of small interfering RNA's and micro-RNA's. *Virus Research* 102, 3–9.

Daddario-DiCaprio, K.M., Geisbert, T.W., Stroher, U., Geisbert, J.B., Grolla, A., and Fritz, E., 2006. Post-exposure protection against Marburg hemorrhagic fever with recombinant vesicular stomatitis virus vectors in non-human primates: An efficacy assessment. *Lancet* 367, 1399–1404.

Delany, J.R., Pentella, M.A., Rodriguez, J.A., Shah, K.J., Baxley, K.B., and Holmes, D.B., 2011. Guidelines for Biosafety Laboratory Competency. *Morbidity and Mortality Weekly Report* 60, 1–6.

Dellinger, R.P., Levy, M.M., Carlet, J.M. et al., 2008. Surviving Sepsis Campaign: International guidelines for management of severe sepsis and septic shock: 2008. *Critical Care Medicine* 36, 296–327.

Dickman, P., Keith, K., Comer, C., Abraham, G., Gopal, R., and Marvi, E., 2009. Report of the International Conference on Risk Communication Strategies for BSL-4 Laboratories, Tokyo, October 3–5, 2007. *Biosecurity and Bioterrorism* 7, 227–233.

Dung, N.M., Day, N., Tam, D.T., Loan, H.T., and Chau, H., 1999. Fluid replacement in dengue shock syndrome: A randomized, double blind comparison of four intravenous fluid regimens. *Clinical Infectious Diseases* 29, 787–794.

Eichacker, P.Q. and Natanson, C., 2007. Increasing evidence that the risks of rhAPC may outweigh its benefits. *Intensive Care Medicine* 33, 396–399.

Emond, R., Evans, b., Bowen, E., and Lloyd, G., 1977. A case of Ebola virus infection. *British Medical Journal* 2, 541–544.

Enria, D.A., Briggiler, A.M., and Sanchez, Z., 2008. Treatment of Argentine hemorrhagic fever. *Antiviral Research* 78, 132–139.

Ergonul, O., 2008. Treatment of Crimean-Congo hemorrhagic fever. *Antiviral Research* 78, 125–131.

Ergonul, O., Celikbas, A., Dokuzoguz, B., Eren, S., Baykam, and N., Esener, H., 2004. Characteristics of patients with Crimean-Congo hemorrhagic fever in a recent outbreak in Turkey and impact of oral ribavirin therapy. *Clinical Infectious Diseases* 39, 284–287.

Ergonul, O., Zeller, H., Celikbas, A., and Dokuzoguz, B., 2007. The lack of Crimean-Congo hemorrhagic fever virus antibodies in health care workers in an endemic region. *International Journal of Infectious Diseases* 11, 48–51.

Feldmann, H., Jones, S., Daddario-DiCaprio, K.M., Geisbert, J.B., Stroher, U., and Grolla, A., 2007. Effective post exposure treatment of Ebola infection. *PLoS Pathogens* 3, 0054–0061.

Fisher-Hoch, S.P., Khan, J., Rehman, S., Mirza, S., Khurshid, M., and McCormick, J.B., 1995. Crimean-Congo hemorrhagic fever treated with oral ribavirin. *The Lancet* 346, 472–475.

Fleming, D. and Hunt, D., 2006. *Biological Safety*. ASM Press, Washington, DC.

Garbutt, M., Leibscher, R., Wahl-Jensen, V., Jones, S., Moller, P., and Wagner, R., 2004. Properties of replication competent vesicular stomatitis virus vectors expressing glycoproteins of filoviruses and arenaviruses. *Journal of Virology* 78, 5458–5465.

Gear, J., 1989. Clinical aspects of African viral hemorrhagic fevers. *Reviews of Infectious Diseases* 11, S777–S782.

Geisbert, T.W., Daddario-DiCaprio, K.M., Geisbert, J.B., Young, H.A., and Formenty, P., 2007. Marburg virus Angola infection of rhesus macaques: Pathogenesis and treatment with recombinant nematode anticoagulant c2. *Journal of Infectious Disease* 196, S372–S381.

Geisbert, T.W., Daddario-DiCaprio, K.M., Williams, K., Geisbert, J.B., and Fernando, L., 2008. Recombinant vesicular stomatitis virus vector mediates post exposure protection against Sudan Ebola hemorrhagic fever in nonhuman primates. *Journal of Virology* 82, 5664–5668.

Geisbert, T.W., Hensley, L.E., Jahrling, P.B., Larsen, T., Geisbert, J.B., and Paragas, J., 2003. Treatment of Ebola virus infection with a recombinant inhibitor of factor VIIa/tissue factor: A study in rhesus monkeys. *The Lancet* 362, 1953–1958.

Geisbert, T.W., Hensley, L.E., Kagan, E., Yu, E., and Geisbert, J.B., 2006. Postexposure of guinea pigs against a lethal Ebola virus challenge is conferred by RNA interference. *Journal of Infectious Disease* 206, 1650–1657.

Geisbert, T.W., Jones, S., Fritz, E., Shurtleff, A.C., Geisbert, J.B., and Liebscher, R., 2005. Development of a new vaccine for the prevention of Lassa fever. *PLoS Medicine* 2, 0537–0545.

Geisbert, T.W., Lee, A.C., Robbins, M., Geisbert, J.B., and Honko, A., 2010. Postexposure protection of non-human primates against a lethal Ebola virus challenge with RNA interference: A proof of concept study. *The Lancet* 375, 1896–1905.

Graci, J.D. and Cameron, C.E., 2006. Mechanisms of action of ribavirin against distinct viruses. *Reviews in Medical Virology* 16, 37–48.

Gubler, D., 2005. The emergence of epidemic dengue fever and dengue hemorrhagic fever in the Americas: A case of failed public health policy. *Pan American Journal of Public Health* 17, 221–224.

Gunther, S., Feldmann, H., Geisbert, T.W. et al., 2011. Management of accidental exposure to Ebola virus in the biosafety level 4 laboratory, Hamburg, Germany. *Journal of Infectious Diseases* 204, S785–S790.

Haas, W.H., Breuer, T., Pfaff, G., Schmitz, H., Kohler, P., and Asper, M., 2003. Imported Lassa fever in Germany: Surveillance and management of contact persons. *Clinical Infectious Diseases* 36, 1254–1258.

Helmick, C., Webb, P.A., Scribner, C., Krebs, J., and McCormick, J.B., 1986. No evidence for increased risk of Lassa fever infection in hospital staff. *Lancet* 2, 1202–1205.

Hensley, L.E., Stevens, E.L., Yan, B., Geisbert, J.B., and Macias, W.L., 2007. Recombinant human activated protein c for the post exposure treatment of Ebola hemorrhagic fever. *Journal of Infectious Disease* 196, S390–399.

Huggins, J.W., 1989. Prospects for treatment of viral hemorrhagic fevers with ribavirin, a broad-spectrum antiviral drug. *Reviews of Infectious Diseases* 11, S750–S761.

Huggins, J.W., Jahrling, P.B., and Kende, M., 1984. Efficacy of ribavirin and tributylribavirin against virulent RNA infections, in: Smith, R. (Ed.), *Clinical Applications of Ribavirin.* Academic Press, New York, pp. 49–63.

Ibrahim, E.H., 2000. The influence of inadequate antimicrobial treatment of bloodstream infections on patient outcomes in the ICU setting. *Chest* 118, 146–155.

Ikegami, T. and Makino, S., 2009. Rift valley fever vaccines. *Vaccine* 27 Suppl 4, D69–D72.

Ippolito, G., Nisii, C., DiCaro, A., Brown, D., and Gopal, R., 2009. European perspective of 2 person rule for biosafety level 4 laboratories. *Emerging Infectious Diseases* 15, 1858.

Jahrling, P.B., Marty, A., and Geisbert, T.W., 2007. *Viral Hemorrhagic Fevers.* Department of Defense, Office of the Surgeon General, US Army; Borden Institute, Washington, DC, pp. 271–310.

Jeffs, B., Roddy, P., Weatherill, D., de la Rosa, O., and Dorion, C., 2007. The Medecins Sans Frontieres intervention in the Marburg hemorrhagic fever epidemic, Uige, Angola, 2005. Lessons learned in the hospital. *The Journal of Infectious Diseases* 196, S154–S161.

Kaufman, S.G. and Berkelman, R., 2007. Biosafety "behavioral based" training for high biocontainment laboratories: Bringing theory into practice for biosafety training. *Journal of the American Biological Safety Association* 12, 178–184.

Kilgore, P., Ksiazek, T., Rollin, P., Mills, J., and Villagra, M., 1997. Treatment of Bolivian hemorrhagic fever with ribavirin. *Clinical Infectious Diseases* 24, 718–722.

Kortepeter, M., Bausch, D.G., and Bray, M., 2011a. Basic clinical and laboratory features of filoviral hemorrhagic fever. *Journal of Infectious Disease* 204(Suppl 3), S810–S816.

Kortepeter, M., Lawler, J.V., Honko, A., Bray, M., Johnson, J., and Purcell, B., 2011b. Real-time monitoring of cardiovascular function in rhesus macaques infected with Zaire ebolavirus. *Journal of Infectious Disease* 204(Suppl 3), S1000–S1010.

Kortepeter, M., Martin, J., Rusnak, J., Cieslak, T., Wakefield, K., Anderson, E., and Ranadive, M., 2008. Managing potential laboratory exposure to Ebola virus by using a patient biocontainment care unit. *Emerging Infectious Diseases* 14, 881–887.

Kunz, S. and de la Torre, J., 2005. Novel antiviral strategies to combat human arenavirus infections. *Current Molecular Medicine* 5, 735–751.

Lawler, A., 2005. Boston University under fire for pathogen mishap. *Science* 307, 501.

LeDuc, J., Anderson, K., Bloom, M., Carrion, R., Feldmann, H., and Fitch, J., 2009. Potential impact of a 2 person security rule on biosafety level 4 laboratory workers. *Emerging Infectious Diseases* 15(7), e1–e6.

Lee, A., Agnelli, G., and Buller, H., 2001. Dose-response study of recombinant factor VIIa/tissue factor inhibitor recombinant nematode anticoagulant protein c2 in prevention of postoperative venous thromboembolism in patients undergoing total knee replacement. *Circulation* 104, 74–78.

Loew-Baselli, A., Poellabauer, E.M., Fritsch, S., Koska, M., Vartian, N., and Himly, C., 2006. Immunogenicity and safety of FSME-Immun 0.5 ml using a rapid immunization schedule. *International Journal of Medical Microbiology* 296, S213–S214.

Mackler, N., Wilkerson, W., and Cinti, S., 2007. Will first-responders show up for work during a pandemic? Lessons from a smallpox vaccination survey of paramedics. *Disaster Management and Response: DMR: An Official Publication of the Emergency Nurses Association* 5, 45–48.

Maltezou, H., Andonova, L., Andraghetti, R., Bouloy, M., Ergonul, O., and Jongejan, F., 2010. Crimean-Congo hemorrhagic fever in Europe: Current situation calls for preparedness. *Eurosurveillance* 15, 19504.

Martin, G.S., Mannino, D.M., Eaton, S., and Moss, M., 2003. The epidemiology of sepsis in the United States from 1979 through 2000. *The New England Journal of Medicine* 348, 1546–1554.

Masterson, L., Steffen, C., Brin, M., Kordick, M.F., and Christos, S., 2009. Willingness to respond: Of emergency department personnel and their predicted participation in mass casualty terrorist events. *The Journal of Emergency Medicine* 36, 43–49.

McCormick, J.B., King, I.J., Webb, P.A., Johnson, K.M., O'Sullivan, R., and Smith, E.S., 1987. A case-control study of the clinical diagnosis and course of Lassa fever. *The Journal of Infectious Diseases* 155, 445–455.

McCormick, J.B., King, I.J., Webb, P.A., Scribner, C.L., Craven, R.B., and Johnson, K.M., 1986. Lassa fever. Effective therapy with ribavirin. *The New England Journal of Medicine* 314, 20–26.

Monath, T.P., 2008. Treatment of yellow fever. *Antiviral Research* 78, 116–124.

Moons, A., Peters, R., and Bijsterveld, N., 2003. Recombinant nematode anticoagulant protein c2, an inhibitor of the tissue factor/factor VIIa complex, in patients undergoing elective coronary angioplasty. *Journal of the American College of Cardiology* 41, 2147–2153.

Morens, D.M. and Fauci, A.S., 2008. *Dengue and Dengue Hemorrhagic Fever.* A potential threat to public health in the United States. *Journal of the American Medical Association* 299, 214–216.

Morse, S., 1995. Factors in the emergence of infectious diseases. *Emerging Infectious Diseases* 1, 7–15.

Mupapa, K., Mukundu, W., Bwaka, M.A., Kipasa, M., and DeRoo, A., 1999. Ebola hemorrhagic fever and pregnancy. *Journal of Infectious Disease* 179, S11–S12.

Orlinger, K.K., Hofmeister, Y., Fritz, R., Holzer, G.W., Falkner, F.G., Unger, B., Loew-Baselli, A., Poellabauer, E.M., Ehrlich, H.J., Barrett, P.N., and Kreil, T.R., 2011. A tick-borne encephalitis virus vaccine based on the European prototype strain induces broadly reactive cross-neutralizing antibodies in humans. *The Journal of Infectious Diseases* 203, 1556–1564.

Paddock, C., 2009. CDC confirms first case of Marburg fever in Colorado. *Medical News Today*, MediLexicon, Intl., February 9, 2009. http://www.medicalnewstoday.com/articles/138304.php

Paweska, J.T., 2009. Nosocomial outbreak of novel arenavirus infection, Southern Africa. *Emerging Infectious Diseases* 15(10), 1598–1602.

Peters, C.J., 1997. Emergence of Rift Valley fever, in: Saluzzo, J. and Dodet, B. (Eds.), *Factors in the Emergence of Arbovirus Diseases*. Elsevier, Paris, France, pp. 253–264.

Peters, C.J., Jahrling, P.B., and Liu, C., 1987. Experimental studies of arenaviral hemorrhagic fevers. *Current Topics in Microbiology and Immunology* 134, 5.

Peters, C.J., Johnson, E., and McKee, K., 1991. Filoviruses and management of viral hemorrhagic fever, in: Belshe, R.B. (Ed.), *Textbook of Human Virology*, 2nd edn. Mosby Year Book Inc, St Louis, MO, pp. 699–712.

Peters, C.J., Reynolds, J., Slone, T., Jones, D., and Stephen, E., 1986. Prophylaxis of Rift Valley fever with antiviral drugs, immune serum, an interferon inducer, and a macrophage activator. *Antiviral Research* 6, 285–297.

Peters, C.J. and Zaki, S.R., 2006. Overview of viral hemorrhagic fevers, in: Guerrant, R.L., Walker, D.H. and Weller, P. (Eds.), *Tropical Infectious Diseases*, 2nd edn. Elsevier Churchill Livingstone, Philadelphia, PA, pp. 726–733.

Price, M., Fisher-Hoch, S.P., Craven, R., and McCormick, J.B., 1988. A prospective study of maternal and fetal outcome in acute Lassa fever infection during pregnancy. *British Medical Journal* 297, 584–587.

ProMED, 2009. Crimean-Congo hemorrhagic fever, fatal: Imported, ex Afghanistan, ProMED 20090919.3286.

Reed, D. and Mohamadzadeh, M., 2007. Status and challenges of filovirus vaccines. *Vaccine* 25, 1923–1934.

Richards, G.A., Murphy, S., Johnson, R., Mer, M., Zinman, C., and Taylor, R., 2000. Unexpected Ebola virus in a tertiary setting: Clinical and epidemiological aspects. *Critical Care Medicine* 28, 240–244.

Richardson, J.S., Ennis, J., Turner, J., and Kobinger, G.P., 2011. Post-exposure efficacy of an adenoviral-based Ebola virus vaccine in mouse, guinea pig and non-human primate animal models, Vaccine and International Society for Vaccines Annual Global Congress, Seattle, WA, p. Abstract 03.06.

Richardson, J.S., Yao, M.K., Tran, K.N., Croyle, M.A., Strong, J.E., Feldmann, H., and Kobinger, G.P., 2009. Enhanced protection against Ebola virus mediated by an improved adenovirus-based vaccine. *PloS one* 4, e5308.

Risi, G.F., 2010. Preparing a community hospital to manage work-related exposures to infectious agents in biosafety level 3 and 4 laboratories. *Emerging Infectious Diseases* 16, 373–378.

Rivers, E., Nguyen, B., Havstad, S., Ressler, J., Muzzin, A., and Knoblich, B., 2001. Early goal-directed therapy in the treatment of severe sepsis and septic shock. *The New England Journal of Medicine* 345, 1368–1377.

Roddy, P., Colebunders, R., Jeffs, B., Palma, P.P., Van Herp, M., and Borchert, M., 2011. Filovirus hemorrhagic Fever outbreak case management: A review of current and future treatment options. *The Journal of Infectious Diseases* 204 Suppl 3, S791–S795.

Salerno, R. and Gaudioso, J.M., 2007. *Laboratory Biosecurity Handbook* (illustrated edn). CRC Press, Boca Raton, FL.

Sanchez, A., Geisbert, T.W., and Feldmann, H., 2007. Filoviridae: Marburg and Ebola viruses, in: Knipe, D. and Howley, P.M., (Eds.), *Field's Virology*, 5th edn. Lippincott Williams & Wilkins, Philadelphia, PA, pp. 1409–1448.

Sidwell, R., Huffman, J., Khare, G., Allen, L., and Witkowski, J., 1972. Broad spectrum antiviral activity of Virazole: 1-beta-D-ribofuranosyl-1,2,4-triazole-3-carboxamide. *Science* 177, 705–706.

Sidwell, R.W. and Smee, D.F., 2003. Viruses of the Bunwa-and Togaviridae families: Potential as bioterrorism agents and means of control. *Antiviral Research* 57, 101–111.

Siegel, J.D., Rhinehart, E., Jackson, M., and Chiarella, L., 2007. *2007 Guidelines for Isolation Precautions: Preventing Transmission of Infectious Agents in Health Care Settings*. Centers for Disease Control and Prevention, http://www.cdc.gov/ncidod/dhqp/pdf/isolation2007, accessed on October 2011.

Skvorc, C. and Wilson, D.E., 2011. Developing a behavioral health screening program for BSL-4 laboratory workers at the National Institutes of Health. *Biosecurity and Bioterrorism: Biodefense Strategy, Practice, and Science* 9, 23–29.

Slenczka, W., 1999. The Marburg virus outbreak of 1967 and subsequent episodes. *Current Topics in Microbiology and Immunology* 235, 49–75.

Solomon, T. and Thomson, G., 2009. Viral haemorrhagic fevers, in: Cook, G.C. and Zumla, A.I. (Eds.), *Manson's Tropical Diseases*, 22nd edn. Saunders Elsevier, China, pp. 763–785.

Sullivan, N.J., Geisbert, T.W., Geisbert, J.B., Zu, L., and Yang, Z., 2003. Accelerated vaccination for Ebola hemorrhagic fever in non-human primates. *Nature* 424, 681–684.

Timen, A., Koopmans, M.P., Vossen, A.C., van Doornum, G.J., Gunther, S., and van den Berkmortel, F., 2009. Response to imported case of Marburg hemorrhagic fever, the Netherlands. *Emerging Infectious Diseases* 15, 1171–1175.

Van de Wal, B., Joubert, J., Van Eeden, S., and King, J., 1985. A nosocomial outbreak of Crimean-Congo hemorrhagic fever at Tygerberg hospital. Part 3: Preventive and prophylactic measures. *South African Medical Journal* 68, 718–728.

VanEeden, P., Van Eeden, S., Joubert, J., King, J., VandeWal, B., and Michell, W., 1985. A nosocomial outbreak of Crimean-Congo haemorrhagic fever at Tygerberg hospital. Part II. Management of patients. *South African Medical Journal* 68, 718–721.

Vlasuk, G., and Rote, W., 2002. Inhibition of factor VIIa/tissue factor with nematode anticoagulant protein c2: From unique mechanism to a promising new clinical anticoagulant. *Trends in Cardiovascular Medicine* 12, 325–331.

Warren, T.K., Warfield, K.L., Wells, J. et al., 2010. Advanced antisense therapies for postexposure protection against lethal filovirus infections. *Nature Medicine* 16, 991–994.

Whitehouse, C.A., 2004. Crimean-Congo hemorrhagic fever. *Antiviral Research* 64, 145–160.

WHO, 1975. *Technical Guides for Diagnosis, Treatment, Surveillance, Prevention and Control of Dengue Hemorrhagic Fever*. World Health Organization, Geneva, Switzerland.

WHO, 1978. Ebola hemorrhagic fever in Zaire, 1976. *Bulletin of the World Health Organization* 56, 271–290.

WHO, 1992. Global health situation and projections-estimates. World Health Organization, Geneva, Switzerland.

WHO, 2011a. Crimean-Congo hemorrhagic fever fact sheet, http://www.who.int/mediacentre/factsheets/fs208/en, accessed on October 2011.

WHO, 2011b. Dengue and Dengue hemorrhagic fever fact sheet, www.who.int/mediacentre/factsheets/fs117/en/, accessed on October 2011.

Wills, B.A., Dung, N.M., Loan, H.T., Tam, D.T., Thuy, T.T., and Minh, L.T., 2005. Comparison of three fluid solutions for resuscitation in dengue shock syndrome. *The New England Journal of Medicine* 353, 877–889.

deWit, E., Feldmann, H., and Munster, V.J., 2011. Tackling Ebola: New insights into prophylactic and therapeutic intervention strategies. *Genome Medicine* 3, 1–10.

Woolhouse, M., 2006. Where do emerging pathogens come from? *Microbe* 1, 511–515.

Yen, J.Y., Garamszegi, S., Geisbert, J.B., Rubins, K.H., Geisbert, T.W., Honko, A., Xia, Y., Connor, J.H., and Hensley, L.E., 2011. Therapeutics of Ebola hemorrhagic fever: Whole-genome transcriptional analysis of successful disease mitigation. *Journal of Infectious Diseases* 204, S1043–S1052.

OTHER RESOURCES

BIOSAFETY, BIOCONTAINMENT, BIOSECURITY

Biosafety in Microbiological and Biomedical Laboratories, 5th edn, HHS publication Number 21-1112 www.cdc.gov/biosafety/publications/bmbl5/index.htm

Fleming, D. and Hunt, D., 2006. *Biological Safety*. ASM Press, Washington, DC.

Laboratory Biosafety Manual, 3rd edn, World Health Organization, Geneva, Switzerland. http://www.who.int/csr/resources/publications/biosafety/Biosafety7.pdf

Salerno, R. and Gaudioso, J., *Laboratory Biosecurity Handbook*. CRC Press, CRC Press, Boca Raton, FL

WEB RESOURCES

American Biological Safety Association: www.absa.org

Public Health Agency of Canada: http://www.phac-aspc.gc.ca/lab-bio/res/psds-ftss/index-eng.php

U. S. Department of Labor: Occupational Safety and Health Administration: www.OSHA.gov

BIOSAFETY TRAINING

Emory University: http://www.sph.emory.edu/CPHPR/biosafetytraining/index.html

National Biosafety and Biocontainment Training Program: www.nbbtp.org

University of Texas Medical Branch: http://www.utmb.edu/biosafetytraining/

9 BSL4 Workforce Preparedness in Hemorrhagic Fever Outbreaks

Sean G. Kaufman

CONTENTS

9.1 INTRODUCTION

The biosafety level 4 (BSL4) laboratory is designed for work with exotic and dangerous agents, offering the scientists working with such agents and the environment where the laboratory is located the highest possible level of protection. The scientists and environment are protected by extensive engineering controls and workplace practices, which include multiple levels of redundancy. There are two types of BSL4 laboratories: the cabinet laboratory and the suit laboratory. This chapter will focuses solely on the suit laboratory where individuals wear positive pressure suits when inside the BSL4 laboratory.

9.2 PROFESSIONAL ETIQUETTE AND SCREENING POTENTIAL CANDIDATES

The BSL4 laboratory environment is the maximum level of containment requiring the highest level of professionalism. Individuals who work in BSL4 laboratories must be aware of not only themselves but their environment as well. These individuals will assist without being asked, will listen more than they speak, and will have a healthy respect for the environment they work in. Individuals who work in BSL4 laboratories are not adrenaline junkies, thrill seekers, and risk takers, nor do they tend to be impulsive. They typically are calculated, have a complete understanding of the risks associated with the work they do, are keenly aware of the environment they work in, and are cool and collective under day-to-day pressure.

The proper screening of potential candidates for BSL4 work is critical. If an individual is not comfortable with a BSL4 suit, goes into the laboratory, and begins to panic, they not only put themselves at risk but they put others at risk as well. Screening of potential candidates includes providing them with an opportunity to get comfortable with the suit, observing them do day-to-day work, and observing them respond to emergency situations with actions that minimize risk rather than increase risk. Screening is only the first step and does not include training. Just because someone may be comfortable in the suit and responds appropriately does not mean they can work in a BSL4 laboratory environment. Only effective training programs will be able to determine whether someone can safely work in a BSL4 laboratory environment.

9.3   PROFESSIONAL PREREQUISITES FOR IDENTIFYING BSL4 WORKFORCE

The BSL4 laboratory requires a workforce with a unique set of skills and attitudes. There has been much debate on how an organization may determine if or when someone is ready to work in the BSL4 laboratory. For example, should individuals have to work in a BSL2 (laboratories that house agents typically spread through the blood-borne/fecal–oral route) and BSL3 (laboratories that house agents typically spread through the aerosol route) prior to being considered for BSL4 laboratory work? The answer is not a simple one.

Individuals who have safely worked in a BSL2 laboratory, mastering specific laboratory practices while demonstrating an extensive commitment to laboratory procedures, may be adequately prepared to work safely in BSL4 laboratories. However, behavioral variances within BSL2 laboratories are vast and specific attention to detail of the workplace is typically minimal. This poses a serious challenge, as a variety of safety behaviors may be appropriate in BSL2 laboratories; however, minimal variation of safety practices is ideal for BSL4 laboratories. It cannot be assumed that work performed in BSL3 prepares an individual to work in BSL4. It is therefore recommended that candidates be screened thoroughly and trained properly regardless of their experiences in BSL2 or BSL3 laboratories.

9.4   BEHAVIORAL EXPECTATIONS OF THOSE WORKING
IN BSL4 LABORATORIES

Individuals gain experience in one laboratory and move to another. During this transition, staff must learn new procedures and adhere to additional policies. If an individual is screened and selected to work in a BSL4, they must have a clear set of behavioral expectations for working in the BSL4 laboratory. These behavioral expectations include the following five items:

1. I will follow all standard operating procedures (SOPs) to the best of my ability.
2. I will ensure others follow the SOPs to the best of their ability.
3. I will report all laboratory near misses and incidents.
4. I will report any symptoms that match the agents I'm working with in the laboratory environment.
5. I will report any new medical conditions (pregnancy, diabetes, cancer, chronic asthmatic conditions, new medications that may cause seizures).

The purpose of a safety training program is to set laboratory staff up for successful safety outcomes. The list of behavioral expectations is derived from incidents which led to laboratory illnesses where staff failed to follow one or more of the listed expectations. Prior to entering any high-containment laboratory, staff should be asked to sign a behavioral contract providing them with a clear understanding of these minimal expectations while producing maximum safety outcomes. Failure to agree to follow any of these behavioral expectations will lead to an increased risk for all staff working in the laboratory environment.

9.5   INTEGRATING LEVELS OF LEARNERS INTO BSL4 TRAINING PROGRAMS

Safety is an outcome, which begins with a plan, and however is directly influenced by an individual's behavior. Therefore, we must make sure individuals who work in laboratory environments understand the risks they face and the benefit of following SOPs, have access to needed resources, have the skills needed to behave safely, and believe their behaviors lead to increased personal safety (see Figure 9.1). These five things are needed for sustained behavior to occur. If an individual is asked to follow an SOP and does not understand why the procedure was needed, the likelihood of following the procedure is minimal. If an individual understands the need to behave but does not have the adequate skills to do so, they cannot follow the procedure. If an individual does not believe the procedure to be effective or efficient, they will not follow the procedure. Training programs must integrate these principles of sustained behavior into all levels of learners.

Individuals who work in laboratory environments range in the amount of professional experience, education, and laboratory practices. Training programs developed to prepare laboratory workers to work in BSL4 laboratories must be prepared to handle these differences. The three levels of learners are listed as novices, practitioners, and experts. A real-life example of the three levels of learners is a picture of the child, parent, and grandparent. The child is a novice who wants to know why they have to do what they are asked to do. The parent is a practitioner who wants to know how they get their child to do what they are asking them to do. The grandparent is an expert who wants to know when and where they can offer their assistance and guidance without having the parent or child completely disengage. All levels of learners require different trainings. Imagine training grandparent like a child! It will not work and can sometimes seem offensive to those who have substantial experience in the laboratory environment.

Novices are individuals who have worked less than one year in the BSL4 laboratory. Training programs developed for these individuals must answer the question why things are done the way they are done. Practitioners are individuals who have worked more than a year and less than 10 years in the BSL4 laboratory. Training programs for these individuals must teach the skills needed to work safely in the BSL4 laboratory. Experts are individuals who work 10 years or more in the BSL4 laboratory. Training programs for these individuals must include leadership, communication skills, team building, and opportunities to integrate their experiences aimed at solving existing laboratory safety issues. One training program will never work for all three levels of learners.

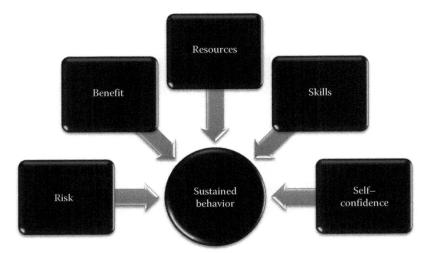

FIGURE 9.1   Before implementing any standard operating procedure, laboratory leaders must ask what it takes to adequately prepare someone for a sustained behavioral practice. Leaders must ensure that all laboratory workers being asked to follow an SOP have the five items needed for sustained behavior.

Therefore, the best practice to prepare individuals to work in BSL4 laboratories is to train them at the appropriate learning level.

9.6 BEHAVIORAL EVOLUTION AND DEVELOPMENT OF TRAINING PROGRAMS

Safety is the intended outcome of a plan. However, behavior is the bridge that fills the gap between an outcome and a plan. Understanding how behavior evolves is critical for developing effective training programs for individuals working in BSL4 laboratories.

SOPs or standard operating behaviors are one and the same. The goal of the SOP is to ensure two or more people do the same thing, in the same way, to achieve the same result. An SOP does not tell someone what to think nor does it tell someone what to feel. It is a procedure aimed at ensuring consistent behaviors, leading to a predictable outcome, among multiple individuals. Therefore, it is important for us to understand how behavior evolves and integrate lessons learned from this concept into BSL4 training programs.

When a new behavior is learned, full attention and awareness are given to that behavior (green = increased awareness). Unfortunately, in a short period of time the individual behaving forms habits (clumping behaviors) and believes they have mastered the behavior (yellow = perceived mastery). This decrease in awareness and attention to detail leads to an incident, accident, or near miss (red = incident/accident/near miss) (Figure 9.2a). Friedrich Nietzsche once stated, "That which does not kill us makes us stronger." If an accident occurs and the individual evolves, a change in behavior takes place. This change in behavior leads to new behaviors that brings the person back to an increased state of awareness (as a result of their new behavior). However, since the person has gained valuable experience they are at a new level of mastery, typically better than before (green = mastery/increased awareness + experience) (Figure 9.2b). Unfortunately, as time passes individuals are challenged with complacency or a state of decreased awareness and attention to detail (yellow = complacency). This again leads to an incident, accident, or near miss (red = incident/accident/near miss). This provides evidence that someone is never "trained" and demonstrates a great need for an individual to remain in a constant state or professional development throughout their career (Figure 9.2c).

Behavioral evolution can be integrated into training programs to prevent incidents and accidents from occurring in BSL4 laboratories. We know that humans will overestimate their skills and abilities. We know complacency leads to a lack of attention to detail, which puts individuals working in BSL4 laboratory environments at increased risks. Both overconfidence and complacency must be addressed when preparing or sustaining individuals who work in BSL4 laboratory environments. Accidents, which occur as a result of overconfidence and complacency, are similar because they deal with a reduction in the amount of attention to detail. For example, a teenager and an adult who get in a car accident may have very different levels of experience, but the cause of the accident may be one in the same.

It is important to make sure individuals working in BSL4 laboratories know their capabilities and understand the consequences of complacency. Training programs should include exercises, which would allow for individuals to learn from being overconfident and complacent while in a controlled environment.

9.7 PHASES OF EFFECTIVE TRAINING PROGRAMS FOR BSL4 LABORATORY ENVIRONMENTS

There are four phases of laboratory training that should be completed before staff are allowed to work alone in a BSL4 laboratory. The following is a description of each stage.

9.7.1 PHASE 1: BSL4 RISK PROGRAM

Many believe that working in the BSL4 laboratory is dangerous work. The top infectious disease killers of the world—meaning those that kill the most people—are found in BSL2 (HIV and influenza)

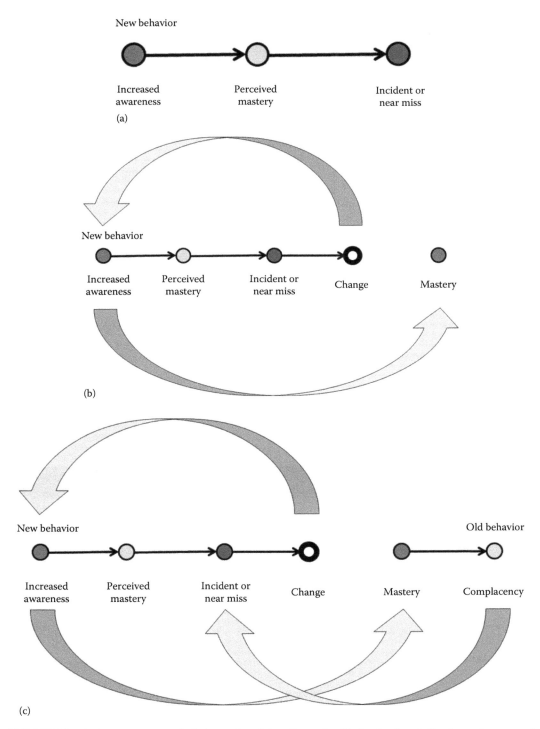

FIGURE 9.2 As long as someone is living and evolving, the cycle of behavioral evolution is continuous and never ending. It is natural for us to become unaware of details and complacent with repetitive behaviors, leading to preventable incidents and accidents over long periods of time. The goal of effective training programs is to keep someone in a state of high awareness, increasing the individual's ability to notice problems prior to an incident or accident occurring. Please see text for descriptions of (a)–(c).

and BSL3 (tuberculosis). It is true that most pathogens worked with in BSL4 are extremely fatal and typically have no known cures or prophylactic treatment. However, the number of controls put in place to protect both the individual and the environment offers an enormous amount of redundancy and allows this work to be done safely.

This phase of training should include teaching an individual the differences between agents worked with in BSL1, BSL2, BSL3, and BSL4. Individuals should know these differences and understand that most agents worked with in BSL4 are typically spread through fecal–oral or blood-borne routes of transmission. Individuals should understand the likelihood of fatality should laboratory-acquired illness occur. Individuals should be made aware of the process, which will occur, should an exposure happen. Individuals should have an opportunity to ask questions and receive information they need to determine whether or not the risk of working in a BSL4 laboratory is acceptable to them personally and professionally.

If the individual is willing to work in the BSL4 laboratory after completing the risk program described earlier, having them sign a contract listing the five behavioral expectations (as discussed earlier in this chapter) should be the final step of this phase of training. Individuals refusing to agree to those five behavioral expectations should not be allowed to work in the BSL4 laboratory environment.

9.7.2 Phase 2: BSL4 Safety Training

Individuals choosing to work in the BSL4 laboratory should be required to complete a comprehensive safety training program inside a laboratory, which has been decontaminated and poses no risk to participants. This training program should teach the essential skills needed to work safely in the BSL4 environment.

9.7.2.1 Primary Controls of Biosafety

Utilizing the four primary controls of biosafety (Figure 9.3), a comprehensive training program can be developed and implemented.

Four Primary Controls of Biosafety			
Engineering	**Personal Protective Equipment**	**Standard Operating Procedures**	**Administrative**
Directional airflow	Gloves	Donning/doffing PPE	Training
Biosafety cabinets	Eye protection	Transporting pathogens	SOP compliance
Interlocked doors	Respirators (N95, PAPR)	Handling needle sticks	SOP evaluation/validation
Hands free sinks	Tyvek suit	Cleaning a spill	Medical surveillance
Self closing doors	Laboratory scrubs/shoes	Laboratory decontamination	Incident surveillance
HEPA filters	Booties	Emergency evacuation	Background checks

FIGURE 9.3 The primary controls of biosafety offer a variety of solutions for working safely with biological agents. If an organization has limitations in the engineering and PPE controls, adding additional SOPs and administrative controls could assist with the offset while allowing safe work practices to continue.

9.7.2.1.1 Engineering Controls

These are systems put in place to ensure the biological agent remains within containment. Examples of BSL4 engineering controls include, but are not limited to, interlocked door systems, double HEPA filter exhaust systems, single HEPA filter supply systems, a sealed environment, double door autoclaves, and liquid waste management decontamination system. It is important to train individuals working in BSL4 laboratories about how the engineering controls protect them and the environment while ensuring the highest level of containment.

Most BSL4 laboratories require a systems checklist to be completed before entering the laboratory on a daily basis. Training individuals to properly complete this systems checklist is a good way to familiarize staff working in the BSL4 laboratory with existing engineering controls. Staff must understand how the laboratory offers them and the environment multiple redundancies of containment. Failure to understand how engineering controls work will lead to behaviors that may cause breaches and breakdowns in overall containment.

9.7.2.1.2 PPE Controls

These are items worn to protect appropriate portals of entry and allow individuals to keep what they were working with in the laboratory—inside the laboratory. Examples of BSL4 PPE controls include scrubs, socks, gloves, and BSL4 suit (including gloves and boots). It is important to train individuals how to properly don and doff their PPE.

Training individuals to wear their PPE properly requires them to interact with the PPE directly. For example, teaching someone how to swim without a swimming pool would not yield good results. Asking someone to read an SOP on how to don PPE properly is insufficient. It is best to provide individuals with multiple opportunities to don and doff PPE in a clean and controlled environment. Failure to both practice and receive feedback on this process will lead to behaviors that could cause breaches and breakdowns in containment.

9.7.2.1.3 SOP Controls

These are procedures that aim to standardize behaviors and ensure consistent outcomes among different individuals. Examples of SOPs include but are not limited to donning and doffing PPE, disposing of waste, working in a biosafety cabinet, responding to a spill outside the biosafety cabinet, and evacuating a laboratory during an emergency situation. It is important to train individuals and ensure adequate understanding with proper demonstration of the SOP occurs.

Individuals should be trained to evaluate and validate their SOPs. Evaluation of SOPs includes ensuring the language is understood (by the user), terminology is clarified (by the author), and what is being asked of the user is physically possible. There are two forms of validation—internal validation and external validation. Internal validation ensures that the behavior produced by the SOP matches the author's intent. External validation ensures that those behaviors produce consistent outcomes among users. Failure to evaluate and validate SOPs may produce variances in practices, which may lead to breaches and breakdowns in containment.

9.7.2.1.4 Administrative Controls

These are measures put into place to offer additional redundancy for laboratory staff and containment. Administrative controls include training, vaccination programs, SOP compliance programs, background checks, and medical and incident surveillance programs. It is important to effectively train all staff working in laboratory environments, ensure they have the most up-to-date and available vaccinations, ensure the highest level of compliance with SOPs, report any incidents that are near misses so that they may not be repeated, and report any clinical symptoms, which may match the agents they are working with in the laboratory so that breaches may be discovered and individuals immediately treated.

Administrative controls are programs put in place to ensure that individuals do what they are expected to do, and should something go wrong in the laboratory, additional redundancy is in place to protect laboratory staff and ensure containment. During any training program, staff should understand that laboratory leadership will expect the highest level of compliance with existing SOPs, reporting any incidents or accidents, which occur in the laboratory, and reporting any illness, which may be associated with the laboratory. Laboratory leadership must also make sure individuals working in the BSL4 laboratory have been properly screened for security purposes.

At a minimum, safety skills should include the ability to list and describe the engineering controls found in the BSL4 laboratory, properly inspecting and assembling the BSL4 suit, donning and donning (putting on PPE) and doffing (taking off PPE) in accordance to the SOP, handling spills inside and outside the BSC, demonstrating proper emergency evacuations, responding to needlesticks or animal bites, and evacuating an unconscious individual. If done properly, the safety program will identify individuals who may not be capable of working in a BSL4 laboratory and allow individuals to self-identify whether or not working in the BSL4 laboratory is something they want to do.

9.7.3   PHASE 3: BSL4 AGENT-SPECIFIC TRAINING

It is critical for all individuals working in a laboratory environment to know the agents they are working with and the clinical symptoms should they become ill. Failure to understand how BSL4 agents are transmitted and the symptoms of clinical presentation could inhibit a proper response or treatment plan should a laboratory exposure occur. All individuals working in the laboratory environment should have a minimum understanding of the agents they may potentially come in contact with.

9.7.4   PHASE 4: BSL4 JOB-SPECIFIC MENTORING

Job-specific mentoring is the last step for preparing individuals to work in the BSL4 laboratory. SOPs will never replace a good mentoring program, which ensures individuals can complete tasks properly. Leaders within organizations should identify critical job functions and only when individuals can demonstrate these functions with comfort and great expertise should that individual be allowed to work alone in a BSL4 laboratory.

All four of these phases of training should be completed before working in a BSL4 laboratory. Individuals must understand risk, must be given the skills to protect themselves against those risks, must understand the agents they may come in contact with, and must have mastery of job-specific skills. These phases of training provide a well-rounded program, which adequately prepares individuals to work in BSL4 laboratories. These phases also allow organizations to determine whether or not the individual is suited and capable for work in the BSL4 laboratory.

9.8   BSL4 WORKFORCE PREPAREDNESS RECOMMENDATIONS

One could argue whether there is a need for so many BSL4 laboratories. BSL4 laboratories are not just built for the diseases we know today, but they are also built for the diseases to come. These unknown pathogens may pose a significant risk to scientists and the general public requiring the highest level of containment. When we encounter an unknown pathogen with a high fatality rate, we must be able to offer the highest level of containment and protection to the scientist and general public. This is the reason for the BSL4 laboratory.

Not all individuals are suited for work in the BSL4 laboratory. The BSL4 laboratory environment requires a keen attention to detail of both work practices and others working within the laboratory environment. Individuals who wish to work in BSL4 laboratories should have great attention to detail, be aware of those working around them, and have a healthy respect for the risks they face in the BSL4 laboratory environment. BSL4 workforce preparedness and readiness begins with proper screening of potential candidates.

If an individual has been deemed suited to work in the BSL4 laboratory, experience is also an important consideration. Several years of experience in a well-run BSL2 laboratory may provide an individual with proper prerequisite skills and experiences to work safely in a BSL4 laboratory. It is preferable, however, that individuals have several years of experience within BSL3 laboratories before being considered to work in a BSL4 laboratory. Attention to detail and compliance of SOPs are more likely to occur at BSL3, providing a similar experience to what is required for work at BSL4 levels.

Different individuals will have different experience levels and will require different levels of training. Some programs must be designed to explain why things are done when they are done. Other programs must focus on providing and increasing the skills needed to work safely in a BSL4 laboratory. Finally, experts will need leadership and management training so they may properly allocate resources and motivate staff to sustain safe laboratory behaviors. One training approach will not work for all. All levels of learners must be considered for effective BSL4 workforce preparedness and readiness.

We know that as individuals learn new behaviors, they begin to overestimate their actual skills and abilities. We also know that individuals who work for sustained periods of time in the same environment, doing the same thing, become complacent. Integrating behavioral evolution into BSL4 workforce preparedness and readiness programs will decrease the likelihood of laboratory accidents and incidents. Both the concepts of behavioral evolution and different levels of learners must be integrated into the four phases of effective training for the BSL4 workforce.

Once individuals have been selected to work in a BSL4 laboratory, risks associated with the work must be discussed and individuals must have the opportunity to determine whether working in a BSL4 laboratory is an acceptable risk to them personally and professionally. If they determine the risk to be acceptable, they should then be asked to sign an agreement of behavioral expectations. Only after accepting the risk in signing the behavioral expectations agreement should individuals progress to BSL4 safety training.

During safety training, individuals learn about engineering controls of the BSL4 laboratory, learn how to assemble and work with PPE found in BSL4 laboratories, evaluate and validate SOPs, and integrate multiple administrative controls to ensure containment and safety of laboratory staff. Before completing this phase, individuals must successfully demonstrate proper emergency response, laboratory evacuation, donning and doffing of PPE, and response to unexpected laboratory events (spills, needlesticks, animal bites).

After successful completion of safety training, individuals should be taught about the specific agents they may be working with inside the laboratory environment. This provides an opportunity to discuss routes of transmission, clinical presentation, and treatment plans should a laboratory exposure occur.

The last and most important step of BSL4 workforce preparedness and readiness training programs is mentoring. There will never be an SOP, which will be capable of replacing the benefit of mentoring. Individuals have different levels of skills and abilities and the mentor will be able to customize their learning experience by fine-tuning their weaknesses while recognizing their strengths. Mentoring should continue until the individual demonstrates all critical workplace functions in a safe and successful manner.

Sustaining readiness AND ensuring BSL4 staff remain suitable for the work they do is a challenge WHICH organizations will need continue to encounter. Annual reviews of staff performance and perceptions of the RISKS they encounter in the laboratory ARE important measures for ensuring sustainable readiness. Organizations must implement controls to frequently assess the readiness levels of staff currently working in the BSL4 laboratory. Assuming a staff member who once demonstrated appropriate levels of readiness will always do so may lead to catastrophic results. Training is an ongoing process, never ending and always striving for the highest level awareness among BSL4 laboratory staff.

10 Major Strategies for Prevention and Control of Viral Hemorrhagic Fevers

Matthew L. Boulton and Eden V. Wells

CONTENTS

10.1 INTRODUCTION

The viral hemorrhagic fevers (VHFs) are caused by a group of taxonomically and pathophysiologically diverse viruses that are collected under a single generalized heading based on their common clinical attribute of having the potential to cause profuse hemorrhage and involve multiple organ systems (LeDuc, 2008). Most, although not all, VHFs are relatively rare diseases but attract disproportionate attention from the media and the public because of their well-publicized capacity for causing significant morbidity and mortality in infected individuals (Crowcroft et al., 2002). In an era of terrorism and terrorist attacks such as the downing of the World Trade Towers on September 11, 2002, in New York City, the VHFs have also attracted increased concern because of their potential for biological weaponization, which could cause widespread death and serious illness through intentional release (Borio et al., 2002).

10.2 GENERAL CHARACTERISTICS OF VIRAL HEMORRHAGIC FEVERS

The VHFs are caused by four different taxonomic families of viruses, the Arenaviridae, the Bunyaviridae, the Filoviridae, and the Flaviviridae, as shown in Table 10.1, which summarizes the taxonomy of the VHF, their associated diseases, geographic distribution, distinctive clinical features, incubation, and mortality statistics (Borio et al., 2002; United States Centers for Disease Control and Prevention [USCDC], 2011b; World Health Organization [WHO], 2001).

Despite the significant variability found among the viruses included in these four distinct families, those included as VHFs have several defining characteristics, which are important for our

TABLE 10.1
Viral Hemorrhagic Fevers

Virus Genus	Virus	Disease	Geographic Distribution	Distinctive Clinical Features	Incubation (Days)	Mortality
Arenaviridae	*Old World*					
Arenavirus	Lassa	Lassa fever	West Africa	Fever, retrosternal pain, sore throat, back pain, cough, abdominal pain, vomiting, diarrhea, conjunctivitis, facial swelling, proteinuria, mucosal bleeding. Neurological problems have been described. Deafness develops during recovery in 30% of cases.	7–21	1% overall; 15%–20% if hospitalized
	LCMV	Lymphocytic choriomeningitis	Europe, Americas, Australia, and Japan	Biphasic illness—initial phase with fever, malaise, myalgias, headache, nausea, and vomiting. After a few days of recovery, meningitis or encephalitis may occur; hydrocephalus, myelitis, and myocarditis may occur.	8–13	<1%
	New World					
	Guanarito	Venezuelan hemorrhagic fever	South America	In general—initial fever, fatigue, dizziness, muscle aches, weakness; severe cases exhibit bleeding (skin, mucosal, and/or internal). Severe cases may exhibit neurological abnormalities, shock, coma, delirium, and seizures.	7–21	15%–30% (Borio et al. 2002)
	Junin	Argentine hemorrhagic fever				
	Machupo	Bolivian hemorrhagic fever				
	Sabia	Brazilian hemorrhagic fever				
Bunyaviridae	Hantaan, Dobrava,	HFRS	Eastern Asia, Balkans, Scandinavia, Western Europe	Sudden intense headaches, back and abdominal pain, fever, chills, nausea, and blurred vision. Flushing of face, inflammation or redness of the eyes, or rash may occur. Later severe symptoms may include shock, vascular leakage, and renal failure.	7–14 (up to 56)	<1%–15%
Hantavirus	Saaremaa	Korean hemorrhagic fever, epidemic hemorrhagic fever, and nephropathia epidemica				
	Seoul					
	Puumala					
	Sin Nombre hantavirus most common in the United States	HPS	The Americas	Early symptoms: fatigue, fever, muscle aches, headaches, dizziness, chills, nausea, vomiting, diarrhea, and abdominal pain	Not known; 7–35 range	38%
				Late symptoms (4–10 days later): coughing and shortness of breath		

Genus/Family	Disease	Virus	Symptoms	Distribution	Incubation	Case fatality
Phlebovirus	RVF	RVF virus	No symptoms or mild illness associated with fever and liver abnormalities. Illness can progress in some patients to hemorrhagic fever, encephalitis, or eye disease. Vision loss is a common complication (1%–10%).	Africa, Saudi Arabia, Yemen	2–6[3]	<1%
Nairovirus	CCHF	CCHF	Sudden onset headache, high fever, back pain, joint pain, stomach pain, vomiting. Reddened eyes, flushed face, red throat, and palatal petechiae common. Jaundice and mood and sensory perception changes may occur. Bruising and bleeding begin around day 4.	Eastern and southern Europe, Mediterranean, China and central Asia, Middle East, and Indian subcontinent	5–12 (World Health Organization 2001)	9%–50%
Filoviridae *Filovirus*	Ebola hemorrhagic fever	Ebola virus (human illness sub-types: Zaire, Sudan, Ivory Coast, Bundibugyo)	Fever, headache, joint and muscle aches, sore throat, and weakness. Subsequent diarrhea, vomiting, and stomach pain. Rash, red eyes, hiccups, and internal and external bleeding may occur.	Africa	2–21	50%–90% (Borio et al. 2002)
	Marburg hemorrhagic fever	Marburg virus	Fever, chills, headache, and myalgia; about day 5, a truncal maculopapular rash develops. Nausea, vomiting, chest pain, sore throat, abdominal pain, and diarrhea develop, and increasingly severe symptoms include jaundice, pancreatitis, severe weight loss, delirium, shock, liver failure, massive hemorrhaging, and multiorgan dysfunction.	Africa	5–10	23%–25%
Flaviviridae *Flavivirus*	Yellow fever	Yellow fever virus	Acute febrile period followed by a brief period of remission
In 15% a severe toxic phase follows fever remission with jaundice, hemorrhage, shock, and organ failure.	Africa South America	3–6	20%–50% of severe cases
	Dengue fever DHF Dengue shock syndrome (DSS)	Dengue virus (DENV 1, DENV 2, DENV 3, DENV 4)	Fever, severe headache, severe pain behind the eyes, joint pain, muscle and bone pain, rash, mild bleeding Initial signs of dengue fever; as fever wanes in 2–7 days, persistent vomiting, severe abdominal pain, and difficulty breathing may occur, with thrombocytopenia, hemorrhage, and capillary leak causing ascites and effusions. DSS and death may follow if unable to correct vascular fluid loss.	Asia, Africa, Pacific, the Americas, Caribbean	4–7	DSS: <1%–10%

(continued)

TABLE 10.1 (continued)
Viral Hemorrhagic Fevers

Virus Genus	Virus	Disease	Geographic Distribution	Distinctive Clinical Features	Incubation (Days)	Mortality
	Kyasanur Forest disease virus	Kyasanur forest disease	Karnataka State, India	Biphasic illness—fever, headache, muscle pain, cough, dehydration, gastrointestinal symptoms, and bleeding. May have low blood pressure, thrombocytopenia, anemia, and neutropenia. Second wave of symptoms at beginning of third week with fever and meningoencephalitis	3–8	3%–5%
	Omsk hemorrhagic fever virus	Omsk hemorrhagic fever	Western Siberia	Fever, cough, conjunctivitis, soft palate rash, hyperemia of face and trunk, lymphadenopathy, and splenomegaly; pneumonia and central nervous system signs and symptoms may occur (Borio et al. 2002).	3–8	0.5%–3%

Source: U.S. Centers for Disease Control and Prevention, CDC Special Pathogens Branch, Viral hemorrhagic fevers, 2011b, http://www.cdc.gov/ncidod/dvrd/spb/mnpages/dipagges/vhf.htm.

understanding and control of these pathogens, including the following (Cleri et al., 2006; LeDuc, 2008; USCDC, 2011b):

1. They are all caused by enveloped RNA viruses.
2. Most are zoonotic illnesses and naturally reside either in a mammalian reservoir host like the deer mouse or in an arthropod vector like mosquitoes and ticks. The reservoir for a few VHFs such as Ebola and Marburg remains largely unknown although there are suspect candidates.
3. Related to this, humans are not the reservoir for any of the VHF viruses although once infected, persons have been known to transmit to other persons in the case of a few of these viruses.
4. The initial clinical presentation can be somewhat indistinct and therefore difficult to distinguish from illness caused by non-VHF viruses and also difficult to differentiate one from another.
5. Once infected, these viruses are capable of causing serious illness and death in infected humans.
6. The epidemiology of many of the VHFs is not well understood and most are characterized sporadic outbreaks occurring at infrequent intervals making them difficult to predict and harder to control.
7. The major outbreaks of some of the VHFs (e.g., Ebola, Lassa fever, and Marburg viruses) have been hospital based and nosocomially spread.
8. With the notable exception of yellow fever and the less common Argentine hemorrhagic fever, there are no vaccines available for prevention of the VHF nor are there standard drug therapies or treatments for most. Ribavirin, an antiviral drug, has been somewhat effective for some of the arenaviruses and bunyaviruses.

Although the VHFs infect humans, that infection is incidental to their otherwise inconspicuous life cycle since all require, and are maintained in, nonhuman vertebrate or arthropod hosts where they cause chronic infection. Conversely, human infections with the VHF are always acute in nature and generally without associated clinical sequelae (Peters, 2006). In addition to the secondary transmission represented by spread through person-to-person contact, VHFs can also be acquired through contact with contaminated body fluids, infected cadavers, and through inadequate infection control practices in hospital or clinical settings due to inadequate sterilization or reuse of contaminated needles and syringes. The role of animal or arthropod vectors in the transmission of VHF (i.e., zoonotic transmission) varies by specific virus. Zoonotic transmission has been documented via several routes but can occur from human consumption of contaminated raw milk or contaminated meat or the care and slaughter of infected animals, via inhalation or contact with other materials, which are contaminated with the saliva or excrement of infected rodents, through nonspecific exposure to other reservoir mammalian reservoir species like bats, and from the bite of mosquitoes or ticks or when a person crushes a tick (LeDuc, 2008; USCDC, 2011b). Although zoonotic transmission drives the overall epidemiology of the VHF, some of the largest documented outbreaks of illness have occurred when person-to-person spread has been the predominant transmission modality, often in a hospital or other clinical setting (Allaranga et al., 2010).

Following exposure, the incubation of the VHFs is variable but can be as long as 21 days creating the risk of travelers returning home who are infected but not yet communicable. The clinical signs and symptoms of VHF are myriad and nonspecific, making it more challenging to differentiate them from other endemic fevers and also difficult to distinguish one VHF from another (USCDC, 2011b). Generally, clinical onset is marked by abrupt onset of fever with myalgias, arthralgias, fatigue, headache, elevated liver enzymes, proteinuria, and profound prostration. In more severe cases of illness, there are often varying degrees of coagulopathy, which can be accompanied by petechiae or ecchymoses and sometimes followed by hemorrhagic phenomena, such as bleeding under the skin or from internal organs or from body orifices like the eyes, mouth, and ears. Depending on the extent of capillary endothelial damage, pulmonary edema, renal failure, severe shock, and coma can ensue potentially leading to death (Speed et al., 1996; USCDC, 2011b, 2012).

Once infected, the treatment of the VHFs usually consists of general supportive care since there are no known effective treatments or cures. Ribavirin, an antiviral, has been used with limited success primarily in reducing illness-associated mortality with some of the arenaviruses (i.e., Lassa fever and New World hemorrhagic fevers) and the bunyaviruses (i.e., hemorrhagic fever with renal syndrome, Hantavirus, and Crimean–Congo hemorrhagic fever [CCHF]) (Borio et al., 2002; USCDC, 2011b). However, ribavirin has no therapeutic role with the other VHFs and its use is further limited by that fact it is a moderately expensive antiviral agent that is not readily available in many developing nations where VHF transmission typically occurs. Convalescent-phase plasma has been utilized with some success in the treatment of New World hemorrhagic fevers (i.e., Argentine hemorrhagic fever) but, again, has no therapeutic value with the other VHFs (USCDC, 2011b, 2012).

The VHFs are spreading geographically and the number of viruses recognized as having the capacity to cause hemorrhagic fevers is increasing (Peters, 2006; USCDC, 2011a). This is likely due in part to progressive improvement in the recognition of the VHF-associated illness by larger numbers of clinicians but also, perhaps more importantly, to the increasing sophistication of global public health surveillance and reporting of emerging pathogens to public health authorities. Despite this, the quality of public health surveillance infrastructure in developing countries, which are often this site of newly emergent infectious disease like VHFs, lags far behind that of the United States and other Western nations. This makes it more likely that individual cases of illness could be missed or go unreported to public health authorities or even that an outbreak of disease would not be recognized especially early on when there are relatively few cases. A lack of reporting or delays in reporting of VHFs can contribute significantly to increases in related morbidity and mortality making earlier identification and intervention to interrupt transmission particularly important for public health authorities. Deficiencies in animal-based surveillance are even more pronounced, which has significant implications for the spread of VHFs given the importance of zoonotic transmission. Many developing nations simply lack the infrastructure to systematically monitor disease status in potential animal hosts or arthropod vectors so that the first indication of disease is often a cluster of human cases. Other factors contributing to the apparent increase and geographic spread of the VHFs are the dramatic rise in the international movement of people, animals, and products, all of which can facilitate global transmission, and the inherent capacity of the viruses themselves to mutate and change in occupying new ecological niches (Peters, 2006).

10.3 GENERAL PRINCIPLES OF PREVENTION AND CONTROL OF VIRAL HEMORRHAGIC FEVERS

Given the potential severity of human illness caused by the VHF viruses and the lack of effective treatments or cures once infected, the use of primary preventive measures to avoid infection is extremely important. Table 10.2 summarizes the major VHF and associated disease; the potential animal, arthropod, and other sources of human infection; whether person-to-person transmission occurs; recommended treatment (if any); infection prevention and control measures to reduce nosocomial transmission; the availability of a vaccine; and recommended vector control measures (Borio et al., 2002; USCDC, 2005, 2011b).

Although the VHFs are geographically distributed all around the world, the risk of acquisition is typically associated with those areas or countries that are inhabited by the specific mammalian hosts or arthropod vectors involved in their transmission. Some VHFs are found in a fairly confined or well-demarcated geographic area such as the New World arenaviruses. However, this is less true for those VHFs that are more widely distributed such as the CCHF or Hantavirus, both of viral family Bunyaviridae, which involve rodent hosts that range widely over large geographic areas (USCDC, 2011b). Generally, international travelers have a very low risk for acquiring VHF since most,

TABLE 10.2

Clinical Presentation and Public Health Control of VHFs

Virus Genus	Disease	Source of Human Infection		Person-to-Person Transmission	Diagnostics	Treatment	Infection Prevention Measures (CDC 2011) (Hospital Isolation, United States)	Vaccine	Vector Control
		Common	Less Common						
Arenaviridae *Arenavirus*	Lassa fever	Old World rats and mice	Nosocomial	Yes	Elisa (IgM, IgG, Lassa antigen) RT-PCR Immunohistochemistry Viral isolation[b]	Supportive, ribavirin	Standard contact droplet, with eye protection Airborne (if indicated)[a]	None currently	Rodent control
	New World hemorrhagic fevers	New World rats and mice	Nosocomial (Machupo)	Yes[3]	Elisa PCR Immunohistochemistry Viral isolation[b]	Supportive, ribavirin[3] Convalescent-phase plasma used with success in Argentine hemorrhagic fever patients		Argentine hemorrhagic fever vaccine (*not FDA approved*)	
Bunyaviridae *Hantavirus*	Hemorrhagic fever with renal syndrome	Rodent	—	No	Elisa (IgM, IgG) RT-PCR Immunohistochemistry	Supportive, ribavirin	Standard contact droplet, with eye protection	None currently	Rodent control
	HPS	Rodent (deer mouse)	—	No	Elisa (IgM, IgG) RT-PCR Immunohistochemistry Western blot Rapid immunoblot strip assay (RIBA)	Supportive only	Airborne (if indicated)[a]	None currently	

(continued)

TABLE 10.2 (continued)
Clinical Presentation and Public Health Control of VHFs

Virus Genus	Disease	Source of Human Infection		Person-to-Person Transmission	Diagnostics	Treatment	Infection Prevention Measures (CDC 2011) (Hospital Isolation, United States)	Vaccine	Vector Control
		Common	Less Common						
Phlebovirus	RVF	Mosquito (*Aedes*)	Contact with infected animal tissues; laboratory aerosolization	No	Elisa (IgM, IgG) RT-PCR Immunohistochemistry Viral isolation[b]	Supportive, ribavirin[3] (promising use in animal studies)	Standard contact droplet, with eye protection Airborne (if indicated)[a]	None currently; human live, attenuated vaccine, MP-12 in laboratory trials in domestic animals	Decrease contact with mosquitoes and to blood or tissues of infected animals
Nairovirus	CCHF	Tick (*Ixodid*)	Contact with infected animal blood; nosocomial	Yes	Elisa (IgM, IgG) PCR Immunohistochemistry Viral isolation[b]	Supportive, ribavirin	Standard contact droplet, with eye protection Airborne (if indicated)[a]	None widely available; an inactivated, mouse-brain-derived vaccine is used on a small scale in Eastern Europe	Avoidance of ticks; avoidance of exposure to blood and tissues of infected animals or humans
Filoviridae *Filovirus*	Ebola hemorrhagic fever	Unknown; fruit bat reservoir	Nosocomial	Yes	Elisa (IgM, IgG) PCR Immunohistochemistry Virus isolation[b]	Supportive	Standard contact droplet, with eye protection Airborne (if indicated)[a]	None currently	Avoid contact with potentially infected blood or secretions of any patient or deceased individual

	Marburg hemorrhagic fever	Unknown; fruit bat reservoir	Nosocomial	Yes	Elisa (IgM, IgG) PCR Immunohistochemistry[b] Virus isolation	Supportive		None currently	Preventive measures against transmission from animal host/fruit bat are not established
Flaviviridae *Flavivirus*	Yellow fever	Mosquito (*Aedes* or *Haemagogus*)	—	No	Elisa (IgM, IgG) PCR Immunohistochemistry Viral isolation[b]	Supportive	Standard contact droplet, with eye protection Airborne (if indicated)[a]	Vaccine (preexposure only)—yellow fever live, attenuated 17D vaccine	Avoid exposures to mosquito bites; mosquito control
	Dengue fever DHF	Mosquito (*Aedes*)	Organ transplants, blood transfusions	Vertical reported	Elisa (IgM, IgG) PCR Immunohistochemistry Viral isolation[b]	Supportive	Standard contact droplet, with eye protection Airborne (if indicated)[a]	Vaccines currently under investigation for probable use in travelers but unlikely to be a solution to hyperendemic dengue transmission that leads to DHF	Avoid exposures to mosquito bites; mosquito control of *Aedes aegypti*
	Kyasanur Forest disease	Tick, or contact with infected animal	—	No	Elisa (IgM, IgG) Virus isolation[b]	Supportive	Standard contact droplet, with eye protection Airborne (if indicated)[a]	None currently	Reduce exposures to ticks

(*continued*)

TABLE 10.2 (continued)
Clinical Presentation and Public Health Control of VHFs

Virus Genus	Disease	Source of Human Infection		Person-to-Person Transmission	Diagnostics	Treatment	Infection Prevention Measures (CDC 2011) (Hospital Isolation, United States)	Vaccine	Vector Control
		Common	Less Common						
	Omsk hemorrhagic fever	Tick or contact with the blood, feces, or urine of an infected sick or dead muskrat	Milk of infected goats or sheep; isolation from aquatic animals and water	No	Elisa (IgM, IgG) Virus isolation[b]	Supportive	Standard contact droplet, with eye protection Airborne (if indicated)[a]	None currently	Reduce exposure to ticks Reduce potential exposure to sick muskrats

Source: U.S. Centers for Disease Control and Prevention, CDC Special Pathogens Branch, Viral hemorrhagic fevers, 2011b, http://www.cdc.gov/ncidod/dvrd/spb/mnpages/dipagges/vhf.htm.

[a] Airborne transmission of VHF is a hypothetical possibility (unproven) during aerosol-generating procedures, such as aerosolized or nebulized medication administration, diagnostic sputum induction, bronchoscopy, airway suctioning, endotracheal intubation, positive pressure ventilation via face mask, and high-frequency oscillatory ventilation (USCDC, 2005).

[b] Biosafety level (BSL)–4 laboratories are recommended for the isolation of the majority of hemorrhagic fever viruses, with few exceptions (e.g., dengue and yellow fevers), that require BSL-2 or BSL-3 standards (USCDC, 2011b).

although by no means all, of the viruses and their mammalian or arthropod vectors tend to be somewhat limited in their geographic range (USCDC, 2012). Certain travelers to these implicated areas, especially when outbreaks of VHF are occurring, are at greater risk principally because of increased occupational risk including health-care professionals working either in clinical settings such as hospitals or clinics and community-based health workers and persons in regular contact with animals or performing animal-based research (USCDC, 2011b, 2012). Avoiding travel to those areas known to be characterized by any transmission of VHF, especially during outbreaks, is perhaps the most effective primary preventive measure depending on the specific disease in question. However, this may not be a viable solution when travel is absolutely required or when the geographic extent of a given VHF is so far-reaching as to render any attempts to stay out of its known range virtually impossible. When undertaking travel to a region associated with VHF transmission, it is always highly advisable to assiduously avoid any areas where active outbreaks are underway as that significantly increases the risk of acquisition for traveler and resident alike. The traveler should also make every attempt to avoid contact with host or vector species associated with transmission, which often entails staying away from rodents and from livestock in areas characterized by endemic transmission of certain Bunyaviridae (CCHF and rift valley fever [RVF]) (Cleri et al., 2006; USCDC, 2011b). The use of insect repellents and insecticide-treated bed nets are generally advisable when traveling in many parts of the world, especially the tropics, so as to reduce the potential exposure to all vector-borne illnesses, including the VHF. Travelers should not consume native bushmeat, which can be contaminated and has been a well-documented source of hemorrhagic fever viruses (LeDuc, 2008; USCDC, 2011b, 2012).

Although the direct risk of acquisition of VHF is greatest in those countries or areas, which harbor implicated mammalian hosts or arthropod vectors, sometimes animals or persons initially infected in those areas then travel or are transported elsewhere. This can result in disease in an area of world or in a setting not typically associated with VHF. There are a number of instances in which animals infected with VHF have been transported to other countries only to result in transmission to humans there, and lab workers have been at particular risk, sometimes causing severe illness or death (Borio et al., 2002; LeDuc, 2008; USCDC, 2011a, 2012). Infected persons also sometimes unknowingly travel, and in cases where it involves those agents capable of person-to-person spread, then additional cases may result from secondary transmission as has been the case with Marburg, Ebola, Lassa, CCHFs, and some of the less common New World arenaviruses (Allaranga et al., 2010; USCDC, 2011a). Person-to-person transmission typically occurs through close contact with infected persons or their contaminated body fluids or indirectly through contact with objects contaminated by their body fluids such as needles. Health-care workers are at particular risk since persons infected with VHFs become quite ill and require hospitalization. Because these persons may have traveled, they may be hospitalized without a definitive diagnosis putting health-care workers at risk for both direct and indirect transmission. A number of outbreaks of VHF have been predominated by nosocomial transmission with large numbers of serious illness and death among health-care workers taking care of infected persons (Speed et al., 1996; USCDC, 2011a, 2011b, 2012; WHO, 2001). Patients with suspected VHF should be placed in strict isolation and all health-care workers should observe standard contact and body fluid precautions to reduce transmission risk. Since cadavers have been known to be a source of infection, direct contact should be avoided with persons who have died from certain VHF like Ebola virus.

Persons traveling to areas with known transmission of yellow fever or Argentine hemorrhagic fever or to countries with vaccination entry requirements for either of these hemorrhagic fevers should be vaccinated well beforehand. The yellow fever vaccine is approved for persons 9 months to 59 years old and is a live, attenuated, viral vaccine, which requires a single dose although persons with ongoing risk of exposure should receive a booster shot every 10 years (USCDC, 2012). The less commonly used, and more recently developed, Argentine hemorrhagic fever vaccine is also a live, attenuated, viral vaccine requiring a single dose although this vaccine has not been approved by the U.S. Food and Drug Administration and is not commonly available in the United States. No booster doses with the Argentine hemorrhagic fever vaccine are recommended (Ambrosio et al., 2011).

An inactivated vaccine derived from mouse brain has been developed for prevention of CCHF (Bunyaviridae), which is in small-scale use in parts of Eastern Europe, but, again, is not approved for use in the United States. A vaccine for use in livestock has been developed for Rift Valley fever (RVF) (Bunyaviridae) and is effective in preventing animal epidemics but has not diminished sporadic, endemic transmission in humans. An experimental human vaccine for RVF has been developed but is not approved and not available in the United States. Finally, a promising vaccine for dengue is currently in clinical trials but will require additional testing before it could be considered for approved use in humans (Borio et al., 2002; USCDC, 2005, 2011b).

10.4 PREVENTION, TREATMENT, AND CONTROL OF VIRAL HEMORRHAGIC FEVERS

As noted, Table 10.2 provides a summary of prevention, treatment, and control measures for the specific hemorrhagic fever viruses. The information is grouped into the four families of viruses known to cause VHF and includes disease-specific information on the availability of a vaccine, effective treatments (if any), principal arthropod vector or mammalian host and control of their populations, whether person-to-person transmission occurs, and appropriate infection control measures to reduce the risk of nosocomial transmission.

10.5 ARENAVIRIDAE-ASSOCIATED VIRAL HEMORRHAGIC FEVERS

The arenavirus-associated hemorrhagic fevers encompass both Old World and New World groups including the widely recognized Lassa fever in the former along with the much rarer Lujo hemorrhagic fever and lymphocytic choriomeningitis. The group referred to as the New World hemorrhagic fevers comprises the Venezuelan, Argentinean, Bolivian, and Brazilian hemorrhagic fevers. The transmission of both Old World and New World arenaviruses to humans is usually from infected rodents. Rodents may transmit VHFs to humans via inhalation of aerosols from rodent urine or fomites, ingestion of rodent-contaminated food, or direct contact of broken skin with rodent feces or urine or saliva. Control and reduction of rodent populations through poisoning, trapping, and other control measures can assist in diminishing the risk of transmission to humans as can minimizing any human contact with rodents. Public health education messages should encourage the public to avoid all contact with rodents and ensure appropriate storage of food in the home to maintain as hygienic an environment as possible to prevent rodent infestations. There is no vaccine available for any of the arenaviruses, but in the event of infection, ribavirin has been a somewhat effective treatment. The recommended loading dose of ribavirin for treatment of Lassa fever is 30 mg/kg intravenously (IV) (maximum 2 g) once, followed by 16 mg/kg IV (maximum, 1 g per dose) every 6 h for 4 days, followed by 8 mg/kg IV (maximum, 500 mg per dose) every 8 h for 6 days. Children can be provided the same ribavirin regimen although dosing needs to be adjusted according to weight. Both Old World and New World arenaviruses can be transmitted from person to person; so VHF precautions should be instituted for all hospitalized patients and all health-care workers need to carefully observe standard contact, droplet, and airborne infection control precautions to prevent nosocomial transmission. At this point, airborne transmission of VHFs is a hypothetical possibility during aerosol-generating procedures, such as aerosolized or nebulized medication administration, diagnostic sputum induction, bronchoscopy, airway suctioning, endotracheal intubation, positive pressure ventilation via face mask, and high-frequency oscillatory ventilation.

10.6 BUNYAVIRIDAE-ASSOCIATED VIRAL HEMORRHAGIC FEVERS

The Bunyaviridae family represents a more diverse group of pathogens in terms of the arthropod vectors and mammalian hosts involved in transmission to humans and with regard to the extent of their geographic range, which is quite large in the case of Hantavirus. Both of these issues speak to an increased level of complexity with regard to appropriate prevention and control issues because a larger geographic area is involved and a greater number of animal vectors need to be addressed. In addition to Hantavirus, which causes both hantavirus pulmonary syndrome (HPS) and hemorrhagic fever with renal syndrome (HFRS), this family of viruses also causes the less common CCHF and Rift Valley hemorrhagic fever, which, unlike Hantavirus, both have arthropod vectors. Hantavirus-related hemorrhagic fevers are typically transmitted by the inhalation of aerosolized excrement of rodents. As such, public health messaging should focus on undertaking efforts to reduce human contact with rodents including the secure storage of foodstuffs as noted earlier and maintenance of a hygienic home environment ensuring no food remnants are present, which could attract rodents into the home. This public health messaging can also be combined with systematic efforts to reduce the rodent population through poisoning or trapping. RVF is primarily transmitted through the bite of mosquitoes (genus *Aedes*) so reducing the probability of being bitten should be the focus of primary prevention efforts. Environmental control measures that may prove useful include elimination of any standing water especially around dwellings as these can provide breeding grounds for mosquitoes and the selective use of insecticides and larvicides to knock down mosquito populations. Public health education messages should also focus on reduction of personal risk through regular use of DEET-containing mosquito repellents, ensuring protective clothing that covers as much exposed skin surface as possible, and the use of insecticide-treated bed nets to lessen risk of being bitten during sleep. CCHF is largely transmitted through the bite of infected ticks (genus *Ixodes*), so use of repellents, protective clothing, and steering clear of known habitats for ticks can help significantly reduce risk of a bite. RVF and CCHF are both less commonly transmitted through human contact with the blood or tissues of infected animals, sometimes in laboratory settings, so avoidance of such should be emphasized with the public. There is no vaccine currently available to prevent hantavirus-related illness nor for RVF although a live, attenuated vaccine is in clinical trials using animals for the latter. An inactivated vaccine is in small-scale use in some parts of Eastern Europe for CCHF, but this vaccine is not approved for use by the U.S. Food and Drug Administration and is not otherwise widely available. In the event of infection, therapy for HPS consists of supportive care, whereas ribavirin has met with some success in the treatment of HFRS and CCHF. Hospitalized patients with HFRS or CCHF should be provided a ribavirin loading dose of 30 mg/kg IV (maximum 2 g) once, followed by 16 mg/kg IV (maximum, 1 g per dose) every 6 h for 4 days, followed by 8 mg/kg IV (maximum, 500 mg per dose) every 8 h for 6 days. Pregnant woman may receive the same ribavirin dosing as adults as should children, except the child's dosing should be adjusted according to weight. Ribavirin has shown promise in the treatment of RVF in animal models but is currently not used in the treatment of infected humans. No person-to-person transmission has been seen with HPS, HFRS, or RVF although it has been documented for CCHF resulting in nosocomial transmission. Nonetheless, VHF precautions should be instituted for all hospitalized patients and all health-care workers need to carefully observe standard contact, droplet, and airborne infection control precautions with eye protection to prevent nosocomial transmission particularly for CCHF since such transmission has been previously documented. As noted earlier, the risk of airborne transmission remains a theoretical risk but should be instituted for hospitalized patients with any of these VHFs as indicated.

10.7 FILOVIRIDAE-ASSOCIATED VIRAL HEMORRHAGIC FEVER

This family of VHFs includes the Ebola and Marburg viruses, which have been associated with some of the most widely publicized and deadly outbreaks of VHFs. Unfortunately, we probably understand their epidemiology the least well, and uncertainties remain about their animal reservoir making prevention and control more difficult. Fruit bats, which are known to carry filoviruses, remain a candidate host reservoir, but their exact role in transmission to humans has not been fully elucidated. Consequently, no specific preventive measures have been recommended to lessen transmission risk to humans from animal/environmental reservoirs, nor have control campaigns been directed at their reducing fruit bat populations. No vaccine exists for either Ebola or Marburg. Both of these pathogens are capable of person-to-person transmission and some of the largest outbreaks associated with them have been in health-care workers through nosocomial transmission as seen in the outbreak among health-care workers in Kikwit in 1995 (Francesconi et al., 2003; USCDC, 1998; Figures 10.1 and 10.2).

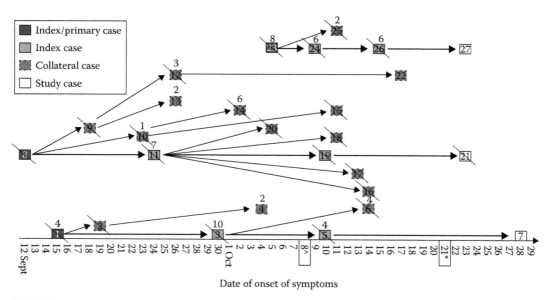

FIGURE 10.1 Chains of transmission relative to 27 Ebola cases, Gulu District, Uganda (September–October 2000). (From Francesconi, P. et al., *Emerg Infect Dis.*, 9(11), 1420, 2003.)

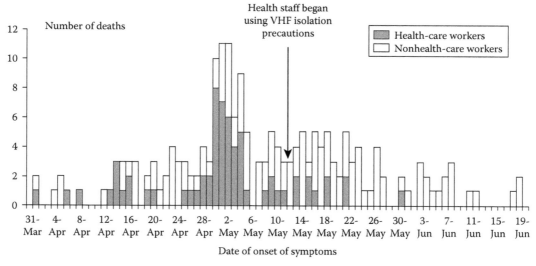

FIGURE 10.2 Epidemic curve of the 1995 Ebola hemorrhagic fever outbreak in Kikwit. (From U.S. Centers for Disease Control and Prevention and World Health Organization, *Infection Control for Viral Haemorrhagic Fevers in the African Health Care Setting*, Centers for Disease Control and Prevention, Atlanta, GA, 1998, pp. 1–198.)

Ebola also has the distinction of documented transmission from cadavers so that contact should be avoided with the blood and secretions not only of infected persons but also from the deceased. There are no effective treatments for either of these filoviruses and hospitalized patients should receive supportive care. VHF precautions should be stringently instituted for all hospitalized patients and all health-care workers need to carefully observe standard contact, droplet, and airborne infection control precautions with eye protection to prevent nosocomial transmission. As noted earlier, the risk of airborne transmission remains a theoretical risk, but caution would probably dictate instituting airborne precautions for all hospitalized patients with either of these VHFs to further minimize the risk of nosocomial spread.

10.8 FLAVIVIRIDAE-ASSOCIATED VIRAL HEMORRHAGIC FEVERS

These represent a diverse group of VHFs and include the well-known and widely spread yellow fever and dengue fever (and dengue hemorrhagic fever [DHF]), in addition to the more localized and lesser known Kyasanur Forest disease and Omsk hemorrhagic fever. All four of these entities generally require transmission to humans through an arthropod vector, mosquitoes in the case of yellow fever and dengue fever and ticks in Kyasanur and Omsk. As previously discussed, public health messages should focus on reduction of individual risk through use of protective clothing, insect repellents, and bed nets, combined with environmental control activities that can diminish arthropod populations. Persons residing in Kyasanur Forest disease-affected areas should also be advised to avoid contact with infected animals as transmission has been documented less frequently via this route. Similarly, persons at potential risk of exposure to Omsk hemorrhagic fever should be advised to avoid contact with infected animals in addition to the blood, feces, or urine of sick or dead muskrats and the milk of infected sheep or goats, which should always be kept away from aquatic animals and standing water to reduce risk of acquisition. No person-to-person transmission has been documented for any of the Flavivirus-associated hemorrhagic fevers although vertical transmission from mother to neonate has been observed with dengue fever. Fortunately, an effective live, attenuated vaccine is available for preexposure prophylaxis against yellow fever, which should always be recommended for any person living in or traveling to a dengue endemic area and is often an entry requirement for international travelers for by most yellow fever endemic countries. A vaccine for dengue fever is currently in clinical trials and is likely to be approved for use in travelers although it probably will not greatly impact the hyperendemic dengue transmission leading to DHF. There is no vaccine available for either Kyasanur Forest disease or Omsk hemorrhagic fever, and ribavirin has no known therapeutic role in the treatment of any of the Flavivirus-associated hemorrhagic fevers.

As such, hospitalized patients should receive supportive care and VHF precautions should be instituted. Health-care workers in these settings need to carefully observe standard contact and droplet infection control precautions with eye protection to prevent nosocomial transmission and implement airborne precautions, as indicated, to reduce the theoretical risk of transmission via this route.

10.9 CONTROL OF VIRAL HEMORRHAGIC FEVERS IN HOSPITALS

Natives of, or travelers to, areas of endemic VHF transmission must undertake preventive measures to avoid exposure to VHF's natural hosts, ill animals or persons, or the contaminated tissues of VHF victims, whether animal or human. Further, health-care workers or laboratorians working with VHF-infected patients or tissues also need to be aware of the risks for exposure. While the majority of VHFs are not endemic to the United States, travelers have carried VHF, such as Marburg and Lassa, presenting with VHF syndromes to hospitals in the United States and Europe (USCDC, 2011b, 2012). The key to prevention, then, is understanding the routes of transmission, the natural hosts, and the potential for secondary spread of VHFs. Experts from the World Health Organization (WHO) and Centers for Disease Control and Prevention (CDC) have developed guidelines for the control of VHFs in the African health-care setting in 1998, which are still in use (USCDC, 1998); the CDC also updated hospital infection prevention guidelines for U.S. hospitals in 2005 (USCDC, 2005). In the United States, the Healthcare Infection Control Practices Advisory Committee (HICPAC) references the CDC isolation recommendations, and these remain the most recent at the time of this publication (Siegel et al., 2007).

VHF transmission from infected patients is primarily through direct contact or droplet exposure; airborne transmission has not been definitively established for VHFs (Siegel et al., 2007; USCDC, 1998, 2005). Person-to-person transmission of VHF has been documented with Ebola, Marburg, Lassa fever, and some arenaviruses (such as Machupo); this is especially relevant for family members and caregivers, as well as health-care providers, as nosocomial transmission events have occurred in the hospital setting with significant morbidity and mortality for both the patients and the health-care workers. HPS-causing Hantaviruses in the United States (Sin Nombre, New York, and Black Creek) have not been documented to be transmitted person to person.

For suspected VHF cases hospitalized in U.S. hospitals, current recommendations are for standard, contact, and droplet precautions, with eye protection (Siegel et al., 2007; USCDC, 2005). Health-care workers or visitors need to ensure that mucosal membranes are barrier protected in addition to the important standard protective measures for skin. Everyone entering a suspected VHF patient's room should be, minimally, wearing a gown and gloves, and face shields, or surgical masks and eye protection, should be worn by anyone within 3 ft of the patient. Additional barrier protection may be needed depending upon the circumstances of the patient (e.g., excessive bleeding, coughing, vomiting). Care to prevent percutaneous injuries with needles or other instrumentation is vital.

While not documented, there is a theoretical risk of airborne transmission of VHF in a hospitalized patient who has severe pneumonia or is undergoing particular procedures. Airborne precautions may be chosen by hospitals for suspect VHF patients with serious lung involvement or who undergo aerosol-generating procedures, such as sputum induction, nebulized treatments, bronchoscopy, airway suctioning, endotracheal intubation, positive pressure ventilation (face mask), or high-frequency oscillatory ventilation, to prevent nosocomial transmission (USCDC, 2005).

Recommendations for the health-care control of VHF in African settings emphasize the need for, and the education about, standard precautions at all times, along with the regular use of hand washing. Further, the ability to identify a suspected VHF case is paramount, along with planning for the potential isolation and care of suspected patients. The WHO- and CDC-developed recommendations, *Infection Control for Viral Haemorrhagic Fevers in the African Health Care Setting* (USCDC, 1998), identify VHF isolation precautions that have been shown to reduce VHF transmission in the health-care setting for any suspected VHF patient. These include

1. Isolating the patient
2. Wearing protective clothing in the isolation area, in the cleaning and laundry areas, and laboratory; wearing a scrub suit, gown, apron, two pairs of gloves, mask, head cover, eyewear, and rubber boots
3. Cleaning and disinfecting spills, waste, and reusable equipment safely
4. Cleaning and disinfecting soiled linens and laundry safely
5. Using safe disposal methods for nonreusable supplies and infectious waste
6. Providing information about the risk of VHF transmission to health facility staff, reinforcing use of VHF isolation precautions with all health facility staff
7. Providing information to families and the community about prevention of VHFs and care of patients

This infection control manual provides clear, step-by-step processes for infection control in the health-care setting in Africa, including cleaning and disinfection instructions, figures illustrating the appropriate donning and doffing procedures, and planning guidance.

There are instances in which a health-care worker may be exposed to secretions inadvertently during the care of a VHF patient. While ribavirin has been shown to be useful in the treatment of some VHFs, such as Lassa fever and HFRS (USCDC, 2011b), there is some debate as to its use for postexposure prophylaxis for some VHFs. While some recent studies support the use of ribavirin for Lassa fever, and possibly other arenaviruses and CCHF virus in the postexposure setting (Bausch et al., 2010), CDC has not issued recommendations, nor is it FDA approved, for use as VHF post-exposure prophylaxis at the time of this publication. Regardless, any person with skin or mucosal exposure to a VHF patient's secretions, tissues, or body fluids must immediately wash the affected skin surfaces with soap and water, and mucous membranes, including eyes, should be flushed generously with water and/or eyewash solution. They should then be monitored for twice daily for 21 days after exposure; any fever within 21 days of exposure necessitates an infectious disease specialist consultation (USCDC, 2005).

Exposures to bodily secretions, potentially infected tissues, and surfaces must be prevented if a VHF (or suspected VHF) patient dies, and the handling of any human remains should be minimized. The 2005 CDC recommendations note that the body should not be embalmed, and it should be wrapped in sealed leak-proof material. CDC also recommends cremation or prompt burial in a sealed casket (USCDC, 2005).

REFERENCES

Allaranga Y., Kone M.L., Formenty P. et al., 2010. Lessons learned during active epidemiological surveillance of Ebola and Marburg viral hemorrhagic fever epidemics in Africa. *East Afr J Public Health*. 7(1), 32–38.

Ambrosio A., Saavedra M., Mariana M. et al., 2011. Argentine hemorrhagic fever vaccines. *Hum Vaccin Immunother*. 7(6), 694–700.

Bausch D., Hadi C., Khan S., and Lertora J., 2010. Review of the literature and proposed guidelines for the use of oral ribavirin as post exposure prophylaxis for Lassa fever. *Clin Infect Dis*. 51(12), 1435–1441.

Borio L., Inglesby T., Peters C.J. et al., 2002. Hemorrhagic fever viruses as biological weapons: Medical and public health management. *JAMA*. 287(18), 2391–2405.

CDC. Interim guidance for managing patients with suspected viral hemorrhagic fever in U.S. hospitals. May 19, 2005 update to previous recommendations (**MMWR** 1995 June 30; 44(25): 475–479). Available at: http://www.cdc.gov/HAI/pdfs/bbp/VHFinterimGuidance05_19_05.pdf. Accessed December 20, 2011.

Cleri D.J., Ricketti A.J., Porwancher R.B., Ramos-Bonner L.S., and Vernaleo J.R., 2006. Viral hemorrhagic fevers: Current status of endemic disease and control strategies for control. *Infect Dis Clin North Am*. 20, 359–393.

Crowcroft N.S., Morgan D., and Brown D., 2002. Editorial: Viral haemorrhagic fevers in Europe-effective control requires a coordinated response. *Eurosurveillance*. 7(3): pii = 343. Available online at: http://www.eurosurveillance.org/ViewArticle.aspx?ArticleId=343. Accessed January 11, 2012.

Francesconi P., Yoti Z., Declich S. et al., 2003. Ebola hemorrhagic fever transmission and risk factors of contacts, Uganda. *Emerg Infect Dis*. 9(11), 1420–1437.

LeDuc J.W., 2008. Viral diseases transmitted primarily by arthropod vectors: The epidemiology of viral hemorrhagic fevers, in: Wallace, R.B. (Ed), *Maxcy-Rosenau-Last Public Health and Preventive Medicine*, 15th edn. McGraw Hill Medical, New York, pp. 352–361.

Peters C.J., 2006. Emerging infections: Lessons from the viral hemorrhagic fevers. *Trans Am Clin Climatol Assoc*. 11, 189–196.

Siegel J.D., Rhinehart E., Jackson M., Chiarello L., and the Healthcare Infection Control Practices Advisory Committee, 2007. *2007 Guideline for Isolation Precautions: Preventing Transmission of Infectious Agents in Healthcare Settings*. U.S. Centers for Disease Control and Prevention, Atlanta, GA. Available online at: http://www.cdc.gov/ncidod/dhqp/pdf/isolation2007.pdf. Accessed December 30, 2011.

Speed B.R., Gerrard M.P., Kennett M.L., Catton M.G., and Harvey B.H., 1996. Reviews: Viral haemorrhagic fevers: Current status, future threats. *Med J Aust*. 164, 79–83.

U.S. Centers for Disease Control and Prevention and World Health Organization, 1998. *Infection Control for Viral Haemorrhagic Fevers in the African Health Care Setting*. Centers for Disease Control and Prevention, Atlanta, GA. Available online at: http://www.cdc.gov/ncidod/dvrd/spb/mnpages/vhfmanual.htm. Accessed December 20, 2011.

U.S. Centers for Disease Control and Prevention, 2005. *Interim Guidance for Managing Patients with Suspected Viral Hemorrhagic Fever in U.S. Hospitals.* Available online at: http://www.cdc.gov/HAI/pdfs/bbp/VHFinterimGuidance05_19_05.pdf. Accessed on January 1, 2012.

U.S. Centers for Disease Control and Prevention. CDC Special Pathogens Branch, 2011a. Viral hemorrhagic fever outbreak postings. Available online at: http://www.cdc.gov/ncidod/dvrd/spb/outbreaks/index.htm. Accessed on December 8, 2011.

U.S. Centers for Disease Control and Prevention. CDC Special Pathogens Branch, 2011b. Viral hemorrhagic fevers. Available online at: http://www.cdc.gov/ncidod/dvrd/spb/mnpages/dipagges/vhf.htm. Accessed December 2, 2011.

U.S. Centers for Disease Control and Prevention, 2012. Infectious diseases related to travel: Viral hemorrhagic fevers, in: Brunette, G.W. (Ed.), *CDC Health Information for International Travel 2012.* Oxford University Press, New York. Available online at: http://wwwnc.cdc.gov/travel/yellowbook/2012/chapter-3-infectious-diseases-related-to-travel/viral-hemorrhagic-fevers.htm. Accessed December 2, 2011.

World Health Organization, 2001. Crimean-Congo haemorrhagic fever. Fact Sheet N208. Available at: http://www.who.int/mediacentre/factsheets/fs208/en/. Accessed January 11, 2012.

11 Practical Considerations for Collection and Transportation of Hemorrhagic Fever Samples

Manfred Weidmann and Mandy Elschner

CONTENTS

11.1 INTRODUCTION

In preparation of an outbreak sortie, the organization of a good shipment chain is of utmost importance. If the infectious agent has already been identified in initial samples and sent to specialized centers, the choice of assays to take along may be easier to determine. If however the agent is unidentified, the assays taken along may not be able to identify the agent on site. Therefore, organizing a shipment chain for patient samples is a very good investment to allow an easy transfer of the samples to a specialized center.

First of all, it is a good idea to contact carriers, which take the responsibility as shippers to transport dangerous goods, e.g., World Courier (http://www.worldcourier.com/) and Marken (http://www.marken.com/) to find out if and from where they can provide transportation services. The next question to be solved is if there is a local source of dry ice. Consider the shipper, hospitals, universities, breweries (carbonation of drinks is often performed using dry ice), and companies using dry ice blasting machines. To find these companies, contact dry ice providers (http://www.dryicedirectory.com/; http://www.yara.com/products_services/industrial_solutions/carbon_dioxide_dry_ice/industrial_processes/blast_cleaning.aspx). You should have an initial set of transport boxes as described below to be able to send your samples, as local transport from the outbreak site to an airport will mostly be entirely dependent on them.

Local rules and regulations on shipment of human and animal samples should be made available from the local public health authorities. Contact your local embassy to ask for support. The type of samples you might want to take (blood, CSF, urine, saliva) will determine the type of sampling material you will have to take along (syringes, swaps). Think of labels to mark your samples with and how you want to run documentation of ongoing sample collection.

Think of how samples will be taken: Which type of samples? What quantities? Which biosafety measures? Which type of sampling environment would you prefer for sampling, and how can you generate this type of environment on site to avoid background contamination? How can the samples be stored before shipment? How will they be shipped? Without answering these questions, your hard work on site may be doomed to fail and you may be doomed too. For detailed advice, check the CDC guide "Infection Control for Viral Haemorrhagic Fevers in the African Health Care Setting" (1998) and the detailed account of a mobile field laboratory sortie during the Marburg outbreak in Angola 2005 (Grolla et al., 2011). A mobile laboratory of course reduces the necessity of shipments. There are however only few mobile laboratories.

11.2 INACTIVATION OF SAMPLES FROM HEMORRHAGIC FEVER PATIENTS

11.2.1 SAMPLES FOR SEROLOGY

Many guides recommend the use of sealed sterile dry tubes (vacutainers type) to contain blood samples in the original receptacle. This approach already solves most of the problems mentioned earlier. Additionally, centrifugation and generation of aerosols can be omitted by letting blood clot in a tube on the bench. Virus antigen preparations and virus culture supernatants of Rift Valley Fever virus and Crimean Congo hemorrhagic fever can be inactivated at 60°C/60 min (Saluzzo et al., 1988). This thermal inactivation is also effective for Lassa virus, Ebola virus, and Marburg virus (Mitchell and McCormick, 1984). More recently, work done in Marburg Germany has confirmed this for the filoviruses. Additionally, this report showed that the detergents NP40 1% and Triton X-100 1% reduce EBOV activity as measured by $TCID_{50}$ by 4,5 log steps while SDS 1% and Michrochem 3% reduce it by only 4 log steps at room temperature (Becker, 2009). Fortunately, chemical inactivation of sera by Triton X-100 and SDS 1% has been shown not to have an effect on serological testing (Dobler et al., 2004).

11.2.2 SAMPLES FOR MOLECULAR BIOLOGY

Samples from vacutainers need to be inactivated before use. Using chaotropic guanidine isothiocyanate (GITC) buffers such as the lysing buffers in many nucleic acid extraction kits has been shown to inactivate enveloped viruses such as alphaviruses, bunyaviruses, filoviruses, and flaviviruses (Blow et al., 2004). If kits are not available, simply add four volumes of GITC buffer to one volume of the sample. Using vacutainers already containing an inactivating agent would be a good alternative. The cationic surfactant tetradecyltrimethyl ammonium oxalate has been used to extract flavivirus RNA from whole blood (Hamel et al., 1995). The PAXgene vacutainers system contains this ingredient (Carrol et al., 2007), but the likely inactivating potential of the buffer used has not yet been tested for viral hemorrhagic fever (VHF) viruses.

In a recent review, the use of dried blood spot cards for viral diagnostics was surveyed (Snijdewind et al., 2012). Viral RNA, DNA antibodies, and antigens remain stable on these easy-to-use cards but have not yet been used for acute VHF patient samples. The use of these cards for highly infectious samples of Hepatitis B virus, Hepatitis C virus, and Dengue virus. Virus-infected patients however indicates that handling is safe and could be considered an alternative for samples of VHF patients.

11.2.3 SKIN BIOPSIES FOR FILOVIRUS DIAGNOSTICS

The CDC has developed a skin biopsy kit, which can be ordered on demand (annex 12 in [1998]).

11.3 SAMPLE SHIPMENT ACCORDING TO IATA-DANGEROUS GOODS REGULATIONS

For air shipments, the "International Air Transport Association-Dangerous Goods Regulations" (IATA-DGR) (2012) are updated annually after consultation of the UN Committee of Experts, the International Atomic Energy Agency, and the International Civil Aviation Organization (ICAO). International, regulations for road shipments are described in the ADR (http://www.unece.org/trans/danger/publi/adr/adr2011/11ContentsE.html) and for transport by rail in the RID (http://eur-lex.europa.eu/LexUriServ/LexUriServ.do?uri=CELEX:31996L0049:EN:NOT).

TABLE 11.1
Infectious Substances Must Be Assigned to a UN Number

Proper Shipping Name/Description	UN Code	Packaging Instruction
Infectious substance, affecting humans (liquid)	UN 2814	PI 620
Infectious substance, affecting humans (solid)	UN 2814	PI 620
Carbon dioxide, solid; dry ice	UN 1845	PI 904

Infectious VHF samples fall into the dangerous goods (DG) class 6.2 for infectious substances, category A. This category refers to an infectious substance that is transported in a form that, when exposure to it occurs, is capable of causing permanent disability, life-threatening disease, or fatal disease in otherwise healthy humans or animal. Therefore, VHF patient samples and infectious animal samples have to be shipped using the shipping names UN 2814 "Infectious substance, affecting humans" (Table 11.1).

The IATA-DGR define patient samples as follows: human or animal materials, collected directly from humans or animals, including, but not limited to, excreta, secreta, blood and its components, tissue and tissue fluid swabs, and body parts being transported for purposes such as research, diagnosis, investigational activities, disease treatment and prevention.

VHF patient samples should be packed according to packaging instructions PI620 (road, rail, air) and PI 904 (dry ice) using UN-type approved packaging material.

Basically, an inner primary receptacle for the samples is protected by an outer rigid container with absorbent material buffering the space between the containers. The secondary container should be externally disinfected by wiping with a surface-active agent, e.g., hypochlorite (1000 ppm available chlorine). In a matrochka manner, all of this goes into a third pack.

In more detail, the inner package consists of

- Watertight primary receptacle(s) (O-ring tubes, vacutainers)
- Watertight secondary packaging (see Figure 11.1)
- Absorbent material in sufficient quantity to absorb the entire contents placed between the primary receptacle and the secondary packaging (e.g., tissue paper)

An itemized list of contents must be enclosed between the secondary and the outer packaging in an extra plastic bag. Do not place it inside the secondary container.

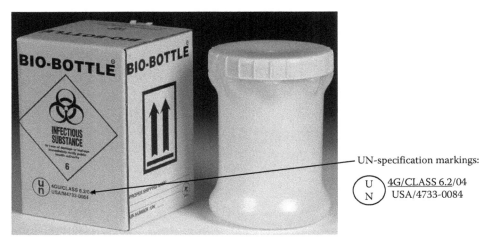

UN-specification markings:

U
N   4G/CLASS 6.2/04
     USA/4733-0084

FIGURE 11.1   Package for infectious substances, category A (UN 2814). Specifications must be visible on the outer package.

The outer package consists of a rigid outer packaging, smallest external dimension 100 mm. It should be marked with the proper shipping name (i.e., "Infectious substance, affecting humans" UN 2814), and the appropriate diamond-shaped warning label for infectious substances (50 mm × 50 mm, Figure 11.2) and the UN specifications.

For refrigerated shipping, refrigerants (ice, dry ice) must be placed outside the secondary package, i.e., into an overpack (Figure 11.3) with one or more complete packages (Figure 11.1). Interior support must be provided to secure the secondary packages in the original position after the ice or dry ice has been filled in. If ice is used, the overpack must be leak-proof. If dry ice is used, the outer package must permit the release of carbon dioxide gas to avoid pressure buildup due to evaporated CO_2.

In order not to meet all formal requirements, the outside markings need to include

- Proper shipping name and UN number
- For dry ice, UN 1845 and net weight in kilogram (kg) used
- Shipper's name, address, and telephone number
- Receiver's name, address, and telephone number
- Name and telephone number of responsible person (available 24 h until shipment arrives)
- Overpack
- UN specification marking
- Infectious substance label (minimum size 50 mm × 50 mm)
- Package orientation label (only used when liquid content in the primary container exceeds 50 mL)
- For dry ice, label class 9 dangerous good

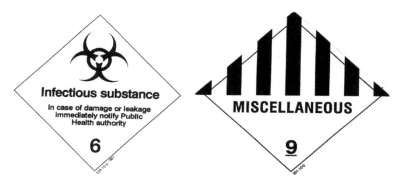

FIGURE 11.2 Labels.

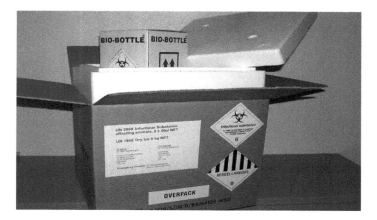

FIGURE 11.3 Overpack markings.

11.4 DOCUMENTATION

For each package, an air waybill of the respective company has to be filled out for transport on road to the airport. For UN codes 2814 or 2900, the column for handling information needs to be filled in with "Dangerous Goods as per attached Shipper's Declaration." This shipper's declaration has to be filled out according to the IATA-DGR regulations as shown in the example in Figure 11.4. According to the IATA-DGR/ICAO-TI regulations, shippers of infectious substances need to be trained and certified every 2 years. The training includes the classification, packaging, labeling, and documentation and authorized to sign the shipper's declaration.

SHIPPER'S DECLARATION FOR DANGEROUS GOODS

Shipper
Dr. Sabine Mustermann
Institut fof Mikrobiology
Mustermannstr. 1
10115 Berlin, Germany
phone +49 1234 56789

Air Waybill No
Page 1 of 1 Pages
Shippers Reference Number
(optional)

Consignee
Dr. Robert Mustermeier
Institut of Virology
Mustermeierstr. 2
20095 Hamburg, Germany
phone +49 3456 7890

WORLD COURIER

Two completed and signed copies of this Declaration must be handed to the operator.

TRANSPORT DETAILS

This shipment is within the limitations prescribed for: (delete non-applicable)
PASSENGER AND CARGO AIRCRAFT | CARGO AIRCRAFT ONLY

Airport of Departure

Airport of Destination:

WARNING

Failure to comply in all respects with the applicable Dangerous Goods Regulations may be in breach of the applicable law subject to legal penalties.

Shipment type: (delete non-applicable)
NON-RADIOACTIVE | RADIOACTIVE

NATURE AND QUANTITY OF DANGEROUS GOODS

Dangerous Good Identification						
UN or ID No.	Proper Shipping Name	Class or Division (Subsidiary Risk)	Pack-ing Group	Quantity and type of packing	Packing Instr.	Authorization
UN 2814	Infectious substance, affecting humans (Ebola) (liquid)	6.2	-	1 x Fiberboard x 2ml	620	
UN1845	Dry Ice	9	III	8 KG Overpack used	904	

Additional Handling Information
24 Hour emergency Contact Number: Dr. Sabine Mustermann+49 3841 12124

I hereby declare that the contents of this consignment are fully and accurately described above by the proper shipping name, and are classified, packaged, marked and labelled/placarded, and are in all respects in proper condition for transport according to applicable international and national governmental regulations. I declare that all of the applicable air transport requirements have been met.

Name/Title of Signatory
Dr. Sabine Mustermann, Microbiologist
Place and Date
Berin, 01.02.2012
Signature
(see warning above) Mustermann

FIGURE 11.4 Example of a shipper's declaration for dangerous goods.

REFERENCES

1998. Infection Control for Viral Haemorrhagic Fevers in the African Health Care Setting. In: W. H. Organization, C. C. F. D. C. A. Prevention, and U. S. D. O. H. H. S. P. H. Service (Eds.).

Becker, S. 2009. Inactivation of filoviruses in clinical samples and establishment of monoclonal antibodies against Marburg virus. Research Report for project FV M SAB1 5A016 (University of Marburg).

Blow, J. A., D. J. Dohm, D. L. Negley, and C. N. Mores. 2004. Virus inactivation by nucleic acid extraction reagents. *J Virol Methods* 119:195–198.

Carrol, E. D., F. Salway, S. D. Pepper, E. Saunders, L. A. Mankhambo, W. E. Ollier, C. A. Hart, and P. Day. 2007. Successful downstream application of the Paxgene Blood RNA system from small blood samples in paediatric patients for quantitative PCR analysis. *BMC Immunol* 8:20.

Council Directive 96/49/EC of 23 July 1996 on the approximation of the laws of the Member States with regard to the transport of dangerous goods by rail. http://eur-lex.europa.eu/LexUriServ/LexUriServ.do?uri=CELEX:31996L0049:EN:NOT

Dobler, G., Essbauer, S., Pfeffer, M., Wölfel, W., and E. J. Finke. 2004. Efficiency of various inactivation methods for alphaviruses, flaviviruses, and bunyaviruses. *Medical Biodefense Meeting*, October 20–21, 2004, Munich, Germany.

European Agreement concerning the International Carriage of Dangerous Goods by Road (ADR). http://www.unece.org/trans/danger/publi/adr/adr2011/11ContentsE.html

Grolla, A., S. M. Jones, L. Fernando, J. E. Strong, U. Stroher, P. Moller, J. T. Paweska, F. Burt, P. Pablo Palma, A. Sprecher, P. Formenty, C. Roth, and H. Feldmann. 2011. The use of a mobile laboratory unit in support of patient management and epidemiological surveillance during the 2005 Marburg Outbreak in Angola. *PLoS Negl Trop Dis* 5:e1183.

Hamel, A. L., M. D. Wasylyshen, and G. P. Nayar. 1995. Rapid detection of bovine viral diarrhea virus by using RNA extracted directly from assorted specimens and a one-tube reverse transcription PCR assay. *J Clin Microbiol* 33:287–291.

http://www.dryicedirectory.com/

http://www.marken.com/

http://www.worldcourier.com/

http://www.yara.com/products_services/industrial_solutions/carbon_dioxide_dry_ice/industrial_processes/blast_cleaning.aspx

International Air Transport Association (IATA). 2012 IATA Dangerous Goods Regulations Manual 2012 (DGR), 53rd edn., http://www.iata.org/ps/publications/dgr/Pages/manuals.aspx

Mitchell, S. W. and J. B. McCormick. 1984. Physicochemical inactivation of Lassa, Ebola, and Marburg viruses and effect on clinical laboratory analyses. *J Clin Microbiol* 20:486–489.

Saluzzo, J. F., B. Leguenno, and G. Van der Groen. 1988. Use of heat inactivated viral haemorrhagic fever antigens in serological assays. *J Virol Methods* 22:165–172.

Snijdewind, I. J., J. J. van Kampen, P. L. Fraaij, M. E. van der Ende, A. D. Osterhaus, and R. A. Gruters. 2012. Current and future applications of dried blood spots in viral disease management. *Antiviral Res* March, 93(3):309–321.

12 Laboratory Diagnosis of Viral Hemorrhagic Fevers

Miroslav Fajfr and Daniel Růžek

CONTENTS

12.1 INTRODUCTION

Viral hemorrhagic fevers (VHFs) are caused by wide and various groups of viral pathogens. They belong to four viral families: *Filoviridae*, *Arenaviridae*, *Flaviviridae*, and *Bunyaviridae*. Although they all belong to RNA viruses, many biological differences can be found between each representative. As already described elsewhere in this book, VHFs belong to one of the most dangerous human diseases not only for individuals but, due to the possibility of their spreading widely, also for populations. Some VHFs exhibit very high mortality, up to 85%–90%. Rapid diagnosis is of high importance, especially for the initiation of adequate treatment of the infected persons, as well as the establishment of necessary epidemiological countermeasures. However, the development or routine usage of laboratory techniques for the diagnosis of VHF pathogens is complicated mostly because work with the causative agents of VHFs is limited to facilities with the highest biosafety level (BSL), BSL-4. In this chapter, we summarize the available ways for the detection and identification of known VHF causative agents. The chapter is divided into four sections. The first section is focused on non-laboratory diagnosis, i.e., clinical and especially epidemiological data, which are essential for the subsequent choice of appropriate laboratory methods and techniques. The second section discusses sampling and available direct and indirect methods of laboratory diagnosis. The third section summarizes the published laboratory methods and techniques for the detection of the most important representatives of viruses causing VHFs. The final section discusses pitfalls associated with these pathogens, mostly with respect to biosecurity issues. We wish to demonstrate here that although laboratory diagnosis of VHFs is not simple, the correct choice of appropriate methods leads to quick results without much difficulty.

12.2 CLINICAL AND EPIDEMIOLOGICAL DIAGNOSIS

Clinical and epidemiological data play a crucial role in the selection of appropriate laboratory tests. VHFs do not have uniform clinical manifestations. Most of the VHFs start with nonspecific febrile prodromal signs, like headache, fever, prostration, and in some cases also with early maculopapular rash. After this nonspecific phase, a typical manifestation of hemorrhagic fever appears, and this can be more or less specific for each pathogen. This underlines the fact that the information about symptoms is very important and helpful for the laboratory because it can help with the selection of appropriate laboratory tests. In particular, the information about bleeding manifestation, its localization and frequency, can be very helpful, e.g., while Ebola hemorrhagic fever exhibits bleeding symptoms in more than 30%, bunyaviral hemorrhagic fever is usually without bleeding; bleeding into gastrointestinal tract is very common in filoviral infections or in yellow fever, but it is rare in the New World arenaviral hemorrhagic fevers. The occurrence of meningoencephalitis is common in bunyaviral hemorrhagic fever, but relatively rare in other VHFs. Jaundice can appear in yellow fever, Lassa hemorrhagic fever, Rift Valley fever, and also filoviral hemorrhagic fever. Conjunctivitis and conjunctiva hemorrhage are common in the New World hemorrhagic fevers. Renal damage and renal syndrome are very common in hemorrhagic fever caused by hantaviruses. An onset of maculopapular rash is seen particularly in filoviral VHFs.

Epidemiological data, especially geographical distribution of these diseases (summarized in Table 12.1), can be very helpful for diagnosis, because in most VHFs, the outbreaks are clearly geographically localized. Therefore, it is essential to obtain records of the patient's travel history. Human filoviral VHFs occur in sub-Saharan Africa (except laboratory-based infections, like the Marburg virus infection in Marburg, Germany, in 1967). Lassa virus (Old World *Arenaviridae*) is naturally endemic in West Africa. New World arenaviral hemorrhagic fevers have natural territory in South America, particularly in Brazil, Venezuela, Bolivia, and Argentina. However, some viruses causing VHFs have large areas of occurrence, e.g., yellow fever virus (YFV) circulates in large areas of the tropical parts of Africa and America, and dengue virus has a large geographical distribution, covering South East Asia, South Pacific, Central and South America, Australia, and some parts of Africa (according to the distribution of the natural vector of the virus, *Aedes aegypti*). Kyasanur Forest disease appears mainly in rural India, and Alkhurma virus (a virus closely related to Kyasanur Forest disease virus [KFDV]) occurs in the Arabian Peninsula and in Egypt only. Omsk hemorrhagic fever virus (OHFV) has been observed in the Siberian region of Russia. Crimean–Congo hemorrhagic fever virus (CCHFV) has an endemic area in Eastern and South-Eastern Europe, Africa, and Asia. Rift Valley fever virus (RVFV) occurs in Africa (particularly in the sub-Saharan part) and the Arabian Peninsula. The causative agents of hemorrhagic fever with renal syndrome (HFRS) have quite diverse distributions. Hantaan virus has an endemic area in Korea, China, and Russia, Dobrava-Belgrade virus in the Balkans, Puumala virus in Europe (mostly in Nordic countries), and Seoul virus can occur virtually worldwide (Howard, 2005). This information, however, has diagnostic importance only in cases of naturally occurring outbreaks of these diseases. Since VHFs belong to Class A of Biological Weapons, there is a risk of their misuse in wars and military conflicts or in terrorist attacks. In these cases, the information about the locality of naturally occurring infections has no relevance. Moreover, in the case of a bioterrorist attack, it can be difficult to recognize VHFs for medical staff in the areas outside their natural occurrence.

TABLE 12.1
Geographical Distribution of Viral Hemorrhagic Fever Pathogens

Viral Family	Viral Name	Disease	Global Distribution
Filoviridae	Ebola virus Zaire, Sudan, Côte dívoire, Bundibugyo	Ebola hemorrhagic fever	Africa continent
	Marburg disease	Marburg disease	Africa continent
Flaviviridae	YFV	Yellow fever	Africa, Caribbean, South America
	Dengue virus 1–4	Dengue	Americas, Asia, Oceania
	OHFV	Omsk hemorrhagic fever	Siberia, Asia
	KFDV	Kyasanur Forest disease	Asia (India)
	Alkhumra virus	Alkhumra disease	Saudi Arabia, Egypt
Arenaviridae	Lassa virus	Lassa fever	West Africa
	Lujo virus	Lujo hemorrhagic fever	Africa
	Junin virus	Argentine hemorrhagic fever	Argentina
	Machupo virus	Bolivian hemorrhagic fever	Bolivia
	Sabia virus	Brazilian hemorrhagic fever	Brazil
	Guanarito virus	Venezuelan hemorrhagic fever	Venezuela
	Whitewater Arroyo virus	Whitewater Arroyo disease	United States
Bunyaviridae	RVFV	Rift Valley fever	East Africa, Arabian Peninsula
	Congo–Crimean hemorrhagic fever virus	Congo–Crimean hemorrhagic fever	East Africa, Easter Europe, Asia
	Virus Hantaan	HFRS	Asia (eastern and central), Europe
	Virus Seoul		Asia and Europe
	Virus Dobrava/Belgrade		Europe, former Soviet Union
	Virus Puumala	Nephropathia endemica	Eastern Europe, Scandinavia, former Soviet Union

The simultaneous onset of multiple patients with a similar clinical course can be the only indicator pointing to the possible bioterrorist misuse of VHF agents (Borio et al., 2002; Drosten et al., 2002).

In endemic areas of VHFs, the clinicians are aware of the symptoms of VHFs, but in other countries it can be a problem. The clinicians should think about VHFs in cases of patients with severe prostration or life-threatening disease, acute onset of disease and rapid progress, bleeding manifestation in cases where there are no predisposing factors for them, and in naturally occurring outbreaks, and travel history into endemic territories.

12.3 SAMPLING

Sampling and any manipulation with the clinical materials from patients or animals with VHFs has to be done extremely cautiously, with the purpose of preventing laboratory-based infections, which usually have a very severe course. Types of samples suitable for specific procedures are discussed in another part of this chapter. For collecting samples, it is very important to perform all procedures less invasively, because several standard invasive methods can cause serious hemorrhages in patients with VHFs.

- Blood samples—blood collection is performed using a sealed collection system (e.g., Vacutainer system). Blood samples can be used for virus culture (whole blood, serum, or plasma), for serological tests (serum), and/or for polymerase chain reaction (PCR) reactions (whole blood, serum, or plasma). For virus cultivation, it is recommended to store the sample at 4°C or deep-freeze it (at −80°C or less). For other examinations, blood or blood components can be stored at −20°C or lower for a longer time. For successful virus cultivation from blood, the sample should be collected during the acute phase (within the first 7 days after incubation period). For serology, it is necessary to collect paired sera (acute and convalescent phases) for monitoring antibody production. Generally, blood samples are suitable for direct diagnosis of all VHFs. Some viruses (like hantaviruses), however, are rapidly cleared from the blood. In field conditions, drying of whole blood spotted on filter paper can be very convenient. According to some studies, viral RNA is stable in filter paper for up to 9 weeks. This simple method can save samples for further PCR detection if freezing is not available (Prado et al., 2005).

- Cerebrospinal fluid (CSF)—collection of CSF is suitable only for viruses present in high concentration in this material. However, in case of intracranial hemorrhage, the lumbar puncture can be very dangerous to the patient and must be indicated very carefully. CSF is used mainly for the detection of some arenaviruses (like Lassa virus; on the other hand, the New World arenaviruses (NWAs) are not isolatable from the CSF despite causing CNS symptoms) and flaviviruses (RVFV or dengue virus).
- Urine—for the direct diagnosis of some VHFs, collection of urine represents mainstream conditions to minimize the risk of contamination of the medical staff. A high viral load in urine is particularly seen in hemorrhagic fevers with renal syndrome. Some other viruses can be also detected in urine, like Lassa virus or less frequently NWAs and sometimes also Marburg virus.
- Tissue samples—mostly all the viruses causing VHFs can be present in several organs (especially parenchymatous tissues). Tissue samples are collected mostly from autopsy (liver, anterior eye fluid, gastrointestinal tissues, etc.), but some tissues samples can be collected also from live patients (semen, vaginal secretions, exudates). Biopsies including fine needles are strictly forbidden due to the high risk of massive hemorrhage, which could be fatal.
- Swab samples—swabs are easy to collect and can be very useful for direct viral detection in VHFs (tissue culture, antigen detection, or PCR). Some viruses can be isolated from saliva or throat swabs (dengue virus), or conjunctiva swabs (NWAs). For virus culture, the swab sample is transported in an appropriate transporting viral medium, which protects the virus from inactivation. For PCR or antigen detection, the swabs can be dried, which makes transportation easier.

12.4   LABORATORY DIAGNOSIS

12.4.1   INDIRECT METHODS

This mode of VHF diagnosis is based on the identification and measurement of virus-specific antibodies. While IgM antibodies are basic serological markers of acute infections or virus reactivation, the presence of IgG antibodies indicates previous infection. Virus-specific IgA antibodies are involved in mucosal antiviral immunity. Serological diagnosis of VHFs is quite intricate, since several VHFs exhibit such a rapid course that there is not enough time for the production of measurable levels of antibodies. Acute serological markers, in particular IgM antibodies, are produced mostly in 5–14 days after incubation period; in case of the most serious VHFs (like Ebola virus Zaire, or Marburg virus Angola strain infections), patients die before virus-specific IgM antibodies are produced (Drosten et al., 2003). Measurement of IgG antibodies is used for epidemiological surveillance during and especially after VHF outbreaks. Examination of paired sera for virus-specific IgG is done for the determination of antibody production dynamics, giving information about the phase of the infection. Acute infection can be confirmed by the evidence of a fourfold rise of the IgG antibodies titer in paired sera. Another method for the determination of the infection stage using IgG antibodies is based on the measurement of antiviral antibody avidity, which can decide between acute and late stages of the infection.

For serological examinations, methods like enzyme immunoassays especially in enzyme-linked variant (ELISA), indirect fluorescence methods (IFA) and virus neutralization tests are ordinarily used. The principle of ELISA and IFA is similar, specific monoclonal antibodies against viral antigens. The principle of virus neutralization testing is the inhibition of the viral cytopathogenic effect on cell culture by specific antibodies. For the diagnosis of some VHFs, a classical complement fixation reaction (CFR) is still used.

The main problem of serological diagnosis of VHFs is the preparation of specific antigens. If full-virion antigens are used, all procedures for the preparation of serological kits have to be done under BSL-4 conditions. Recombinant viral proteins or virus-like particles as targets for antibodies overcome these limitations, i.e., their preparation does not require a BSL-4 facility. This method is used for filoviruses, Lassa virus, RVFV, dengue virus, and YFV diagnostic ELISA tools (Ksiazek et al., 1999a,b; Bausch et al., 2000; Koraka et al., 2002; Balmaseda et al., 2003; Schilling et al., 2004; Emmerich et al., 2006; Paweska et al., 2007). Analogous limitations are met when a method of virus neutralization is performed. This method requires the using of viable viruses, which is limited to BSL-4 laboratories. The laboratory must also have adequate cell lines for virus multiplication and experienced laboratory staff.

In summary, indirect VHF diagnostic methods have advantages, but also several limitations. Despite these limitations, for several VHFs, indirect diagnosis is the only option. For example, dengue virus or hantaviruses have very fast clearance from the human body, and IgM antibodies are then the only marker of acute infection. Another application of indirect detection methods includes serological surveys, like the detection of IgG antibodies against dengue virus in endemic areas, since they have a predictive value for the possible development of hemorrhagic forms of dengue fever (antigen–antibody enhancement) (Drosten et al., 2003). Taken together, indirect diagnostics has still, in the era of amplification methods, its irreplaceable role in the diagnosis of VHFs.

12.4.2 Direct Methods

12.4.2.1 Electron Microscopy

Electron microscopy (EM) has been used in the detection and identification of viruses for several decades. The first electron micrograph of a human viral pathogen (poxvirus) was published in 1938, and after the introduction of negative staining, it became a common diagnostic method. A wide variety of viral agents can be detected by EM, especially in cases when specific reagents like antibodies, antigens, or nucleic acid primers and/or probes are not available. As demonstrated during the last SARS outbreak, EM still has an irreplaceable role in the identification of new viral pathogens. EM has several unique characteristics that make this method important in viral identification, especially in sporadic infections. The main advantage of this diagnostic tool is its non-specificity, which means that EM is able to identify virus as well as bacteria or other pathogens especially in negative contrast (catchall method). EM is also a very rapid method of identification; results could be available in 30 min (classical negative stain preparation may be made and a result could be obtained within 5 min of the arrival of the specimen to the laboratory). Another advantage of this method is its possibility to analyze various samples, from clinical specimen to tissues or infected cell culture (Hazelton and Gelderblom, 2003). EM has, however, also some limitations. The biggest limitation is the need of a high virus concentration in the sample for its capture ($>10^7$ viral particles/mL serum or 10^5/mL after specimen ultracentrifugation). The sample can be concentrated by direct centrifugation to the EM grid with the Beckman Airfuge, which increases the yield of viral particles by three or more orders of magnitude. But this method is not always available (Hazelton and Gelderblom, 2003). EM is an expensive method with high acquisition as well as operating costs. An experienced operator is required for a reliable EM analysis (Drosten et al., 2003). Therefore, the success of EM diagnostics depends greatly on the quality of the sample collected as well as on the method of preparation and skills of the microscopist (Drosten et al., 2003; Hazelton and Gelderblom, 2003). Since EM is based on morphology identification of pathogens, it is able to identify only the viral family. Better virus identification can be achieved by the immunological alternative of EM (immunoelectron microscopy).

As mentioned earlier, the basic method of EM is based on negative staining. This approach is the most suitable for liquid samples. A small volume of the sample is placed on a special EM grid covered with a special membrane (Formvar, Collodion, Butvar, or Pioloform). For negative staining, various chemicals can be used; mostly 0.05%–2% uranyl acetate (UA), 1%–2% phosphotungstic acid (PTA), or 0.05%–5% ammonium molybdate (AM). UA not only stains but also fixes the samples for a long time (unlike the other two stains). AM is better for the resolution of small details in the sample. PTA can downgrade some viruses, but has the ability to increase better resolution of surface structures, like envelopes or viral spikes. For interpretation of the results, a highly experienced investigator is needed to prevent misinterpretation. High toxicity and even radioactivity are some of the disadvantages of the negative staining approach. With respect to avoiding these disadvantages, some other methods were developed. Very good results are obtained from immunological modifications using the EM. Monoclonal antibodies could be used for virus-specific immunoaggregation and can be added directly to liquid samples and then centrifuged, or can be applied directly to the EM grids. This approach increases the sensitivity and is helpful especially for the detection of viruses in low concentrations. However, this method does not have the "catchall" ability. For virus visualization inside the host cells, a method called "immunolocalization" can be used. This is based on virus-recognizing antibodies and gold-labeled secondary antibodies (Goldsmith and Miller, 2009). For research work, tomography EM (also called electron tomography—ET) can be used. This EM modification serves for 3D modeling of viral structures or viral particles. But this method is not suitable for routine applications for its exactingness. The ET method was used, e.g., in research of filovirus budding (Welsch et al., 2010).

12.4.2.2 Cell Cultures and Animal Models

Animal models: Suitable animal models allow investigation of virus interaction with a live organism. In case of the availability of suitable animals, pathogenesis, clinical manifestations, organ or tissue damage, host antibody response, or effectiveness of curative interventions can be investigated. There are some rules for animal choosing in accordance with the desired process modeling. For most investigations, mammals are used because of their basic similarity with humans. Based on their high biological similarity to humans, nonhuman primates are frequently used for infectious disease investigation as the best model. For VHFs, rhesus monkeys (*Macaca mulatta*), cynomolgus monkeys (*Macaca fascicularis*), or chimpanzee (*Pan troglodytes*) are most frequently used. However, usage of these animals brings also many disadvantages, including the need of large menageries (especially in BSL-4 conditions), difficult laboratory work with large animals, and last but not least also ethical problems and negative public opinion. Therefore, only a very limited number of facilities use nonhuman primates for VHF research. Nonhuman primates have been used for the research of filoviral infections (both MARV and EBOV), arenaviral infections (Lassa fever, Bolivian or Argentine hemorrhagic fever), flaviviral infections (dengue fever, Omsk hemorrhagic fever), and bunyaviral infection (Puumala hantavirus infection or RVHF).

Recently, guinea pigs (*Cavia porcellus*) have been commonly used for the study of filoviral infections. Use of this model is not as ethically problematic as the work with nonhuman primates. However, viruses must be adapted for infection of these nonnatural hosts, and this could lead to some changes in viral properties and influence clinical manifestations. In bunyaviral and arenaviral research, rodents are generally frequently used, because rodents represent a natural reservoir of these viruses, and viruses are well adapted to these hosts. But infection of the natural reservoir hosts is usually without any symptoms, so these animals are used mostly only for virus propagations and only in some cases for the investigation of pathologic changes in the rodent organs.

Hamsters *Calomys callosus* (for Machupo research, RVHF), mice *Mus musculus* (for hantaviral, RVHF, or New World arenaviral infection research), rats (RVHF research), and some others are most often used. For VHF research, special inbred rodent strains can be used. These can represent highly susceptible animals for VHFs, and each strain can have different manifestations and organ pathology. Frequently used are also mice with specific immunological depletions, like SCID mice (severe combined immunodeficient), athymic BALB/c mice, or AG129 mice (INF-α/β and γ receptor-deficient) (Yauch and Shresta, 2008; Gowen et al., 2010; do Valle et al., 2010). These hybrid mice are used mostly for immunological studies or treatment development. They also could be used for the examination of organs and tissue changes after virus infection. Some VHF viruses can be cultured in their arthropod vectors (mosquitoes or ticks). Use of this cultivation is however limited to facilities equipped with an insectarium under BSL-3 or -4 conditions.

The use of cell or tissue cultures is still a "gold standard" in virus detection and identification. But it is important to keep in mind that some viruses do not grow in cell cultures or grow only after special manipulation. Moreover, viruses may not survive the collection and transportation conditions to the lab and are not then cultivable. For VHF VERO cells, a cell lineage from cynomolgus macaque kidney tissues is universally used. E6 lineage is the most sensitive and is also frequently used in VHF diagnosis or research. It is also possible to use similar mammalian cell or tissue lines, like porcine kidney cells (Jayekeerthi et al., 2006). Virus growth in cell lines could be detected by the appearance of the cytopathic effect (CPE). But CPE detection has very low virus specificity because it has similar features in different virus infections. Moreover, classical virus cultivation takes a long time. Shell vial cultivations represent a method, which can shorten the cultivation time from weeks to a few days. In this case, the virus (sample) is centrifuged onto the cell line's monolayer.

For exact virus identification, it is necessary to use some accurate method, like EM (virus family morphology), or virus-specific antigen detection by immunofluorescence. It is also possible to use PCR amplification for virus detection (Drosten et al., 2003). Another complication, which limits the use of cell cultures in VHF diagnosis, is the requirement to perform all work under BSL-3 or BSL-4 conditions.

12.4.2.3 Antigen Detection

These methods are based on specific viral antigen detection by labeled monoclonal antibodies. Antigens capture ELISA (enzyme labeled monoclonal antibodies) and direct immunofluorescences (monoclonal antibodies marked by immunofluorescence stain) are the most commonly used methods for antigen detection. Both methods can be used in field conditions with basic equipment (ELISA reader or fluorescence microscope), which increases rapid detection capabilities during VHF outbreaks. For their relative simplicity and quickness, these methods have been mainstays of rapid detection methods during VHF outbreaks, including the Kikwit EBOV outbreak (Ksiazek et al., 1999a). The main pitfall of this method is specific antibodies production, which requires BSL-4 conditions and appropriate laboratory animals as antibody producers. This problem can be solved by using recombinant proteins for specific monoclonal antibody inductions in standard animal producers (like BALB/c mice for monoclonal antibody production and rabbits for polyclonal antibodies) (Niikura et al., 2001; Ikegami et al., 2003; Saijo et al., 2005). Recombinant viral proteins can be prepared by standard techniques of molecular biology in an *Escherichia coli* expression system or in a baculovirus system. Because of relatively frequent changes in outer antigenic structure, virus nucleoproteins are preferred targets. Use of artificial proteins instead of live viruses means lower biohazard security but is also more demanding for the preparation of proteins. Methods of antigen detection such as multiple desk test or single strip tests could be used and give rapid results. But some publications showed lower sensitivity if compared to nucleic acid amplification tests (Bausch et al., 2000; Leroy et al., 2000). Antigen detection methods are ideal especially for first rapid screening in outbreak areas and also as an addition to other diagnostic procedures.

12.4.2.4 Amplification Methods

As already mentioned earlier, detection and identification of VHF causative agents can provide clinical benefits only if the methods used are highly specific, sensitive, and rapid. All these prerequisites can be fulfilled with the use of nucleic acid amplification methods. Especially PCR has become one of the most used direct detection methods not only for VHF. The principle of the nucleic acid amplification method is the detection of short and pathogen-specific sequences of nucleic acid.

The most common clinical material for VHF diagnostics by amplification method is serum or plasma. However, the highest PCR efficiency is ensured when whole blood is collected in EDTA tubes (Howard, 2005). Collection of CSF can be useful if neurological symptoms are present. Generally, volumes of 100 µL of the clinical sample are used and the purified nucleic acids are resuspended in 50–100 µL.

There are a lot of variations of the basic principle of PCR (denaturation–annealing–elongation). The detection process consists of several steps—nucleic acid extraction (from various samples), reverse transcription (in case of RNA viruses—all members of VHF), amplification, and visualization of specific amplicon. Classical PCR was introduced in the 1990s but had some disadvantages. Reaction with product visualization in agar is time consuming. Nested PCR, a modification that uses two pairs of specific primers, is more specific but still time consuming. Real-time PCR with product visualization in real time using a probe system markedly shortens the time needed for analysis. Single tube PCR allows performing reverse transcription and PCR itself in one tube, capillary, or desk well without other manipulations. Recently, real-time PCR has become the most used nucleic acid amplification method for VHF diagnosis.

Separate detection of each virus under different PCR conditions represents another disadvantage of PCR diagnosis of VHF. An ideal situation is based on simultaneous detection of various pathogens in one run under identical reaction conditions. Such a method was used, e.g., by Drosten et al., who developed a system of single-step PCR for six viral pathogens using the same reaction conditions allowing sample examination for all pathogens in one reaction cycle.

Multiplex systems allow detection of more pathogens in one tube. Multiplex PCR can be used as classical PCR (products detection on agarose gel) or in real-time format (by using several fluorescent dyes, each for a different pathogen). Multiplex systems have a crucial advantage in the short diagnosis time because they allow multiple pathogen identification, but the designing and optimization of basic reaction parameters are quite difficult (Drosten et al., 2003). Reaction conditions allow only very minor differences in melting temperature (T_m) of all primers used in the reaction. The difficulty of designing primers with similar T_m increases with the number of pathogens simultaneously detected. Therefore, multiplex PCR is not often used in pathogen detection. Only a few protocols are available for VHF. Examples include detection of all four serotypes of dengue viruses (Sánches-Seco et al., 2006) or for a small group of VHF pathogens (Jabado et al., 2006). MassTag PCR, a novel modification allowing simultaneous detection of several viral pathogens, has been designed as a multiplex assay that uses primers labeled by a library of 64 distinct mass tags. After PCR reaction, the incorporated mass tags are released from products using UV light. Each tag (and also each primer) is then identified by mass spectrometry. The pathogen identification is based on the presence of two specific mass tags (one from each primer), which means that theoretically 32 different pathogens can be detected. This method was successfully used for the detection of 22 respiratory pathogens. Palacios et al. applied MassTag PCR for multiplex detection of 10 VHF pathogens. According to the authors, this method is rapid, sensitive, and not expensive. The disadvantage lies in the need for mass spectrometry for the analysis (Palacios et al., 2006).

Sequencing allows detailed genetic characterization of every pathogen including those causing VHF by determination of nucleotide sequences in a fragment or the whole genome. Standard sequencing applies basic biochemical dideoxy-termination principles developed by Sanger, which needs single-stranded DNA as a template for complementary strain synthesis using specific dideoxy nucleotides. Recently, new methods have been introduced such as new-generation sequencing like pyro-sequencing, which are faster than older ones. But even new technological sequencing is still not suitable for routine diagnosis.

12.4.2.5 Biosensors

Biosensors are a relatively new trend in diagnostic systems. They are defined as complex sensing systems that use biological reaction for pathogen detection. They are composed of a biological or biochemical sensing system, using enzyme, antibody, microorganism, or nucleic acid detection techniques, and a conversion system, which is able to convert results of the sensing system into electronic signals. Biosensors have high sensitivity and specificity because their composition allows combinations of several diagnostic tests in one panel. This diagnostic system could be very helpful for pathogen detection in field conditions. For VHF, immune sensors have been developed, e.g., for dengue virus detection (Fang et al., 2010). Wide application of biosensors for pathogen detection is limited by the relatively high price and difficulty in developing and production.

12.4.3 Filovirus Detection

Virus presentation—Filoviruses are pantropic, which means that they could affect nearly every organ in the human body. However, the viruses have an affinity to organs well supplied by the blood circulation, like liver or spleen. Necrotic lesions can be seen in a large number of organs: pancreas, gonads, adrenals, hypophysis, thyroid, kidney, skin, and mucosa. Large amounts of virus particles are detectable not only in blood, but also in semen, vaginal secretion, and tears. In some clinical material, the virus can be detectable for a relatively long time. Marburg virus can be cultured from a convalescent secretion 1–3 months after acute disease. Data from the Kikwit Ebola outbreak had shown the presence of Ebola virus nucleic acids in semen or vaginal secretion for 1 month, but cultivation attempts were not successful (Peters and LeDuc, 1999; Howard, 2005).

Cell culture—The most frequent cell culture for filovirus cultivation is the Vero cell line, especially E6 clone and MA-104 or SW13 cell cultures (all cell lines from monkey kidney). Viruses can also be cultivated in human endothelial cell lines, like HMEC-1 or HUVEC. All filoviruses are cytopathic for the cells in culture but do not cause much extensive CPE. MARV and ZEBOV have usually a faster replication period than SEBOV and REBOV, which results in the earlier appearance of CE (from 48 h after cell infection). CE can be seen by microscopy, or using some additional staining, e.g., immunofluorescence or enzymatic. According to published data, plaque reduction assay (CPE identification) was 4–10× less sensitive than fluorescence antigen detection methods, and the optimal time for test evaluation was 7 days of incubation (Moe et al., 1981). A very good experience was also with viral inclusion detection in culture cell cytoplasm, which can be seen from 3 to 5 days after the inoculation. Another method for virus detection in cell culture is the use of PCR, which can be positive from 3 h after infection (optimal results about 18 h). Methods of laboratory detection of filoviruses are summarized in Table 12.2.

Electron microscopy—This is very useful for filovirus detection. This method is applicable for virus identification from cell cultures, human fluid samples, or in modification also for virus detection in human tissues. It is able to detect characteristic firoviral morphology by using standard EM methods such as negative staining and transmission EM. The characteristic filamentous morphology of filovirus is unmistakable, but it is necessary to have a relatively large amount of viral particles in the sample to get a positive result. Also differentiation between two filoviral genera is possible but mostly very difficult for very slight differences, and it needs a well-experienced researcher. Also commonly used is the monoclonal antibody staining, which is able to differentiate all kinds of filoviruses in immunoelectron microscopy.

Antigen detection—Viral antigens are detectable very early, e.g., in blood samples from day 3 after symptom onset. Some antigen-capture ELISA systems were developed for EBOV (Zaire, Sudan, Reston, and Ivory Coast) and MARV (Lake Victoria) detection. Published methods use as

TABLE 12.2

Laboratory Diagnosis of *Filoviridae* Family

Test	Samples	Principle	Notes
Indirect immunofluorescence assay (IFA)	Blood, serum	Detection of virus specific antibodies (IgM, IgG)	Easy method, subjective interpretation, lower sensitivity than ELISA
ELISA	Blood, serum	Detection of virus-specific antibodies (IgM, IgG)	Easy method, sensitive and specific, faster initiation of Ab response than IFA (IgM 2nd–9th day, disappear 30th–168th day of illness), IgG antibodies from the second week of illness, cross-reaction inside viral family
Immunoblot	Blood, serum	Detection of virus-specific antibodies (IgM, IgG)	Protein-specific, difficult interpretation, mostly used as a confirmatory method
Virus neutralization test	Blood, serum	Detection of virus-neutralizing antibodies	Specific method, work with live viruses (extremely cautious work in BSL-4 lab)
Electron microscopy	Blood, tissues, exudates, cell culture	Detection of viral particle morphology	Fast method, only filovirus family-specific morphology (for specificity extent possibility of immunostaining), extensive equipment, lower sensitivity
Antigen-capture ELISA	Blood, tissues, exudates	Detection of viral antigens NP, VP40, or GP	Rapid and sensitive method, in case of strip variant suitable for field conditions as an effective tool in early diagnosis
Fluorescence assay (FA)	Blood, tissues, exudates, cell culture	Detection of viral antigens NP, VP40, or GP	Rapid method, subjective interpretation, necessity of fluorescence microscopy, suitable for using in cell culture
Immunochromatography	Blood, tissues, exudates	Detection of viral antigens	Rapid and very simply test, lower sensitivity and specificity, appropriate for field conditions
RT-PCR	Blood, tissues, exudates, cell culture	Detection of viral nucleic acids, specific genome sequences	Rapid (real-time RT-PCR up to 2 h), sensitive, and specific method (minimal cross-reactions), relatively high costs, special laboratory equipment, after nucleic acid isolation, there is no BSL-4 necessity, relatively many published homemade PCR protocols that include multiplex
Virus culture	Blood, tissues, exudates	Culture of viruses (cell lines or animals)	Long-term method (MARV and ZEBOB from the second day after infection, the rest later), without further identification (PCR, FA) low specificity, afford virus for other examinations, necessity of BSL-4 lab

target antigens mostly filoviral NP, VP40, or GP. Some research groups use artificial recombinant viral proteins for the production of monoclonal antibodies (MAbs), and the produced MAbs have very good affinity against real viral proteins. This method allows the development of novel antigen detection methods in countries without BSL-4 laboratories necessary for working with live viruses. For every new method, it is necessary to test it using real human samples. Some methods based on the principle of antigen-capture ELISA were used during Ebola hemorrhagic fever outbreaks in the Democratic Republic of Congo in 1995, in Gabon in 1996, and in Uganda in 2000 and were evaluated as an effective tool in VHF early diagnosis (Ksiazek et al., 1999a; Saijo et al., 2006).

For EBOV, immunohistochemistry assay can be used as well. This method is developed for rapid detection of EBOV in tissues, especially in skin biopsies. The assay uses monoclonal antibody targeting of viral glycoproteins or nucleoprotein. The great advantage of this method is the ability to detect the virus in standard formalin-fixed, which means that BSL-4 conditions are not necessary during the examination (Zaki et al., 1999). A very similar method exists also for Marburg virus, which is used for tissue samples, too (Geisbert and Jaax, 1998).

Antibody detection—According to published data, the first onset of EBOV IgM-specific antibodies is between 2 and 9 days after incubation period with a peak in IgM titers around day 18 after clinical symptoms onset, and this serological marker disappears in survivals between 30 and 168 days. EBOV-specific IgG antibodies then appear between 6 and 18 days after clinical manifestation onset and persist for a long time (Ksiazek et al., 1999a). In Marburg hemorrhagic fever, the situation is very similar to EBOV. IgM-specific antibodies appear between 4 and 7 days after incubation period, and IgM antibody titers reach a maximum 1–2 weeks later. IgG antibodies are developed a little later after IgM antibodies and persist also for a long time (Wulff and Johnson, 1979). Because of the delayed production of antibodies, IgM detection is mostly used in epidemiological investigations for surveillance of contacts during filoviral outbreaks (Peters and LeDuc, 1999).

There are various serological diagnostic tools for filoviruses. The indirect fluorescence antibody test is commonly used, especially for EBOV. This method has good sensitivity especially during early convalescent periods (Ksiazek et al., 1999a). Other serological methods are ELISA, RIA, or Western blot. Some different kinds of ELISA for filovirus IgM or IgG antibodies were developed, but none are commercially available (Saijo et al., 2001; Ikegami et al., 2003). Detection of nucleoproteins as antigens in ELISA exhibits high cross-reactivity within the filovirus family, so this method is not very suitable for filoviral species-specific humoral response, but it can be used for genus specificity differentiation. Therefore, another method using virus structural glycoproteins as an ELISA target antigen was developed. These structural glycoproteins exhibit high diversity between each filovirus species (Nakayama et al., 2010). This method shows very limited cross-reactivity between each filoviral species and is more suitable for antibody detection than ELISA using NP antigen.

Amplification methods—PCR has been used for diagnosis of filoviral infection for more than 20 years. All VHF viruses are RNA viruses, so their detection requires reverse transcription of the viral RNA into cDNA prior to PCR amplification. Conventional reverse transcription PCR (RT-PCR) with product visualization on ethidium bromide agarose is sensitive and more specific than other diagnostic tools used, but takes a relatively long time. Another method, nested PCR, is a reaction with two sets of primers, which provides more specific results. Real-time RT-PCR with fluorogenic probes is recently commonly used for its speed (results up to 1 h including RNA extraction time), and very high sensitivity and specificity (Gibb et al., 2001). Because of the high mutation frequencies of filoviruses, many genetically diverse isolates of each filovirus species can be found. Five or more internal mismatches or a single mismatch at the 3′ position in the sequence of the primers can drastically reduce the amplification efficiency. Since sequence information on VHF viruses is often limited, it is difficult to design appropriate primer sets. Therefore, not all PCRs are able to detect all virus isolates, and some assays require revisions when sequences of new isolates are available (Drosten et al., 2003). A new system developed by Panning et al. (2007) is based on a mixture of five primers and three probes. But even this new method is not able to detect the new Ebola virus strain, which emerged in 2007/2008. Filoviruses are also included in several developed multiplex systems for VHV detection (Briese et al., 2005; Palacios et al., 2006; Trombley et al., 2010).

12.4.4 Arenavirus Detection

Virus presentation—Arenaviruses can be found in various human organs. A massive viral load in blood is common during acute illness. In milder cases, the virus is present in blood from days to several weeks, but in severe illness, it can be for 4 or more weeks (especially in Lassa fever).

Viruses are present also in the parenchyma organs, especially liver (both Old World and New World arenaviruses cause multifocal hepatocellular necrosis). Arenaviruses are often found also in lung tissues, and also in pleural exudates or transudates. Some viruses from this family cause neurological damage and can be found in CSF (Lassa). Throat swabs can be helpful for virus isolation during the acute phases (mostly during the first week of illness). This is true for Lassa virus, Junin virus, and to a lesser extent also Machupo virus. Long-term presentation (up to 2 months after illness onset) of Lassa virus in urine is diagnostically valuable. Some viruses can be found in breast milk of animals and also humans (confirmed for Lassa and Junin virus infection) and possibly in vaginal secretions. Virus can be readily isolated from conjunctiva and conjunctival secretions. Methods of laboratory diagnosis of arenaviruses are summarized in Table 12.3.

TABLE 12.3
Laboratory Diagnosis of *Arenaviridae* Family

Test	Samples	Principle	Notes
Indirect immunofluorescence assay (IFA)	Blood, serum	Detection of virus-specific antibodies (IgM, IgG)	Easy method, subjective interpretation, lower sensitivity, first antibodies can be detectable from the second week of illness, high cross-reactivity inside viral family
ELISA	Blood, serum	Detection of virus-specific antibodies (IgM, IgG)	Easy method, sensitive and specific, high cross-reactivity among each viruses (especially in Tacaribe virus complex), IgM antibodies from 4th to 40th day
Virus neutralization test	Blood, serum	Detection of virus-neutralizing antibodies	Specific method, work with live viruses (cautious work in BSL-4 lab), difficult evaluation
Complement fixation test	Blood, serum	Detection of complement fixation antibodies	Complement fixating antibodies detectable from the third week of illness, not always developed
Electron microscopy	Blood, tissues, exudates, cell culture	Detection of viral particle morphology	Viral morphology highly pleiomorphic (for specificity extent necessity of immunostaining), extensive equipment, experienced laboratory workers, lower sensitivity
Antigen-detected ELISA	Blood, tissues, exudates	Detection of viral antigens NP, GP	Rapid and sensitive method, in case of strip variant suitable for field conditions, cross-reaction in GP targeting monoclonal antibodies, method detected NP developed for Junin virus
Fluorescence assay (FA)	Blood, tissues, exudates, cell culture	Detection of viral antigens NP, GP	Rapid method, subjective interpretation, necessity of IF microscope, cross-reaction in GP targeting monoclonal antibodies
RT-PCR	Blood, tissues, exudates, cell culture	Detection of viral nucleic acids	Rapid (real-time RT-PCR), sensitive and specific method, relatively high costs, require special laboratory equipment, after nucleic acid isolation, there is no BSL-4 necessity. Able to detect virus from the third day of illness, enough published homemade PCR protocols
Virus culture	Blood, tissues, exudates	Detection of living viruses (mammalian cell lines or animals)	Long-term method, without further identification (PCR, FA) low specificity, afford virus for other examinations, mostly without clear CE in cell lines, using mostly only for Lassa virus

Cell culture—Arenaviruses readily grow in Vero cell culture or other mammalian cell lines, but this method of diagnosis is used mainly only for Lassa virus, because NWAs are mostly detected by other methods. First CPE can be seen after days, but often it is not very clearly presented. For virus identification, immunofluorescence or PCR is recommended.

Electron microscopy—Arenaviruses do not have such specific morphology as filoviruses; the particles are highly pleiomorphic. It is known that arenaviral particles are unstable to chemical fixation. For morphological examination, it is better to use other methods of sample preparation, like cryotechniques. Arenaviruses can also produce defective particles, which can interfere during EM. Immunoelectron microscopy with specific monoclonal antibodies is, therefore, the method of choice.

Antibody detection—Serological tests are widely used in arenavirus diagnosis, but there are differences in which method fits better to each member of this viral family. For Lassa virus, there is very good experience with indirect immunofluorescence; first antibodies are detectable from the second week after illness onset. Complement fixating antibodies against Lassa virus can be detected from the third week of disease, and this method is very helpful for early diagnosis (a disadvantage is the need for paired sera). A virus neutralization test method, suitable only for specialized laboratories because of the use of live viruses, is useful mostly for NWA diagnostics. In vitro measurement of virus-neutralizing antibodies against Lassa virus is very difficult. ELISA tests are available for several members of this virus family. For Lassa virus, reverse ELISA method capable of the detection of both IgM and IgG antibodies can be used (Emmerich et al., 2006). Comparison of ELISA and plaque reduction neutralization test in Junin virus diagnostics showed no significant difference in sensitivity and specificity of both these tests (Riera et al., 1997). The main weakness of the serological tests lies in the high antigenic relationships among the NWAs. This results in a frequent cross-reaction in enzymatic or immunofluorescent serological tests (Howard, 2005). Viral antigens used in serological methods have been traditionally prepared as lysates of infected culture cells, which required a BSL-4 laboratory. Recently, the antigens for ELISA are mainly prepared using recombination techniques (Ure et al., 2008), which allow the work with these antigens without any special requirements for biological safety. Saijo et al. (2007) used a baculovirus system for the production of recombinant nucleoproteins (rNP) of Lassa virus, lymphocytic choriomeningitis virus, and Junin virus. They prepared polyclonal antibodies by standard rabbit immunization and monoclonal antibodies by using BALB/c mice and purified rNP. Antigens made by this technique can be used in indirect immunofluorescence, ELISA (IgG, IgM), and antigen-capture ELISA. These methods are highly sensitive and specific and suitable for use in clinical diagnostics.

Amplification method—All methods mentioned earlier (with the exception of EM) are not much useful for the diagnosis of acute disease, because they provide results after several days or weeks. Antigen detection methods have sensitivity and specificity sufficient for rapid field diagnostics, but not for verification tests. PCR is commonly used as a confirmation method. For Lassa virus detection, this method has been used for more than 20 years. RT-PCR has the ability to detect a virus as early as on day 3 after illness onset, whereas antibodies are detectable in only 52% patients at this point in time (Demby et al., 1994). Various RT-PCR protocols have been developed for the NWAs, from conventional RT-PCR in the 1990s to real-time RT-PCR in the present day. PCR is a very sensitive diagnostic tool, with a detection limit of 10–100 RNA copies. Recently, multiplex RT-PCR systems were developed with the capacity to detect more VHF pathogens in one reaction (Vieth et al., 2007). A new diagnostic PCR panel with artificial viral genomic segments was developed. This approach eliminates the need for live viruses in the analysis. This method was optimized for diagnostics of the Clade B NWAs (by conventional PCR) and for the detection of Junin virus and Guanarito virus (by real-time RT-PCR). The method is fully operational and gives similar results to methods using real virus particles (Vieth et al., 2005).

12.4.5 FLAVIVIRUS DETECTION

Virus presentation—Flaviviruses are a very wide group of viruses with different tropisms. YFV has tropism to hepatic tissue. But the virus can be detected also from blood, other parenchyma organs, urine, and also from the central nervous system. OHFV is pantropic, but it has a marked affinity to hematopoietic tissues and nervous system. Virus can be isolated also from blood or liver tissue. Dengue viruses can be simply detected in saliva samples (Yap et al., 2011). Methods of laboratory diagnosis of flaviviruses are summarized in Table 12.4.

Cell culture—Flaviviruses can be easily grown in Vero cell culture. However, viruses can also be effectively isolated in special invertebrate cell cultures. Commonly used cultures include C6–36 cell line derived from *Aedes albopictus*, AP-61 line derived from *Aedes pseudoscuttelaris*, and

TABLE 12.4
Laboratory Diagnosis of *Flaviviridae* Family

Test	Samples	Principle	Notes
ELISA, indirect immunofluorescence assay (IFA)	Blood, serum	Detection of virus-specific antibodies (IgM, IgG, IgA)	Easy method, sensitive and specific, IgM antibodies detectable from the third to fifth day, persist up to 3 months, IgG from the second week, for DENV and YFV available commercial ELISA kits, for DENV, also possibility of IgA antibodies measurement (from the first to sixth day of illness), problem with cross-reaction at some ELISA methods
Virus neutralization test	Blood, serum	Detection of virus-neutralizing antibodies	Rapid test, positive from the first week of illness, allow serotyping (dengue virus), necessity of live viruses
Complement fixation test	Blood, serum	Detection of complement fixation virus antibodies	Recently not so often used method (OHFV), difficult evaluations, needed of fourfold titer change in paired sera
Electron microscopy	Blood, tissues, exudates, cell culture	Detection of viral particle morphology	Only flavivirus family-specific morphology, low sensitivity and specificity without using immunoelectron microscopy
Antigen-capture ELISA	Blood, tissues, exudates	Detection of viral antigens NS1	Dengue NS1 protein can be detected earlier than IgM immune response, rapid and sensitive method for all four types of DENV, commercial tests available
Fluorescence assay	Blood, tissues, exudates, cell culture	Detection of viral antigens	Rapid method with subjective interpretation, used for DENV or OHFV
RT-PCR	Blood, tissues, exudates, cell culture	Detection of viral nucleic acids	Rapid method with high sensitivity and specificity, excellent for KFDV and YFV detection, a lot of published RT-PCR protocols also for DENV (there is a problem with four serotypes, often used multiplex PCR for detection of all of them). OHFV has short-term viremia, PCR not so successful from blood samples
Virus culture	Blood, tissues, exudates	Culture viruses in tissue lines (mammalian or insect)	Long-term, low sensitivity and specificity (without extensive detection), results after 3–7 days (IF or enzymatic staining), more successfully using an animal model (intracerebral inoculation of sucking mice)

TRA-248 cell line from the tick, *Toxorhynchites ambionensis*. Viruses can be detected on day 3 as soon as after inoculation by monoclonal antibodies or RT-PCR (Howard, 2005).

Electron microscopy—The morphology of flaviviruses is very similar for all members of the family. The particles consist of an isometric core structure, which is surrounded by a marked lipid envelope. Without additional methods like immunoelectron microscopy, this diagnostic technique has low sensitivity and specificity. For research purposes, other modifications of EM are also available. For example, the structure of dengue envelope proteins was investigated using cryo-electron microscopy, which allowed three-dimensional viewing of examined proteins (Gadkari and Srinivasan, 2010).

Antibody detection—Serological methods represent a gold standard in diagnosis of flaviviral infections. Hemagglutination tests are positive within the first week of illness, and the presence of antibodies is long-lasting. This method is still used, e.g., for Omsk hemorrhagic fever diagnosis. The complement fixation tests are not widely used for lower sensitivity and specificity and the need of fourfold titer change in paired sera. Both methods mentioned earlier were replaced recently by ELISA, which is able to detect IgM or IgG antibodies. For the best results, investigation of paired serum samples is preferred, i.e., of one serum from the acute stage and a second from the convalescent period. IgM antibodies are detectable from the third to fifth day and persist up to 3 months (dengue virus), and IgG antibodies are detectable with 1 or 2 weeks' delay and persist for long time periods. Virus neutralization tests can detect virus-neutralizing antibodies during the first week of illness, but they can be detectable for years. The neutralization test has one main disadvantage, which is the need for a live virus in the experiment. Measurement of neutralizing antibodies allows the determination of virus serotypes (characteristically for dengue viruses). Other serological methods show relatively frequent cross-reactions among different flaviviruses, which decrease diagnostic information of serology in this viral family (Howard, 2005; Růžek et al., 2010). For example, when comparing the serological tests for yellow fever and dengue viruses, the cross-reactivity of ELISA was observed in 46% of the tests and virus neutralization test in 80% (Houghton-Trivino et al., 2008). A new ELISA test detecting anti-dengue IgA antibodies was shown to be highly sensitive (particularly in second and other reinfections) and rapid (result in hours). Anti-dengue antibodies are detectable from the first day of illness (in a small percentage of patients). On day 6, the antibodies are detectable in more than 90% of patients (Yap et al., 2011). More recently, chimeric tetravalent or noncovalent antigens for anti-dengue antibodies that cover all serotypes with high specificity are used in ELISA tests (Batra et al., 2011).

Antigen detection—Dengue virus ELISA with monoclonal antibodies against conservative antigens of the nonstructural protein 1 (NS1; genome region of NS1 has been also used in PCR) is highly specific (no cross-reaction) and able to detect antibodies against all four serotypes (Ding et al., 2011). NS1 proteins are detectable in the first few days of infection and are present earlier than IgM antibodies. This method of detection of NS1 antigen is suitable for rapid diagnosis of all serotypes of dengue viruses (Datta and Wattal, 2010).

Amplification method—KFDV and YFV can be detected by RT-PCR. These viruses are genetically stable in comparison to other flaviviruses. Methods from conventional RT-PCR to real-time RT-PCR using SYBR Green I or TaqMan probes are available and detect the viruses from blood or tissue samples. For dengue virus serotypes, the primers target conservative genome sequences, and then it is possible to detect all serotypes, but primer and probe sequences must include degenerated bases that decrease the sensitivity. Another possibility is the use of four primer sets and probes for each dengue virus serotype. This variant can distinguish the serotype of dengue virus but for each sample, four reactions are necessary. Multiplex (fourplex) RT-PCR, which incorporates all four reactions into one reaction tube, has also been developed (Johnson et al., 2005; Lai et al., 2007). All hemorrhagic flaviviruses (with the exception of OHFV) are included in the multiplex PCR panel for the detection of 10 VHF pathogens (Palacios et al., 2006) or for the detection of 8 flaviviruses (Chao et al., 2007). For OHFV, real-time RT-PCR protocols using SYBR Green 1 or TaqMan probes in singleplex or multiplex systems are available. Detection of OHFV by PCR is not standardly used in the diagnosis of Omsk hemorrhagic fever because of fast virus clearance from the blood. Virus is present in blood only in the first nonspecific phase of diseases, but not in the second one, and therefore it can give false-negative results (as is common also in other tick-borne flavivirus diseases). According to the published data, the PCR method is more suitable for surveillance samples of ticks or wild animals or for postmortem samples than in clinical diagnosis (Růžek et al., 2010).

12.4.6 BUNYAVIRUS DETECTION

Virus presentation—Bunyaviruses are also a very diverse group of viral pathogens with many differences. Also virus presentations are partially different in each bunyavirus infection. Methods of laboratory diagnosis of bunyaviruses are summarized in Table 12.5. CCHFV has an affinity to endothelial cells and therefore it can be isolated from various tissues (liver, kidney, central nervous system, adrenals, and generally in the lymphoid system). For diagnostic purposes, blood samples are the most important, i.e., for serology and also, in the first stage, for direct detection. Viruses causing HFRS can be isolated from urine samples. RVFV can be isolated from blood,

TABLE 12.5

Laboratory Diagnosis of *Bunyaviridae* Family

Test	Samples	Principle	Notes
Indirect immunofluorescence assay (IFA)	Blood, serum	Detection of virus-specific antibodies (IgM, IgG)	Easy method with subjective interpretation. IgM detectable from the seventh to ninth day and disappear after 4 months, IgG from the second week and remain up to 5 years
ELISA	Blood, serum	Detection of virus-specific antibodies (IgM, IgG)	Easy method, sensitive and specific, IgM detectable earlier than by IFA (fifth to seventh day of illness), but other characteristics are the same as in IFA. Cross-reaction common in Hantavirus genera members
Immunoblot	Blood, serum	Detection of virus-specific antibodies (IgM, IgG)	Specific method, for its specificity and sensitivity, it is used as confirmatory test, not commonly available
Virus neutralization test	Blood, serum	Detection of virus-neutralizing antibodies	Can give inconsistent results depending on the actual immune response of infected persons
Electron microscopy (EM)	Blood, tissues, exudates, cell culture	Detection of viral particle morphology	Bunyaviral EM morphology is quite unspecific and pleiomorphic, needed appropriate methods
Antigen-capture ELISA	Blood, tissues, exudates	Detection of viral antigens NP, NSP, or GP	Rapid techniques, usable also for virus detection from tissues, used for RVFV or Hantaan virus detection
Fluorescence assay	Blood, tissues, exudates, cell culture		
RT-PCR	Blood, tissues, exudates, cell culture	Detection of viral nucleic acids	Sensitive and specific methods, mostly used for the confirmation of serological tests, usage of RT-PCR-complicated rapid virus clearance (RVFV, Hantaviruses) and high genetic instability (Hantaviruses, CCHFV), minimum cross-reaction inside viral family, possibility of using saliva as suitable sample
Virus culture	Blood, tissues, exudates	Virus culture on mammalian or insect tissue lines or mammals model	Long-term method with low sensitivity, viruses are detectable after 1–5 days (by PCR, IF) (CPE) after inoculation, higher viral harvest from mammals model

liver, spleen, and other tissues. Moreover, this virus has a higher affinity to the nervous system than other bunyaviruses and can be therefore isolated from CSF or from brain tissue. In bunyavirus infections, the viremia is usually very short (especially in the case of hantaviruses and RVFV) taking only several days, when the virus can be isolated directly from blood samples.

Virus culture—Methods of culturing bunyavirus vary according to their natural reservoir hosts and vectors. They are generally cultured in mammalian cells like Vero E6 (simian kidney cell line) or BHK-21 (hamster kidney cell line). For viruses transmitted by mosquitoes or ticks, an invertebrate cell line can be used. Viruses can be detected by immunofluorescence after 1–5 days after inoculations. But this method has low sensitivity and, for clinical diagnosis, can be suitable only in severe cases with high virus loads. For virus isolation, it is more sensitive if the sample is inoculated to animal models than to cell lines. Intracerebral inoculation of the sample to suckling mice is a very sensitive method. This method is more useful for clinical diagnosis than cell culture, but it takes more time (from 6 to 9 days) (Howard, 2005).

Antibody detection—Because of very short viremia in bunyavirus infections (up to 7 days after disease onset, but often only days), serology represents the main diagnostic approach. Indirect immunofluorescence (IFA) and ELISA tests are the most commonly used. IgM-specific antibodies can be detected by IFA from seventh to ninth day after disease onset with maximum titer in second and third weeks and disappear after about 4 months. IgG antibodies are detectable by IFA starting from the second week of illness and can be detectable for at least 5 years. IgM antibodies can be detected by ELISA a little earlier, from fifth to seventh day of illness, and also can be detectable for at least 3 or 4 months. In the case of IgG antibody detection, there are no big differences between IFA and ELISA. The virus neutralization test provides, e.g., in CCHFV infection, very inconsistent results, some infected patients develop very low titers, but some relatively high. Therefore, the virus neutralization test cannot be used for virus detection as a single method without confirmation by another (Mardani and Keshtkar-Jahromi, 2007). Mostly, all serological tests especially for diagnosis in the hantavirus group have low specificity, which decreases their diagnostic value. On the other hand, at least three antigens should be used for serological diagnosis of HFRS, because only 80% of HFRS can be diagnosed if only one antigen is used in the assay (Howard, 2005). Use of immunoblotting can improve the specificity of the diagnosis of hantavirus infections (Scubert et al., 2001).

Antigen detection—For RVFV, several techniques able to detect viral antigens have been developed. Mostly, IgG antibodies reacting with viral nucleoprotein or a mixture of nonstructural proteins and glycoprotein are used. Antigens can be detected by ELISA or IF methods, both suitable for virus detection from clinical material (mostly from blood from patients in viremic phases or from various tissues) or from cell culture (Pépin et al., 2010).

Amplification method—RT-PCR is one of the most commonly used methods in bunyavirus diagnosis. RT-PCR is a very sensitive and specific diagnostic tool, which can also be very fast in real-time modification. Despite the rapid clearance of bunyaviruses from blood up to 7 days after illness onset (which limits other direct diagnosis methods), RT-PCR can be positive for a longer time and in some cases up to convalescent phase (Burt et al., 1998). If the assay is properly designed, there should be no cross-reaction with other bunyaviruses. But based on the segmented genome, bunyaviruses are genetically very instable, and there are many genetically different strains of each virus, which complicates the design of RT-PCR primers and/or probes. Methods from conventional RT-PCR to, recently, the most commonly used real-time RT-PCR have been used in the multiplex format in some cases.

Most RT-PCR assays target sequences within the small or medium genomic segment. But the genetic variability in these genomic segments should be taken into account. For that reason, targeting of the L segment is preferred by some authors. For RT-PCR detection of bunyaviruses, blood or tissue samples are mostly used, which are collected by an invasive sampling method. Some studies, however, have shown that the use of saliva (not invasive sampling) provides identical results as the use of blood samples (Pettersson et al., 2008).

12.5 PITFALLS OF WORKING WITH VIRAL HEMORRHAGIC FEVER CAUSATIVE PATHOGENS

12.5.1 PATHOGEN BIOHAZARD CLASSIFICATION

The BSL expresses the degree of laboratory security, safety, and equipment necessary for work with various biological agents. Not entirely correct, the term "Biosafety level" (BSL) is used also for the classification of biological agents (e.g., BSL-4 pathogens); more correct would be the use of the term "Biohazard level" (BHL) or "Biological risk group" (BRG). All biological agents are divided into four groups or levels according to the possibility of prophylaxis and treatment, and human risk. The laboratory BSL corresponds with the level of biological agent (4-BSL for 4-BHL). Risk Group 1 includes pathogens not associated with human disease in a healthy adult. Risk Group 2 contains biological agents able to cause human disease, but rarely serious, and prophylaxis and treatment are commonly available. Risk Group 3 includes pathogens causing serious or lethal human disease, but prophylaxis or therapy is available (pathogens with high individual risk but low community risk). In the most dangerous Risk Group 4 are pathogens causing human diseases for which prophylaxis or therapy does not exist or are not usually available (high risk both for individuals and for the community).

Pathogens causing VHFs belong to BRG-2, BRG-3, and BRG-4, according to CDC classifications. Members of viral families *Filoviridae* and *Arenaviridae* belong to the highest Risk Group 4. An exception from the arenaviruses is Junin virus for which effective vaccination (Candid#1) is available, and therefore, this virus can be assigned to Risk Group 3. In accordance with data regarding candidate vaccines against other causative agents of VHF, it is possible that some other pathogens could be assigned to lower groups. From the last two viral families, OHFV, KFDV/Alkhumra virus, and CCHFV are in Risk Group 4. Other members of the *Bunyaviridae* family causing VHF belong to Risk Group 3, and also Yellow fever from the flaviviruses is within this group. And at the lowest level, dengue viruses are assigned to Risk Group 2. But there is a rule that if there are pathogens used in aerosol (for research purposes) or there are real threats of aerosol development, these pathogens are reclassified into a risk group one level higher (Chasewood and Wilson, 2009). Risk classifications of VHF causative agents are in Table 12.6.

Other criteria for pathogen classifications are by its possibility of misuse during warfare or terrorist attack. According to CDC criteria, the agents are divided into three categories A, B, and C. Category A represents the most risky pathogens causing severe diseases, which can be easily disseminated and difficult to prevent or treat. Category B agents have moderate morbidity and mortality, and are moderately easy to disseminate. Category C includes newly emerging pathogens or pathogens with the potential to be abused. The majority of VHF causative agents is placed by CDC in A category (NWAs, Lassa virus, Ebola virus, and Marburg virus) and represents pathogens with national security risk. Some other authors have classified as high risk, besides CDC Category A pathogens, RVFV, YFV, KFDV, and OHFV (Borio et al., 2002).

12.5.2 LABORATORY BIOSAFETY CLASSIFICATION

The basic principle of laboratory biosafety is strict adherence to standard microbiological practices and techniques. Persons working with dangerous infectious agents must know the potential hazards and must be trained well in the practices and techniques required for handling and working with these agents. Biosafety procedures are implemented into standard operational manuals of each laboratory and are adopted for use on laboratory equipment, personnel situation, or use of biological agents. Biosafety procedures have been created to minimize or eliminate potential exposure of laboratory personnel or environment by biological hazards. As mentioned earlier according to CDC guidelines, there are four levels of biosafety from BSL-1 to BSL-4.

TABLE 12.6

VHF Causative Agents: Different Biological Hazard Classifications

Name of Pathogen (Disease Caused by It)	Agents Usable for Biological Weapons (WHO—1970)[a]	Critical Biological Agents for Biological Attack (CDC)[b]	Threat Agents, according to Working Group on Civilian Biodefense[c]	Biological Hazard Classification (CDC)[d]
Chikungunya virus (Chikungunya disease)	X			2/3
CCHFV (Crimean–Congo hemorrhagic fever)	X			4
Dengue (dengue fever)	X			2/3
Ebola virus (Ebola hemorrhagic fever)		Cat A	X	4
Dobrava/Belgrade virus ((HFRS)		Cat C		3
Guanarito virus (Venezuelan hemorrhagic fever)		Cat A	X	4
Hantaan virus (HFRS)	X	Cat C		3
Junin virus (Argentine hemorrhagic fever)		Cat A	X	4(3)
KFDV (Kyasanur Forest disease)/Alkhurma virus			X	4
Lassa virus (Lassa fever)		Cat A	X	4
Machupo virus (Bolivian hemorrhagic fever)		Cat A	X	4
Marburg virus (Marburg hemorrhagic fever)		Cat A	X	4
OHFV (Omsk hemorrhagic fever)			X	4
Puumala virus ((HFRS)		Cat C		3
RVFV (Rift Valley fever)	X		X	3
Sabia virus (Brazil hemorrhagic fever)		Cat A	X	4
Seoul virus (HFRS)		Cat C		3
YFV (Yellow fever)	X	Cat C	X	3

[a] World Health Organization, Health aspects of chemical and biological weapons: Report of a WHO group of consultants, Geneva, Switzerland, 1970.

[b] Centers for Disease Control and Prevention: Biological and chemical terrorism: Strategic plan for preparedness and response, *MMWR*, 49(RR-4):1–26, 2000.

[c] Borio L., Inglesby T., Peters C.J. et al. Hemorrhagic fever viruses as biological weapons (Medical and Public Health Management). *JAMA*, 287(18):2391–2405, 2002.

[d] Centers for Disease Control and Prevention: Biosafety in Microbiological and Biomedical Laboratories, 5th edn., 2009, pp. 246–265.

BSL-1—The extent of laboratory biosafety for working with Risk Group 1 biological agents. Laboratory safety procedures are based on basic microbiological practices without special primary or secondary barriers against microbes.

BSL-2—Laboratory procedures, equipment, or facility construction sufficient for standard clinical, diagnostic, training, or other laboratories working with moderate risk agents (agents

BRG-2). Personnel must minimize work developing an infectious aerosol (e.g., centrifugation must be done in special tight cups) and to minimize accidental percutaneous or mucous exposures of infectious materials. The laboratory staff must be extremely cautious while working with sharp instruments like needles or scalpels. All work is under the close supervision of a scientist competent in handling infectious agents and associated procedures. Access to the laboratory is restricted. Basic personal protective equipment is used, such as face protection masks (if necessary), splash shields, gowns (ideally one-off use), and gloves (while working in the lab). Under this BSL, there are basic secondary barriers like waste decontamination or hand disinfection to prevent environmental contamination. Eyewash stations must be at their disposal. It is recommended to have biological safety cabinets (BSCs) Class II (for infectious aerosol reduction).

BSL-3—Laboratory biosafety procedures allow work with high-risk pathogens from Risk Group 3. All personnel must be trained in appropriate infectious agent handling and procedures. It is recommended to vaccinate all personnel with appropriate vaccines. The laboratory must be specially designed and engineered. A great emphasis is put on primary and secondary barriers. All waste (cultures, stocks, consumable laboratory materials, and so on) should be decontaminated by an effective method within the laboratory. If the decontamination must be done outside, the waste must be packed into durable, leak-proof containers, and transport must be safe. There is controlled and restricted access to the laboratory while working with infectious agents. All manipulation with infectious materials must be done in BSCs Class II or III with HEPA filtered exhausters. Personal protective equipment are in general the same as in BSL-2 laboratories with the emphasis on one-off usage (special suits only for the laboratory), and it is also recommended using two pairs of gloves. If personnel use contact lenses in the laboratory, they must use eye protection. The entrance into the laboratory must be through two self-closing door, between which is the space for changing clothing. All windows in BSL-3 laboratories must be sealed. The laboratory must be also equipped with an air system with HEPA filters on the system exhaust. Airflow must be directional and controlled by sensors and protected against reverse flow (it means to not allow air to flow away from the laboratory).

BSL-4—Biological protection sufficient for working with the most dangerous pathogens, like the majority of VHF causative agents. There are two kinds of BSL-4 laboratories, i.e., simple cabinet laboratory where all manipulation with infectious substances is done only in the Class III BSCs, and a suit laboratory where work is done in underpressurized rooms and laboratory staff must wear overpressurized protective suits. All personnel must have specific training in handling extremely hazardous infectious agents and procedures in BSL-4 conditions. Entrance into the BSL-4 laboratory is restricted: only approved and well-experienced laboratory staff can work in this lab. Entrance must be through a secure, locked door with the documentation of each person entering the lab. All personal clothing including undergarments must be changed in the clothing changing room before entering the lab. All laboratory staff must be under periodic medical surveillance. For supplies and material entry into the lab, there must be double-door autoclaves, fumigation chambers, or airlocks to minimize environmental contamination. Inside the lab, only necessary equipment should be situated. Class III BSC must have an HEPA filter on the air entry and two HEPA filters in series on the cabinet exhaust: the same principle must also be in the air system of the suit laboratory. In the cabinet laboratory, personnel wear personal protective equipment as in BSL-3 laboratories, but in suit labs, workers must wear a one-piece overpressurized protective suit with supplied air. All workers must wear at least two pairs of gloves. Exiting of laboratory staff from an underpressurized laboratory in suit variant of BSL-4 labs is possible only through a decontamination shower (of the surface of the protective suits), and in both variants, all personnel must have a personal shower before exiting the laboratory tract (in the clothing changing room). The BSL-4 laboratory must be situated in a separate building or a fully isolated zone of the building. All BSL-4 laboratory engineering (decontamination systems, power supplies, air distribution systems, and so on) must be doubled, and standby systems must be automatically activated in an emergency. All windows must be unbreakable and sealed. Suit BSL-4 laboratories must also contain sealed doors. All laboratory space is permanently controlled by a supervisor and also recorded for follow-up control. Also all engineering systems are permanently monitored. Decontamination of all waste (solid or liquid) must be documented and performed by validated methods (physically and biologically). Also the interior of cabinets or laboratory rooms must be subjected to sufficient decontamination.

12.5.3 RESEARCH AND DIAGNOSTIC PROBLEMS OF VHFs

Many problems with research in VHF fields are related to the necessity of high BSLs. Not many BSL-4 laboratories are available (e.g., in Europe, there are only eight BSL-4 laboratories according to the ENVID database), and only a few of them are suitable for work with large laboratory animals, like monkeys. The disadvantage, which is tightly related to the facts mentioned earlier, is the strict restriction on the transportation of VHF pathogens. Therefore, research institutes working with these agents must obtain many certificates and permissions before obtaining viruses. To simplify this problem, it is possible, e.g., to work with recombinant proteins (no special permission needed) or with full/partial genomes (not so strict restrictions). But the consequence of this simplification is only the theoretical evaluation of specificity, which can be different from the real wild viruses. This relates to the enormous instability of genomic sequences of RNA viruses in general, which results in various genome sequence variants of one virus, and of course also in some differentiation in the chemical structure of antigens, both of which can interfere with development methods. Another problem is the frequency of detection of VHF viruses in laboratories and laboratory staff experience with these diagnostics. According to the results of multinational (28 laboratories from 17 countries) quality tests of highly infectious disease (orthopoxvirus, EBOV, MARV, LASV) rapid detection, only 45.8% laboratories detected low-quantity virus, and only 66.7% labs detected a high quantity of viruses (Niedrig et al., 2004). A couple of highly specialized world reference laboratories responsible for VHF diagnosis and experienced enough to give reliable results would be of benefit. Unfortunately, there are many laboratories across the world now doing diagnostics of VHFs, but only some of them can give reliable results (e.g., laboratories CDC, WHO, or Institute of Tropical Diseases Hamburg, Germany).

12.6 CONCLUSIONS

Diagnostics of VHFs is very difficult due to biosafety, limitations of identification methods, and many other problems. There is not available any simple method for detection of all VHF causative agents. For accurate detection, it is necessary to combine all available diagnostic approaches including clinical and epidemiological data. Only then, a well-prepared laboratory can provide a reliable diagnosis of the VHF causative agents.

ACKNOWLEDGMENTS

We acknowledge financial support by grant OVUVZU2008002 of the Ministry of Defense of the Czech Republic, and financial support by the Czech Science Foundation project Nos. P302/10/P438 and P502/11/2116, and grant Z60220518 from the Ministry of Education, Youth, Sports of the Czech Republic, and the AdmireVet project No. CZ.1.05./2.1.00/01.006 (ED006/01/01).

REFERENCES

Balmaseda, A., Guzman, M.G., and Hammond, S., 2003. Diagnosis of dengue virus infection by detection of specific Immunoglobulin M (IgM) and IgA antibodies in serum and saliva. *Clin. Diagn. Lab. Immunol.* 10(2), 317–322.

Batra, G., Nemani, S.K., Tyagi, P., Swaminathan, and S., Khanna, N., 2011. Evaluation of envelope domain III-based single chimeric tetravalent antigen and monovalent antigen mixtures for the detection of anti-dengue antibodies in human sera. *BMC Infect. Dis.* 11, 64.

Bausch, D.G., Rollin, P.E., Demby, A.H., Coulibaly, M., Kanu, J., Conteh, A.S., Wagoner, K.D., McMullan, L.K., Bowen, M.D., Peters, C.J., and Ksiazek, T.G., 2000. Diagnosis and clinical virology of Lassa fever as evaluated by enzyme-linked immunosorbent assay, indirect fluorescent-antibody test, and virus isolation. *J. Clin. Microbiol.* 38, 2670–2677.

Borio, L., Inglesby, T., Peters, C.J., Schmaljohn, A.L., Hughes, J.M., Jahrling, P.B., Ksiazek, T., Johnson, K.M., Meyerhoff, A., O'Toole, T., Ascher, M.S., Bartlett, J., Breman, J.G., Eitzen, E.M. Jr, Hamburg, M., Hauer, J., Henderson, D.A., Johnson, R.T., Kwik, G., Layton, M., Lillibridge, S., Nabel, G.J., Osterholm, M.T., Perl, T.M., Russell, P., Tonat, K., and Working Group on Civilian Biodefense, 2002. Hemorrhagic fever viruses as biological weapons (medical and public health management). *JAMA* 287(18), 2391–2405.

Briese, T., Palacios, G., Kokoris, M., Jabado, O., Liu, Z., Renwick, N., Kapoor, V., Casas, I., Pozo, F., Limberger, R., Perez-Brena, P., Ju, J., and Lipkin, W.I., 2005. Diagnostic system for rapid and sensitive differential detection of pathogens. *Emerg. Infect. Dis.* 11(2), 310–313.

Burt, F.J., Leman, P.A., Smith, J.F., and Swanepoel, R., 1998. The use of a reverse transcription-polymerase chain reaction for the detection of viral nucleic acid in the diagnosis of Crimean-Congo hemorrhagic fever. *J. Virol. Methods* 70(2), 129–137.

Chao, D.Y., Davis, B.S., and Chang, G.J., 2007. Development of multiplex Real-Time reverse transcriptase PCR assays for detecting eight medically important flaviviruses in mosquitoes. *J. Clin. Microbiol.* 45(2), 584–589.

Chasewood, L.C. and Wilson, D.E., 2009. *Biosafety in Microbiological and Biomedical Laboratories*, 5th edn., U.S. Health department, Diane Publishing, Washington, DC.

Datta, S. and Wattal, C., 2010. Dengue NS1 antigen detection: A useful tool in early diagnosis of dengue virus infection. *Indian J. Med. Microbiol.* 28(2), 107–110.

Demby, A.H., Chamberlain, J., Brown, D.W., and Clegg, C.S., 1994. Early diagnosis of Lassa fever by reverse transcription-PCR. *J. Clin. Microbiol.* 32(12), 2898–2903.

Ding, X., Hu, D., Chen, Y., Di, B., Jin, J., Pan, Y., Qiu, L., Wang, Y., Wen, K., Wang, M., and Che, X., 2011. Full serotype- and group-specific NS1 capture enzyme-linked immunosorbent assay for rapid differential diagnosis of dengue virus infection. *Clin. Vaccine Immunol.* 18(3), 430–434.

do Valle, T.Z., Billecocq, A., Guillemot, L., Alberts, R., Gommet, C., Geffers, R., Calabrese, K., Schughart, K., Bouloy, M., Montagutelli, X., and Panthier, J.J., 2010. A new mouse model reveals a critical role for host innate immunity in resistance to Rift Valley fever. *J. Immunol.* 185(10), 6146–6156.

Drosten, C., Göttig, S., Schilling, S., Asper, M., Panning, M., Schmitz, H., and Günther S., 2002. Rapid detection and quantification of RNA of Ebola and Marburg viruses, Lassa virus, Crimean-Congo hemorrhagic fever virus, Rift Valley fever virus, dengue virus, and yellow fever virus by real-time reverse transcription-PCR. *J. Clin. Microbiol.* 40(7), 2323–2330.

Drosten, C., Kümmerer, B.M., Schmitz, H., and Günther, S., 2003. Molecular diagnostics of viral hemorrhagic fevers. *Antiviral Res.* 57(1–2), 61–87.

Emmerich, P., Thome-Bolduan, C., Drosten, C., Gunther, S., Ban, E., Sawinsky, I., and Schmitz, H., 2006. Reverse ELISA for IgG and IgM antibodies to detect Lassa virus infections in Africa. *J. Clin. Virol.* 37(4), 277–281.

Fang, X., Tan, O.K., Tse, M.S., and Ooi, E.E., 2010. A label-free immunosensors for diagnosis of Dengue infection with simple electrical measurements. *Biosens. Bioelectron.* 25(5), 1137–1142.

Gadkari, R.A. and Srinivasan, N., 2010. Prediction of protein–protein interaction in dengue virus coat protein guided by low resolution cryoEM structures. *BMC Struct. Biol.* 10, 17.

Geisbert, T.W. and Jaax, N.K., 1998. Marburg hemorrhagic fever: Report of a case studied by immunohistochemistry and electron microscopy. *Ultrastruct. Pathol.* 22(1), 3–17.

Gibb, T.R., Norwood, D.A. Jr, Woollen, N., and Henchal, E.A., 2001. Development and evaluation of a fluorogenic 5′ nuclease assay to detect and differentiate between Ebola virus subtypes Zaire and Sudan. *J. Clin. Microbiol.* 39, 4125–4130.

Goldsmith, C.S. and Miller, S.E., 2009. Modern uses of electron microscopy for detection of viruses. *Clin. Microbiol. Rev.* 22(4), 522–563.

Gowen, B.B., Wong, M.H., Larson, D., Ye, W., Jung, K.H., Sefing, E.J., Skirpstunas, R., Smee, D.F., Morrey, J.D., and Schneller, SW., 2010. Development of a new Tacaribe arenavirus infection model and its use to explore antiviral activity of a novel aristeromycin analog. *PLoS. One* 5(9):e12760.doi:10.1371/journal.pone.0012760.

Hazelton, P.R. and Gelderblom, H.R., 2003. Electron microscopy for rapid diagnosis of infectious agents in emergent situations. *Emerg. Infect. Dis.* 9(3), 294–303.

Houghton-Triviño, N., Montaña, D., and Castellanos, J., 2008. Dengue-yellow fever sera cross-reactivity; challenges for diagnosis. *Rev. Salud. Publica(Bogota)* 10(2), 299–307.

Howard, C.R. 2005. Viral hemorrhagic fever, 1st edn., Elsevier Inc., San Diego, CA.

Ikegami, T., Niikura, M., Saijo, M., Miranda, M.E., Calaor, A.B., Hernandez, M., Acosta, L.P., Manalo, D.L., Kurane, I., Yoshikawa, Y., and Morikawa, S. 2003. Antigen capture enzyme-linked immunosorbent assay for specific detection of Reston Ebola virus nucleoprotein. *Clin. Diagn. Lab. Immunol.* 10(4), 552–557.

Jabado, O.J., Palacios, G., Kapoor, V., Hui, J., Renwick, N., Zhai, J., Briese, T., and Lipkin, W.I., 2006. Greene SCRPrimer: A rapid comprehensive tool for designing degenerate primers from multiple sequence alignments. *Nucleic Acids Res.* 34(22), 6605–6611.

Jayakeerthi, R.S., Potula, R.V., Shrinivasan, S., and Badrinath, S., 2006. Shell vial culture assay for rapid diagnosis of Japanese encephalitis, West Nile and Dengue-2 viral encephalitis. *Virol. J.* 3:2 doi:10.1186/1743–422X-3-2.

Johnson, B.W., Russell, B.J., and Lanciotti, R.S., 2005. Serotype-specific detection of dengue viruses in a fourplex real-time reverse transcriptase PCR assay. *J. Clin. Microbiol.* 43(10), 4977–4983.

Koraka, P., Zeller, H., Niedrig, M., Osterhaus, A.D., and Groen, J., 2002. Reactivity of serum samples from patients with a flavivirus infection measured by immunofluorescence assay and ELISA. *Microbes Infect.* 4(12), 1209–1215.

Ksiazek, T.G., Rollin, P.E., Williams, A.J., Bressler, D.S., Martin, M.L., Swanepoel, R., Burt, F.J., Leman, P.A., Khan, A.S., Rowe, A.K., Mukunu, R., Sanchez, A., and Peters, C.J., 1999a. Clinical virology of Ebola hemorrhagic fever (EHF): virus, virus antigen, and IgG and IgM antibody findings among EHF patients in Kikwit, Democratic Republic of the Congo, 1995. *J. Infect. Dis.* 179(Suppl. 1), S177–S187.

Ksiazek, T.G., Westm, C.P., Rollin, P.E., Jahrling, P.B., and Peters, C.J., 1999b. ELISA for the detection of antibodies to Ebola viruses. *J. Infect. Dis.* 179(Suppl. 1), S192–S198.

Lai, Y.L., Chung, Y.K., Tan, H.C., Yap, H.F., Yap, G., Ooi, E.E., and Ng, L.C., 2007. Cost-effective real-time reverse transcriptase PCR (RT-PCR) to screen for Dengue virus followed by rapid single-tube multiplex RT-PCR for serotyping of the virus. *J. Clin. Microbiol.* 45(3), 935–941.

Leroy, E.M., Baize, S., Lu, C.Y., McCormick, J.B., Georges, A.J., Georges-Courbot, M.C., Lansoud-Soukate, J., and Fisher-Hoch, S.P., 2000. Diagnosis of Ebola haemorrhagic fever by RT-PCR in an epidemic setting. *J. Med. Virol.* 60(4), 463–467.

Mardani, M. and Keshtkar-Jahromi, M., 2007. Crimean-Congo hemorrhagic fever. *Arch. Iran. Med.* 10(2), 204–214.

Moe, A.B., Lambert, R.D., and Lupton, H.W., 1981. Plaque assay for Ebola virus. *J. Clin. Microbiol.* 13(4), 791–793.

Nakayama, E., Yokoyama, A., Miyamoto, H., Igarashi, M., Kishida, N., Matsuno, K., Marzi, A., Feldmann, H., Ito, K., Saijo, M., and Takada, A., 2010. Enzyme-linked immunosorbent assay for detection of filovirus species-specific antibodies. *Clin. Vaccine Immunol.* 17(11), 1723–1728.

Niedrig, M., Schmitz, H., Becker, S., Günther, S., ter Meulen, J., Meyer, H., Ellerbrok, H., Nitsche, A., Gelderblom, H.R., and Drosten, C., 2004. First international quality assurance study on the rapid detection of viral agents of bioterrorism. *J. Clin. Microbiol.* 42(4), 1753–1755.

Niikura, M., Ikegami, T., Saijo, M., Kurane, I., Miranda, M.E., and Morikawa, S. 2001. Detection of Ebola viral antigen by enzyme-linked immunosorbent assay using a novel monoclonal antibody to nucleoprotein. *J. Clin. Microbiol.* 39(9), 3267–3271.

Palacios, G., Briese, T., Kapoor, V., Jabado, O., Liu, Z., Venter, M., Zhai, J., Renwick, N., Grolla, A., Geisbert, T.W., Drosten, C., Towner, J., Ju, J., Paweska, J., Nichol, S.T., Swanepoel, R., Feldmann, H., Jahrling, P.B., and Lipkin, W.I., 2006. Masstag polymerase chain reaction for differential diagnosis of viral hemorrhagic fevers. *Emerg. Infect. Dis.* 12(4), 692–695.

Panning, M., Laue, T., Olschlager, S., Eickmann, M., Becker, S., Raith, S., Courbot, M.C., Nilsson, M., Gopal, R., Lundkvist, A., Caro, A., Brown, D., Meyer, H., Lloyd, G., Kummerer, B.M., Gunther, S., and Drosten, C. 2007. Diagnostic reverse-transcription polymerase chain reaction kit for filoviruses based on the strain collections of all European biosafety level 4 laboratories. *J. Infect. Dis.* (Suppl. 2), S199–S204.

Paweska, J.T., Jansen van Vuren, P., and Swanepoel, R. 2007. Validation of an indirect ELISA based on a recombinant nucleocapsid protein of Rift Valley fever virus for the detection of IgG antibody in humans. *J. Virol. Methods* 146(1–2), 119–124.

Pepin, M., Bouloy, M., Bird, B.H., Kemp, A., and Paweska, J., 2010. Rift Valley fever virus (Bunyaviridae: Phlebovirus): an update on pathogenesis, molecular epidemiology, vectors, diagnostics and prevention. *Vet. Res.* 41, 61 doi:10.1051/vetres/2010033.

Peters, C.J. and LeDuc, J.W., 1999. An introduction to Ebola: the virus and the disease. *J. Infect. Dis.* 179(Suppl. 1), 9–16.

Pettersson, L., Klingström, J., Hardestam, J., Lundkvist, A., Ahlm, C., and Evander, M., 2008. Hantavirus RNA in saliva from patients with hemorrhagic fever with renal syndrome. *Emerg. Infect. Dis.* 14(3), 406–411.

Prado, I., Rosario, D., Bernardo, L., Alvarez, M., Rodríguez, R., Vázquez, S., and Guzmán, M.G., 2005. PCR detection of dengue virus using dried whole blood spotted on filter paper. *J. Virol. Methods* 125(1), 75–81.

Riera, L.M., Feuillade, M.R., Saavedra, M.C., and Ambrosio, A.M. 1997. Evaluation of an enzyme immunosorbent assay for the diagnosis of Argentine haemorrhagic fever. *Acta Virol.* 41(6), 305–310.

Růžek, D., Yakimenko, V.V., Karan, L.S., and Tkachev, S.E., 2010. Omsk haemorrhagic fever. *Lancet* 376, 2104–2113.

Saijo, M., Georges-Courbot, M.C., Marianneau, P., Romanowski, V., Fukushi, S., Mizutani, T., Georges, A.J., Kurata, T., Kurane, I., and Morikawa, S., 2007. Development of recombinant nucleoprotein-based diagnostic systems for Lassa fever. *Clin. Vaccine Immunol.* 14(9), 1182–1189.

Saijo, M., Niikura, M., Ikegami, T., Kurane, I., Kurata, T., and Morikawa, S., 2006. Laboratory diagnostic systems for Ebola and Marburg hemorrhagic fevers developed with recombinant proteins. *Clin. Vaccine Immunol.* 13(4), 444–451.

Saijo, M., Niikura, M., Morikawa, S., Ksiazek, T.G., Meyer, R.F., Peters, C.J., and Kurane, I., 2001. Enzyme-linked immunosorbent assay for detection of antibodies to Ebola and Marburg viruses using recombinant nucleoproteins. *J. Clin. Microbiol.* 39(1), 1–7.

Saijo, M., Tang, Q., Shimayi, B., Han, L., Zhang, Y., Asiguma, M., Tianshu, D., Maeda, A., Kurane, I., and Morikawa, S., 2005. Antigen-capture enzyme-linked immunosorbent assay for the diagnosis of Crimean-Congo hemorrhagic fever using a novel monoclonal antibody. *J. Med. Virol.* 77(1), 83–88.

Sánchez-Seco, M.P., Rosario, D., Hernández, L., Domingo, C., Valdés, K., Guzmán, M.G., and Tenorio, A., 2006. Detection and subtyping of Dengue 1–4 and Yellow fever viruses by means of a multiplex RT-nested-PCR using degenerated primers. *Trop. Med. Int. Health* 11(9), 1432–1441.

Schilling, S., Ludolfs, D., Van An, L., and Schmitz, H. 2004. Laboratory diagnosis of primary and secondary dengue infection. *J. Clin. Virol.* 31(3), 179–184.

Scubert, J., Tollmann, F., and Weissbrich, B., 2001. Evaluation of a panreactive hantavirus enzyme immunoassay and a hantavirus immunoblot for the diagnosis of nephropathia epidemica. *J. Clin. Virol.* 21(1), 63–73.

Trombley, A.R., Wachter, L., Garrison, J., Buckley-Beason, V.A., Jahrling, J., Hensley, L.E., Schoepp, R.J., Norwood, D.A., Goba, A., Fair, J.N., and Kulesh, D.A., 2010. Short report: Comprehensive panel of Real-time TaqMan™ polymerase chain reaction assays for detection and absolute quantification of filoviruses, arenaviruses and New World Hantaviruses. *Am. J. Trop. Med. Hyg.* 82(5), 954–960.

Ure, A.E., Ghiringhelli, P.D., Possee, R.D., Morikawa, S., and Romanowski, V. 2008. Argentine hemorrhagic fever diagnostic test based on recombinant Junín virus N protein. *J. Med. Virol.* 80(12), 2127–2133.

Vieth, S., Drosten, C., Charrel, R., Feldmann, H., and Günther, S., 2005. Establishment of conventional and fluorescence resonance energy transfer-based real-time PCR assays for detection of pathogenic New World arenaviruses. *J. Clin. Virol.* 32, 229–235.

Vieth, S., Drosten, C., Lenz, O., Vincent, M., Omilabu, S., Hass, M., Becker-Ziaja, B., Ter Meulen, J., Nichol, S.T., Schmitz, H., and Günther, S., 2007. RT-PCR assay for detection of Lassa virus and related Old World arenaviruses targeting the *L* gene. *Trans. R. Soc. Trop. Med. Hyg.* 101(12), 1253–1264.

Welsch, S., Kolesnikova, L., Krähling, V., Riches, J.D., Becker, S., and Briggs, J.A., 2010. Electron tomography reveals the steps in filovirus budding. *PLoS Pathog.* 6(4): e1000875. doi:10.1371/journal.ppat.1000875.

Wulff, H. and Johnson, K.M., 1979. Immunoglobulin M and G responses measured by immunofluorescence in patients with Lassa or Marburg virus infections. *Bull. World Health Organ.* 57(4), 631–635.

Yap, G., Sil, B.K., and Ng, L.C., 2011. Use saliva for early dengue diagnosis. *PLoS Negl. Trop. Dis.* 5(5), e1046.

Yauch, L.E. and Shresta, S., 2008. Mouse model of dengue virus infection and disease. *Antiviral Res.* 80(2), 87–93.

Zaki, S.R., Shieh, W.J., Greer, P.W., Goldsmith, C.S., Ferebee, T., Katshitshi, J., Tshioko, F.K., Bwaka, M.A., Swanepoel, R., Calain, P., Khan, A.S., Lloyd, E., Rollin, P.E., Ksiazek, T.G., and Peters, C.J., 1999. A novel immunohistochemical assay for the detection of Ebola virus in skin: implications for diagnosis, spread, and surveillance of Ebola hemorrhagic fever. Commission de Lutte contre les Epidemies a Kikwit. *J. Infect. Dis.* 179(Suppl. 1), S36–S47.

13 Vaccine Research Efforts for Hemorrhagic Fever Viruses

Shannon S. Martin and Mary Kate Hart

CONTENTS

13.1 INTRODUCTION

Viral hemorrhagic fevers (VHFs) represent a serious public health problem with recurrent outbreaks in endemic regions primarily located in South America, Eastern Europe, Africa, and Asia. The VHFs are from four virus families, *Flaviviridae*, *Filoviridae*, *Arenaviridae*, and *Bunyaviridae*. Many of the VHFs are listed as category A priority pathogens by the Centers for Disease Control and Prevention (CDC) and the National Institute of Allergy and Infectious Diseases (NIAID). Currently, an approved vaccine, licensed by the U.S. Food and Drug Administration (FDA), is not available for any of the VHFs. Development of vaccines for the VHFs faces numerous challenges.

TABLE 13.1

Advantages and Disadvantages Associated with the Vaccine Platforms

Vaccine Platform	Advantages	Disadvantages
DNA	• Simple to construct • Amenable to large-scale production • Unable to revert to virulent phenotype • Long shelf life, more temperature stable than other vaccine types	• Poorly immunogenic • Requires multiple vaccinations • Potential for host genome integration • Potential for antibiotic resistance • Potential for induction of autoimmunity
Live attenuated	• Induces strong cellular and humoral immune responses • Induction of durable/memory immune response • Low manufacturing costs	• Reversion to virulent phenotype • Environmental release concerns • Risk of disease in immunocompromised
Virus vector mediated	• Strong and durable immune response	• Preexisting immunity to vector • Development of immunity to vector limits boosting. • Cell tropism may not support optimal production of immunogen.
Viruslike particles	• Delivery of immunogenic proteins in native conformation	• Multiple immunizations
Recombinant proteins	• Eliminates exposure to virus—safety • Easily manipulated to induce optimal immune response	• Multiple immunizations • Poorly immunogenic • Requirement for adjuvant
Inactivated	• Inexpensive	• Poorly immunogenic • Short duration of protection

The high pathogenicity of the VHFs restricts any manipulation to biosafety level 3 (BSL3) and, for the majority, BSL4 containment facilities, which limit the number of research institutions that can support research programs using the virulent forms of these viruses. Challenges facing vaccine development further include a limited understanding of the pathogenesis of the VHFs in humans and the limited availability of relevant animal models necessary to evaluate vaccine immunogenicity and efficacy. Licensure of vaccines for the majority of VHFs will likely require regulatory strategies involving the FDA Animal Rule, further increasing the complexity of the obstacles the vaccine developers face.

Due to the multitude of vaccine efforts for each of the VHF viruses, particularly dengue virus (DENV), this review focuses on vaccine efforts targeting the VHF viruses reported in the last 5 years. A variety of vaccine platforms have been investigated for each of the VHFs including live-attenuated, recombinant protein, DNA, viruslike particles (VLPs), virus-vectored, and inactivated virus vaccines. The development of vaccines based on these platforms and novel approaches to improve performance is described as applicable for VHF in the following sections. Table 13.1 presents the advantages and disadvantages of each platform.

13.2 FLAVIVIRUSES

13.2.1 Dengue Virus

Dengue is the most important vector-borne viral disease in terms of morbidity and mortality with an estimated 3.6 billion people from 110 countries at risk of infection. Disease caused by DENV infection ranges from asymptomatic to severe life-threatening disease generally referred to as dengue hemorrhagic fever (DHF) and dengue shock syndrome (DSS). An estimated 36 million infections with DENV occur worldwide annually, with approximately 2.1 million severe cases (DHF/DSS) and 20,000 deaths (Guzman and Isturiz, 2010). Epidemiological studies determined that the risk for more severe dengue illness is higher following a second, heterotypic DENV infection than for a primary DENV infection. While infection with one type of DENV is widely believed to provide lifelong immunity against that virus serotype, it does not provide cross-protection against infections from the other DENV types. The vast majority of DHF cases occur following a second infection with a DENV serotype different from the first infection. Despite the impact of this disease on the global human population, no vaccine is licensed for general use.

The development of a safe, immunogenic, and efficacious vaccine for DENV is hindered by a number of challenges. The general consensus of vaccine developers is that a DENV vaccine should induce an immune response that prevents disease caused by all four serotypes simultaneously because of the concern that a dengue vaccine may induce immune responses that potentiate a more severe dengue disease upon subsequent exposure (Durbin et al., 2011a). The availability of relevant animal models demonstrating similar disease manifestations as humans following infection for evaluating vaccine efficacy also presents a challenge for investigators (Cassetti et al., 2010). Mice infected with DENV often fail to develop significant viremias and signs of disease. Alternative mouse models such as humanized mice (Bente et al., 2005; Kuruvilla et al., 2007; Mota and Rico-Hesse, 2009) and genetically modified mice have been developed, and immune-competent models challenged with mouse-adapted DENV strains are used for early assessments of efficacy (Johnson and Roehrig, 1999; Shresta et al., 2006; Williams et al., 2009). Nonhuman primates (NHPs) are also used; however, NHPs do not develop overt signs of DENV disease, although they do become viremic following challenge. Since development of viremia is also a hallmark of DENV infection in humans, vaccine candidates are considered protective if viremia can be reduced or prevented in the NHP model. A complete understanding of the mechanisms of protective immunity continues to elude investigators. Significant evidence exists indicating that neutralizing antibodies play a major role in protection; however, the quantity of neutralizing antibody needed for protection has not been established. Other immune mechanisms, such as cytotoxic T-cell responses, are believed to play a role in protection, but additional research is needed to further define and validate immune correlates of protection for vaccine development (Hombach et al., 2007). Finally, clinical development of a dengue vaccine is a challenging goal since the disease is prevalent in 110 countries and multiple genetic variants of each DENV exist such that extensive clinical trials in multiple countries are likely needed before a vaccine can be licensed. The development of vaccines for DENV faces additional challenges including achievement of a balanced immune response for all four serotypes, replication interference between different serotypes of DENV in tetravalent formulations, and prevention of transmissibility by mosquitoes.

A number of vaccine candidates are currently undergoing preclinical evaluation, and five vaccines have advanced to clinical trials (Danko et al., 2011; Osorio et al., 2011; Durbin et al., 2011b; Guy et al., 2011b; Coller et al., 2011). The following sections will discuss the status of vaccines in clinical and preclinical development. Due to the sheer number of vaccines for DENV in preclinical development, discussion of all vaccine candidates in preclinical development is beyond the scope of this review. Table 13.2 presents the most recently reported DENV vaccine candidates and their stage of development, while the following sections highlight selected reports describing the generation and testing of tetravalent vaccine candidates.

13.2.1.1 Live-Attenuated Virus Vaccines

Two live-attenuated virus vaccines are currently in clinical development. A live-attenuated tetravalent vaccine, generated by serial passage of a DENV from each serotype in primary dog kidney (pdk) cells (Eckels et al., 2003), advanced to clinical testing in 2003. A Phase 2 clinical study was recently completed that identified a tetravalent formulation (Formulation 17) that provides a more

TABLE 13.2

Vaccine Efforts for Development of DENV

Vaccine Candidates by Platform	DENV Subtypes Targeted	Stage of Development	Reference
Live-attenuated virus vaccine			
Attenuated by insertion of genomic mutations	DENV1–4 (tetravalent)	Phase 1	Reviewed in Durbin et al. (2011a)
Attenuated by serial passage in pdk cells	DENV1–4	Phase 2	Sun et al. (2009)
Attenuated DENV/DENV chimeric virus	DENV2	Preclinical attenuation and immunogenicity in mice	Reviewed in Schmitz et al. (2011)
Recombinant protein vaccines			
DEN-80E	DENV1	Phase 1	Reviewed in Coller et al. (2011)
DEN-80E	DENV1–4 (tetravalent)	Preclinical immunogenicity and efficacy in NHPs	Coller et al. (2011)
Lipidated consensus DENV envelope protein domain III (LcEDIII)	DENV1–4 (tetravalent)	Preclinical efficacy in mice	Chiang et al. (2011)
Domain III of the DENV 1 envelope protein (pD1) + Freund adjuvant	DENV1	Preclinical efficacy in NHPs	Bernardo et al. (2008)
Aggregated DENV2 EDIII-capsid chimeric protein	DENV2	Preclinical immunogenicity in mice	Valdes et al. (2009)
Fusion of DENV2 EDIII to p64K formulated with outer membrane vesicle or capsular polysaccharide of *Neisseria meningitidis*	DENV2	Preclinical immunogenicity and efficacy in NHPs	Valdes et al. (2009)
Mutated transmembrane domain of E	DENV2	Preclinical immunogenicity in mice	Smith et al. (2011)
DENV1/3 and DENV2/4 EDIII/STF2D (flagellin) fusion protein	DENV1–4 (tetravalent) (administered as two bivalents (DENV1/3 and DEV2/4)	Preclinical immunogenicity in mice	Reviewed in Schmitz et al. (2011)
Expression of consensus EDIII (cEDIII) in *E. coli*; adjuvanted with aluminum phosphate	DENV1–4 (tetravalent)	Preclinical immunogenicity in mice	Leng et al. (2009)
Expression of cEDIII-Ag473	DENV1–4 (tetravalent)	Preclinical immunogenicity in mice	Chen et al. (2009)
Expression of the EDIII region of the four DENV by *P. pastoris*	DENV1–4 (tetravalent)	Preclinical immunogenicity in mice	Etemad et al. (2008)

Virus-vectored vaccines

Description	Serotype	Status	Reference
ChimeriVax™-TDV	DENV1–4 (tetravalent)	Phase 2	clinicaltrials.gov (NCT0875524; NCT00842530)
Adenovirus-mediated delivery of DENV1/2 prM/E and DENV3/4 prM/E	DENV1–4 (tetravalent) (administered as two bivalents (DENV1/2 and DEV3/4))	Preclinical immunogenicity and efficacy in mice and NHPs	Holman et al. (2007); Raviprakash et al. (2008)
Adenovirus-mediated delivery of an EDIII fusion protein	DENV1–4 (tetravalent)	Preclinical immunogenicity in mice	Khanam et al. (2009)
YF 17D containing DENV prM/E	DENV1	Preclinical immunogenicity in mice and NHPs	Mateu et al. (2007); Trindade et al. (2008)
Measles vector expressing DENV1–4 EDIII/ectodomain of M fusion protein	DENV1–4 (tetravalent)	Preclinical immunogenicity in mice	Brandler et al. (2010)
Alphavirus replicon expression of EDIII, E85, or prM/E	DENV2	Preclinical immunogenicity in mice	White et al. (2007)
West Nile-vectored DENV2 prM/E	DENV2	Preclinical immunogenicity and efficacy in mice	Suzuki et al. (2009)

DNA vaccine

Description	Serotype	Status	Reference
EDI, II, III (pE1D2)	DENV2	Phase 2	Danko et al. (2011)
	DENV2	Preclinical efficacy in mice	Azevedo et al. (2011)
Synthetic consensus (SynCon™) DIII of envelope protein	DENV1–4 (tetravalent)	Induction of neutralizing antibodies mice	Ramanathan et al. (2009)

VLP

Description	Serotype	Status	Reference
Expression of prM/E VLP in mammalian cells	DENV1–4 (monovalent and tetravalent)	Induction of neutralizing antibodies and specific cytotoxic T-cells	Zhang et al. (2011)
Expression of prM/E of DENV2 VLP in insect cells	DENV2	Induction of antibodies in mice	Kuwahara and Konishi (2010)

Whole virus inactivated vaccines

Description	Serotype	Status	Reference
Psoralen-inactivated DENV1	DENV1	Preclinical immunogenicity and efficacy in NHPs	Maves et al. (2011)

balanced tetravalent neutralizing antibody response than formulations evaluated in Phase 1 studies (Sun et al., 2009). Formulation 17 is now the lead candidate and will be evaluated in human volunteers spanning a broad age range in endemic areas. In contrast to the attenuation strategy used by Eckels and associates (Eckels et al., 2003), the second live-attenuated DENV vaccine in clinical trials was generated using recombinant DNA technology to introduce defined attenuating mutations into full-length clones of DEN viruses representing each of the four serotypes to generate monovalent vaccines that will be combined for use in a tetravalent formulation. These were recently evaluated in Phase 1 clinical studies for safety, immunogenicity, and viral kinetics for identification of monovalents to include in tetravalent formulation (Durbin et al., 2011a). The results of these studies identified six monovalents for further evaluation.

13.2.1.2 Nonreplicating Vaccines

The strategy for development of nonreplicating DENV vaccines largely relies on generating immune responses to the envelope (E) glycoprotein due to its being a target of neutralizing antibodies. The E protein contains three distinct ectodomains (EDI, EDII, and EDIII). The EDIII domain is involved in receptor binding and contains critical epitopes that can elicit cross-neutralizing antibodies to all four serotypes as well as epitopes that elicit serotype-specific neutralizing antibodies (Sukupolvi-Petty et al., 2007; Crill et al., 2009; Rajamanonmani et al., 2009). Proper folding of the recombinant E protein is required to maintain the conformation of its neutralizing epitopes and generally requires the coexpression of the premembrane (prM) protein (Allison et al., 2001) to facilitate proper folding and trafficking through the secretory pathway (Lorenz et al., 2002; Mukhopadhyay et al., 2005). Proper processing of the E protein is also dependent on a variety of posttranslational events such as facilitation of disulfide bond formation, proteolytic cleavage, and glycosylation. Selection of the most appropriate expression system is critical for achieving an optimally immunogenic protein. A variety of expression systems have been evaluated for production of recombinant E proteins including *E. coli* (Mason et al., 1990), baculovirus vectors (Kelly et al., 2000), yeast (Etemad et al., 2008), mammalian cells (Coller et al., 2011), insect cells (Clements et al., 2010), and plant-based systems (Yap and Smith, 2010).

13.2.1.2.1 Recombinant Protein Vaccines

Recombinant protein vaccine approaches offer a variety of advantages over other platforms for the development of DENV vaccines as they provide one of the safest approaches toward avoiding the issue of viral interference. Administration of recombinant proteins may more easily achieve the objective of a balanced immune response to each serotype as adjustment of dosages of recombinant proteins is more easily made.

Recombinant proteins comprising the full-length prM protein and 80% of the E protein (DEN-80E) for each of the DENV serotypes were recently generated for development into DENV vaccine candidates (Coller et al., 2011). The production of high-yield and high-quality recombinant proteins has challenged DENV vaccine developers for years. The Drosophila S2 cell expression system was selected for expression of DEN-80E and provided DEN-80E developers a product suitable for evaluation in the clinic. As with nearly all recombinant protein vaccines, immunogenicity was quite low, and induction of robust immune responses requires a formulation containing adjuvant. A series of preclinical studies demonstrated that DEN-80E formulated with adjuvants results in a potent immunogen capable of inducing high-titer neutralizing antibody responses. DEN-80E is the only DENV recombinant protein vaccine to advance to clinical development (Coller et al., 2011). A Phase 1 clinical trial was recently completed in which the two dosages of DEN-80E adjuvanted with aluminum hydroxide were evaluated. The results of this study are not yet available (NCT00936429).

A variety of recombinant protein vaccines are currently in preclinical development. A tetravalent chimeric vaccine (rEDIII-T) was developed by fusion of the EDIII region of the four DENV serotypes using flexible pentaglycyl peptide linkers (Etemad et al., 2008). The proteins were expressed in the methylotrophic yeast, *Pichia pastoris*. There are many advantages associated with the use of the *P. pastoris* expression system including the following: it supports high-level expression of transgenes, is capable of posttranslational modifications characteristic of higher eukaryotes, is amenable to scale-up, and is relatively inexpensive. *P. pastoris* is also an expression system that works well with disulfide-rich proteins such as the tetravalent EDIII chimera. Vaccination of mice with rEDIII-T formulated with adjuvant stimulates the production of neutralizing antibodies to all four DENV serotypes. This initial study supported further investigation of rEDIII-T as a vaccine candidate; however, the authors reported failure of chimeric protein secretion into the growth medium and the inability to purify the protein under nondenaturing conditions, both being issues that will need to be resolved for this construct to be a viable vaccine candidate.

A tetravalent protein, expressed in *Escherichia coli*, was generated by fusing the consensus sequence of the EDIII protein (cEDIII) for all four DENV serotypes to the lipoprotein signal sequence, Ag473 from *Neisseria meningitidis*, which serves as an intrinsic adjuvant (Chen et al., 2009). Immunogenicity testing in mice showed EDIII-Ag473 induces a neutralizing antibody response to all four DENV serotypes that is greater in magnitude to that induced by cEDIII when formulated with alum. Additional preclinical characterization studies on cEDIII-Ag473 in mice demonstrated the ability of a single vaccination with cEDIII to stimulate neutralizing antibodies such that a rapid anamnestic response is triggered upon challenge with wild-type virus (Chiang et al., 2011). The ability to induce a strong memory response after one dose of vaccine is a very attractive feature for a DENV vaccine. Further, the delivery of a "built-in" adjuvant is also attractive as it eliminates the need to conduct formulation studies with adjuvants.

13.2.1.2.2 DNA Vaccines

One of the most significant challenges for any DNA vaccine is the induction of a sufficient immune response; therefore, formulating DNA vaccines with adjuvant is often necessary. In addition to efforts to improve immunogenicity of DENV DNA vaccines with adjuvants, DENV DNA vaccine developers focused on modifying the E protein to direct its translocation to the endosomal/lysosomal compartment (Lu et al., 2003). Targeting the E protein to this compartment colocates the E protein and major histocompatibility complex (MHC) class II, which enhances MHC class II presentation of the E protein and production of neutralizing antibodies.

DENV DNA vaccines, consisting of the prM and E genes, were generated for DENV serotypes 1–3 and induced neutralizing antibodies to the respective DENV serotype in mice and NHPs but only provided marginal protection against homologous challenge in NHPs (Kochel et al., 2000; Raviprakash et al., 2000; Putnak et al., 2003; Lu et al., 2003). Later efficacy studies conducted in NHPs with modified preparations and delivery methods demonstrated that the induced neutralizing antibodies could prevent the development of viremia (Raviprakash et al., 2003). A Phase 1 clinical study was conducted to evaluate the safety and immunogenicity of the prototype, DENV DNA vaccine, D1ME[100] (Beckett et al., 2011). Two dosages of D1ME[100], administered using the needle-free Biojector® 2000, were evaluated for safety and immunogenicity in human volunteers. The safety profile of D1ME[100] was found to be acceptable; however, the immune responses stimulated were low. Studies are planned to further evaluate approaches to enhance immunogenicity.

A tetravalent DNA vaccine based on the EDIII domain was recently reported (Ramanathan et al., 2009). This DNA vaccine contains a gene with a single open reading frame with built-in cleavage sites for the expression of the EDIII gene from each of the four serotypes. The EDIII sequence for each serotype is based on consensus sequences for each serotype and is human codon optimized. Evaluation of immunogenicity was conducted in mice. The DNA was administered using an electroporation-enhanced transfection system that is currently being evaluated in Phase 1 clinical studies for other vaccines. Vaccination with the tetravalent chimeric EDIII expressing plasmid induced neutralizing antibody responses to all four serotypes. This study encourages future evaluation of the tetravalent DNA vaccine in a clinical setting; however, additional testing will be needed to support the transition to the clinic.

13.2.1.2.3 *Viruslike Particle*

VLP vaccine candidates for DENV have been developed based on the coexpression of prM and E for and are undergoing preclinical testing. These DENV VLPs utilize different expression systems including baculovirus (Kuwahara and Konishi, 2010), yeast (Qing et al., 2010), and mammalian cells (Zhang et al., 2011). Of these reports, the VLP generated by Zhang et al. (2011) is the only tetravalent VLP and was generated by coadministering four monovalent VLPs generated using plasmids optimized for prM/E expression. The VLPs were expressed in 293 T-cells. Following administration of the tetravalent VLP, recipient mice developed a specific immunoglobulin G (IgG) and neutralizing antibodies against all four serotypes. The level of neutralizing antibodies to each serotype was balanced. Interestingly, vaccination with the tetravalent VLP-induced IgG titers greater than those induced by each of DENV 1–4 virions and neutralizing antibodies was capable of blocking DENV 1–4 infection. Further, splenocytes derived from mice vaccinated with the tetravalent VLP produced significantly higher levels of IFN-γ, indicating induction of a cellular immune response. Together, the results of this study support further evaluation of the tetravalent VLP-based vaccine candidate.

13.2.1.3　Virus-Vectored Vaccines

Two virus-vectored DENV tetravalent vaccines have advanced to clinical trials. The ChimeriVax™ dengue tetravalent vaccine (ChimeriVax™-DTV) was generated using recombinant technology to construct four chimeric monovalent viruses comprising genes from the attenuated 17D yellow fever (YF) vaccine and the prM and E proteins of each DENV serotypes (Bray and Lai, 1991). The immunogenicity and efficacy of the ChimeriVax-DEN1–4 was evaluated in NHPs demonstrating induction of robust neutralizing antibody responses to all four serotypes and prevention of viremia following challenge (Guirakhoo et al., 2004). The results of the ChimeriVax™-DTV Phase 1 clinical trials demonstrated the vaccine is well tolerated and immunogenic for all four serotypes in human volunteers with ages ranging from 2 years to adult. Together these studies prompted the initiation of ChimeriVax™-DTV Phase 2 clinical studies (Guy et al., 2011a). Two Phase 2 clinical trials are currently ongoing for ChimeriVax™-DTV, one evaluating the immunogenicity of the vaccine in healthy subjects in 4 age cohorts (NCT00875524) and the second evaluating safety and efficacy in children (NCT00842530).

Two recent reports describe the use of adenovirus subtype 5 (Ad5) vectors for delivery of DENV proteins. A tetravalent DENV vaccine was developed that involves codelivery of two Ad5 bivalent constructs. Each construct contains the genes for expression of the DENV prM and E: construct one contains these genes for DENV1 and 2 (CAdVax-Den 12), and construct two contains these genes for DENV3 and 4 (CAdVax-Den 34) (Holman et al., 2007). Vaccination of mice with CAdVax-DenTV (combined) stimulated humoral and cellular immune responses to all four DENV serotypes after the second vaccination (Holman et al., 2007). Short- and long-term efficacy of CAdVax-DenTV was evaluated in NHPs at 4 and 24 weeks post-vaccination (Raviprakash et al., 2008). At both time points, post-challenge viremias were completely prevented in NHPs challenged with DENV1 and DENV3; however, NHPs were viremic following challenge with DENV2 and DENV4 at both time points, but the viremias were significantly reduced at the 24-week time point. The basis of the second Ad5-mediated DENV vaccine recently reported is an in-frame fusion protein of the EDIII-encoding sequences of the four DENV serotypes (Khanam et al., 2009). This approach circumvents the complexities associated with codelivery of monovalent or bivalent constructs. Vaccination of mice with the Ad5 expressing the EDIII fusion protein stimulates a tetravalent immune response involving both humoral- and cell-mediated responses. The results of both of these studies provide potential for future use of Ad5 as a vector for DENV vaccines. Interestingly, in both studies, antibodies to the Ad5 vector were developed that were expected to suppress the production of neutralizing antibodies following the second vaccination, but this was not observed.

13.3 FILOVIRUSES

Filovirus outbreaks occurred in Germany, Yugoslavia, and the United States following import of NHPs and occur in Africa, possibly as a result of transmission from bats to primates (Pourrut et al., 2007, 2009; Towner et al., 2007; Leroy et al., 2009). Vaccine efforts focus on several Ebola viruses identified as human pathogens: Ebola (Zaire), Bundibugyo, Sudan and Tai Forest (Ivory Coast), and on the Marburg viruses Marburg and Ravn. Vaccines for filoviruses will likely require the use of the Animal Rule for FDA approval, as outbreaks do not occur often. Challenges facing vaccine developers include the requirement for handling the viruses under BSL4 containment and the lack of a widely accepted correlate for immunity. The Animal Rule regulatory path establishes an effective human dosage based on vaccine-induced efficacy in animals and extrapolations from animal immune responses that correlate with efficacy. The human and animal vaccine-induced responses are compared and used to predict the likely benefit in humans. For filoviruses, a generally accepted correlate of immunity is not available. Protection has been conferred in mice using passive transfer of monoclonal antibodies or adoptive transfer of cytotoxic T-cells (Wilson et al., 2000; Wilson and Hart, 2001; Olinger et al., 2005) and in macaques using convalescent sera (Dye, 2012). In mice, post-challenge analysis indicated that both antibodies and cytotoxic T lymphocytes (CTLs) were present in survivors, regardless of what the animals were provided (Olinger et al., 2005). The challenge facing vaccine sponsors lies in the ability of the vaccine platform to safely induce both responses and to identify an assay that meets the Animal Rule requirements for predicting protection.

Mouse, guinea pig, and NHP models are available for Ebola viruses and Marburg viruses. Some animal models/viruses required the human virus to be adapted to the animals before pathogenicity was observed, which complicates interpretation of study results. Additional challenges facing sponsors include safety associated with replication-competent vectors, reduced immune responses due to preexisting immunity, and insufficient immune responses induced by recombinant or replication-deficient vectors. Significant progress has been made to demonstrate the immunogenicity of these approaches; however, some of the safety and efficacy issues remain to be resolved.

13.3.1 Nonreplicating Vaccines

13.3.1.1 DNA Vaccines

A Phase 1 clinical trial was conducted for the DNA-based vaccine (Martin et al., 2006). In the first study, placebo or three DNA plasmids encoding the nucleoprotein of Ebola virus and the glycoproteins from Ebola (formerly Zaire) and Sudan were administered by intramuscular (IM) injection at one of three dosages (2, 4, or 8 mg). Twenty-one subjects received the vaccine candidate, and six subjects received placebo. Subjects were 18–44 years of age and healthy. The vaccine was safe and immunogenic, inducing both antibody and T-cell responses (measured by ELISpot and intracellular cytokine staining).

13.3.2 Virus-Vectored Vaccines

13.3.2.1 Replication-Competent Virus-Vectored Vaccines

13.3.2.1.1 Recombinant Vesicular Stomatitis Virus (rVSV) Platform

Live-attenuated filovirus vaccines are not feasible due to safety concerns. However, replication-competent rVSV vaccines, in which the glycoprotein of the VSV is replaced with the filovirus glycoprotein, are under evaluation as potential vaccine candidates. One IM immunization with the rVSV for Marburg or Ebola protected cynomolgus macaques from challenge with the homologous filovirus (Jones et al., 2005), and cross-protection against other strains of Ebola or Marburg was observed after immunization with a single component (Daddario-DiCaprio et al., 2006; Falzarano et al., 2011a). Combined administration of rVSVs expressing Zaire, Sudan, and Marburg glycoproteins in a proof-of-concept study induced protective immunity to challenge with any of these viruses (Geisbert et al., 2009). Protection was also induced following vaccination by oral and intranasal routes (Qiu et al., 2009).

The rVSV platform induces high titers of antibodies to filoviruses in vaccinated animals as measured by ELISA and low levels of neutralizing antibodies in about half of the animals (Falzarano et al., 2011b). This platform provided efficacy from Marburg and Sudan Ebola viruses but less effectively against Zaire Ebola virus, in which onset of symptoms occurs sooner.

No published toxicology evaluations are available for the recombinant VSVs, making it difficult to assess their safety profile in humans. The rVSV filovirus construct expressing the Zaire Ebola virus glycoprotein was administered to one laboratory worker following a potential laboratory exposure. The subject developed fever, myalgia, and headache after the vaccination, but it is unclear whether the patient was actually infected (Falzarano et al., 2011b).

13.3.2.1.2 Recombinant Parainfluenza Virus 3 Platform

Modified parainfluenza 3, expressing the Zaire Ebola virus glycoprotein, was given intranasally and induced protective immunity in guinea pigs (Bukreyev et al., 2006, 2007), with significant survival induced in rhesus macaques following one dose and improved survival observed after two administrations (Bukreyev et al., 2007). The impact of preexisting immunity to the parainfluenza virus platform remains to be determined, although efforts to modify the vector to avoid this interference are underway (Bukreyev et al., 2009).

13.3.2.2 Replication-Deficient Virus-Vectored Vaccines

13.3.2.2.1 Venezuelan Equine Encephalitis Virus (VEEV) Replicon Particles (VRP)

This platform comprises a replication-deficient VEEV virus expressing filovirus antigens. The replicon has the nonstructural VEEV genes from VEE 3014 and a filovirus gene in place of the structural VEEV genes. It is packaged into particles using helper constructs that express the VEEV structural proteins (Pushko et al., 1997). Once administered, the particles infect cells and undergo a single round of replication during which the foreign gene product is made. VRP vaccination induces protective monoclonal antibodies to the glycoprotein and protective CTL responses to the glycoprotein and other Ebola virus proteins in mice (Wilson et al., 2000; Wilson and Hart, 2001; Olinger et al., 2005). VRP also protect guinea pigs and cynomolgus macaques from Marburg virus challenge (Hevey et al., 1998). This was the first platform to induce protection in NHPs, and the platform is being further evaluated in early advanced development efforts as a potential vaccine for filoviruses.

13.3.2.2.2 Adenovirus-Vectored Vaccines

Used for gene therapy trials in humans, the replication-defective adenovirus serotype 5 platform expresses the filovirus glycoprotein or nucleoprotein. Two adenovirus vectors were studied, one which encodes only a single filovirus gene per construct and a second (CAdVax) encodes more than one gene. The adenovirus platform induces protection in mice, guinea pigs, and NHPs (Sullivan et al., 2003, 2006; Wang et al., 2006a,b; Pratt et al., 2010) and boosted DNA-primed immune responses to filoviruses (Sullivan et al., 2000).

A Phase 1 clinical trial was conducted for the single-gene adenovirus platform (Ledgerwood et al., 2010). The trial coadministered two adenovirus constructs, one expressing the Ebola (Zaire) glycoprotein and the second expressing the Sudan glycoprotein. Thirty one subjects were given the adenovirus-vectored constructs, and eight subjects received placebo by IM injection. This platform was well tolerated in subjects and induced detectable B- and T-cell immune responses. Two dosages (2×10^{10} and 2×10^{9} particles) were evaluated, with the higher dosage more effectively inducing seroconversion to both antigens. Preexisting immunity to the adenovirus vector resulted in reduced antibody responses following vaccination.

13.3.2.2.3 VLP or Protein-Based Vaccines

Most of the published studies for filovirus VLPs use a platform comprising some viral proteins (the VP40 matrix protein, glycoprotein, and nucleoprotein) but lack the genetic material needed for replication. This represents a safety advantage over replication-competent approaches, and this platform also does not have concerns regarding preexisting immunity. VLPs for Ebola or Marburg viruses induced protective immunity in small animals and macaques (reviewed in Warfield and Aman, 2011). Vaccination induced antibody and T-cell-mediated immune responses in animals. Improved immune responses were observed in animals administered VLPs formulated with an adjuvant. The inclusion of an adjuvant may complicate the regulatory pathway for VLPs, as the successful adjuvant-containing studies used RIBI or QS21 (Warfield and Aman, 2011). Of note, the approved adjuvant alum was not tested with VLPs.

Replicons derived from Kunjin virus (a flavivirus that is a subtype of West Nile virus) expressing various forms of Ebola virus glycoprotein were transfected into a packaging cell line, which produced VLPs containing the replicons. Guinea pigs received intraperitoneal vaccination with VLPs on days 0 and 20 and were challenged on day 40. Replicons encoding a full-length (wild-type or mutated) glycoprotein protected a significant number of guinea pigs from death (Reynard et al., 2011).

Recently, an Ebola virus glycoprotein Fc fusion protein was tested in mice. Vaccination with the protein formulated in complete Freund's adjuvant induced both antibody and T-cell responses to the glycoprotein. Challenge virus was administered 2 weeks after the final vaccination, with survival observed in approximately 90% of the experimentally treated mice (Konduru et al., 2011).

13.4 ARENAVIRUSES

Twenty-two known viral species compose the *Arenaviridae* family and are classified phylogenetically into Old World and New World complexes (Charrel et al., 2008). The Old World complex viruses that cause human disease include Lassa virus (LASV), Lujo virus, and lymphocytic choriomeningitis virus (LCMV). LASV and Lujo virus can lead to infections that progress to hemorrhagic fevers, while LCMV does not. Junin virus (JUNV), Machupo virus (MACV), Guanarito virus (GTOV), Sabia virus (SABV), and Whitewater Arroyo virus (WWAV) comprise the New World complex. JUNV, MACV, GTOV, and SABV are the etiologic agents of hemorrhagic fever syndromes in South America, while WWAV has been identified as the etiologic agent of two fatalities in North America (Clegg, 2002).

Vaccine strategies for arenaviruses reported in the past 5 years have focused on reassortant technology, replication-competent virus vectors, VLP, and plasmid DNA. In addition to these approaches, recent efforts have been made to develop multivalent epitope-based vaccines that induce protective CD8+ T-cell responses specific to multiple arenaviruses. Epitopes were identified using a two-pronged approach involving the identification of epitopes reactive across a group of seven arenaviruses and identification of a pool of epitopes capable of stimulating cross-reactive CD8+ T-cell responses (Kotturi et al., 2009). Identified HLA-restricted, cross-reactive epitopes were subsequently evaluated in mice and found to provide cross-protection against heterologous arenavirus infection (Botten et al., 2010). These studies provide proof-of-concept data to support further evaluation of a multivalent vaccination strategy based on cross-reactive epitopes.

13.4.1 LASSA FEVER VIRUS

Annually, 100,000–300,000 infections and 5,000 deaths occur in western Africa due to LASV infection (Fisher-Hoch and McCormick, 2004). Symptoms of LASV infection include fever, headache, malaise, and sore throat. Hemorrhagic symptoms are observed in 15%–20% of patients (Fisher-Hoch and McCormick, 2004). Ribavirin is an effective therapy for LASV when administered within a week of infection; however, access to ribavirin in endemic regions is limited (McCormick et al., 1987). Licensure of an LASV vaccine faces many obstacles. For example, the high pathogenicity of LASV requires the virus be handled in BSL4 containment laboratories, which reduces the number of investigators with access to the virus for evaluation. The availability of relevant animal models also presents a challenge. Strain 13 guinea pigs are the most commonly used small animal model although mice are used in proof-of-concept studies. NHPs are considered a better model of human disease and are commonly used to evaluate vaccine efficacy; however, the cost, space require-ments, particularly in BSL4 containment facilities, and ethical concerns associated with usage of NHPs constrain their use. A great deal of genetic diversity has been reported among LASV strains, with up to 27% variation in the NP gene sequence (Bowen et al., 2000) raising the question of whether multiple LASV vaccines will need to be developed. Finally, many questions remain unanswered regarding the relationship between viral pathogenesis and the immune responses required for protection. The absence of reports of repeat LASV infections among survivors sug-gests the immunological responses generated to primary infection protect survivors and that upon subsequent exposure, an anamnestic response is sufficient to protect against disease. These reports suggest that production of a vaccine that generates the appropriate immune response has the poten-tial to be successful (Grant-Klein et al., 2011). Cell-mediated immune responses appear to play a significant role in protection against LASV, whereas the role of neutralizing antibodies in protec-tion has not been well defined (Baize et al., 2009; Grant-Klein et al., 2011). A growing number of studies indicate that both major LASV antigens, the glycoproteins (GP1 and GP2) and the NP, are important for induction of long-lasting cell-mediated immunity (McCormick and Fisher-Hoch, 2002; Fisher-Hoch and McCormick, 2004).

13.4.1.1   Live-Attenuated Virus Vaccines

Live-attenuated vaccines developed for LASV used reassortant technology. The ML29 vaccine can-didate was generated to encode the NP and glycoprotein complex (GPC) of LASV (Josiah SL/76/H) and the Z protein (zinc binding protein) and RNA polymerase of Mopeia virus (MOPV-AN201410). The immunogenicity, efficacy, and safety of ML29 reassortants were evaluated in marmosets following high- and low-dose vaccination (Lukashevich et al., 2008). Viral replication was detected in target tissues at levels directly related to the vaccine dosages. Notably, damage occurred to the blood–brain barrier following vaccination with high dosages. Immunization of marmosets with ML29 stimulated cell-mediated immune responses that were protective against a lethal challenge with the LASV Josiah strain. Although this study demonstrated the immunogenicity and efficacy of ML29, future ML29 studies will need to address safety concerns before this candidate is likely to advance. Further, while natural reassortants between arenaviruses have not been reported, the potential for natural reassortants arising between vaccine strains and virulent strains is possible and will also need to be addressed.

13.4.1.2   Nonreplicating Vaccines

13.4.1.2.1   VLPs

The first generation LASV VLP-based vaccine candidates express the major immunological deter-minants of LASV including the GPC, NP, and Z protein (Branco et al., 2010). Two VLPs were developed, one containing the Z protein, GPC, and NP and the other containing the Z protein and GPC. The Z protein was included as it is involved in formation and release of VLP from infected and transfected cells (Bausch et al., 2008). The LASV VLPs were generated by expression in the mammalian cell line, HEK-293 T/17 cells, due to its high transfection rates and capacity to support expression of recombinant proteins. The use of this expression system provided high yields of the LASV VLP. The immunogenicity of the VLPS was evaluated in mice using a three-dose vaccina-tion schedule. Following the first immunization, low levels of IgG were detected in a small percent-age of animals. Both the magnitude and frequency of IgG increased with each additional vaccine boost. A comparison of the two VLPs revealed the Z+GPC+NP VLP was more immunogenic than the VLP containing only Z+GPC demonstrating the advantage of including the NP protein. These proof-of-concept evaluations of immunogenicity in mice support additional evaluation of immunogenicity and efficacy in in vivo models, and such studies are currently ongoing.

13.4.1.3 Virus-Vectored Vaccines

13.4.1.3.1 Replication-Competent Virus-Vectored Vaccines

Recently, the YF 17D vaccine was pursued as a vector due to its superior safety profile and extensive testing in humans. Jiang et al. reported a 17D-based vaccine candidate comprised of two constructs: one construct contained the LASV-GP1 and the other contained the LASV-GP2 (Jiang et al., 2011). CBA/J mice were used to evaluate vaccine immunogenicity to assess the capacity of the vaccine to induce protective CD8+ T-cell responses that are known to play a prominent role in LASV protective immunity. Coimmunization of mice with the YF17D/LASV-GP1 and LASV-GP-2 candidates stimulated an antigen-specific CD8+ T-cell response. YF17D/LASV-GP1 and LASV-GP-2 were evaluated for protective efficacy in strain 13 guinea pigs. Five of six vaccinated guinea pigs survived lethal LASV challenge. However, all surviving guinea pigs experienced clinical signs associated with Lassa fever, and LASV was detected in tissues tested.

13.4.2 Junin Virus

The JUNV is the etiologic agent of Argentine hemorrhagic fever (AHF), causing 300–1000 cases of each year in endemic regions (Enria et al., 2008). Infection with JUNV typically manifests as a non-specific febrile illness that may progress to AHF and ultimately death. A live-attenuated vaccine, Candid 1, was developed by the U.S. Army and field tested in Argentina and is currently available for use in endemic regions to prevent AHF (Barrera Oro and McKee, 1991; Maiztegui et al., 1998). Currently, Candid 1 maintains an investigational new drug status with the FDA. Licensure of Candid 1 has not been obtainable due to its high reactogenicity in humans, safety profile in pregnant women, and lack of understanding of the exact basis of attenuation (Seregin et al., 2010). A protective immune response to JUNV has been demonstrated in patients and animals to be antibody mediated with the GCP as the primary protective antigen.

13.4.2.1 Virus-Vectored Vaccines

13.4.2.1.1 Replication-Deficient Virus-Vectored Vaccines

A replicon system based on the live-attenuated VEEV virus vaccine, TC-83, expressing the GPC was engineered to evaluate the protective capacity of neutralizing antibodies to GPC in animals following JUNV challenge (Seregin et al., 2010). The nucleotide sequence of the gene encoding the glycoprotein precursor of Candid 1 JUNV was codon optimized (GPCopt) to enhance the expression in mammalian cells. The immunogenicity of the TC-83rep/GCPopt replicon was assessed in guinea pigs. Following two doses of the vaccine, all vaccinated guinea pigs developed detectable neutralizing antibody titers to JUNV. The protective efficacy of the candidate vaccine was evaluated in guinea pigs on day 28 following the second vaccination. Animals were challenged with a dose of JUNV Romero that was lethal in 100% of control animals. All guinea pigs vaccinated were protected from death. The vaccinated animals experienced a mild decrease in body weight following challenge but did not experience any other overt signs of disease.

13.5 BUNYAVIRUSES

13.5.1 Rift Valley Fever Virus

Rift Valley fever (RVF) is a mosquito-borne disease that affects both humans and livestock. RVFV traditionally caused recurrent outbreaks affecting humans in sub-Saharan Africa but continues to spread throughout Africa. Infection of humans with RVFV is associated with benign fever that can lead to retinal vasculitis, encephalitis, neurologic deficits, hepatic necrosis, or fatal hemorrhagic fever. Historically less than 2% of infected individuals develop fatal hemorrhagic fever; however, recent outbreaks have been associated with a 20%–30% case fatality rate (LaBeaud et al., 2008). In addition to being categorized as a high-priority pathogen by NIAID and the CDC, RVFV is also considered a select agent by the U.S. Department of Agriculture because of its large impact on animal health and its potential for rapid spread. The use of RVFV vaccines for agricultural use presents a challenge for vaccine developers due to the requirement to differentiate vaccinated from naturally exposed animals.

A formalin-inactivated RVFV vaccine, TSI-GSD-200, is currently administered to personnel at risk of exposure to RVFV (Pittman et al., 1999). Approximately 90% of the vaccinees initially responded to the immunization with neutralization antibody titers considered to be protective, whereas the remaining 10% of vaccinees fail to achieve a protective titer (Pittman et al., 1999). Currently, a regular booster is the administration to maintain maximum protection after the series of primary immunization. TSI-GSD 200 is not approved for general use by the FDA, which may be attributed to the hypersensitivity reactions in vaccinees and the suboptimal immunogenic effect (Kark et al., 1982).

13.5.1.1 Live-Attenuated Virus Vaccines

A live-attenuated vaccine, MP-12, was developed for both human and veterinary use by performing the serial passage of the ZH548 strain in the presence of the chemical mutagen, 5-fluorouracil (5FU) (Caplen et al., 1985). MP-12 is a temperature-sensitive mutant (Saluzzo and Smith, 1990) that carries mutations in all three RNA segments. An MP-12 Phase 1 clinical trial was conducted to evaluate its potential for use as a human vaccine for RVFV. MP-12 was administered into more than 100 human volunteers and was shown to be safe and immunogenic (reviewed in Ikegami and Makino, 2009). The safety, immunogenicity, and genetic stability of MP-12 were recently evaluated in a Phase II clinical evaluation in human volunteers (clinicaltrials.gov NCT00415051). Recently conducted studies in NHPs demonstrated the ability of vaccination with MP-12, using various routes of vaccination-induced protective immune responses that provided protection against lethal RVFV pulmonary challenges, thereby providing further support for the continued evaluation of this vaccine in the clinic (Morrill and Peters, 2011a,b).

The risk of reversion to a virulent phenotype is always a concern associated with live-attenuated vaccines prepared by serial passage in cell culture or those that rely on point mutations for attenuation. Using a reverse genetic approach, Bird and colleagues (Bird et al., 2008) created two rationally designed RVFV vaccine candidates in which (1) the entire NSs gene was deleted or (2) the NSs and NSm genes were deleted providing a live-attenuated vaccine with greater resistance to reversion. The NSs and NSm genes were specifically targeted for deletion because of their involvement in host pathogenesis. The immunogenicity and efficacy of the two vaccine candidates were evaluated in a rat RVFV model, and following one dose of vaccine, all animals developed high neutralizing antibody titers and were protected against challenge dosages 500 times the lethal dose 50 (LD_{50}). In many RVFV vaccines that are developed for dual use, vaccination of humans and livestock, it is important that immune responses generated by RVFV vaccines can be differentiated from immune response induced by natural exposure when used in the agriculture setting. The deletion of NSs allows for this distinction as vaccinated animals do not develop anti-NSs antibodies, whereas animals exposed to wild-type virus do. The codevelopment of the NSs-deleted vaccine candidates along with the assays for detection of anti-NSs antibodies will provide both a vaccine and screening tool.

13.5.1.2 Nonreplicating Vaccines

13.5.1.2.1 VLP

The successful development of VLP for other virus members of the *Bunyaviridae* family including Uukuniemi virus and Hantaan virus (HTNV) suggested that development of VLP for RVFV may result in similar successes. Initial efforts to generate RVFV VLP used a baculovirus expression system that coexpressed the glycoproteins Gn and Gc and the nucleoprotein. This system was capable of efficiently generating RVFV VLPs and had the potential to produce large amounts of VLP. Subsequent efforts involved transfection of mammalian cells with plasmid constructs expressing N (nucleoprotein), L (encodes the RNA-dependent RNA polymerase), and the viral glycoproteins, together with the reporter minireplicon construct expressing Renilla luciferase (Habjan et al., 2009). Immunogenicity and efficacy studies in mice revealed the RVFV VLP is highly immunogenic when high dosages are administered and a high percentage of vaccinated mice are protected following challenge (Naslund et al., 2009).

In 2010, Mandell and associates (Mandell et al., 2010) reported the generation of a chimeric RVF VLP (RVF chimVLP). The RVF chimVLP was generated by transient transfection of HEK-293-gag cells with plasmids encoding the RVFV Gn and Gn with or without the N protein. The HEK293-gag cells are HEK293 that constitutively express Moloney murine leukemia virus (MoMLV) gag protein. The MoMLV gag protein was included as it was previously shown that inclusion of retroviral gag can increase uniformity, and quantity (Rovinski et al., 1992; Szecsi et al., 2006), and stability (Hammonds et al., 2003) of VLPs. Further modifications to optimize codon usage of the RVFV glycoprotein gene were made to the RVF chimVLP to enhance expression in mammalian cells. These VLPs with and without N were shown to be immunogenic in mice based on the induction of neutralizing antibodies and antigen-specific secretion of immune-related cytokines from splenocytes derived from vaccinated mice. The protective efficacy of the RVF VLP was evaluated in mice and rats, showing partial protection in mice and 100% protection in rats. The generation of the RVF VLP in the presence and absence of the N protein indicates the N protein is not necessary for formation of the VLP. This finding contradicts the report by Habjan and associates (Habjan et al., 2009); however, Mandell and colleagues (Mandell et al., 2010) suggest the difference is attributed to differences in the expression systems used in the studies.

Generation of RVF VLPs by de Boer and colleagues (de Boer et al., 2010) provided additional evidence that the N protein was not necessary for production of VLP. In this study, RVF VLPs comprising only the RVF Gn and Gc proteins using a *Drosophila* insect protein expression system were generated. One of the objectives of this study was to evaluate the enhancement of neutralizing antibodies in vivo following vaccination with RVF VLPs alone or in the presence of Stimune. Vaccinated mice developed a neutralizing antibody response following one dose of vaccine that was enhanced following a second dose. Following the first and second dose, mice that received RVF VLPs plus Stimune generated antibody response significantly higher than the group of mice receiving RVF VLPs alone. However, when vaccinated mice were challenged 3 weeks following the second vaccination, all mice, regardless of vaccine formulation received, survived challenge and no overt signs of disease were reported. The RVF VLPs reported to date support further development of VLPs for RVF.

13.5.1.2.2 *DNA Vaccines*

In 2009, Lagerqvist and colleagues generated a series of RVFV DNA vaccine candidates that encoded either the N protein, the Gn/Gc proteins (combined), the Gn protein alone, or the Gc protein alone (Lagerqvist et al., 2009). Vaccination of mice with the construct encoding the N protein induced a humoral and lymphocyte proliferative response, whereas induction of neutralizing antibodies was only observed following vaccination with constructs encoding the glycoproteins. Although vaccination reduced the clinical signs of disease following challenge, vaccination with these constructs did not prevent death in all animals. Although this study demonstrates the potential of DNA vaccines for RVFV, further development is needed.

The use of the molecular adjuvant C3d has been shown to significantly enhance antibody responses against DNA vaccine-delivered antigens (Ross et al., 2000; Green et al., 2003). Bhardwaj and associate*s* investigated the use of C3d with RVFV DNA vaccines by fusing the C3d to the RVFV Gn glycoprotein (Gn-C3d) (Bhardwaj et al., 2010). Vaccinating mice with Gn-C3d stimulated the production of antibodies that could neutralize wild-type RVFV, and following challenge, all Gn-C3d-vaccinated mice survived challenge with no overt signs of disease, whereas 60% of mice vaccinated with Gn alone succumbed to challenge.

13.5.1.3 Virus-Vectored Vaccines

13.5.1.3.1 Replication-Competent Virus-Vectored Vaccines

Newcastle disease virus (NDV), a paramyxovirus, was evaluated as an in vivo expression system for the RVFV Gn and Gc. The utility of NDV-vectored vaccines in mammals has an inherent safety feature in that although NDV can infect mammalian cells, mammalian species are not a natural host, and therefore the spread of the virus is limited (Kortekaas et al., 2010). A prime-boost vaccination strategy was employed to evaluate the immunogenicity and efficacy of NDV-GnGc in mice. All vaccinated mice survived a lethal RVFV challenge 3 weeks after the final vaccination with no overt signs of disease reported. Based on these promising results in mice, the NDV-GnGc vaccine was evaluated in lambs and found to induce a neutralizing antibody response after the first vaccination that was enhanced after the second vaccination.

13.5.1.3.2 Replication-Deficient Virus-Vectored Vaccines

Sindbis virus (SINV) replicons have been investigated for development of RVFV vaccines with efforts focusing on expression of RVFV Gn, Gc, and M proteins. These replicons, when evaluated in mice, induced immune responses that were protective against both systemic and mucosal RVFV challenge. The SINV replicons were then evaluated in sheep and found to induce neutralizing antibody titers (Heise et al., 2009).

The use of adenovirus vectors has been explored as an alternative vaccine vector for RVFV. Holman and associates generated an adenovirus expressing the RVFV glycoproteins (CAdVax-RVF), and upon vaccination of mice, strong immune responses were induced that provided protection against subsequent challenge with lethal dosages of wild-type RVFV (Heise et al., 2009). With adenovirus vectors, the issue of preexisting immunity to the vector is always a concern; however, the presence of preexisting immunity to the vector did not interfere with the induction of a strong immune response to the RVFV glycoproteins.

13.5.2 HANTAVIRUS

Hemorrhagic fever with renal syndrome (HFRS), characterized by fever, hemorrhage, and nephritis, can be caused by one of four hantaviruses, HTNV, Seoul virus (SEOV), Puumala virus (PUUV), and Dobrava virus (DOBV). Approximately 100,000–150,000 cases of HFRS are reported annually worldwide with the majority of HFRS occurring in Asia (Lee, 1989).

Development of neutralizing antibodies to Gn and Gc is thought to play a major role in protective immunity (Schmaljohn, 1990; Hooper et al., 2001; Klingstrom et al., 2008); however, cytotoxic T-cells have been identified from convalescent human samples and may play a role in protective immunity and viral pathogenesis (Van Epps et al., 2002; Kilpatrick et al., 2004; Tuuminen et al., 2007). A formalin-inactivated, suckling mouse brain–derived vaccine, Hantavax, has been marketed for use in Korea. Although it provides good protective immunity in vaccinated people, the duration of protective antibody titers is less than optimal. Research efforts to develop highly effective vaccines that induce long-term protective immunity are ongoing. Unfortunately, progress in vaccine development is hindered by the lack of relevant animal models for assessment of vaccine efficacy.

13.5.2.1 Nonreplicating Vaccines

13.5.2.1.1 DNA Vaccines

A limited number of DNA vaccines for HTNV have been reported. Most recently, an approach involving administration of two DNA plasmids, alone or in combination, expressing the M segment of HTNV or PUUV was reported (Spik et al., 2008). The authors found that when administered separately neutralizing antibodies were generated against both HTNV and PUUV; however, when administered together, neutralizing antibodies to only PUUV were detected suggesting interference at the protein–protein level. Interestingly, this interference was not detected when the HTNV and PUUV were administered at the same time but different sites. These results suggest coadministration of Hantavirus DNA vaccines in one site may not be feasible and that further investigations are needed to understand this phenomenon.

13.5.2.1.2 VLP

Similar to other viruses, hantaviruses possess the intrinsic ability to assemble into VLPs independent of the viral RNA. Li C et al. generate recombinant Hantaan viruslike particle (HTN-VLP) in CHO cells and evaluated the immunogenicity in mice following IM or subcutaneous (SC) vaccination (Li et al., 2010). The HTN-VLP were found to induce specific humoral and cellular immune responses, and more importantly induced neutralizing antibody titers were comparable to titers induced by the inactivated bivalent vaccine. This study provides proof-of-concept for Hantavirus VLP-based vaccines and warrants further investigation.

13.6 CONCLUSION

In general, a number of traditional and novel vaccine approaches are being evaluated for efficacy against the VHFs. These are primarily in the preclinical phase of development although a few vaccines are in clinical testing. Significant but not insurmountable obstacles will challenge their continued development. These include the demonstration of safety and appropriate immunogenicity in humans. Vaccines pursuing licensure using the Animal Rule will also need appropriate animal models and correlates for establishing an effective dosage in humans.

REFERENCES

Allison, S. L., Schalich, J., Stiasny, K., Mandl, C. W., Heinz, F. X., 2001. Mutational evidence for an internal fusion peptide in flavivirus envelope protein E. *J Virol*. 75, 4268–4275.

Azevedo, A. S., Yamamura, A. M., Freire, M. S., Trindade, G. F., Bonaldo, M., Galler, R., Alves, A. M., 2011. DNA vaccines against dengue virus type 2 based on truncate envelope protein or its domain III. *PLoS One*. 6, e20528.

Baize, S., Marianneau, P., Loth, P., Reynard, S., Journeaux, A., Chevallier, M., Tordo, N., Deubel, V., Contamin, H., 2009. Early and strong immune responses are associated with control of viral replication and recovery in Lassa virus-infected cynomolgus monkeys. *J Virol*. 83, 5890–5903.

Barrera Oro, J. G., McKee, K. T. Jr., 1991. Toward a vaccine against Argentine hemorrhagic fever. *Bull Pan Am Health Organ*. 25, 118–126.

Bausch, D. G., Sprecher, A. G., Jeffs, B., Boumandouki, P., 2008. Treatment of Marburg and Ebola hemorrhagic fevers: A strategy for testing new drugs and vaccines under outbreak conditions. *Antiviral Res*. 78, 150–161.

Beckett, C. G., Tjaden, J., Burgess, T., Danko, J. R., Tamminga, C., Simmons, M., Wu, S. J., Sun, P., Kochel, T., Raviprakash, K., Hayes, C. G., Porter, K. R., 2011. Evaluation of a prototype dengue-1 DNA vaccine in a Phase 1 clinical trial. *Vaccine*. 29, 960–968.

Bente, D. A., Melkus, M. W., Garcia, J. V., Rico-Hesse, R., 2005. Dengue fever in humanized NOD/SCID mice. *J Virol*. 79, 13797–13799.

Bernardo, L., Izquierdo, A., Alvarez, M., Rosario, D., Prado, I., Lopez, C., Martinez, R. et al., 2008. Immunogenicity and protective efficacy of a recombinant fusion protein containing the domain III of the dengue 1 envelope protein in non-human primates. *Antiviral Res*. 80, 194–199.

Bhardwaj, N., Heise, M. T., Ross, T. M., 2010. Vaccination with DNA plasmids expressing Gn coupled to C3d or alphavirus replicons expressing gn protects mice against Rift Valley fever virus. *PLoS Negl Trop Dis*. 4, e725.

Bird, B. H., Albarino, C. G., Hartman, A. L., Erickson, B. R., Ksiazek, T. G., Nichol, S. T., 2008. Rift valley fever virus lacking the NSs and NSm genes is highly attenuated, confers protective immunity from virulent virus challenge, and allows for differential identification of infected and vaccinated animals. *J Virol.* 82, 2681–2691.

de Boer, S. M., Kortekaas, J., Antonis, A. F., Kant, J., van Oploo, J. L., Rottier, P. J., Moormann, R. J., Bosch, B. J., 2010. Rift Valley fever virus subunit vaccines confer complete protection against a lethal virus challenge. *Vaccine.* 28, 2330–2339.

Botten, J., Whitton, J. L., Barrowman, P., Sidney, J., Whitmire, J. K., Alexander, J., Kotturi, M. F., Sette, A., Buchmeier, M. J., 2010. A multivalent vaccination strategy for the prevention of Old World arenavirus infection in humans. *J Virol.* 84, 9947–9956.

Bowen, M. D., Rollin, P. E., Ksiazek, T. G., Hustad, H. L., Bausch, D. G., Demby, A. H., Bajani, M. D., Peters, C. J., Nichol, S. T., 2000. Genetic diversity among Lassa virus strains. *J Virol.* 74, 6992–7004.

Branco, L. M., Grove, J. N., Geske, F. J., Boisen, M. L., Muncy, I. J., Magliato, S. A., Henderson, L. A. et al., 2010. Lassa virus-like particles displaying all major immunological determinants as a vaccine candidate for Lassa hemorrhagic fever. *Virol J.* 7, e279.

Brandler, S., Ruffie, C., Najburg, V., Frenkiel, M. P., Bedouelle, H., Després, P., Tangy, F., 2010. Pediatric measles vaccine expressing a dengue tetravalent antigen elicits neutralizing antibodies against all four dengue viruses. *Vaccine.* 41, 6730–6739.

Bray, M., Lai, C. J., 1991. Construction of intertypic chimeric dengue viruses by substitution of structural protein genes. *Proc Natl Acad Sci USA.* 88, 10342–10346.

Bukreyev, A., Marzi, A., Feldmann, F., Zhang, L., Yang, L., Ward, J. M., Dorward, D. W., Pickles, R. J., Murphy, B. R., Feldmann, H., Collins, P. L., 2009. Chimeric human parainfluenza virus bearing the Ebola virus glycoprotein as the sole surface protein is immunogenic and highly protective against Ebola virus challenge. *Virology.* 383, 348–361.

Bukreyev, A., Rollin, P. E., Tate, M. K., Yang, L., Zaki, S. R., Shieh, W. J., Murphy, B. R., Collins, P. L., Sanchez, A., 2007. Successful topical respiratory tract immunization of primates against Ebola virus. *J Virol.* 81, 6379–6388.

Bukreyev, A., Yang, L., Zaki, S. R., Shieh, W. J., Rollin, P. E., Murphy, B. R., Collins, P. L., Sanchez, A., 2006. A single intranasal inoculation with a paramyxovirus-vectored vaccine protects guinea pigs against a lethal-dose Ebola virus challenge. *J Virol.* 80, 2267–2279.

Caplen, H., Peters, C. J., Bishop, D. H., 1985. Mutagen-directed attenuation of Rift Valley fever virus as a method for vaccine development. *J Gen Virol.* 66(Pt 10), 2271–2277.

Cassetti, M. C., Durbin, A., Harris, E., Rico-Hesse, R., Roehrig, J., Rothman, A., Whitehead, S., Natarajan, R., Laughlin, C., 2010. Report of an NIAID workshop on dengue animal models. *Vaccine.* 28, 4229–4234.

Charrel, R. N., de, L., X, Emonet, S., 2008. Phylogeny of the genus Arenavirus. *Curr Opin Microbiol.* 11, 362–368.

Chen, H. W., Liu, S. J., Liu, H. H., Kwok, Y., Lin, C. L., Lin, L. H., Chen, M. Y. et al., 2009. A novel technology for the production of a heterologous lipoprotein immunogen in high yield has implications for the field of vaccine design. *Vaccine.* 27, 1400–1409.

Chiang, C. Y., Liu, S. J., Tsai, J. P., Li, Y. S., Chen, M. Y., Liu, H. H., Chong, P., Leng, C. H., Chen, H. W., 2011. A novel single-dose dengue subunit vaccine induces memory immune responses. *PLoS One.* 6, e23319.

Clegg, J. C., 2002. Molecular phylogeny of the arenaviruses. *Curr Top Microbiol Immunol.* 262, 1–24.

Clements, D. E., Coller, B. A., Lieberman, M. M., Ogata, S., Wang, G., Harada, K. E., Putnak, J. R. et al., 2010. Development of a recombinant tetravalent dengue virus vaccine: Immunogenicity and efficacy studies in mice and monkeys. *Vaccine.* 28, 2705–2715.

Coller, B. A., Clements, D. E., Bett, A. J., Sagar, S. L., Ter Meulen, J. H., 2011. The development of recombinant subunit envelope-based vaccines to protect against dengue virus induced disease. *Vaccine.* 29, 7267–7275.

Crill, W. D., Hughes, H. R., Delorey, M. J., Chang, G. J., 2009. Humoral immune responses of dengue fever patients using epitope-specific serotype-2 virus-like particle antigens. *PLoS One.* 4, e4991.

Daddario-DiCaprio, K. M., Geisbert, T. W., Geisbert, J. B., Stroher, U., Hensley, L. E., Grolla, A., Fritz, E. A., Feldmann, F., Feldmann, H., Jones, S. M., 2006. Cross-protection against Marburg virus strains by using a live, attenuated recombinant vaccine. *J Virol.* 80, 9659–9666.

Danko, J. R., Beckett, C. G., Porter, K. R., 2011. Development of dengue DNA vaccines. *Vaccine.* 29, 7261–7266.

Durbin, A. P., Kirkpatrick, B. D., Pierce, K. K., Schmidt, A. C., Whitehead, S. S., 2011a. Development and clinical evaluation of multiple investigational monovalent DENV vaccines to identify components for inclusion in a live attenuated tetravalent DENV vaccine. *Vaccine.* 29, 7242–7250.

Durbin, A. P., Whitehead, S. S., Shaffer, D., Elwood, D., Wanionek, K., Thumar, B., Blaney, J. E., Murphy, B. R., Schmidt, A. C., 2011b. A single dose of the DENV-1 candidate vaccine rDEN1Delta30 is strongly immunogenic and induces resistance to a second dose in a randomized trial. *PLoS Negl Trop Dis.* 5, e1267.

Dye, J. M., Herbert, A. S., Kuehne, A. I., Barth, J. F., Muhammad, M. A., Zak, S. E., Ortiz, R. A., Prugar, L. I., Pratt, W. D., 2012. Post-exposure transfer of immunoglobulin G from convalescent survivors protects rhesus macaques from Marburg virus. *Proc Natl Acad Sci USA.* 109, 5034–5039.

Eckels, K. H., Dubois, D. R., Putnak, R., Vaughn, D. W., Innis, B. L., Henchal, E. A., Hoke, C. H., Jr., 2003. Modification of dengue virus strains by passage in primary dog kidney cells: Preparation of candidate vaccines and immunization of monkeys. *Am J Trop Med Hyg.* 69, 12–16.

Enria, D. A., Briggiler, A. M., Sanchez, Z., 2008. Treatment of Argentine hemorrhagic fever. *Antiviral Res.* 78, 132–139.

Etemad, B., Batra, G., Raut, R., Dahiya, S., Khanam, S., Swaminathan, S., Khanna, N., 2008. An envelope domain III-based chimeric antigen produced in Pichia pastoris elicits neutralizing antibodies against all four dengue virus serotypes. *Am J Trop Med Hyg.* 79, 353–363.

Falzarano, D., Feldmann, F., Grolla, A., Leung, A., Ebihara, H., Strong, J. E., Marzi, A. et al., 2011a. Single immunization with a monovalent vesicular stomatitis virus-based vaccine protects nonhuman primates against heterologous challenge with Bundibugyo ebolavirus. *J Infect Dis.* 204(Suppl 3), S1082–S1089.

Falzarano, D., Geisbert, T. W., Feldmann, H., 2011b. Progress in filovirus vaccine development: Evaluating the potential for clinical use. *Expert Rev Vaccines.* 10, 63–77.

Fisher-Hoch, S. P., McCormick, J. B., 2004. Lassa fever vaccine. *Expert Rev Vaccines.* 3, 189–197.

Geisbert, T. W., Geisbert, J. B., Leung, A., Daddario-DiCaprio, K. M., Hensley, L. E., Grolla, A., Feldmann, H., 2009. Single-injection vaccine protects nonhuman primates against infection with Marburg virus and three species of Ebola virus. *J Virol.* 83, 7296–7304.

Grant-Klein, R. J., Altamura, L. A., Schmaljohn, C. S., 2011. Progress in recombinant DNA-derived vaccines for Lassa virus and filoviruses. *Virus Res.* 162, 148–161. doi:10.1016/j.virusres.2011.09.005.

Green, T. D., Montefiori, D. C., Ross, T. M., 2003. Enhancement of antibodies to the human immunodeficiency virus type 1 envelope by using the molecular adjuvant C3d. *J Virol.* 77, 2046–2055.

Guirakhoo, F., Pugachev, K., Zhang, Z., Myers, G., Levenbook, I., Draper, K., Lang, J. et al., 2004. Safety and efficacy of chimeric yellow fever-dengue virus tetravalent vaccine formulations in nonhuman primates. *J Virol.* 78, 4761–4775.

Guy, B., Barrere, B., Malinowski, C., Saville, M., Teyssou, R., Lang, J., 2011a. From research to phase III: Preclinical, industrial and clinical development of the Sanofi Pasteur tetravalent dengue vaccine. *Vaccine.* 29, 7229–7241.

Guy, B., Barrere, B., Malinowski, C., Saville, M., Teyssou, R., Lang, J., 2011b. From research to phase III: Preclinical, industrial and clinical development of the Sanofi Pasteur tetravalent dengue vaccine. *Vaccine.* 29, 7229–7241.

Guzman, A., Isturiz, R. E., 2010. Update on the global spread of dengue. *Int J Antimicrob Agents.* 36(Suppl 1), S40–S42.

Habjan, M., Penski, N., Wagner, V., Spiegel, M., Overby, A. K., Kochs, G., Huiskonen, J. T., Weber, F., 2009. Efficient production of Rift Valley fever virus-like particles: The antiviral protein MxA can inhibit primary transcription of bunyaviruses. *Virology.* 385, 400–408.

Hammonds, J., Chen, X., Ding, L., Fouts, T., De Vico, A., zur Megede, J., Barnett, S., Spearman, P., 2003. Gp120 stability on HIV-1 virions and Gag-Env pseudovirions is enhanced by an uncleaved Gag core. *Virology.* 314, 636–649.

Heise, M. T., Whitmore, A., Thompson, J., Parsons, M., Grobbelaar, A. A., Kemp, A., Paweska, J. T. et al., 2009. An alphavirus replicon-derived candidate vaccine against Rift Valley fever virus. *Epidemiol Infect.* 137, 1309–1318.

Hevey, M., Negley, D., Pushko, P., Smith, J., Schmaljohn, A., 1998. Marburg virus vaccines based upon alphavirus replicons protect guinea pigs and nonhuman primates. *Virology.* 251, 28–37.

Holman, D. H., Wang, D., Raviprakash, K., Raja, N. U., Luo, M., Zhang, J., Porter, K. R., Dong, J. Y., 2007. Two complex, adenovirus-based vaccines that together induce immune responses to all four dengue virus serotypes. *Clin Vaccine Immunol.* 14, 182–189.

Hombach, J., Cardosa, M. J., Sabchareon, A., Vaughn, D. W., Barrett, A. D., 2007. Scientific consultation on immunological correlates of protection induced by dengue vaccines report from a meeting held at the World Health Organization November 17–18, 2005. *Vaccine.* 25, 4130–4139.

Hooper, J. W., Larsen, T., Custer, D. M., Schmaljohn, C. S., 2001. A lethal disease model for hantavirus pulmonary syndrome. *Virology.* 289, 6–14.

Ikegami, T., Makino, S., 2009. Rift valley fever vaccines. *Vaccine.* 27(Suppl 4), D69–D72.

Jiang, X., Dalebout, T. J., Bredenbeek, P. J., Carrion, R., Jr., Brasky, K., Patterson, J., Goicochea, M., Bryant, J., Salvato, M. S., Lukashevich, I. S., 2011. Yellow fever 17D-vectored vaccines expressing Lassa virus GP1 and GP2 glycoproteins provide protection against fatal disease in guinea pigs. *Vaccine*. 29, 1248–1257.

Johnson, A. J., Roehrig, J. T., 1999. New mouse model for dengue virus vaccine testing. *J Virol*. 73, 783–786.

Jones, S. M., Feldmann, H., Stroher, U., Geisbert, J. B., Fernando, L., Grolla, A., Klenk, H. D. et al., 2005. Live attenuated recombinant vaccine protects nonhuman primates against Ebola and Marburg viruses. *Nat Med*. 11, 786–790.

Kark, J. D., Aynor, Y., Peters, C. J., 1982. A rift Valley fever vaccine trial. I. Side effects and serologic response over a six-month follow-up. *Am J Epidemiol*. 116, 808–820.

Kelly, E. P., Greene, J. J., King, A. D., Innis, B. L., 2000. Purified dengue 2 virus envelope glycoprotein aggregates produced by baculovirus are immunogenic in mice. *Vaccine*. 18, 2549–2559.

Khanam, S., Pilankatta, R., Khanna, N., Swaminathan, S., 2009. An adenovirus type 5 (AdV5) vector encoding an envelope domain III-based tetravalent antigen elicits immune responses against all four dengue viruses in the presence of prior AdV5 immunity. *Vaccine*. 27, 6011–6021.

Kilpatrick, E. D., Terajima, M., Koster, F. T., Catalina, M. D., Cruz, J., Ennis, F. A., 2004. Role of specific CD8+ T cells in the severity of a fulminant zoonotic viral hemorrhagic fever, hantavirus pulmonary syndrome. *J Immunol*. 172, 3297–3304.

Klingstrom, J., Stoltz, M., Hardestam, J., Ahlm, C., Lundkvist, A., 2008. Passive immunization protects cynomolgus macaques against Puumala hantavirus challenge. *Antivir Ther*. 13, 125–133.

Kochel, T. J., Raviprakash, K., Hayes, C. G., Watts, D. M., Russell, K. L., Gozalo, A. S., Phillips, I. A., Ewing, D. F., Murphy, G. S., Porter, K. R., 2000. A dengue virus serotype-1 DNA vaccine induces virus neutralizing antibodies and provides protection from viral challenge in Aotus monkeys. *Vaccine*. 18, 3166–3173.

Konduru, K., Bradfute, S. B., Jacques, J., Manangeeswaran, M., Nakamura, S., Morshed, S., Wood, S. C., Bavari, S., Kaplan, G. G., 2011. Ebola virus glycoprotein Fc fusion protein confers protection against lethal challenge in vaccinated mice. *Vaccine*. 29, 2968–2977.

Kortekaas, J., de Boer, S. M., Kant, J., Vloet, R. P., Antonis, A. F., Moormann, R. J., 2010. Rift Valley fever virus immunity provided by a paramyxovirus vaccine vector. *Vaccine*. 28, 4394–4401.

Kotturi, M. F., Botten, J., Sidney, J., Bui, H. H., Giancola, L., Maybeno, M., Babin, J., Oseroff, C., Pasquetto, V., Greenbaum, J. A., Peters, B., Ting, J., Do, D., Vang, L., Alexander, J., Grey, H., Buchmeier, M. J., Sette, A., 2009. A multivalent and cross-protective vaccine strategy against arenaviruses associated with human disease. *PLoS Pathog*. 5, e1000695.

Kuruvilla, J. G., Troyer, R. M., Devi, S., Akkina, R., 2007. Dengue virus infection and immune response in humanized RAG2(-/-)gamma(c)(-/-) (RAG-hu) mice. *Virology*. 369, 143–152.

Kuwahara, M., Konishi, E., 2010. Evaluation of extracellular subviral particles of dengue virus type 2 and Japanese encephalitis virus produced by *Spodoptera frugiperda* cells for use as vaccine and diagnostic antigens. *Clin Vaccine Immunol*. 17, 1560–1566.

LaBeaud, A. D., Muchiri, E. M., Ndzovu, M., Mwanje, M. T., Muiruri, S., Peters, C. J., King, C. H., 2008. Interepidemic Rift Valley fever virus seropositivity, northeastern Kenya. *Emerg Infect Dis*. 14, 1240–1246.

Lagerqvist, N., Naslund, J., Lundkvist, A., Bouloy, M., Ahlm, C., Bucht, G., 2009. Characterisation of immune responses and protective efficacy in mice after immunisation with Rift Valley Fever virus cDNA constructs. *Virol J*. 6, e6.

Ledgerwood, J. E., Costner, P., Desai, N., Holman, L., Enama, M. E., Yamshchikov, G., Mulangu, S. et al., 2010. A replication defective recombinant Ad5 vaccine expressing Ebola virus GP is safe and immunogenic in healthy adults. *Vaccine*. 29, 304–313.

Lee, H. W., 1989. Hemorrhagic fever with renal syndrome in Korea. *Rev Infect Dis*. 11(Suppl 4), S864–S876.

Leng, C. H., Liu, S. J., Tsai, J. P., Li, Y. S., Chen, M. Y., Liu, H. H., Lien, S. P., Yueh, A., Hsiao, K. N., Lai, L. W., Liu, F. C., Chong, P., Chen, H. W., 2009. A novel dengue vaccine candidate that induces cross-neutralizing antibodies and memory immunity. *Microbes Infect*. 11, 288–295.

Leroy, E. M., Epelboin, A., Mondonge, V., Pourrut, X., Gonzalez, J. P., Muyembe-Tamfum, J. J., Formenty, P., 2009. Human Ebola outbreak resulting from direct exposure to fruit bats in Luebo, Democratic Republic of Congo, 2007. *Vector Borne Zoonotic Dis*. 9, 723–728.

Li, C., Liu, F., Liang, M., Zhang, Q., Wang, X., Wang, T., Li, J., Li, D., 2010. Hantavirus-like particles generated in CHO cells induce specific immune responses in C57BL/6 mice. *Vaccine*. 28, 4294–4300.

Lorenz, I. C., Allison, S. L., Heinz, F. X., Helenius, A., 2002. Folding and dimerization of tick-borne encephalitis virus envelope proteins prM and E in the endoplasmic reticulum. *J Virol*. 76, 5480–5491.

Lu, Y., Raviprakash, K., Leao, I. C., Chikhlikar, P. R., Ewing, D., Anwar, A., Chougnet, C. et al., 2003. Dengue 2 PreM-E/LAMP chimera targeted to the MHC class II compartment elicits long-lasting neutralizing antibodies. *Vaccine*. 21, 2178–2189.

Lukashevich, I. S., Carrion, R., Jr., Salvato, M. S., Mansfield, K., Brasky, K., Zapata, J., Cairo, C., Goicochea, M., Hoosien, G. E., Ticer, A., Bryant, J., Davis, H., Hammamieh, R., Mayda, M., Jett, M., Patterson, J., 2008. Safety, immunogenicity, and efficacy of the ML29 reassortant vaccine for Lassa fever in small nonhuman primates. *Vaccine*. 26, 5246–5254.

Maiztegui, J. I., McKee, K. T., Jr., Barrera Oro, J. G., Harrison, L. H., Gibbs, P. H., Feuillade, M. R., Enria, D. A. et al., 1998. Protective efficacy of a live attenuated vaccine against Argentine hemorrhagic fever. AHF Study Group. *J Infect Dis*. 177, 277–283.

Mandell, R. B., Koukuntla, R., Mogler, L. J., Carzoli, A. K., Freiberg, A. N., Holbrook, M. R., Martin, B. K. et al., 2010. A replication-incompetent Rift Valley fever vaccine: Chimeric virus-like particles protect mice and rats against lethal challenge. *Virology*. 397, 187–198.

Martin, J. E., Sullivan, N. J., Enama, M. E., Gordon, I. J., Roederer, M., Koup, R. A., Bailer, R. T., Chakrabarti, B. K. et al., 2006. A DNA vaccine for Ebola virus is safe and immunogenic in a phase I clinical trial. *Clin Vaccine Immunol*. 13, 1267–1277.

Mason, P. W., Zugel, M. U., Semproni, A. R., Fournier, M. J., Mason, T. L., 1990. The antigenic structure of dengue type 1 virus envelope and NS1 proteins expressed in *Escherichia coli*. *J Gen Virol*. 71(Pt 9), 2107–2114.

Mateu, G. P., Marchevsky, R. S., Liprandi, F., Bonaldo, M. C., Coutinho, E. S., Dieudonne, M., Caride, E., Jabor, A. V., Freire, M. S., Galler, R., 2007. Construction and biological properties of yellow fever 17D/dengue type 1 recombinant virus. *Trans R Soc Trop Med Hyg*. 101, 289–298.

Maves, R. C., Ore, R. M., Porter, K. R., Kochel, T. J., 2011. Immunogenicity and protective efficacy of a psoralen-inactivated dengue-1 virus vaccine candidate in Aotus nancymaae monkeys. *Vaccine*. 29, 2691–2696.

McCormick, J. B., Fisher-Hoch, S. P., 2002. Lassa fever. *Curr Top Microbiol Immunol*. 262, 75–109.

McCormick, J. B., King, I. J., Webb, P. A., Johnson, K. M., O'Sullivan, R., Smith, E. S., Trippel, S., Tong, T. C., 1987. A case-control study of the clinical diagnosis and course of Lassa fever. *J Infect Dis*. 155, 445–455.

Morrill, J. C., Peters, C. J., 2011a. Mucosal immunization of rhesus macaques with Rift Valley Fever MP-12 vaccine. *J Infect Dis*. 204, 617–625.

Morrill, J. C., Peters, C. J., 2011b. Protection of MP-12-vaccinated rhesus macaques against parenteral and aerosol challenge with virulent rift valley fever virus. *J Infect Dis*. 204, 229–236.

Mota, J., Rico-Hesse, R., 2009. Humanized mice show clinical signs of dengue fever according to infecting virus genotype. *J Virol*. 83, 8638–8645.

Mukhopadhyay, S., Kuhn, R. J., Rossmann, M. G., 2005. A structural perspective of the flavivirus life cycle. *Nat Rev Microbiol*. 3, 13–22.

Naslund, J., Lagerqvist, N., Habjan, M., Lundkvist, A., Evander, M., Ahlm, C., Weber, F., Bucht, G., 2009. Vaccination with virus-like particles protects mice from lethal infection of Rift Valley Fever Virus. *Virology*. 385, 409–415.

Olinger, G. G., Bailey, M. A., Dye, J. M., Bakken, R., Kuehne, A., Kondig, J., Wilson, J., Hogan, R. J., Hart, M. K., 2005. Protective cytotoxic T-cell responses induced by Venezuelan equine encephalitis virus replicons expressing Ebola virus proteins. *J Virol*. 79, 14189–14196.

Osorio, J. E., Huang, C. Y., Kinney, R. M., Stinchcomb, D. T., 2011. Development of DENVax: A chimeric dengue-2 PDK-53-based tetravalent vaccine for protection against dengue fever. *Vaccine*. 29, 7251–7260.

Pittman, P. R., Liu, C. T., Cannon, T. L., Makuch, R. S., Mangiafico, J. A., Gibbs, P. H., Peters, C. J., 1999. Immunogenicity of an inactivated Rift Valley fever vaccine in humans: A 12-year experience. *Vaccine*. 18, 181–189.

Pourrut, X., Delicat, A., Rollin, P. E., Ksiazek, T. G., Gonzalez, J. P., Leroy, E. M., 2007. Spatial and temporal patterns of Zaire ebolavirus antibody prevalence in the possible reservoir bat species. *J Infect Dis*. 196(Suppl 2), S176–S183.

Pourrut, X., Souris, M., Towner, J. S., Rollin, P. E., Nichol, S. T., Gonzalez, J. P., Leroy, E., 2009. Large serological survey showing cocirculation of Ebola and Marburg viruses in Gabonese bat populations, and a high seroprevalence of both viruses in *Rousettus aegyptiacus*. BMC. *Infect Dis*. 9, e159.

Pratt, W. D., Wang, D., Nichols, D. K., Luo, M., Woraratanadharm, J., Dye, J. M., Holman, D. H., Dong, J. Y., 2010. Protection of nonhuman primates against two species of Ebola virus infection with a single complex adenovirus vector. *Clin Vaccine Immunol*. 17, 572–581.

Pushko, P., Parker, M., Ludwig, G. V., Davis, N. L., Johnston, R. E., Smith, J. F., 1997. Replicon-helper systems from attenuated Venezuelan equine encephalitis virus: Expression of heterologous genes in vitro and immunization against heterologous pathogens in vivo. *Virology*. 239, 389–401.

Putnak, R., Fuller, J., VanderZanden, L., Innis, B. L., Vaughn, D. W., 2003. Vaccination of rhesus macaques against dengue-2 virus with a plasmid DNA vaccine encoding the viral pre-membrane and envelope genes. *Am J Trop Med Hyg*. 68, 469–476.

Qing, M., Liu, W., Yuan, Z., Gu, F., Shi, P. Y., 2010. A high-throughput assay using dengue-1 virus-like particles for drug discovery. *Antiviral Res*. 86, 163–171.

Qiu, X., Fernando, L., Alimonti, J. B., Melito, P. L., Feldmann, F., Dick, D., Stroher, U., Feldmann, H., Jones, S. M., 2009. Mucosal immunization of cynomolgus macaques with the VSVDeltaG/ZEBOVGP vaccine stimulates strong Ebola GP-specific immune responses. *PLoS One*. 4, e5547.

Rajamanonmani, R., Nkenfou, C., Clancy, P., Yau, Y. H., Shochat, S. G., Sukupolvi-Petty, S., Schul, W., Diamond, M. S., Vasudevan, S. G., Lescar, J., 2009. On a mouse monoclonal antibody that neutralizes all four dengue virus serotypes. *J Gen Virol*. 90, 799–809.

Ramanathan, M. P., Kuo, Y. C., Selling, B. H., Li, Q., Sardesai, N. Y., Kim, J. J., Weiner, D. B., 2009. Development of a novel DNA SynCon tetravalent dengue vaccine that elicits immune responses against four serotypes. *Vaccine*. 27, 6444–6453.

Raviprakash, K., Ewing, D., Simmons, M., Porter, K. R., Jones, T. R., Hayes, C. G., Stout, R., Murphy, G. S., 2003. Needle-free Biojector injection of a dengue virus type 1 DNA vaccine with human immunostimulatory sequences and the GM-CSF gene increases immunogenicity and protection from virus challenge in Aotus monkeys. *Virology*. 315, 345–352.

Raviprakash, K., Porter, K. R., Kochel, T. J., Ewing, D., Simmons, M., Phillips, I., Murphy, G. S., Weiss, W. R., Hayes, C. G., 2000. Dengue virus type 1 DNA vaccine induces protective immune responses in rhesus macaques. *J Gen Virol*. 81, 1659–1667.

Raviprakash, K., Wang, D., Ewing, D., Holman, D. H., Block, K., Woraratanadharm, J., Chen, L., Hayes, C., Dong, J. Y., Porter, K., 2008. A tetravalent dengue vaccine based on a complex adenovirus vector provides significant protection in rhesus monkeys against all four serotypes of dengue virus. *J Virol*. 82, 6927–6934.

Reynard, O., Mokhonov, V., Mokhonova, E., Leung, J., Page, A., Mateo, M., Pyankova, O. et al., 2011. Kunjin virus replicon-based vaccines expressing Ebola virus glycoprotein GP protect the guinea pig against lethal Ebola Virus infection. *J Infect Dis*. 204(Suppl 3), S1060–S1065.

Ross, T. M., Xu, Y., Bright, R. A., Robinson, H. L., 2000. C3d enhancement of antibodies to hemagglutinin accelerates protection against influenza virus challenge. *Nat Immunol*. 1, 127–131.

Rovinski, B., Haynes, J. R., Cao, S. X., James, O., Sia, C., Zolla-Pazner, S., Matthews, T. J., Klein, M. H., 1992. Expression and characterization of genetically engineered human immunodeficiency virus-like particles containing modified envelope glycoproteins: Implications for development of a cross-protective AIDS vaccine. *J Virol*. 66, 4003–4012.

Saluzzo, J. F., Smith, J. F., 1990. Use of reassortant viruses to map attenuating and temperature-sensitive mutations of the Rift Valley fever virus MP-12 vaccine. *Vaccine*. 8, 369–375.

Schmaljohn, C. S., 1990. Nucleotide sequence of the L genome segment of Hantaan virus. *Nucleic Acids Res*. 18, 6728.

Schmitz, J., Roehrig, J., Barrett, A., Hombach, J., 2011. Next generation dengue vaccines: A review of candidates in preclinical development. *Vaccine*. 29, 7276–7284.

Seregin, A. V., Yun, N. E., Poussard, A. L., Peng, B. H., Smith, J. K., Smith, J. N., Salazar, M., Paessler, S., 2010. TC83 replicon vectored vaccine provides protection against Junin virus in guinea pigs. *Vaccine*. 28, 4713–4718.

Shresta, S., Sharar, K. L., Prigozhin, D. M., Beatty, P. R., Harris, E., 2006. Murine model for dengue virus-induced lethal disease with increased vascular permeability. *J Virol*. 80, 10208–10217.

Smith, K. M., Nanda, K., Spears, C. J., Ribeiro, M., Vancini, R., Piper, A., Thomas, G. S., Thomas, M. E., Brown, D. T., Hernandez, R., 2011. Structural mutants of dengue virus 2 transmembrane domains exhibit host-range phenotype. *Virol J*. 8, e289.

Spik, K. W., Badger, C., Mathiessen, I., Tjelle, T., Hooper, J. W., Schmaljohn, C., 2008. Mixing of M segment DNA vaccines to Hantaan virus and Puumala virus reduces their immunogenicity in hamsters. *Vaccine*. 26, 5177–5181.

Sukupolvi-Petty, S., Austin, S. K., Purtha, W. E., Oliphant, T., Nybakken, G. E., Schlesinger, J. J., Roehrig, J. T., Gromowski, G. D., Barrett, A. D., Fremont, D. H., Diamond, M. S., 2007. Type- and subcomplex-specific neutralizing antibodies against domain III of dengue virus type 2 envelope protein recognize adjacent epitopes. *J Virol*. 81, 12816–12826.

Sullivan, N. J., Geisbert, T. W., Geisbert, J. B., Shedlock, D. J., Xu, L., Lamoreaux, L., Custers, J. H. et al., 2006. Immune protection of nonhuman primates against Ebola virus with single low-dose adenovirus vectors encoding modified GPs. *PLoS Med.* 3, e177.

Sullivan, N. J., Geisbert, T. W., Geisbert, J. B., Xu, L., Yang, Z. Y., Roederer, M., Koup, R. A., Jahrling, P. B., Nabel, G. J., 2003. Accelerated vaccination for Ebola virus haemorrhagic fever in non-human primates. *Nature.* 424, 681–684.

Sullivan, N. J., Sanchez, A., Rollin, P. E., Yang, Z. Y., Nabel, G. J., 2000. Development of a preventive vaccine for Ebola virus infection in primates. *Nature.* 408, 605–609.

Sun, W., Cunningham, D., Wasserman, S. S., Perry, J., Putnak, J. R., Eckels, K. H., Vaughn, D. W. et al., 2009. Phase 2 clinical trial of three formulations of tetravalent live-attenuated dengue vaccine in flavivirus-naive adults. *Hum Vaccin.* 5, 33–40.

Suzuki, R., Winkelmann, E. R., Mason, P. W., 2009. Construction and characterization of a single-cycle chimeric flavivirus vaccine candidate that protects mice against lethal challenge with dengue virus type 2. *J Virol.* 83, 1870–1880.

Szecsi, J., Boson, B., Johnsson, P., Dupeyrot-Lacas, P., Matrosovich, M., Klenk, H. D., Klatzmann, D., Volchkov, V., Cosset, F. L., 2006. Induction of neutralising antibodies by virus-like particles harbouring surface proteins from highly pathogenic H5N1 and H7N1 influenza viruses. *Virol J.* 3, e70.

Towner, J. S., Pourrut, X., Albarino, C. G., Nkogue, C. N., Bird, B. H., Grard, G., Ksiazek, T. G., Gonzalez, J. P., Nichol, S. T., Leroy, E. M., 2007. Marburg virus infection detected in a common African bat. *PLoS One.* 2, e764.

Trindade, G. F., Marchevsky, R. S., Fillipis, A. M., Nogueira, R. M., Bonaldo, M. C., Acero, P. C., Caride, E., Freire, M. S., Galler, R., 2008. Limited replication of yellow fever 17DD and 17D-Dengue recombinant viruses in rhesus monkeys. *An Acad Bras Cienc.* 80, 311–321.

Tuuminen, T., Kekalainen, E., Makela, S., Ala-Houhala, I., Ennis, F. A., Hedman, K., Mustonen, J., Vaheri, A., Arstila, T. P., 2007. Human CD8+ T cell memory generation in Puumala hantavirus infection occurs after the acute phase and is associated with boosting of EBV-specific CD8+ memory T cells. *J Immunol.* 179, 1988–1995.

Valdes, I., Hermida, L., Martin, J., Menendez, T., Gil, L., Lazo, L., Castro, J. et al., 2009. Immunological evaluation in nonhuman primates of formulations based on the chimeric protein P64k-domain III of dengue 2 and two components of *Neisseria meningitidis.* *Vaccine.* 27, 995–1001.

Van Epps, H. L., Terajima, M., Mustonen, J., Arstila, T. P., Corey, E. A., Vaheri, A., Ennis, F. A., 2002. Long-lived memory T lymphocyte responses after hantavirus infection. *J Exp Med.* 196, 579–588.

Wang, D., Hevey, M., Juompan, L. Y., Trubey, C. M., Raja, N. U., Deitz, S. B., Woraratanadharm, J. et al., 2006a. Complex adenovirus-vectored vaccine protects guinea pigs from three strains of Marburg virus challenges. *Virology.* 353, 324–332.

Wang, D., Raja, N. U., Trubey, C. M., Juompan, L. Y., Luo, M., Woraratanadharm, J., Deitz, S. B. et al., 2006b. Development of a cAdVax-based bivalent Ebola virus vaccine that induces immune responses against both the Sudan and Zaire species of Ebola virus. *J Virol.* 80, 2738–2746.

Warfield, K. L., Aman, M. J., 2011. Advances in virus-like particle vaccines for filoviruses. *J Infect Dis.* 204(Suppl 3), S1053–S1059.

White, L. J., Parsons, M. M., Whitmore, A. C., Williams, B. M., de Silva, A., Johnston, R. E., 2007. An immunogenic and protective alphavirus replicon particle-based dengue vaccine overcomes maternal antibody interference in weanling mice. *J Virol.* 81, 10329–10339.

Williams, K. L., Zompi, S., Beatty, P. R., Harris, E., 2009. A mouse model for studying dengue virus pathogenesis and immune response. *Ann N Y Acad Sci.* 1171(Suppl 1), E12–E23.

Wilson, J. A., Hart, M. K., 2001. Protection from Ebola virus mediated by cytotoxic T lymphocytes specific for the viral nucleoprotein. *J Virol.* 75, 2660–2664.

Wilson, J. A., Hevey, M., Bakken, R., Guest, S., Bray, M., Schmaljohn, A. L., Hart, M. K., 2000. Epitopes involved in antibody-mediated protection from Ebola virus. *Science.* 287, 1664–1666.

Yap, Y. K., Smith, D. R., 2010. Strategies for the plant-based expression of dengue subunit vaccines. *Biotechnol Appl Biochem.* 57, 47–53.

Zhang, S., Liang, M., Gu, W., Li, C., Miao, F., Wang, X., Jin, C. et al., 2011. Vaccination with dengue virus-like particles induces humoral and celluar immune responses in mice. *Virol J.* 8, e333.

Part II

Specific Infections

14 Pathogenic Old World Arenaviruses

*Jason Botten, Benjamin King, Joseph Klaus,
and Christopher Ziegler*

CONTENTS

14.1 INTRODUCTION

Arenaviruses are enveloped, negative-sense RNA viruses that are maintained in nature primarily in rodent reservoirs. Phylogenetically, the arenaviruses are organized into Old or New World groups (Clegg, 2002). At least nine arenaviruses are known to cause severe human diseases. Of the New World viruses, Junin (JUNV), Machupo (MACV), Guanarito (GTOV), Sabia (SABV), and Chapare viruses are etiologic agents of hemorrhagic fever (HF) syndromes in South America while White Water Arroyo virus (WWAV) was linked to fatal hemorrhagic infections in North America (Buchmeier et al., 2007; Delgado et al., 2008). In the Old World group, Lassa (LASV) (Frame et al., 1970) and Lujo (Briese et al., 2009; Paweska et al., 2009) viruses are responsible for severe HF syndromes in Africa, while lymphocytic choriomeningitis (LCMV), which is found throughout the world, can cause aseptic meningitis in immunocompetent individuals (Armstrong and Lillie, 1934).

LCMV is also a potent teratogen to the developing fetus (Barton et al., 2002) and is highly lethal in immunosuppressed individuals (Fischer et al., 2006). LASV, LCMV, JUNV, GTOV, and MACV are NIAID Category A agents, and with the exception of LCMV, each of these viruses is also on the select agent list of potential bioterrorism threat agents and requires biosafety level (BSL)-4 containment for safe handling. There are no U.S. Food and Drug Administration (FDA)-approved vaccines for the prevention of arenavirus disease, and therapeutic treatment is limited to the use of ribavirin and/or immune plasma for a subset of the pathogenic arenaviruses. Accordingly, there is a clear need to develop effective vaccines and antiviral therapies to combat these devastating infections. In this chapter, we focus on the pathogenic Old World arenaviruses and discuss the following themes: genomic organization and replication, structure/function of viral proteins, ecology of infection in reservoir rodents, arenavirus phylogeny and evolution, epidemiology and clinical disease, the basis for pathogenesis during human infection, adaptive immunity to infection, and progress toward the development of vaccines and antivirals for prevention and treatment of arenavirus disease.

14.2   GENOMIC ORGANIZATION AND CODING STRATEGY

The arenaviruses have an RNA genome that is encoded in an ambisense fashion and consists of two single-stranded RNA segments, the ~3 kb small (S) and ~7 kb large (L) segments (see Figure 14.1). The arenavirus proteome consists of four proteins: the viral RNA-dependent RNA polymerase (L) (~250 kDa) and small really interesting new gene (RING) finger protein Z (~11 kDa), which are encoded on the L segment, and the S segment encoded nucleoprotein (NP) (~63 kDa) and glycoprotein precursor (GPC) (~75 kDa), which is post-translationally modified to yield a 58 amino acid stable signal peptide (SSP) and the envelope glycoproteins GP1 (~45 kDa) and GP2 (~35 kDa) (Southern, 1996). The SSP, GP1, and GP2 subunits form a mature glycoprotein (GP) complex that mediates attachment and entry into permissive host cells. The two open reading frames (ORFs) present on each of the genomic RNAs do not overlap and are separated by a noncoding, intergenic region (IGR). The IGR sequences have the potential to form stable secondary structures in the form of hairpins and contain the sites of transcriptional termination (Meyer et al., 2002). Each RNA segment is flanked by untranslated regions (UTRs) at the 3′ and 5′ termini. The first 19 bases of the 3′ and 5′ UTRs are complementary to one another and facilitate the formation of panhandle structures (Meyer et al., 2002). The terminal UTRs, together with the IGR, constitute the minimal

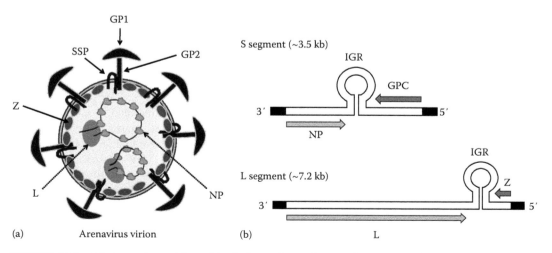

FIGURE 14.1 (a) Arenavirus virion with viral proteins and genomic segments depicted. (b) Arenavirus genome and encoded proteins. The protein coding regions within each segment are separated by a noncoding IGR that forms a hairpin structure. Each segment is flanked by 5′ and 3′ UTRs (black rectangles).

cis-acting signals required for replication and transcription in reverse genetic systems for LCMV and Tacaribe virus (TCRV) (Buchmeier et al., 2007). Although arenaviruses are classified as negative-sense viruses, each RNA segment encodes viral proteins in an ambisense fashion. As a result of this coding strategy, the generation of subgenomic viral mRNAs for NP and L is accomplished via transcription of the negative-sense, genomic (vRNA) templates (similar to other negative-sense viruses), while the transcription of GPC and Z requires the positive-strand, antigenomic (vcRNA) templates (in a pseudo-positive-sense virus manner). Therefore, dependent upon the availability of the appropriate transcriptional templates, this coding strategy allows for differential temporal regulation of viral transcription and protein expression (i.e., NP and L vs. GPC and Z) during infection.

14.3 ARENAVIRUS REPLICATION

Arenavirus replication begins with the attachment and entry of mature virions into permissive host cells. In the setting of human infection, this can include epithelial cells, endothelial cells, and monocytes/macrophages (Buchmeier et al., 2007; Dylla et al., 2008; Gomez et al., 2003; Lee et al., 2008). Viral attachment is mediated through an interaction between GP and either α-dystroglycan (α-DG) for the Old World viruses LASV and certain immunosuppressive strains of LCMV (Cao et al., 1998) or transferrin receptor 1 for a subset of the pathogenic New World arenaviruses (Radoshitzky et al., 2007). Additional receptors utilized by LASV include dendritic cell–specific intercellular adhesion molecule-3-grabbing nonintegrin (DC-SIGN) and liver and lymph node sinusoidal endothelial calcium-dependent lectin (LSECtin), Axl receptor tyrosine kinase (Axl), and protein tyrosine kinase receptor 3 (Tyro3) (Shimojima et al., 2012). Upon receptor attachment, the virions are taken into the cell via endocytosis and delivered to acidified endosomes where low pH induces membrane fusion between viral and endosomal membranes (see Section 14.4.4 for more details regarding entry) (Rojek and Kunz, 2008). Viral ribonucleoproteins (RNPs), which consist of L or S segment RNAs encapsidated by NP and associated with L, are then released into the cytoplasm where viral genome replication and transcription take place (Meyer et al., 2002) (see Figure 14.2 for overview of replication strategy). The genetic material contained in each virion includes the L and S viral RNAs (vRNAs), as well as significant levels of Z mRNA (Buchmeier et al., 2007). Kinetic studies investigating the temporal appearance of S segment-derived proteins (NP and GPC) have demonstrated that NP expression precedes that of GPC after initial infection (Franze-Fernandez et al., 1987; Southern, 1996). This observation confirms the requirement for genome replication to take place in order to generate the S segment viral complementary RNAs (vcRNA) template needed to facilitate transcription and expression of GP. The initial transcription and expression of NP and L allow replication of the L and S vRNAs to take place. From the newly formed replicative intermediates (S and L vcRNA), transcription and expression of GPC and Z are carried out. When all of the viral proteins are expressed, the L and S vRNAs are encapsidated by NP and L to form RNP complexes (Meyer et al., 2002). These RNPs are then packaged with Z, SSP, GP1, and GP2, allowing mature virions to bud from the plasma membrane. The budding process, which is largely driven by Z, is covered in more detail in Section 14.4.3. Arenaviruses establish noncytopathic infections in both rodents and humans (Buchmeier et al., 2007).

14.4 VIRAL PROTEOME

14.4.1 NUCLEOPROTEIN

The arenavirus NP consists of between 558 and 570 amino acids and has a molecular weight ranging from 60 to 68 kDa (Buchmeier, 2002). It is the most abundant viral protein produced during both acute and persistent infection. The primary role of NP has classically been thought to be encapsidation of the vRNA and vcRNA, though additional roles are beginning

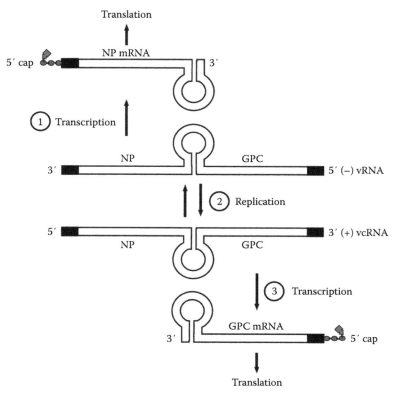

FIGURE 14.2 Arenavirus replication and transcription. This figure depicts the initial events of viral replication and transcription for the viral S segment immediately following the release of viral RNPs into the cytoplasm of a newly infected cell. (1) The viral L protein transcribes NP and L mRNAs from the incoming S and L vRNAs, respectively. (2) Newly formed NP and L proteins then read through the IGR to replicate full-length L and S vcRNAs. (3) The S and L vcRNAs then serve as templates for the transcription of the GPC and L mRNAs, respectively, permitting translation of GPC and Z protein.

to be appreciated (Buchmeier et al., 2007). The crystal structure of the LASV NP reveals the existence of two independently folding domains located at the N-terminus (residues 1–338) and the C-terminus (residues 364–561) separated by a linker region (Figure 14.3) (Qi et al., 2010). Structural characterization of the LASV NP identified a basic cleft in the N-terminal domain that is responsible for the binding of single-stranded viral RNA (Hastie et al., 2011b). NP also is

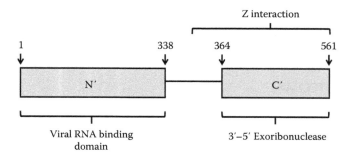

FIGURE 14.3 Representative arenavirus NP protein. Known functions are mapped to associated domains. Amino acid residues shown as determined for Lassa virus NP. (From Qi, X. et al., *Nature*, 468, 779, 2010.)

able to oligomerize, and it is likely that the self-interaction between RNA-bound NP monomers could provide the structural basis for its ability to encapsidate arenaviral vRNA and vcRNA (Brunotte et al., 2011a; Hastie et al., 2011b).

NP of all members of the arenavirus family has been shown to suppress the induction of type I interferon (IFN) expression, with the exception of TCRV NP, which has no anti-IFN activity, and Mopeia virus (MOPV) NP, which seems to exhibit only weak anti-IFN activity (Martinez-Sobrido et al., 2006, 2007; Ortiz-Riano et al., 2011; Pannetier et al., 2004). The ability of LCMV NP to inhibit the induction of type I IFN was mapped to the C-terminal domain (Martinez-Sobrido et al., 2009). Structural analysis of the C-terminal domain of LASV NP revealed a high level of structural homology with the DEDD superfamily of exonucleases. This domain exhibits 3′–5′ exoribonuclease activity with specificity for double-stranded RNA (dsRNA) (Hastie et al., 2011a; Qi et al., 2010). Thus, NP appears to inhibit the induction of type I IFN through its C-terminal exonuclease domain by limiting the abundance of viral dsRNA, which is a potent ligand for the cytoplasmic innate immune sensors retinoic acid–inducible gene I (RIG-I) and melanoma differentiation-associated gene 5 (MDA5), which upregulate the expression of IFN-β upon ligand binding.

NP and L have been shown to interact, though the structural basis and functional consequences of this interaction are unknown (Kerber et al., 2011). NP and Z have also been shown to interact (Eichler et al., 2004). This interaction with Z is critical for the packaging of NP into virus-like particles (Groseth et al., 2010; Schlie et al., 2010). Thus, the Z–NP interaction is likely important for the incorporation of viral RNP into budding virions.

14.4.2 Viral RNA-Dependent RNA Polymerase

The L protein (~2,200 amino acids; ~250 kDa) serves as the arenavirus RNA-dependent RNA polymerase (RdRp) (Lukashevich et al., 1997). L and NP represent the minimal *trans*-acting factors for transcription and replication of vRNA and vcRNA (Buchmeier et al., 2007). L is well conserved throughout the arenavirus family and shares sequence similarity to the RdRps encoded by bunyaviruses and orthomyxoviruses (Brunotte et al., 2011b; Vieth et al., 2004). Sequence analysis of the L gene reveals four highly conserved regions among all arenaviruses. These conserved regions have been hypothesized to fold into four independent domains, putative domains I (residues 1–250), II (500–900), III (1,000–1,650), and IV (1,750–1,900), which are separated by flexible linker regions of more variable sequence (Figure 14.4) (Brunotte et al., 2011b; Vieth et al., 2004). L has been shown to oligomerize, and this ability to self-associate seems to be important for its function (Sanchez and de la Torre, 2005).

Domain I of L seems to play a role in the cap-snatching mechanism employed by arenaviruses to prime their nascent mRNAs. Domain I of LCMV L exhibits extensive structural homology to the

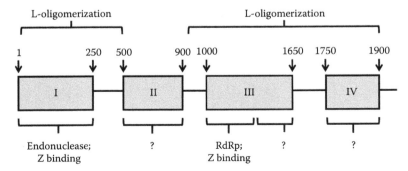

FIGURE 14.4 Representative arenavirus L protein. Known functions are mapped to associated domains. No known function has been determined for putative domains II, IV, or the C terminal region of domain III. Amino acid residues shown as determined for Lassa virus L. (From Vieth, S. et al., *Virology*, 318, 153, 2004.)

N-terminus of the influenza PA protein. PA, a subunit of the influenza virus heterotrimeric RdRp, has *bona fide* "cap-snatching" endonuclease activity (Morin et al., 2010). Functional characterization of domain I of both LASV and LCMV L demonstrated that these domains indeed possess endonuclease activity and can bind mRNA (Lelke et al., 2010; Morin et al., 2010).

The N-terminal region of domain III (residues 1040–1550) is the putative RdRp domain of L (Hass et al., 2008; Sanchez and de la Torre, 2005; Vieth et al., 2004). Using purified MACV L protein, electron microscopy showed the presence of a central ring–shaped domain with three appendages attached by flexible linkers. It has been hypothesized that the central ring domain of L is the RdRp due to its similarity to the ring-like structure of other viral RdRps (Kranzusch et al., 2010). Indeed, it was demonstrated that purified MACV L protein can bind either the 3′ or 5′ termini of the viral genome.

The L protein has been shown to interact with Z. This interaction is necessary for Z's ability to inhibit the transcription and the replication of the viral genome (Jacamo et al., 2003; Kranzusch and Whelan, 2011). The MACV Z protein was shown to directly inhibit the catalytic RdRp activity of MACV L protein. It has been proposed that the interaction with Z protein locks the L protein in a viral promoter bound, catalytically inactive complex that may play a critical role in the regulation of the viral life cycle and perhaps aid in the packaging of functional RNP into budding viral particles (Kranzusch and Whelan, 2011).

14.4.3 Z PROTEIN

The Z protein is the matrix protein of arenaviruses. Z ranges in size from 90 to 103 amino acids with nearly two-thirds of the protein consisting of a central zinc binding, RING domain. The RING domain, which is central to many of Z's interactions and functions, is composed of a $C–X_2–C–X_9–C–X_2–H–X_2–C–X_2–C–X_{10}–C–X_2–C$ sequence that coordinates two Zn^{2+} molecules at two sites (Figure 14.5) (Kentsis et al., 2002). Similar to other viral matrix proteins, Z is an authentic budding protein that drives the formation and release of arenavirus particles at the plasma membrane. Z also plays a role in modulating the activity of the arenavirus polymerase. Finally, Z influences the cellular environment by repressing translation of cellular mRNAs and acting as a type I IFN antagonist.

Z plays a critical role in the generation of infectious viral particles by driving the viral budding process at the plasma membrane. Z itself is capable of forming virus-like particles independent of other viral proteins. Arenavirus Z proteins, with the exception of TCRV Z, contain one or two C-terminal late domains that are required for efficient budding. Arenaviral late domains are proline-rich sequences of the type PTAP, PSAP, or PPPY (Urata and de la Torre, 2011). In addition to late domains, budding of Z requires myristoylation of the strictly conserved N-terminal glycine residue at position 2, which facilitates Z's interaction with the plasma membrane (Capul et al., 2011; Perez et al., 2004).

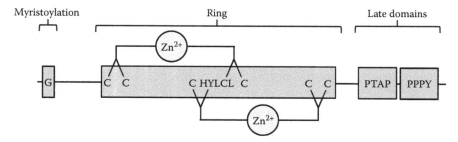

FIGURE 14.5 LASV Z protein. The myristoylated glycine residue at position 2 is required for interaction with the plasma membrane, budding, and recruitment of GP. The RING domain binds two Zn^{2+} molecules and is required to repress the viral L protein and inhibit cellular mRNA translation. The late domains PTAP and PPPY interact with proteins in the vacuolar protein sorting pathway and are required for budding.

One function of matrix proteins is to recruit other viral proteins into enveloped particles. Accordingly, the arenavirus Z protein can recruit GP and NP independently into virus particles, although the precise mechanism by which these proteins are recruited into viral particles is currently unclear (Capul et al., 2011; Groseth et al., 2010; Levingston Macleod et al., 2011). Another open question is whether a conserved recruitment mechanism is utilized by all arenaviruses.

It is becoming increasingly clear that Z has a central role in modulating viral replication as well as the cellular environment. First, as described in Section 14.4.2, Z can potently inhibit viral genome replication and transcription in a dose-dependent manner by repressing the RdRp activity of the viral L protein (Kranzusch and Whelan, 2011). Z is also able to repress translation of cellular mRNAs with a 5′ 7-methyl guanosine cap by impairing the translation initiation factor eIF-4E's ability to bind the mRNA cap (Kentsis et al., 2001; Volpon et al., 2010). The Z protein also has a role in evading the host innate immune response. In a mechanism employed by New World arenaviruses but not Old World arenaviruses, Z suppresses expression of the type I IFN, IFN-β, by binding RIG-I, which recognizes viral RNA and induces the transcription of IFN-β through downstream mediators (Fan et al., 2010).

14.4.4 GLYCOPROTEIN

The arenavirus envelope GP complex, which consists of the subunits SSP, GP1, and GP2, decorates the surface of virions, and is critical for mediating attachment and entry of virions into permissive host cells (Buchmeier et al., 1978; Buchmeier and Oldstone, 1979; Oldstone, 1992). As such, GP largely defines arenavirus tropism on the basis of its ability to interact with specific cellular surface receptors. Recent studies have shown that GP may also play a role in directing polarized assembly and release of viral particles (Schlie et al., 2010). Electron micrographs of virions revealed that GP spikes, which resemble multimeric club-like projections, are 5–10 nM in length and evenly spaced in the host-derived viral membrane (Neuman et al., 2005).

GP synthesis and maturation require an intricate interplay of GP with host cellular machinery along the exocytic pathway. Key maturation events include myristoylation, N-linked glycosylation, coordinated folding and multimerization, sequential proteolysis, and carbohydrate maturation (see Figure 14.6) (Nunberg and York, 2012). GP biogenesis begins with co-translational translocation of the nascent GPC into the rough ER by virtue of the unusually long (58 amino acids) SSP, which, following cleavage by the host signal peptidase (SPase), assumes a bitopic membrane orientation and is myristoylated (Agnihothram et al., 2007; Eichler et al., 2003; York et al., 2004). Following SSP cleavage, GPC is glycosylated at multiple asparagine residues; these modifications are critical to the functionality of the GP as well as its early maturation events, including proper folding necessary for ER exit (Bonhomme et al., 2011; Wright et al., 1990). For LASV, there are 11 predicted sites of N-linked glycosylation: 7 on GP1 and 4 on GP2 (Eichler et al., 2006). While in the ER, the GP selectively induces an ER stress response, thereby enhancing the folding capacity of the compartment (Pasqual et al., 2011a). The SSP and GPC form a non-covalently associated complex in the ER via multi-domain interactions between the SSP and the GP2 domain within GPC (Nunberg and York, 2012). Formation of this complex masks a distal ER retention signal located within the cytoplasmic tail of GP2 and facilitates ER exit and anterograde movement of the SSP–GPC complex to the Golgi apparatus (Agnihothram et al., 2006). For most arenavirus GPs, including those encoded by JUNV, LCMV, MACV, and GTOV, a second proteolytic event takes place presumably in the Golgi, whereby the cellular site 1 protease (S1P) recognizes a conserved tetrapeptide R[X]L[X] motif and cleaves GPC into the mature GP1 (40–46 kDa) and GP2 (35–38 kDa) subunits (Buchmeier et al., 2007), which form non-covalent GP1–GP2 heterodimers and remain associated with the SSP (Beyer et al., 2003; Rojek et al., 2008). LASV GPC is unique in that it is cleaved by S1P earlier in the exocytic pathway in either the ER or cis-Golgi (Lenz et al., 2001). High-mannose N-linked glycans on GP1 and GP2 are further processed to a complex form in the Golgi apparatus as evidenced by their acquisition of endoglycosidase H resistance (Kunz et al., 2003; Wright et al., 1990). The mature tripartite complex (SSP–GP1–GP2) then traffics to the cell

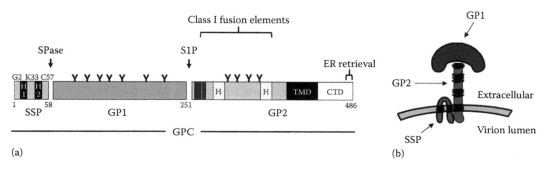

FIGURE 14.6 (a) LASV GP. Proteolytic sites within the GPC are indicated by black arrows. The SSP is cleaved after amino acid 58 by the signal peptidase (SPase) while cleavage of GPC into GP1 and GP2 occurs at amino acid 251 via site 1 protease (S1P). Highlighted features within SSP are (i) the conserved myristoylated glycine at position 2 (G2), (ii) two membrane-spanning hydrophobic domains (H1 and H2) important for SSP topology, and (iii) residues K33 (critical pH sensor required for fusion activation) and C57 (required for association with GPC). The GP1 subunit contains the receptor binding domain and seven N-linked glycosylation additions indicated by (Y). Highlighted GP2 features include (i) class I fusion elements, including two sequential fusion peptides at the N terminus (dark gray boxes) as well as two heptad repeats indicated by (H), (ii) four N-linked sugars as indicated (Y), (iii) the transmembrane domain (TMD), and (iv) the C-terminal domain (CTD). Although not shown, the TMD and CTD both contain SSP binding sites. (b) Structural representation of mature arenavirus GP complex in virions. Arrows indicate the GP1 ectodomain, the GP2 transmembrane stalk, and the SSP. Note that the structural representation is not drawn to scale.

membrane, where assembly and budding take place in a polarized fashion (Kunz et al., 2003; Schlie et al., 2010). The GP presumably plays a role in the assembly and release process via interactions of the SSP with Z (Capul et al., 2007).

Old World arenaviruses such as LASV and certain immunosuppressive strains of LCMV initiate entry into permissive host cells via attachment of GP1 to one of several surface receptors, including α-DG, DC-SIGN, LSECtin, Axl, and Tyro3 (Cao et al., 1998; Shimojima et al., 2012). Entry is mediated via a mechanism of clathrin- and caveolin-independent endocytosis whereby virions traffic, via ESCRT machinery, through the multivesicular body network for delivery to late endosomes (Pasqual et al., 2011b). Upon acidification of the endosomes, a pH-dependent molecular rearrangement of the GP complex is triggered that leads to shedding of the GP1 ectodomain (Di Simone et al., 1994) and transformation of GP2 from an SSP-dependent, pre-fusion metastable form (York and Nunberg, 2006) to a post-fusion hairpin structure (Igonet et al., 2011; Thomas et al., 2011). Following GP1 ejection, a fusion peptide on GP2 can insert into the target host membrane (Klewitz et al., 2007). Heptad repeats in GP2 are then reorganized into a six-helix structure, which brings host and viral membranes into close apposition to facilitate membrane fusion and release of viral RNPs into the cytoplasm (Eschli et al., 2006; York et al., 2005).

14.5 ECOLOGY OF ARENAVIRUS INFECTION

Arenaviruses are maintained in nature primarily in rodent reservoirs. The lone exception is TCRV, which was isolated from the *Artibeus* sp. of bat (Downs et al., 1963). Each arenavirus is principally associated with a single rodent species, although infection of additional species can occur (Salazar-Bravo et al., 2002). Typically, each virus can establish a persistent, largely asymptomatic infection in its reservoir rodent (Childs and Peters, 1993; Keenlyside et al., 1983; Salazar-Bravo et al., 2002). Infected rodents are thought to shed infectious virus into the environment via urine, feces, and/or saliva (Buchmeier et al., 2007). Viral maintenance within the rodent population is achieved through vertical transmission from mother to pup and/or via horizontal transmission events involving exposure of naive rodents to infected rodents (possibly through aggressive encounters) or their infectious excreta (saliva, urine, feces) (Banerjee et al., 2008; Buchmeier et al., 2007; Childs and Peters, 1993). Our current understanding of the natural history of arenavirus infection in reservoir species in natural settings is limited, and many critical questions remain to be answered regarding the precise mechanisms of viral transmission and maintenance as well as the characteristics of infection in reservoir rodents, including the impact of infection on host fitness.

14.6   ARENAVIRUS PHYLOGENY AND EVOLUTION

There are currently 23 species of arenaviruses (Family *Arenaviridae*; Genus *Arenavirus*) recognized by the International Committee on Taxonomy of Viruses (ICTV) (http://www.ictvonline.org//virusTaxonomy.asp). Arenaviruses are traditionally classified into two groups or complexes based on their serologic, genetic, and geographic relationships (Bowen et al., 1997). The Old World viruses are maintained in rodents belonging to the subfamily *Murinae*, while the New World viruses, with the exception of TCRV and Tamiami viruses, have been isolated from rodents in the subfamilies *Neotominae* (North American viruses) or *Sigmodontinae* (South American viruses), respectively (Emonet et al., 2009). The geographic distribution of each virus is determined by the range of its respective host. The Old World complex consists of LCMV, which has a worldwide distribution, and the African viruses Lassa Mobala, Mopeia and Ippy (Emonet et al., 2009) (see Table 14.1). It should be noted that considerable genetic diversity exists among LASV isolates found within the region endemic for Lassa fever, and phylogenetically four genetic lineages exist (Bowen et al., 2000). Isolates from the Western end of the endemic zone (Sierra Leone, Liberia, and Guinea) are fairly conserved and compose a single lineage, while those from the Far Eastern end in Nigeria are more diverse, constituting the remaining three lineages. Over the past 5 years, at least eight new arenaviruses that group with the Old World complex have been reported. Whether these new isolates constitute new arenaviral species awaits approval from the ICTV. Potential new species that are pathogenic include Dandenong virus (DANV), an LCMV-like virus that was isolated from a cluster of fatal infections among solid-organ transplant recipients in Australia (note that the index case in this outbreak may have contracted infection in Yugoslavia) (see Section 14.7.2) (Palacios et al., 2008), and Lujo virus, which recently caused an outbreak of LASV-like disease in Africa (see Section 14.7.3) (Briese et al., 2009; Paweska et al., 2009). The natural reservoirs of these viruses are currently not known. The remaining new viruses, which have not been associated with human disease at this time, are the African viruses Morogoro, Merino Walk, Menekre, Gbagroube, Luna, and Kodoko (Coulibaly-N'Golo et al., 2011; Emonet et al., 2009; Ishii et al., 2011; Palacios et al., 2010).

The New World complex consists of viruses found in North and South America. This group is also referred to as the Tacaribe complex and can be divided into four clades (A, B, C, and A/Rec) (Charrel et al., 2011; Emonet et al., 2009). Clade A consists of the South American viruses Flexal, Pirital, Pichinde, Parana, and Allpahuayo. Clade C is composed of the South American viruses Oliveros and Latino. Clade B consists of eight South American viruses, five of which are etiologic agents of HF syndromes in South America (JUNV, Argentine HF; MACV, Bolivian HF; GTOV, Venezuelan HF; SABV, Brazilian HF; and Chapare virus, isolated from a fatal human case of HF in Bolivia in 2003 [Buchmeier et al., 2007; Delgado et al., 2008]). The remaining clade B members are Cupixi, Amapari, and TCRV viruses. Clade A/Rec, which consists of North American viruses, is unique because it has been hypothesized that viruses in this clade are the product of recombination events between ancestral viruses of clades A and B (Charrel et al., 2001). The only pathogenic virus in this clade is WWAV, which was linked to several fatalities in the Southwestern United States (Byrd et al., 2000); the other species are Bear Canyon and Tamiami viruses. Similar to the Old World complex, several potentially novel arenavirus species have been isolated in the Americas over the past 9 years. In North America, new viruses include Skinner Tank, Big Brushy Tank, Tonto Creek, and Catarina, which all group in the A/Rec clade (Charrel et al., 2011).

There are several mechanisms that have been proposed to explain arenavirus diversity and genetic distribution (for review see Clegg, 2002; Emonet et al., 2009). One potential mechanism is

TABLE 14.1

Old World Arenaviruses and Selected New World Arenaviruses

Virus	Acronym[b]	Lineage	Rodent Host	Human Disease	Geographic Distribution
Lassa	LASV	OW	*Mastomys* sp.	Lassa fever	West Africa
Lymphocytic choriomeningitis	LCMV	OW	*Mus musculus, Mus domesticus*	Aseptic meningitis, congenital deformities, and high lethality in immunosuppressed patients	Worldwide
Dandenong*[a]		OW	Unknown	High lethality in immunosuppressed patients	Yugoslavia or Australia
Lujo*		OW	Unknown	Febrile illness with high lethality	Zambia, South Africa
Mobala	MOBV	OW	*Praomys* sp.	None[c]	Central African Republic
Mopeia	MOPV	OW	*Mastomys* sp.	None	Mozambique, Zimbabwe
Ippy	IPPYV	OW	*Arvicanthis niloticus*	None	Central African Republic
Kodoko*		OW	*Mus Nannomys minutoides*	None	Guinea, West Africa
Morogoro*		OW	*Mastomys* sp.	None	Tanzania, Africa
Merino Walk*		OW	*Myotomys unisulcatus*	None	Eastern Cape, South Africa
Luna virus*		OW	*Mastomys* sp.	None	Zambia, Africa
Menekre*		OW	*Hylomyscus* sp.	None	Cote d'Ivoire, Africa
Gbagroube*		OW	*Mus Nannomys setulosus*	None	Cote d'Ivoire, Africa
Junin	JUNV	NW, clade B	*Calomys musculinus*	Argentine HF	Argentina
Machupo	MACV	NW, clade B	*Calomys callosus*	Bolivian HF	Bolivia
Sabia	SABV	NW, clade B	Unknown	Brazilian HF	Brazil
Guanarito	GTOV	NW, clade B	*Zygodontomys brevicauda, Sigmodon alstoni*	Venezuelan HF	Venezuela
White Water Arroyo	WWAV	NW, clade Rec/A	*Neotoma albigula*	HF	Southwestern United States
Chapare		NW, clade B	Unknown	HF	Bolivia

[a] Newly described viruses that are not officially recognized by International Committee on Taxonomy of Viruses at this time are denoted with an asterisk (*).

[b] Acronyms are assigned by the International Committee for the Taxonomy of Viruses.

[c] Not known to cause human disease at this time.

co-evolution of arenaviruses with their respective rodent hosts. There are several lines of evidence to support this mechanism, including the fact that arenaviruses are able to establish persistent, asymptomatic infections in their hosts. This would be an advantageous setting for arenaviral quasi-species to evolve and adapt to the host rodent (Clegg, 2002). Another line of evidence is that most arenaviruses are associated with only one or two rodent species (Salazar-Bravo et al., 2002). Further, genetically related arenaviruses are maintained in rodents that have analogous genetic similarities. Another potential mechanism for evolution involves interspecies transmission of arenaviruses. In this setting, two unique arenavirus species would co-infect a common host such that individual cells would be co-infected with both viral species. This would permit the following outcomes: (i) a genetic recombination event resulting in gene transfer between the viruses or (ii) a genome segment swapping event resulting in the creation of new viral reassortants containing one parental genome segment and one from the other viral species (Clegg, 2002; Emonet et al., 2009). While viral reassortants have been generated in the laboratory (Kirk et al., 1980; Riviere et al., 1985), there is currently no evidence that such an event has taken place in nature (Emonet et al., 2009). Interestingly, it has been hypothesized that an intra-segmental recombination event between ancestral viruses of clades A and B led to the generation of the clade A/Rec virus WWAV (Charrel et al., 2001). At present, there is not enough evidence to fully support any of these mechanisms (Emonet et al., 2009).

14.7 EPIDEMIOLOGY, CLINICAL DISEASE, AND PATHOGENESIS OF OLD WORLD ARENAVIRUSES

14.7.1 LASV

Of the pathogenic arenaviruses, LASV is by far the most significant cause of morbidity and mortality. LASV was identified in 1969 during an outbreak of Lassa fever in the village of Lassa in northern Nigeria (Frame et al., 1970). Each year, it is estimated that ~200,000 cases of Lassa fever occur in several West African countries including Nigeria, Sierra Leone, Liberia, and Guinea (McCormick et al., 1987b). Lassa fever cases originating in endemic African regions are also routinely exported to locations around the world (Gunther and Lenz, 2004). Mortality following LASV infection in hospitalized patients is between 15% and 20% (McCormick et al., 1987a) while an estimated 5% of all infections are thought to be fatal (McCormick and Fisher-Hoch, 2002). Mortality rates can exceed 50% during nosocomial outbreaks of disease (Fisher-Hoch et al., 1995). Lassa fever is also highly lethal in pregnant women, especially during the third trimester, where mortality rates can reach 30%, with fetal and neonatal losses of 88% (Price et al., 1988). Deafness is a common complication of LASV infection and occurs in ~30% of cases (Cummins et al., 1990). Serosurveys conducted within endemic regions have demonstrated that, in certain locations, 20%–50% of the population can show evidence of a previous LASV infection (Richmond and Baglole, 2003). There are no approved vaccines for the prevention of LASV; ribavirin is partially effective provided it is administered within the first week following the onset of symptoms (McCormick, 1986; McCormick et al., 1986a).

LASV is primarily maintained in the multimammate mouse, *Mastomys natalensis* (Lecompte et al., 2006; Monath et al., 1974), but has also been isolated from related *Mastomys* species, including *M. huberti* and *M. erythroleucus* (McCormick et al., 1987b). It should be noted that the exact taxonomy of the *Mastomys* genus is not fully understood (Salazar-Bravo et al., 2002); continued taxonomic refinement will be required to more precisely define the range of LASV reservoir species in nature. *Mastomys* rodents, which are peridomestic, are commonly found in human dwellings in the endemic region (McCormick et al., 1987b; Robbins et al., 1983). Infected animals can shed infectious virus into the environment via urine (McCormick et al., 1987b; Walker et al., 1975). Therefore, infected rodents can deposit infectious virus on many surfaces (e.g., floors, counters, tables, dishes, and bedding) as well as into food stores and sources of water. Humans can contract infection via direct contact with infected mice or their infectious excreta (Fraser et al., 1974; Keenlyside et al., 1983). Likely routes of infection include viral contact with the mucosa (e.g., ingestion of virus via contaminated food or water) and direct inoculation of virus into cuts or abrasions from contaminated surfaces (McCormick and Fisher-Hoch, 2002). Virus may also be transmitted through aerosols created from dried excreta (e.g., via sweeping excreta), albeit rarely (Keenlyside et al., 1983). Another high-risk activity that is common in the endemic region is catching and eating *Mastomys* rodents (Ter Meulen et al., 1996). LASV can also be spread from person to person in household or hospital settings. High-risk activities in hospital settings include close contact with blood, secretions, or tissues from an infected individual or re-use of virus-contaminated needles (Fisher-Hoch et al., 1995; McCormick and Fisher-Hoch, 2002); in household settings, close contact with secretions from acutely ill individuals (Keenlyside et al., 1983) or sexual intercourse with convalescent individuals poses a high risk of infection (virus can be shed in urine for upwards of 2 months following recovery from acute disease) (Emond et al., 1982). It does not appear, however, that aerosol transmission is an efficient means of transmission from person to person in hospital settings (Fisher-Hoch et al., 1985). Accordingly, standard barrier practices are highly effective at limiting nosocomial transmission.

Lassa fever is often difficult to diagnose as the broad spectrum of symptoms exhibited is not readily distinguishable from those presented in other common febrile illnesses seen within the endemic region (McCormick et al., 1987a). Following an incubation period of 7–18 days, a wide range of symptoms can present, including fever, sore throat, cough, severe headache, and dizziness (Frame, 1989; Gunther and Lenz, 2004; McCormick and Fisher-Hoch, 2002; McCormick et al., 1987a). Other common symptoms include nausea, diarrhea, vomiting, and abdominal pain. Predictors of severe disease include high viremia, facial and pulmonary edema, bleeding from mucosal surfaces, high serum aspartate transaminase levels, and neurological symptoms ranging from confusion to seizures (Johnson et al., 1987; McCormick et al., 1986a, 1987a). In severe cases of disease leading to death, health deteriorates rapidly following ~6–8 days of prodromal illness (McCormick and Fisher-Hoch, 2002). Features of severe disease include pulmonary edema and adult respiratory distress syndrome, hypovolemic shock, and neurological complications including severe encephalopathy (Cummins et al., 1992; Fisher-Hoch et al., 1985; Frame, 1989; Knobloch et al., 1980; McCormick et al., 1987a). Death typically occurs ~12 days following the onset of symptoms. The duration of illness in patients that recover is generally 2 weeks, and virus is cleared from the blood ~3 weeks following the onset of disease (Demby et al., 1994; Johnson et al., 1987; Mertens et al., 1973; Trappier et al., 1993). Sensorineural deafness is also a common complication seen late in disease or early into convalescence (Cummins et al., 1990).

Pathogenesis during severe Lassa fever is associated with several features including uncontrolled virus replication, lymphopenia, platelet dysfunction, neutrophilia, hemorrhage, hypotension, vascular leakage, and shock (Fisher-Hoch et al., 1988; Johnson et al., 1987; McCormick et al., 1987a). Vascular leakage can manifest as effusions and edema in the peritoneal, pleural, and cardiac cavities (Edington and White, 1972; McCormick and Fisher-Hoch, 2002). Unlike other HFs such as Ebola, disseminated intravascular coagulation and thrombocytopenia do not appear to play a role in Lassa fever pathogenesis (Fisher-Hoch et al., 1988; Kunz, 2009b; Moraz and Kunz, 2011). Evidence of a "cytokine storm" (e.g., elevated levels of proinflammatory cytokines in patient sera) has not been documented (Mahanty et al., 2001). Hemorrhage is also unlikely to contribute to disease severity as it is mild and limited to mucosal surfaces (Kunz, 2009b; McCormick and Fisher-Hoch, 2002; Moraz and Kunz, 2011). The role, if any, that neutrophilia plays is currently unknown. It is common to observe high viral titers in many tissues during severe infection, including the heart, lung, spleen, kidney, and adrenal gland (Walker et al., 1982). However, the corresponding histopathological changes observed in these tissues, which includes acute myocarditis, splenic necrosis (in the marginal zone and periarteriolar lymphocytic sheath), and multifocal hepatocellular necrosis in the liver, are modest and not sufficient to explain the profound capillary leakage and shock that ultimately lead to death (Edington and White, 1972; Frame et al., 1970; McCormick et al., 1986b; Walker et al., 1982). In particular, vascular lesions are mild and inflammatory infiltrates modest

(Walker et al., 1982). LASV infection of endothelial cells *in vitro* does not appear to cause cellular damage (Lukashevich et al., 1999). These findings, coupled with the lack of circulating proinflammatory cytokines (Mahanty et al., 2001), suggest that capillary leakage, while not the result of overt endothelial cell destruction, may occur as a direct result of viral infection of the endothelium and/or indirectly following endothelial cell exposure to soluble mediators released from infected cells or inflammatory cells (Kunz, 2009b). As an example, it has been shown that a soluble factor found in the plasma of acutely ill Lassa fever patients can inhibit platelet aggregation as well as neutrophil degranulation *in vitro* (Cummins et al., 1989; Roberts et al., 1989). It is possible that a soluble mediator such as this may target endothelial cells and impair barrier function.

The uncontrolled viral replication observed during severe Lassa fever is thought to occur due to a virus-induced state of immunosuppression that may impact both the innate and adaptive immune responses. There are several potential explanations for how this immunosuppressive state might be induced. As described in Section 14.4.1, LASV NP is able to subvert the production of type I IFNs and thereby impair the innate immune response to infection (Martinez-Sobrido et al., 2007). Additionally, a feature of the LASV genome, an overhanging 5′ ppp-nucleotide found in the double-stranded panhandles that form between the 3′ and 5′ UTRs, can act as a RIG-I decoy and inhibit the production type I IFNs (Marq et al., 2011). Another relevant observation is that antigen-presenting cells such as macrophages and dendritic cells, which are prominent targets of LASV, fail to become activated or produce proinflammatory cytokines following infection (Baize et al., 2004; Lukashevich et al., 1999; Mahanty et al., 2003). Once infected, these cells are also impaired in their ability to activate T cells (Mahanty et al., 2003). The pronounced lymphopenia that occurs during severe LASV infection could certainly result in depressed or suboptimal adaptive T or B cell responses (Fisher-Hoch et al., 1988). While the mechanism responsible for the observed lymphopenia is unknown, it is likely not a result of direct infection of T and B cells by LASV as these cells are refractory to infection due to an absence of the proper cellular receptor (Reignier et al., 2006). Finally, the fact that neutralizing antibodies (nAbs) are typically not produced during acute infection could certainly contribute to uncontrolled viral replication (Johnson et al., 1987). The failure to generate nAbs is likely not attributable to lymphopenia as virus-specific, nonneutralizing antibodies are generated early during infection (see Section 14.8.1).

14.7.2 LCMV

The outcome of human LCMV infection can range from subclinical infection to debilitating febrile disease with central nervous system involvement (Buchmeier and Zajac, 1999). LCMV is a significant cause of aseptic meningitis, accounting for ~8%–9% of diagnosed cases in previous studies (Adair et al., 1953; Barton and Hyndman, 2000; Meyer et al., 1960). Vertical transmission of LCMV from mother to fetus can lead to severe teratogenic consequences including chorioretinitis, microcephaly, or hydrocephalus (Buchmeier et al., 2007). Maternal infection during the first trimester is also associated with increased incidence of spontaneous abortion (Barton and Mets, 2001). The mortality rate for children born with congenital LCMV can approach 30%; nearly two-thirds of surviving children can exhibit neurological disorders such as mental retardation and seizures (Wright et al., 1997). Infection of immunosuppressed individuals following transplantation of LCMV-infected tissues can result in high rates of mortality, as evidenced by the fact that, in two independent outbreaks, seven of eight recipients succumbed to infection (Fischer et al., 2006). Likewise, DANV, which is genetically similar to LCMV, also led to three fatalities following transplantation of DANV-infected tissues to immunosuppressed recipients (Palacios et al., 2008). There are no licensed vaccines available for LCMV. Ribavirin is an effective antiviral *in vitro* (Gessner and Lother, 1989), but to date, trials of antiviral agents have not been conducted in humans.

The primary reservoir for LCMV in nature is the common house mouse (*Mus musculus* or *Mus domesticus*), a peridomestic rodent that has a worldwide distribution (Lehmann-Grube, 1971). Infection of the reservoir rodent results in an asymptomatic, life-long infection, during which infectious virus is persistently shed in urine, feces, and/or saliva (Childs and Peters, 1993). Viral maintenance in the reservoir population is thought to occur through vertical transmission from infected dams to their pups (Childs and Peters, 1993). LCMV can be introduced into pet hamster breeding colonies where a similar pattern of asymptomatic maintenance and viral shedding has been observed (Buchmeier and Zajac, 1999). Human infection likely occurs as a result of direct contact with infected rodents, or by exposure to excreta or infectious aerosols (Buchmeier et al., 2001; Childs and Peters, 1993). As described earlier, viral transmission can also occur following solid organ transplantation of LCMV-infected tissues (Fischer et al., 2006; Palacios et al., 2008). In the United States, ~5% of adults are seropositive for LCMV-specific antibodies (Childs et al., 1992).

14.7.3 Lujo Virus

In 2008, a previously unrecognized arenavirus caused an outbreak of severe febrile illness in South Africa. The new virus, called Lujo, was highly lethal, leading to death in four of five patients (Briese et al., 2009; Paweska et al., 2009). The index case contracted infection from an unknown source in Zambia and was flown to South Africa for medical treatment. In South Africa, nosocomial transmission led to the infection of the remaining four individuals (medical workers and a member of the cleaning staff). Excluding the index case, the incubation period for the remaining patients ranged from 7 to 13 days (Paweska et al., 2009). Initial symptoms for each patient included a nonspecific, flu-like prodromal phase with fever, headache, and myalgia. By 1 week after disease onset, the severity of the original symptoms had increased while diarrhea and pharyngitis developed. In the four fatal cases, disease severity lessened for a short time ~1 week following disease onset, then rapidly progressed to fatal illness characterized by respiratory distress, neurologic symptoms, and circulatory collapse. All patients had thrombocytopenia upon hospital admission. The single survivor was the only patient to receive ribavirin. Phylogenetically, Lujo virus groups with the Old World arenavirus complex, but is genetically distinct from LASV and LCMV (Briese et al., 2009).

14.8 ADAPTIVE IMMUNITY TO OLD WORLD ARENAVIRUSES

14.8.1 Adaptive Immunity to LASV

The protective role of T and B cells during viral infection is well established. Generally, CD4+ T cells are thought to control viral infection through multiple mechanisms, including enhancement of B and CD8+ T cell responses, production of inflammatory cytokines, cytotoxicity against infected cells, and promotion of memory responses. Similarly, CD8+ T cells control viral infection through several mechanisms, including direct cytotoxicity of infected cells and production of cytokines such as IFN-γ and TNF-α. Finally, one mechanism by which B cells control infection is through the production of nAbs to help control primary infection as well as provide sterile immunity against reinfection. In most cases, the humoral and cell-mediated immune responses act synergistically to provide protection. LASV infection in humans or experimental animal models, however, is atypical in that it resolves prior to the appearance of nAbs (Fisher-Hoch and McCormick, 2001; Johnson et al., 1987; Peters et al., 1987). Low-titer nAbs, if formed, generally appear weeks to months following viral clearance. This phenomenon is also seen during experimental infection of mice with LCMV (Bruns et al., 1983; Buchmeier et al., 1980). Additionally, immune plasma does not protect LASV patients against disease (Fisher-Hoch and McCormick, 2001; McCormick, 1986). Based largely on these observations, it is thought that CD8+ and CD4+ T cell responses are primarily responsible for providing protective immunity during human LASV infection (Fisher-Hoch and McCormick, 2001). At present, our current understanding of the human T cell response to LASV infection is limited to the fact that LASV-specific, memory CD4+ T cells are detectable in convalescent patients up to 6 years after infection (ter Meulen et al., 2000, 2004).

More direct evidence confirming the importance of T cells for protective immunity against LASV has been obtained through experimental animal infection models. First, it was shown that guinea pigs or nonhuman primates vaccinated with recombinant vaccinia viruses expressing LASV GPC were protected against subsequent lethal LASV challenge in the absence of detectable nAbs (Auperin et al., 1988; Fisher-Hoch et al., 2000; Morrison et al., 1989). Second, in two separate studies, transfer of immune-splenocytes (from either guinea pigs or CBA mice that had been successfully vaccinated against LASV) to naive recipients demonstrated that splenocytes, but not serum, were capable of providing protection against a subsequent lethal LASV challenge in recipient animals (Lukashevich et al., 2005; Peters et al., 1987). Last, it was recently demonstrated that CD8+ T cells, but not CD4+ T cells, were critical for clearing a primary LASV virus infection in a newly developed murine model of LASV infection (Flatz et al., 2010b). Interestingly, this study also reported a direct role for CD8+ T cells in the pathogenesis of disease, presumably through activation of monocyte/macrophage lineage cells. Collectively, these observations illustrate the importance of the antiviral T cell response during LASV infection.

In contrast to the delay observed in the formation of nAbs, nonneutralizing antibodies specific for viral antigens can be detected early during human LASV infection or in animal infection models for LCMV (Johnson et al., 1987; Lehmann-Grube, 1971). This suggests that the failure to produce nAbs is not due to insufficient B cell numbers or functionality. Interestingly, it was shown that the poor nAb kinetics seen following LCMV infection in mice could be completely attributed to the LCMV GP as opposed to the rest of the viral backbone (Pinschewer et al., 2004). One possible explanation for why this occurs is that structural features inherent to GP, such as its heavy glycan shielding, could mask neutralizing epitopes on GP1. These findings are significant for vaccine development as they suggest that even the most immunogenic delivery vectors may not be capable of improving the poor nAb kinetics seen following immunization with LCMV or LASV GPC.

14.8.2 Adaptive Immunity to LCMV

Murine models of LCMV infection represent a valuable tool for the study of the adaptive immune response to Old World arenaviruses. Infection of immunocompetent mice by intraperitoneal inoculation with LCMV strain Armstrong results in an asymptomatic infection that is cleared ~8 days following infection. Protective immunity in this infection setting is well defined (Botten and Kotturi, 2007). Virus-specific CD8+ cytolytic T cells (CTLs) are the primary effector cells required for viral clearance (Lehmann-Grube et al., 1993; Oldstone, 2002; Wille et al., 1989). They establish protective immunity mainly through perforin-mediated lysis of virus-infected cells (Kagi et al., 1994; Walsh et al., 1994). Interferon (IFN)-γ production is also important considering that IFN-γ knockout mice cannot clear an LCMV strain Armstrong infection following intravenous inoculation (Bartholdy et al., 2000). Immunization of mice with individual CD8+ T cell epitopes is sufficient to confer protection against infection with LCMV and other arenaviruses (Botten et al., 2006, 2007; Klavinskis et al., 1989; Kotturi et al., 2007; Rodriguez-Carreno et al., 2005; Schulz et al., 1991).

The importance of CD4+ T cells in providing protective immunity varies depending upon the context of a particular infection. For example, during the primary infection described earlier, CD4+ T cells are not required for protective immunity as virus-specific CTL will develop in CD4+ T cell–deficient mice and clear infection with the same kinetics seen in wild-type mice (Matloubian et al., 1994). However, in this same context, CD4+ T cell help is critical for programming CTLs to develop into memory cells capable of effectively responding to a secondary challenge (Janssen et al., 2003; Shedlock and Shen, 2003). This CD4+ T cell help is required during, not after, initial CTL priming (Janssen et al., 2003; Shedlock and Shen, 2003). Additionally, CD4+ T cells are required during memory maintenance for the health and survival of memory CD8+ T cell populations (Sun et al., 2004).

14.9   HUMAN (HLA)-RESTRICTED T CELL EPITOPE DISCOVERY FROM THE PATHOGENIC ARENAVIRUSES

As described earlier, the natural history of the human T cell response to arenavirus infection is largely unknown. Profiling the phenotypic characteristics of antiviral CD8+ and CD4+ T cells during acute infection and convalescence is necessary to define correlates of protective or immuno-pathogenic T cell responses. Key reagents required for such studies are epitopes that can be used to interrogate T cells from patients. In 2004, only six HLA-restricted T cell epitopes had been identified from the arenaviruses; all were CD4+ T cell epitopes from LASV (ter Meulen et al., 2000, 2004). To provide diagnostic reagents for future human T cell studies, our group conducted a large-scale epitope discovery program to systematically define epitopes restricted by common human (HLA) MHC class I and class II molecules from seven of the pathogenic arenaviruses (LASV, LCMV, JUNV, MACV, GTOV, SABV, and WWAV). As is discussed in Section 14.10, the other main goal of our program was to test whether a subset of the identified epitopes could be used to develop a universal vaccination strategy designed to elicit a T-cell-mediated immune response that would provide broad coverage against a variety of arenaviruses and across different ethnic populations. With that in mind, we deliberately biased our epitope discovery toward those peptides with the highest degree of amino acid conservation among the targeted viruses. The most direct method to identify new epitopes typically involves screening peripheral blood mononuclear cells from patients with a history of arenavirus infection. However, this approach was not feasible due to several logistical obstacles, including the paucity of samples available from previously infected patients and the BSL-4 requirement associated with the majority of the targeted viruses. To overcome these limitations, we developed a novel approach for epitope discovery that featured the use of bioinformatic prediction algorithms to identify candidate epitope sequences from arenavirus proteins and HLA transgenic mice—which express human MHC class I or class II molecules and can generate "human T cell responses"—for epitope validation (for detailed reviews of this approach, see [Botten et al., 2006, 2010a; Botten and Kotturi, 2007; Kotturi et al., 2010]). Using this approach, we were able to identify 10 HLA-A*0201- and 10 HLA-A*1101-restricted CD8+ T cell epitopes (Botten et al., 2006, 2007; Kotturi et al., 2009b) and 37 HLA-DRB1*0101-restricted CD4+ T cell epitopes (Kotturi et al., 2010) from the seven antigenically distinct arenaviruses. The epitopes identified provide human population coverage of ~60% (class I epitopes) and ~80% (class II epitopes); nearly 100% coverage should be possible by applying the same approach to additional HLA supertype families (Bui et al., 2006; Sette and Sidney, 1999). As discussed in greater detail later, a subset of the identified HLA-A*0201-restricted epitopes from LASV and LCMV are also legitimate vaccine determinants as immunization of HLA-A*0201 mice with these peptides protected against a subsequent viral challenge (Botten et al., 2006, 2007). It should also be noted that a number of tools and reagents generated during the course of our studies are openly available to the scientific community, including a database containing 333 manually annotated protein sequences derived from the seven targeted arenaviruses (Bui et al., 2007) as well as a panel of recombinant vaccinia viruses that express 24 of the 28 ORFs encoded by these viruses (Botten et al., 2010a) (BEI Resources; http://www.beiresources.org; BEIR NR-15486 to NR-15509). Additionally, our team comprehensively mapped class I and class II epitopes for LCMV in common mouse backgrounds (H-2b for class I; H-2b and H2d for class II) to permit detailed studies of cell-mediated immunity during experimental infection (Dow et al., 2008; Kotturi et al., 2007; Mothe et al., 2007).

One important consideration regarding our studies is the fact that there is an incomplete overlap in the epitopes recognized by HLA transgenic mice and humans (Kotturi et al., 2009a). Our group recently had an opportunity to directly measure the degree of overlap by screening T lymphocytes from persons previously exposed to LCMV for their ability to recognize the LCMV-derived, CD8+ T cell epitopes described earlier. There was excellent overlap as four of the five HLA-A*0201-restricted epitopes identified in the HLA*0201 mice were also recognized by HLA-A*0201-positive donors (Kotturi et al., 2011). Importantly, two of the recognized epitopes were the same peptides that protected HLA-A*0201 mice against LCMV challenge. This finding further illustrates the relevance of these peptides for vaccine design (Botten et al., 2007). This study also advanced our understanding of memory phase, LCMV-specific T cells in humans. One finding was that LCMV-specific T cell responses are long-lived (4–8 years post-exposure). Interestingly, increasing severity of LCMV disease seems to correlate with an increase in the magnitude of memory CD8+ T cell responsiveness. Last, virus-specific CD8+ and CD4+ T cells undergo distinct memory differentiation programs.

14.10   VACCINATION STRATEGIES FOR ARENAVIRUSES

There are currently no FDA-approved vaccines for the prevention of arenavirus disease in humans. The only vaccine that has been used in humans is the attenuated Candid#1 strain of JUNV, which has investigational new drug status in the United States. Despite having been shown to be safe and efficacious after being used in greater than 200,000 individuals in Argentina (Maiztegui et al., 1998), concerns regarding its phenotypic stability have restricted its use to persons with the high risk of exposure to JUNV (e.g., agricultural workers) (Contigiani et al., 1993). A clear need exists for the development of licensed vaccines for the prevention of arenavirus disease.

Pathogen heterogeneity represents a significant challenge to vaccine development. This is particularly relevant for the pathogenic arenaviruses, which display a high degree of both interspecies and intraspecies heterogeneity. The low incidence of disease observed for the majority of these antigenically diverse viruses combined with the low socioeconomic status of the countries where they are endemic makes it unlikely that individual vaccines will be developed for each of the nine pathogenic arenavirus species. Therefore, engineering multivalent vaccines capable of providing cross-protection against multiple arenavirus species is desirable. Our research team recently focused on the development of such multivalent vaccination strategies. Our approach was to capitalize on the known importance of the cellular immune response for providing protective immunity (see Section 14.8) and accordingly develop epitope-based vaccines that would be capable of inducing protective CD8+ T cell responses specific to multiple arenaviruses (Botten et al., 2010a). Using the HLA-A*0201-restricted epitopes described in Section 14.9, we tested two strategies to induce multivalent protection: inclusion of protective HLA-restricted epitopes corresponding to each of several targeted arenaviruses in a single vaccine or alternatively, to reduce the complexity of the vaccine, inclusion of highly conserved, cross-reactive epitopes that each has the capacity to induce cross-protection against multiple arenaviruses. For the first approach, immunization of HLA-A*0201 mice with a cocktail of 14 epitopes derived from the GPCs of six different arenaviruses led to significant simultaneous CD8+ T cell responses to each epitope and protected mice from challenge with recombinant vaccinia viruses expressing the GPCs from LASV, LCMV, or SABV (Kotturi et al., 2009b). For the second approach, HLA-A*0201 mice were immunized with a single peptide derived from LASV GPC (YL*I*SIFLHL) that is highly conserved with the corresponding peptide sequence found in LCMV (YL*V*SIFLHL). Immunization with this peptide led to nearly identical recognition of either the LASV or LCMV peptide by in vivo CTL assay and provided cross-protection against heterologous challenge with LCMV (Botten et al., 2010b). Collectively, our findings provide proof-of-concept that this cell-mediated vaccination strategy may be able to protect against multiple, antigenically distinct arenavirus species while providing broad population coverage.

There are several additional vaccination approaches that have been developed to target LASV. These include the use of (i) various recombinant vectors that express LASV antigens such as Yellow fever 17D virus (Jiang et al., 2011), vesicular stomatitis virus (Geisbert et al., 2005), vaccinia virus (Fisher-Hoch et al., 2000), alphavirus (Pushko et al., 2001), salmonella (Djavani et al., 2001), or DNA (Rodriguez-Carreno et al., 2005), (ii) an arenavirus reassortant between LASV and MOPV (Lukashevich et al., 2005), (iii) LASV-like particles (Branco et al., 2010), (iv) a replication-defective LCMV vector (Flatz et al., 2010a), and (v) stable, live-attenuated arenaviruses that express a foreign GP to maintain attenuation (Bergthaler et al., 2006). There are several characteristics that a

LASV vaccine will have to possess to be effective. First, due the high incidence of HIV within LASV-endemic regions, the vaccine will need to be safe for immunocompromised individuals. This requirement unfortunately precludes the use of the highly effective vaccinia virus–based vaccines. Second, the vaccine will likely need to be capable of inducing protective immunity in a single immunization due to the logistical difficulties that would be associated with providing a booster vaccination within endemic regions. Last, the vaccine will have to be capable of inducing long-lasting immunity as persons living in the endemic region will be at continual risk for exposure to LASV during their lives.

14.11   ANTIVIRALS

The only drug with proven efficacy for the treatment of Lassa fever in humans is the guanosine analogue ribavirin. One limitation of ribavirin is that to be effective, it must be given within the first week after disease onset (McCormick, 1986; McCormick et al., 1986a). A second concern relates to the side effects it causes, including thrombocytosis, severe anemia, and birth defects (Enria et al., 1987; McKee et al., 1988). Immune plasma therapy, while effective for treatment of Argentine HF (Enria et al., 1984; Maiztegui et al., 1979), does not protect against Lassa fever (Fisher-Hoch and McCormick, 2001; McCormick, 1986). Therefore, there is a clear need for additional antiviral compounds. Many groups have recently focused on the identification of novel small molecules that will disrupt arenavirus replication. Common themes include targeting (i) specific steps in the viral life cycle such as virion attachment and fusion (Cashman et al., 2011; Nunberg and York, 2012), (ii) critical cellular proteins such as S1P, which is required for proteolytic cleavage of GPC (Kunz, 2009a), or (iii) genetic elements of the viral genome such as the promoter regions using antisense oligonucleotides (Muller and Gunther, 2007; Neuman et al., 2011). Two reviews provide an excellent overview of recent discoveries related to novel antiviral compounds (Emonet et al., 2011; Lee et al., 2011). Finally, a recent siRNA-based study identified 54 human cellular proteins that are critically required for LCMV replication and therefore represent novel antiviral targets (Panda et al., 2011).

14.12   CONCLUSION

Over the past 9 years, Lujo, Chapare, and Dandenong viruses have all emerged as previously unrecognized arenavirus species that cause severe human disease. In addition, while not associated with human disease at this time, 10 additional arenaviruses have been identified during this same time period in both the Old World ($N = 6$) and New World ($N = 4$). These findings suggest that many additional species of arenaviruses exist in nature and may represent a significant threat to human health in the future. Moving forward, a high priority must be placed on the development of licensed vaccines for the prevention of arenavirus disease. New vaccines should minimally target LASV due to its significant impact on human health, but ideally target multiple pathogenic arenavirus species. Studies of the human T cell response during arenavirus infection will also be important to define the role of T cells in protection or enhancement of disease. Defining the phenotypic properties of protective T cells will ultimately help to guide in the selection of new vaccines. Another key priority will be to develop effective antivirals for the treatment of arenavirus infection. Furthermore, the mechanisms by which LASV causes human disease, particularly capillary leakage and shock, need to be defined to better understand the pathogenesis of disease and guide the development of new targeted therapies. Finally, it will be important to continue to improve our understanding of the molecular and cellular biology of arenavirus replication to elucidate novel ways to combat infection. In particular, little is known regarding the interactions that arenavirus proteins have with host cellular proteins or the importance of these interactions for viral replication and disease pathogenesis. Identification of novel arenavirus protein–host protein interactions that are critical for viral replication and/or pathogenesis will advance our understanding of the basic biology of these highly pathogenic viruses and provide new targets for the development of antivirals.

ACKNOWLEDGMENTS

Due to space limitations, we were regretfully unable to fully cite all of the original publications related to the topics discussed. Therefore, in some instances, we instead cite recent reviews, which contain links to the original publications. We apologize in advance to the many colleagues whose original publications were not cited in this chapter. This work was supported by National Institutes of Health grants AI50840, AI065359, AI074862, AI27028, AI077607, P20RR021905, 1R21AI88059-1A1, T32 AI055402-06A1T32, AI07354, F32 AI056827, and contract N01-AI-40023.

REFERENCES

Adair, C.V., Gauld, R.L., Smadel, J.E., 1953. Aseptic meningitis, a disease of diverse etiology: Clinical and etiologic studies on 854 cases. *Ann. Int. Med.* 39, 675–704.

Agnihothram, S.S., York, J., Nunberg, J.H., 2006. Role of the stable signal peptide and cytoplasmic domain of G2 in regulating intracellular transport of the Junin virus envelope glycoprotein complex. *J. Virol.* 80, 5189–5198.

Agnihothram, S.S., York, J., Trahey, M., Nunberg, J.H., 2007. Bitopic membrane topology of the stable signal peptide in the tripartite Junin virus GP-C envelope glycoprotein complex. *J. Virol.* 81, 4331–4337.

Armstrong, C., Lillie, R.D., 1934. Experimental lymphocytic choriomeningitis of monkeys and mice produced by a virus encountered in studies of the 1933 St. Louis encephalitis epidemic. *Publ. Health Rep.* 49, 1019–1027.

Auperin, D.D., Esposito, J.J., Lange, J.V., Bauer, S.P., Knight, J., Sasso, D.R., McCormick, J.B., 1988. Construction of a recombinant vaccinia virus expressing the Lassa virus glycoprotein gene and protection of guinea pigs from a lethal Lassa virus infection. *Virus Res.* 9, 233–248.

Baize, S., Kaplon, J., Faure, C., Pannetier, D., Georges-Courbot, M.C., Deubel, V., 2004. Lassa virus infection of human dendritic cells and macrophages is productive but fails to activate cells. *J. Immunol.* 172, 2861–2869.

Banerjee, C., Allen, L.J., Salazar-Bravo, J., 2008. Models for an arenavirus infection in a rodent population: Consequences of horizontal, vertical and sexual transmission. *Math. Biosci. Eng.* 5, 617–645.

Bartholdy, C., Christensen, J.P., Wodarz, D., Thomsen, A.R., 2000. Persistent virus infection despite chronic cytotoxic T-lymphocyte activation in gamma interferon-deficient mice infected with lymphocytic choriomeningitis virus. *J. Virol.* 74, 10304–10311.

Barton, L.L., Hyndman, N.J., 2000. Lymphocytic choriomeningitis virus: Reemerging central nervous system pathogen. *Pediatrics* 105, E35.

Barton, L.L., Mets, M.B., 2001. Congenital lymphocytic choriomeningitis virus infection: Decade of rediscovery. *Clin. Infect. Dis.* 33, 370–374.

Barton, L.L., Mets, M.B., Beauchamp, C.L., 2002. Lymphocytic choriomeningitis virus: Emerging fetal teratogen. *Am. J. Obstet. Gynecol.* 187, 1715–1716.

Bergthaler, A., Gerber, N.U., Merkler, D., Horvath, E., de la Torre, J.C., Pinschewer, D.D., 2006. Envelope exchange for the generation of live-attenuated arenavirus vaccines. *PLoS Pathogens* 2, e51.

Beyer, W.R., Popplau, D., Garten, W., von Laer, D., Lenz, O., 2003. Endoproteolytic processing of the lymphocytic choriomeningitis virus glycoprotein by the subtilase SKI-1/S1P. *J. Virol.* 77, 2866–2872.

Bonhomme, C.J., Capul, A.A., Lauron, E.J., Bederka, L.H., Knopp, K.A., Buchmeier, M.J., 2011. Glycosylation modulates arenavirus glycoprotein expression and function. *Virology* 409, 223–233.

Botten, J., Alexander, J., Pasquetto, V., Sidney, J., Barrowman, P., Ting, J., Peters, B. et al., 2006. Identification of protective Lassa virus epitopes that are restricted by HLA-A2. *J. Virol.* 80, 8351–8361.

Botten, J.W., Kotturi, M.F., 2007. Adaptive immunity to Lymphocytic choriomeningitis virus: New insights into antigenic determinants. *Future Virology* 2, 495–508.

Botten, J., Sidney, J., Mothe, B.R., Peters, B., Sette, A., Kotturi, M.F., 2010a. Coverage of related pathogenic species by multivalent and cross-protective vaccine design: Arenaviruses as a model system. *Microbiol. Mol. Biol. Rev.* 74, 157–170.

Botten, J., Whitton, J.L., Barrowman, P., Sidney, J., Whitmire, J.K., Alexander, J., Kotturi, M.F., Sette, A., Buchmeier, M.J., 2010b. A multivalent vaccination strategy for the prevention of Old World arenavirus infection in humans. *J. Virol.* 84, 9947–9956.

Botten, J., Whitton, J.L., Barrowman, P., Sidney, J., Whitmire, J.K., Alexander, J., Ting, J.P., Bui, H.H., Sette, A., Buchmeier, M.J., 2007. HLA-A2-restricted protection against lethal lymphocytic choriomeningitis. *J. Virol.* 81, 2307–2317.

Bowen, M.D., Peters, C.J., Nichol, S.T., 1997. Phylogenetic analysis of the Arenaviridae: Patterns of virus evolution and evidence for cospeciation between arenaviruses and their rodent hosts. *Mol. Phylogenet. Evol.* 8, 301–316.

Bowen, M.D., Rollin, P.E., Ksiazek, T.G., Hustad, H.L., Bausch, D.G., Demby, A.H., Bajani, M.D., Peters, C.J., Nichol, S.T., 2000. Genetic diversity among Lassa virus strains. *J. Virol.* 74, 6992–7004.

Branco, L.M., Grove, J.N., Geske, F.J., Boisen, M.L., Muncy, I.J., Magliato, S.A., Henderson, L.A. et al., 2010. Lassa virus-like particles displaying all major immunological determinants as a vaccine candidate for Lassa hemorrhagic fever. *J. Virol.* 7, 279.

Briese, T., Paweska, J.T., McMullan, L.K., Hutchison, S.K., Street, C., Palacios, G., Khristova, M.L. et al., 2009. Genetic detection and characterization of Lujo virus, a new hemorrhagic fever-associated arenavirus from southern Africa. *PLoS Pathogens* 5, e1000455.

Brunotte, L., Kerber, R., Shang, W., Hauer, F., Hass, M., Gabriel, M., Lelke, M. et al., 2011a. Structure of the Lassa virus nucleoprotein revealed by X-ray crystallography, small-angle X-ray scattering, and electron microscopy. *J. Biol. Chem.* 286, 38748–38756.

Brunotte, L., Lelke, M., Hass, M., Kleinsteuber, K., Becker-Ziaja, B., Gunther, S., 2011b. Domain structure of Lassa virus L protein. *J. Virol.* 85, 324–333.

Bruns, M., Cihak, J., Muller, G., Lehmann-Grube, F., 1983. Lymphocytic choriomeningitis virus. VI. Isolation of a glycoprotein mediating neutralization. *Virology* 130, 247–251.

Buchmeier, M.J., 2002. Arenaviruses: protein structure and function. *Curr. Top. Microbiol. Immunol.* 262, 159–173.

Buchmeier, M.J., Bowen, M.D., Peters, C.J., 2001. *Arenaviridae*: The viruses and their replication, in: Knipe, D.M., Howley, P.M., Griffin, D.E., Lamb, R.A., Martin, M.A., Roizman, B., Straus, S.E. (Eds.), *Fields Virology*. Lippincott Williams & Wilkins, Philadelphia, PA, pp. 1635–1668.

Buchmeier, M.J., Elder, J.H., Oldstone, M.B., 1978. Protein structure of lymphocytic choriomeningitis virus: Identification of the virus structural and cell associated polypeptides. *Virology* 89, 133–145.

Buchmeier, M.J., Oldstone, M.B., 1979. Protein structure of lymphocytic choriomeningitis virus: Evidence for a cell-associated precursor of the virion glycopeptides. *Virology* 99, 111–120.

Buchmeier, M.J., de la Torre, J.C., Peters, C.J., 2007. *Arenaviridae*: The viruses and their replication, in: Knipe, D.M., Howley, P.M., Griffin, D.E., Lamb, R.A., Martin, M.A., Roizman, B., Straus, S.E. (Eds.), *Fields Virology*, 5th edn. Wolters Kluwer Health/Lippincott Williams & Wilkins, Philadelphia, PA, pp. 1791–1827.

Buchmeier, M.J., Welsh, R.M., Dutko, F.J., Oldstone, M.B., 1980. The virology and immunobiology of lymphocytic choriomeningitis virus infection. *Adv. Immunol.* 30, 275–331.

Buchmeier, M.J., Zajac, A.J., 1999. *Lymphocytic Choriomeningitis Virus*. John Wiley & Sons Ltd, Chichester, U.K.

Bui, H.H., Botten, J., Fusseder, N., Pasquetto, V., Mothe, B., Buchmeier, M.J., Sette, A., 2007. Protein sequence database for pathogenic arenaviruses. *Immunome Res.* 3, 1.

Bui, H.H., Sidney, J., Dinh, K., Southwood, S., Newman, M.J., Sette, A., 2006. Predicting population coverage of T-cell epitope-based diagnostics and vaccines. *BMC Bioinform.* 7, 153.

Byrd, R.G., Cone, L.A., Commess, B.C., Williams-Herman, D., Rowland, J.M., Lee, B., Fitzgibbons, M.W. et al., 2000. Fatal illness associated with a new world arenavirus- California, 1999–2000. *Morb. Mortal. Wkly. Rep.* 49, 709–711.

Cao, W., Henry, M.D., Borrow, P., Yamada, H., Elder, J.H., Ravkov, E.V., Nichol, S.T., Compans, R.W., Campbell, K.P., Oldstone, M.B., 1998. Identification of alpha-dystroglycan as a receptor for lymphocytic choriomeningitis virus and Lassa fever virus. *Science* 282, 2079–2081.

Capul, A.A., Perez, M., Burke, E., Kunz, S., Buchmeier, M.J., de la Torre, J.C., 2007. Arenavirus Z-glycoprotein association requires Z myristoylation but not functional RING or late domains. *J. Virol.* 81, 9451–9460.

Capul, A.A., de la Torre, J.C., Buchmeier, M.J., 2011. Conserved residues in Lassa fever virus Z protein modulate viral infectivity at the level of the ribonucleoprotein. *J. Virol.* 85, 3172–3178.

Cashman, K.A., Smith, M.A., Twenhafel, N.A., Larson, R.A., Jones, K.F., Allen, R.D., 3rd, Dai, D. et al., 2011. Evaluation of Lassa antiviral compound ST-193 in a guinea pig model. *Antiviral Res.* 90, 70–79.

Charrel, R.N., Coutard, B., Baronti, C., Canard, B., Nougairede, A., Frangeul, A., Morin, B. et al., 2011. Arenaviruses and hantaviruses: From epidemiology and genomics to antivirals. *Antiviral Res.* 90, 102–114.

Charrel, R.N., de Lamballerie, X., Fulhorst, C.F., 2001. The Whitewater Arroyo virus: Natural evidence for genetic recombination among Tacaribe serocomplex viruses (family Arenaviridae). *Virology* 283, 161–166.

Childs, J.C., Peters, C.J., 1993. Ecology and epidemiology of arenaviruses and their hosts, in: Salvato, M.S. (Ed.), *The Arenaviridae*. Plenum Press, New York, pp. 331–373.

Childs, J.E., Glass, G.E., Korch, G.W., Ksiazek, T.G., Leduc, J.W., 1992. Lymphocytic choriomeningitis virus infection and house mouse (*Mus musculus*) distribution in urban Baltimore. *Am. J. Trop. Med. Hyg.* 47, 27–34.

Clegg, J.C., 2002. Molecular phylogeny of the arenaviruses. *Curr. Top. Microbiol. Immunol.* 262, 1–24.

Contigiani, M., Medeot, S., Diaz, G., 1993. Heterogeneity and stability characteristics of Candid 1 attenuated strain of Junin Virus. *Acta Virol.* 37, 41–46.

Coulibaly-N'Golo, D., Allali, B., Kouassi, S.K., Fichet-Calvet, E., Becker-Ziaja, B., Rieger, T., Olschlager, S. et al., 2011. Novel arenavirus sequences in Hylomyscus sp. and Mus (Nannomys) setulosus from Cote d'Ivoire: Implications for evolution of arenaviruses in Africa. *PLoS One* 6, e20893.

Cummins, D., Bennett, D., Fisher-Hoch, S.P., Farrar, B., Machin, S.J., McCormick, J.B., 1992. Lassa fever encephalopathy: Clinical and laboratory findings. *J. Trop. Med. Hyg.* 95, 197–201.

Cummins, D., Fisher-Hoch, S.P., Walshe, K.J., Mackie, I.J., McCormick, J.B., Bennett, D., Perez, G., Farrar, B., Machin, S.J., 1989. A plasma inhibitor of platelet aggregation in patients with Lassa fever. *Br. J. Haematol.* 72, 543–548.

Cummins, D., McCormick, J.B., Bennett, D., Samba, J.A., Farrar, B., Machin, S.J., Fisher-Hoch, S.P., 1990. Acute sensorineural deafness in Lassa fever [see comments]. *JAMA* 264, 2093–2096.

Delgado, S., Erickson, B.R., Agudo, R., Blair, P.J., Vallejo, E., Albarino, C.G., Vargas, J. et al., 2008. Chapare virus, a newly discovered arenavirus isolated from a fatal hemorrhagic fever case in Bolivia. *PLoS Pathogens* 4, e1000047.

Demby, A.H., Chamberlain, J., Brown, D.W., Clegg, C.S., 1994. Early diagnosis of Lassa fever by reverse transcription-PCR. *J. Clin. Microbiol.* 32, 2898–2903.

Di Simone, C., Zandonatti, M.A., Buchmeier, M.J., 1994. Acidic pH triggers LCMV membrane fusion activity and conformational change in the glycoprotein spike. *Virology* 198, 455–465.

Djavani, M., Yin, C., Lukashevich, I.S., Rodas, J., Rai, S.K., Salvato, M.S., 2001. Mucosal immunization with Salmonella typhimurium expressing Lassa virus nucleocapsid protein cross-protects mice from lethal challenge with lymphocytic choriomeningitis virus. *J. Hum. Virol.* 4, 103–108.

Dow, C., Oseroff, C., Peters, B., Nance-Sotelo, C., Sidney, J., Buchmeier, M., Sette, A., Mothe, B.R., 2008. Lymphocytic choriomeningitis virus infection yields overlapping CD4+ and CD8+ T-cell responses. *J. Virol.* 82, 11734–11741.

Downs, W.G., Anderson, C.R., Spence, L., Aitken, T., Greenhal, A.H., 1963. Tacaribe virus, a new agent isolated from Artibeus bats and mosquitoes in Trinidad, West Indies. *Am. J. Trop. Med. Hyg.* 12, 640–646.

Dylla, D.E., Michele, D.E., Campbell, K.P., McCray, P.B., Jr., 2008. Basolateral entry and release of New and Old World arenaviruses from human airway epithelia. *J. Virol.* 82, 6034–6038.

Edington, G.M., White, H.A., 1972. The pathology of Lassa fever. *Trans. R. Soc. Trop. Med. Hyg.* 66, 381–389.

Eichler, R., Lenz, O., Garten, W., Strecker, T., 2006. The role of single N-glycans in proteolytic processing and cell surface transport of the Lassa virus glycoprotein GP-C. *Virol. J.* 3, 41.

Eichler, R., Lenz, O., Strecker, T., Eickmann, M., Klenk, H.D., Garten, W., 2003. Identification of Lassa virus glycoprotein signal peptide as a trans-acting maturation factor. *EMBO Rep.* 4, 1084–1088.

Eichler, R., Strecker, T., Kolesnikova, L., ter Meulen, J., Weissenhorn, W., Becker, S., Klenk, H.D., Garten, W., Lenz, O., 2004. Characterization of the Lassa virus matrix protein Z: Electron microscopic study of virus-like particles and interaction with the nucleoprotein (NP). *Virus Res.* 100, 249–255.

Emond, R.T., Bannister, B., Lloyd, G., Southee, T.J., Bowen, E.T., 1982. A case of Lassa fever: Clinical and virological findings. *Br. Med. J. (Clin. Res. Ed).* 285, 1001–1002.

Emonet, S.E., Urata, S., de la Torre, J.C., 2011. Arenavirus reverse genetics: new approaches for the investigation of arenavirus biology and development of antiviral strategies. *Virology* 411, 416–425.

Emonet, S.F., de la Torre, J.C., Domingo, E., Sevilla, N., 2009. Arenavirus genetic diversity and its biological implications. *Infect. Genet. Evol.* 9, 417–429.

Enria, D.A., Briggiler, A.M., Fernandez, N.J., Levis, S.C., Maiztegui, J.I., 1984. Importance of dose of neutralising antibodies in treatment of Argentine haemorrhagic fever with immune plasma. *Lancet* 2, 255–256.

Enria, D.A., Briggiler, A.M., Levis, S., Vallejos, D., Maiztegui, J.I., Canonico, P.G., 1987. Tolerance and antiviral effect of ribavirin in patients with Argentine hemorrhagic fever. *Antiviral Res.* 7, 353–359.

Eschli, B., Quirin, K., Wepf, A., Weber, J., Zinkernagel, R., Hengartner, H., 2006. Identification of an N-terminal trimeric coiled-coil core within arenavirus glycoprotein 2 permits assignment to class I viral fusion proteins. *J. Virol.* 80, 5897–5907.

Fan, L., Briese, T., Lipkin, W.I., 2010. Z proteins of New World arenaviruses bind RIG-I and interfere with type I interferon induction. *J. Virol.* 84, 1785–1791.

Fischer, S.A., Graham, M.B., Kuehnert, M.J., Kotton, C.N., Srinivasan, A., Marty, F.M., Comer, J.A. et al., 2006. Transmission of lymphocytic choriomeningitis virus by organ transplantation. *N. Engl. J. Med.* 354, 2235–2249.

Fisher-Hoch, S.P., Hutwagner, L., Brown, B., McCormick, J.B., 2000. Effective vaccine for Lassa fever. *J. Virol.* 74, 6777–6783.

Fisher-Hoch, S.P., McCormick, J.B., 2001. Towards a human Lassa fever vaccine. *Rev. Med. Virol.* 11, 331–341.

Fisher-Hoch, S., McCormick, J.B., Sasso, D., Craven, R.B., 1988. Hematologic dysfunction in Lassa fever. *J. Med. Virol.* 26, 127–135.

Fisher-Hoch, S.P., Price, M.E., Craven, R.B., Price, F.M., Forthall, D.N., Sasso, D.R., Scott, S.M., McCormick, J.B., 1985. Safe intensive-care management of a severe case of Lassa fever with simple barrier nursing techniques. *Lancet* 2, 1227–1229.

Fisher-Hoch, S.P., Tomori, O., Nasidi, A., Perez-Oronoz, G.I., Fakile, Y., Hutwagner, L., McCormick, J.B., 1995. Review of cases of nosocomial Lassa fever in Nigeria: The high price of poor medical practice. *BMJ* 311, 857–859.

Flatz, L., Hegazy, A.N., Bergthaler, A., Verschoor, A., Claus, C., Fernandez, M., Gattinoni, L. et al., 2010a. Development of replication-defective lymphocytic choriomeningitis virus vectors for the induction of potent CD8+ T cell immunity. *Nat. Med.* 16, 339–345.

Flatz, L., Rieger, T., Merkler, D., Bergthaler, A., Regen, T., Schedensack, M., Bestmann, L. et al., 2010b. T cell-dependence of Lassa fever pathogenesis. *PLoS Pathogens* 6, e1000836.

Frame, J.D., 1989. Clinical features of Lassa fever in Liberia. *Rev. Infect. Dis.* 11(Suppl 4), S783–S789.

Frame, J.D., Baldwin, J.M., Jr., Gocke, D.J., Troup, J.M., 1970. Lassa fever, a new virus disease of man from West Africa. I. Clinical description and pathological findings. *Am. J. Trop. Med. Hyg.* 19, 670–676.

Franze-Fernandez, M.T., Zetina, C., Iapalucci, S., Lucero, M.A., Bouissou, C., Lopez, R., Rey, O., Daheli, M., Cohen, G.N., Zakin, M.M., 1987. Molecular structure and early events in the replication of Tacaribe arenavirus S RNA. *Virus Res.* 7, 309–324.

Fraser, D.W., Campbell, C.C., Monath, T.P., Goff, P.A., Gregg, M.B., 1974. Lassa fever in the Eastern Province of Sierra Leone, 1970–1972. I. Epidemiologic studies. *Am. J. Trop. Med. Hyg.* 23, 1131–1139.

Geisbert, T.W., Jones, S., Fritz, E.A., Shurtleff, A.C., Geisbert, J.B., Liebscher, R., Grolla, A. et al., 2005. Development of a new vaccine for the prevention of Lassa fever. *PLoS Med.* 2, e183.

Gessner, A., Lother, H., 1989. Homologous interference of lymphocytic choriomeningitis virus involves a ribavirin-susceptible block in virus replication. *J. Virol.* 63, 1827–1832.

Gomez, R.M., Pozner, R.G., Lazzari, M.A., D'Atri, L.P., Negrotto, S., Chudzinski-Tavassi, A.M., Berria, M.I., Schattner, M., 2003. Endothelial cell function alteration after Junin virus infection. *Thromb. Haemost.* 90, 326–333.

Groseth, A., Wolff, S., Strecker, T., Hoenen, T., Becker, S., 2010. Efficient budding of the Tacaribe virus matrix protein z requires the nucleoprotein. *J. Virol.* 84, 3603–3611.

Gunther, S., Lenz, O., 2004. Lassa virus. *Crit. Rev. Clin. Lab. Sci.* 41, 339–390.

Hass, M., Lelke, M., Busch, C., Becker-Ziaja, B., Gunther, S., 2008. Mutational evidence for a structural model of the Lassa virus RNA polymerase domain and identification of two residues, Gly1394 and Asp1395, that are critical for transcription but not replication of the genome. *J. Virol.* 82, 10207–10217.

Hastie, K.M., Kimberlin, C.R., Zandonatti, M.A., MacRae, I.J., Saphire, E.O., 2011a. Structure of the Lassa virus nucleoprotein reveals a dsRNA-specific 3' to 5' exonuclease activity essential for immune suppression. *Proc. Natl. Acad. Sci. USA* 108, 2396–2401.

Hastie, K.M., Liu, T., Li, S., King, L.B., Ngo, N., Zandonatti, M.A., Woods, V.L., Jr., de la Torre, J.C., Saphire, E.O., 2011b. Crystal structure of the Lassa virus nucleoprotein-RNA complex reveals a gating mechanism for RNA binding. *Proc. Natl. Acad. Sci. USA* 108, 19365–19370.

Igonet, S., Vaney, M.C., Vonrhein, C., Bricogne, G., Stura, E.A., Hengartner, H., Eschli, B., Rey, F.A., 2011. X-ray structure of the arenavirus glycoprotein GP2 in its postfusion hairpin conformation. *Proc. Natl. Acad. Sci. USA* 108, 19967–19972.

Ishii, A., Thomas, Y., Moonga, L., Nakamura, I., Ohnuma, A., Hang'ombe, B., Takada, A., Mweene, A., Sawa, H., 2011. Novel arenavirus, zambia. *Emerg. Infect. Dis.* 17, 1921–1924.

Jacamo, R., Lopez, N., Wilda, M., Franze-Fernandez, M.T., 2003. Tacaribe virus Z protein interacts with the L polymerase protein to inhibit viral RNA synthesis. *J. Virol.* 77, 10383–10393.

Janssen, E.M., Lemmens, E.E., Wolfe, T., Christen, U., von Herrath, M.G., Schoenberger, S.P., 2003. CD4+ T cells are required for secondary expansion and memory in CD8+ T lymphocytes. *Nature* 421, 852–856.

Jiang, X., Dalebout, T.J., Bredenbeek, P.J., Carrion, R., Jr., Brasky, K., Patterson, J., Goicochea, M., Bryant, J., Salvato, M.S., Lukashevich, I.S., 2011. Yellow fever 17D-vectored vaccines expressing Lassa virus GP1 and GP2 glycoproteins provide protection against fatal disease in guinea pigs. *Vaccine* 29, 1248–1257.

Johnson, K.M., McCormick, J.B., Webb, P.A., Smith, E.S., Elliott, L.H., King, I.J., 1987. Clinical virology of Lassa fever in hospitalized patients. *J. Infect. Dis.* 155, 456–464.

Kagi, D., Ledermann, B., Burki, K., Seiler, P., Odermatt, B., Olsen, K.J., Podack, E.R., Zinkernagel, R.M., Hengartner, H., 1994. Cytotoxicity mediated by T cells and natural killer cells is greatly impaired in perforin-deficient mice. *Nature* 369, 31–37.

Keenlyside, R.A., McCormick, J.B., Webb, P.A., Smith, E., Elliott, L., Johnson, K.M., 1983. Case control study of mastomys-natalensis and humans in Lassa virus infected households in Sierra-Leone. *Am. J. Trop. Med. Hyg.* 32, 829–837.

Kentsis, A., Dwyer, E.C., Perez, J.M., Sharma, M., Chen, A., Pan, Z.Q., Borden, K.L., 2001. The RING domains of the promyelocytic leukemia protein PML and the arenaviral protein Z repress translation by directly inhibiting translation initiation factor eIF4E. *J. Mol. Biol.* 312, 609–623.

Kentsis, A., Gordon, R.E., Borden, K.L., 2002. Self-assembly properties of a model RING domain. *Proc. Natl. Acad. Sci. USA.* 99, 667–672.

Kerber, R., Rieger, T., Busch, C., Flatz, L., Pinschewer, D.D., Kummerer, B.M., Gunther, S., 2011. Cross-species analysis of the replication complex of Old World arenaviruses reveals two nucleoprotein sites involved in L protein function. *J. Virol.* 85, 12518–12528.

Kirk, W.E., Cash, P., Peters, C.J., Bishop, D.H., 1980. Formation and characterization of an intertypic lymphocytic choriomeningitis recombinant virus. *J. Gen. Virol.* 51, 213–218.

Klavinskis, L.S., Whitton, J.L., Oldstone, M.B., 1989. Molecularly engineered vaccine which expresses an immunodominant T-cell epitope induces cytotoxic T lymphocytes that confer protection from lethal virus infection. *J. Virol.* 63, 4311–4316.

Klewitz, C., Klenk, H.D., ter Meulen, J., 2007. Amino acids from both N-terminal hydrophobic regions of the Lassa virus envelope glycoprotein GP-2 are critical for pH-dependent membrane fusion and infectivity. *J. Gen. Virol.* 88, 2320–2328.

Knobloch, J., McCormick, J.B., Webb, P.A., Dietrich, M., Schumacher, H.H., Dennis, E., 1980. Clinical observations in 42 patients with Lassa fever. *Tropenmed. Parasitol.* 31, 389–398.

Kotturi, M.F., Assarsson, E., Peters, B., Grey, H., Oseroff, C., Pasquetto, V., Sette, A., 2009a. Of mice and humans: How good are HLA transgenic mice as a model of human immune responses? *Immunome Res.* 5, 3.

Kotturi, M.F., Botten, J., Maybeno, M., Sidney, J., Glenn, J., Bui, H.H., Oseroff, C. et al., 2010. Polyfunctional CD4+ T cell responses to a set of pathogenic arenaviruses provide broad population coverage. *Immunome Res.* 6, 4.

Kotturi, M.F., Botten, J., Sidney, J., Bui, H.H., Giancola, L., Maybeno, M., Babin, J. et al., 2009b. A multivalent and cross-protective vaccine strategy against arenaviruses associated with human disease. *PLoS Pathogens* 5, e1000695.

Kotturi, M.F., Peters, B., Buendia-Laysa, F., Jr., Sidney, J., Oseroff, C., Botten, J., Grey, H., Buchmeier, M.J., Sette, A., 2007. The CD8+ T-cell response to lymphocytic choriomeningitis virus involves the L antigen: Uncovering new tricks for an old virus. *J. Virol.* 81, 4928–4940.

Kotturi, M.F., Swann, J.A., Peters, B., Arlehamn, C.L., Sidney, J., Kolla, R.V., James, E.A. et al., 2011. Human CD8 and CD4 T cell memory to lymphocytic choriomeningitis virus infection. *J. Virol.* 85, 11770–11780.

Kranzusch, P.J., Schenk, A.D., Rahmeh, A.A., Radoshitzky, S.R., Bavari, S., Walz, T., Whelan, S.P., 2010. Assembly of a functional Machupo virus polymerase complex. *Proc. Natl. Acad. Sci. USA* 107, 20069–20074.

Kranzusch, P.J., Whelan, S.P., 2011. Arenavirus Z protein controls viral RNA synthesis by locking a polymerase-promoter complex. *Proc. Natl. Acad. Sci. USA* 108, 19743–19748.

Kunz, S., 2009a. Receptor binding and cell entry of Old World arenaviruses reveal novel aspects of virus-host interaction. *Virology* 387, 245–249.

Kunz, S., 2009b. The role of the vascular endothelium in arenavirus haemorrhagic fevers. Thromb. *Haemost.* 102, 1024–1029.

Kunz, S., Edelmann, K.H., de la Torre, J.C., Gorney, R., Oldstone, M.B., 2003. Mechanisms for lymphocytic choriomeningitis virus glycoprotein cleavage, transport, and incorporation into virions. *Virology* 314, 168–178.

Lecompte, E., Fichet-Calvet, E., Daffis, S., Koulemou, K., Sylla, O., Kourouma, F., Dore, A. et al., 2006. Mastomys natalensis and Lassa fever, West Africa. *Emerg. Infect. Dis.* 12, 1971–1974.

Lee, A.M., Pasquato, A., Kunz, S., 2011. Novel approaches in anti-arenaviral drug development. *Virology* 411, 163–169.

Lee, A.M., Rojek, J.M., Spiropoulou, C.F., Gundersen, A.T., Jin, W., Shaginian, A., York, J., Nunberg, J.H., Boger, D.L., Oldstone, M.B., Kunz, S., 2008. Unique small molecule entry inhibitors of hemorrhagic fever arenaviruses. *J. Biol. Chem.* 283, 18734–18742.

Lehmann-Grube, F., 1971. *Lymphocytic Choriomeningitis Virus. Virology Monographs*. Springer Verlag, New York.

Lehmann-Grube, F., Lohler, J., Utermohlen, O., Gegin, C., 1993. Antiviral immune responses of lymphocytic choriomeningitis virus-infected mice lacking CD8+ T lymphocytes because of disruption of the beta 2-microglobulin gene. *J. Virol.* 67, 332–339.

Lelke, M., Brunotte, L., Busch, C., Gunther, S., 2010. An N-terminal region of Lassa virus L protein plays a critical role in transcription but not replication of the virus genome. *J. Virol.* 84, 1934–1944.

Lenz, O., ter Meulen, J., Klenk, H.D., Seidah, N.G., Garten, W., 2001. The Lassa virus glycoprotein precursor GP-C is proteolytically processed by subtilase SKI-1/S1P. *Proc. Natl. Acad. Sci. USA* 98, 12701–12705.

Levingston Macleod, J.M., D'Antuono, A., Loureiro, M.E., Casabona, J.C., Gomez, G.A., Lopez, N., 2011. Identification of two functional domains within the arenavirus nucleoprotein. *J. Virol.* 85, 2012–2023.

Lukashevich, I.S., Djavani, M., Shapiro, K., Sanchez, A., Ravkov, E., Nichol, S.T., Salvato, M.S., 1997. The Lassa fever virus L gene: nucleotide sequence, comparison, and precipitation of a predicted 250 kDa protein with monospecific antiserum. *J. Gen. Virol.* 78, 547–551.

Lukashevich, I.S., Maryankova, R., Vladyko, A.S., Nashkevich, N., Koleda, S., Djavani, M., Horejsh, D., Voitenok, N.N., Salvato, M.S., 1999. Lassa and Mopeia virus replication in human monocytes/macrophages and in endothelial cells: different effects on IL-8 and TNF-alpha gene expression. *J. Med. Virol.* 59, 552–560.

Lukashevich, I.S., Patterson, J., Carrion, R., Moshkoff, D., Ticer, A., Zapata, J., Brasky, K., Geiger, R., Hubbard, G.B., Bryant, J., Salvato, M.S., 2005. A live attenuated vaccine for Lassa fever made by reassortment of Lassa and Mopeia viruses. *J. Virol.* 79, 13934–13942.

Mahanty, S., Bausch, D.G., Thomas, R.L., Goba, A., Bah, A., Peters, C.J., Rollin, P.E., 2001. Low levels of interleukin-8 and interferon-inducible protein-10 in serum are associated with fatal infections in acute Lassa fever. *J. Infect. Dis.* 183, 1713–1721.

Mahanty, S., Hutchinson, K., Agarwal, S., McRae, M., Rollin, P.E., Pulendran, B., 2003. Cutting edge: Impairment of dendritic cells and adaptive immunity by Ebola and Lassa viruses. *J. Immunol.* 170, 2797–2801.

Maiztegui, J.I., Fernandez, N.J., de Damilano, A.J., 1979. Efficacy of immune plasma in treatment of Argentine haemorrhagic fever and association between treatment and a late neurological syndrome. *Lancet* 2, 1216–1217.

Maiztegui, J.I., McKee, K.T., Jr., Barrera Oro, J.G., Harrison, L.H., Gibbs, P.H., Feuillade, M.R., Enria, D.A. et al., 1998. Protective efficacy of a live attenuated vaccine against Argentine hemorrhagic fever. AHF Study Group. *J. Infect. Dis.* 177, 277–283.

Marq, J.B., Hausmann, S., Veillard, N., Kolakofsky, D., Garcin, D., 2011. Short double-stranded RNAs with an overhanging 5′ ppp-nucleotide, as found in arenavirus genomes, act as RIG-I decoys. *J. Biol. Chem.* 286, 6108–6116.

Martinez-Sobrido, L., Emonet, S., Giannakas, P., Cubitt, B., Garcia-Sastre, A., de la Torre, J.C., 2009. Identification of amino acid residues critical for the anti-interferon activity of the nucleoprotein of the prototypic arenavirus lymphocytic choriomeningitis virus. *J. Virol.* 83, 11330–11340.

Martinez-Sobrido, L., Giannakas, P., Cubitt, B., Garcia-Sastre, A., de la Torre, J.C., 2007. Differential inhibition of type I interferon induction by arenavirus nucleoproteins. *J. Virol.* 81, 12696–12703.

Martinez-Sobrido, L., Zuniga, E.I., Rosario, D., Garcia-Sastre, A., de la Torre, J.C., 2006. Inhibition of the type I interferon response by the nucleoprotein of the prototypic arenavirus lymphocytic choriomeningitis virus. *J. Virol.* 80, 9192–9199.

Matloubian, M., Concepcion, R.J., Ahmed, R., 1994. CD4+ T cells are required to sustain CD8+ cytotoxic T-cell responses during chronic viral infection. *J. Virol.* 68, 8056–8063.

McCormick, J.B., 1986. Clinical, epidemiologic, and therapeutic aspects of Lassa fever. *Med. Microbiol. Immunol. (Berl)* 175, 153–155.

McCormick, J.B., Fisher-Hoch, S.P., 2002. Lassa fever. *Curr. Top. Microbiol. Immunol.* 262, 75–109.

McCormick, J.B., King, I.J., Webb, P.A., Johnson, K.M., O'Sullivan, R., Smith, E.S., Trippel, S., Tong, T.C., 1987a. A case-control study of the clinical diagnosis and course of Lassa fever. *J. Infect. Dis.* 155, 445–455.

McCormick, J.B., King, I.J., Webb, P.A., Scribner, C.L., Craven, R.B., Johnson, K.M., Elliott, L.H., Belmont-Williams, R., 1986a. Lassa fever. Effective therapy with ribavirin. *N. Engl. J. Med.* 314, 20–26.

McCormick, J.B., Walker, D.H., King, I.J., Webb, P.A., Elliott, L.H., Whitfield, S.G., Johnson, K.M., 1986b. Lassa virus hepatitis: a study of fatal Lassa fever in humans. *Am. J. Trop. Med. Hyg.* 35, 401–407.

McCormick, J.B., Webb, P.A., Krebs, J.W., Johnson, K.M., Smith, E.S., 1987b. A prospective study of the epidemiology and ecology of Lassa fever. *J. Infect. Dis.* 155, 437–444.

McKee, K.T., Jr., Huggins, J.W., Trahan, C.J., Mahlandt, B.G., 1988. Ribavirin prophylaxis and therapy for experimental argentine hemorrhagic fever. Antimicrob. *Agents Chemother.* 32, 1304–1309.

Mertens, P.E., Patton, R., Baum, J.J., Monath, T.P., 1973. Clinical presentation of Lassa fever cases during the hospital epidemic at Zorzor, Liberia, March–April 1972. *Am. J. Trop. Med. Hyg.* 22, 780–784.

Meyer, H.M., Johnson, R.T., Crawford, E.P., Dascomb, H.E., Rogers, N.G., 1960. Central nervous system syndromes of "viral" etiology. *Am. J. Med.* 29, 334–347.

Meyer, B.J., de la Torre, J.C., Southern, P.J., 2002. Arenaviruses: genomic RNAs, transcription, and replication. *Curr. Top. Microbiol. Immunol.* 262, 139–157.

Monath, T.P., Newhouse, V.F., Kemp, G.E., Setzer, H.W., Cacciapuoti, A., 1974. Lassa virus isolation from Mastomys natalensis rodents during an epidemic in Sierra Leone. *Science* 185, 263–265.

Moraz, M.L., Kunz, S., 2011. Pathogenesis of arenavirus hemorrhagic fevers. *Expert Rev. Anti. Infect. Ther.* 9, 49–59.

Morin, B., Coutard, B., Lelke, M., Ferron, F., Kerber, R., Jamal, S., Frangeul, A. et al., 2010. The N-terminal domain of the arenavirus L protein is an RNA endonuclease essential in mRNA transcription. *PLoS Pathogens* 6, e1001038.

Morrison, H.G., Bauer, S.P., Lange, J.V., Esposito, J.J., McCormick, J.B., Auperin, D.D., 1989. Protection of guinea pigs from Lassa fever by vaccinia virus recombinants expressing the nucleoprotein or the envelope glycoproteins of Lassa virus. *Virology* 171, 179–188.

Mothe, B.R., Stewart, B.S., Oseroff, C., Bui, H.H., Stogiera, S., Garcia, Z., Dow, C. et al., 2007. Chronic lymphocytic choriomeningitis virus infection actively down-regulates CD4+ T cell responses directed against a broad range of epitopes. *J. Immunol.* 179, 1058–1067.

Muller, S., Gunther, S., 2007. Broad-spectrum antiviral activity of small interfering RNA targeting the conserved RNA termini of Lassa virus. *Antimicrob. Agents Chemother.* 51, 2215–2218.

Neuman, B.W., Adair, B.D., Burns, J.W., Milligan, R.A., Buchmeier, M.J., Yeager, M., 2005. Complementarity in the supramolecular design of arenaviruses and retroviruses revealed by electron cryomicroscopy and image analysis. *J. Virol.* 79, 3822–3830.

Neuman, B.W., Bederka, L.H., Stein, D.A., Ting, J.P., Moulton, H.M., Buchmeier, M.J., 2011. Development of peptide-conjugated morpholino oligomers as pan-arenavirus inhibitors. *Antimicrob. Agents Chemother.* 55, 4631–4638.

Nunberg, J.H., York, J., 2012. The curious case of arenavirus entry, and its inhibition. *Viruses* 4, 83–101.

Oldstone, M.B., 2002. Biology and pathogenesis of lymphocytic choriomeningitis virus infection. *Curr. Top. Microbiol. Immunol.* 263, 83–117.

Oldstone, M.B., Borrow, P., 1992. Characterization of lymphocytic choriomeningitis virus-binding protein(s): a candidate cellular receptor for the virus. *J. Virol.* 66, 7270–7281.

Ortiz-Riano, E., Cheng, B.Y., de la Torre, J.C., Martinez-Sobrido, L., 2011. The C-terminal region of lymphocytic choriomeningitis virus nucleoprotein contains distinct and segregable functional domains involved in NP-Z interaction and counteraction of the type I interferon response. *J. Virol.* 85, 13038–13048.

Palacios, G., Druce, J., Du, L., Tran, T., Birch, C., Briese, T., Conlan, S. et al., 2008. A new arenavirus in a cluster of fatal transplant-associated diseases. *N. Engl. J. Med.* 358, 991–998.

Palacios, G., Savji, N., Hui, J., Travassos da Rosa, A., Popov, V., Briese, T., Tesh, R., Lipkin, W.I., 2010. Genomic and phylogenetic characterization of Merino Walk virus, a novel arenavirus isolated in South Africa. *J. Gen. Virol.* 91, 1315–1324.

Panda, D., Das, A., Dinh, P.X., Subramaniam, S., Nayak, D., Barrows, N.J., Pearson, J.L. et al., 2011. RNAi screening reveals requirement for host cell secretory pathway in infection by diverse families of negative-strand RNA viruses. *Proc. Natl. Acad. Sci. USA* 108, 19036–19041.

Pannetier, D., Faure, C., Georges-Courbot, M.C., Deubel, V., Baize, S., 2004. Human macrophages, but not dendritic cells, are activated and produce alpha/beta interferons in response to Mopeia virus infection. *J. Virol.* 78, 10516–10524.

Pasqual, G., Burri, D.J., Pasquato, A., de la Torre, J.C., Kunz, S., 2011a. Role of the host cell's unfolded protein response in arenavirus infection. *J. Virol.* 85, 1662–1670.

Pasqual, G., Rojek, J.M., Masin, M., Chatton, J.Y., Kunz, S., 2011b. Old world arenaviruses enter the host cell via the multivesicular body and depend on the endosomal sorting complex required for transport. *PLoS Pathogens* 7, e1002232.

Paweska, J.T., Sewlall, N.H., Ksiazek, T.G., Blumberg, L.H., Hale, M.J., Lipkin, W.I., Weyer, J. et al., 2009. Nosocomial outbreak of novel arenavirus infection, southern Africa. *Emerg. Infect. Dis.* 15, 1598–1602.

Perez, M., Greenwald, D.L., de la Torre, J.C., 2004. Myristoylation of the RING finger Z protein is essential for arenavirus budding. *J. Virol.* 78, 11443–11448.

Peters, C.J., Jahrling, P.B., Liu, C.T., Kenyon, R.H., McKee, K.T., Jr., Barrera Oro, J.G., 1987. Experimental studies of arenaviral hemorrhagic fevers. *Curr. Top. Microbiol. Immunol.* 134, 5–68.

Pinschewer, D.D., Perez, M., Jeetendra, E., Bachi, T., Horvath, E., Hengartner, H., Whitt, M.A., de la Torre, J.C., Zinkernagel, R.M., 2004. Kinetics of protective antibodies are determined by the viral surface antigen. *J. Clin. Invest.* 114, 988–993.

Price, M.E., Fisher-Hoch, S.P., Craven, R.B., McCormick, J.B., 1988. A prospective study of maternal and fetal outcome in acute Lassa fever infection during pregnancy. *BMJ* 297, 584–587.

Pushko, P., Geisbert, J., Parker, M., Jahrling, P., Smith, J., 2001. Individual and bivalent vaccines based on alphavirus replicons protect guinea pigs against infection with Lassa and Ebola viruses. *J. Virol.* 75, 11677–11685.

Qi, X., Lan, S., Wang, W., Schelde, L.M., Dong, H., Wallat, G.D., Ly, H., Liang, Y., Dong, C., 2010. Cap binding and immune evasion revealed by Lassa nucleoprotein structure. *Nature* 468, 779–783.

Radoshitzky, S.R., Abraham, J., Spiropoulou, C.F., Kuhn, J.H., Nguyen, D., Li, W., Nagel, J. et al., 2007. Transferrin receptor 1 is a cellular receptor for New World haemorrhagic fever arenaviruses. *Nature* 446, 92–96.

Reignier, T., Oldenburg, J., Noble, B., Lamb, E., Romanowski, V., Buchmeier, M.J., Cannon, P.M., 2006. Receptor use by pathogenic arenaviruses. *Virology* 353, 111–120.

Richmond, J.K., Baglole, D.J., 2003. Lassa fever: Epidemiology, clinical features, and social consequences. *BMJ* 327, 1271–1275.

Riviere, Y., Ahmed, R., Southern, P.J., Buchmeier, M.J., Oldstone, M.B., 1985. Genetic mapping of lymphocytic choriomeningitis virus pathogenicity: Virulence in guinea pigs is associated with the L RNA segment. *J. Virol.* 55, 704–709.

Robbins, C.B., Krebs, J.W., Jr., Johnson, K.M., 1983. Mastomys (rodentia: muridae) species distinguished by hemoglobin pattern differences. *Am. J. Trop. Med. Hyg.* 32, 624–630.

Roberts, P.J., Cummins, D., Bainton, A.L., Walshe, K.J., Fisher-Hoch, S.P., McCormick, J.B., Gribben, J.G., Machin, S.J., Linch, D.C., 1989. Plasma from patients with severe Lassa fever profoundly modulates f-met-leu-phe induced superoxide generation in neutrophils. *Br. J. Haematol.* 73, 152–157.

Rodriguez-Carreno, M.P., Nelson, M.S., Botten, J., Smith-Nixon, K., Buchmeier, M.J., Whitton, J.L., 2005. Evaluating the immunogenicity and protective efficacy of a DNA vaccine encoding Lassa virus nucleoprotein. *Virology* 335, 87–98.

Rojek, J.M., Kunz, S., 2008. Cell entry by human pathogenic arenaviruses. *Cell Microbiol.* 10, 828–835.

Rojek, J.M., Lee, A.M., Nguyen, N., Spiropoulou, C.F., Kunz, S., 2008. Site 1 protease is required for proteolytic processing of the glycoproteins of the South American hemorrhagic fever viruses Junin, Machupo, and Guanarito. *J. Virol.* 82, 6045–6051.

Salazar-Bravo, J., Ruedas, L.A., Yates, T.L., 2002. Mammalian reservoirs of arenaviruses. *Curr. Top. Microbiol. Immunol.* 262, 25–63.

Sanchez, A.B., de la Torre, J.C., 2005. Genetic and biochemical evidence for an oligomeric structure of the functional L polymerase of the prototypic arenavirus lymphocytic choriomeningitis virus. *J. Virol.* 79, 7262–7268.

Schlie, K., Maisa, A., Freiberg, F., Groseth, A., Strecker, T., Garten, W., 2010. Viral protein determinants of Lassa virus entry and release from polarized epithelial cells. *J. Virol.* 84, 3178–3188.

Schulz, M., Zinkernagel, R.M., Hengartner, H., 1991. Peptide-induced antiviral protection by cytotoxic T cells. *Proc. Natl. Acad. Sci. USA* 88, 991–993.

Sette, A., Sidney, J., 1999. Nine major HLA class I supertypes account for the vast preponderance of HLA-A and -B polymorphism. *Immunogenetics* 50, 201–212.

Shedlock, D.J., Shen, H., 2003. Requirement for CD4 T cell help in generating functional CD8 T cell memory. *Science* 300, 337–339.

Shimojima, M., Stroher, U., Ebihara, H., Feldmann, H., Kawaoka, Y., 2012. Identification of cell surface molecules involved in dystroglycan-independent Lassa virus cell entry. *J. Virol.* 86, 2067–2078.

Southern, P.J., 1996. Arenaviridae: The viruses and their replication, in: Fields, B.N., Knipe, D.M., Howley, P.M. et al. (Eds.), *Fields Virology*. Lippencott-Raven, Philadelphia, PA, pp. 1505–1519.

Sun, J.C., Williams, M.A., Bevan, M.J., 2004. CD4+ T cells are required for the maintenance, not programming, of memory CD8+ T cells after acute infection. *Nat. Immunol.* 5, 927–933.

Ter Meulen, J., Badusche, M., Kuhnt, K., Doetze, A., Satoguina, J., Marti, T., Loeliger, C. et al., 2000. Characterization of human CD4(+) T-cell clones recognizing conserved and variable epitopes of the Lassa virus nucleoprotein. *J. Virol.* 74, 2186–2192.

Ter Meulen, J., Badusche, M., Satoguina, J., Strecker, T., Lenz, O., Loeliger, C., Sakho, M., Koulemou, K., Koivogui, L., Hoerauf, A., 2004. Old and New World arenaviruses share a highly conserved epitope in the fusion domain of the glycoprotein 2, which is recognized by Lassa virus-specific human CD4+ T-cell clones. *Virology* 321, 134–143.

Ter Meulen, J., Lukashevich, I., Sidibe, K., Inapogui, A., Marx, M., Dorlemann, A., Yansane, M.L., Koulemou, K., Chang-Claude, J., Schmitz, H., 1996. Hunting of peridomestic rodents and consumption of their meat as possible risk factors for rodent-to-human transmission of Lassa virus in the Republic of Guinea. *Am. J. Trop. Med. Hyg.* 55, 661–666.

Thomas, C.J., Casquilho-Gray, H.E., York, J., DeCamp, D.L., Dai, D., Petrilli, E.B., Boger, D.L. et al., 2011. A specific interaction of small molecule entry inhibitors with the envelope glycoprotein complex of the Junin hemorrhagic fever arenavirus. *J. Biol. Chem.* 286, 6192–6200.

Trappier, S.G., Conaty, A.L., Farrar, B.B., Auperin, D.D., McCormick, J.B., Fisher-Hoch, S.P., 1993. Evaluation of the polymerase chain reaction for diagnosis of Lassa virus infection. *Am. J. Trop. Med. Hyg.* 49, 214–221.

Urata, S., de la Torre, J.C., 2011. Arenavirus budding. *Adv. Virol.* 2011, 180326.

Vieth, S., Torda, A.E., Asper, M., Schmitz, H., Gunther, S., 2004. Sequence analysis of L RNA of Lassa virus. *Virology* 318, 153–168.

Volpon, L., Osborne, M.J., Capul, A.A., de la Torre, J.C., Borden, K.L., 2010. Structural characterization of the Z RING-eIF4E complex reveals a distinct mode of control for eIF4E. *Proc. Natl. Acad. Sci. USA* 107, 5441–5446.

Walker, D.H., McCormick, J.B., Johnson, K.M., Webb, P.A., Komba-Kono, G., Elliott, L.H., Gardner, J.J., 1982. Pathologic and virologic study of fatal Lassa fever in man. *Am. J. Pathol.* 107, 349–356.

Walker, D.H., Wulff, H., Lange, J.V., Murphy, F.A., 1975. Comparative pathology of Lassa virus infection in monkeys, guinea-pigs, and Mastomys natalensis. *Bull. World Health Organ.* 52, 523–534.

Walsh, C.M., Matloubian, M., Liu, C.C., Ueda, R., Kurahara, C.G., Christensen, J.L., Huang, M.T., Young, J.D., Ahmed, R., Clark, W.R., 1994. Immune function in mice lacking the perforin gene. *Proc. Natl. Acad. Sci. USA* 91, 10854–10858.

Wille, A., Gessner, A., Lother, H., Lehmann-Grube, F., 1989. Mechanism of recovery from acute virus infection. VIII. Treatment of lymphocytic choriomeningitis virus-infected mice with anti-interferon- gamma monoclonal antibody blocks generation of virus-specific cytotoxic T lymphocytes and virus elimination. *Eur. J. Immunol.* 19, 1283–1288.

Wright, K.E., Spiro, R.C., Burns, J.W., Buchmeier, M.J., 1990. Post-translational processing of the glycoproteins of lymphocytic choriomeningitis virus. *Virology* 177, 175–183.

Wright, R., Johnson, D., Neumann, M., Ksiazek, T.G., Rollin, P., Keech, R.V., Bonthius, D.J., Hitchon, P., Grose, C.F., Bell, W.E., Bale, J.F., Jr., 1997. Congenital lymphocytic choriomeningitis virus syndrome: A disease that mimics congenital toxoplasmosis or Cytomegalovirus infection. *Pediatrics* 100, E9.

York, J., Agnihothram, S.S., Romanowski, V., Nunberg, J.H., 2005. Genetic analysis of heptad-repeat regions in the G2 fusion subunit of the Junin arenavirus envelope glycoprotein. *Virology* 343, 267–274.

York, J., Nunberg, J.H., 2006. Role of the stable signal peptide of Junin arenavirus envelope glycoprotein in pH-dependent membrane fusion. *J. Virol.* 80, 7775–7780.

York, J., Romanowski, V., Lu, M., Nunberg, J.H., 2004. The signal peptide of the Junin arenavirus envelope glycoprotein is myristoylated and forms an essential subunit of the mature G1-G2 complex. *J. Virol.* 78, 10783–10792.

15 Lassa Fever

Daniel G. Bausch, Lina M. Moses, Augustine Goba, Donald S. Grant, and Humarr Khan

CONTENTS

15.1   INTRODUCTION

Viral hemorrhagic fever (VHF) is an acute systemic illness classically involving fever, a constellation of initially nonspecific signs and symptoms, and a propensity for bleeding and shock. The syndrome may be caused by more than 25 different viruses from four taxonomic families. One of the most important hemorrhagic fever viruses is Lassa virus (LASV), the cause of Lassa fever (LF), a rodent-borne disease that is endemic in West Africa (Figure 15.1) (Enria et al., 2011). LF was first recognized in Nigeria in 1969 and named after the village Lassa in

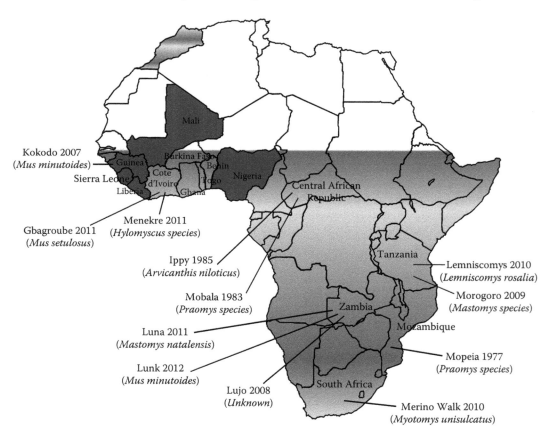

FIGURE 15.1 Geographic distribution of Old World arenaviruses. The virus name and date of discovery of each virus are shown, with the known or suspected rodent reservoir listed in parentheses. Countries where LF (virus discovered 1969, reservoir Mastomys natalensis) has been definitively shown are depicted in dark gray. Indirect evidence, such as anecdotal reports or seroprevalence data, exists for the other named countries in West Africa. The distribution of rodents of the genus Mastomys is shown in light gray. With the exception of Lassa and lymphocytic choriomeningitis viruses, most Old World arenaviruses have been isolated on single or very few occasions, and the precise distribution of the virus beyond the place of first isolation is unknown. Not shown on the map: Lymphocytic choriomeningitis viruses (discovered 1933, reservoir Mus musculus), which has a worldwide distribution, and Dandenong virus (discovered 2009, unknown reservoir), which is thought to be found in Eastern Europe. Only Lassa, Lujo, Dandenong, and lymphocytic choriomeningitis viruses are known to be human pathogens.

northeastern Nigeria, from which the first recognized case came (Frame et al., 1970). However, hospital records and anecdotal reports describe cases consistent with LF from this area as far back as 1956 (Bond et al., 2013).

Because of their high pathogenicity, risk of secondary spread (although this is often overestimated), and tendency to cause public panic and social disruption, the use of LASV and other hemorrhagic fever viruses as bioweapons is a concern (Bausch and Peters, 2009). LASV is classified as biosafety level 4 (BSL-4) or "high containment." Laboratory accidents resulting in LASV infection have been reported before, but not since, the BSL-4 system was implemented.

Much of our knowledge about LF comes from field studies conducted in West Africa in the 1970s and 1980s. In subsequent decades, civil unrest in the areas of West Africa where LF is of highest endemicity severely curtailed research, often preventing confirmation of initial findings. Consequently, some assumptions have become *de facto* dogma over the years, despite limited objective data. Renewed peace in many areas of West Africa in recent years, especially in Sierra Leone and Liberia, combined with increased funding driven primarily by industrialized countries' concerns about hemorrhagic fever viruses as bioweapons have rejuvenated research in LF, offering the opportunity to revisit many pertinent questions (Khan et al., 2008; Schuler, 2005). Furthermore, research advances have put new "tools" at our disposition, including more sensitive and specific diagnostics, molecular methods for accurate species designation of rodents, reverse genetics (or so-called mini-replicon systems) to explore the biology and pathogenesis, and various animal models.

15.2 BIOLOGICAL PROPERTIES

15.2.1 TAXONOMY

LASV is a member of the *Arenaviridae* family, which is serologically, phylogenetically, and geographically divided into an Old World (or Lymphocytic Choriomeningitis) Complex, which includes LASV, and a New World (or Tacaribe) Complex (Figures 15.1 and 15.2). Arenaviruses are phylogenetically closely related to other segmented negative-strand RNA viruses such as the *Bunyaviridae* and *Orthomyxoviridae*, with which they share basic features of the replication cycle (Buchmeier et al., 2007). Based on the phylogeny of the conserved amino acid sequences of the viral RNA polymerase, *Arenaviridae* are most closely related to the Nairovirus genus of *Bunyaviridae*. Although Old World arenaviruses continue to be discovered, especially in Africa, only the recently discovered Lujo virus is known to cause VHF (see Chapter 16) (Figure 15.1) (Paweska et al., 2009).

15.2.2 PHYSICAL PROPERTIES

The *Arenaviridae* family derives its name from the Latin "arenosus" for "sandy," referring to the grainy appearance of internal electron-dense particles seen on electron microscopy, which are host cell ribosomes (Figure 15.3). Surface glycoproteins are seen as club-shaped projections or spikes protruding from the envelope membrane. Virions are pleomorphic, ranging in size from 60 to 300 nM, and have a lipid envelope.

15.2.3 GENE ORGANIZATION, REPLICATION, AND TRANSCRIPTION

Arenaviruses are bisegmented viruses with a ~11 kb genome comprising two single-stranded RNA segments denoted small (S) and large (L) of 3.4 and 7.2 kb, respectively (Buchmeier et al., 2007). The S RNA encodes the viral nucleoprotein (NP) and a precursor glycoprotein protein (GPC) that is post-translationally cleaved into GP1 and GP2 by the cellular subtilase SK1-1/S1P. Proteolytic processing of GPC is necessary for LASV infectivity. GP1 is a peripheral membrane protein while GP2

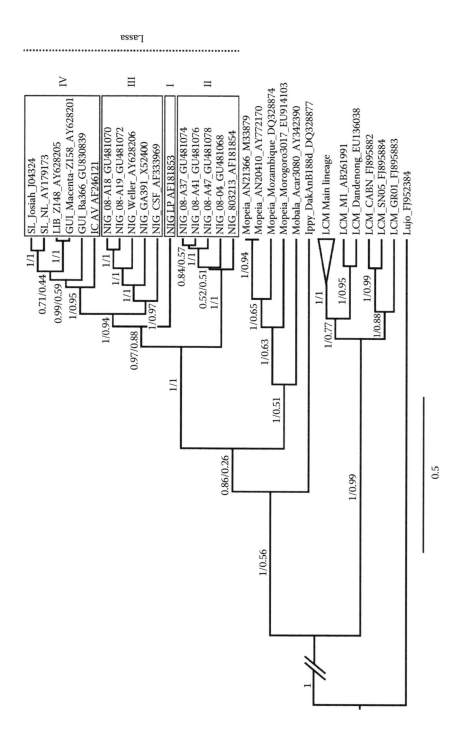

FIGURE 15.2 Phylogenetic analysis of the Old World Arenavirus Complex. The phylogram is based on complete nucleotide sequences of the GP gene (1473 nucleotides). Similar results (not shown) were found with complete nucleotide sequencing of the LASV NP and L genes. The origin of LASV strains is indicated by the following prefixes: **SL**, Sierra Leone; **LIB**, Liberia; **GUI**, Guinea; **IC**, Ivory Coast; **NIG**, Nigeria; **LCM**, Lymphocytic choriomeningitis virus. Recently discovered nonpathogenic arenaviruses are not shown. (Adapted with permission from Ehichioya, D.U. et al., *J. Clin. Microbiol.*, 49, 1157, 2011.)

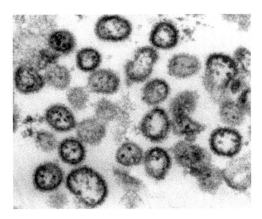

FIGURE 15.3 Electron micrograph of LASV. The sandy appearance from the ribosomes inside virions is evident, as well as the surface glycoprotein spikes. Magnification approximately × 55,000. (Micrograph courtesy of F. A. Murphy, University of Texas Medical Branch, Galveston, TX.)

is transmembrane. Both are involved in receptor binding and cell entry. The L RNA encodes the viral polymerase (L protein) and a small zinc-binding protein (Z), which appears to play a regulatory role in virus replication, particle formation, and budding, as well as having a structural function as a matrix protein. The genes on the two RNA segments are separated by an intergenic region that folds into a stable secondary structure.

Arenavirus genes are oriented in both negative and positive senses on the two RNA segments, a coding strategy called ambisense. Through this mechanism, GPC and NP gene expression are independently regulated. Viral RNA must be transcribed before GPC can be expressed. Replication and transcription of the genome take place in the cytoplasm and require the association of viral proteins with the viral RNA in the form of ribonucleoprotein (RNP) complexes. The NP and L proteins, together with virus RNA, are the minimal components of the RNP complex and are sufficient for genome replication and transcription. Purified RNPs are competent for RNA synthesis *in vitro* and can initiate virus replication after transfection into cells. Naked RNA is not infectious.

During genome replication, a full-length copy of the genome is synthesized yielding the corresponding antigenomic S and L RNAs. Due to the ambisense coding strategy, both genomic and antigenomic RNA serve as templates for transcription of viral mRNA. The transcripts contain a cap but are not polyadenylated. The viral RNA species that are packaged into virions are defined as the genomic RNAs; however, smaller amounts of antigenomic RNA and Z gene mRNA are also packaged. In addition to viral RNA, ribosomal RNA is present within virions. Virions mature by budding from the plasma membrane.

15.2.4 PHYLOGENETICS

There is considerable sequence heterogeneity of LASVs across West Africa, with four recognized lineages—three in Nigeria and one in the area comprising Sierra Leone, Liberia, Guinea, and Ivory Coast (Figure 15.2) (Bowen et al., 2000). There is also considerable genetic heterogeneity within lineages, especially in Nigeria. The largest variation is in the L and Z genes, with mean differences of 26% and 20% at the nucleotide and amino acid levels, respectively. The structural genes NP and GPC are more conserved, with mean differences of about 20% and 8% at the nucleotide and amino acid levels, respectively (Vieth et al., 2004). Phylogenetic analyses suggest that Nigerian strains are ancestral to those found further west in Africa, with limited recombination in the evolution of LASV.

15.2.5 Receptor

α-Dystroglycan, which is expressed in a wide range of tissues in humans and rodents, is at least one essential receptor for LASV, likely binding to GP1 (Oldstone and Campbell, 2011). GP2 is thought to then mediate fusion of the viral envelope with the cellular membrane and thus entry of the virus into the host cell. Interestingly, LASV and other nonpathogenic arenaviruses bind α-dystroglycan with equal affinity. The transmembrane proteins Axl, Tyro3, dendritic cell-specific intercellular adhesion molecule 3-grabbing nonintegrin (DC-SIGN), and liver and lymph node sinusoidal endothelial calcium-dependent lectin (LSECtin) may also serve as LASV receptors independent of α-dystroglycan (Shimojima et al., 2012).

15.3 CLINICAL PRESENTATION

LF is seen in both genders and all age groups, with a spectrum ranging from mild disease to shock, multi-organ system failure, and death. Most patients present with nonspecific signs and symptoms difficult to distinguish from a host of other febrile illnesses common in the tropics. The incubation period is usually about 1 week (range 3–21 days). Illness typically begins with the gradual onset of fever and constitutional symptoms, including general malaise, anorexia, headache, chest or retrosternal pain, sore throat, myalgia, arthralgia, lumbosacral pain, and dizziness (Bausch et al., 2001; McCormick et al., 1987a). The pharynx may be erythemic or even exudative, a finding that has at times led to misdiagnosis of streptococcal pharyngitis. In one study in Sierra Leone, pharyngitis was the most sensitive (70%) indicator of LF, although not as specific (60%) (McCormick et al., 1987a). Gastrointestinal signs and symptoms occur early in the course of disease and may include nausea, vomiting, epigastric and abdominal pain and tenderness, and diarrhea. LF has sometimes been mistaken for acute appendicitis or other abdominal emergencies. A morbilliform, maculopapular, or petechial skin rash almost always occurs in fair-skinned persons but, for unclear reasons, rarely in blacks. Conjunctival injection or hemorrhage is frequent but is not accompanied by itching, discharge, or rhinitis (Figure 15.4a). A dry cough, sometimes accompanied by a few scattered rales on auscultation, may be noted, but prominent pulmonary symptoms are uncommon early in the course of the disease. Jaundice is not typical and should suggest another diagnosis.

In severe cases, patients progress to vascular instability, which may be manifested by subconjunctival hemorrhage, facial flushing, edema, bleeding, hypotension, shock, and proteinuria. Swelling in the face and neck and bleeding are particularly specific signs but are not very sensitive—seen in less than 20% of cases (Figure 15.4b). Despite the term "hemorrhagic fever," clinically discernible hemorrhage is seen in less than 20% of cases and never in the first few days of illness. Hematemesis, melena, hematochezia, metrorrhagia, petechiae, epistaxis, and bleeding from the gums and venipuncture sites may develop, but hemoptysis and hematuria are infrequent (Figure 15.4c). Significant internal bleeding from the gastrointestinal tract may occur even in the absence of external hemorrhage.

Neurological complications, including disorientation, tremor, ataxia, seizures, and coma, may be seen, particularly in the late stages, and usually portend a fatal outcome (Solbrig and McCormick, 1991). Laboratory studies show arenavirus tropism for Schwann cells (though not neurons), which abundantly express α-dystroglycan (Rambukkana et al., 2003). LASV can be isolated from the cerebrospinal fluid (CSF) of some but not all patients with neurological manifestations, without apparent correlation between disease severity and virus or antibody titer (Cummins et al., 1992; Johnson et al., 1987). In one unusual case, LASV was isolated from the CSF, but not the blood, of a patient with encephalopathy after the febrile stage of disease (Gunther et al., 2001). The cellular and chemistry profile in CSF is often normal.

Pregnant women with LF often present with spontaneous abortion and vaginal bleeding, with LASV found at high concentrations in placenta and fetal tissues (Walker et al., 1982). Anasarca has

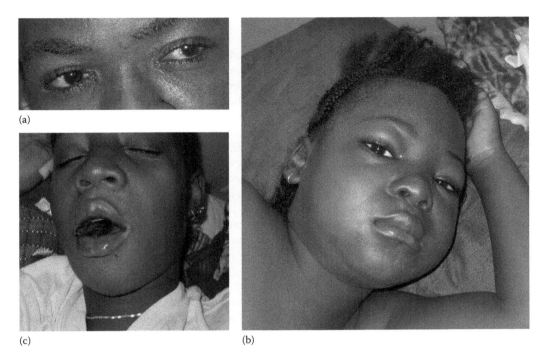

FIGURE 15.4 **(See color insert.)** Clinical manifestations of LF. Beginning clockwise from the top left are examples of (a) conjunctival injection, (b) facial and neck swelling, and (c) oral mucosal bleeding. (Photos by Daniel Bausch, Donald Grant, and Victor Lungi, respectively.)

been described in a single report of children with LF (termed the "swollen baby syndrome") but may have been related to aggressive rehydration (Monson et al., 1987). One instance of polyserositis with pleural and pericardial effusions and ascites 6 months after infection was reported (Hirabayashi et al., 1988). LASV could not be recovered from the effusion fluid, but lymphocytes and high levels of antibody were noted, suggesting an immune-mediated mechanism.

Typical clinical laboratory findings for LF include early leukopenia and lymphocytopenia, sometimes with atypical lymphocytes, followed later by leukocytosis with a left shift; mild-to-moderate thrombocytopenia, hemoconcentration; elevated aspartate aminotransferase (AST), alanine aminotransferase (ALT), and amylase; electrolyte perturbations; and proteinuria (Bausch, 2011). Unlike classic viral hepatitides, in LF the AST is typically much higher than the ALT, suggesting that its source is not exclusively the liver. Since a broad range of tissues can release AST, it should be considered a marker of systemic organ damage. Radiographic and electrocardiographic findings are generally nonspecific and correlate with the physical examination (Cummins et al., 1989a; Ketai et al., 2003).

15.4 DIAGNOSIS

15.4.1 Clinical Diagnosis

The nonspecific presentation of LF and other VHFs makes them extremely difficult to diagnose clinically, especially early in the course of disease before hemorrhage and other more severe manifestations develop. The differential diagnosis includes a broad array of febrile illnesses common in West Africa, including malaria, typhoid fever, and rickettsial infections (Bausch, 2011). A diagnosis of LF should be considered in patients with a clinically compatible syndrome who, within 3 weeks prior to disease onset, (1) live in or traveled to West Africa, especially in areas where LF is known to be endemic (Figure 15.1); (2) had potential direct contact with blood or bodily fluids of a person with LF during their acute illness (this group most often comprises healthcare workers); (3) worked in a laboratory or animal facility where LASV is handled; or (4) had sex with someone recovering from LF in the last 3 months (see Section 15.6). Recognized direct contact with rodents in West Africa should heighten suspicion but is rare even among confirmed cases. Acts of bioterrorism must be considered if LF is strongly suspected in a patient without any of the aforementioned risk factors, especially if clusters of cases are seen. It should be noted that even persons who meet the earlier criteria most commonly have a disease other than LF, so alternative diagnoses should always be aggressively sought, especially malaria and typhoid fever.

Consultation with infectious disease specialists with experience diagnosing and treating patients with VHF should be sought when the diagnosis is suspected. The inclusion of VHF in the differential diagnosis has the potential to induce considerable anxiety in patients, hospital staff, and the general community. Knowledge that imported LF is rare and that routinely practiced barrier nursing is protective in the vast majority of cases should offer reassurance (see Sections 15.6 and 15.8). All confirmed cases of LF should be reported immediately to government health authorities.

15.4.2 LABORATORY DIAGNOSIS

Prompt laboratory testing for LF is imperative. Unfortunately, no commercial assays are currently available, posing a major impediment to both patient diagnosis and research. Various "in-house" assays have been developed and are performed in a few specialized laboratories. Common diagnostic modalities include cell culture, serologic testing by enzyme-linked immunosorbent assay (ELISA) or immunofluorescent antibody assay (IFA), and the reverse transcriptase polymerase chain reaction (RT-PCR), each with its unique advantages and disadvantages.

The most definitive and perhaps sensitive diagnostic technique is cell culture. Serum is the most reliable sample to test, but virus can be variably isolated from throat washings, urine, CSF, breast milk, and various other tissues (Johnson et al., 1987). LASV grows well on a variety of monkeys, human, rodents, and other cell lines. Vero cells are most commonly used, in which most cells are productively infected in 4–9 days with peak infectivity titers approaching 10^8/mL. Cytopathogenic effect may be minimal even in infected cells, so all cultures should be harvested and examined by IFA using LASV-specific antiserum (Buchmeier et al., 2007).

A particular advantage of cell culture is that it enables detection of virtually any strain of LASV or, with appropriate antiserum, other viruses. However, the time required for LASV propagation and the need for a BSL-4 laboratory render cell culture more of a confirmatory test and research tool. Laboratory animal systems for propagating LASV in suckling mice, hamsters, guinea pigs, and monkeys also exist, but have little role in the modern-day diagnosis of LF. Passage of LASV at high multiplicity of infection leads to the production of defective interfering particles that can suppress replication both in cell culture and in animals, but the significance of this observation relative to natural infection is unknown (Meyer and Southern, 1997). Cell culture at higher dilutions may yield a positive result from a specimen found negative at lower dilutions.

The ELISA for LASV-specific antigen and IgM antibody is the usual mainstay of LF diagnosis (Bausch et al., 2000). Since antigen and IgM antibody detected by ELISA are not typically seen at the same time in the course of the disease, both components are necessary. ELISA assays are of high throughput and can be performed with inactivated specimens using standard equipment present in many diagnostic laboratories. Although extensive standardization and validation of ELISA tests for LF have not been conducted, they appear to have sensitivities and specificities over 90% when antigen and IgM antibody assays are conducted in tandem (Niedrig et al., 2004). Detection of antibody through IFA can also be a valuable tool, but it is not as routinely sensitive or specific and is more subjective in its interpretation, varying with the experience of the technician (Bausch et al., 2000; VanderWaals et al., 1986). Potential cross-reaction with other Old World arenaviruses is a possibility with any serologic assay.

The lack of commercially available reagents is a major drawback to both ELISA and IFA; both techniques traditionally rely on antigen produced through LASV propagation in cell culture in BSL-4 laboratories, which poses a major barrier to production given the scarcity of such laboratories due to their high construction and maintenance costs and real or perceived safety concerns. Various recombinant-protein and virus-like particle-based assays are being developed that may eventually relieve this bottleneck as well as further improve sensitivity and specificity (Branco et al., 2008; Saijo et al., 2007).

RT-PCR is becoming an increasingly valuable tool for the diagnosis of LF and can detect LASV in over 80% in the first 10 days of illness (Olschlager et al., 2010). The technique can be performed in hours, appears to be at least as sensitive as cell culture, and has the significant advantage of not requiring any reagent that must be produced in a BSL-4. Since high levels of LASV (or cDNA copy number or ELISA antigen as surrogates) correlate with death, quantitative RT-PCR also provides prognostic value (Bausch et al., 2000; McCormick et al., 1987a). Sequence heterogeneity of LASVs across West Africa has traditionally posed a challenge to PCR-based diagnostics due to primer–target mismatch, but recent development of assays targeting conserved portions of the LASV GPC or arenavirus L genes may have resolved this problem (Olschlager et al., 2010). Real-time and multiplex PCR assays have been developed that may vastly improve the rapidity and ease of the diagnosis of LF and the many diseases in its differential diagnosis (Palacios et al., 2006; Trombley et al., 2010).

Due to the extreme sensitivity of RT-PCR, contamination and false-positive results are a real concern, especially when the assay is being performed in more rudimentary facilities in West Africa where separate spaces for pre- and post-PCR procedures and the routine use of positive and negative controls are not always possible. In the worst case, outbreaks or even bioterrorism could be falsely suspected. The use of one-step assays, sequencing of PCR products to distinguish them from reference strains, targeting different portions of the genome, and use of multiple supporting diagnostic methods can minimize the risk of false positives (Drosten et al., 2002). False-negative RT-PCR results in VHF are also a concern, which may be due to viral heterogeneity and primer mismatch or inhibition by substances circulating in blood, perhaps released by tissue damage or during the RNA extraction process (Vieth et al., 2007). Appropriate inhibition controls must be included in all assays.

Post-mortem diagnosis of LF may be established by pathology examination with immunohistochemical staining of formalin-fixed tissue, but the assay does not appear to be as reliable for LF as for some of the other VHFs (Zaki et al., 1999).

15.5 PATHOGENESIS AND PATHOLOGY

Knowledge of the pathogenesis of LF is based on limited data from humans combined with extrapolation from more extensive observations made in animal models. There appears to be considerable overlap in the pathogenesis of VHF and septic shock. Insufficient effective circulating intravascular volume leads to hypotension, cellular dysfunction, and multi-organ system failure.

As with all VHFs, microvascular instability and impaired hemostasis are the hallmarks of LF. However, these are essentially "downstream" results of infection for which the "upstream" determinants are poorly understood. The pathogenesis of LF is thought to relate primarily to disruption of cellular function, as opposed to extensive cell death; patients often die without significant bleeding, and histopathological lesions (on the few cases in which autopsies have been performed) are usually not severe enough to account for death (Walker et al., 1982).

A long-standing mystery of LF is the apparent extreme range of clinical severity. The reasons for this variation are unknown, but may relate to heterogeneity in the virulence of infecting LASV strains, route and dose of inoculation, genetic predisposition, underlying co-infections and/or pre-morbid conditions (such as malaria, malnutrition, and diabetes), or misclassification of reinfection as new infection due to waning of antibody (see later text). Most studies in nonhuman primates use challenge doses designed to produce uniformly fatal disease and thus are not particularly illustrative regarding the pathogenesis of the full spectrum of LF. Interestingly, in a monkey model of LF using the WE strain of lymphocytic choriomeningitis virus (LCMV), intravenous (IV) inoculation resulted in fatal VHF while monkeys inoculated via the gastric route mostly had an attenuated infection with no disease (Lukashevich et al., 2002).

Field and laboratory data suggest variation in virulence among the various lineages and strains of LASV; the four different lineages of LASV roughly correspond with geographic areas of low and high incidence of LF, suggesting a relationship between lineage and virulence (Figures 15.1 and 15.2). In laboratory experiments, virulence of LASV strains in guinea pigs roughly correlated with the severity of disease noted in the humans from whom the viruses were isolated (Jahrling et al., 1985b). Interestingly, strains isolated from pregnant women and infants were benign in guinea pigs, suggesting that host factors such as immunosuppression play a significant role in human disease. Strains of LASV from Nigeria are sometimes said to be more virulent than those from areas further west in Africa, but specific data are lacking to support or refute this theory. The virulence factors of the LASV genome are not known, although for LCMV, they are suspected to map to the L segment (Buchmeier et al., 2007). On the host side, there is evidence that three human genes, LARGE, DMD, and IL-21, have apparently undergone positive selection in populations in endemic areas for LF in Nigeria, suggesting a protective effect (Andersen et al., 2012).

After inoculation, LASV first replicates in dendritic cells and other local tissues, with subsequent migration to regional lymph nodes and dissemination via the lymph and blood monocytes to a broad range of tissues and organs, including the liver, spleen, pancreas, lymph nodes, adrenal glands, lungs, heart, and endothelium (Hensley et al., 2011). Migration of tissue monocyte/macrophages results in secondary infection of permissive parenchymal cells. The broad organ distribution of LASV is not surprising considering that the primary receptor, α-dystroglycan, is expressed on most tissues. The lowest titers are in the central nervous system, presumably due to protection afforded by the blood–brain barrier (Walker et al., 1982). Although lymphocytes remain free of infection, they may be destroyed through apoptosis (Rodas et al., 2009). Infection of mesothelial cells can explain the serous effusions sometimes seen in LF. LASV infection of hormone-secreting cells may relate to the more severe pathophysiology of the disease noted in pregnant women.

Although the data are mixed, severe LF appears to result from an insufficient or suppressed immune response; most studies show that LASV infection of dendritic and peripheral blood mononuclear cells does not result in significant secretion of proinflammatory cytokines, up-regulation of costimulatory molecules, or significant T-cell proliferation (Baize et al., 2004; Mahanty et al., 2003). Antibody titers to LASV are significantly lower in fatal cases relative to survivors (Bausch et al., 2000).

Studies suggest that interferon (IFN) plays a role in controlling LASV infection, although this conclusion has not always been supported from *in vitro* and animal model studies (Asper et al., 2004). Absent or diminished inflammatory cytokine responses (including TNF-α, IL-8, and IFN IP-10) are often noted in both *in vitro* and *in vivo* experiments and have been correlated with a poor outcome in humans (Mahanty et al., 2001). In contrast, early and strong cytokine and cellular immune responses correlated with survival in monkey models and in case reports on humans. The lack of an immune response may reflect active down-regulation induced, at least in part, by counteraction of the type I IFN response by the LASV NP. Based on data from guinea pig models, cardiac inotropy may be directly or indirectly inhibited from a yet-to-be identified soluble mediator in the serum (Cummins et al., 1989b).

Impaired hemostasis in LF may result from endothelial cell, platelet, and/or coagulation factor dysfunction. Serum from patients with LF inhibits both neutrophil function and platelet aggregation, and consequently hemorrhage may occur even when platelet counts are not drastically low (Roberts et al., 1989). The synthesis of cell surface tissue factor triggers the extrinsic coagulation pathway.

In contrast to some other VHFs, based on animal studies, disseminated intravascular coagulopathy does not appear to be part of the pathogenesis of LF, although this finding bears confirmation (Fisher-Hoch, 1987).

Tissue damage may ultimately be mediated through direct necrosis of infected cells or indirectly through apoptosis of immune cells. Gross pathologic findings include pulmonary edema, pleural effusion, ascites, and hemorrhage from the gastrointestinal mucosa (Edington and White, 1972). Microscopic lesions include hepatocellular and splenic necrosis, renal tubular injury with interstitial nephritis, interstitial pneumonitis, and myocarditis. Inflammatory cell infiltrates are usually mild. The liver is consistently the most affected organ, which is in keeping with the finding of elevated hepatic transaminases in severe cases.

15.5.1 Virus and Antibody Dynamics

Unchecked viremia appears to be central to the pathogenesis of LF. Virus titer correlates with death, reaching up to 10^8 $TCID_{50}$/mL in fatal cases (Johnson et al., 1987). Viremia is usually present at patient presentation (presumably starting with disease onset, although patients are rarely available for testing at this time), peaking between days 4 and 9 and clearing within 2–3 weeks in survivors. Clearance of virus from a few immunologically protected sites such as the kidney and gonads may be delayed for weeks to months (Johnson et al., 1987; Lunkenheimer et al., 1990). LASV has been isolated from human urine up to 67 days after the onset of illness. Excretion may be intermittent (Zweighaft et al., 1977). No data exist on the duration of LASV in the semen.

In survivors, IgM antibodies, primarily directed at the NP and to a lesser extent GPC and Z proteins, begin to appear after about a week and progressively increase as virus clears, lasting at least some months (Bausch et al., 2000). There are conflicting reports regarding the timing of IgM antibody appearance; this discrepancy may reflect differences in the production of reagents and their target epitopes, variations in assay sensitivity and specificity, and experience of the persons conducting the assay (Bausch et al., 2000; Johnson et al., 1987). LASV antigen and IgM antibodies are often seen together when using IFA but are largely mutually exclusive when using ELISA. Recent studies in which IgM antibody responses are reported to be not consistent or helpful in the diagnosis of acute LF are based on a recombinant assay whose specificity has not been validated and contradict published data and years of field experience to the contrary (Branco et al., 2011). IgG antibody begins to appear 2–3 weeks after onset and recovery and lasts for years, including being found in persons over 40 years after acute disease who left the endemic area for LF and had no opportunity for re-exposure (Bond et al., 2013).

Fatal disease is occasionally seen in persons who have already cleared LASV from the blood and produced a strong IgM antibody response. The pathogenesis of these unusual cases is poorly understood but may relate to persistent LASV in immunologically protected areas outside the blood, most notably in the central nervous system behind the blood–brain barrier, perhaps facilitated by an immunocompromised state in some patients, such as HIV/AIDS, severe malnutrition, or diabetes. Other cases with sudden deterioration after apparent stabilization may involve acute complicating events, such as pericardial tamponade (Hirabayashi et al., 1988).

Cell-mediated immunity is thought to be the primary arm of recovery in LF (Buchmeier et al., 2007). The initial IgM and IgG antibodies produced in LF are generally not neutralizing (Jahrling et al., 1985a). Neutralizing antibodies may be produced, but usually months after recovery and often at a low titer. The continued increase in antibody titer months after infection suggests a sustained B-cell response that might be attributable to low-level virus persistence; in vaccine experiments in monkeys, replication-competent virus was cleared within 14 days after LASV challenge, but RNA could be detected up to 112 days, suggesting low-level viral persistence or the presence of defective interfering particles (Fisher-Hoch et al., 2000). Delayed clearance of LASV from immunologically protected sites could explain the finding.

15.6 EPIDEMIOLOGY

15.6.1 Geographic Distribution and Incidence

LF is endemic exclusively in West Africa (Figure 15.1). After first recognition in Nigeria in 1969, the disease was noted in Sierra Leone, Liberia, and Guinea in the wake of hospital outbreaks and field studies in the early 1970s (Enria et al., 2011). Subsequent antibody prevalence studies and case reports indicate that LASV is present in varying degrees in other West African countries (Fichet-Calvet and Rogers, 2009; Richmond and Baglole, 2003). However, the risk of exposure to LASV varies significantly in a given country and often among regions or even villages within endemic areas. The highest incidence of disease appears to be in areas of eastern Sierra Leone, northern Liberia, southeastern Guinea, and central and southern Nigeria (Figure 15.1) (Bausch et al., 2001; Bloch, 1978; Ehichioya et al., 2010; McCormick et al., 1987b). The reasons for the extreme heterogeneity in incidence are not clear, especially considering that the rodent reservoir for LASV is often readily found in areas where little or no human LF has been recognized. Varied intensity of surveillance may contribute to the apparent heterogeneous distribution, but cannot completely explain it.

An annual incidence of 300,000–500,000 LASV infections with up to 5000 deaths is often quoted in the scientific literature, but these figures are extrapolations from surveillance in the 1970s and 1980s in eastern Sierra Leone, where LF is clearly hyperendemic (McCormick et al., 1987b). Estimating the true incidence and mortality of LF is extremely difficult due to the non-specific clinical presentation; logistical impediments presented by civil unrest (Fair et al., 2007), unstable governments with underdeveloped surveillance systems (Allan et al., 1999), extensive human migration, and perturbation of the physical landscape; and lack of reagents and laboratories for laboratory confirmation in West Africa (Khan et al., 2008).

The incidence of LF is consistently highest during the dry season (Bausch et al., 2001). The reasons for this are not clear but may relate to greater stability of LASV in lower humidity (Stephenson et al., 1984) and seasonal fluctuations of numbers and prevalence of LASV infection in rodents (Fichet-Calvet et al., 2007). Interestingly, in one hospital in Guinea, the seasonal fluctuation in the incidence of LF mirrored the general pattern of other febrile diseases and hospital admissions, suggesting that non-biological factors may be involved (Bausch et al., 2001); cultural, economic, and other logistical impediments such as poor road conditions, the need to attend to crops, and lack of funds prior to the seasonal harvest often limit the ability of persons to seek medical care during the rainy season.

15.6.2 Maintenance in Nature

LASV is maintained in nature in *Mastomys natalensis* (the "multimammate rat"), presumably due to coevolution of the virus and rodent (Figure 15.5) (Bowen et al., 1997). On rare occasions, LASV has been isolated from other rodent species, including *Mastomys erythroleucus* (the "Guinea multimammate mouse"), *Rattus rattus* (the "roof rat"), and *Mus minutoides* (the "Southern African Pygmy mouse"), although the animal species could not be definitively confirmed in all cases (McCormick et al., 1987b; Safronetz et al., 2010; Wulff et al., 1975) (L. Moses, manuscript in preparation). The finding of LASV in species other than *M. natalensis* is usually considered to be due to spillover infection (i.e., incidental transient infection of a non-reservoir host). These animals are not thought to play a role in LASV maintenance. Another issue may be the difficulty identifying the species; at least five morphometrically identical species of *Mastomys* exist in sub-Saharan Africa that, until recently, could only be distinguished through karyotyping, a technically cumbersome and perhaps error-prone technique. A reliable PCR-based assay to genotypically identify *Mastomys* at the species level has now been developed and become the standard (Lecompte et al., 2006). Studies using this technique are limited but, to date, have not shown LASV in *M. erythroleucus*

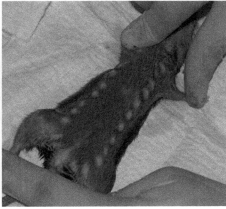

FIGURE 15.5 *Mastomys natalensis*, the reservoir for LASV. The animal is commonly known as the "multi-mammate rat" due to the female's multiple and prominent mammary glands, as seen in the right-hand panel. (Photo by Daniel Bausch.)

(Lecompte et al., 2006; Safronetz et al., 2010). Humans are dead-end hosts who play no role in the natural maintenance of LASV.

Despite the occurrence of *Mastomys* species throughout sub-Saharan Africa, LASV has not been found in rodents outside of West Africa (Figure 15.1). The reasons for this are unclear but may relate to bottlenecks in dispersal of the virus, reservoir, or both. Landscape features are the most likely barriers to migration. The competency of *Mastomys* species found outside of West Africa as reservoirs of LASV has not been investigated.

M. natalensis is almost always found in close association with humans in rural villages and surrounding cultivated fields and, less commonly, in grasslands and at the forest edge (Fichet-Calvet et al., 2007). In highly endemic regions, over 50% of all rodents caught in houses are *M. natalensis*, with a prevalence of LASV infection ranging as high as 80% (Keenlyside et al., 1983). Consumption of rodents and poor quality housing, which may reflect ease of rodent access to the home, have been shown to be risk factors for LF (Bonner et al., 2007; ter Meulen et al., 1996). In Sierra Leone, the incidence of LF is consistently high in diamond-mining areas, presumably due to increased contact with rodent excreta in the soil, although poor hygiene in mining camps and consumption of rodents may again play roles. Foreign military personnel, peacekeepers, and aid workers in rural settings are occasionally infected, sometimes importing LASV back to their countries of origin (ter Meulen et al., 2001). *M. natalensis* is not typically found in large urban centers, so the risk of rodent transmission of LASV to humans in these environments is negligible.

Studies on the transmission and maintenance of LASV in *M. natalensis* are limited. Consequently, many conclusions are based on extrapolations from laboratory experiments using LCMV in laboratory mice as a model. It is an open question how well this system actually represents LASV. In addition, the age of the animal, host genotype, and route of inoculation affect the outcome of infection (Peters et al., 1987).

Available data suggest that LASV is maintained in *M. natalensis* via chronic asymptomatic infection and vertical transmission from dam to offspring; in the only published laboratory investigation of LASV transmission in *Mastomys* species (reported to be *M. natalensis*), LASV virus persistence was achieved when animals were infected as neonates while most animals inoculated as adults cleared the virus (Walker et al., 1975). Neonates did not produce LASV antibody. It is important to note that some laboratory strains of rodents identified as *M. natalensis* were later determined to be *M. coucha*, although it is unclear which species was used for the aforementioned study (Kruppa et al., 1990).

In wild-caught *Mastomys* species, LASV antibody and antigen are usually mutually exclusive (Demby et al., 2001). Antibody-positive animals usually outnumber antigen-positive rodents, with a J-shaped curve of antibody prevalence (high at birth, decreasing in early adolescence, and then gradually rising as animals age). This pattern is consistent with the transmission of maternal antibody to offspring, which is then lost as animals wean, to be regained by animals exposed to LASV in adulthood. Although no pathology related to LASV infection was observed in *Mastomys* infected in the laboratory, LASV-infected wild *M. natalensis* were smaller in size and weight and had higher frequencies of myocardial and perivascular inflammatory lesions compared to their uninfected counterparts (Demartini et al., 1975).

15.6.3 Rodent-to-Human Transmission

Transmission of LASV to humans is believed to occur via exposure to rodent excreta, either from direct inoculation to the mucous membranes or from inhalation of aerosols produced when rodents urinate (McCormick et al., 1987b). The relative frequency of these modes of transmission is unknown. Experimental data illustrate that arenavirus infection may also occur by the oral route, perhaps through a gastric portal (Rai et al., 1997). LASV may also be contracted when rodents are trapped and prepared for consumption, a common practice in some parts of West Africa (ter Meulen et al., 1996). Since LASV is easily inactivated by heating, eating cooked rodent meat should pose no danger. It is not known whether LASV can be transmitted through a rodent bite, although the virus has been found in rodent saliva (Walker et al., 1975). The infectious dose for LASV is unknown but is thought to be low.

Transmission of LASV through aerosolized rodent urine or virus-contaminated dust particles is often referred to in the scientific literature, but there are few data to support or refute this mode of transmission. Although household clusters of LF cases from communities in West Africa are occasionally noted, single cases are much more common, suggesting that aerosol transmission to humans is not common. However, infectious and moderately stable aerosols of LASV and other arenaviruses have been artificially produced in the laboratory so the possibility of primary aerosol transmission cannot be discarded (Bausch and Peters, 2009). Regardless of their role in natural infection, the artificial production of infectious aerosols has obvious implications for the potential use of LASV as a bioweapon.

15.6.4 Human-to-Human Transmission

Empiric evidence suggests that transmission of LASV between humans occurs through direct contact with infected blood or bodily fluids. Data on the precise modes of transmission are lacking, but presumably result from oral or mucous membrane exposure, most often in the context of providing care to a sick family member (community) or patient (nosocomial transmission). Funeral rituals that entail the touching of the corpse prior to burial may also result in transmission of hemorrhagic fever viruses, although this has not been specifically recognized with LASV (Boumandouki et al., 2005). Despite delayed clearance of LASV from the kidney and gonads, secondary transmission during convalescence has not been noted, with the exception of rare reports of sexual transmission occurring months after recovery from acute disease (Bausch et al., 2000). Although aerosol spread of LASV was speculated to occur in the first recognized outbreaks of LF, extensive field experience since then has not suggested aerosol transmission between humans in natural settings (Carey et al., 1972). Viremia and infectivity of persons with LF generally parallels the clinical state, with highest infectivity late in the course of severe disease, especially when bleeding. The risk of transmission during the incubation period or from asymptomatic persons is negligible.

Contrary to popular concept, secondary attack rates for LASV are generally low, probably on the order of 5% or less as long as strict barrier nursing practices are observed (see Section 15.8). Tertiary transmission is unusual. In over 25 imported cases with at least 1500 cumulative identified contacts reported since 1969, only a single putative and asymptomatic instance of secondary transmission of LASV has been noted (Haas et al., 2003). Large outbreaks are almost always fueled by nosocomial transmission, usually in resource-poor regions where barrier nursing practices may not be maintained (Enria et al., 2011).

Based on limited data, mild or asymptomatic LASV infection appears to be frequent; in a study performed in a hyperendemic area of Sierra Leone in the 1980s examining rates of antibody conversion detected by IFA, less than 20% of persons who recently developed antibody reported a recent febrile illness (McCormick et al., 1987b). However, since antibody reversion (i.e., loss of antibody after infection) was also frequent, some of the seroconversions and asymptomatic infections may have actually represented re-exposure in persons with pre-existing immunity from past LASV exposure. The possibility of cross-reacting antibodies from previous infection with other non-pathogenic arenaviruses in West Africa, such as Kodoko, Gbagroube, and Menekre viruses, cannot be excluded (Figure 15.1). Studies on the rate of asymptomatic transmission bear repetition with newer more sensitive and specific diagnostic modalities (Bausch et al., 2000).

15.7 PROGNOSIS AND TREATMENT

15.7.1 PROGNOSIS

The clinical course of LF unfolds rapidly, with death in fatal cases usually occurring within 2–3 weeks after onset. The case-fatality rate in hospitalized cases is usually around 20%, although the overall mortality may be less than 5% when mild or asymptomatic transmission in the community is taken into account (McCormick et al., 1987b). Common indicators of a poor prognosis include shock, bleeding, neurological manifestations, viremia $>10^8$ $TCID_{50}$/mL (or cDNA copy number or ELISA antigen as surrogates), AST > 150 IU/L, and pregnancy, especially during the third trimester when maternal and fetal mortality approach 100% (McCormick et al., 1987a).

Convalescence from LF may be prolonged, with persistent myalgia, arthralgia, anorexia, weight loss, and alopecia up to a year after infection. Cerebellar ataxia has also been occasionally reported (Solbrig and McCormick, 1991). The psychological effects of LF may also be significant and often overlooked, with some patients experiencing depression or posttraumatic stress, as well as social stigmatization.

Sensorineural deafness is the only recognized permanent sequela. It is reported to occur in as many as 25% of cases, although this seems like a significant overestimate from our experience in Sierra Leone and Guinea over the last 15 years (Cummins et al., 1990). Deafness typically presents during convalescence and is unassociated with the severity of the acute illness or level of viremia, suggesting an immune-mediated pathogenesis. Deafness may be uni- or bilateral, and is permanent in approximately two-thirds of cases. Auditory patterns resemble idiopathic nerve deafness (Liao et al., 1992).

15.7.2 GENERAL SUPPORTIVE MEASURES

Patients with LF should generally be treated in intensive care units since severe microvascular instability, often complicated by vomiting, diarrhea, and decreased fluid intake, may require continuous monitoring and aggressive fluid replacement. Overaggressive and unmonitored rehydration may lead to significant third-spacing and pulmonary edema. Because of the risk of bleeding at insertion sites, intravascular hemodynamic devices are contraindicated, instead monitoring hemodynamic status by blood pressure cuff or other noninvasive means.

Fluid and blood pressure management guidelines for septic shock are recommended for VHF due to the common elements in the pathogenesis of these two conditions, although there are no efficacy data on their use in VHF (Bausch, 2011). Crystalloids (Ringers lactate or normal saline) and, if necessary, vasopressors should be infused to maintain central venous pressure between 8 and 12 mm Hg or mean arterial blood pressure above 65 mm Hg (88.4 cm H_2O). Early use of vasopressors, especially dopamine and norepinephrine, may diminish the risk of fluid overload. Dobutamine should be added if the earlier measures and blood transfusion fail to maintain the target blood pressure and adequate organ perfusion. Transfusions, preferably with packed red blood cells, should be used to maintain a hematocrit over 30% while avoiding volume overload, especially in patients in West Africa where chronic anemia may be frequent due to problems such as malaria and malnutrition. Although disseminated intravascular coagulopathy is not frequently noted in LF, the possibility merits checking the relevant laboratory parameters if bleeding and thrombocytopenia persist, with transfusion of platelets and/or fresh frozen plasma as required. Vitamin K (10 mg on 2 consecutive days) may be given, especially if underlying malnutrition or liver disease is suspected.

Patients should be immediately covered with appropriate broad-spectrum antibacterial and/or antiparasitic therapy, with specific consideration of antimalarial agents for malaria and doxycycline for tick-borne rickettsial diseases, until a diagnosis of LF can be confirmed. Secondary bacterial infection should be suspected if patients have persistent or new fever after about 2 weeks of illness. Acetaminophen (500–1000 mg q4–6 h), tramadol (50–100 mg q4–6 h), opiates, or other analgesics should be used for pain control. Salicylates and nonsteroidal anti-inflammatory drugs should not be used, and intramuscular and subcutaneous injections should be avoided due to the risk of bleeding. Prophylactic therapy for stress ulcers with H_2 receptor antagonists (e.g., ranitidine 50 mg IV every 8 h) is appropriate. Seizures can usually be managed with benzodiazepines or phenytoin, with careful attention to possible respiratory depression. The use of sedatives and neuromuscular blocking agents should be minimized, but haloperidol (0.5–5 mg two or three times daily) or a benzodiazepine (e.g., lorazepam 1–10 mg orally daily in two to three divided doses) may be used. Impaired gas exchange is not typically a prominent feature of LF, especially in the absence of iatrogenic pulmonary edema. Intubation and mechanical ventilation should be avoided if possible because of the risk of barotrauma and pleural-pulmonary hemorrhage. Uterine evacuation appears to lower maternal mortality and should be considered in pregnant patients, although performed with extreme caution as this can be considered a high-risk procedure with regard to potential nosocomial transmission (Price et al., 1988).

15.7.3 ANTIVIRAL THERAPY

Intravenous administration of the nucleoside analogue drug ribavirin should be given to all patients with LF (Table 15.1). Ribavirin has been shown to decrease mortality in severe LF from 55% to 5% when begun within the first 6 days of illness (McCormick et al., 1986). The mechanism of action is unknown, although lethal mutagenesis is suspected (Bausch et al., 2010). Although few data are available, oral ribavirin may also be effective in some cases, but less so than the IV form, most likely because the serum concentration achieved through oral administration is on the borderline of the mean inhibitory concentration of ribavirin for LASV (4–40 μM) (Bausch et al., 2010; McCormick et al., 1986). Absorption of oral ribavirin from the gut may also pose a barrier given the vomiting and diarrhea often present in LF. Until more data are available on the efficacy of oral ribavirin, the entire treatment course of ribavirin should be administered IV.

Patent issues and high cost (up to $1000 per patient from most sources in Europe and North America) have historically severely limited the availability of IV ribavirin. However, the patent is now expired, and the World Health Organization has applied to add the drug to the list of essential medicines, which will hopefully significantly lower cost and improve availability.

TABLE 15.1
Ribavirin Therapy for LF

Indication	Route	Dose	Interval
Treatment	IV[a]	30 mg/kg (maximum 2 g)[b]	Loading dose, followed by
	IV[a]	15 mg/kg (maximum 1 g)[b]	Every 6 h for 4 days, followed by
	IV[a]	7.5 mg/kg (maximum 500 mg)[b]	Every 8 h for 6 days
Prophylaxis	PO	35 mg/kg (maximum 2.5 g)[b]	Loading dose, followed by
	PO	15 mg/kg (maximum 1 g)[b]	Every 8 h for 10 days

Notes: IV, intravenous; PO, oral administration.

[a] The drug should be diluted in 150 mL of 0.9% saline and infused slowly.

[b] Reduce the dose in persons known to have significant renal insufficiency (creatinine clearance of less than 50 mL/min).

Meanwhile, many countries in Africa import the drug from less expensive makers in China and Russia (Hadi et al., 2010).

Major adverse effects due to short-term ribavirin are rare. The primary adverse effect is a dose-dependent, mild-to-moderate hemolytic anemia that infrequently necessitates transfusion and disappears with cessation of treatment (Bausch et al., 2010; McCormick et al., 1986). Rigors may occur when ribavirin is infused too rapidly. Relative contraindications include severe anemia or hemoglobinopathy, coronary artery disease, renal insufficiency, decompensated liver disease, breast-feeding, and known hypersensitivity. Although findings of teratogenicity and fetal loss in laboratory animals have rendered that ribavirin technically contraindicated in pregnancy (pregnancy category X), its use must still be considered as a life-saving measure given the extremely high maternal and fetal mortality associated with LF in pregnancy.

Hemoglobin, hematocrit, and bilirubin levels should be checked at initiation of ribavirin therapy and then every few days, with consideration of transfusion of packed red blood cells if significant anemia develops. Because of the long terminal half-life (~24 h) and large volume of distribution, ribavirin may still have effect for a time even after cessation, particularly in red blood cells where it accumulates.

A number of experimental therapies for LASV and other arenaviruses have shown activities *in vitro* and *in vivo*, including small molecules, nucleoside analogues, inhibitors of S-adenosyl-l-homocysteine hydrolase, and tyrosine kinase inhibitors, but are not yet ready for clinical application (Gowen and Bray, 2011).

15.7.4 CONVALESCENT IMMUNE PLASMA

Convalescent plasma has been used in LF with apparent benefit, but results from animal studies suggest that it is only efficacious if it contains a high titer of neutralizing antibody, which is not always the case even in survivors (see Section 15.5), and there is a close antigenic match between the infecting viruses of the donor and the recipient. In Argentine hemorrhagic fever caused by the New World arenavirus Junín, convalescent plasma has been associated with a convalescent-phase neurologic syndrome characterized by fever, cerebellar signs, and cranial nerve palsies in 10% of treated patients. The neurologic syndrome has not been reported in patients with LF treated with convalescent plasma, although opportunities for observation and possible detection of this syndrome have been limited. Given the logistical challenges inherent in the use of immune plasma, including the risk of concomitant transmission of other blood-borne pathogens and lack of an existing bank of immune plasma for this purpose, this therapy should be reserved for severe and refractory cases unresponsive to ribavirin or when ribavirin is not available.

15.7.5 Immune Modulators

Recently, there has been renewed interest in the use of immunomodulating drugs in the treatment of septic shock and, by extension due to the common elements of their pathogenesis, VHF. However, trials of various immune modulators in septic shock, including ibuprofen, corticosteroids, anti-TNF-α, nitric oxide inhibitors, statins (HMG-CoA reductase inhibitors), and interleukins have not shown conclusive benefit. Ribavirin combined with IFN alfacon-1, a consensus IFN, diminished mortality and disease severity in a hamster arenavirus model (Gowen et al., 2006). Although approved for clinical use in humans, IFN aflacon-1 has not been tested in human LF, perhaps in part due to its high cost, systemic toxicity, and need for repeated doses. These problems can potentially be overcome through the delivery of a recombinant, replication-deficient type 5 human adenovirus that encodes and elicits the production of IFN alfacon-1 from infected cells.

A number of other immunomodulating approaches are being explored, including those that enhance immune recognition of infected cells and dampen immune responses through the blockage of toll-like receptors (Gowen and Bray, 2011). Statin drugs, which appear to have immunomodulatory, anti-inflammatory, antimicrobial, and vasculature-stabilizing properties, were included in the successful treatment of a case of Lujo virus infection along with the antioxidant and free radical scavenger N-acetylcysteine.

15.7.6 Coagulation Modulators

A growing body of literature suggests that disturbances in the procoagulant–anticoagulant balance play an important role in the mediation of septic shock.

Recombinant activated protein C (24 µg/kg/h constant infusion) is efficacious for some patients with septic shock and reduces mortality in Ebola virus–infected monkeys but should still be considered experimental for VHF in humans (Bernard et al., 2001). At first glance, the major adverse effect of activated protein C—serious bleeding (including intracranial hemorrhage) that has been reported in up to 5% of treated patients—would seem to contraindicate its use in VHF. However, the mechanism of the drug may not be via direct anticoagulation, but rather through modulation of inflammation. Conceivably, early use could mitigate the pathogenic processes in VHF that ultimately result in hemorrhage with no additional risk of bleeding due to the drug itself. Furthermore, the infrequency of bleeding in LF relative to most other VHFs might make it a logical candidate for trials with activated protein C (Bausch et al., 2001; McCormick et al., 1987a).

An experimental recombinant inhibitor of the tissue factor/factor VIIa coagulation pathway, rNAPc2, also decreases mortality in Ebola virus–infected monkeys and has completed a Phase I trial in humans. Somewhat paradoxically since it would have the opposite effect of rNAPc2, recombinant factor VIIa itself was included in the successful treatment of the aforementioned case of Lujo virus infection. Data on the use of other anticoagulants, such as heparin sulfate, antithrombin III, recombinant factor VIIa, and tissue-factor pathway inhibitor to treat VHF and LF are either nonexistent or inconclusive.

15.7.7 Management of Convalescence

Since the clinical status of patients with LF correlates with the level of viremia and infectivity, patients who have recovered from their acute illness can safely be assumed to have cleared their viremia and can be discharged from the hospital without concern of subsequent transmission at home. Because of the potential delayed clearance in the urine and semen, abstinence or condom use is recommended for 3 months after acute illness. Transmission through toilet facilities has not been noted, but simple precautions to avoid contact with potentially infectious excretions in this setting are prudent, including separate toilet facilities and regular hand washing. Breast-feeding should be avoided during convalescence unless there is no other way to support the baby. Clinical management during convalescence includes the use of warm packs, acetaminophen, nonsteroidal anti-inflammatory drugs, cosmetics, hair-growth stimulants, anxiolytics, antidepressants, nutritional supplements, and nutritional and psychological counseling as indicated.

15.8 PREVENTION AND CONTROL

15.8.1 CASE IDENTIFICATION AND PATIENT ISOLATION

Infection control of LF relies on classic public health principles of identification and isolation of cases and monitoring of their contacts. However, the early nonspecific presentation of LF poses a serious challenge to case identification. Fortunately, the low secondary attack rate affords a measure of reassurance even when cases go unrecognized as long as proper barrier nursing is maintained. Furthermore, since mild cases are not very infectious, missed or delayed diagnosis of these patients is unlikely to pose a problem from an infection control standpoint. All patients with a clinically compatible syndrome should be presumed infectious and kept under "VHF isolation precautions" (see later text) until a specific diagnosis is made (CDC and WHO, 1998). If available, placement in a negative airflow room is prudent, but hermetically sealed isolation chambers are not required and may have severe adverse psychological effects on both patient and staff. Access to the patient should be limited to a small number of designated staff and family members with specific instructions and training on the implementation of VHF isolation precautions.

15.8.2 PERSONAL PROTECTIVE EQUIPMENT AND NURSING PRECAUTIONS

Although routine strict barrier nursing is protective in most cases, specific VHF isolation precautions (surgical mask, double gloves, gown, protective apron, face shield, and shoe covers) are advised for added security. Positive-airway pressure masks and other small particle aerosol precautions should be used when performing procedures that may generate aerosols, such as endotracheal intubation. Items that were in direct contact with the patient should be decontaminated using household bleach or other disinfectants (see later text). A 1:100 (1%) solution of ordinary 5% chlorine bleach can be used for reusable items, and a 1:10 (10%) solution for disinfecting excreta, corpses, and items to be discarded.

15.8.3 CONTACT TRACING

Persons with unprotected direct contact with a patient during the symptomatic phase of LF should be monitored daily for the evidence of disease for 3 weeks (the longest possible incubation period) after their last contact. Contacts should check their temperature daily and record the results in a log. Despite the lack of evidence for transmission during the incubation period, it is usually recommended that exposed persons avoid close contact or activities with household members that might result in exposure to bodily fluids, such as sex, kissing, and sharing of utensils. Hospitalization or other confinement of asymptomatic persons is not warranted, but persons who develop fever or other signs and symptoms consistent with LF should immediately be isolated until the diagnosis can be excluded.

15.8.4 VACCINES

A number of vaccine platforms have been explored in animal models of LF over the years, including inactivated LASV, attenuated arenaviruses, recombinant vaccinia viruses, RNA replicon vectors derived from an attenuated strain of Venezuelan equine encephalitis virus, recombinant salmonella typhimurium, LASV protein subunits, naked DNA, recombinant yellow fever 17D viruses, reassortant Lassa/Mopeia viruses, virus-like particles, and recombinant vesicular stomatitis virus

(Enria et al., 2011). The LASV GP appears to be the most important immunogenic protein. The recombinant vesicular stomatitis virus platform is perhaps the most promising candidate, providing 100% protection after a single dose in a monkey model of LF. Furthermore, this vaccine platform has been effective when given by the nasal or oral route, potentially facilitating its use in epidemics, as well as for postexposure prophylaxis for various VHFs, although LF was not among those tested (Geisbert et al., 2010). Although there are concerns over the safety of an unproven live vector vaccine that might be administered in populations with a high frequency of immunosuppressed persons due to HIV/AIDS or malnutrition, preliminary studies in immunocompromised monkeys have not indicated problems (Geisbert et al., 2008).

15.8.5 POSTEXPOSURE PROPHYLAXIS

Postexposure prophylaxis with oral ribavirin should be considered for persons with direct unprotected contact with blood or bodily fluids from a person with confirmed LF (Bausch et al., 2010). Persons who develop manifestations of LF should be immediately converted to IV ribavirin. Prophylaxis should never be given if the only exposure was during the incubation period.

15.8.6 ENVIRONMENTAL CLEANUP

There are few data available on the airflow dynamics or viability of LASV or other hemorrhagic fever viruses once released into the environment. Considering that the lipid envelope of these viruses is generally easily disrupted, their longevity when shed naturally in body fluids, which would then dry, appears to be on the order of hours to days, varying with temperature and humidity (Bond et al., 2012). However, hemorrhagic fever viruses have been isolated from samples kept for weeks at ambient temperatures if stored hydrated in a biological buffer, such as blood or serum. The survival of LASV in rodent urine is probably affected by factors such as the animal's diet and urinary pH and protein.

There is little real concern of LASV or other hemorrhagic fever viruses seeping into groundwater or posing any long-term risk through casual exposures in the outdoor environment. When contamination may have recently occurred, such as in homes of persons with LF, barrier precautions and decontamination as described earlier are warranted and effective. The lipid envelope of hemorrhagic fever viruses renders them easily inactivated by various methods, including heating to 60°C for 1 h, ultraviolet light, gamma irradiation, surfactant nanoemulsions, and use of disinfectants containing phenolic compounds, hypochlorite, quaternary amines, and acidic or basic pH (Bond et al., 2012). Some of these inactivation techniques, such as heat and gamma irradiation, preserve the LASV proteins for serologic testing. Workers cleaning areas where rodents may inhabit should let the area aerate before entering and use mops soaked in 10% bleach to avoid propelling dust potentially contaminated with rodent excreta and LASV into the air.

15.8.7 RESERVOIR CONTROL

Measures to prevent human contact with rodents are important in the control of LF. Since *M. natalensis* often colonizes human dwellings, where most infections are thought to occur, prevention is best achieved by improving "village hygiene," such as eliminating unprotected storage of garbage, foodstuffs, and water, and, when possible, plugging holes that allow rodents to enter homes. Rodent trapping or poisoning is generally not thought to be an effective long-term control strategy because animals from surrounding fields will likely soon recolonize the area. A single trapping session in houses in Sierra Leone did not diminish the incidence of human LASV infection (Keenlyside et al., 1983). It is unknown whether a more sustained effort could reduce rodent populations sufficiently to break transmission cycles for LASV.

15.9 CONCLUSIONS AND FUTURE PERSPECTIVES

Significant progress has been made in our understanding of VHF in recent years, mostly in the realm of the laboratory and basic sciences, especially regarding pathogenesis and the development of vaccines from studies in animal models. However, the primary emphasis to date has been on Ebola and Marburg viruses, which are considered greater bioterrorism threats, despite a much lower incidence of disease compared to LF. A detailed description of research advances and future directions in LF is published elsewhere (Khan et al., 2008), but priority areas are highlighted here.

15.9.1 DIAGNOSTIC ASSAYS

The absence of rapid, low-cost, easy-to-use, sensitive, and specific diagnostic assays for LF poses a major barrier to both research and patient care. Various recombinant-protein and virus-like particle-based assays are currently being developed that will hopefully soon relieve this bottleneck, although interpretation and validation of these assays are proving challenging.

15.9.2 THERAPIES AND VACCINES

Guidelines for clinical management of LF are patterned on those for septic shock, but the efficacy of these measures remains unknown. Detailed and ideally controlled clinical observations, including intensive hemodynamic monitoring and measuring of serum electrolytes, should be made on the efficacy of aggressive rehydration and administration of blood products and pressor agents. Priorities in the investigation of specific antiviral therapies include exploring the efficacy of alternative dosing regimens of ribavirin, including shorter treatment courses and those that allow conversion to oral therapy after patients stabilize on IV, both of which would diminish costs and adverse effects. Although evaluation of the efficacy of oral ribavirin as postexposure prophylaxis is also a theoretical research priority, the low secondary attack rate of LF makes it unlikely that enough subjects could ever be enrolled for a conclusive study to be performed (Bausch et al., 2010).

Subsequent steps would be to examine promising new drugs, such as small interfering RNAs and novel nucleoside analogues that would be given, at least initially, in conjunction with ribavirin. Transcriptome profiles from monkey models of arenavirus infection may aid in identifying key genes, some of which are the targets of drugs already in clinical use. Rather than targeting LASV itself, ideal therapies would mediate the underlying pathogenesis of LF and other VHFs and thus have utility regardless of the specific etiologic virus. A number of promising vaccine candidates exist for LF, some of which could soon be ready for Phase 1 clinical trials. However, finding economic backing for VHF vaccine development is proving to be a major hurdle.

15.9.3 PATHOGENESIS

The pathogenesis of LF remains largely a mystery. It is still unclear whether severe LF is the result of an overactive or underactive inflammatory response, and what the determinants are of the observed variation in disease severity. Experiments in nonhuman primate models are proving illustrative, but correlation with detailed clinical observations and laboratory testing in humans is essential. The role of human host genotype should not be overlooked.

15.9.4 BASIC SCIENCE INVESTIGATION

Research priorities here include identifying the role of any co-receptors for LASV. Reverse genetics approaches may play an important role, including allowing the safe shipment of noninfectious LASV RNA to laboratories registered to work with hemorrhagic fever viruses, where infectious recombinant LASV could be reconstituted under BSL-4 conditions (Hass et al., 2004).

15.9.5 Field Studies on Epidemiology and Control

Debate lingers regarding various aspects of the epidemiology of LF, including the precise modes of LASV transmission between rodents and humans, and the true frequency of infection to disease in humans. Furthermore, little attention has been paid to community control, including the efficacy of measures such as regular rodent trapping or use of rodenticides, perhaps timed to coincide with the breeding season, rodent-proofing of homes, and the introduction of predators, such as cats. Interdisciplinary programs bridging human and animal health and conservation offer further opportunities to understand ecosystem regulation of LASV and how human relationship with the environment enhances or reduces LF risk.

It remains to be seen whether the more stable research environment in West Africa and the availability of modern research tools will translate to a better life for those at risk of LF in Africa as well as increased safety for persons in other areas of the world where bioterrorism or imported cases are the primary concerns. The current global economic crisis that threatens to blunt research funding, the lack of economic incentive for pharmaceutical and other companies to invest in research and development for VHFs because they are primarily endemic in resource-poor regions of the world, and the relatively weak research and training infrastructure in Africa that continues to suffer from the "brain drain" of African scientists to more lucrative jobs on other continents will be major challenges.

ACKNOWLEDGMENTS

The authors thank Andrew Bennett, Cecilia Gonzales, Ricardo Hora, and Landon Vom Steeg for assistance in preparing the manuscript and Mike Bowen, Jim Mills, and John Schieffelin for critical review.

REFERENCES

Allan, R., Mardell, S., Ladbury, R., Pearce, E., Skinner, K., 1999. The progression from endemic to epidemic Lassa fever in war-torn West Africa, in: Saluzzo, J.F., Dodet, B. (Eds.), *Factors in the Emergence and Control of Rodent-Borne Viral Diseases*. Elsevier SAS, Annecy, France, pp. 197–125.

Andersen, K.G., Shylakhter, I., Tabrizi, S., Grossman, S.R., Happi, C.T., Sabeti, P.C., 2012. Genome-wide scans provide evidence for positive selection of genes implicated in Lassa fever. *Philos Trans R Soc Lond B Biol Sci* 367, 868–877.

Asper, M., Sternsdorf, T., Hass, M., Drosten, C., Rhode, A., Schmitz, H., Gunther, S., 2004. Inhibition of different Lassa virus strains by alpha and gamma interferons and comparison with a less pathogenic arenavirus. *J Virol* 78, 3162–3169.

Baize, S., Kaplon, J., Faure, C., Pannetier, D., Georges-Courbot, M.C., Deubel, V., 2004. Lassa virus infection of human dendritic cells and macrophages is productive but fails to activate cells. *J Immunol* 172, 2861–2869.

Bausch, D.G., 2011. Viral hemorrhagic fevers, in: Arend, W.P., Armitage, J.O., Clemmons, D.R., Drazen, J.M., Griggs, R.C., Landry, D.W., Levinson, W., Rustgi, A.K., Scheld, W.M. (Eds.), *Goldman's Cecil Medicine*, 24th edn., Elsevier Saunders, Philadelphia, PA, p. 2704, Chapter 389.

Bausch, D.G., Demby, A.H., Coulibaly, M., Kanu, J., Goba, A., Bah, A., Conde, N., Wurtzel, H.L., Cavallaro, K.F., Lloyd, E., Baldet, F.B., Cisse, S.D., Fofona, D., Savane, I.K., Tolno, R.T., Mahy, B., Wagoner, K.D., Ksiazek, T.G., Peters, C.J., Rollin, P.E., 2001. Lassa fever in Guinea: I. Epidemiology of human disease and clinical observations. *Vector Borne Zoonotic Dis* 1, 269–281.

Bausch, D.G., Hadi, C.M., Khan, S.H., Lertora, J.J., 2010. Review of the literature and proposed guidelines for the use of oral ribavirin as postexposure prophylaxis for Lassa fever. *Clin Infect Dis* 51, 1435–1441.

Bausch, D.G., Peters, C.J., 2009. The viral hemorrhagic fevers, in: Lutwick, L.I., Lutwick, S.M. (Eds.), *Beyond Anthrax: The Weaponization of Infectious Diseases*. Humana Press, New York, pp. 107–144.

Bausch, D.G., Rollin, P.E., Demby, A.H., Coulibaly, M., Kanu, J., Conteh, A.S., Wagoner, K.D., McMullan, L.K., Bowen, M.D., Peters, C.J., Ksiazek, T.G., 2000. Diagnosis and clinical virology of Lassa fever as evaluated by enzyme-linked immunosorbent assay, indirect fluorescent-antibody test, and virus isolation. *J Clin Microbiol* 38, 2670–2677.

Bernard, G.R., Vincent, J.L., Laterre, P.F., LaRosa, S.P., Dhainaut, J.F., Lopez-Rodriguez, A., Steingrub, J.S., Garber, G.E., Helterbrand, J.D., Ely, E.W., Fisher, C.J., Jr., 2001. Efficacy and safety of recombinant human activated protein C for severe sepsis. *New Engl J Med* 344, 699–709.

Bloch, A., 1978. A serological survey of Lassa fever in Liberia. *Bull World Health Organ* 56, 811–813.

Bond, N., Moses, L., Peterson, A., Mills, J., Bausch, D., 2012. Environmental aspects of the viral hemorrhagic fevers, in: Friis, R. (Ed.), *Praeger Handbook of Environmental Health*. Praeger Publishing Company, Santa Barbara, CA, pp. 133–161.

Bond, N., Schieffelin, J.S., Moses, L.M., Bennett, A.J., Bausch, D.G., 2013. A historical look at the first reported cases of Lassa fever: IgG antibodies 40 years after acute infection. *Am J Trop Med Hyg* 88, 241–244.

Bonner, P.C., Schmidt, W.-P., Belmain, S.R., Oshin, B., Baglole, D., Borchert, M., 2007. Poor housing quality increases risk of rodent infestation and Lassa fever in refugee camps of Sierra Leone. *Am J Trop Med Hyg* 77, 169–175.

Boumandouki, P., Formenty, P., Epelboin, A., Campbell, P., Atsangandoko, C., Allarangar, Y., Leroy, E.M., Kone, M.L., Molamou, A., Dinga-Longa, O., Salemo, A., Kounkou, R.Y., Mombouli, V., Ibara, J.R., Gaturuku, P., Nkunku, S., Lucht, A., Feldmann, H., 2005. Clinical management of patients and deceased during the Ebola outbreak from October to December 2003 in Republic of Congo (Frenc). *Bull Soc Pathol Exot* 98, 218–223.

Bowen, M.D., Peters, C.J., Nichol, S.T., 1997. Phylogenetic analysis of the Arenaviridae: patterns of virus evolution and evidence for cospeciation between arenaviruses and their rodent hosts. *Mol Phylogenet Evol* 8, 301–316.

Bowen, M.D., Rollin, P.E., Ksiazek, T.G., Hustad, H.L., Bausch, D.G., Demby, A.H., Bajani, M.D., Peters, C.J., Nichol, S.T., 2000. Genetic diversity among Lassa virus strains. *J Virol* 74, 6992–7004.

Branco, L.M., Grove, J.N., Boisen, M.L., Shaffer, J.G., Goba, A., Fullah, M., Momoh, M., Grant, D.S., Garry, R.F., 2011. Emerging trends in Lassa fever: redefining the role of immunoglobulin M and inflammation in diagnosing acute infection. *Virol J* 8, 478.

Branco, L.M., Matschiner, A., Fair, J.N., Goba, A., Sampey, D.B., Ferro, P.J., Cashman, K.A., Schoepp, R.J., Tesh, R.B., Bausch, D.G., Garry, R.F., Guttieri, M.C., 2008. Bacterial-based systems for expression and purification of recombinant Lassa virus proteins of immunological relevance. *Virol J* 5, 74.

Buchmeier, M.J., de la Torre, J.C., Peters, C.J., 2007. Arenaviridae: The viruses and their replication, in: Knipe, D.M., Howley, P.M. (Eds.), *Fields' Virology*, 5th edn. Wolters Kluwer Health/Lippincott Williams & Wilkins, Philadelphia, PA, pp. 1635–1668.

Carey, D.E., Kemp, G.E., White, H.A., Pinneo, L., Addy, R.F., Fom, A.L., Stroh, G., Casals, J., Henderson, B.E., 1972. Lassa fever. Epidemiological aspects of the 1970 epidemic, Jos, Nigeria. *Trans R Soc Trop Med Hyg* 66, 402–408.

CDC and WHO, 1998. Infection control for viral haemorrhagic fevers in the African health care setting. Centers for Disease Control and Prevention, Atlanta.

Cummins, D., Bennett, D., Fisher-Hoch, S.P., Farrar, B., Machin, S.J., McCormick, J.B., 1992. Lassa fever encephalopathy: clinical and laboratory findings. *J Trop Med Hyg* 95, 197–201.

Cummins, D., Bennett, D., Fisher-Hoch, S.P., Farrar, B., McCormick, J.B., 1989a. Electrocardiographic abnormalities in patients with Lassa fever. *J Trop Med Hyg* 92, 350–355.

Cummins, D., Fisher-Hoch, S.P., Walshe, K.J., Mackie, I.J., McCormick, J.B., Bennett, D., Perez, G., Farrar, B., Machin, S.J., 1989b. A plasma inhibitor of platelet aggregation in patients with Lassa fever. *Br J Haematol* 72, 543–548.

Cummins, D., McCormick, J.B., Bennett, D., Samba, J.A., Farrar, B., Machin, S.J., Fisher-Hoch, S.P., 1990. Acute sensorineural deafness in Lassa fever. *JAMA* 264, 2093–2096.

Demartini, J.C., Green, D.E., Monath, T.P., 1975. Lassa virus infection in *Mastomys natalensis* in Sierra Leone. *Bull World Health Organ* 52, 651–663.

Demby, A.H., Inapogui, A., Kargbo, K., Koninga, J., Kourouma, K., Kanu, J., Coulibaly, M., Wagoner, K.D., Ksiazek, T.G., Peters, C.J., Rollin, P.E., Bausch, D.G., 2001. Lassa fever in Guinea: II. Distribution and prevalence of Lassa virus infection in small mammals. *Vector Borne Zoonotic Dis* 1, 283–297.

Drosten, C., Gottig, S., Schilling, S., Asper, M., Panning, M., Schmitz, H., Gunther, S., 2002. Rapid detection and quantification of RNA of Ebola and Marburg viruses, Lassa virus, Crimean-Congo hemorrhagic fever virus, Rift Valley fever virus, dengue virus, and yellow fever virus by real-time reverse transcription-PCR. *J Clin Microbiol* 40, 2323–2330.

Edington, G.M., White, H.A., 1972. The pathology of Lassa fever. *Trans R Soc Trop Med Hyg* 66, 381–389.

Ehichioya, D.U., Hass, M., Olschlager, S., Becker-Ziaja, B., Onyebuchi Chukwu, C.O., Coker, J., Nasidi, A., Ogugua, O.O., Gunther, S., Omilabu, S.A., 2010. Lassa fever, Nigeria, 2005–2008. *Emerg Infect Dis* 16, 1040–1041.

Ehichioya, D.U., Hass, M., Becker-Ziaja, B., Ehimuan, J., Asogun, D.A., Fichet-Calvet, E., Kleinsteuber, K., Lelke, M., ter Meulen, J., Akpede, G.O., Omilabu, S.A., Günther, S., Olschläger, S., 2011. Current molecular epidemiology of Lassa virus in Nigeria. *J Clin Microbiol* 49, 1157–1161.

Enria, D.A., Mills, J.N., Bausch, D., Shieh, W., Peters, C.J., 2011. Arenavirus Infections, in: Guerrant, R.L., Walker, D.H., Weller, P.F. (Eds.), *Tropical Infectious Diseases: Principles, Pathogens, and Practice*, 3rd edn. Churchill Livingstone, Philadelphia, PA, pp. 449–461.

Fair, J., Jentes, E., Inapogui, A., Kourouma, K., Goba, A., Bah, A., Tounkara, M., Coulibaly, M., Garry, R.F., Bausch, D.G., 2007. Lassa virus-infected rodents in refugee camps in Guinea: A looming threat to public health in a politically unstable region. *Vector Borne Zoonotic Dis* 7, 167–171.

Fichet-Calvet, E., Lecompte, E., Koivogui, L., Soropogui, B., Dore, A., Kourouma, F., Sylla, O., Daffis, S., Koulemou, K., ter Meulen, J., 2007. Fluctuation of abundance and Lassa virus prevalence in Mastomys natalensis in Guinea, West Africa. *Vector Borne Zoonotic Dis* 7, 119–128.

Fichet-Calvet, E., Rogers, D.J., 2009. Risk maps of Lassa fever in West Africa. *PLoS Negl Trop Dis* 3, e388.

Fisher-Hoch, S., 1987. Pathophysiology of shock and haemorrhage in viral haemorrhagic fevers. Southeast Asian. *J Trop Med Public Health* 18, 390–391.

Fisher-Hoch, S.P., Hutwagner, L., Brown, B., McCormick, J.B., 2000. Effective vaccine for Lassa fever. *J Virol* 74, 6777–6783.

Frame, J.D., Baldwin, J.M., Jr., Gocke, D.J., Troup, J.M., 1970. Lassa fever, a new virus disease of man from West Africa. I. Clinical description and pathological findings. *Am J Trop Med Hyg* 19, 670–676.

Geisbert, T.W., Bausch, D.G., Feldmann, H., 2010. Prospects for immunisation against Marburg and Ebola viruses. *Rev Med Virol* 20, 344–357.

Geisbert, T.W., Daddario-Dicaprio, K.M., Lewis, M.G., Geisbert, J.B., Grolla, A., Leung, A., Paragas, J., Matthias, L., Smith, M.A., Jones, S.M., Hensley, L.E., Feldmann, H., Jahrling, P.B., 2008. Vesicular stomatitis virus-based Ebola vaccine is well-tolerated and protects immunocompromised nonhuman primates. *PLoS Pathog* 4, e1000225.

Gowen, B.B., Bray, M., 2011. Progress in the experimental therapy of severe arenaviral infections. *Future Microbiol* 6, 1429–1441.

Gowen, B.B., Smee, D.F., Wong, M.H., Pace, A.M., Jung, K.H., Bailey, K.W., Blatt, L.M., Sidwell, R.W., 2006. Combinatorial ribavirin and interferon alfacon-1 therapy of acute arenaviral disease in hamsters. *Antivir Chem Chemother* 17, 175–183.

Gunther, S., Weisner, B., Roth, A., Grewing, T., Asper, M., Drosten, C., Emmerich, P., Petersen, J., Wilczek, M., Schmitz, H., 2001. Lassa fever encephalopathy: Lassa virus in cerebrospinal fluid but not in serum. *J Infect Dis* 184, 345–349.

Haas, W.H., Breuer, T., Pfaff, G., Schmitz, H., Kohler, P., Asper, M., Emmerich, P., Drosten, C., Golnitz, U., Fleischer, K., Gunther, S., 2003. Imported Lassa fever in Germany: Surveillance and management of contact persons. *Clin Infect Dis* 36, 1254–1258.

Hadi, C.M., Goba, A., Khan, S.H., Bangura, J., Sankoh, M., Koroma, S., Juana, B., Bah, A., Coulibaly, M., Bausch, D.G., 2010. Ribavirin for Lassa fever postexposure prophylaxis. *Emerg Infect Dis* 16, 2009–2011.

Hass, M., Golnitz, U., Muller, S., Becker-Ziaja, B., Gunther, S., 2004. Replicon system for Lassa virus. *J Virol* 78, 13793–13803.

Hensley, L.E., Smith, M.A., Geisbert, J.B., Fritz, E.A., Daddario-DiCaprio, K.M., Larsen, T., Geisbert, T.W., 2011. Pathogenesis of Lassa fever in cynomolgus macaques. *Virol J* 8, 205.

Hirabayashi, Y., Oka, S., Goto, H., Shimada, K., Kurata, T., Fisher-Hoch, S.P., McCormick, J.B., 1988. An imported case of Lassa fever with late appearance of polyserositis. *J Infect Dis* 158, 872–875.

Jahrling, P.B., Frame, J.D., Rhoderick, J.B., Monson, M.H., 1985a. Endemic Lassa fever in Liberia. IV. Selection of optimally effective plasma for treatment by passive immunization. *Trans R Soc Trop Med Hyg* 79, 380–384.

Jahrling, P.B., Frame, J.D., Smith, S.B., Monson, M.H., 1985b. Endemic Lassa fever in Liberia. III. Characterization of Lassa virus isolates. *Trans R Soc Trop Med Hyg* 79, 374–379.

Johnson, K.M., McCormick, J.B., Webb, P.A., Smith, E.S., Elliott, L.H., King, I.J., 1987. Clinical virology of Lassa fever in hospitalized patients. *J Infect Dis* 155, 456–464.

Keenlyside, R.A., McCormack, J.B., Webb, P.A., Smith, E., Elliott, L., Johnson, K.M., 1983. Case-control study of *Mastomys natalensis* and humans in Lassa virus-infected households in Sierra Leone. *Am J Trop Med Hyg* 32, 829–837.

Ketai, L., Alrahji, A.A., Hart, B., Enria, D., Mettler, F., Jr., 2003. Radiologic manifestations of potential bioterrorist agents of infection. *AJR Am J Roentgenol* 180, 565–575.

Khan, S., Goba, A., Chu, M., Roth, C., Healing, T., Marx, A., Fair, J., Guttieri, M., Ferro, P., Imes, T., Monagin, C., Garry, R., Bausch, D., 2008. New opportunities for field research on the pathogenesis and treatment of Lassa fever. *Antiviral Res* 78, 103–115.

Kruppa, T.F., Iglauer, F., Ihnen, E., Miller, K., Kunstyr, I., 1990. Mastomys natalensis or Mastomys coucha. Correct species designation in animal experiments. *Trop Med Parasitol* 41, 219–220.

Lecompte, E., Fichet-Calvet, E., Daffis, S., Koulemou, K., Sylla, O., Kourouma, F., Dore, A., Soropogui, B., Aniskin, V., Allali, B., Kouassi Kan, S., Lalis, A., Koivogui, L., Gunther, S., Denys, C., ter Meulen, J., 2006. Mastomys natalensis and Lassa fever, West Africa. *Emerg Infect Dis* 12, 1971–1974.

Liao, B.S., Byl, F.M., Adour, K.K., 1992. Audiometric comparison of Lassa fever hearing loss and idiopathic sudden hearing loss: evidence for viral cause. *Otolaryngol Head Neck Surg* 106, 226–229.

Lukashevich, I.S., Djavani, M., Rodas, J.D., Zapata, J.C., Usborne, A., Emerson, C., Mitchen, J., Jahrling, P.B., Salvato, M.S., 2002. Hemorrhagic fever occurs after intravenous, but not after intragastric, inoculation of rhesus macaques with lymphocytic choriomeningitis virus. *J Med Virol* 67, 171–186.

Lunkenheimer, K., Hufert, F.T., Schmitz, H., 1990. Detection of Lassa virus RNA in specimens from patients with Lassa fever by using the polymerase chain reaction. *J Clin Microbiol* 28, 2689–2692.

Mahanty, S., Bausch, D.G., Thomas, R.L., Goba, A., Bah, A., Peters, C.J., Rollin, P.E., 2001. Low levels of interleukin-8 and interferon-inducible protein-10 in serum are associated with fatal infections in acute Lassa fever. *J Infect Dis* 183, 1713–1721.

Mahanty, S., Hutchinson, K., Agarwal, S., McRae, M., Rollin, P.E., Pulendran, B., 2003. Cutting edge: impairment of dendritic cells and adaptive immunity by Ebola and Lassa viruses. *J Immunol* 170, 2797–2801.

McCormick, J.B., King, I.J., Webb, P.A., Johnson, K.M., O'Sullivan, R., Smith, E.S., Trippel, S., Tong, T.C., 1987a. A case-control study of the clinical diagnosis and course of Lassa fever. *J Infect Dis* 155, 445–455.

McCormick, J.B., King, I.J., Webb, P.A., Scribner, C.L., Craven, R.B., Johnson, K.M., Elliott, L.H., Belmont-Williams, R., 1986. Lassa fever. Effective therapy with ribavirin. *New Engl J Med* 314, 20–26.

McCormick, J.B., Webb, P.A., Krebs, J.W., Johnson, K.M., Smith, E.S., 1987b. A prospective study of the epidemiology and ecology of Lassa fever. *J Infect Dis* 155, 437–444.

Meyer, B.J., Southern, P.J., 1997. A novel type of defective viral genome suggests a unique strategy to establish and maintain persistent lymphocytic choriomeningitis virus infections. *J Virol* 71, 6757–6764.

Monson, M.H., Cole, A.K., Frame, J.D., Serwint, J.R., Alexander, S., Jahrling, P.B., 1987. Pediatric Lassa fever: a review of 33 Liberian cases. *Am J Trop Med Hyg* 36, 408–415.

Niedrig, M., Schmitz, H., Becker, S., Gunther, S., ter Meulen, J., Meyer, H., Ellerbrok, H., Nitsche, A., Gelderblom, H.R., Drosten, C., 2004. First international quality assurance study on the rapid detection of viral agents of bioterrorism. *J Clin Microbiol* 42, 1753–1755.

Oldstone, M.B., Campbell, K.P., 2011. Decoding arenavirus pathogenesis: Essential roles for alpha-dystroglycan-virus interactions and the immune response. *Virology* 411, 170–179.

Olschlager, S., Lelke, M., Emmerich, P., Panning, M., Drosten, C., Hass, M., Asogun, D., Ehichioya, D., Omilabu, S., Gunther, S., 2010. Improved detection of Lassa virus by reverse transcription-PCR targeting the 5′ region of S RNA. *J Clin Microbiol* 48, 2009–2013.

Palacios, G., Briese, T., Kapoor, V., Jabado, O., Liu, Z., Venter, M., Zhai, J., Renwick, N., Grolla, A., Geisbert, T.W., Drosten, C., Towner, J., Ju, J., Paweska, J., Nichol, S.T., Swanapoel, R., Feldmann, H., Jahrling, P.B., Lipkin, W.I., 2006. MassTag polymerase chain reaction for differential diagnosis of viral hemorrhagic fevers. *Emerg Infect Dis* 12, 692–695.

Paweska, J.T., Sewlall, N.H., Ksiazek, T.G., Blumberg, L.H., Hale, M.J., Lipkin, W.I., Weyer, J., Nichol, S.T., Rollin, P.E., McMullan, L.K., Paddock, C.D., Briese, T., Mnyaluza, J., Dinh, T.H., Mukonka, V., Ching, P., Duse, A., Richards, G., de Jong, G., Cohen, C., Ikalafeng, B., Mugero, C., Asomugha, C., Malotle, M.M., Nteo, D.M., Misiani, E., Swanepoel, R., Zaki, S.R., 2009. Nosocomial outbreak of novel arenavirus infection, southern Africa. *Emerg Infect Dis* 15, 1598–1602.

Peters, C.J., Jahrling, P.B., Liu, C.T., Kenyon, R.H., McKee, K.T., Jr., Barrera Oro, J.G., 1987. Experimental studies of arenaviral hemorrhagic fevers. *Curr Top Microbiol Immunol* 134, 5–68.

Price, M.E., Fisher-Hoch, S.P., Craven, R.B., McCormick, J.B., 1988. A prospective study of maternal and fetal outcome in acute Lassa fever infection during pregnancy. *BMJ* 297, 584–587.

Rai, S.K., Micales, B.K., Wu, M.S., Cheung, D.S., Pugh, T.D., Lyons, G.E., Salvato, M.S., 1997. Timed appearance of lymphocytic Choriomeningitis virus after gastric inoculation of mice. *Am J Pathol* 151, 633–639.

Rambukkana, A., Kunz, S., Min, J., Campbell, K.P., Oldstone, M.B., 2003. Targeting Schwann cells by nonlytic arenaviral infection selectively inhibits myelination. *Proc Natl Acad Sci USA* 100, 16071–16076.

Richmond, J.K., Baglole, D.J., 2003. Lassa fever: epidemiology, clinical features, and social consequences. *BMJ* 327, 1271–1275.

Roberts, P.J., Cummins, D., Bainton, A.L., Walshe, K.J., Fisher-Hoch, S.P., McCormick, J.B., Gribben, J.G., Machin, S.J., Linch, D.C., 1989. Plasma from patients with severe Lassa fever profoundly modulates f-met-leu-phe induced superoxide generation in neutrophils. *Br J Haematol* 73, 152–157.

Rodas, J.D., Cairo, C., Djavani, M., Zapata, J.C., Ruckwardt, T., Bryant, J., Pauza, C.D., Lukashevich, I.S., Salvato, M.S., 2009. Circulating natural killer and gammadelta T cells decrease soon after infection of rhesus macaques with lymphocytic choriomeningitis virus. *Mem Inst Oswaldo Cruz* 104, 583–591.

Safronetz, D., Lopez, J.E., Sogoba, N., Traore, S.F., Raffel, S.J., Fischer, E.R., Ebihara, H., Branco, L., Garry, R.F., Schwan, T.G., Feldmann, H., 2010. Detection of Lassa virus, Mali. *Emerg Infect Dis* 16, 1123–1126.

Saijo, M., Georges-Courbot, M.C., Marianneau, P., Romanowski, V., Fukushi, S., Mizutani, T., Georges, A.J., Kurata, T., Kurane, I., Morikawa, S., 2007. Development of recombinant nucleoprotein-based diagnostic systems for Lassa fever. *Clin Vaccine Immunol* 14, 1182–1189.

Schuler, A., 2005. Billions for biodefense: federal agency biodefense budgeting, FY2005-FY2006. *Biosecur Bioterror* 3, 94–101.

Shimojima, M., Stroher, U., Ebihara, H., Feldmann, H., Kawaoka, Y., 2012. Identification of cell surface molecules involved in dystroglycan-independent Lassa virus cell entry. *J Virol* 86, 2067–2078.

Solbrig, M.V., McCormick, J.B., 1991. Lassa fever: central nervous system manifestations. *J Trop Geograph Neurol* 1, 23–30.

Stephenson, E.H., Larson, E.W., Dominik, J.W., 1984. Effect of environmental factors on aerosol-induced Lassa virus infection. *J Med Virol* 14, 295–303.

ter Meulen, J., Lenz, O., Koivogui, L., Magassouba, N., Kaushik, S.K., Lewis, R., Aldis, W., 2001. Short communication: Lassa fever in Sierra Leone: UN peacekeepers are at risk. *Trop Med Int Health* 6, 83–84.

ter Meulen, J., Lukashevich, I., Sidibe, K., Inapogui, A., Marx, M., Dorlemann, A., Yansane, M.L., Koulemou, K., Chang-Claude, J., Schmitz, H., 1996. Hunting of peridomestic rodents and consumption of their meat as possible risk factors for rodent-to-human transmission of Lassa virus in the Republic of Guinea. *Am J Tropical Med Hyg* 55, 661–666.

Trombley, A.R., Wachter, L., Garrison, J., Buckley-Beason, V.A., Jahrling, J., Hensley, L.E., Schoepp, R.J., Norwood, D.A., Goba, A., Fair, J.N., Kulesh, D.A., 2010. Comprehensive panel of real-time TaqMan polymerase chain reaction assays for detection and absolute quantification of filoviruses, arenaviruses, and New World hantaviruses. *Am J Trop Med Hyg* 82, 954–960.

Van der Waals, F.W., Pomeroy, K.L., Goudsmit, J., Asher, D.M., Gajdusek, D.C., 1986. Hemorrhagic fever virus infections in an isolated rainforest area of central Liberia. Limitations of the indirect immunofluorescence slide test for antibody screening in Africa. *Trop Geogr Med* 38, 209–214.

Vieth, S., Drosten, C., Lenz, O., Vincent, M., Omilabu, S., Hass, M., Becker-Ziaja, B., ter Meulen, J., Nichol, S.T., Schmitz, H., Gunther, S., 2007. RT-PCR assay for detection of Lassa virus and related Old World arenaviruses targeting the L gene. *Trans R Soc Trop Med Hyg* 101, 1253–1264.

Vieth, S., Torda, A.E., Asper, M., Schmitz, H., Gunther, S., 2004. Sequence analysis of L RNA of Lassa virus. *Virology* 318, 153–168.

Walker, D.H., McCormick, J.B., Johnson, K.M., Webb, P.A., Komba-Kono, G., Elliott, L.H., Gardner, J.J., 1982. Pathologic and virologic study of fatal Lassa fever in man. *Am J Pathol* 107, 349–356.

Walker, D.H., Wulff, H., Lange, J.V., Murphy, F.A., 1975. Comparative pathology of Lassa virus infection in monkeys, guinea-pigs, and Mastomys natalensis. *Bull World Health Organ* 52, 523–534.

Wulff, H., Fabiyi, A., Monath, T.P., 1975. Recent isolations of Lassa virus from Nigerian rodents. *Bull World Health Organ* 52, 609–613.

Zaki, S.R., Shieh, W.J., Greer, P.W., Goldsmith, C.S., Ferebee, T., Katshitshi, J., Tshioko, F.K., Bwaka, M.A., Swanepoel, R., Calain, P., Khan, A.S., Lloyd, E., Rollin, P.E., Ksiazek, T.G., Peters, C.J., 1999. A novel immunohistochemical assay for the detection of Ebola virus in skin: implications for diagnosis, spread, and surveillance of Ebola hemorrhagic fever. Commission de Lutte contre les Epidemies a Kikwit. *J Infect Dis* 179 Suppl 1, S36–S47.

Zweighaft, R.M., Fraser, D.W., Hattwick, M.A., Winkler, W.G., Jordan, W.C., Alter, M., Wolfe, M., Wulff, H., Johnson, K.M., 1977. Lassa fever: Response to an imported case. *New Engl J Med* 297, 803–807.

16 Lujo Virus Hemorrhagic Fever

Janusz T. Paweska, Petrus Jansen van Vuren,
and Jacqueline Weyer

CONTENTS

16.1 INTRODUCTION

The emergence of infectious diseases usually encompasses any of three situations: (1) appearance of a known agent in a new geographic area or population or increase of incidence of the disease, (2) evolution of a pathogen to increased pathogenicity or transmissibility, and (3) recognition of a previously unknown agent (Fauci, 2005; Taylor et al., 2001). It is estimated that 12% of human infectious diseases (of bacterial, fungal, and viral nature) can be considered emerging, with majority having an animal source. In many instances, the hosts of emerging zoonotic viruses have proven to be small mammals. Some of the major factors responsible for the emergence of infectious diseases include ecological changes, such as those due to agricultural or economic development or anomalies in climate, microbial adaptation and change, and the breakdown of public health measures. Increasing international travel, trafficking in wildlife, political instability, and bioterrorism have made also emerging infectious diseases a global concern. In this context, viral hemorrhagic fevers (VHFs) warrant specific emphasis because of their high rates of severe illness and death and the potential for rapid dissemination by human-to-human transmission. VHFs are caused by infection with several negative-stranded RNA viruses belonging to four families: *Arenaviridae*, *Bunyaviridae*, *Filoviridae*, and *Flaviviridae*. Although clinical management of VHFs is primarily supportive, early diagnosis is needed in order to timely implement infection control measures to combat further spread of the disease. This is especially important if VHF pathogens are encountered out of their traditional geographic areas. Of the Old World (OW) arenaviruses, only lymphocytic choriomeningitis virus (LCMV) and Lassa fever virus (LASV) were known to cause disease in humans until September 2008. During September and October 2008, five VHF cases caused by a novel arenavirus, Lujo virus (LUJV), occurred in South Africa. LUJV is the first hemorrhagic fever–associated arenavirus from Africa identified in four decades and from Southern Africa. It was originally isolated in South Africa during a nosocomial outbreak characterized by high case-fatality rate of 80% (4/5 cases). The primary case originating from Zambia was airlifted in critical condition to South Africa for treatment (Briese et al., 2009; Keeton, 2008; Paweska et al., 2009). The successful international collaboration during this highly fatal outbreak highlighted the importance of global cooperation in outbreak response to emerging and highly dangerous pathogens.

16.1.1 General Characteristics of Arenaviruses

The family *Arenaviridae* consists of a single genus *Arenavirus* that currently comprises 23 recognized members by the International Committee for Taxonomy of Viruses (ICTV, 2009; Salvato et al., 2005; http://ictvonline.org/virus) with at least 12 recently described species awaiting classification (Briese et al., 2009; Cajimat et al., 2007a, 2008; Charrel et al., 2008; Coulibaly-N'Golo et al., 2011; De Bellocq et al., 2010; Delgado et al., 2008; Günter et al., 2009; Inizan at al., 2010; Ishii at al., 2011; Lecompte et al., 2007; Palacios et al., 2008, 2010). The type species of the family *Arenaviridae* is LCMV, isolated in 1933 in North America from a human case with aseptic meningitis (Armstrong and Lillie, 1934). Arenaviruses are grouped into two major complexes based on serological cross-reactions, genetic, and geographic relationships, namely, the New World (NW) or Tacaribe (TCRV) complex and the OW or Lassa–lymphocytic choriomeningitis complex (Bowen et al., 1997; Gonzalez et al., 2007; Moncayo et al., 2001). The NW complex includes viruses indigenous to the Americas. The OW complex includes viruses occurring in Africa and the ubiquitous LCMV. The geographic distribution of each arenavirus is largely determined by the range of its natural vertebrate reservoir hosts, mostly rodents. LCMV is the only arenavirus to have a worldwide distribution due to its association with the ubiquitous *Mus musculus*. The NW arenaviruses are further divided into three clades A, B, and C (Archer et al., 2002; Bowen et al., 1996; Charel et al., 2003) with clade B containing all the South American VHF-associated viruses. Chronic infections with excretion of arenaviruses are well documented and are an important mechanism for their maintenance and perpetuation in nature.

Historically, the evolution of arenaviruses has been closely tied with that of its rodent reservoir (Bowen et al., 1997). Each arenavirus appears to be uniquely associated with particular rodent species, the only exception to date being the TCRV virus with a fruit bat reservoir. Genomic data from the increasing number of newly discovered arenaviruses challenge this hypothesis and suggest that host switching and lineage extinction also contribute to the phylogeny of the arenaviruses (Coulibaly-N'Golo et al., 2011). While NW arenaviruses are associated with rodents in the *Sigmodontinae* subfamily of the family *Cricetidae*, OW arenaviruses are associated with rodents in the *Murinae* subfamily of the family *Muridae*. Rodent-to-human infection occurs most frequently through contact with infected rodent excreta, commonly via inhalation of dust or aerosolized virus-containing materials, or fomites soiled with rodent feces or urine, or ingestion of contaminated foods (McCormick et al., 1987; Stephenson et al., 1984). Rodent consumption has also been evoked as possible risk behavior (Ter Meulen et al., 1996). Person-to-person transmission occurs via direct contact with infected blood, urine, or pharyngeal secretions. Severe outbreaks are mostly reported in health-care facilities, often as a result of inadequate implementations of standard precautions (Bajani et al., 1997; Frame et al., 1970). Poor quality and lack of hygiene in individual housing and of the immediate surrounding environment increase risk of rodent infestation and contracting Lassa fever (Bonner et al., 2007). The recent study in communities of the forest region of Guinea underlines the importance of person-to-person transmission of Lassa fever via close contact in the same household or nosocomial exposure (Kerneis et al., 2009). Transmission of arenaviruses may also occur during transplantation of infected organs (Fisher et al., 2006; Palacios et al., 2008). Arenavirus genome consists of two RNA segments, a small (S) and a large (L) segment, which together total about 11 kb, each encoding for two proteins in ambisense coding strategy. The S segment encodes for the nucleocapsid protein (NP) and the glycoprotein precursor (GPC), subsequently cleaved into the envelope proteins GP1 and GP2. As with other class I fusion proteins, the GP1 associates with cellular receptors. The GP2 is a transmembrane protein that mediates fusion of the viral and cellular membranes after internalization of the virus into acidified endosomes. The L segment encodes for the viral RNA-dependent RNA polymerase (L protein) and a zinc-binding matrix protein (Z protein). It is a small protein of about 11 kDa, has a ring finger motif, and binds zinc (Buchmeier et al., 2007). It has multiple functions, including downregulation of RNA replication and the synthesis of mRNAs, and it is also required for budding of virions (Schlie et al., 2010; York and Nunberg, 2007).

16.1.2 Host Cellular Receptors: Determinants of Pathogenesis and Zoonotic Transmission

To date, two different cell surface molecules have been implicated as cellular receptors for arenaviruses. OW arenaviruses (LASV, certain strains of LCMV) and NW clade C viruses use α-dystroglycan (α-DG) to enter cells (Spiropoulou et al., 2002). This cellular receptor plays a critical role in cell-mediated assembly of basement membranes. Viruses with a high binding affinity for α-DG replicate preferentially in the white pulp of the spleen and infect large numbers of lymphocytes. The ability of these cells to act as antigen-presenting cells results in impairment of host immune responses. Although immunosuppression may be important for the establishment of persistent infection in rodent hosts, in which the infection is subclinical, in humans it may lead to serious illness. Immunosuppressive strains of LCMV and highly pathogenic LASV have been shown to bind to α-DG with very high affinity, which is crucial for their ability to infect dendritic and Schwann cells to cause a generalized immunosuppression (Smelt et al., 2001). The existence of a high degree of conservation in the receptor binding characteristics makes α-DG a promising target for the development of novel drugs for inhibition of LASV replication (Kunz et al., 2005).

The pathogenic NW clade B arenaviruses Junín virus (JUNV), Machupo virus (MACV), Guanarito virus (GTOV), and Sabiá virus (SABV) use both human transferrin receptor 1 (TfR1) and the TfR1 orthologs of their reservoir rodent species. This molecule has a number of properties, which favor their replication and disease pathology in humans (Flanagan et al., 2008; Radoshitzky et al., 2008). TfR1 is rapidly endocytosed into acidic compartments, expressed on endothelial cells, and upregulated on activated lymphocytes. In contrast to NW pathogenic arenaviruses, the nonpathogenic NW arenaviruses use only the TfR1 orthologs of their animal hosts. For example, the Ampari (AMAV) and TCRV viruses use only the TfR1 orthologs of their respective rodent or bat animal hosts (*Neacomys spinosus, Artibeus jamaicensis*), but not human TfR1. However, mutation of one residue can convert human TfR1 into an efficient receptor for TCRV; similarly, replacement of four residues with those present in *N. spinosus* TfR1 converts human TfR1 into an efficient receptor for AMAV (Abraham et al., 2009). These recent findings underscore the biological role of TfR1 cellular receptor in arenavirus replication and evolution. While pathogenic NW arenaviruses specifically adapted to the TfR1 orthologs of their animal hosts, they identify key commonalities between these orthologs and human TfR1 necessary for efficient zoonotic transmission to humans. Full genome sequencing of LUJV revealed that it lacks the sequences to support the use of α-DG and TfR1 receptors, which further supports the novelty of this newly described arenavirus (Briese et al., 2009).

TABLE 16.1

Known Arenaviruses Causing Viral Hemorrhagic Fevers in Humans

Virus Name and Abbreviation	Year of First Reported Outbreak	Natural Host	Transmission	World Distribution
Lassa (LASV)[OW]	1969	*Mastomys sp.*	Urine, saliva	West Africa (Nigeria, Sierra Leone, Liberia, Guinea)
Lujo (LUJV)[OW]	2008	Unknown	Unknown	Sub-Saharan Africa
Guanarito (GTOV)[NW]	1990	*Zygodontomys brevicauda*	Urine, saliva	Venezuela
Junín (JUNV)[NW]	1958	*Calomys musculinus*	Urine, saliva	Argentina
Machupo (MACV)[NW]	1963	*Calomys callosus*	Urine, saliva	Bolivia
Sabiá (SABV)[NW]	1990	Unidentified rodents	Unknown	Brazil
Chapare	2003	Unknown	Unknown	Bolivia

Note: OW, Old World arenavirus; NW, New World arenavirus.

16.2 VIRAL HEMORRHAGIC FEVERS CAUSED BY ARENAVIRUSES

Several members of the family *Arenaviridae* can cause severe hemorrhagic fevers in humans (Table 16.1), thus representing a serious public health problem in endemic areas of Africa and South America (Aguilar et al., 2009; Charrel and de Lamballerie, 2003; Richmond and Baglole, 2003; Tesh, 2002).

Together with OW LASV, NW clade B arenaviruses are considered as category A potential agents for bioterrorism and are restricted to biosafety level 4 (BSL-4) containment. Importation of these viruses from endemic into nonendemic countries carries potentially severe implication, including interruption of normal life, commerce or social structure of the community, widespread anxiety about severe symptoms and high fatality rates, fear of epidemic or pandemic spread, and concerns about travel-associated and health-care-acquired infections. Management of VHF cases could be very challenging for laboratory, medical, and public health systems and requires highly organized and specialized public health measures in controlling these diseases (Bannister, 2010; Borio et al., 2002; Franz et al., 1997). It is noteworthy that two highly pathogenic arenaviruses, namely, LUJV and Dandenong, have been discovered in the past 5 years. The International Health Regulations are designed to protect against the international spread of these dangerous infections (WHO, 2005).

16.2.1 Clinical Presentation and Pathogenesis

Asymptomatic arenavirus infections appear to be frequent, but two major types of clinical presentations include neurological and hemorrhagic fever symptoms. The incubation period is about 10 days (3–21 days). LCM viruses cause aseptic meningitis or meningoencephalitis with an overall case fatality <1%. Fetal infections can result in congenital abnormalities or death. Immunosuppressed recipients of organ transplants can develop fatal hemorrhagic fever–like disease. The Lassa VHF usually presents as a nonspecific illness with symptoms including fever, headache, dizziness, asthenia, sore throat, pharyngitis, cough, retrosternal and abdominal pain, and vomiting. In severe forms, facial edema is associated with hemorrhagic conjunctivitis, moderate bleeding (from nose, gums, vagina, etc.), and exanthema. Neurological signs may develop and progress to confusion, convulsion, coma, and death. Severe prognosis is associated with a high viremia, elevated aspartate aminotransferase (AST) liver enzymes, bleeding, encephalitis, and edema (Monat and Casals, 1975). There is a very high risk of fetal mortality in pregnant women during the third trimester of pregnancy. Case-fatality rates range from 5% to 20% for hospitalized cases. The most consistently observed lesions in fatal cases of Lassa fever include hepatocellular, adrenal, and splenic necrosis and adrenal cytoplasmic inclusions. LASV replicates to high titers of liver, lung, spleen, kidney, heart, placenta, and mammary gland (Walker et al., 1982). Clinical symptoms of infection by arenaviruses in South America are similar to Lassa fever in Africa. The nucleoprotein of LASV is the main immunogen, being produced in excess of other proteins and inducing a dominating humoral response. There is, however, no correlation between this response and protection against viral replication and disease development. Immunization of monkeys with the glycoproteins of LASV does not induce detectable neutralizing antibody responses, yet it protects them against lethal challenge, which suggests involvement of cellular immunity (Fisher-Hoch et al., 2000; Geisbert et al., 2005). Severe or fatal Lassa fever is likely not only as a result of uncontrolled viral replication and destruction of host cells by viral lysis but also because of immunopathogenic effects as a result of altered host immune responses, such as an uncontrolled inflammatory response. As with any viral infection, type I interferons play an important role in fighting arenavirus infection of the host. And also, as for most if not all viruses, arenaviruses employ certain strategies to counteract the type I interferon response to allow evasion of host immunity. Studies on the role of inflammatory mediators in the pathogenesis of Lassa fever indicate that low levels of interleukin-8 and interferon-inducible protein 10 in serum are associated with fatal infections (Mahanty et al., 2001). Type I interferon induces the production of tetherin, an antiviral molecule, which has been shown to inhibit the release of arenaviruses, including LASV, from host cells. Studies on LCMV have implicated the nucleoprotein of arenaviruses as the pathogenesis marker as a result of its ability to counteract type I interferon, specifically inhibiting the interferon regulatory factor 3 (IRF3) (Borrow et al., 2010).

16.2.2 TREATMENT AND PROPHYLACTIC VACCINATION

Based on the danger of arenaviruses for human health, the increased emergence of new viral species in recent years and the lack of effective tools for their control or prevention, the search for novel antiviral compounds effective against these pathogenic agents is a continuous demanding effort (Khan et al., 2008). Unfortunately, therapeutic options for all VHFs are limited; thus, early differential diagnosis has implications for containment and clinical management (Gowen and McBray, 2011). A number of vaccines have been developed for VHFs, but only yellow fever vaccine is widely available for public. Early treatment with immune plasma was effective in JUNV and LASV infections (Enria et al., 2008). The nucleoside analog ribavirin may be helpful if given early in the course of Lassa fever, Crimean–Congo hemorrhagic fever, or hemorrhagic fever with renal syndrome and is recommended in postexposure prophylaxis and early treatment of arenavirus infections (Bannister, 2010) but not anymore in treatment of Rift Valley fever. Ribavirin has been shown to be an effective treatment for Lassa fever, especially when started within the first 6 days of illness. One available vaccine is licensed in Argentina for JUNV (Maiztegui et al 1998). There is currently no vaccine for Lassa fever and most of the VHFs caused by arenaviruses, but several candidates are under development with successful trials in primates (Geisbert et al., 2005). Recent identification of HLA-restricted CD8[+] T cell epitopes that are either cross-reactive or species specific has a promise in the development of a multivalent vaccine strategy against arenaviruses pathogenic for humans (Kotturi et al., 2009).

16.3 OUTBREAK OF LUJO VIRUS HEMORRHAGIC FEVER IN SOUTH AFRICA

16.3.1 HISTORY OF THE OUTBREAK

Epidemiological description of the LUJV outbreak is drawn from literature available to date (Bateman, 2008; Paweska et al., 2009; Richards et al., 2009; Sewlall, 2011) and information available to authors of the chapter for whom the outbreak was a challenging experience.

The first case, a 36-year-old woman, lived in the outskirts of Lusaka, Zambia, on an agricultural smallholding and worked as tourist guide for safari trips. Apparently, on the August 30, 2008, she was cut on the shin by broken glass and 2 days later developed severe headache and malaise. On September 4, she attended a wedding in South Africa. Four of 110 guests who attended the wedding developed diarrhea and vomiting on September 6. While taking medication for suspected influenza, on return to Lusaka 3 days later, she had an attack of diarrhea and vomiting. Her symptoms worsened with time, and from the 8th to 11th of September, she developed fever, severe chest pain, myalgia, sore throat, followed by intensive skin rash, and swelling of the face. She was initially treated for food poisoning and influenza and then for suspected allergic reaction to antibiotic treatment (cephalosporin). Due to a progressive course of illness, resulting in deterioration and generalized tonic–clonic seizures, on September 12, she was moved from a clinic to a hospital in Lusaka where she was intubated for a deteriorating Glasgow Coma Scale (GSC) score prior to air evacuation by the Speciality Emergency Services (SES) to the Morningside Medi-Clinic in Johannesburg, South Africa. She arrived at the Morningside Medi-Clinic Sandton, Johannesburg, intubated, heavily sedated, and unresponsive (brain-dead) with a macular rash on her trunk and upper arm. The attending physician, Dr Nivesh Sewall, observed a classic tick bite (eschar) on her right foot, which prompted treatment for rickettsiosis. For him, the patient appeared to suffer from tick bite fever with end-stage multiple organ failure. The patient arrived from Lusaka without any accompanying clinical notes or laboratory results, placing the attending physician and other hospital staff in a challenging situation (Bateman, 2008). Despite intensive care, including hemodialysis, she died on September 14, with severe cerebral edema, generalized inflammatory capillary leak, and multiorgan failure, but no specific diagnosis could be made. Although she was markedly thrombocytopenic, no bleeding or overt hemorrhage was noted, but she displayed diffused erythematous macular rash. Neither autopsy was performed on the first case nor clinical specimens taken in South Africa for microbiological testing. Later on during the outbreak investigation, blood samples collected from case 1 on September 12 were traced back by a member of Special Pathogens Unit of the National Institute for the Communicable Diseases of the National Health Laboratory Services (SPU NICD/NHLS) in a hospital in Lusaka and shipped to CDC, Atlanta, USA.

During the medical evacuation of the first case, the attending physicians and a paramedic were involved in nebulization, suctioning, and manual ventilation of the patient. They had worn only nonsterile examination gloves, without masks. The paramedic recalled handling some of the patient's secretions, including diarrheal stool, but denied having contact with mucous membranes. On September 21, he developed a progressive febrile illness with increasing malaise and gastrointestinal symptomatology. He was admitted to a hospital in Lusaka on September 24 and 3 days later evacuated to the same private hospital in Johannesburg as the first case. On September 28, the epidemiological connection between these two cases was recognized by the same attending physician as for first case, Dr Nivesh Sewall. In consultation with NICD-NHLS, the clinical syndrome of VHF was recognized. Immediate tracing of contacts and barrier nursing procedure were instituted, and blood samples collected from the second case on September 29 and 30 were submitted for laboratory testing to the SPU of the NICD-NHLS. Despite intensive screening of samples for all known African VHF agents (Ebola HF, Marburg HF, Lassa fever, Rift Valley fever, and Crimean–Congo HF) and using various assay targeting specific antibody, antigens, and nucleic acids, all tests yielded negative results. They were also negative for tick bite fever, malaria, and bacterial culture. In the meantime, the condition of the second case deteriorated rapidly and he died on October 2, 5 days after hospital admission.

An intensive care unit nurse at Morningside Medi-Clinic (case 3—34-year-old woman), who cared for and cleaned the body of the first case, was on leave when she developed flu-like symptoms on September 23. She was admitted to a private hospital, the Sir Albert Robinson Hospital in the West Rand, Johannesburg, on October 1. Her condition deteriorated and she died 4 days later, on October 5. Similarly as for the second case, laboratory testing of blood samples collected from her on only one occasion, October 3 (11th day of illness), gave negative results. The fourth victim (case 4—38-year-old woman) was Morningside Medi-Clinic contract cleaner who was involved in the terminal cleaning of the cubicle of the first case. She became ill on September 27, admitted to a provincial hospital on October 5, and then transferred to the

Charlotte Maxeke Johannesburg Academic Hospital where she was treated for a chronic illness but eventually she died the following day.

At this point of the outbreak, the mode of nosocomial transmission of the unknown deadly VHF agent had been realized. Calming fears of the hospital staff in four hospitals in Johannesburg, which cared for the patients, the Johannesburg's public community in general, but also responding to increasing international interest became an additional formidable challenge to medical, laboratory, and public health authorities in South Africa. After endless hours of local and international consultations, the possibility of the outbreak being caused by unknown highly dangerous VHF agent had been seriously taken into consideration. Where did this frightening disease come from? How did it spread? Is the South African public health and laboratory system prepared to control it? Can it be stopped? Can this newly emerged unknown pathogen escape out of Africa? These were some of the questions and fears that the South African doctors and scientists while risking their own lives were asked to answer or directly face (Keeton, 2008).

Because of the nonspecific nature and high mortality of VHFs, the diagnosis and etiology are often not determined until autopsy. Eventually, decision was made to take tissue biopsies, particularly liver samples because of high tropism of VHF agents to this organ (Figure 16.1). For a number of administrative and logistic reasons, taking of biopsies was delayed until October 9. While laboratory tests, including virus isolation from collected biopsies, continued in South Africa, blood, liver, and skin samples from cases 2 and 3 were shipped the same day to the Centers for Disease Control and Prevention, Atlanta, USA, for virological testing and RNA extracts from postmortem liver biopsies and one serum sample to Columbia University, New York, USA, for unbiased high-throughput pyrosequencing.

The next day, on October 10, the Department of Anatomical Pathology of the University of the Witwatersrand, South Africa, reported hepatocyte necrosis and skin vasculitis compatible with VHF histopathological lesions. On October 11, the Infectious Disease Pathology Branch, CDC, reported detection of antigen in liver and skin sections by immunohistochemical staining using

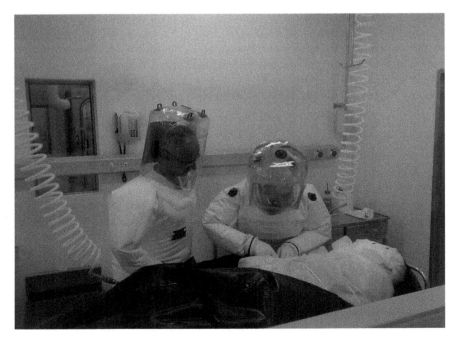

FIGURE 16.1 **(See color insert.)** SPU NICD/NHLS staff members taking biopsies from a victim of a fatal nosocomial outbreak in Johannesburg in BSL-4 facility. September 9, 2008, Sandringham-Johannesburg, South Africa.

monoclonal antibody broadly cross-reactive for OW arenaviruses. This was the first indication of an etiological diagnosis. Consequently, generic RT-PCR procedures for the detection of all known OW arenaviruses were applied to liver extracts and produced positive results at SPU NICD/NHLS and at the Special Pathogens Branch, CDC. Partial nucleotide sequencing of PCR products and phylogenetic analysis performed by the SPU NICD/NHLS indicated that the etiological agent was not LASV but likely a novel, previously unreported OW arenavirus.

Nucleic acid extracts from serum and tissue samples subjected to unbiased pyrosequencing yielded sequence fragments representing about 50% of a prototypic arenavirus genome within 72 h of samples receipt at the Columbia University. Subsequently, full genome sequence was generated by PCR amplification of remaining fragments using specific primers complementary to sequence obtained by pyrosequencing and a universal primer targeting the conserved arenaviral termini. Phylogenetic analyses based on full genome sequences confirmed the presence of a new member of the family *Arenaviridae* (Briese et al., 2009), provisionally named LUJV in recognition of its geographic origin and geographic location of the first recognized outbreak (*Lu*saka, Zambia, and *Jo*hannesburg, South Africa). The initial negative RT-PCR results in the blood samples of cases 2 and 3 were likely due to too low viral loads, but eventually an arenavirus was isolated in Vero cells from blood samples from the two cases.

Case 5 was a nurse who became sick on October 9 and was hospitalized on the following day. She had attended case 2 (paramedic) and was involved in the insertion of a central venous line on September 27. At this time, the cause of the outbreak was recognized and on October 11 she was subjected to treatment with ribavirin along with supportive therapy, but the hepatitis and thrombocytopenia worsened and she became confused. It was realized that oral therapy was unlikely to achieve the minimum inhibitory concentrations necessary to eradicate the virus rapidly. Consequently, a source of intravenous ribavirin was urgently sought but could only be obtained after one week from the Netherlands at an excessive price of 45,000 US dollars to supply a 10 day course, which started on October 17. This highlights pure opportunism by a pharmaceutical company, which took advantage of the urgency of this dramatic situation. It has been suggested that ribavirin should be available to the WHO for distribution at reasonable cost when urgently needed (Richards et al., 2009). On initiation of intravenous ribavirin treatment, there was a prompt improvement and the patient was discharged from the hospital on December 2 after arenaviral RNA was not detected in blood and urine on three consecutive occasions. Schedule for oral and intravenous administration of ribavirin and virological and other laboratory findings in patients during this outbreak are described in more detail by Paweska et al. (2009).

16.3.2 Source and Mode of Lujo Virus Transmission

The natural reservoir hosts of LUJV remain unknown likewise the source and the route of infection of the first (index) case. She kept domestic pets and horses on her agricultural holding near Lusaka and there was evidence of rodent activity in the stables. Rodents could contaminate objects left unattended on the ground by excretion of virus in urine, including broken glass, which apparently resulted in a deep cut to her foot about 10 days before medical evacuation to South Africa (Paweska et al., 2009). These circumstances might be of potential epidemiological relevance as she developed first symptoms within the incubation period of arenavirus infection after the cut. However, Dr Nivesh Sewlall, who admitted and treated the first arenavirus victim, did not note any cut, laceration, or sutures during examination of her soles and feet upon her arrival at the Morningside Medi-Clinic in Johannesburg on September 12 (Bateman, 2008). The natural reservoirs of arenavirus in Africa are rodents of the family *Muridae*, especially *Mastomys natalensis*. Up to now, only nonpathogenic arenaviruses have been found in areas surrounding Zambia. A study on the prevalence of arenaviruses among *M. natalensis* rodents in Zambia was conducted less than 1 year after the LUJV outbreak, from May to August 2009, including areas surrounding the cities of Lusaka, Namwala, and Mfuwe, but not specifically on the farm of the index case. Nevertheless, of the total of 263 rodents captured, five were positive for an arenavirus infection in kidneys, of which 17% of the 23 rodents captured near Lusaka and 4% of the 24 captured in Namwala were positive, but none of the 143 rodents captured in Mfuwe were positive for arenavirus. Phylogenetic analysis of the four Zambian arenavirus isolates showed distinct sequences between OW and NW arenaviruses. The novel Lusaka and Namwala strains, collectively designated Luna virus (LUNV), are genetically different from the LUJV but closely related to nonpathogenic arenaviruses that have been found from central to eastern Africa (Ishii et al., 2011).

It still remains rather intriguing that no nosocomial transmission took place in clinics in Lusaka either among health-care workers or other contacts of the index case. Despite the fact that the first four patients in hospitals in Johannesburg were initially managed without special infection control measures (use of only surgical gloves and impervious aprons in wards but no masks, goggles, or visors), it appears that the LUJV was transmitted to a very limited number of hospital workers and only within a single private hospital. Tracing and clinical monitoring of contacts of the 5 known patients or fomites in Zambia (at least 34 contacts) and South Africa (more than 240 contacts, including 110 guests of the wedding attended by case 1) did not identify any additional cases. Of the 94 people on the Morningside Medi-Clinic contact list in Johannesburg, all but one passed safely the recommended 21-day observation period (maximum incubation time for all know VHF infections), with six cautionary admissions due to temperature fluctuations. However, antibody surveys in all known contacts in Zambia and South Africa would have to be conducted to assess the possibility of less severe infections and to determine the distribution and prevalence of the virus in human and rodent populations in sub-Saharan Africa, particularly in Zambia. Fortunately, nosocomial transmission of LUJV outbreak occurred in a single health-care setting in Johannesburg, despite the fact that three victims infected at Morningside Medi-Clinic were hospitalized in three different hospitals in Johannesburg (Bateman, 2008; Paweska et al., 2009; Sewlall, 2011).

The low level of LUJV secondary spread from the five outbreak patients might indicate that aerosol transmission of the virus is not very efficient and infection with the virus requires very close, unprotected skin or mucosal contact with either patient's excreta, vomitus, blood, or direct, unprotected exposure to contaminated fomites (e.g., bedding, utensils, bedpans, hospital equipment) or might result from accidentally acquired infection, e.g., by a needlestick. No apparent incidents occurred during the outbreak that could result in a specific exposure. However, during outbreak of this nature, contact with infected patients or fomites is usually inadvertent, and specific exposure to infected secretions or tissues might be unnoted or not recalled as being significant at the time. VHF outbreaks are unpredictable and rare and usually their recognition is delayed. Therefore, constant vigilance and attention to standard precautions, although basic, should be very strict and correctly applied to the daily, routine medical practice. This should apply not only to hospitals and clinics but also during medical air evacuation procedures (Bateman, 2008; Sewlall, 2011).

No genome sequence differences were found between LUJV detected in primary clinical material directly obtained from liver biopsy (cases 2 and 3) and serum (cases 1 and 2) and virus isolates from Vero cell cultures at SPU NICD/NHLS and CDC. The lack of sequence variation is consistent with the epidemiological data, indicating an initial natural exposure of the index case, followed by a chain of nosocomial transmission among subsequent cases (Briese et al., 2009).

16.3.3 Medical Evacuation of LUJV Patient: Lessons Learned

One of the weaknesses recognized shortly after the outbreak started in Johannesburg was lack of prior consultation by air evacuation services (SES) with South African Port Control officers employed by the South African Department of Health and timely consultations of these officials with their own local medical experts. Although airport health officials checked the SES flight, which brought the index case from Lusaka to Johannesburg, the list of symptoms concerned (fever, severe headache, abnormal sweating, difficulties in breathing, coughing, vomiting, diarrhea) covers more common diseases (malaria, flu, and tick bite) being one of the reasons why case 1 was not recognized as a potential communicable threat. However, it is worrying that the attending doctor at Morningside Medi-Clinic in Johannesburg was not given any documentation related to the patient disease history (Bateman, 2008).

International transfer of patients must follow strict international health regulations to minimize the possibility of rapid spread of infectious diseases by air transport. This is of paramount importance considering the number of patients evacuated from neighboring countries to South Africa. More than 50 foreign febrile patients are admitted weekly to hospitals only in Johannesburg (Bateman, 2008).

16.3.4 CLINICAL PRESENTATION

Clinical presentation and histological lesions in the course of infection with LUJV are similar to those of other arenavirus VHFs (De Manzione et al., 1998; Elnser et al., 1973; Frame et al., 1970; Maiztequi 1975; McCormick et al., 1987; Salas et al., 1991; Tesh, 2002; Walker et al., 1982). Severe bleeding or hemorrhage was not a prominent clinical feature. Initial symptoms were nonspecific and included fever, headache, and myalgia followed by gastrointestinal symptoms and sore throat with increasing severity. Terminal features included severe respiratory distress, neurological signs, and circulatory collapse. Summary of major symptoms and signs in patients infected with LUJV drawn from limited literature on the outbreak (Bateman, 2008; Paweska et al., 2009; Richards et al., 2009; Sewlall, 2011) is given in Table 16.2.

16.3.5 IMMUNOLOGICAL RESPONSES TO INFECTION WITH LUJV

Humans infected with an arenavirus usually develop a transient IgM and delayed but persistent IgG response. IgM antibody to LASV can be detected in patients within a few days after hospitalization (Wulff and Johnson, 1979). In rhesus monkeys experimentally infected with LASV, IgM is

TABLE 16.2

Prevalence of Major Signs and Symptoms among Five Patients Confirmed to Be Infected with LUJV during a Nosocomial Outbreak in Johannesburg, South Africa, 2008

Signs and Symptoms	Occurrence (%)
Fever	100
Headache	100
Myalgia	100
Sore throat	100
Diarrhea	100
Respiratory distress	100
Neurological signs	100
Morbilliform rash	60
Neck and facial swelling	60
Chest pain	40
Petechial rash	20
Bleeding	20
Thrombocytopenia	100
Increased aspartate aminotransferase	80
Leukocytosis	80

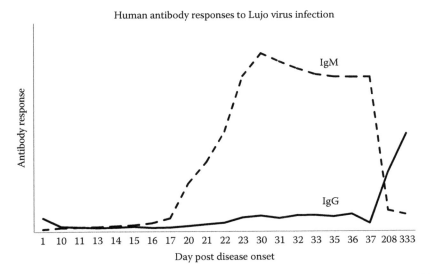

FIGURE 16.2 Dynamics of immune response in a Lujo hemorrhagic fever patient as measured by ELISA. The dotted line indicates the IgM response and the solid line the IgG response.

detectable 10 days postinfection, with peak after 36 days followed by gradual decrease, but still detectable low level for at least 18 months. The IgG response is slightly delayed, detectable 3–9 days after the appearance of IgM with peak on week 10, and still detectable at high levels for at least 18 months (Niklasson et al., 1984). A recent study showed that IgM can be detected in patients for up to at least 2 years after infection (Branco et al., 2011). Lassa fever patients that fail to develop an IgM and/or IgG response within the first week of onset usually suffer severe infection (Bausch et al., 2000; Wulff and Johnson, 1979).

In four fatal cases infected with LUJV, IgM and IgG could not be demonstrated during acute course of infection, 10–13 days from disease onset to death (NICD/NHLS, unpublished data). This corresponds to observations made in fatal Lassa fever cases (Bausch et al., 2000). Therefore, we could only monitor humoral responses using serial bleeds from one patient who survived. The IgM was first detected on day 20, with peak on day 30, followed by decreasing but still detectable level on day 333 after onset of the disease (Figure 16.2). The long persistence of IgM in a Lujo patient corresponds to recent findings with Lassa fever and some of the South American arenaviruses. Possible explanations for the sustained IgM response include inhibition of antibody class switching, presence of IgM + memory B cells, or decreased T-helper cell (CD4+) function (Branco et al., 2011). The Lujo patient developed a delayed IgG response, detectable between day 37 and 208, but with still increasing level by day 333 post-onset (Figure 16.1). It remains unknown whether the patient survived solely due to humoral immune response or because of ribavirin treatment or combination of both. Ribavirin suppresses pro-inflammatory cytokines and Th-2 cytokines, involved in humoral immunity, while maintaining normal Th-1 cytokine expression that favors cellular immunity (Ning et al., 1998). Infection with the NW Machupo arenavirus also induces a late humoral response in patients, with antibodies rarely being detectable within the first 2 weeks post-onset (Peters et al., 1973).

16.3.6 Cross-Reactivity of LUJV with Other Arenaviruses

Our data show that LUJV glycoprotein 1 (GP1) strongly cross-reacts with immune serum to LASV, weakly with Mobala and LCM immune sera, but does not cross-react with sera against Junín, Mopeia, and Ippy (Figure 16.3). It has been shown that GP1 of LASV, compared to NP and GP2, is less

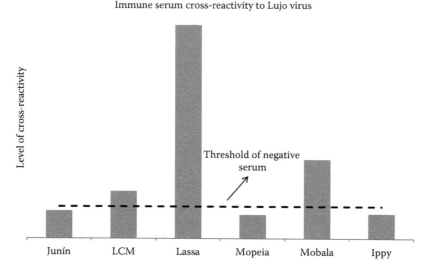

FIGURE 16.3   Cross-reactivity of LUJV GP1 with immune sera against other arenaviruses.

cross-reactive to immune serum against other OW and NW arenaviruses, which might explain the low cross-reactivity of LUJV with some of these viruses (Branco et al., 2008). The cross-reactivity of LUJV NP and GP2 with antisera to other arenaviruses remains to be investigated. Host immune responses to the different structural antigens of LUJV require further studies. Whether humoral responses to either of the LUJV antigens correlate with protection against disease is unknown.

16.3.7   Molecular Biology and Phylogeny of Lujo Virus

A new species of arenavirus is designated based on the combination of two or more of the following features: the specific host species of the virus, defined geographic distribution of the virus, the pathogenicity of the disease, antigenic cross-reaction patterns, and/or significant amino acid divergence compared to other arenaviruses (Salvato et al., 2005). The classification of arenaviruses based on genetic information alone, which has become the case with many virus groups, is complicated (Charrel et al., 2008), and common similarity or divergence cutoff values for all of the arenaviruses cannot be established. On a genetic level, the interspecies variation for the NW complex is much higher than for the OW complex. In addition, only five OW species have been classified opposed to 18 NW species. The interspecies level of sequence divergence of the OW arenaviruses is much higher though than with the NW viruses. Example of evolution through recombination or reassortment events is also described for a number of the NW viruses, but this has not yet been observed for the OW viruses (Charrel et al., 2008). Although only 23 arenaviral species have been officially recognized by the International Committee on the Taxonomy of Viruses (2009), this group of viruses is rapidly expanding with the description of several additional putative species in the recent past. The latter includes Kodoko, Morogoro, Pinhal, Skinner Tank, Catarina, Dandenong, Merino Walk, Menekre, Gbagroube, Lujo, and Luna viruses (Briese et al., 2009; Cajimat et al., 2007a; Coulibaly-N'Golo et al., 2011; Günther et al., 2009; Ishii et al., 2011; Palacios et al., 2008, 2010). Full genome phylogenetic analyses of LUJV showed it to be unique and branching off the root of the OW arenaviruses, suggesting that it represents a highly novel genetic lineage, very distinct from previously characterized virus species and clearly separating from the LCMV lineage. The virus G1 glycoprotein sequence is highly diverse and almost equidistant from that of other OW and NW arenaviruses, indicating a potential distinctive receptor tropism. This seems to be further supported by lack of recognizable homology to other arenavirus G1 sequences, making use of one of the two known arenavirus receptors, the α-DG, which binds OW arenaviruses and NW clade C viruses, or TfR1, which binds pathogenic NW arenaviruses JUNV, MACV, GTOV, and SABV. No evidence of genome segment reassortment was found with other OW arenaviruses based on S and L segment nucleotide sequences. In addition, a phylogenetic tree constructed from deduced L-polymerase amino acid sequence also showed LUJV near the root of the OW arenaviruses, distinct from characterized species, and separate from the LCMV branch. A distant relationship to OW arenaviruses was also deduced from the analysis of Z protein sequence. The NP gene sequence of LUJV was shown to differ from other arenaviruses from 36% (IPPYV) to 43% (TAMV) at the nucleotide level and from 41% (MOBV/LASV) to 55% (TAMV) at the amino acid sequence level. This degree of divergence is considerably higher than proposed threshold values both within (<10%–12%) and between (>21.5%) OW arenavirus species, giving the LUJV phylogenetically a particular position. A number of protein motifs potentially relevant to LUJV biology have been identified, including a unique Y77REL motif, which interacts with the clathrin adaptor protein 2 (AP2) complex. No evidence of recombination or reassortment could be established by Briese and coworkers, as has been the case with the evolution of some NW species, specifically Skinner Tank and Whitewater Arroyo viruses (Cajimat et al., 2008; Charrel et al., 2001). Study on principal host relationships and evolutionary history of the North American arenaviruses suggests that genetic recombination between arenaviruses with significantly different phylogenetics did not play a role in their evolution (Cajimat et al., 2007b). Increasing genomic data from newly discovered arenaviruses seem to challenge the concept that the evolution of arenaviruses is solely tied with that of its specific rodent reservoir. New data suggest that host switching and lineage extinction also might contribute to the phylogeny of the arenaviruses (Coulibaly-N'Golo et al., 2011). In this context, the unique phylogenetic position of LUJV prompts various hypotheses regarding the likely reservoir host, including murinae or nonmurinae hosts. However, to date only one arenavirus has been linked with nonrodent species, namely, the TCRV virus associated with *Artibeus* species of fruit bat (Downs et al., 1963). The multimammate rat (*M. natalensis*) is abundantly distributed in the region where the index case of the LUJV-associated outbreak lived and worked. Recent pathogen discovery expeditions in the area following the LUJV outbreak could not find LUJV in *M. natalensis* (Ishii et al., 2011) (Figure 16.4).

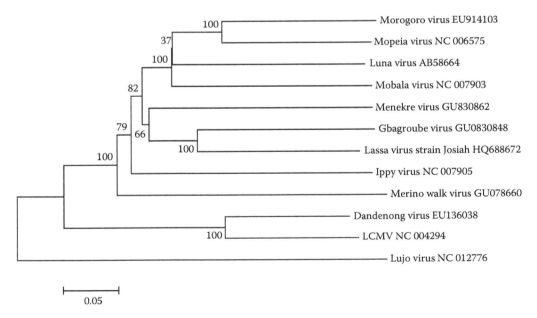

FIGURE 16.4 Phylogeny of currently described OW arenaviruses based on complete S segment sequences (NCBI accession numbers provided). Kodoko virus is not included due to limited sequencing information. This unrooted neighbor-joining tree was constructed using 1000 bootstrap replicates.

16.4 CONCLUSIONS

Compared to other arenaviruses causing VHF syndrome, the mortality rate in patients infected with LUJV appears to be unusually high. Since only one out of the five patients who survived infection was treated with ribavirin, it is unknown whether the surviving patient responded to treatment or presented a milder infection or combination of both. Therapeutic effectiveness of ribavirin in LUJV infection needs to be further investigated. Symptoms and signs in the course of infection with LUJV are similar to those of other VHFs; severe bleeding disorders were not a prominent clinical feature in Lujo patients. The virus seems to be more readily transmissible through human contact than most VHF agents, but this is also true for some other highly pathogenic arenaviruses. Likewise most of the arenaviruses, the LUJV is possibly rodent-borne pathogen, but the precise animal vector has not been identified to date, and it remains unknown how the primary patient had become infected in Zambia. Further monitoring of rodent populations in sub-Saharan Africa is highly recommended to determine the geographic distribution of the virus as well as to investigate the possibility for occurrence of other pathogenic arenaviruses. Longitudinal reservoir ecology studies are essential component of any integrated public health response to established, reemerging, or emerging zoonotic disease. The development of molecular and serological diagnostic tools specific for LUJV will aid greatly future surveillance or pathogen discovery programs, with the ultimate goal of identifying the reservoir host and mapping areas at risk of LUJV infection. Research is also required into the cell receptors that LUJV utilizes for entry into human cells, which appear to be different from those recognized to date for other highly pathogenic arenaviruses. Their identification is vital for future developments of therapeutic interventions. Lastly, a very challenging question is the mechanism of LUJV emergence. Was it one-off interspecies jump event driven by a combination of specific conditions, which will remain uncovered, or was it the first recognized event of already well-established potential for LUJV zoonotic transmission, which consequence is to be witnessed again in a near future?

REFERENCES

Abraham, J., Kwong, J.A., Albarino, C.G., Lu, J.G., Radoshitzky, S.R., Salazar-Bravo, J., Farzanan, M., Spiropoulou, C.F., and Choe, H., 2009. Host-species transferrin receptor 1ortholods are cellular receptors for nonpathogenic New World Clade B arenaviruses. *PLos Pathog.* 5(4), e1000358.

Aguilar, P.V., Camargo, W., Vargas, J., Guevara, C., Felices, V., Laguana-Torres, V.A., Tesh, R., and Ksiazek, T.G., 2009. Reemergence of Bolivian hemorrhagic fever, 2007–2008. *Emerg. Infect. Dis.* 15, 1526–1528.

Archer, A.M. and Rico-Hesse, R., 2002. High genetic divergence and recombination in Arenaviruses from the Americas. *Virology* 304, 274–281.

Armstrong, C. and Lillie, R.D., 1934. Experimental lymphocytic choriomeningitis of monkeys and mice produced by a virus encountered in studies of the 1933 St. Louis encephalitis epidemic. *Public Health Rep.* 49, 1019–1027.

Bajani, M.D., Tomori, O., Rollin, P.E., Harry, T.O., Bukbuk, N.D., Wilson, L., Childs, J.E., and Peters, C.J., 1997. A survey for antibodies to Lassa virus among health workers in Nigeria. *Trans. R. Soc. Trop. Med. Hyg.* 91, 379–381.

Bannister, B., 2010. Viral hemorrhagic fevers imported into non-endemic countries: Risk assessment and management. *Br. Med. Bull.* 95, 193–225.

Barry, M., Russi, M., Armstrong, L., Geller, D., Tesh, R., Dembty, L., Gonzalez, J.P., Khan, A.S., and Peters, C.J., 1995. Brief report: Treatment of laboratory-acquired Sabia virus infection. *New Engl. J. Med.* 333, 294–296.

Bateman, C., 2008. Arenavirus deaths—Emergency air services tighten up. *South Afr. Med. J.* 98, 910–914.

Bausch, D.G., Rollin, P.E., Demby, A.H., Coulibaly, M, Kanu, J., Conteh, A.S., Wagoner, K.D. et al., 2000. Diagnosis and clinical virology of Lassa fever as evaluated by enzyme-linked immunosorbent assay, indirect fluorescent-antibody test, and virus isolation. *J. Clin. Microbiol.* 38(7), 2670–2677.

Bonner, P.C., Schmidt, W.P., Belmain, S.R., Oshin, B., and Borchert, M., 2007. Poor housing quality increases risk of rodent infestation and Lassa fever in refugee camps of Sierra Leone. *Am. J. Trop. Med. Hyg.* 77, 169–175.

Borio, L., Inglesby, T., Peters, C.J., Schmaljohn, A.L., Hughers, J.M., Jahrling, P.B., Ksiazek, T. et al. and for the Working Group on Civilian Biodefense. 2002. Hemorrhagic fever viruses as biological weapons: Medical and public health management. *J. Am. Med. Ass.* 287, 2391–2405.

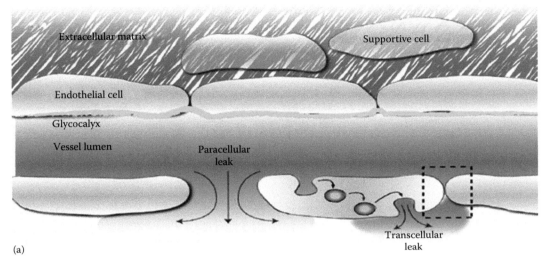

(a)

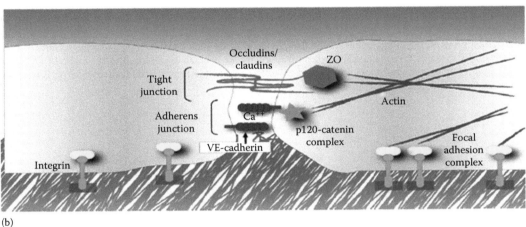

(b)

FIGURE 5.1 Microvascular constituents and leak pathways. (a) The blood vessel lumen is lined by ECs, which comprise the primary component of the microvascular permeability barrier. Vessel integrity is further influenced by the interaction of ECs with the extracellular matrix (ECM), glycocalyx, and supporting cells. In certain physiological and pathological settings, microvascular leak pathways can be initiated via paracellular and transcellular mechanisms. (b) The principal interendothelial junctions are the TJ and AJ, which interface with the cell's cytoskeleton via specific adaptor proteins. The TJ is composed of claudin and occludin proteins, which interact with the zona occludens proteins (labeled ZO in the figure) along their cytoplasmic surface. The calcium-dependent homotypic interaction of endothelial VE-cadherin of the AJ is relayed to the actin cytoskeleton by multiple adaptor proteins. The p120-catenin complex is shown here. The importance of the different junctions in preventing leak varies across different vascular beds. The cytoskeleton also engages with integrin proteins that anchor ECs to the surrounding ECM. (Reproduced from Steinberg, B.E. et al., *Antiviral Res.*, 93, 2, 2012. With permission.)

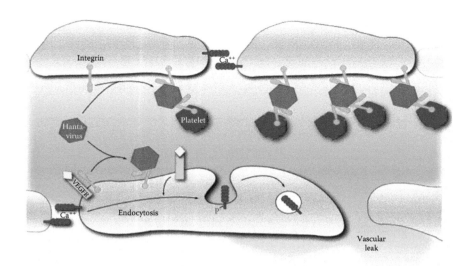

FIGURE 5.2 Binding and entry of Hantavirus onto the integrins of ECs. Hantavirus binds to the β_3 integrin of the $\alpha_v\beta_3$ integrins on the EC membrane to gain entry by means of endocytosis. The binding of hantavirus to the $\alpha_v\beta_3$ integrins impairs its function of reducing the VEGFR2-mediated internalization of VE-cadherin. The β_3 integrin, in its inactive conformation, cannot mediate the VEGF signaling through its receptor, which leads to phosphorylation and internalization of VE-cadherin by endocytosis. VE-cadherin maintains the AJ integrity. Upon VEGF stimulation, VE-cadherin is phosphorylated and internalized reducing the adherence between ECs and aiding in angiogenesis and leukocyte extravasation and ultimately causes paracellular leakage. Also, ECs covered by hantavirus also bind to quiescent platelets by binding to the integrin on them. This can directly lead to thrombocytopenia and coagulation defects. (Reproduced from Steinberg, B.E. et al., *Antiviral Res.*, 93, 2, 2012. With permission.)

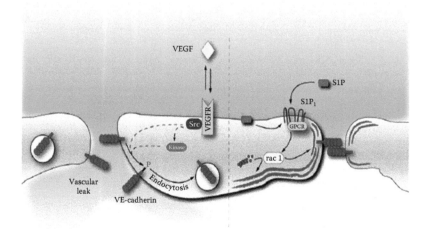

FIGURE 5.4 Barrier disruptive and protective signaling pathways at the AJ. VEGF signaling through its receptor leads to the activation of Src family kinases and downstream phosphorylation of the VE-cadherin, which is subsequently endocytosed. Removal of VE-cadherin from the interendothelial junction results in paracellular leak (Carr et al., 2003). Sphingosine-1-phosphate (S1P) is produced within ECs as well as in the circulation. Binding to the S1P1 receptor (S1P1) leads to G-protein-coupled receptor activation and downstream stabilization of the cortical actin network and maintenance of intercellular junctional proteins. (Reproduced from Steinberg, B.E. et al., *Antiviral Res.*, 93, 2, 2012. With permission.)

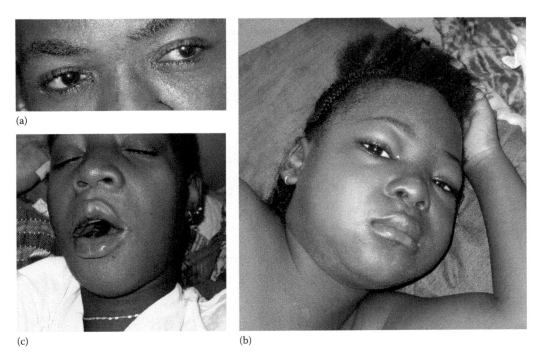

FIGURE 15.4 Clinical manifestations of LF. Beginning clockwise from the top left are examples of (a) conjunctival injection, (b) facial and neck swelling, and (c) oral mucosal bleeding. (Photos by Daniel Bausch, Donald Grant, and Victor Lungi, respectively.)

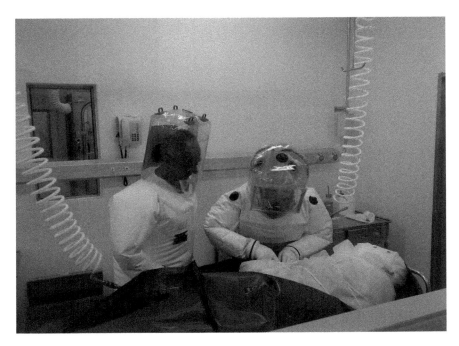

FIGURE 16.1 SPU NICD/NHLS staff members taking biopsies from a victim of a fatal nosocomial outbreak in Johannesburg in BSL-4 facility. September 9, 2008, Sandringham-Johannesburg, South Africa.

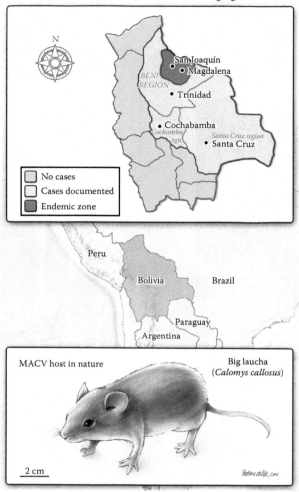

Bolivia, July 1–September 30, 1994. Cities and regions where alleged cases were documented. Endemic zone is highlighted.

FIGURE 19.1 Geography and ecology of Machupo virus.

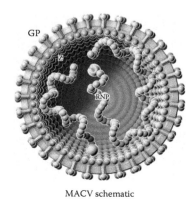

MACV schematic

FIGURE 19.2 Schematic of a Machupovirion. GP, glycoprotein; RNP, ribonucleoprotein complex.

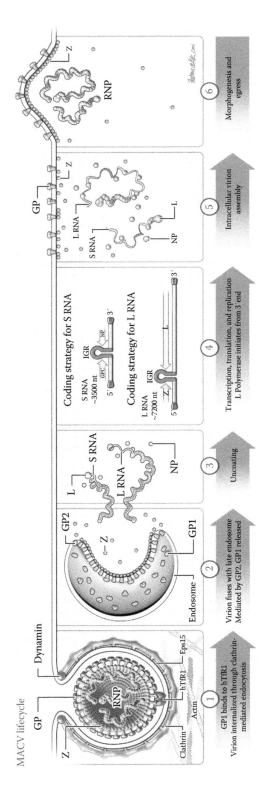

FIGURE 19.3 The Machupo virus cellular life cycle: (1) Virion uptake; (2) virus-cell membrane fusion; (3) uncoating; (4) transcription, translation, and replication; (5) virion assembly; and (6) virion budding. GP, glycoprotein; hTfR1, human transferrin receptor 1; IGR, intergenic region; L, RNA-dependent RNA polymerase; NP, nucleoprotein; Z, matrix protein.

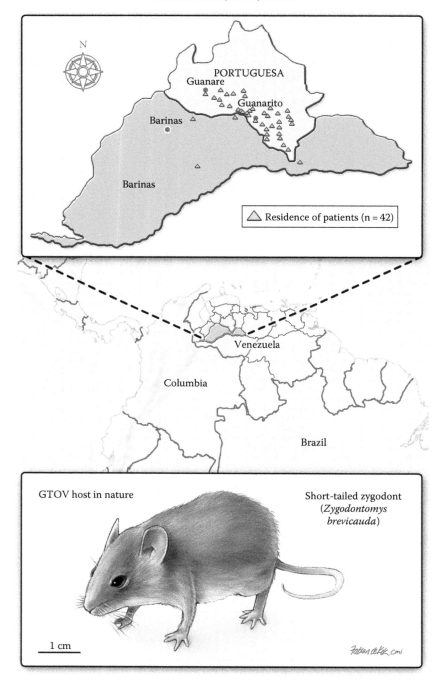

FIGURE 20.1 Geography and ecology of Guanarito virus.

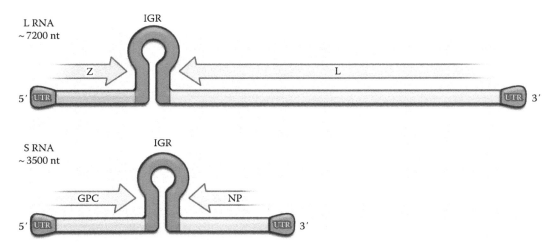

FIGURE 20.2 Schematic of the Guanarito virus genome. GPC, glycoprotein precursor gene; IGR, intergenic region; L, RNA-dependent RNA polymerase; NP, nucleoprotein; UTR, untranslated region; Z, matrix protein.

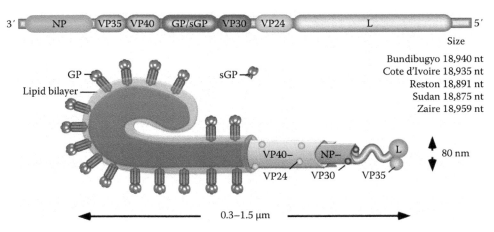

FIGURE 24.1 Schematic representation of *Ebola virus* genome and structure. Upper panel. The arrangement and the relative sizes of *Ebola virus* genes are shown. Gene sizes are similar between all family members with some differences in the intragenic regions. The size of genomes for family members is indicated at right. Lower panel. The structure of the virion is shown. Particles are pleiomorphic and range in length from 300 to 1500 nm. The diameter of particles is relatively constant at 80 nm, except at ends where the virus filament can loop back (as shown at left) as well as form ring structures. The virus is enveloped, being coated with a lipid bilayer derived from the host cell membrane at time of budding. The homotrimeric GP is anchored in the bilayer by a transmembrane domain with the surface portion being exposed to the medium. Following an RNA-editing event, a truncated, secreted form of GP is generated (sGP). Underlying the lipid envelope is the matrix, which is composed mostly of VP40 and a lesser amount of VP24. The matrix tightly encases the nucleocapsid that is made mostly of NP with some VP30. The nucleocapsid coats the vRNA that interacts with the polymerase complex composed of L and the accessory protein VP35.

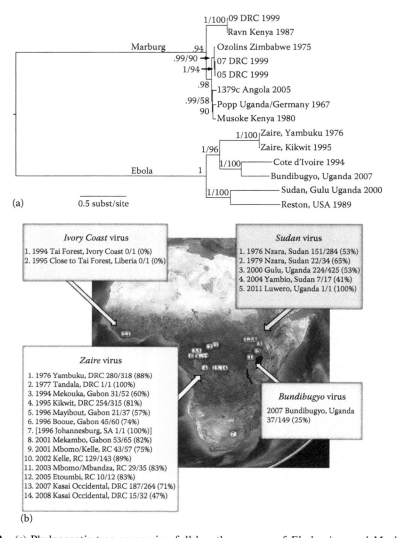

FIGURE 24.2 (a) Phylogenetic tree comparing full-length genomes of *Ebola virus* and *Marburg virus* by Bayesian analysis. Posterior probabilities greater than 0.5 and maximum likelihood bootstrap values greater than 50 are indicated at the nodes. (Reproduced from Towner, J.S. et al., *PLOS Pathog.*, 4, e1000212, 2008. With permission.) (b) Viruses without borders: outbreaks of the African ebolaviruses occur close to the equator primarily across Gabon, RC, DRC, Southern Sudan, and Western Uganda. Note how the species appear to have clustered within tremendously diverse biomes with *Ivory Coast* western most, *Zaire* central, and *Sudan* eastern central with no intermingling, suggesting local ecological niches have been self-limiting to a degree so far. Insets show starting year of outbreak, village, country, fatalities/total cases, and mortality rate (occasionally, these numbers vary a little between reports). Outbreak data based on Leroy et al. (2009) and CDC/WHO Ebola websites (note that CDC is missing the Etoumbi 2005 outbreak at the time of writing) and locations were placed on a satellite image. The #7 *Zaire virus* outbreak is bracketed since a doctor who became infected in Gabon traveled to South Africa and infected a nurse who subsequently died. *Reston virus* has been imported in primate shipments to Reston, Virginia (1989, 1990), Philadelphia, Pennsylvania (1989), Alice, Texas (1989, 1990, and 1996), and Sienna (Italy 1992). *Reston virus* outbreaks in primate exporting facilities in the Philippines were detected in 1989 and 1996 and in swine in 2008. Laboratory-acquired infections have been reported to have occurred in Salisbury (UK, 1976), USAMRIID (2004), Vector (2004), and Hamburg (2009), which was immediately countered with a recombinant VSV-based vaccine (Tuffs 2009). All laboratorians except the vector case recovered. (The NASA satellite map of Africa used to create Figure 24.2b was obtained from http://www.zonu.com/detail-en/2009-11-08-10937/Africa-satellite-map.html.)

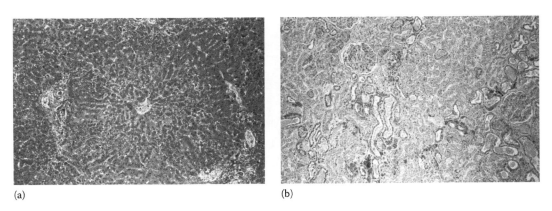

(a)   (b)

FIGURE 25.2 Histopathology of liver and spleen. (a) Liver section from a patient that succumbed to MVD. Note bleeding into the sinusoids. (b) Kidney section from same patient. Glomerular fibrin thrombosis and tubular necrosis are present. (Obtained from the CDC Public Health Image Library, http://phil.cdc.gov/phil/home.asp. Content providers: CDC/J. Lyle Conrad, 1975.)

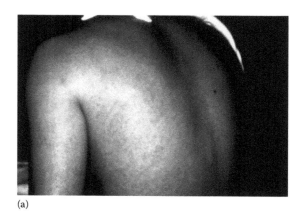

(a)

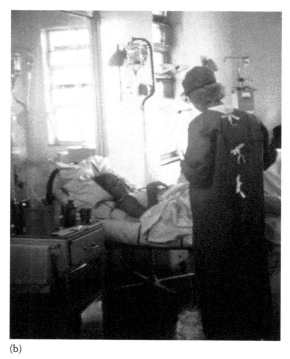

(b)

FIGURE 25.3 (a) Characteristic rash found in MVD patient. This individual survived infection. (b) Same patient in isolation ward in hospital in South Africa. The doctor is wearing personal protective equipment to minimize the chance of infection. (Obtained from the CDC Public Health Image Library, http://phil.cdc.gov/phil/home.asp. Content providers: CDC/J. Lyle Conrad, 1975.)

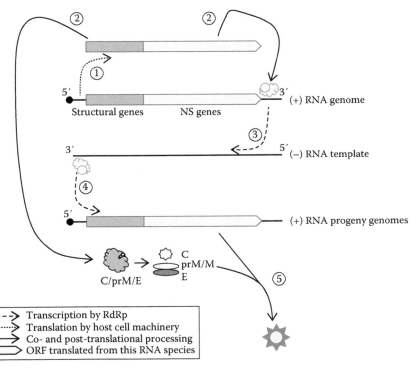

FIGURE 27.8 Diagrammatic representation of YFV replication in a permissive cell. The replication cycle is depicted as a series of temporally regulated steps: (1) translation of the polyprotein; (2) co- and post-translational processing to produce the structural proteins (C, prM, and E) and the NS proteins (NS1, NS2A, NS2B, NS3, NS4A, NS4B, and NS5); (3) synthesis of complementary, negative-sense RNA by the RdRp NS5 and other viral replicase components; (4) synthesis of progeny genomes by transcription of the negative strand; and (5) packaging of progeny genomes into nucleocapsids and bud intracellularly to acquire envelope.

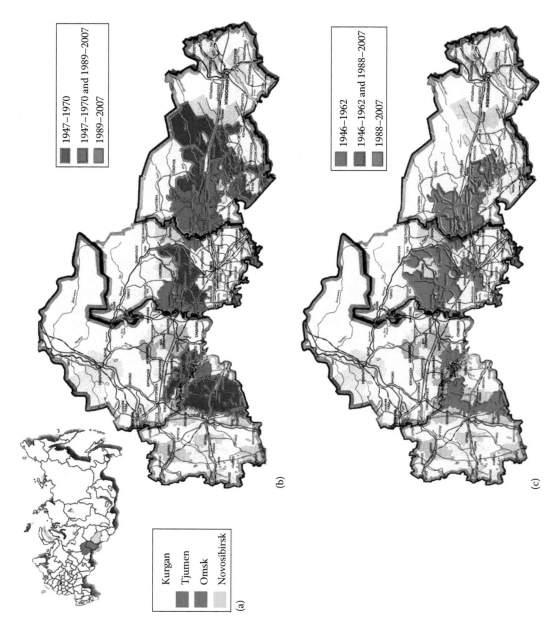

FIGURE 30.1 (a) OHF endemic administrative regions of Russia, i.e., Kurgan, Tjumen, Omsk, and Novosibirsk regions. (b) Geographical distribution of epizootics of OHF in the muskrat population. (c) Geographical distribution of areas of OHF human morbidity since 1946.

Borrow, P.L., Martinez-Sobrido, J.C., and de la Torre, L., 2010. Inhibition of the type I interferon antiviral response during Arenavirus infection. *Viruses* 2, 2443–2480.

Bowen, M.D., Peters, C.J., and Nichol, S.T., 1996. The phylogeny of New World (Tacaribe complex) arenaviruses. *Virology* 219, 285–290.

Bowen, M.D., Peters, C.J., and Nichol, S.T., 1997. Phylogenetic analysis of the arenaviridae: Patterns of virus evolution ad evidence for cospeciation between arenaviruses and their rodent hosts. *Mol. Phylogenet. Evol.* 8(3), 301–316.

Branco, L.M., Grove, J.N., Boisen, M.L., Shaffer, J.G., Goba, A., Fullah, M., Momoh, M., Grant, D.S., and Garry, R.F., 2011. Emerging trends in Lassa fever: Redefining the role of immunoglobulin M and inflammation in diagnosing acute infection. *Virol. J.* 8, 478.

Branco, L.M., Matschiner, A., Fair, J.N., Goba, A., Sampey, D.B., Ferro, P.J., Cashman, K.A. et al., 2008. Bacterial-based systems for expression and purification of recombinant Lassa virus proteins of immunological relevance. *Virol. J.* 5, 74.

Briese, T., Paweska, J.T., McMullan, L.K., Hutchison, S.K., Street, C., Palacios, G., Khristova, M.L. et al., 2009. Genetic detection and characterization of Lujo virus, a new hemorrhagic fever-associated arenavirus from southern Africa. *PLoS Pathog.* 5, e1000455.

Buchmeier, M., de la Torre, J.C., and Peters, C.J., 2007. Arenaviridae: The viruses and their replication. In Knipe, D.M. and Howley, P.M. (eds). *Fields Virology*, 5th edn. Philadelphia, PA: Lippincott Williams & Wilkins, pp. 1791–1827.

Cajimat, M.N.B., Milazzo, M.L., Borchert, J.N., Abbott, K.D., Bradley, R.D., and Fulhorst, C.F., 2008. Diversity among Tacaribe serocomplex viruses (family Arenaviridae) naturally associated with the Mexican woodrat (*Neotoma mexicana*). *Virus Res.* 133, 211–217.

Cajimat, M.N.B., Milazzo, M.L., Bradley, R.D., and Fulhorst, C.F., 2007a. Catarina virus, an arenaviral species principally associated with *Neotoma micropus* (southern plains woodrat) in Texas. *Am. J. Trop. Med. Hyg.* 77, 732–736.

Cajimat, M.N.B., Milazzo, M.L., Hess, B.D., Rood, M.P., and Fulhorst, C.F., 2007b. Principal host relationships and evolutionary history of the North American arenaviruses. *Virology* 367, 235–243.

Charrel, R.N. and de Lamballerie, X., 2003. Arenaviruses other than Lassa virus. *Antiviral Res.* 57, 89–100.

Charrel, R.N., de Lamballerie, X., and Emonet, S., 2008. Phylogeny of the genus Arenavirus. *Curr. Opin. Microbiol.* 11, 362–368.

Charrel, R.N., de Lamballerie, X., and Fulhorst, C.F., 2001. The Whitewater Arroyo virus: Natural evidence for genetic recombination among Tacaribe serocomplex viruses (Family Arenaviridae). *Virology* 283, 161–166.

Coulibaly-N'Golo, D., Allali, B., Kouassi, S.K., Fichet-Calvet, E., Becker-Ziaja, B., Riegre, T., Ölchläger, S. et al., 2011. Novel arenavirus sequences in Hylomyscus sp. and Mus (Nannomys) setulosus from Côte d'Ivoire: Implications for evolution of arenaviruses in Africa. *PLoS One* 6(6), e20893.

De Bellocq, J.G., Borremans, A.K., Katakweba, R. A., Makundi, R., Baird, S.J.E., Becker-Ziaja, B., Günter, S., and Leirs, H., 2010. Sympatric occurrence of 3 arenaviruses, Tanzania. *Emerg. Infect. Dis.* 16, 692–694.

De Manzione, N., Salas, R.A., Paredes, H., Godoy, O., Rojas, L., Araoz, F., Fulhorst, C.F. et al., 1998. Venezuelan hemorrhagic fever: Clinical and epidemiological studies of 165 cases. *Clin. Infect. Dis.* 26, 308–313.

Delgado, S., Erickson, B.R., Agudo, R., Blair, P.J., Vallejo, E., Albarion, C.G., Vargas, J., Comer, J.A., Rollin, P.E., Ksiazek, T.G., Olson, J.G., and Nichol, S.T., 2008. Chapare virus, a newly discovered arenavirus isolated from a fatal hemorrhagic fever case in Bolivia. *PLoS Pathog.* 4(4), e1000047.

Downs, W.G., Anderson, C.R., Spence, L.P., Aitken, A.T.H., and Greehhall, A.A., 1963. Tacaribe virus, a new agent isolated from Artibeus bats and mosquitoes in Trinidad, West Indies. *Am. J. Trop. Med. Hyg.* 12, 640–646.

Elsner, B., Schwarz, E., Mando, O.G., Maiztegui, J., and Vilches, A., 1973. Pathology of 12 cases of Argentine hemorrhagic fever. *Am. J. Trop. Med. Hyg.* 22, 229–236.

Enria, D.A., Briggiler, A.M., and Sáchez, Z., 2008. Treatment of Argentine hemorrhagic fever. *Antiviral Res.* 78, 132–139.

Fauci, A.S., 2005. Emerging and reemerging infectious diseases: The perpetual challenge. *Academic Med.* 80(12), 1079–1085.

Fisher, S.A., Graham, M.B., Kuehnert, M.J., Kotton, C. N., Marty, A.F.M. Comer, J.A., Guarner, J. et al. and the LCMV in Transplant Recipients Investigation Team, 2006. Transmission of lymphocytic choriomeningitis virus by organ transplantation. *N. Engl. J. Med.* 354, 2235–2249.

Fisher-Hoch, S.P., Hutwagner, L., Brown, L.B., and McCormick, J.B., 2000. Effective vaccine for Lassa fever. *J. Virol.* 74, 6777–6783.

Flanagan, M.L., Oldenburg, J., Reignier, T., Holt, N., Hamilton, G.A., Martin, V.K., and Cannon, P.M., 2008. New World clade B arenaviruses can use transferrin receptor 1 (TfR1)-dependent and-independent entry pathways and glycoproteins from human pathogenic strains are associated with the use of TfR1. *J. Virol.* 82, 938–948.

Frame, J.D., Baldwin, J.M., and Troup, J.M., 1970. Lassa fever, a new virus disease of man from West Africa. I. Clinical description and pathological findings. I. *Am. J. Trop. Med. Hyg.* 19, 670–676.

Franz, D.R., Jahrling, P.B., Friedlander, A.M., McClain, D.J., Hoover, D.L., Bryne, W.R., Pavlin, J.A., Christopher, G.W., and Eitzen, E.M., 1997. Clinical recognition and management of patients exposed to biological warfare agents. *J. Am. Med. Ass.* 278, 399–411.

Geisbert, T.W., Jones, S., Fritz, E.A., Schurtleft, A.C., Geisbert, J.B., Libscher, R., Grolla, A. et al., 2005. Development of a new vaccine for the prevention of Lassa fever. *PLoS Med.* 2(6), e183.

Gonzalez, J.P., Emonet, S., de Lamballerie, X., and Charrel, R., 2007. Arenaviruses. *Curr. Top. Microbiol. Immunol.* 315, 253–288.

Gowen, B.B. and Bray, M., 2011. Progress in the experimental therapy of severe arenaviral infections. *Future Microbiol.* 6, 1429–1441.

Günter, S., Hoofd, G., Charrel, R., Röser, C., Becker-Ziaja, B., Lloyd, G., Sabuni, C. et al., 2009. Mopeia virus-related arenavirus in Natal multimammate mice, Morogoro, Tanzania. *Emerg. Infect. Dis.* 15, 2008–2012.

Inizan, C.C., Cajimat, M.N.B., Milazzo, M.L., Barragan-Gomez, A., Bradlye, R.D., and Fulhorst, C.F., 2010. Genetic evidence for a Tacaribe Serocomplex Virus, Mexico. *Emerg. Infect. Dis.* 16, 1007–1010.

International Committee on the Taxonomy of Viruses. Updates since the 8th Report, 2009. http://ictvonline. org/. Accessed October 3, 2011.

Ishii, A., Thomas, Y., Moonga, L., Nakamura, I., Ohnuma, A., Hang'omebe, B., Takada, A., Mweene, A., and Sawa, H., 2011. Novel arenavirus, Zambia. *Emerg. Infect. Dis.* 17, 1921–1924.

Keeton, C., 2008. South African doctors move quickly to contain new virus. *Bull World Health Organ.* 86, 912–913.

Kerneis, S., Koivogui, L., Magassouba, N.F., Koulemou, K., Lewis, R., Aplogan, A., Grais, R.F., Guerin, P.J., and Fichet-Calvet, E., 2009. Prevalence and risk factors of Lassa seropositivity in inhabitants of the forest region of Guinea; a cross-sectional study. *PLoS Negl. Trop. Dis.* 3(11), e548.

Khan, S.H., Goba, A., Chu, M., Roth, C., Healing, T., Marx, A., Marx, J. et al. and Mano River Union Lassa Fever Network., 2008. New opportunities for field research on the pathogenesis and treatment of Lassa fever. *Antiviral Res.* 78, 103–115.

Kotturi, M.F., Botten, J., Sidney, J., Bui, H.H., Giancola, L., Maybeno, M., Babin, J. et al., 2009. A multivalent and cross-protective vaccine strategy against arenaviruses associated with human disease. *PLos Pathog.* 5(12), e1000695.

Kunz, S., Rojek, J.M., Perez, M., Spiropoulou, C.F., and Oldstone, M.B.A., 2005. Characterization of the interaction of Lassa fever virus with its cellular receptor α-dystroglycan. *J. Virol.* 79, 5979–5987.

Lecompte, E., Ter Meulen, J., Emonet, S., Daffis, S., and Charrel, R.N., 2007. Genetic identification of Kodoko virus, a novel arenavirus of the American pigmy mouse (Mus Nannomys minutoides) in West Africa. *Virology* 364, 178–183.

Mahanty, S., Bausch, D.G., Thomas, R.L., Goba, A., Bah, A., Peters, C.J., and Rollin, P.E., 2001. Low levels of interleukin-8 and interferon-inducible protein-10 in serum are associated with fatal infections in acute Lassa fever. *J. Infect. Dis.* 183, 1713–1721.

Maiztegui, J.I., 1975. Clinical and epidemiological patterns of Argentine hemorrhagic fever. *Bull. World Health Organ.* 52, 567–575.

Maiztegui, J.I., McKee, K.T. Jr., Barrera-Oro, J.G., Harrison, L.H., Gibbs, P.H., Feuillade, M.R., Enria, D.A. et al., 1998. Protective efficacy of a live attenuated vaccine against Argentine hemorrhagic fever. *J. Infect. Dis.* 177, 277–283.

McCormick, J.B., King, I.J., Webb, P.A., Johnson, K.M., O'Sullivan, R., Smith, E.S., Trippel, S., and Tong, T.C., 1987. A case-control study of the clinical diagnosis and course of Lassa fever. *J. Infect. Dis.* 155, 445–455.

Monat, T.P. and Casals, J., 1975. Diagnosis of Lassa fever and the isolation and management of patients. *Bull. World Health Organ.* 52, 707–715.

Moncayo, A.C., Hice, C.L., Watts, D.M., Travassos de Rosa, A.P., Guzman, H., Russell, K.L., Calampa, C. et al., 2001. Allpahuayo virus: A newly recognized arenavirus (arenaviridae) from arboreal rice rats (oecomys bicolor and oecomys paricola) in northeastern Peru. *Virology* 284, 277–286.

Niklasson, B.S., Jahrling, P.B., and Peters, C.J., 1984. Detection of Lassa virus antigens and Lassa virus-specific immunoglobulins G and M by enzyme-linked immunosorbent assay. *J. Clin. Microb.* 20(2), 239–244.

Ning, Q., Brown, D., Parodo, J., Cattral, M., Gorczynski, R., Cole, F.L., Ding, J.W. et al., 1998. Ribavirin inhibits viral-induced macrophage production of TNF, IL-1, the procoagulant fgl2 prothrombinase and preserves Th1 cytokine production but inhibits Th2 cytokine response. *J. Immunol.* 160, 3487–3493.

Palacios, G., Druce, J., Du, L., Tran, T., Birch, C., Briese, T., Conlan, S. et al., 2008. A new arenavirus in a cluster of fatal transplant-associated diseases. *N. Engl. J. Med.* 358, 991–998.

Palacios, G., Savji, N., Hui, J., Travassos de Rosa, A., Popov, V., Briese, T., Tesh, R., and Lipkin, W.I., 2010. Genomic and phylogenetic characterization of Merino Walk virus, a novel arenavirus isolated in South Africa. *J. Gen. Virol.* 91, 1315–1324.

Paweska, J.T., Sewlall, N.H., Ksiazek, T.G., Blumberg, L.H., Hale, M.J., Lipkin, W.I., Weyer, J. et al. and Members of the Outbreak Control and Investigation Teams, 2009. Nosocomial outbreak of novel arenavirus infection, Southern Africa. *Emerg. Infect. Dis.* 15(10), 1598–1602.

Peters, C.J., Webb, P.A., Johnson, K.M., 1973. Measurement of antibodies to Machupo virus by the indirect fluorescent technique. *Proc. Soc. Exp. Biol. Med.* 142, 526–531.

Radoshitzky, S.R., Kuhn, J.H., Spiropoulou, C.F., Albarino, C.G., Nguyen, D.P., Salazar-Bravo, J., Dorfman, T. et al., 2008. Receptor determinants of zoonotic transmission of New World hemorrhagic fever arenaviruses. *Proc. Natl. Acad. Sci. USA* 105, 2664–2669.

Richards, G.A., Sewlall, N.H., and Duse, A., 2009. Availability of drugs for formidable communicable diseases. *Lancet* 373, 545–546.

Richmond, J.K. and Baglole, D.J., 2003. Lassa fever: Epidemiology, clinical features, and social consequences. *Br. Med. J.* 327, 1271–1275.

Salas, R., Pacheco, M.E., Ramos, B., Taibo, M.E., Jaimes, E., Vasquez, C., Querales, J. et al., 1991. Venezuelan heamorrhagic fever. *Lancet* 338, 1033–1036.

Salvato, M.S., Clegg, J.C.S., Bowen, M.D., Buchmeier, M.J., Charrel, R.N., Gonzalez, J.P., Lukashevich, I.S., Peters, C.J., Rico-Hesse, R., and Romanowski, V., 2005. Family Arenaviridae. In Fauquet, C.M., Mayo, M.A., Maniloff, J., Desselberger, U., and Ball, L.A. (eds). *Virus Taxonomy: Eighth Report of the International Committee on Taxonomy of Viruses.* Elsevier Academic Press, Amsterdam, the Netherlands, pp. 725–733.

Schlie, K., Maisa, A., Freiberg, F., Groseth, A., Strecker, T., and Garten, W., 2010. Viral protein determinants of Lassa virus entry and release from polarized epithelial cells. *J. Virol.* 84, 3178–3188.

Sewlall, N., 2011. Arenavirus outbreak with nosocomial transmission: Infection control and the lessons learned. *South Afr. Anaesth. Analg.* 17, 54–56.

Smelt, S.C., Borrow, P., Kunz, S., Cao, W., Tishon, A., Lewicki, H., Campbell, K.P., and Oldstone, M.B., 2001. Differences in affinity of binding of lymphocytic choriomeningitis virus strains to the cellular receptor α-dystroglycan correlate with viral tropism and disease kinetics. *J. Virol.* 75, 448–457.

Spiropoulou, C.F., Kunz, S., Rollin, P.E., Campbell, K.P., and Oldstone, M.B., 2002. New World arenaviruses clade C, but not clade A and B viruses, utilizes α-dystroglycan as its major receptor. *J. Virol.* 76, 5140–5146.

Stephenson, E.H., Larson, E.W., and Dominik, J.W., 1984. Effect of environmental factors on aerosol-induced Lassa virus infection. *J. Med. Virol.* 14, 295–303.

Taylor, L.H., Latham, S.M., and Woolhouse, M.E., 2001. Risk factors for human disease emergence. *Phil. Trans. R. Soc.* 356, 983–989.

Ter Meulen, J., Lukashevich, I., Sidibe, K., Inapogui, A., Marx, M., Dorlemann, A., Yansane, M.L., Koulemou, K., Chang-Claude, J., and Schmitz, H., 1996. Hunting of peridomestic rodents and consumption of their meat as possible risk factors for rodent-to-human transmission of Lassa virus in the Republic of Guinea. *Am. J. Trop. Med. Hyg.* 55, 661–666.

Tesh, R.B., 2002. Viral hemorrhagic fevers of South America. *Bimédica* 22, 287–295.

Walker, D.H., McCormick, J.B., Johnson, K.M., Webb, P.A., Komba-Kono, G., Elliott, L.H., and Gardner, J.J., 1982. Pathologic and virologic study of fatal Lassa fever in man. *Am. Ass. Pathol.* 107, 349–356.

World Health Organization. 2005. International Health Regulations, 2nd edn. ISBN 9789241580410.

Wulff, H. and Johnson, K.M., 1979. Immunoglobulin M and G responses measured by immunofluorescence in patients with Lassa or Marburg virus infections. *Bull. World Health. Organ.* 57, 631–635.

York, J. and Nunberg, J.H., 2007. A novel zinc-binding domain is essential for formation of the functional Junin virus envelope glycoprotein complex. *J. Virol.* 81, 13385–13391.

17 Receptor Determinants of Zoonosis and Pathogenesis of New World Hemorrhagic Fever Arenaviruses

Stephanie Jemielity, Michael Farzan, and Hyeryun Choe

CONTENTS

17.1 INTRODUCTION

Five rodent-borne New World (NW) arenaviruses are known to cause recurrent hemorrhagic fever outbreaks in South America. Each of these viruses uses the ubiquitously expressed transferrin receptor 1 (TfR1) to enter human cells. In contrast, closely related but nonpathogenic NW arenaviruses use TfR1-independent mechanisms to enter human cells, although they do use the TfR1 orthologs of their rodent host species as receptors. Mutagenesis studies showed that two key human TfR1 (hTfR1) residues are essential for the entry of pathogenic NW arenaviruses: tyrosine 211 and asparagine 348. These residues are also conserved in the TfR1 orthologs of all known NW arenaviral host species, suggesting that fortuitous similarities between human and host-species TfR1 facilitate arenaviral zoonoses. Less is known about the critical residues on the virus side. Most of the arenavirus glycoprotein (GP) residues in direct contact with hTfR1 are poorly or moderately conserved, suggesting that at given positions variable GP residues can interact with the hTfR1 GP-binding site. This variability implies that the TfR1-binding regions of arenaviral GP molecules have multiple pathways to adapt to the TfR1 ortholog of a new host. In addition, properties of TfR1 may contribute to the severity of South American hemorrhagic fevers. For example, feedback mechanisms triggered by the early immune response to infection may upregulate hTfR1 expression and thereby amplify viral replication.

17.2 ROLE OF RECEPTORS IN THE ZOONOTIC TRANSMISSION OF VIRUSES

The plasma membrane of the cell is the first barrier that must be breached by an infecting virus. To do so, viruses require suitable host-cell receptors. Most viruses need a specific receptor to efficiently attach to the plasma membrane, and many depend on receptor-mediated endocytosis to reach an environment suitable for cytosolic delivery of their genetic materials. Although receptors are not the only host factors necessary to support a successful infectious cycle, recent outbreaks by SARS coronavirus and H5N1 bird flu viruses show that receptor utilization can be a critical determinant of viral zoonoses and the emergence of new human pathogens (Li et al., 2005; Watanabe et al., 2011, 2012). Here we discuss the role of receptor usage in the zoonotic transmission and pathobiology of the NW hemorrhagic fever arenaviruses.

17.3 NW ARENAVIRUSES AND THEIR ENTRY INTO TARGET CELLS

NW arenaviruses can be divided into three phylogenetically distinct clades (A, B, C) endemic to South America, as well as clade A/B recombinants endemic to North America (Charrel et al., 2008; Clegg, 2002). Together with Old World (OW) arenaviruses, they constitute the Arenaviridae family, whose members share a number of biological and epidemiological characteristics (reviewed in Buchmeier et al., 2007; Charrel and de Lamballerie, 2010; Jay et al., 2005; Rojek and Kunz, 2008). Arenaviruses chronically infect their host animals, in most cases rodent species, which are thus capable of shedding viruses over a prolonged period of time. There are seven arenaviruses known to cause hemorrhagic fevers in humans; the OW arenaviruses Lassa (LASV) and Lujo (LUJV) in Africa (see Chapter 21) and the NW arenaviruses Machupo (MACV), Junín (JUNV), Gunarito (GTOV), Sabiá (SABV), and Chapare (CHAV) in South America (see Chapters 2 through 4). All five pathogenic NW arenaviruses belong to clade B, which also includes several nonpathogenic viruses. The significance of these nonpathogenic clade B viruses in future zoonoses is discussed in the following. For a brief overview of arenaviruses, their distribution, host animals, and associated diseases, see Table 17.1.

The single-stranded ambisense RNA genome of arenaviruses consists of two segments and encodes four proteins: a nucleoprotein (NP), a matrix protein (Z), an RNA polymerase (L), and an envelope GP (Buchmeier et al., 2007; Rojek and Kunz, 2008). Among these proteins, GP alone is present on the host-cell-derived viral membrane and mediates the entry of viruses into cells. GP is initially produced as a precursor protein GPC, which, during viral maturation in the host cell, is proteolytically cleaved into three parts: the stable signal peptide (SSP), the attachment protein GP1, and the fusion protein GP2. On mature virions, GP1 and GP2 form heterodimers that are assembled into trimers. When the receptor-binding domain of GP1 engages the host-cell receptor, the virion is endocytosed. Progressive acidification within the endosome then induces a conformational change in GP1 and GP2, which in turn triggers the fusion between viral and endosomal membranes, thus facilitating the entry of the viral ribonucleoprotein complexes into the cytoplasm. One unusual feature of arenaviruses is that the SSP remains associated with the mature GP1, complexes on the virions (York et al., 2004) and may be involved in controlling the fusion machinery (York and Nunberg, 2006). Viral entry is completed with the unpacking of RNA, which allows the virally encoded RNA polymerase to initiate viral replication in the cytoplasm, followed by virus assembly and budding at the cell membrane.

17.3.1 NW Arenavirus Receptors

Receptor usage within the Arenaviridae family does not strictly follow the phylogenetic boundaries (Table 17.1). While OW and clade C NW arenaviruses use a dystroglycan as a receptor (Cao et al., 1998; Reignier et al., 2006; Spiropoulou et al., 2002), all five pathogenic clade B viruses, i.e., MACV, JUNV, GTOV, SABV, and CHAV, use human transferrin receptor 1 (hTfR1) to enter human cells (Helguera et al., in press; Hentze et al., 2004). The nonpathogenic clade B arenaviruses

TABLE 17.1

Overview of New and Old World Arenaviruses

Virus	Abbr. Used	Serotype/Clade	Human Disease	Host Species; Region	Human Receptor	Host-Species Receptor	References
Pichinde	—	NW/A	—	*Oryzomys albigularis*; Columbia	? not α–DG	?	Spiropoulou et al. (2002)
Paraná	—	NW/A	—	*Oryzomys buccinatus*; Paraguay	? not α–DG	?	Spiropoulou et al. (2002)
Pirital	—	NW/A	—	*Sigmodon alstoni*; Venezuela	? not α–DG	?	Spiropoulou et al. (2002)
Tamiami	—	NW/A/B	—	*Sigmodon hispidus* (cotton rat); Florida, USA	? not α–DG	?	Spiropoulou et al. (2002)
Bear Canyon	—	NW/A/B	—	*Peromyscus californicus*; California, USA	?	?	
Whitewater Arroyo	—	NW/A/B	Unclear (tr)	*Neotoma micropus* (woodrat); Texas, USA	? not TfR1, not α–DG	?	Reignier et al. (2008)
Junín	JUNV	NW/B	Argentine HF	*Calomys musculinus* (corn mouse); Argentina	hTfR1	TfR1	Radoshitzky et al. (2007, 2008)
Machupo	MACV	NW/B	Bolivian HF	*Calomys callosus* (vesper mouse); Bolivia	hTfR1	TfR1	Radoshitzky et al. (2007, 2008)
Tacaribe	TCRV	NW/B	- (tr)	*Artibeus jamaicensis* (bat); Trinidad	? not hTfR1	TfR1	Abraham et al. (2009) and Flanagan et al. (2008)
Guanarito	GTOV	NW/B	Venezuelan HF	*Zygodontomys brevicauda* (cane mouse); Venezuela	hTfR1	TfR1	Radoshitzky et al. (2007, 2008)
Amapari	AMAV	NW/B	—	*Neacomys spinosus* (bristly mouse); Brazil	? not hTfR1	TfR1	Abraham et al. (2009) and Flanagan et al. (2008)
Cupixi	CPXV	NW/B	—	*Oryzomys goeldii* (rice rat); Brazil	?	?	—
Sabiá	SABV	NW/B	Brazilian HF	Unknown; Brazil	hTfR1	Likely TfR1	Radoshitzky et al. (2007)

(continued)

TABLE 17.1 (continued)
Overview of New and Old World Arenaviruses

Virus	Abbr. Used	Serotype/ Clade	Human Disease	Host Species; Region	Human Receptor	Host-Species Receptor	References
Chapare	CHAV	NW/B	HF	Unknown; Bolivia	hTfR1	Likely TfR1	Helguera et al. (in press)
Oliveros	—	NW/C	—	*Necromys benefactus*; Argentina	α-DG	Likely α-DG	Spiropoulou et al. (2002)
Latino	—	NW/C	—	*Calomys callosus* (vesper mouse); Bolivia	α-DG	Likely α-DG	Spiropoulou et al. (2002)
Lassa	LASV	OW	Lassa HF	*Mastomys natalensis* (rat); West Africa	α-DG	Likely α-DG	Cao et al. (1998)
Lymphocytic choriomeningitis	LCMV	OW	Meningitis, birth defects	*Mus musculus* (house mouse); ubiquitous	α-DG/ unknown	α-DG	Cao et al. (1998), Kunz et al. (2004), and Reignier et al. (2006)
Lujo	LUJV	OW	HF	Unknown; South Africa	?	?	—

Listed are only selected representative viruses from the Arenaviridae, which currently includes over 30 putative members, 23 of which have been officially classified by the International Committee on Taxonomy of Viruses. Along with virus abbreviations (Abbr.), phylogenetic affiliation and human diseases caused by these viruses, respective host species and their distribution, as well as receptor usage are indicated. Note that all clade B NW arenaviruses use TfR1 as a host-species receptor, but only the pathogenic clade B members are also capable of using human TfR1 (hTfR1). In contrast, clade C NW arenaviruses and OW arenaviruses use alpha-dystroglycan (α-DG) as a receptor. Clade A/B viruses likely arose through recombination between clade A and clade B viruses. Although their GP proteins are derived from the clade B progenitor, the receptors they use are still unknown.

Notes: ?, unknown; —, none; (tr), transmission to humans reported.

Amapari (AMAV) and Tacaribe (TCRV) also enter human cells but in a TfR1-independent manner (Abraham et al., 2009; Flanagan et al., 2008). Their human receptors, as well as those for clade A and A/B viruses, remain unknown.

Although pathogenic and nonpathogenic clade B viruses use different mechanisms to enter human cells, both groups of viruses efficiently use the TfR1 orthologs of their host species (Abraham et al., 2009; Radoshitzky et al., 2008). Consistent with the coevolution of these viruses with their host species, each of these viruses uses its respective host-species TfR1 more efficiently than those of other viruses.

The expression pattern of TfR1 is compatible with its role as a major receptor for pathogenic NW arenaviruses. TfR1 is expressed ubiquitously because of its essential roles in iron hemostasis and cell growth (Anderson and Vulpe, 2009; Daniels et al., 2006). It is particularly highly expressed in rapidly dividing cells, including those of the intestinal epithelium, liver, kidney tubules, and lymphatic tissues, as well as in macrophages and activated peripheral blood mononuclear cells (Anderson and Vulpe, 2009; Daniels et al., 2006; Pallone et al., 1987; Tacchini et al., 2008). This expression pattern is consistent with histopathological findings from autopsies of fatal human cases and experimentally infected primates; most of the tissues were infected with high virus titers, including spleen, lymph nodes, bone marrow, lung, liver, kidney, and blood cells of the monocyte–macrophage lineage (Gonzalez et al., 1980, 1983; Maiztegui et al., 1975; McLeod et al., 1978).

Several groups have suggested that there may be a second receptor for pathogenic NW arenaviruses (Cuevas et al., 2011; Flanagan et al., 2008), based on the observation that these viruses are capable of entering mouse cells, in spite of the fact that mouse TfR1 does not support the entry of NW arenaviruses (Flanagan et al., 2008; Radoshitzky et al., 2008). Nevertheless, this hypothetical alternative receptor appears to play a minor role, at least in vitro, because in several cell lines the majority of entry is hTfR1 dependent (Helguera et al., in press; Radoshitzky et al., 2007). One interesting possibility is that this alternative receptor may also be used as the entry gateway for the nonpathogenic clade B NW arenaviruses AMAV, TCRV, and CPXV (cupixi virus). Alternatively, this second receptor may function as a coreceptor, analogous to the coreceptors of HIV-1, that some viruses can utilize in the absence of TfR1.

17.3.2 OTHER HOST FACTORS

In addition to receptors, many other host factors are required for a virus to productively infect a cell. For instance, various cellular molecules are known to enhance the entry of a wide range of viruses because they promote viral attachment by nonspecifically binding to components on the virion surface, such as high mannose-bearing or negatively charged carbohydrates. These molecules include DC-SIGN, DC-SIGNR, and heparan sulfate proteoglycans (Bartlett and Park, 2010; Helenius, 2007; Lozach et al., 2007). More recently, it has become apparent that phosphatidylserine (PS)-binding molecules may similarly enhance the attachment and internalization of viruses by binding to PS exposed on the viral membrane (Morizono et al., 2011). We have examined PS-binding proteins such as Axl and T-cell immunoglobulin and mucin domain-containing proteins (Tim family members) for their ability to enhance NW arenavirus entry. As expected, transient expression of at least three human PS receptors—Axl, Tim1, and Tim4—increased the entry of AMAV and TCRV GP-bearing pseudoviruses by more than an order of magnitude and also moderately increased the entry of MACV and JUNV pseudoviruses (SJ et al., unpublished data). Whether these PS receptors are sufficient to complete the entry of these viruses in the absence of the receptors or whether they are responsible for the aforementioned entry of pathogenic NW arenaviruses into murine cells remains to be seen.

17.4 CRITICAL TfR1 RESIDUES IN NW ARENAVIRUS ZOONOSES

The physiological role of TfR1 is to facilitate iron uptake into cells. TfR1 binds two molecules of iron-bound transferrin (holo-transferrin), the main iron carrier in mammalian blood, and transports them to endosomes, where low pH promotes release of the iron. Iron is then transported to

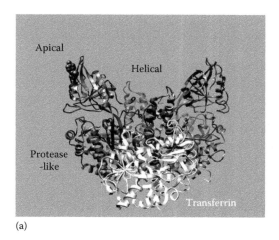
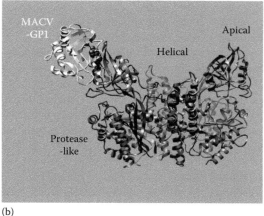

(a)   (b)

FIGURE 17.1 Ribbon diagrams of the hTfR1 dimer in complex with human transferrin or MACV GP1. (a) Human transferrin (in white) binds hTfR1 encompassing the protease-like and helical domains (PDB ID: 1SUV). (From Cheng, Y. et al., *Cell*, 116, 565, 2004.) (b) MACV GP1 (in white) binds the tip of the hTfR1 apical domains (PDB ID: 3KAS). (From Abraham, J. et al., *Nat. Struct. Mol. Biol.*, 17, 438, 2010.)

the cytosol, and TfR1 bound to cargo-free transferrin (apo-transferrin) is recycled back to the cell surface (Aisen, 2004; Andrews et al., 1999; Hentze et al., 2004). Tfr1 is a type II dimeric membrane protein composed of three distinct extracellular domains: a protease-like domain, a helical domain that mediates homodimerization, and an apical domain (Figure 17.1). While transferrin binds the first two domains (Cheng et al., 2004) (Figure 17.1a), pathogenic NW arenaviruses interact only with the tip of the apical domain (Figure 17.1b), as demonstrated first through studies with human–murine TfR1 chimeras, and then crystallographically (Abraham et al., 2010; Radoshitzky et al., 2008). Further evidence indicated that this TfR1 binding site is shared by all pathogenic NW arenaviruses. Specifically, two independent monoclonal antibodies targeting the hTfR1 apical domain efficiently neutralized all pathogenic arenaviruses (Helguera et al., in press; Radoshitzky et al., 2007).

Several apical domain residues critical for clade B arenaviruses' usage of hTfR1 have been identified. A mapping experiment using human–mouse TfR1 chimeras narrowed the arenavirus binding sites to two small regions in the apical domain (Radoshitzky et al., 2008). Based on amino acid sequence alignments, the most interesting residue in the first region was tyrosine 211 (Y211), which is conserved in all virus-permissive TfR1 orthologs, but not in three of four nonpermissive TfR1 orthologs (Figure 17.2). In the second region, asparagine 348 (N348) stood out, because it is highly exposed and located adjacent to tyrosine 211 in the TfR1 structure. While the residue 348 in most of the TfR1 molecules is an asparagine, it is replaced with a lysine in the nonpermissive mouse TfR1 ortholog. Binding and entry experiments with Y211T, N348K, and similar mutants confirmed that these residues are indeed essential for hTfR1 use by MACV, JUNV, and GTOV pseudoviruses (Radoshitzky et al., 2008). The sequence alignment of TfR1 orthologs also showed a loss of a potential N-glycosylation site at hTfR1 residue 204 compared to host-species TfR1 molecules (Figure 17.2). Further mutagenesis studies indicated that lack of glycosylation at this position is beneficial for the entry of most of the pathogenic NW arenaviruses (Radoshitzky et al., 2008).

Unlike residues that facilitate hTfR1 usage, some residues attenuate its efficiency as a receptor for pathogenic NW arenaviruses. In particular, arginine at residue 208 (R208) displays such a characteristic; when R208 was mutated to glycine, this R208G mutant hTfR1 allowed for more efficient entry of MACV, JUNV, and GTOV than wild-type hTfR1 (Abraham et al., 2010). This finding is interesting, because all of the host-species TfR1 orthologs have uncharged residues at this position,

```
                   204      211          293          348       370
                    |        |            |            |         |           M J G S C
H.sapiens      IIVDKNGRLVYLVENP...IVNAELS...RAAAEKLFGNME...TSESKNVK          + + + + +
M.mulatta      IIVDKNGGLVYLVENP...IVKADLS...RAAAEKLFGNME...TSENKSVK          + +
C.jacchus      TITGTNSEFVYLVENP...IVDADVS...RAAAERLFENME...TSQNKSVK          + +
F.catus        TIVGTNSGMVYLVESP...ITNAEIP...RANAEKLFGNME...TSRNWNVK          + + +
C.callosus     TIINASNGV-YLLESP...VVKADLS...REAADNLFQNME...LTSDKTVK          + + -
C.musculinus   TIINASGGS-YPLENP...VVKADLS...RETADKLFQKME...LTSDKTVK          - + -
Z.brevicauda   TIINTSGGL-YLLENP...VVKADLS...READKLFQNME...LTSNKNVN           + + +
C.familiaris   TIVDMESDLVYLAESP...IVNARIP...RAAAEKLFENME...TSSNKNVN          - - -
M.musculus     TIVQSNGNL-DPVESP...VVEADLA...RAAAEKLFGKME...LSQNQNVK          - - -
R.norvegicus   TIN-SGSNI-DPVEAP...VVEADLQ...RAAAEKLFKNME...LSQNQNVK          - - -
C.griseus      TIINVNGDS-DLVENP...VVEAELS...RKAAEKLFQNME...SSQGINVN          - - -
```

FIGURE 17.2 Alignment of TfR1 amino acid sequences and respective viral usage. All 16 hTfR1 residues shown by the co-crystal structure to be in direct contact with MACV GP1 are marked in bold and underlined (Abraham et al., 2010). TfR1 key residues, tyrosine 211 and asparagine 348, experimentally shown to be essential for hTfR1 usage by pathogenic NW arenaviruses are marked with black boxes. Loss of N-glycosylation at position 204, which may improve hTfR1's efficiency as a viral receptor, is also indicated (Radoshitzky et al., 2008). In contrast, arginine 208 (boxed in gray) reduces the efficiency of hTfR1 as a receptor for NW arenaviruses (Abraham et al., 2010). The TfR1 usage patterns indicated to the right are based on published data (Helguera et al., in press; Radoshitzky et al., 2007, 2008) with the exception of those for *M. mulatta* and *C. jacchus* (SJ et al., unpublished data). Abbreviations M, J, G, S, and C represent MACV, JUNV, GTOV, SABV, and CHAV, respectively. TfR1 orthologs from human (*H. sapiens*), rhesus macaque (*M. mulatta*), marmoset (*C. jacchus*), cat (*F. catus*), MACV rodent host (*C. callosus*), JUNV rodent host (*C. musculinus*), GTOV rodent host (*Z. brevicauda*), dog (*C. familiaris*), house mouse (*M. musculus*), rat (*R. norvegicus*), and hamster (*C. griseus*) are represented.

consistent with the more efficient entry of pathogenic NW arenaviruses in cells expressing these TfR1 orthologs compared to those expressing hTfR1.

Finally, based on the co-crystal structure (Abraham et al., 2010), there are 12 more hTfR1 residues that directly interact with MACV GP1 in addition to the four residues already discussed. The difference between human and host-species TfR1 for each of these residues could positively or negatively affect the efficiency of hTfR1 as a viral receptor. While this has not been tested experimentally, several of these residues are quite variable among virus-permissive TfR1 orthologs (Figure 17.2), indicating that their influence is likely modest.

17.5 VIRAL GP1 REQUIREMENTS FOR HUMAN TfR1 USAGE AND IMPLICATIONS FOR ZOONOSES

Despite the pivotal role of GP1 in receptor binding, overall GP1 sequence homologies among pathogenic NW arenaviruses are a poor predictor of receptor usage. GP1 protein sequences are considerably divergent among pathogenic NW arenaviruses, with pairwise sequence identities ranging from 21% to 58% and similarity scores ranging from 40% to 78% (Figure 17.3). In fact, this high degree of divergence is true even for the 16 GP1 residues directly involved in hTfR1 binding (Abraham et al., 2010), supporting the idea that arenavirus GPs can establish contacts with hTfR1 using various combinations of residues. This flexibility may contribute to the zoonotic potential of NW arenaviruses by affording viruses multiple ways to acquire specificity for a TfR1 ortholog of a new host while maintaining high affinity for the TfR1 of its original host species. Unfortunately, with a few exceptions (Radoshitzky et al., 2011), most GP1 residues in contact with hTfR1 have not been assessed for their influence on receptor usage.

Another interesting observation is that the nonpathogenic AMAV and TCRV are phylogenetically closer to pathogenic viruses than most pathogenic viruses among themselves (Charrel et al., 2008; Emonet et al., 2009), which is well reflected in the sequence homologies of their GP1 proteins. For instance, AMAV shares a GP1 similarity score of 71% with GTOV versus 43% between GTOV and JUNV. TCRV shares a similarity score of 65% with JUNV versus 42% shared between JUNV

	MACV	CHAV	SABV	GTOV	AMAV	JUNV	TCRV	CPXV	
MACV	100.0	24.2	22.9	25.5	25.0	45.6	41.7	21.1	% identity
CHAV	39.8	100.0	57.7	29.7	26.0	23.4	20.4	28.0	
SABV	41.6	78.1	100.0	28.6	27.0	20.6	17.6	30.5	
GTOV	41.8	53.1	52.6	100.0	52.1	25.3	21.3	52.7	
AMAV	43.3	49.0	48.0	70.7	100.0	26.8	26.4	52.4	
JUNV	60.3	40.3	42.2	43.4	48.5	100.0	45.6	21.1	
TCRV	60.8	38.3	37.7	40.6	43.7	65.3	100.0	19.2	
CPXV	38.3	48.7	49.2	65.4	68.3	44.2	37.9	100.0	
	% similarity								

FIGURE 17.3 Analysis of pairwise sequence identities (%) and similarities (%) for clade B arenavirus GP1 proteins. GP1 amino acid sequences were aligned in ClustalW and pairwise identity and similarity scores calculated using the Gonnet matrix. Only one strain per species was included in the analysis because GP1 sequence variation among various isolates of the same species is not substantial. Black squares mark the highest and lowest identity and similarity scores among pathogenic NW arenaviruses. Black ellipses indicate the highest similarity scores for AMAV and TCRV.

and SABV (Figure 17.3). Such relatedness patterns suggest that pathogenicity might have been acquired or lost several times within clade B arenaviruses and raise the possibility that other strains with a zoonotic potential may emerge. Consistent with this hypothesis, it was shown that only one to four amino acid substitutions in hTfR1 are necessary to allow TCRV and AMAV, respectively, to use hTfR1 efficiently (Abraham et al., 2009). Thus, modest changes in their GP1 proteins may also be sufficient for these viruses to acquire hTfR1 usage and emerge as human pathogens. Of course, the GP1 mutations necessary to do so should not cost viral fitness or impose a replication disadvantage in the host animals, if a virus with such mutations is to survive the competition with other well-adapted viruses. It is also possible that variants already capable of utilizing hTfR1 are circulating undetected in the reservoir species of nonpathogenic clade B viruses.

17.6 ROLE OF HUMAN TfR1 IN DISEASE ENHANCEMENT

The absolute correlation between hTfR1 usage by pathogenic NW arenaviruses and the severe diseases they cause suggests that hTfR1 usage may be a necessary attribute of a virus that causes South American hemorrhagic fever. Although other viral factors are likely to contribute, hTfR1 use may facilitate infection of cell populations that play a decisive role in disease progression. In addition, it may promote a positive feedback mechanism that allows the infection to spiral out of control. There are at least two ways in which such a feedback loop might operate. First, activation of immune cells as a result of virus infection will increase TfR1 expression in those cells and thus viral entry and replication. Second, release of the pro-inflammatory cytokine IL-6 during the acute-phase response induced by virus infection will cause a hepcidin-/ferroportin-mediated sequestration of iron from the plasma. Hepcidin is a liver-produced peptide hormone and its binding to ferroportin, an iron exporter, causes internalization and degradation of ferroportin and thus blocks iron release from iron-storing cells such as duodenal enterocytes, macrophages, and hepatocytes (Ganz, 2011). The resulting hypoferremic condition will in turn upregulate TfR1 expression in the tissues via iron-regulatory proteins (IRPs), thus leading to enhanced viral entry and replication (Figure 17.4).

While this hypothetical feedback loop described earlier still awaits confirmation, its individual steps are well supported by the published literature. Elevated IL-6 levels in the serum are a common response to viral infections (Hoppe and Hulthen, 2007; Linderholm et al., 1996; Ng et al., 2009) and have been reported for Argentine hemorrhagic fever (HF) patients (Marta et al., 1999). IL-6 was further shown to be both necessary and sufficient to induce hepcidin mRNA expression in human cells (Nemeth et al., 2003, 2004), and infusion of human volunteers with IL-6 or other

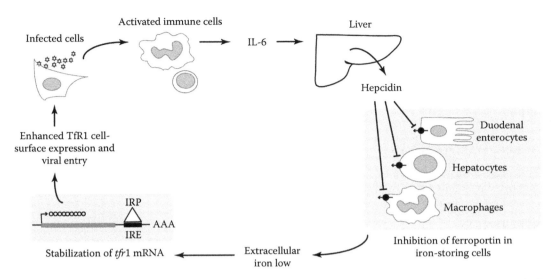

FIGURE 17.4 Hypothetical vicious cycle triggered by NW arenavirus hTfR1 usage. In response to viral infection, cells release cytokines, including IL-6, which induces hepcidin production by the liver. By triggering the uptake and degradation of the iron-exporter ferroportin, hepcidin leads to the sequestration of iron in iron-storing cells, thus lowering iron levels in the plasma and tissues. The concomitant drop in intracellular iron levels results in an IRP-mediated stabilization of *tfr1* mRNA, which upregulates TfR1 cell-surface expression and thus viral entry. IRP, iron-regulatory protein; IRE, iron-responsive element; AAA, polyA tail.

pro-inflammatory molecules caused a rapid increase in urinary hepcidin followed by a significant reduction in serum iron levels (Kemna et al., 2005; Nemeth et al., 2004). It also has been shown that low serum iron levels are strongly correlated with enhanced tissue TfR1 expression (Collins et al., 2005; Fry et al., 2009; Liu et al., 2007; Lu et al., 1995; Pietrangelo et al., 1992; Siddappa et al., 2003). The molecular players behind this correlation are IRPs, which at low intracellular iron concentration bind the multiple iron-responsive elements (IREs) in the 3′ untranslated region of *tfr1* mRNA. IRP-bound *tfr1* mRNA is more stable than unbound mRNA, and the resulting longer mRNA half-life translates into increased TfR1 cell-surface expression (reviewed in Ganz, 2011; Hentze et al., 2004). Together these studies show a clear connection among infection, hepcidin, serum iron levels, and tissue TfR1 expression, supporting the possibility that a destructive feedback loop, triggered by viral hTfR1 usage, contributes to the severity of South American hemorrhagic fevers.

17.7 OUTLOOK

One way to better understand whether TfR1 usage per se is required for severe disease would be to compare disease severity resulting from infection by a wild-type pathogenic arenavirus to that by variants defective for hTfR1 binding. Such an experiment could be carried out in hTfR1 knock-in mice, which are currently under development. Furthermore, to confirm the existence of an IL-6/hepcidin-mediated feedback loop, one could assess whether targeted interruption of the process, for instance, with antimurine IL-6 antibodies or shRNAs against murine hepcidins, alleviates disease severity. If such a mechanism exists for NW arenaviruses, it will be interesting to know whether the course of infection by very divergent viruses that also utilize TfR1 as a receptor (e.g., mouse mammary tumor virus, as well as canine and feline parvoviruses) is similarly amplified. It will also be interesting to ask whether modulation of plasma iron levels during the early stage of infection (by providing iron supplements) can lessen the severity of South American hemorrhagic fevers.

Finally, there are many open questions regarding the emergence of new human pathogens. Can nonpathogenic clade B arenaviruses such as AMAV and TCRV readily acquire hTfR1 use? Or do new pathogenic viruses like CHAV evolve from preexisting pathogenic species? Are hTfR1-utilizing variants of nonpathogenic viruses already circulating in their respective reservoir species? Additional work—greater monitoring, sequencing, and characterization of such variants—will be necessary to limit the risks of arenaviral zoonoses and help anticipating the emergence of new viruses.

ACKNOWLEDGMENTS

This project was supported by NIH grants AI074879 to HC and AI57159 to HC and MF. SJ was supported by a fellowship from the Swiss National Science Foundation.

REFERENCES

Abraham, J., K. D. Corbett, M. Farzan, H. Choe, and S. C. Harrison. 2010. Structural basis for receptor recognition by New World hemorrhagic fever arenaviruses. *Nat Struct Mol Biol* **17**:438–444.

Abraham, J., J. A. Kwong, C. G. Albarino, J. G. Lu, S. R. Radoshitzky, J. Salazar-Bravo, M. Farzan, C. F. Spiropoulou, and H. Choe. 2009. Host-species transferrin receptor 1 orthologs are cellular receptors for nonpathogenic new world clade B arenaviruses. *PLoS Pathog* **5**:e1000358.

Aisen, P. 2004. Transferrin receptor 1. *Int J Biochem Cell Biol* **36**:2137–2143.

Anderson, G. J. and C. D. Vulpe. 2009. Mammalian iron transport. *Cell Mol Life Sci* **66**:3241–3261.

Andrews, N. C., M. D. Fleming, and J. E. Levy. 1999. Molecular insights into mechanisms of iron transport. *Curr Opin Hematol* **6**:61–64.

Bartlett, A. H. and P. W. Park. 2010. Proteoglycans in host-pathogen interactions: Molecular mechanisms and therapeutic implications. *Expert Rev Mol Med* **12**:e5.

Buchmeier, M. J., J. C. de la Torre, and C. J. Peters. 2007. Arenaviridae: The viruses and their replication, pp. 1791–1819. In D. M. Knipe and P. M. Howley (eds.), *Fields Virology*, 5th edn., vol. 2. Lippincott Williams & Wilkins, Philadelphia, PA.

Cao, W., M. D. Henry, P. Borrow, H. Yamada, J. H. Elder, E. V. Ravkov, S. T. Nichol, R. W. Compans, K. P. Campbell, and M. B. Oldstone. 1998. Identification of alpha-dystroglycan as a receptor for lymphocytic choriomeningitis virus and Lassa fever virus. *Science* **282**:2079–2081.

Charrel, R. N. and X. de Lamballerie. 2010. Zoonotic aspects of arenavirus infections. *Vet Microbiol* **140**:213–220.

Charrel, R. N., X. de Lamballerie, and S. Emonet. 2008. Phylogeny of the genus *Arenavirus. Curr Opin Microbiol* **11**:362–368.

Cheng, Y., O. Zak, P. Aisen, S. C. Harrison, and T. Walz. 2004. Structure of the human transferrin receptor-transferrin complex. *Cell* **116**:565–576.

Clegg, J. C. 2002. Molecular phylogeny of the arenaviruses. *Curr Top Microbiol Immunol* **262**:1–24.

Collins, J. F., C. A. Franck, K. V. Kowdley, and F. K. Ghishan. 2005. Identification of differentially expressed genes in response to dietary iron deprivation in rat duodenum. *Am J Physiol Gastrointest Liver Physiol* **288**:G964–G971.

Cuevas, C. D., M. Lavanya, E. Wang, and S. R. Ross. 2011. Junin virus infects mouse cells and induces innate immune responses. *J Virol* **85**:11058–11068.

Daniels, T. R., T. Delgado, J. A. Rodriguez, G. Helguera, and M. L. Penichet. 2006. The transferrin receptor part I: Biology and targeting with cytotoxic antibodies for the treatment of cancer. *Clin Immunol* **121**:144–158.

Emonet, S. F., J. C. de la Torre, E. Domingo, and N. Sevilla. 2009. Arenavirus genetic diversity and its biological implications. *Infect Genet Evol* **9**:417–429.

Flanagan, M. L., J. Oldenburg, T. Reignier, N. Holt, G. A. Hamilton, V. K. Martin, and P. M. Cannon. 2008. New world clade B arenaviruses can use transferrin receptor 1 (TfR1)-dependent and -independent entry pathways, and glycoproteins from human pathogenic strains are associated with the use of TfR1. *J Virol* **82**:938–948.

Fry, M. M., C. A. Kirk, J. L. Liggett, G. B. Daniel, S. J. Baek, J. S. Gouffon, P. M. Chimakurthy, and B. Rekapalli. 2009. Changes in hepatic gene expression in dogs with experimentally induced nutritional iron deficiency. *Vet Clin Pathol* **38**:13–19.

Ganz, T. 2011. Hepcidin and iron regulation, 10 years later. *Blood* **117**:4425–4433.

Gonzalez, P. H., P. M. Cossio, R. Arana, J. I. Maiztegui, and R. P. Laguens. 1980. Lymphatic tissue in Argentine hemorrhagic fever. Pathologic features. *Arch Pathol Lab Med* **104**:250–254.

Gonzalez, P. H., R. P. Laguens, M. J. Frigerio, M. A. Calello, and M. C. Weissenbacher. 1983. Junin virus infection of *Callithrix jacchus*: Pathologic features. *Am J Trop Med Hyg* **32**:417–423.

Helenius, A. 2007. Virus entry and uncoating, pp. 99–118. In D. M. Knipe and P. M. Howley (eds.), *Fields Virology*, 5th edn., vol. 1. Lippincott Williams & Wilkins, Philadelphia, PA.

Helguera, G., S. Jemielity, J. Abraham, S. M. Cordo, M. G. Martinez, J. A. Rodríguez, C. Bregni, J. J. Wang, M. Farzan, M. L. Penichet, N. A. Candurra, and H. Choe. 2012. An antibody recognizing the apical domain of human transferrin receptor 1 efficiently inhibits the entry of all New World hemorrhagic fever arenaviruses. *J Virol* **86**:4024–4028.

Hentze, M. W., M. U. Muckenthaler, and N. C. Andrews. 2004. Balancing acts: Molecular control of mammalian iron metabolism. *Cell* **117**:285–297.

Hoppe, M. and L. Hulthen. 2007. Capturing the onset of the common cold and its effects on iron absorption. *Eur J Clin Nutr* **61**:1032–1034.

Jay, M. T., C. Glaser, and C. F. Fulhorst. 2005. The arenaviruses. *J Am Vet Med Assoc* **227**:904–915.

Kemna, E., P. Pickkers, E. Nemeth, H. van der Hoeven, and D. Swinkels. 2005. Time-course analysis of hepcidin, serum iron, and plasma cytokine levels in humans injected with LPS. *Blood* **106**:1864–1866.

Kunz, S., N. Sevilla, J. M. Rojek, and M. B. Oldstone. 2004. Use of alternative receptors different than alpha-dystroglycan by selected isolates of lymphocytic choriomeningitis virus. *Virology* **325**:432–445.

Li, W., C. Zhang, J. Sui, J. H. Kuhn, M. J. Moore, S. Luo, S. K. Wong et al. 2005. Receptor and viral determinants of SARS-coronavirus adaptation to human ACE2. *EMBO J* **24**:1634–1643.

Linderholm, M., C. Ahlm, B. Settergren, A. Waage, and A. Tarnvik. 1996. Elevated plasma levels of tumor necrosis factor (TNF)-alpha, soluble TNF receptors, interleukin (IL)-6, and IL-10 in patients with hemorrhagic fever with renal syndrome. *J Infect Dis* **173**:38–43.

Liu, C. Y., Y. F. Liu, L. Zeng, S. G. Zhang, and H. Xu. 2007. The expression of TfR1 mRNA and IRP1 mRNA in the placenta from different maternal iron status. *Zhonghua Xue Ye Xue Za Zhi* **28**:255–258.

Lozach, P. Y., L. Burleigh, I. Staropoli, and A. Amara. 2007. The C type lectins DC-SIGN and L-SIGN: Receptors for viral glycoproteins. *Methods Mol Biol* **379**:51–68.

Lu, J., K. Hayashi, and X. Hu. 1995. Transferrin receptor and iron deposition pattern in the hepatic lobules of the iron-deficient and iron-overloaded rats. *Zhonghua Bing Li Xue Za Zhi* **24**:75–77.

Maiztegui, J. I., R. P. Laguens, P. M. Cossio, M. B. Casanova, M. T. de la Vega, V. Ritacco, A. Segal, N. J. Fernandez, and R. M. Arana. 1975. Ultrastructural and immunohistochemical studies in five cases of Argentine hemorrhagic fever. *J Infect Dis* **132**:35–53.

Marta, R. F., V. S. Montero, C. E. Hack, A. Sturk, J. I. Maiztegui, and F. C. Molinas. 1999. Proinflammatory cytokines and elastase-alpha-1-antitrypsin in Argentine hemorrhagic fever. *Am J Trop Med Hyg* **60**:85–89.

McLeod, C. G., Jr., J. L. Stookey, J. D. White, G. A. Eddy, and G. A. Fry. 1978. Pathology of Bolivian Hemorrhagic fever in the African green monkey. *Am J Trop Med Hyg* **27**:822–826.

McLeod, C. G., J. L. Stookey, G. A. Eddy, and K. Scott. 1976. Pathology of chronic Bolivian hemorrhagic fever in the rhesus monkey. *Am J Pathol* **84**:211–224.

Morizono, K., Y. Xie, T. Olafsen, B. Lee, A. Dasgupta, A. M. Wu, and I. S. Chen. 2011. The soluble serum protein Gas6 bridges virion envelope phosphatidylserine to the TAM receptor tyrosine kinase Axl to mediate viral entry. *Cell Host Microbe* **9**:286–298.

Nemeth, E., S. Rivera, V. Gabayan, C. Keller, S. Taudorf, B. K. Pedersen, and T. Ganz. 2004. IL-6 mediates hypoferremia of inflammation by inducing the synthesis of the iron regulatory hormone hepcidin. *J Clin Invest* **113**:1271–1276.

Nemeth, E., E. V. Valore, M. Territo, G. Schiller, A. Lichtenstein, and T. Ganz. 2003. Hepcidin, a putative mediator of anemia of inflammation, is a type II acute-phase protein. *Blood* **101**:2461–2463.

Ng, L. F., A. Chow, Y. J. Sun, D. J. Kwek, P. L. Lim, F. Dimatatac, L. C. Ng, E. E. Ooi, K. H. Choo, Z. Her, P. Kourilsky, and Y. S. Leo. 2009. IL-1beta, IL-6, and RANTES as biomarkers of Chikungunya severity. *PLoS One* **4**:e4261.

Pallone, F., S. Fais, O. Squarcia, L. Biancone, P. Pozzilli, and M. Boirivant. 1987. Activation of peripheral blood and intestinal lamina propria lymphocytes in Crohn's disease. In vivo state of activation and in vitro response to stimulation as defined by the expression of early activation antigens. *Gut* **28**:745–753.

Pietrangelo, A., E. Rocchi, G. Casalgrandi, G. Rigo, A. Ferrari, M. Perini, E. Ventura, and G. Cairo. 1992. Regulation of transferrin, transferrin receptor, and ferritin genes in human duodenum. *Gastroenterology* **102**:802–809.

Radoshitzky, S. R., J. Abraham, C. F. Spiropoulou, J. H. Kuhn, D. Nguyen, W. Li, J. Nagel et al. 2007. Transferrin receptor 1 is a cellular receptor for New World haemorrhagic fever arenaviruses. *Nature* **446**:92–96.

Radoshitzky, S. R., J. H. Kuhn, C. F. Spiropoulou, C. G. Albarino, D. P. Nguyen, J. Salazar-Bravo, T. Dorfman, A. S. Lee, E. Wang, S. R. Ross, H. Choe, and M. Farzan. 2008. Receptor determinants of zoonotic transmission of New World hemorrhagic fever arenaviruses. *Proc Natl Acad Sci USA* **105**:2664–2669.

Radoshitzky, S. R., L. E. Longobardi, J. H. Kuhn, C. Retterer, L. Dong, J. C. Clester, K. Kota, J. Carra, and S. Bavari. 2011. Machupo virus glycoprotein determinants for human transferrin receptor 1 binding and cell entry. *PLoS One* **6**:e21398.

Reignier, T., J. Oldenburg, M. L. Flanagan, G. A. Hamilton, V. K. Martin, and P. M. Cannon. 2008. Receptor use by the Whitewater Arroyo virus glycoprotein. *Virology* **371**:439–446.

Reignier, T., J. Oldenburg, B. Noble, E. Lamb, V. Romanowski, M. J. Buchmeier, and P. M. Cannon. 2006. Receptor use by pathogenic arenaviruses. *Virology* **353**:111–120.

Rojek, J. M. and S. Kunz. 2008. Cell entry by human pathogenic arenaviruses. *Cell Microbiol* **10**:828–835.

Siddappa, A. J., R. B. Rao, J. D. Wobken, K. Casperson, E. A. Leibold, J. R. Connor, and M. K. Georgieff. 2003. Iron deficiency alters iron regulatory protein and iron transport protein expression in the perinatal rat brain. *Pediatr Res* **53**:800–807.

Spiropoulou, C. F., S. Kunz, P. E. Rollin, K. P. Campbell, and M. B. Oldstone. 2002. New World arenavirus clade C, but not clade A and B viruses, utilizes alpha-dystroglycan as its major receptor. *J Virol* **76**:5140–5146.

Tacchini, L., E. Gammella, C. De Ponti, S. Recalcati, and G. Cairo. 2008. Role of HIF-1 and NF-kappaB transcription factors in the modulation of transferrin receptor by inflammatory and anti-inflammatory signals. *J Biol Chem* **283**:20674–20686.

Watanabe, Y., M. S. Ibrahim, H. F. Ellakany, N. Kawashita, R. Mizuike, H. Hiramatsu, N. Sriwilaijaroen, T. Takagi, Y. Suzuki, and K. Ikuta. 2011. Acquisition of human-type receptor binding specificity by new H5N1 influenza virus sublineages during their emergence in birds in Egypt. *PLoS Pathog* **7**:e1002068.

Watanabe, Y., M. S. Ibrahim, Y. Suzuki, and K. Ikuta. 2012. The changing nature of avian influenza A virus (H5N1). *Trends Microbiol* **20**:11–20.

York, J. and J. H. Nunberg. 2006. Role of the stable signal peptide of Junin arenavirus envelope glycoprotein in pH-dependent membrane fusion. *J Virol* **80**:7775–7780.

York, J., V. Romanowski, M. Lu, and J. H. Nunberg. 2004. The signal peptide of the Junin arenavirus envelope glycoprotein is myristoylated and forms an essential subunit of the mature G1-G2 complex. *J Virol* **78**:10783–10792.

18 Argentine Hemorrhagic Fever

*Víctor Romanowski, Matías L. Pidre, M. Leticia Ferrelli,
Cecilia Bender, and Ricardo M. Gómez*

CONTENTS

18.1 INTRODUCTION

Argentine hemorrhagic fever (AHF) is a severe viral hemorrhagic syndrome endemic to the agricultural plains of central Argentina. Its incidence is mainly seasonal (Moraz and Kunz, 2011). The clinical symptoms of AHF include hematological, neurological, cardiovascular, renal, and immunological alterations. This emerging disease was first recognized in 1955, and its etiological agent was characterized and designated Junín virus (JUNV) for the geographical site where it was first isolated (Parodi et al., 1958, 1959). JUNV is a rodent-borne virus and belongs to the clade B New World (NW) arenavirus within the *Arenaviridae* family (Salvato et al., 2011).

The population of humans at risk is composed mainly of agricultural workers, believed to become infected with urine, saliva, or blood from infected rodents (Mills et al., 1994). Since its emergence in the 1950s, annual epidemics of the disease have been recorded. The initially high case fatality rate of the disease was markedly reduced, first with adequate supportive measures and, more significantly, with the use of immune plasma (Enria et al., 2008). Former endemic hot spots are currently cooling off; however, there is a steady and progressive geographic expansion of the endemic region into north-central Argentina, and currently almost five million people are considered to be at risk of contracting AHF (Enria et al., 2008). A collaborative effort conducted by the U.S. and Argentine Governments in the 1980s led to the production of a live attenuated JUNV vaccine (Maiztegui et al., 1998). The availability of the live attenuated vaccine contributed to a substantial reduction in the number of AHF cases in the recent years (Enria et al., 2008).

18.2 BIOLOGICAL PROPERTIES

18.2.1 VIRUS STRUCTURE AND GENOME ORGANIZATION

Like other members of the *Arenaviridae* family, JUNV virions are enveloped, pleomorphic, and ~120 nm in diameter; they contain several copies of circular nucleocapsids with helicoidal symmetry and include a variable number of ribosomes (Romanowski and Bishop, 1983; Salvato et al., 2005, 2011). They acquire a lipid envelope during the budding process from the host cell membrane, enclosing nucleocapsids as well as ribosomes at the end of the infectious cycle. Their genome consists of two single-stranded RNA segments: S (small) and L (large) of lengths of about 3.5 kb and 7.3 kb, respectively. Each segment has two non-overlapping open reading frames (ORFs) of opposite sense, which was the origin of the term *ambisense* to describe this type of coding strategy (Auperin et al., 1984). The ORFs of opposite polarity are separated in both RNAs by a noncoding intergenic region predicted to fold into a stable secondary structure (Goñi et al., 2006). The L segment codes for both the 94 amino acid (aa) zinc-binding Z matrix protein that drives virus budding (*ca.* 11 kDa), and the RNA-dependent RNA polymerase L (2210 aa, *ca.* 250 kDa). The S RNA codes for both the nucleocapsid protein N (564 aa) and the glycoprotein (GP) precursor GPC (485 aa) (Goñi et al., 2006; Salvato et al., 2005, 2011). GPC is synthesized as a single polypeptide chain and post-translationally cleaved to yield mature virion GPs G1 (192 aa), G2 (235 aa), and a stable signal peptide (SSP) (58 aa) (York et al., 2004). G1/G2/SSP trimers form the spikes decorating the virus surface (Figure 18.1).

G1 is located at the top of the spike and mediates virus interaction with host cell surface receptors (Radoshitzky et al., 2007) and G2 is similar to other class I viral fusion proteins (York et al., 2005). SSP is generated by signal peptidase cleavage, but unlike conventional signal peptides is stable, unusually long (58 *vs.* the usual 15–25 aa), myristoylated (York et al., 2004), contains two

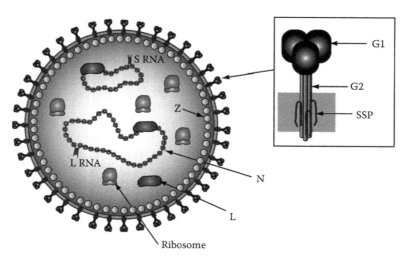

FIGURE 18.1 Schematic representation of a JUNV particle. A lipid envelope acquired during budding from cell is underlined by a matrix-like protein Z and decorated with trimeric GP complex spikes (consisting of transmembrane fusion protein G2 associated with SSP and topped with a trimeric G1 head). The envelope encloses a variable number of nucleocapsids, host cell ribosomes, and one or more L polymerase molecules. Circular nucleocapsids are formed by L and S RNAs associated with multiple copies of the major structural protein N forming helicoidal RNP complexes.

hydrophobic segments that span the lipid bilayer with both N and C termini residing in the cytosol (Figure 18.1), and contribute to G2 fusion activity (York and Nunberg, 2006) through its C-terminal region (York and Nunberg, 2007b).

The nucleocapsid protein N can bind zinc, and according to a recent structural study of LASV (Lassa virus) N protein, the N domain has a structure with a cavity for binding the cap that is required for viral RNA transcription, whereas the C domain contains 3′–5′ exoribonuclease activity involved in suppressing interferon induction (Qi et al., 2010). JUNV is believed to have the same protein distribution estimated for other arenaviruses, N is the most abundant virion protein, followed by G2, G1, Z, and L (~1,500, 650, 650, 450, and 30 molecules per virion, respectively) (Vezza et al., 1977; Gunther and Lenz, 2004). Finally, it is worth mentioning that a protein sequence database for pathogenic arenaviruses has been constructed (Bui et al., 2007).

18.2.2 Virus Entry and Cell Tropism

Specific virus interaction with receptor molecules on the cell membrane is a crucial step in the infectious process: it drives subsequent entry into the host cell, making cell receptors the major determinants of viral cell tropism, host range, and pathogenesis. Until 2005, little was known about the mechanism by which JUNV entered host cells. By using pseudotyped retroviruses, several laboratories confirmed that JUNV, as well as other clade B NW arenaviruses, did not interact with α-dystroglycan, the known receptor for Old World (OW) arenaviruses, to enter the cells (Rojek and Kunz, 2008).

The next big breakthrough came 1 year later when using a proteomic pull-down approach, applying a recombinant receptor-binding G1 moiety of Machupo virus (MACV) as bait, transferrin receptor 1 (TfR1) was identified as the first known JUNV, MACV, GTOV, and SABV cell receptor (Radoshitzky et al., 2007). Authors also showed that expression of human TfR1, but not TfR2, greatly enhanced susceptibility of otherwise somewhat resistant hamster cell lines to JUNV, MACV, Guanarito virus (GTOV), and Sabiá virus (SABV) pseudotypes, but not to LASV or lymphocytic choriomeningitis virus (LCMV). In addition, infection of human cells with JUNV, MACV, GTOV, and SABV, but not with LASV, was efficiently blocked with anti-TfR1 antibodies. Furthermore, iron loading of culture medium also affected infective capacity, since depletion enhanced and supplementation decreased the infection efficiency of JUNV and MACV but not of LASV pseudoviruses. These results provide strong evidence for TfR1 being the major cell receptor for pathogenic human clade B NW arenaviruses. Interestingly, TfR1 is also used by canine and feline parvoviruses and by the mouse mammary tumor virus [reviewed in Rojek and Kunz (2008)]. In contrast, nonpathogenic human clade B NW arenaviruses like Amapari virus (AMAV) and Tacaribe virus (TCRV) attach to cells via a TfR1-independent pathway (Flanagan et al., 2008).

Given the conserved tropism of both pathogenic and nonpathogenic NW clade B arenaviruses for TfR1-expressing cells and the fact that MACV G1 has the greatest structural similarity to the JUNV G1, MACV G1 structure and its interaction with the TfR1 are used here to speculate on the interactions of JUNV G1 and TfR1 (Figure 18.2).

MACV G1 secondary structure consists of seven antiparallel β-strands forming a left-handed sheet, three α-helices, and two additional 3_{10} helices. The secondary structure is stabilized by four disulfide bonds. Two of these disulfide bonds appear to be conserved across the NW HF arenaviruses, a third is also found in the G1 of JUNV, while the fourth stabilizes a MACV-specific insertion. The presence of this additional conserved disulfide bond in both MACV G1 and JUNV G1 reflects the close relationship between these two viruses compared to other NW arenaviruses (Bowden et al., 2009).

A total of 10 potential N-linked glycosylation sites were found on all NW HF arenavirus G1s (MACV, JUNV, GTOV, SABV, and Chapare virus [CHPV]) and decorate solvent-accessible loops on the perimeter of the β-sheet. Two sites are completely conserved among all these viruses and a third site is conserved between MACV and JUNV. It is believed that the area containing the

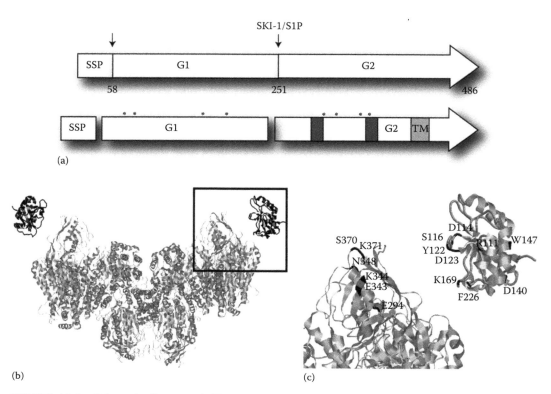

FIGURE 18.2 Schematic diagram of GP precursor (GPC) maturation and interaction with cell surface receptor. (a) GPC is synthesized as a single polypeptide chain of 485 aa length that is proteolytically processed (indicated by arrows) to yield mature virion GPs G1, G2, and SSP. Asterisks indicate potential glycosylation sites; dark gray boxes in G2 are the fusion peptide sequences (located in the ectodomain) and the light gray box indicates the transmembrane domain (followed by the C-terminus located inside the virion). G1/G2/SSP trimers form the spikes decorating the virus surface as indicated in Figure 18.1. (b) Interaction of butterfly-shaped human transferrin receptor 1 (hTfR1) with G1 (two dark G1 monomers are shown approaching hTfR1, for clarity). The structures have been obtained from http://www.rcsb.org/pdb/. (c) Amplified portions of hTfR1 and G1 indicating the amino acid residues relevant for protein–protein interactions (both polypeptides are shown more separated than they are when they actually interact). The actual 3D structure of G1 shown in this figure is that of MACV, assumed to closely match that of JUNV, as indicated in the text. See review by Choe et al. (2011).

carbohydrate cluster is likely to be structurally variable. In contrast, the surfaces involved in receptor binding, as well as interaction with G2, lie outside this region and contain conserved aa residues.

The G1-binding site on human TfR1 (hTfR1) has been attributed to a prominent loop within the apical domain, between the residues 201 and 212. Recent studies using G1 deletion mutants of MACV have established that the 20 N-terminal amino acids of MACV G1 are dispensable for TfR1 binding.

A co-crystal structure of the MACV G1:TfR1 complex provided relevant information about the residues of G1 that contact TfR1. Five interaction motifs were identified at the interface between MACV G1 and the apical domain of TfR1. The ability of a panel of MACV G1 variants in these motifs to bind hTfR1 were tested measuring MACV entry using simian and human cells as experimental systems. These results demonstrate that G1 residues R111, D123, Y122, and F226 are important for hTfR1 binding and cell entry of MACV. R111 residue is not conserved in any of the other NW arenaviruses and is part of interaction motif 1 making prominent contacts with hTfR1 bII-2 strand. F226 is part of interaction motif 3 and has a hydrophobic interaction with hTfR1 as well as a van der Waals contact. D123 and Y122 are conserved between MACV, JUNV, and SABV. D123 forms a hydrogen bond with hTfR1 K344, whereas Y122 forms one with hTfR1 E343 (Abraham et al., 2010; Radoshitzky et al., 2011) (Figure 18.2b and c).

MACV GPC residues D114, S116, D140, K169, and W147 are dispensable for cell attachment and entry of MACV. D114 and S116 form interaction motif 2, which contacts hTfR1 residues N348, S370, and K371. K169, a part of interaction motif 4, contacts K344, and E294 of hTfR1 (Figure 18.2b and c).

Seven N-linked glycans on Asn178 were revealed in the MACV G1-hTfR1 crystal structure. These glycans perhaps play an important role in MACV GP folding, trafficking, GP complex formation, or entry (Radoshitzky et al., 2011).

Although the hTfR1 is definitely a major receptor that allows JUNV infection, there is information on additional or alternative cell surface molecules that seem to promote virus entry. Dendritic cell-specific intercellular adhesion molecule-3-grabbing nonintegrin (DC-SIGN) is a type II transmembrane lectin receptor. DC-SIGN is abundantly expressed on immature dendritic cells (iDCs), one of the principal targets of JUNV (Svajger et al., 2010). Previous reports showed that nonpermissive cells lacking TfR1 became significantly more susceptible for JUNV infection when transfected with a construct expressing the DC-SIGN receptor. In addition, pre-treatment of these genetically modified cells with anti-DC-SIGN or mannan reduces the infection with JUNV. This work validates a direct cell-to-cell transmission of JUNV and supports the role of type C lectins in the viral adsorption, internalization, and intracellular transport (Forlenza et al., 2011; Iula et al., 2011).

After binding to its receptor, a virus can resort to different internalization mechanisms. Briefly, there are three general mechanisms for viral internalization: clathrin-mediated endocytosis, caveolar/rafts pathway, and cholesterol-dependent endocytosis. It has been demonstrated in Vero cells that clathrin-mediated endocytosis is the main route used by JUNV (Martinez et al., 2007) (Figure 18.3). Compounds impairing clathrin-mediated endocytosis reduce virus internalization without affecting virion binding. In contrast, drugs altering lipid-raft micro-domains and, therefore, impairing caveolae-mediated endocytosis do not block virus entry. Finally, direct evidence of JUNV cell entry was obtained using transmission electron microscopy (Martinez et al., 2007). PICV, another NW arenavirus, has also been shown to enter cells through a clathrin-dependent endocytic pathway, trafficked through the dynamin 2 endocytic pathway in which the virus travels through Rab5-mediated early and Rab7-mediated late endosomes (Vela et al., 2008). Similar results have been observed for JUNV (Martinez et al., 2009). JUNV internalization leads to PI3K/Akt signaling pathway activation (Linero and Scolaro, 2009), which requires both intact actin and a dynamic microtubule network (Martinez et al., 2008). It is not known whether other molecules are involved in the process. Because TfR1 is recycled through the early endosome, but JUNV GP complex requires delivery to more acidic late endosomal compartments (York and Nunberg, 2007a), it has been suggested that JUNV manipulates endocytic host cell machinery by modifying the oligomeric state of TfR1 ligand (Rojek and Kunz, 2008).

Later, the fusion of the viral envelope with the endosomal membrane is essential for the progression of the infection cycle. While G1 interacts with the TfR1, G2 is responsible for the fusion process. The principal features of the current model for arenavirus G2 include a hydrophobic N-terminus and two antiparallel helices separated by a glycosylated loop region. In the conformation that precedes the fusion, the N-terminal α-helices form a coiled-coil core structure. Low pH generates a series of rearrangements leading to the post-fusion six-helix bundle conformation. Thereby, the viral envelope and the target membrane are pulled together to give rise to the fusion pore. G2-directed monoclonal antibodies (mAbs) prevent membrane fusion by binding to an intermediate form of the protein on the fusion pathway (York et al., 2010). Nunberg's laboratory showed that small-molecule compounds inhibit arenavirus entry and protect against lethal infection in animal models, acting on the GP spikes through the SSP or interfering with the G2–SSP association, respectively (York et al., 2008).

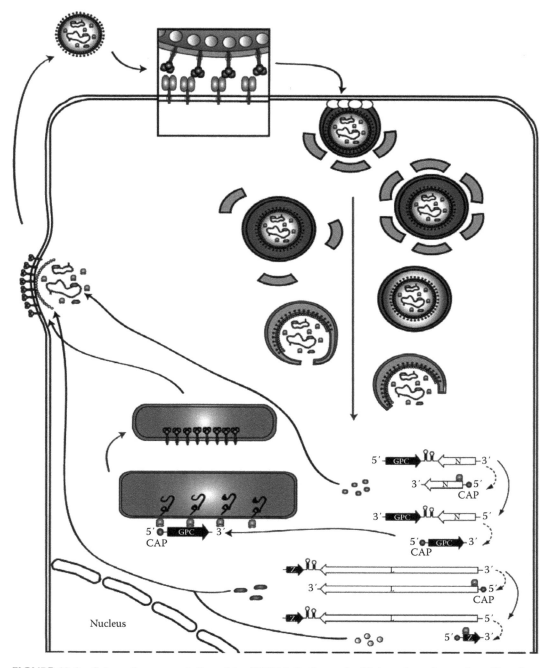

FIGURE 18.3 Schematic representation of the JUNV infection cycle. Virion adsorption to the cell surface is mediated by G1 (head of trimeric GP complex) interaction with the TfR1 cell receptor. After receptor-mediated endocytosis, the drop in pH triggers conformational changes in G2 that drive virion and endosome membrane fusion. Uncoating releases of the viral nucleoprotein, which serves as a template for transcription and replication of the ambisense genomic RNAs (see Figure 18.4 for details). Translation of GPC mRNA and processing of the precursor via the secretory pathway yield the GP complexes inserted in the cell membrane. The Z protein drives the budding process by curving the GP containing membrane GP patches and interacting with the newly formed nucleocapsids. No nuclear phase is required for arenavirus replication.

These studies revealed a crucial role of the SSP in pH-dependent membrane fusion. The SSP has an unusual length of 58 amino acids, contains two distinct hydrophobic domains, and is myristoylated at its N-terminus. Both N- and C-termini of SSP are located in the cytosol and SSP associates non-covalently with the cytoplasmic domain (CTD) and the fusion machinery in the ectodomain of G2 (York et al., 2004). Amino acid substitutions in SSP affect the ability of GPC to mediate cell–cell fusion (Albarino et al., 2010; York and Nunberg, 2006). This dual interaction provides a novel mechanism for fusion of both viral and cellular membranes (Rojek and Kunz, 2008).

18.2.3 TRANSCRIPTION AND REPLICATION

The 3′ and 5′ terminal sequences of 19-nucleotide RNA segments are complementary and very well conserved in all arenaviruses. The base complementarity of these termini is probably the molecular basis for the circular conformation of the nucleocapsids that has been observed (Ghiringhelli et al., 1991). These termini are essential for replication and transcription and are believed to function as a binding site for viral polymerases (reviewed in Albarino et al., 2009; Gunther and Lenz, 2004).

Genome replication and transcription take place in the cytoplasm of infected cells and requires that viral proteins combine with viral RNA to form ribonucleoprotein (RNP) complexes. The L protein mediates viral transcription and replication using RNPs as templates. In TCRV, N and L proteins together with virus RNA are the minimal components of RNP complexes and are sufficient for genome replication and transcription (Lopez et al., 2001). Both N and L proteins are sufficient and necessary for these early steps *in vivo* in reconstructed JUNV transcription-replication (Albarino et al., 2009). Purified RNPs were shown to be competent for RNA synthesis *in vitro* (Fuller-Pace and Southern, 1989). During genome replication, full-length copies of genomic S and L RNAs are synthesized generating the corresponding antigenomic S and L RNAs (Figure 18.4). Due to the *ambisense* coding strategy, both genomic and antigenomic RNAs serve as template for viral mRNA transcription (Figure 18.4). Transcripts contain a cap but are not polyadenylated. The 3′-end sequences of the subgenomic mRNAs fall within the intergenic region suggesting that the stem-loop structure plays a role in transcription termination (Iapalucci et al., 1991; Tortorici et al., 2001). However, elements regulating termination have not yet been well defined. Genomic and antigenomic full-length S RNAs as well as S RNA transcripts often contain non-templated extensions at their 5′ ends. These are thought to be generated at the beginning of replication and transcription and may be important to maintain integrity of genomic terminal segments; the 5′-ends of mRNAs are generated by cap snatching [reviewed in Gunther and Lenz (2004)]. Viral RNA species packaged into virions are defined as genomic RNAs; however, smaller amounts of antigenomic RNA and Z gene mRNA are also included (Salvato and Shimomaye, 1989). Ribosomal RNA is also found inside virions, but its relevance is not known (Salvato et al., 2011).

The functional role of the Z protein in genome replication and transcription is unclear, but it has been assigned both a role in genomic RNA and N mRNA synthesis (Garcin et al., 1993), as well as an inhibitory effect on transcription and replication, but not on encapsidation, via interaction with the L protein (Jacamo et al., 2003).

18.2.4 ASSEMBLY, MATURATION, AND RELEASE

SSP is required for transport of the G1G2 precursor protein GPC from the endoplasmic reticulum (ER). It has been suggested and later demonstrated that JUNV SSP is retained and positioned in the GP complex through interaction with a zinc-binding domain in the cytoplasmic tail of G2 (Agnihothram et al., 2006; York and Nunberg, 2007b). The interaction with SSP appears to mask endogenous dibasic ER retention/retrieval signals in the CTD of G2 (Agnihothram et al., 2006) to

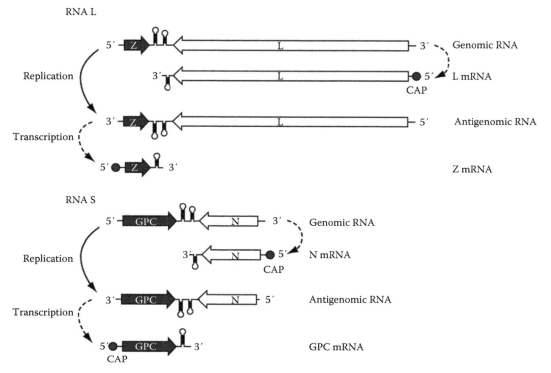

FIGURE 18.4 JUNV genome replication and transcription. The ambisense coding strategy of JUNV consists of two ORFs of opposite polarities in both L and S RNAs represented by rectangles with arrowheads. Genome-sense or positive-sense (Z, GPC) are dark gray, while the negative-sense or antigenomic ORFs (L, N) are white. The intergenic region contains a transcription termination signal for the generation of the subgenomic mRNAs, which include cap snatching.

allow transit through the Golgi compartment, where the G1G2 precursor is cleaved by the cellular SKI-1/S1P protease to form the mature G1 and G2 subunits. As with other class I viral fusion proteins, proteolytic cleavage of the GPC precursor is required to render the GP complex competent for membrane fusion (Albarino et al., 2009; York et al., 2004).

Interestingly, cleavage by SKI-1/S1P is not necessary for JUNV GPC transport to the cell surface, where budding occurs, but it is a prerequisite for GP incorporation into viral particles (Agnihothram et al., 2006). In the absence of LASV GPC cleavage, enveloped non-infectious virus-like particles containing at least N, Z protein, and viral RNA, but devoid of viral GP, are released. Thus, proteolytic processing of GPC is necessary for infectivity. The consensus motif at the cleavage site appears to be preserved in arenaviruses. However, SKI-1/S1P cleaves peptides representing the processing site of GTOV only to negligible levels, indicating that some arenavirus GPCs are indeed processed by a related protease (Gunther and Lenz, 2004).

By using a TCRV–JUNV hybrid system, it has been shown that interaction between Z and N is required for the assembly of both nucleocapsids and GP into infectious budding particles (Casabona et al., 2009). Sequence analysis revealed the presence of so-called late domains, short proline-rich sequences, in the Z protein of LCMV (PPPY) and LASV (PTAP and PPPY). Late domains were previously identified in Gag proteins of a number of retroviruses as well as in matrix proteins of rhabdo- and filoviruses, and are known to interact with cellular proteins such as Tsg101 and Nedd4 promoting virus particle release; their integrity is essential for this function (Gunther and Lenz, 2004).

Z protein has been assigned a major role in virus particle budding, at least for some arenavirus including JUNV (Groseth et al., 2010). Additionally, Z function may be related to cell response to viral infection. It has been recently demonstrated that Z proteins from NW arenaviruses such as JUNV, GTOV, MAVC, and SABV, but not from OW arenaviruses like LCMV or LASV, bind to RIG-I, resulting in the down-regulation of IFNβ response (Fan et al., 2010). Similar findings had been reported earlier for N protein in several arenaviruses including JUNV, after observing inhibition of IFNβ- and IRF3-dependent promoter activation, as well as nuclear translocation of IRF-3 (Martinez-Sobrido et al., 2007; Qi et al., 2010). Confirmation of other JUNV Z protein functions, such as a reported interaction between LCMV or LASV Z protein and promyelocytic leukemia protein or the ribosomal protein P0, will require future studies.

18.2.5 REVERSE GENETICS OF JUNV

The development of a reverse genetics system for JUNV represented an important breakthrough in efforts to understand and combat AHF at the molecular level. This success was achieved by first determining the correct sequence of the virus RNA genome termini that contain critical promoter and encapsidation signals, followed by the development of a JUNV minigenome system to allow screening of functional JUNV L polymerase-encoding clones. These elements, coupled with the construction of an appropriate negative control L polymerase (lacking the SDD core motif), laid the necessary foundation for the infectious virus rescue attempts. The design of the procedure consisting of a simple two-plasmid system (expressing the full-length S and L RNAs) resulted in the generation of a highly efficient and robust virus rescue system (Albarino et al., 2009). This recent development represents a powerful tool to precisely address questions regarding JUNV (and other NW arenaviruses) biology and pathogenicity.

As a first evaluation of the power of this system, it was tested to examine elements of virus GP processing. As indicated earlier in this chapter, the proteolytic processing of GPC leads to the formation of G1/G2/SSP complexes that associate to form the trimeric spikes, which are incorporated in budding virions at the cell surface. By creating a plasmid that expressed a mutant JUNV S RNA sequence in the reverse genetic system, it was unequivocally confirmed that GPC processing by the cell protease SKI-1/S1P was a strict requirement for the production of infectious virus.

Further studies using site-directed mutagenesis and chimeric arenavirus genomes lead to more sophisticated discoveries, e.g., the mutations responsible for the attenuated phenotype of live Candid#1 vaccine and the requirement for specific interactions of homologous GP SSP and G2 CTDs to yield infectious virus (Albarino et al., 2011a,b; Emonet et al., 2011).

18.3   CLINICAL PRESENTATION

AHF incubation period ranges from 6 to 12 days, ending with the onset of fever, usually associated with a flu-like syndrome that may include myalgia, arthralgia, headache, relative bradycardia, conjunctivitis, nausea, vomiting, and diarrhea, with little central nervous system (CNS) or hematological involvement during the first week. The early symptoms of AHF differ from those of acute respiratory infections by an almost constant absence of sore throat, cough, or nasal congestion. At the end of the first week of evolution, there is oliguria and different degrees of dehydration, neurological symptoms are common, and, in female patients, mild-to-moderate metrorrhagia is always present, being in some cases the first symptom of this disease.

In the second week of the disease, around 75% of infected individuals begin to improve, while the remaining 25% manifest neurological disorders or severe bleeding. Overlapping shock and bacterial infections appear 6–12 days after the onset of symptoms. Fever persists, while petechiae in the oral mucosa and the axillary region as well as gingival bleeding can be observed. Less common, more severe hemorrhagic signs may be present including hematemesis, melena, hemoptysis, epistaxis, hematomas, metrorrhagia, and hematuria. CNS involvement can also be present during the second week in the form of hyporeflexia and mental confusion. When severe, it can progress to include areflexia, muscular hypotonia, ataxia, increased irritability and tremors, followed by delirium, generalized seizures, and coma (Molinas et al., 1981).

Clinically apparent disease occurs in almost two-thirds of the infected subjects. Fatality rate is as high as 30% among untreated patients. Immune plasma therapy reduces mortality to less than 1% although this specific therapy is effective only when started during the first week of illness (Moraz and Kunz, 2011).

In early convalescence, 10% of cases treated with immune plasma from convalescent patients develop a late neurological syndrome (LNS). The LNS occurs after a period free of manifestations, differs from the neurological symptoms of the acute period of AHF, and is characterized by fever syndrome and manifestations from the cerebellar trunk (Enria et al., 1998). Patients have a prolonged convalescence. Temporary loss of hair is common; many patients experience fatigue, irritability, and memory changes, but these symptoms are temporary and disappear gradually.

During the second week of illness, patients that are improving start to produce antibodies against JUNV as well as cellular immune response to clear up the virus. Moreover, robust titers of neutralizing antiviral antibodies can be detected in immune plasma from convalescent patients [reviewed in Enria et al. (2008)].

18.4 DIAGNOSIS

AHF is of mandatory reporting in Argentina. At present, the AHF diagnosis to establish specific therapy is based on clinical and laboratory data. During the early phase of illness, the clinical manifestations of AHF are nonspecific and can be confused with several acute febrile conditions. Therefore, if platelet counts below 100,000/mm^3 in combination with white blood cell counts under 2,500/mm^3 are detected, when screening patients in endemic areas, can be considered potentially useful to identify individuals at risk [reviewed in (Enria et al., 2008)].

Seroconversion occurs only late in the course of infection: the serological tests are not useful markers in early stages of the disease. Neutralizing anti-JUNV antibodies (Abs) consisting mainly of the IgG1 subtype are usually present from day 12 onwards [reviewed in Enria et al. (2008)]. Serologic diagnosis can be done by complement fixation, indirect immunofluorescent antibody assays, neutralization tests, and ELISA. Due to its sensitivity and specificity, ELISA is the routine method of choice for the etiologic diagnosis of reported cases retrospectively and for the surveillance of the zoonosis (Morales et al., 2002). More recently, a more accurate ELISA was developed employing recombinant JUNV N protein (Ure et al., 2008). Immunohistochemistry is used to examine organ specimens from autopsy and confirm etiology.

During the acute phase of infection, virus titers are low in blood. Therefore, JUNV antigen detection is not a method of choice for early diagnosis until more sensitive techniques become available. During this phase, virus isolation can be performed from whole blood or peripheral blood mononuclear cells (PBMCs), a useful (reliable, but lengthy and cumbersome) tool to retrospectively confirm the clinical diagnosis in addition to serological tests (Ambrosio et al., 1986).

Since immune plasma therapy is capable of reducing mortality when performed during the first 8 days of infection, the availability of rapid and early diagnostic test is extremely important. In this context, a reverse transcriptase (RT) polymerase chain reaction (PCR)-based assay has been established for rapid diagnosis (Lozano et al., 1995) and has also been successfully applied to establish an etiological diagnosis in subjects who died before the appearance of the specific antibodies. More recently, a real-time RT-PCR assay was described capable of detecting a 0.5 TCID$_{50}$ of cell culture-derived JUNV (Vieth et al., 2005). However, it has not been used yet with clinical or field specimens.

At present, the RT-PCR analysis to detect JUNV genome seems to be the most sensitive, rapid, and early test for the specific diagnosis of the infection.

18.5 PATHOLOGY AND PATHOGENESIS

JUNV may enter the body through skin, respiratory tract, or gastrointestinal mucosa (Figure 18.5). After replication, generalized dissemination occurs, but gross pathology changes are nonspecific. Capillary dilatation ensues with perivascular erythrocyte diapedesis and bleeding; minor edema of the vascular wall has also been observed. Erythroblastopenia with morphologically abnormal erythroid and leukopoietic cell lines and normal megakaryocytes have been described in bone marrow, as well as severe meningeal edema and hemorrhages in Virchow–Robin spaces in the CNS [reviewed in Marta et al. (1998).

Decreased T and B lymphocyte counts and a diminished response to mitogens are expressions of immunosuppression during the acute phase of the disease. Low numbers of null, B, and T cells as well as a lower T4/T8 ratio have been observed during the acute phase of AHF. Null and T8 cell numbers improve after immune plasma infusion, and all cell subsets return to normal in early convalescence. It has been proposed that circulating monocytes (macrophages) are targets for JUNV

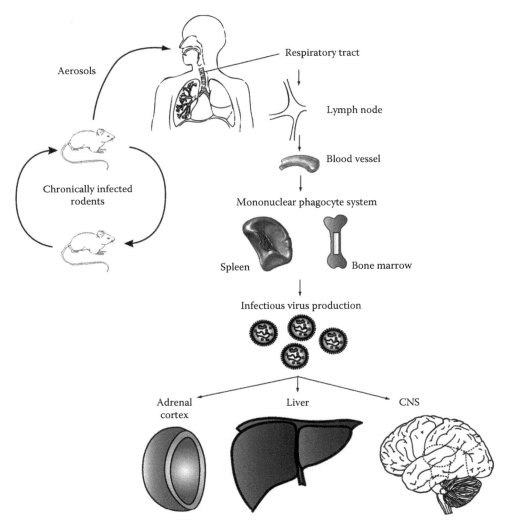

FIGURE 18.5 AHF: JUNV circulation in nature, including rodent reservoirs, major route of infection, and main affected organs.

replication contributing to viral spread during the acute AHF (Marta et al., 1998). However, at this stage as well as in early convalescence, patient PBMCs may exert antibody-dependent cell cytotoxicity, suggesting that JUNV replication in macrophages does not affect their killing capacity (Marta et al., 1998).

Although causes of bleeding are poorly understood, impaired hemostasis, endothelial cell dysfunction, and low platelet counts are considered the major determinants of AHF.

Prolonged kaolin partial thromboplastin time with low levels of factors VIII, IX, and XI and elevated levels of factors V and von Willebrand (vWF) has been found in early disease stages. Fibrinogen is significantly increased in moderate and severe cases, and increased plasma levels of prothrombin fragment 1 + 2 and thrombin–antithrombin complexes suggest that factor Xa and thrombin are also generated (Marta et al., 1998). These results, in combination with patient clinical features and absence of fibrin deposits in microcirculation, rule out disseminated intravascular coagulation as a contributing factor in AHF pathogenesis (Molinas et al., 1981).

In moderate and severe cases, marked increase in tissue plasminogen activator is found in early stages of AHF; functional and antigenic plasminogen levels decrease slightly (Marta et al., 1998). Fibrin split products show raised D-dimer levels in almost all patients on admission, indicating that insoluble fibrin is broken down before being deposited in the microcirculation (Marta et al., 1998). Because a plasma inhibitor of platelet aggregation was found in patients with severe Lassa fever, search for a similar abnormality was conducted in AHF patients. An inhibitor of platelet aggregation and serotonin release was found. However, it was considered to be different from the one present in Lassa fever because it was not found in immunoglobulin G fraction and because its activity was abolished by heat [reviewed in Marta et al. (1998)].

AHF patients show reduced complement hemolytic activity with a drop in antigenic levels of Clq, C3, C5, and functional C2, but with increased C4 antigen and factors B and C3 degradation products (de Bracco et al., 1978), suggesting that the alternative complement pathway is the most affected. To gain further insight into complement activation pathways in AHF patients, levels of C3 (C3b, C3bi, C3c) and C4 activation products (C4b, C4bi, and C4c) were measured using ELISA (Marta et al., 1998). High levels were present in all cases regardless of AHF severity. In addition, antigenic C2 and C4 levels were also elevated, thus confirming that both complement pathways are activated in AHF patients.

In most cells, TfR1 expression is subject to regulation by cellular iron, with an inverse correlation between cytosol iron concentration and TfR1 surface expression levels (Kunz, 2009). As mentioned earlier, iron loading of culture medium affected cell infection levels by JUNV pseudotypes (Radoshitzky et al., 2007) or infectious JUNV (Pozner et al., 2010). Interestingly, a revision of AHF patients' records indicated that anemia, an expected condition for individuals with iron deficiency, was not associated to early or late AHF (Marta et al., 2000).

During the first week after symptom onset, AHF patients show very high serum α-interferon titers (IFNα). Even though these values slowly normalize during the second week of illness in survivors, they remain elevated in severe cases. Interferon levels at admission correlate with outcome and are significantly lower in patients who survive (Levis et al., 1985). This apparent lack of endogenous IFNα efficacy led researchers to consider that circulating IFN might be biologically inactive. This hypothesis was later ruled out after 2′-5′-oligoadenylate synthetase, a marker of IFN biological activity, was found in above-normal range in peripheral blood mononuclear cells (Ferbus et al., 1988).

Other cytokines described as significantly elevated in the serum of acute AHF patients include TNFα, IL-6, IL-8, and IL-10 (Marta et al., 1999) although their individual role in disease pathogenesis has not been studied. For more details on human disease findings, readers should consult the review by Marta et al. (1998).

Pathological lesions in fatal AHF include generalized vasocongestion with multiple hemorrhages in the gastrointestinal mucosa, different organs, such as liver, kidney, and lungs, as well as subcutaneous tissue. The highest virus titers are found in the spleen, lymph nodes, and lungs, and high levels of viral antigen are found in cells of the monocyte/macrophage lineage in peripheral blood, lymphatic tissue, lung, and liver (Marta et al., 1998).

18.6 EPIDEMIOLOGY

18.6.1 Expansion of Agriculture and Emergence of AHF

This disease is endemic, with annual outbreaks from the end of summer until midwinter, coincident with the harvest of maize and with the increase in the population of wild rodents *Calomys musculinus, Calomys laucha, Akodon azarae,* and *Oryzomys flavescens* (Mills et al., 1994). It was assumed that appearance of AHF disease in the mid-1950s was due to human changes made in natural habitats in relation to agricultural practices. Those environmental modifications are thought to have favored the growth of *C. musculinus* population and facilitated its contact with humans. The epidemiological features of AHF are determined by the natural cycle of JUNV and by the behavior of the rodent reservoirs (Mills et al., 1994). AHF mainly affects rural workers from the agricultural region known as the Humid Pampa, in central east Argentina (Enria et al., 2008). There are, however, urban cases in which the origin of infection is not easy to establish. Since the first description of the disease in the 1950s, uninterrupted annual outbreaks have been observed in a progressively expanding region in north-central Argentina, to the point that almost five million individuals are considered today to be at risk for AHF (Enria et al., 2008).

In 1958, cases were limited to an area of approximately 16,000 km^2, with a population at risk estimated to be 270,000 inhabitants. In 1963, cases of AHF were confirmed in the Southeast of the province of Córdoba, and between 1964 and 1967, new areas were affected in the province of Buenos Aires, and later cases were detected in the South of Santa Fe province. At present, the endemoepidemic region spans an area of approximately 150,000 km^2, with a population at risk estimated at more than 5,000,000 people (Enria et al., 2008). Yearly geographic expansion in the latest periods has been smaller than that seen immediately after the first outbreaks of AHF, suggesting an autolimitation in the extension of the endemic area. Nevertheless, rodent studies indicate the possibility of a northward expansion (Enria et al., 2008).

Since 1958, annual outbreaks of AHF have been registered without interruption, with more than 25,000 cases reported. AHF has a focal distribution that could be related to the patchy spatial distribution of the infected rodent, *C. musculinus*. Annual incidence within the endemic area may be as low as 1/100,000, but in the areas of highest activity, it reaches 140/100,000 inhabitants and 355/100,000 adult males. The incidence of AHF also varies in time. In general, it is higher during a period of 5–10 years in newly involved areas and it tends to decline later. Nevertheless, cases continue to be reported in the older locations within the endemic area. This epidemiological pattern is being modified since 1992, through vaccination of high-risk population and, probably, also by viral genetic variation and virus–reservoir population dynamics, affected by diverse environmental factors (Enria et al., 2008; Gomez et al., 2011).

18.6.2 Seasonality and Occupational Risk Factors

Although cases of the disease can be seen throughout the whole year, epidemics occur predominantly during the Southern autumn and winter, with a peak incidence in May, coinciding with the major harvesting season in Argentina.

AHF is four times more prevalent in males than in females, and is more prevalent among rural workers (90%) than in urban populations. Children under the age of 14 account for *ca.* 10% of cases, but disease is uncommon in those younger than 4 and exceptional in those less than 2 years of age.

The seasonal distribution of the illness and the prevalence in adult male rural workers reflect occupational exposure of humans to JUNV rodent hosts. The increase in rodent population densities during late summer and autumn is coincident with an increase in farm labor in the fields. AHF emerged in the 1950s, affecting men working in agriculture-related occupations, mainly those harvesting crops by hand. With the mechanization of agricultural practices that took place in the 1960s–1970s, the group at higher risk comprised those workers driving tractors or other machines in the field.

Seroepidemiological surveys demonstrated the occurrence of inapparent infections with JUNV, but their prevalence is low, ranging from 2% to 4%. These surveys also showed that in different rural areas in which the disease was endemic for more than 10 years, only 10% of the population was immune, thus 90% was still susceptible to infection (Enria et al., 1998).

18.6.3 RODENT RESERVOIRS

All arenaviruses pathogenic for humans are rodent viruses. Although each arenavirus can infect many species of rodents, in every geographical site, there is one species that, due to higher population density and the prevalence and characteristics of infection, is the principal reservoir. *Calomys musculinus* (family Muridae, subfamily Sigmodontinae) has been identified as the principal reservoir of JUNV, although virus has also been isolated from the organs and body fluids of other rodents captured in the endemic area, e.g., *C. laucha* and *A. azarae,* and occasionally from *Mus musculus, Necromys benefactus,* and *Oligoryzomys flavescens* (Sabattini et al., 1977).

Some of these animals develop an acute disease with antibody response and clearance of the virus, while others develop a persistent infection, with low titers or absence of antibodies, chronic viremia, and shedding of virus in urine, feces, and saliva (Vitullo et al., 1987). The chronically infected rodents are usually asymptomatic and exhibit a normal behavior. Field studies of natural populations demonstrated that infection with JUNV among *C. musculinus* was more frequent among males than females and was positively correlated with age and the presence of wounds and scars (Mills et al., 1994). JUNV among rodents may be transmitted via aerosols and bites, as well as sexually.

JUNV-specific mAbs are available and may find interesting applications for the analysis of blood or other tissue samples from patients or wild rodents captured in epidemiological surveillance programs. First mAbs against JUNV were generated against N and G2 (Sanchez et al., 1989), and more recently, new mAbs against JUNV N were obtained by Nakauchi et al. (2009). Three epitopes comprising residues 12–17 (WTQSLR), 72–79 (KEVDRLMS), and 551–558 (PPSLLFLP) are recognized by different mAbs with different degrees of specificity, i.e., ranging from broadly reactive with South American Arenaviruses to JUNV specific (Nakauchi et al., 2009). RT-PCR-based methods have been described using both Arenaviridae family-specific and species-specific that can be employed to detect arenaviruses in rodents captured in the field (Lozano et al., 1997; Vieth et al., 2005).

18.6.4 HUMAN INFECTION

The mechanism of virus transmission from rodents to humans has not yet been well established. Nevertheless, there is strong scientific evidence to support that virus is transmitted by inhaling aerosols of rodent excreta, although viral entry may occur by other routes, such as the conjunctival membranes, other mucous membranes, ingestion, and direct contact with damaged skin. Transmission between humans was reported, even if AHF is usually not contagious from human to human. In patients with AHF, the viremia is present during the entire acute febrile period. Moreover, the virus was occasionally isolated from oral swabs, urine, and breast milk from infected subjects. Sexual transmission of JUNV was reported from convalescent men to women (Briggiler et al., 1987).

18.6.5 VACCINES

A successful rodent control program in endemic areas for the elimination of rodent reservoirs is often not practical. Therefore, since the discovery of JUNV, all efforts to prevent AHF have been directed toward the development of an effective vaccine.

A scientific collaboration between the U.S. and Argentine governments allowed the development of a live attenuated JUNV vaccine, Candid#1 (Maiztegui et al., 1998).

Recently, Albarino et al. (2011a) demonstrated that the major determinant of attenuation in mice of the Candid#1 vaccine strain is located in the G2 glycoprotein transmembrane domain. Nucleotide and amino acid sequence alignments, performed on Candid#1 and XJ ancestor strains, also revealed 6 amino acid substitutions resulting from 12 point mutations throughout the L gene (Goni et al., 2006); some of these changes might contribute to the attenuated phenotype. Candid strain turned out to be safe, immunogenic, and effective in preventing AHF in preclinical studies in mice, guinea pigs, and Rhesus monkeys. Guinea pigs and Rhesus monkeys inoculated with increasing doses of Candid#1 developed neutralizing antibodies and became JUNV resistant if inoculated with highly virulent strains. Candid#1 also protected these animals against MACV, the etiologic agent of the Bolivian hemorrhagic fever (Enria and Barrera Oro, 2002). These studies also showed an absence of neurovirulence, neurotropism, or hemorrhagic manifestations, and the stability of the attenuated strain.

In phase III clinical trials conducted in the period 1988–1990, Candid#1 showed a protective efficacy $\geq$84% and no serious adverse effects. As expected, immune response to Candid#1 boosts preexisting immunity to JUNV but is not changed by previous exposure to LCMV (Enria et al., 2008).

The live attenuated JUNV vaccine, Candid#1, has proven effective during the last two decades in more than 100,000 persons (Enria et al., 2008).

The vaccine has been recently produced in Argentina and tested in a compatible clinical study with 946 healthy volunteers that participated to support the comparability of Candid#1 vaccine manufacturing in the United States and Argentina (Enria et al., 2010). Results presented by Enria et al. (2010) showed that the vaccine produced in Argentina is equivalent to that manufactured in the United States, both in ability to immunize against JUNV (immunogenicity $\geq$ 95.5%) and in absence of promoting any serious adverse effects. Adverse effects of immunization are of low clinical significance, resolving spontaneously or with the use of a symptomatic treatment. Candid#1 is the first effective vaccine against arenaviruses; it is effective to protect against AHF, promoting humoral and cell-mediated responses, and since January 2007, in Argentina, is part of the National Immunization Program in the AHF risk area (Enria et al., 2010). Candid#1 vaccine was used in persons starting at the age of 15. The humoral immune response was shown to persist for years in these subjects. Although the availability of such live attenuated vaccine has led to a substantial reduction in the incidence of AHF disease (Enria and Barrera Oro, 2002), more recent epidemiological data record an increase in the incidence of AHF individuals younger than 15, indicating that more effort is needed to evaluate the use of Candid#1 in children living in risky areas (Enria et al., 2010).

In a recent report, Seregin et al. (2010) used a live attenuated alphavirus-based approach to engineer a replicon system based on a human vaccine TC83 that expresses JUNV's GP heterologous antigen. This vaccine was later tested in guinea pigs for its immunogenicity and efficacy, showing that a single dose of that vaccine is immunogenic and provides partial protection against JUNV, while a double dose provides a complete protection. The protection correlates with neutralizing antibody development and absence of virus dissemination. These data suggested that, in guinea pigs, this experimental vaccine is immunogenic, and the viral GP alone is sufficient to provide an effective immune response against JUNV (Seregin et al., 2010).

18.7 PROGNOSIS AND TREATMENT

Without treatment, after the second week, over 80% of patients improve, although bacterial infection is a frequent complication.

Significant improvement in clinical AHF management has been achieved using immune plasma from convalescents (3,000 units/kg body weight), with mortality rates dropping from almost 30% to less than 1%. Additional administration of ribavirin may enhance these results even further [reviewed in Enria et al. (2008)]. Approximately 10% of cases treated with immune plasma develop LNS. After a symptom-free period, LNS onset is characterized by fever, cerebellar signs, and cranial nerve palsies. LNS has never been registered among AHF patients recovering without specific treatment. Current antiarenaviral therapy is limited to an off-label use of ribavirin (1-β-D-ribofuranosyl-1,2,4-triazole-3-carboxamide), which has had only mixed success in the treatment of severe infections and is associated with significant toxicity in humans (Enria et al., 2008). New potential therapeutic molecules are being tested (Gowen and Bray, 2011).

18.8 CONCLUSIONS AND FUTURE PERSPECTIVES

During the past years, an impressive progress has been made toward our understanding of the basic molecular and cellular biology of JUNV.

In particular, the development of a reverse genetics system for JUNV represented an important breakthrough and provided a powerful tool to precisely address questions regarding JUNV (and other NW arenaviruses) biology and pathogenicity (Albarino et al., 2011a,b; Emonet et al., 2011).

The reverse genetics system paves the way for sophisticated structure–function studies and the molecular dissection of the mechanisms of virus–host interaction *in vitro* and also *in vivo* (Emonet et al., 2011; Moraz and Kunz, 2011). Particular interests are focused on the ability of arenaviruses to subvert the host cell's innate antiviral defenses, the impact of arenavirus infection on the differentiation and function of cells targeted by hemorrhagic arenaviruses *in vivo*, including APCs such as macrophages and DCs, endothelial cells, and megakaryocytes involved in platelet formation (Gomez et al., 2003; Moraz and Kunz, 2011; Pozner et al., 2010). At the same time, novel animal models will provide important new information about the interaction of hemorrhagic arenaviruses with the host's adaptive immune system, in particular virus-induced immunosuppression, and the understanding of the terminal hemorrhagic shock syndrome (Yun et al., 2008).

To study direct effects of virus replication and gene expression that may be responsible for the perturbation of endothelial cell function, future research involving cell culture models for human endothelium, which allow detailed analysis of virus-induced cell biological and biochemical alterations, will be of great importance. Advances in endothelial cell culture, combined with high-resolution confocal microscopy and the ability to measure endothelial cell functions such as transendothelial electrical resistance in live cells will allow monitoring of subtle, virus-induced functional changes as a consequence of JUNV infection (Gowen et al., 2010; Moraz and Kunz, 2011).

The studies on early molecular events of JUNV infection involving viral GP spikes and cell surface receptors as well as virion and cell membrane fusion provided the basis for the development of novel therapeutic strategies (Thomas et al., 2010; York et al., 2008a). All other steps specific for virus entry, processing, and replication are potential therapeutic targets being explored (García et al., 2011; Gowen and Bray, 2011; Lee et al., 2008; Neuman et al., 2011; Sepulveda et al., 2011).

In addition to small antiviral molecules, immune therapy can be regarded as an alternative for AHF patients. Based on the success of immune therapy in controlling AHF mortality, it is possible to design a strategy replacing convalescent immune plasma, of short supply, with controlled humanized neutralizing mAbs appropriately tested for efficacy (Sepulveda et al., 2011; York et al., 2010).

Promising small animal models for AHF will facilitate the validation of mechanistic findings obtained *in vitro* in the systemic context and allow evaluation of novel therapeutic strategies in preclinical studies (Gowen and Bray, 2011; Gowen et al., 2010).

Favipiravir or T-705 is a promising pyrazine derivative (6-fluoro-3-hydroxy-2-pyrazinecarboxamide) with broad antiviral activity against RNA viruses. So far, animal models of acute arenaviral disease have demonstrated that T-705 can be used effectively to treat advanced infections (Gowen et al., 2008). The ability of T-705 to specifically target the viral replication machinery may minimize the possibility of *in vivo* toxicity. In contrast, ribavirin also inhibits cellular IMP dehydrogenase, a key enzyme in guanosine biosynthesis, and thereby perturbs cellular nucleotide pools (Bausch et al., 2011; Mendenhall et al., 2011). Further studies on new potential treatments are needed to block viral replication without causing toxicity and to prevent the increased vascular permeability that is responsible for hypotension and shock.

Rapid and specific diagnostic tools have been developed that can be applied both for early detection of JUNV in AHF suspected patients and in epidemiological surveillance studies, including samples from field rodents.

Finally, the availability of a powerful reverse genetic system for JUNV combined with new detailed knowledge on virus–host interactions has the potential to be utilized in a rational design of novel live attenuated virus vaccines with precisely engineered disruptions of pathogenic properties. The attenuation features of Candid #1 identified by site-directed mutagenesis can be used as starting information to precisely engineer a prototypic live attenuated vaccine incorporating these and other attenuating features, thereby improving on the vaccine identity, efficacy, and safety (Albarino et al., 2011a). Other approaches to safe and effective vaccines cannot be ruled out (Emonet et al., 2011; Seregin et al., 2010).

REFERENCES

Abraham, J., Corbett, K. D., Farzan, M., Choe, H., and Harrison, S. C. (2010). Structural basis for receptor recognition by New World hemorrhagic fever arenaviruses. *Nat. Struct. Mol. Biol.* 17(4), 438–444.

Agnihothram, S. S., York, J., and Nunberg, J. H. (2006). Role of the stable signal peptide and cytoplasmic domain of G2 in regulating intracellular transport of the Junin virus envelope glycoprotein complex. *J. Virol.* 80(11), 5189–5198.

Albarino, C. G., Bergeron, E., Erickson, B. R., Khristova, M. L., Rollin, P. E., and Nichol, S. T. (2009). Efficient reverse genetics generation of infectious Junin viruses differing in glycoprotein processing. *J. Virol.* 83(11), 5606–5614.

Albarino, C. G., Bird, B. H., Chakrabarti, A. K., Dodd, K. A., Flint, M., Bergeron, E., White, D. M., and Nichol, S. T. (2011a). The major determinant of attenuation in mice of the candid1 vaccine for Argentine hemorrhagic fever is located in the g2 glycoprotein transmembrane domain. *J. Virol.* 85(19), 10404–10408.

Albarino, C. G., Bird, B. H., Chakrabarti, A. K., Dodd, K. A., White, D. M., Bergeron, E., Shrivastava-Ranjan, P., and Nichol, S. T. (2010). Reverse genetics generation of chimeric infectious Junin/Lassa virus is dependent on interaction of homologous glycoprotein stable signal peptide and g2 cytoplasmic domains. *J. Virol.* 85(1), 112–122.

Albarino, C. G., Bird, B. H., Chakrabarti, A. K., Dodd, K. A., White, D. M., Bergeron, E., Shrivastava-Ranjan, P., and Nichol, S. T. (2011b). Reverse genetics generation of chimeric infectious Junin/Lassa virus is dependent on interaction of homologous glycoprotein stable signal peptide and G2 cytoplasmic domains. *J. Virol.* 85(1), 112–122.

Ambrosio, A. M., Enria, D. A., and Maiztegui, J. I. (1986). Junin virus isolation from lympho-mononuclear cells of patients with Argentine hemorrhagic fever. *Intervirology* 25(2), 97–102.

Auperin, D. D., Romanowski, V., Galinski, M. S., and Bishop, H. L. (1984). Sequencing studies of Pichinde arenavirus S RNA indicate a novel coding strategy, an ambisense viral S RNA. *J. Virol.* (52), 987–904.

Bausch, D. G., Mendenhall, M., Russell, A., Smee, D. F., Hall, J. O., Skirpstunas, R., Furuta, Y., and Gowen, B. B. (2011). Effective Oral Favipiravir (T-705) Therapy initiated after the onset of clinical disease in a model of arenavirus hemorrhagic fever. *PLoS Negl. Trop. Dis.* 5(10), e1342.

Bowden, T. A., Crispin, M., Graham, S. C., Harvey, D. J., Grimes, J. M., Jones, E. Y., and Stuart, D. I. (2009). Unusual molecular architecture of the machupo virus attachment glycoprotein. *J. Virol.* 83(16), 8259–8265.

de Bracco, M. M., Rimoldi, M. T., Cossio, P. M., Rabinovich, A., Maiztegui, J. I., Carballal, G., and Arana, R. M. (1978). Argentine hemorrhagic fever. Alterations of the complement system and anti-Junin-virus humoral response. *N. Engl. J. Med.* 299(5), 216–221.

Briggiler, A. M., Enria, D. A., Feuillade, M. R., and Maiztegui, J. I. (1987). Contagio interhumano e infección clínica con virus Junin (V J) en matrimonios residentes en el area endémica de Fiebre Hemorráigica Argentina (FHA). *Medicina (B. Aires)* 47, 565.

Bui, H. H., Botten, J., Fusseder, N., Pasquetto, V., Mothe, B., Buchmeier, M. J., and Sette, A. (2007). Protein sequence database for pathogenic arenaviruses. *Immunome Res.* 3, 1.

Casabona, J. C., Levingston Macleod, J. M., Loureiro, M. E., Gomez, G. A., and Lopez, N. (2009). The RING domain and the L79 residue of Z protein are involved in both the rescue of nucleocapsids and the incorporation of glycoproteins into infectious chimeric arenavirus-like particles. *J. Virol.* 83(14), 7029–7039.

Choe, H., Jemielity, S., Abraham, J., Radoshitzky, S. R., and Farzan, M. (2011). Transferrin receptor 1 in the zoonosis and pathogenesis of New World hemorrhagic fever arenaviruses. *Curr. Opin. Microbiol.* 14(4), 476–482.

Emonet, S. E., Urata, S., and de la Torre, J. C. (2011). Arenavirus reverse genetics: New approaches for the investigation of arenavirus biology and development of antiviral strategies. *Virology* 411(2), 416–425.

Enria, D. A., Ambrosio, A. M., Briggiler, A. M., Feuillade, M. R., Crivelli, E., and Study Group on Argentine Hemorrhagic Fever Vaccine. (2010). Candid#1 vaccine against Argentine hemorrhagic fever produced in Argentina. Immunogenicity and safety. *Medicina (B Aires)* 70(3), 215–222.

Enria, D. A. and Barrera Oro, J. G. (2002). Junin virus vaccines. *Curr. Top. Microbiol. Immunol.* 263, 239–261.

Enria, D., Briggiler, A. M., and Feuillade, M. R. (1998). An overview of the epidemiological, ecological and preventive hallmarks of Argentine hemorrhagic fever (Junin virus). *Bull. Inst. Pasteur* 96, 103–114.

Enria, D. A., Briggiler, A. M., and Sanchez, Z. (2008). Treatment of Argentine hemorrhagic fever. *Antiviral Res.* 78(1), 132–139.

Fan, L., Briese, T., and Lipkin, W. I. (2010). Z proteins of New World arenaviruses bind RIG-I and interfere with type I interferon induction. *J. Virol.* 84(4), 1785–1791.

Ferbus, D., Saavedra, M. C., Levis, S., Maiztegui, J., and Falcoff, R. (1988). Relation of endogenous interferon and high levels of 2′-5′ oligoadenylate synthetase in leukocytes from patients with Argentine hemorrhagic fever. *J. Infect. Dis.* 157(5), 1061–1064.

Flanagan, M. L., Oldenburg, J., Reignier, T., Holt, N., Hamilton, G. A., Martin, V. K., and Cannon, P. M. (2008). New world clade B arenaviruses can use transferrin receptor 1 (TfR1)-dependent and -independent entry pathways, and glycoproteins from human pathogenic strains are associated with the use of TfR1. *J. Virol.* 82(2), 938–948.

Forlenza, M. B., Roldán, J. S., Martínez, M. G., Cordo, S. M., and Candurra, N. A. (2011). Interacción del virus Junín con lectinas de tipo C y mecanismos endocíticos involucrados. *Revista Argentina de Microbiología* 43(sup.1; X Congreso Argentino de Virología), 162–163.

Fuller-Pace, F. V. and Southern, P. J. (1989). Detection of virus-specific RNA-dependent RNA polymerase activity in extracts from cells infected with lymphocytic choriomeningitis virus: In vitro synthesis of full-length viral RNA species. *J. Virol.* 63(5), 1938–1944.

García, C. C., Sepúlveda, C. S., and Damonte, E. B. (2011). Novel therapeutic targets for arenavirus hemorrhagic fevers. *Future Virol.* 6(1), 27–44.

Garcin, D., Rochat, S., and Kolakofsky, D. (1993). The Tacaribe arenavirus small zinc finger protein is required for both mRNA synthesis and genome replication. *J. Virol.* 67(2), 807–812.

Ghiringhelli, P. D., Rivera-Pomar, R. V., Lozano, M. E., Grau, O., and Romanowski, V. (1991). Molecular organization of Junin virus S RNA: Complete nucleotide sequence, relationship with other members of the Arenaviridae and unusual secondary structures. *J. Gen. Virol.* 72(Pt 9), 2129–2141.

Gomez, R. M., Jaquenod de Giusti, C., Sanchez Vallduvi, M. M., Frik, J., Ferrer, M. F., and Schattner, M. (2011). Junin virus. A XXI century update. *Microbes Infect.* 13(4), 303–311.

Gomez, R. M., Pozner, R. G., Lazzari, M. A., D'Atri, L. P., Negrotto, S., Chudzinski-Tavassi, A. M., Berria, M. I., and Schattner, M. (2003). Endothelial cell function alteration after Junin virus infection. *Thromb. Haemost.* 90(2), 326–333.

Goni, S. E., Iserte, J. A., Ambrosio, A. M., Romanowski, V., Ghiringhelli, P. D., and Lozano, M. E. (2006). Genomic features of attenuated Junin virus vaccine strain candidate. *Virus Genes* 32(1), 37–41.

Gowen, B. B. and Bray, M. (2011). Progress in the experimental therapy of severe arenaviral infections. *Future Microbiol.* 6(12), 1429–1441.

Gowen, B. B., Julander, J. G., London, N. R., Wong, M. H., Larson, D., Morrey, J. D., Li, D. Y., and Bray, M. (2010). Assessing changes in vascular permeability in a hamster model of viral hemorrhagic fever. *Virol. J.* 7, 240.

Gowen, B. B., Smee, D. F., Wong, M. H., Hall, J. O., Jung, K. H., Bailey, K. W., Stevens, J. R., Furuta, Y., and Morrey, J. D. (2008). Treatment of late stage disease in a model of arenaviral hemorrhagic fever: T-705 efficacy and reduced toxicity suggests an alternative to ribavirin. *PLoS One* 3(11), e3725.

Groseth, A., Wolff, S., Strecker, T., Hoenen, T., and Becker, S. (2010). Efficient budding of the Tacaribe virus matrix protein z requires the nucleoprotein. *J. Virol.* 84(7), 3603–3611.

Gunther, S. and Lenz, O. (2004). Lassa virus. *Crit. Rev. Clin. Lab. Sci.* 41(4), 339–390.

Iapalucci, S., Lopez, N., and Franze-Fernandez, M. T. (1991). The 3′ end termini of the Tacaribe arenavirus subgenomic RNAs. *Virology* 182(1), 269–278.

Iula, L. J., Martínez, M. G., and Cordo, S. M. (2011). La expresión del receptor celular DC-SIGN aumenta la capacidad infectiva del virus Junín. Interacción directa del receptor con la glicoproteína viral. *Revista Argentina de Microbiología* 43(sup.1, X Congreso Argentino de Virología), 42.

Jacamo, R., Lopez, N., Wilda, M., and Franze-Fernandez, M. T. (2003). Tacaribe virus Z protein interacts with the L polymerase protein to inhibit viral RNA synthesis. *J. Virol.* 77(19), 10383–10393.

Kunz, S. (2009). The role of the vascular endothelium in arenavirus hemorrhagic fevers. *Thromb. Haemost.* 102(6), 1024–1029.

Lee, A. M., Rojek, J. M., Spiropoulou, C. F., Gundersen, A. T., Jin, W., Shaginian, A., York, J. et al. (2008). Unique small molecule entry inhibitors of hemorrhagic fever arenaviruses. *J. Biol. Chem.* 283(27), 18734–18742.

Levis, S. C., Saavedra, M. C., Ceccoli, C., Feuillade, M. R., Enria, D. A., Maiztegui, J. I., and Falcoff, R. (1985). Correlation between endogenous interferon and the clinical evolution of patients with Argentine hemorrhagic fever. *J. Interferon Res.* 5(3), 383–389.

Linero, F. N. and Scolaro, L. A. (2009). Participation of the phosphatidylinositol 3-kinase/Akt pathway in Junin virus replication in vitro. *Virus Res.* 145(1), 166–170.

Lopez, N., Jacamo, R., and Franze-Fernandez, M. T. (2001). Transcription and RNA replication of Tacaribe virus genome and antigenome analogs require N and L proteins: Z protein is an inhibitor of these processes. *J. Virol.* 75(24), 12241–12251.

Lozano, M. E., Enria, D., Maiztegui, J. I., Grau, O., and Romanowski, V. (1995). Rapid diagnosis of Argentine hemorrhagic fever by reverse transcriptase PCR-based assay. *J. Clin. Microbiol.* 33(5), 1327–1332.

Lozano, M. E., Posik, D. M., Albarino, C. G., Schujman, G., Ghiringhelli, P. D., Calderon, G., Sabattini, M., and Romanowski, V. (1997). Characterization of arenaviruses using a family-specific primer set for RT-PCR amplification and RFLP analysis. Its potential use for detection of uncharacterized arenaviruses. *Virus Res.* 49(1), 79–89.

Maiztegui, J. I., McKee, K. T., Jr., Barrera Oro, J. G., Harrison, L. H., Gibbs, P. H., Feuillade, M. R., Enria, D. A. et al. (1998). Protective efficacy of a live attenuated vaccine against Argentine hemorrhagic fever. AHF Study Group. *J. Infect. Dis.* 177(2), 277–283.

Marta, R. F., Enria, D., and Molinas, F. C. (2000). Relationship between hematopoietic growth factors levels and hematological parameters in Argentine hemorrhagic fever. *Am. J. Hematol.* 64, 1–6.

Marta, R. F., Montero, V. S., Hack, C. E., Sturk, A., Maiztegui, J. I., and Molinas, F. C. (1999). Proinflammatory cytokines and elastase-alpha-1-antitrypsin in Argentine hemorrhagic fever. *Am. J. Trop. Med. Hyg.* 60(1), 85–89.

Marta, R. F., Montero, V. S., and Molinas, F. C. (1998). Systemic disorders in Argentine hemorrhagic fever. *Bull. Inst. Pasteur* 96, 115–124.

Martinez, M. G., Cordo, S. M., and Candurra, N. A. (2007). Characterization of Junin arenavirus cell entry. *J. Gen. Virol.* 88(Pt 6), 1776–1784.

Martinez, M. G., Cordo, S. M., and Candurra, N. A. (2008). Involvement of cytoskeleton in Junin virus entry. *Virus Res.* 138(1–2), 17–25.

Martinez, M. G., Forlenza, M. B., and Candurra, N. A. (2009). Involvement of cellular proteins in Junin arenavirus entry. *Biotechnol. J.* 4(6), 866–870.

Martinez-Sobrido, L., Giannakas, P., Cubitt, B., Garcia-Sastre, A., and de la Torre, J. C. (2007). Differential inhibition of type I interferon induction by arenavirus nucleoproteins. *J. Virol.* 81(22), 12696–12703.

Mendenhall, M., Russell, A., Juelich, T., Messina, E. L., Smee, D. F., Freiberg, A. N., Holbrook, M. R., Furuta, Y., de la Torre, J. C., Nunberg, J. H., and Gowen, B. B. (2011). T-705 (favipiravir) inhibition of arenavirus replication in cell culture. *Antimicrob. Agents Chemother.* 55(2), 782–787.

Mills, J. N., Ellis, B. A., Childs, J. E., McKee, K. T., Jr., Maiztegui, J. I., Peters, C. J., Ksiazek, T. G., and Jahrling, P. B. (1994). Prevalence of infection with Junin virus in rodent populations in the epidemic area of Argentine hemorrhagic fever. *Am. J. Trop. Med. Hyg.* 51(5), 554–562.

Molinas, F. C., de Bracco, M. M., and Maiztegui, J. I. (1981). Coagulation studies in Argentine hemorrhagic fever. *J. Infect. Dis.* 143(1), 1–6.

Morales, M. A., Calderon, G. E., Riera, L. M., Ambrosio, A. M., Enria, D. A., and Sabattini, M. S. (2002). Evaluation of an enzyme-linked immunosorbent assay for detection of antibodies to Junin virus in rodents. *J. Virol. Methods* 103(1), 57–66.

Moraz, M.-L. and Kunz, S. (2011). Pathogenesis of arenavirus hemorrhagic fevers. *Expert Rev. Anti-infect. Ther.* 9, 49–59.

Nakauchi, M., Fukushi, S., Saijo, M., Mizutani, T., Ure, A. E., Romanowski, V., Kurane, I., and Morikawa, S. (2009). Characterization of monoclonal antibodies to Junin virus nucleocapsid protein and application to the diagnosis of hemorrhagic fever caused by South American arenaviruses. *Clin. Vaccine Immunol.* 16(8), 1132–1138.

Neuman, B. W., Bederka, L. H., Stein, D. A., Ting, J. P. C., Moulton, H. M., and Buchmeier, M. J. (2011). Development of peptide-conjugated morpholino oligomers as pan-arenavirus inhibitors. *Antimicrob. Agents Chemother.* 55(10), 4631–4638.

Parodi, A. S., Greenway, D. J., Rugiero, H. R., Frigerio, M., De La Barrera, J. M., Mettler, N., Garzon, F., Boxaca, M., Guerrero, L., and Nota, N. (1958). Sobre la etiología del brote epidémico de Junín. *Dia. Med.* 30, 2300–2301.

Parodi, A. S., Rugiero, H. R., Greenway, D. J., Mettler, N., Martinez, A., Boxaca, M., and De La Barrera, J. M. (1959). Isolation of the Junin virus (epidemic hemorrhagic fever) from the mites of the epidemic area (*Echinolaelaps echidninus*, Barlese). *Prensa Med. Argent* 46, 2242–2244.

Pozner, R. G., Ure, A. E., Jaquenod de Giusti, C., D'Atri, L. P., Italiano, J. E., Torres, O., Romanowski, V., Schattner, M., and Gomez, R. M. (2010). Junin virus infection of human hematopoietic progenitors impairs in vitro proplatelet formation and platelet release via a bystander effect involving type I IFN signaling. *PLoS. Pathog.* 6(4), e1000847.

Qi, X., Lan, S., Wang, W., Schelde, L. M., Dong, H., Wallat, G. D., Ly, H., Liang, Y., and Dong, C. (2010). Cap binding and immune evasion revealed by Lassa nucleoprotein structure. *Nature* 468(7325), 779–783.

Radoshitzky, S. R., Abraham, J., Spiropoulou, C. F., Kuhn, J. H., Nguyen, D., Li, W., Nagel, J., Schmidt, P. J., Nunberg, J. H., Andrews, N. C., Farzan, M., and Choe, H. (2007). Transferrin receptor 1 is a cellular receptor for New World hemorrhagic fever arenaviruses. *Nature* 446(7131), 92–96.

Radoshitzky, S. R., Longobardi, L. E., Kuhn, J. H., Retterer, C., Dong, L., Clester, J. C., Kota, K., Carra, J., and Bavari, S. (2011). Machupo virus glycoprotein determinants for human transferring receptor 1 binding and cell entry. *PLoS One* 6(7), e21398.

Rojek, J. M. and Kunz, S. (2008). Cell entry by human pathogenic arenaviruses. *Cell Microbiol.* 10(4), 828–835.

Romanowski, V. and Bishop, D. H. (1983). The formation of arenaviruses that are genetically diploid. *Virology* 126(1), 87–95.

Sabattini, M. S., Gonzalez de Rios, L. E., Diaz, G., and Vega, V. R. (1977). Infección natural y experimental de roedores con virus Junin. *Medicina (B. Aires)* 37, 149–159.

Salvato, M. S., Clegg, J. C. S., Buchmeier, M. J., Charrel, R. N., Gonzales, J. P., Lukashevich, I. S., Peters, C. J., Rico-Hesse, R., and Romanowski, V. (2005). Arenaviridae. In *Virus Taxonomy. Eighth Report of the International Committee on Taxonomy of Viruses* (C. M. Fauquet, J. Maniloff, M. A. Mayo, U. Desselberger, and L. A. Ball, Eds.), pp. 725–733. Elsevier/Academic Press, London, U.K.

Salvato, M. S., Clegg, J. C. S., Buchmeier, M. J., Charrel, R. N., Gonzalez, J. P., Lukashevich, I.S., Peters, C. J., and Romanowski, V. (2011). Arenaviridae. In *Virus Taxonomy. Classification and Nomenclature of Viruses: Ninth Report of the International Committee on Taxonomy of Viruses* (A. M. Q. King, M. J. Adams, E. B. Carstens, and E. J. Lefkowitz, Eds.), pp. 715–724. Elsevier-Academic Press, Oxford, U.K.

Salvato, M. S. and Shimomaye, E. M. (1989). The completed sequence of lymphocytic choriomeningitis virus reveals a unique RNA structure and a gene for a zinc finger protein. *Virology* 173(1), 1–10.

Sanchez, A., Pifat, D. Y., Kenyon, R. H., Peters, C. J., McCormick, J. B., and Kiley, M. P. (1989). Junin virus monoclonal antibodies: Characterization and cross-reactivity with other arenaviruses. *J. Gen. Virol.* 70, 1125–1132.

Sepulveda, C. S., Garcia, C. C., Fascio, M. L., D'Accorso, N. B., Docampo Palacios, M. L., Pellon, R. F., and Damonte, E. B. (2011). Inhibition of Junin virus RNA synthesis by an antiviral acridone derivative. *Antiviral Res.* 93(1), 16–22.

Seregin, A. V., Yun, N. E., Poussard, A. L., Peng, B. H., Smith, J. K., Smith, J. N., Salazar, M., and Paessler, S. (2010). TC83 replicon vectored vaccine provides protection against Junin virus in guinea pigs. *Vaccine* 28(30), 4713–4718.

Svajger, U., Anderluh, M., Jeras, M., and Obermajer, N. (2010). C-type lectin DC-SIGN: an adhesion, signalling and antigen-uptake molecule that guides dendritic cells in immunity. *Cell Signal.* 22(10), 1397–1405.

Thomas, C. J., Casquilho-Gray, H. E., York, J., DeCamp, D. L., Dai, D., Petrilli, E. B., Boger, D. L. et al. (2010). A specific interaction of small molecule entry inhibitors with the envelope glycoprotein complex of the Junin hemorrhagic fever arenavirus. *J. Biol. Chem.* 286(8), 6192–6200.

Tortorici, M. A., Albarino, C. G., Posik, D. M., Ghiringhelli, P. D., Lozano, M. E., Rivera, P. R., and Romanowski, V. (2001). Arenavirus nucleocapsid protein displays a transcriptional antitermination activity in vivo. *Virus Res.* 73(1), 41–55.

Ure, A. E., Ghiringhelli, P. D., Possee, R. D., Morikawa, S., and Romanowski, V. (2008). Argentine hemorrhagic fever diagnostic test based on recombinant Junín virus N protein. *J. Med. Virol.* 80(12), 2127–2133.

Vela, E. M., Colpitts, T. M., Zhang, L., Davey, R. A., and Aronson, J. F. (2008). Pichinde virus is trafficked through a dynamin 2 endocytic pathway that is dependent on cellular Rab5- and Rab7-mediated endosomes. *Arch. Virol.* 153(7), 1391–1396.

Vezza, A. C., Gard, G. P., Compans, R. W., and Bishop, D. H. (1977). Structural components of the arenavirus Pichinde. *J. Virol.* 23(3), 776–786.

Vieth, S., Drosten, C., Charrel, R., Feldmann, H., and Gunther, S. (2005). Establishment of conventional and fluorescence resonance energy transfer-based real-time PCR assays for detection of pathogenic New World arenaviruses. *J. Clin. Virol.* 32(3), 229–235.

Vitullo, A. D., Hodara, V. L., and Merani, M. S. (1987). Effect of persistent infection with Junin virus on growth and reproduction of its natural reservoir, Calomys musculinus. *Am. J. Trop. Med. Hyg.* 37(3), 663–669.

York, J., Agnihothram, S. S., Romanowski, V., and Nunberg, J. H. (2005). Genetic analysis of heptad-repeat regions in the G2 fusion subunit of the Junin arenavirus envelope glycoprotein. *Virology* 343(2), 267–274.

York, J., Berry, J. D., Stroher, U., Li, Q., Feldmann, H., Lu, M., Trahey, M., and Nunberg, J. H. (2010). An antibody directed against the fusion peptide of Junin virus envelope glycoprotein GPC inhibits pH-induced membrane fusion. *J. Virol.* 84(12), 6119–6129.

York, J., Dai, D., Amberg, S. M., and Nunberg, J. H. (2008). pH-Induced activation of arenavirus membrane fusion is antagonized by small-molecule inhibitors. *J. Virol.* 82(21), 10932–10939.

York, J. and Nunberg, J. H. (2006). Role of the stable signal peptide of Junin arenavirus envelope glycoprotein in pH-dependent membrane fusion. *J. Virol.* 80(15), 7775–7780.

York, J. and Nunberg, J. H. (2007a). Distinct requirements for signal peptidase processing and function in the stable signal peptide subunit of the Junin virus envelope glycoprotein. *Virology* 359(1), 72–81.

York, J. and Nunberg, J. H. (2007b). A novel zinc-binding domain is essential for formation of the functional Junin virus envelope glycoprotein complex. *J. Virol.* 81(24), 13385–13391.

York, J., Romanowski, V., Lu, M., and Nunberg, J. H. (2004). The signal peptide of the Junin arenavirus envelope glycoprotein is myristoylated and forms an essential subunit of the mature G1-G2 complex. *J. Virol.* 78(19), 10783–10792.

Yun, N. E., Linde, N. S., Dziuba, N., Zacks, M. A., Smith, J. N., Smith, J. K., Aronson, J. F. et al. (2008). Pathogenesis of XJ and romero strains of Junin virus in two strains of guinea pigs. *Am. J. Trop. Med. Hyg.* 79, 275–282.

19 Bolivian Hemorrhagic Fever

Sheli R. Radoshitzky, Fabian de Kok-Mercado,
Peter B. Jahrling, Sina Bavari, and Jens H. Kuhn

CONTENTS

19.1 INTRODUCTION: HISTORY AND EPIDEMIOLOGY OF BOLIVIAN HEMORRHAGIC FEVER

In 1964, Mackenzie et al. described a new hemorrhagic fever in the Beni region in northeastern Bolivia (Figure 19.1) (Mackenzie et al. 1964). The disease, by then named Bolivian hemorrhagic fever (BHF), had been first recognized in 1959 on the Bolivian island of Orobayaya. There, 470 cases were reported in the years up to 1962. Another series of localized outbreaks then occurred between 1962 and 1964, which involved more than 1000 patients, of which 180 died. After 20 years of no reported cases, mainly as a result of rodent control measures (Kuns 1965), an outbreak of 19 cases was reported in 1994. Eight more cases were described in 1999, 18 cases in 2000, and 20 suspected cases between 2007 and 2008 (Aguilar et al. 2009). The latest BHF outbreak occurred at the end of 2011, again in the Beni region. At the time of writing, case numbers were still unclear. Machupo virus (MACV), named after a river close to the outbreak area, was isolated in 1963 from the spleen of a fatal human case studied in the town of San Joaquín and identified as the etiological agent of BHF (Johnson et al. 1965b). Similarly to Argentinian hemorrhagic fever (AHF), the BHF outbreak frequency peaks during the annual harvest (April–July). The case-fatality rate of BHF is approximately 5%–30%. Humans become infected with MACV through contact with infected rodents or inhalation of aerosolized virus from contaminated rodent blood, excreta or secreta (Charrel and de Lamballerie 2003). Direct human-to-human transmission, though possible, is probably not the

339

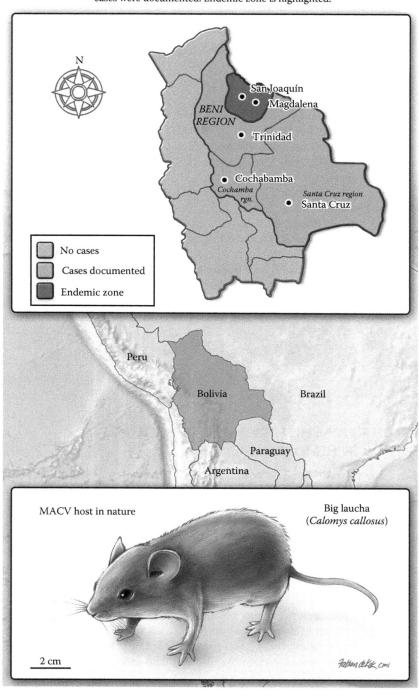

Bolivia, July 1–September 30, 1994. Cities and regions where alleged cases were documented. Endemic zone is highlighted.

FIGURE 19.1   **(See color insert.)** Geography and ecology of Machupo virus.

principal mode of disease dissemination. In fact, only small quantities of MACV can be isolated from human blood or from throat and oral swabs (Johnson 1965).

19.2 CLINICAL COURSE OF BOLIVIAN HEMORRHAGIC FEVER

BHF is clinically similar to the other South American hemorrhagic fevers (Junín/AHF, "BHF," and "Venezuelan hemorrhagic fever")) (Arribalzaga 1955; Molteni et al. 1961; Rugiero et al. 1964; Stinebaugh et al. 1966; de Manzione et al. 1998; Harrison et al. 1999). The disease begins insidiously after an incubation period of 1–2 weeks with fever and malaise, weakness, dehydration, cutaneous hyperesthesia, headache, myalgia, back pain, epigastric pain, and anorexia. After 3–4 days, the symptoms become increasingly severe and indicate multisystem involvement: abdominal pain, nausea and vomiting, and mild diarrhea are typical symptoms. The majority of patients are characterized by thrombocytopenia, leukopenia, proteinuria, and hematuria upon admission (Mackenzie et al. 1964; Stinebaugh et al. 1966; Aguilar et al. 2009). The earliest signs of vascular damage might also appear, such as conjunctivitis (or conjunctival hyperemia), periorbital edema, flushing over the head and upper torso, skin petechiae on the neck and upper thorax and occasionally on the oral mucus membranes, and mild (postural) hypotension. During the second week of illness, the more severe cases (≈30% of patients) begin to display neurologic or hemorrhagic manifestations and superimposed bacterial infections. Neurological signs may be prominent and include intention tremor of the tongue and extremities, irritability, short attention span, delirium, and hallucinations; followed by tetanic-like spasms; and finally, clonic–tonic convulsions and coma. Hemorrhagic manifestations consist of hematemesis, gingival bleeding, epistaxis, metrorrhagia, and melena. Ecchymoses at needle puncture sites develop, and shock supervenes. Death usually occurs 7–12 days after disease onset. Patients who survive begin to improve during the second week of disease, as the appearance of neutralizing antibodies signals the onset of an immune response. Convalescence often lasts several weeks or months with fatigue, dizziness, Beau lines in digital nails, and hair loss.

19.3 TAXONOMY AND PHYLOGENETIC RELATIONSHIPS

The family *Arenaviridae* includes a single genus, *Arenavirus*, currently comprising 25 approved species (Clegg 2002; Oldstone 2002; Charrel et al. 2003; Jay et al. 2005; Lecompte et al. 2007; Delgado et al. 2008; Salvato et al. 2011). Based on their antigenic properties, arenaviruses have been divided into two distinct groups. The Old World arenaviruses (Lassa–lymphocytic choriomeningitis serocomplex) include viruses indigenous to Africa and the ubiquitous lymphocytic choriomeningitis virus (LCMV). The New World arenaviruses (Tacaribe serocomplex) include viruses indigenous to the Americas (Bowen et al. 1996; Rowe et al. 1970; Clegg 2002; Cajimat and Fulhorst 2004).

The sequences of the *NP* gene of all known arenaviruses have provided the basis for their phylogenetic analysis. This analysis supports the previously defined antigenic groupings and further defines virus relationships within them. Sequence data derived from other regions of the genome, where available, are largely consistent with this analysis. The 25 member viruses of the 25 species represent four phylogenetic groups. The Old World complex forms one monophyletic group that is deeply rooted to the three New World lineages (Bowen et al. 1997, 2000; Albarino et al. 1998; Vieth et al. 2004). New World arenaviruses are subdivided into four groups (A, B, C, Rec). MACV is a member of group B New World arenaviruses (Gonzalez et al. 1995; Bowen et al. 1996; Cajimat et al. 2012), which also contains the human pathogenic viruses Guanarito virus (GTOV), Junín virus (JUNV), Sabiá virus (SABV), and Chapare virus (CHPV) and the nonpathogenic Tacaribe virus, Amaparí virus, and Cupixi virus (Bowen et al. 1996; Charrel et al. 2002, 2003). This group has been further subdivided into B1, B2, and B3 sublineages. Interestingly, the four hemorrhagic fever–causing viruses do not group together, but rather are distributed in the three sublineages together with nonpathogenic viruses. MACV belongs to sublineage B1 and is most closely related to JUNV and the nonpathogenic Tacaribe virus. GTOV belongs to sublineage B2, and CHPV virus and SABV to B3 (Charrel and de Lamballerie 2003). It is therefore apparent that during arenavirus evolution, the trait of human pathogenicity has arisen on at least two independent occasions.

19.4 VIRUS STRUCTURE AND LIFE CYCLE

19.4.1 VIRUS PROPERTIES

Like all arenaviruses, MACV particles are enveloped, round, oval, or pleomorphic (60–260 nm in diameter) with short surface projections (Figure 19.2). The particles appear sandy (electron-dense granules) in electron microscopy sections (Murphy et al. 1969, 1970).

19.4.2 GENOMIC ORGANIZATION

The arenavirus genome consists of two single-stranded RNA molecules, designated large (L) and small (S). Each segment encodes two different proteins in two nonoverlapping reading frames of opposite polarities. The L segment (≈7200 bp) encodes the viral RNA-dependent RNA polymerase (L) and a small zinc-binding matrix protein (Z). The S segment (≈3500 bp) encodes the structural viral proteins: the nucleoprotein (NP), which associates with viral RNA in the form of bead-like structures to form nucleocapsids, and the envelope glycoprotein precursor (GPC), which is proteolytically processed into two mature envelope proteins GP1 and GP2 and a stable signal peptide (Buchmeier et al. 1987; Lenz et al. 2001; Beyer et al. 2003; Kunz et al. 2003). Extracted virion RNA is not infectious, and therefore, arenaviruses are considered by some as negative-sense RNA viruses despite the presence of the ambisense coding strategy.

The two open reading frames on each genomic segment are separated by a noncoding intergenic region (IGR) with the potential to form one or more energetically stable stem-loop (hairpin) structures (Auperin et al. 1984; Wilson and Clegg 1991). The IGR serves individual functions in

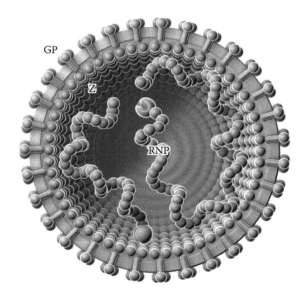

MACV schematic

FIGURE 19.2   (See color insert.) Schematic of a Machupovirion. GP, glycoprotein; RNP, ribonucleoprotein complex.

structure-dependent transcription termination for enhanced gene expression (Meyer and Southern 1993, 1994; Tortorici et al. 2001) and in virus assembly and/or budding (Pinschewer et al. 2005).

The 5' and 3' ends of the L and S RNA segments are noncoding untranslated regions (UTRs) and contain a conserved reverse complementary sequence of 19 or 20 nucleotides at each extremity (Auperin et al. 1982a,b). These termini are thought to base pair and form panhandle structures (Salvato et al. 1989; Harnish et al. 1993), a prediction that is supported by electron microscopic studies of arenaviral nucleocapsids (Young and Howard 1983). The 3' UTR of each segment contains the arenaviral promoter, which functions as a duplex and is composed of two elements: (1) a sequence-specific region at the 3' end of the promoter and (2) a variable complementary region at the 5' end of the promoter where solely base pairing between the 3' and the 5' ends of the segment is important (Perez and de la Torre 2003; Hass et al. 2006). In the case of MACV, RNA template recognition by L requires a sequence-specific motif, 3' N_{1-2}HKUG 5' ($C_2G_3U_4G_5$), located at positions 2–5 in the 3' terminus of the viral genome. This motif is conserved in all experimentally determined arenavirus termini. Furthermore, L–RNA complex formation depends on a single-stranded nature of the RNA, indicating that inter-termini dsRNA interactions must be partially broken for complex assembly to occur (Kranzusch et al. 2011).

19.4.3 VIRUS LIFE CYCLE

19.4.3.1 Virus Entry

Arenaviral cell entry is mediated by the GP spike complex on the virion surface. This complex is a trimer of heterodimers comprising the integral membrane protein GP2, which is noncovalently attached to the highly glycosylated peripheral protein GP1 (Gangemi et al. 1978; Daelli and Coto 1982; Padula and de Martinez Segovia 1984; Wright et al. 1990; Eichler et al. 2006). The complex also includes the signal peptide of the arenavirus GPC (Lenz et al. 2001; Froeschke et al. 2003; York et al. 2004), which is associated noncovalently with the cytoplasmic domain of GP2 (Agnihothram et al. 2006). The arenaviral GP shares many characteristics with class I fusion proteins, and in accordance with class I architecture, the arenaviral GP1 subunit has been shown to mediate receptor binding. In the case of MACV, as well as other pathogenic New World arenaviruses, cell entry occurs after GP1 binds the cellular receptor transferrin receptor 1 (TfR1) (Radoshitzky et al. 2007, 2008; Abraham et al. 2009; Helguera et al. 2012). Following TfR1 binding, the virus is most likely endocytosed through clathrin-coated vesicles (Martinez et al. 2007; Vela et al. 2007) into intracellular endosomal compartments, where GP2-dependent acid-induced membrane fusion occurs (Figure 19.3) (Glushakova and Lukashevich 1989; Glushakova et al. 1990a; Castilla et al. 1994; Kunz et al. 2002; Martinez et al. 2007; Oldenburg et al. 2007). This internalization process is dependent on the cellular clathrin coat–associated protein Eps15 and dynamin (Martinez et al. 2009) and on an intact actin network (Martinez et al. 2008). The activation of the PI3K/AKT signaling pathway seems to be necessary for successful completion of the entry pathway (Linero and Scolaro 2009).

GP2-dependent acid-induced fusion of the virions with cellular membranes occurs in late endosomes (Figure 19.3). The current hypothesis suggests that the putative signal peptide-GP2 interface stabilizes the prefusion state of the GP complex at neutral pH and triggers the conformational changes leading to membrane fusion at acidic pH (York and Nunberg 2006, 2009). Both the myristoylation of the signal peptide and a positively charged amino acid residue, lysine residue 33, are specifically required for its function as fusion modulator (York et al. 2004; York and Nunberg 2006). Exposure to low pH destabilizes the high-energy prefusion state of the GP spike complex resulting in irreversible conformational changes, in which the GP1 subunit is released from the GP spike (Di Simone et al. 1994; Di Simone and Buchmeier 1995). The hydrophobic fusion peptide at the N-terminus of GP2 (Glushakova et al. 1990b, 1992) becomes exposed and inserts into the target membrane. This results in a series of conformational rearrangements in the GP2 subunit leading to the postfusion six-α-helix core structure. Subsequently, the viral envelope and the target

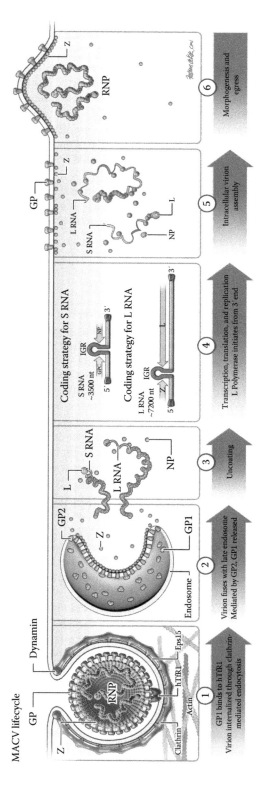

FIGURE 19.3 **(See color insert.)** The Machupo virus cellular life cycle: (1) Virion uptake; (2) virus-cell membrane fusion; (3) uncoating; (4) transcription, translation, and replication; (5) virion assembly; and (6) virion budding. GP, glycoprotein; hTfR1, human transferrin receptor 1; IGR, intergenic region; L, RNA-dependent RNA polymerase; NP, nucleoprotein; Z, matrix protein.

membrane are pulled together and give rise to the fusion pore (Eschli et al. 2006). Viral nucleocapsids are then delivered into the cell cytoplasm, where arenavirus transcription and replication exclusively occurs (Figure 19.3) (Meyer et al. 2002).

19.4.3.2 Virus Transcription and Protein Expression

Arenavirus RNA synthesis is initiated after delivery of the two encapsidated S and L segments, each associated with L, into the cytosol. L initiates from the 3′ end of the genomic template and produces subgenomic, genome-complementary, NP and L mRNAs terminating at nonspecific sites at the IGR. These mRNAs are capped and not polyadenylated (Singh et al. 1987; Southern et al. 1987; Meyer and Southern 1993). The 5′ ends of viral mRNAs contain, in addition to a cap structure, several extra random nontemplated bases, resembling the mRNAs of orthomyxoviruses and bunyaviruses (Garcin and Kolakofsky 1990; Raju et al. 1990; Meyer and Southern 1993). Therefore, it has been inferred that the arenaviral L protein uses a transcription-initiation mechanism involving an oligonucleotide-cap primer of host origin. NP, L, and viral RNA form the transcriptionally active unit, the ribonucleocapsid protein (RNP) complex. Both proteins are the minimal *trans*-acting factors required for arenaviral RNA replication and transcription. Minimal *cis*-acting elements are the promoter sequences residing within the 3′ UTR at the end of each RNA segment, as well as the IGRs (Meyer and Southern 1993; Lee et al. 2000; Lopez et al. 2001; Hass et al. 2004).

The transcription of GPC and Z mRNAs occurs only after one round of arenavirus replication, in which S and L antigenomes are synthesized (Figure 19.3). Consequently, NP mRNA and NP protein accumulate earlier than the GPC mRNA and the GP1 and GP2 glycoproteins. Furthermore, NP is much more abundant in cells than GP1 and GP2 (Meyer et al. 2002; Buchmeier et al. 2006). These observations illustrate how the ambisense coding arrangement of arenaviruses allows for independent regulation of viral gene expression separate from virion production.

GP is translated as a single precursor protein (GPC) into the lumen of the endoplasmic reticulum, where the signal peptide is cleaved off (Eichler et al. 2003a). GPC then undergoes extensive *N*-linked glycosylation (Padula and de Martinez Segovia 1984; Wright et al. 1990) and is thought to oligomerize prior to being proteolytically processed by the subtilase SKI-1/S1P to generate the two mature glycoproteins GP1 and GP2 (Buchmeier et al. 1987; Lenz et al. 2001; Beyer et al. 2003; Kunz et al. 2003; Rojek et al. 2008). In addition to its role as a signal sequence for ER import, the signal peptide of arenaviruses is also specifically required for intracellular trafficking of GPC from the ER to the cell surface (Eichler et al. 2003a; Agnihothram et al. 2006), as well as for the proteolytic maturation of GPC (Eichler et al. 2003b; York and Nunberg 2006). The signal peptide associates with and masks endogenous dibasic ER retention/retrieval signals present in the cytoplasmic tail of GP2 and thereby ensures that only the fully assembled GP spike is transported to the plasma membrane (Agnihothram et al. 2006).

19.4.3.3 Virus Replication

During replication, L reads through the IGR transcription-termination signal and generates uncapped antigenomic and genomic RNA species (Leung et al. 1977) that contain a single nontemplated G on the 5′ end (Garcin and Kolakofsky 1990; Raju et al. 1990). A model for this extra nontemplated G contends that a dinucleotide primer (pppGpC) used in replication initiation first base pairs with positions 2 and 3 of the template RNA and then produces a single G overhang when the nascent RNA slips two bases "backward" on the template (Garcin and Kolakofsky 1992).

19.4.3.4 Virus Assembly and Budding

The newly synthesized full-length antigenomic and genomic RNA species are encapsidated by NP to generate RNP complexes for further mRNA transcription and for production of virus progeny (Raju et al. 1990). It is thought that arenaviral Z plays an important role in regulating the dynamics of arenaviral infection by a direct feedback loop during which increased viral gene expression and higher levels of Z act to suppress RNA synthesis and balance the infection process (Cornu and de la Torre 2001, 2002; Lopez et al. 2001; Cornu et al. 2004). Z inhibits RNA synthesis by locking the L polymerase in a catalytically inactive promoter-bound state (Jacamo et al. 2003; Kranzusch and Whelan 2011). It is plausible that the resulting Z–L–RNA complex may serve as a critical intermediate to ensure genomic RNA is packaged along with L poised to reinitiate viral RNA synthesis in the newly infected cell (Kranzusch and Whelan 2011). The integrity of the RING domain within Z is essential for Z–L interaction and therefore for Z inhibitory activity (Cornu and de la Torre 2002).

MACV buds from the plasma membrane (Murphy et al. 1969, 1970), where the virus RNP core associates with the host derived membrane that is highly enriched with the viral GP spike complex (Figure 19.3). In addition to its role as a potent inhibitor of virus RNA synthesis, Z is also essential for arenavirus assembly and budding (Perez et al. 2003; Strecker et al. 2003; Eichler et al. 2004). Z is strongly membrane associated and accumulates near the inner surface of the plasma membrane where budding takes place (Strecker et al. 2003). Myristoylation of Z is critical for its binding to the viral GP complex, specifically with the signal peptide component, and for its association with lipid membranes and thus for effective virus budding (Perez et al. 2004; Strecker et al. 2006; Capul et al. 2007). Interaction of NP and Z protein was also reported, suggesting that Z might recruit NP, complexed in the virus RNP core, to cellular membranes where virus assembly takes place (Salvato et al. 1992; Salvato 1993). Consistent with its features as a bona fide budding protein, MACV Z contains the canonical proline-rich late (L) domain PSAP motif. L domains are highly conserved and have been shown to mediate interaction with host cell proteins required for virus budding. For example, the cellular multivesicular body pathway proteins Vps4A, Vps4B, and Tsg101 are involved in Lassa virus budding (Urata et al. 2006), and Z was shown to recruit Tsg101 to the plasma membrane (Perez et al. 2003).

MACV, JUNV, GTOV, and SABV Z also bind RIG-I, resulting in the downregulation of the IFN-β response (Fan et al. 2010). MACV and JUNV NP also inhibit IFN-β- and IRF3-dependent promoter activation, as well as nuclear translocation of IRF3 (Martinez-Sobrido et al. 2007).

19.5 PATHOLOGY AND PATHOGENESIS OF BOLIVIAN HEMORRHAGIC FEVER

The pathology and pathogenesis of MACV infections is incompletely characterized. Only few studies have been performed, and the majority of those were done in the 1970s before the advent of sophisticated molecular tools. As such, these studies largely address only the gross pathology and histopathology of humans that succumbed to BHF or of various animals that were infected with MACV with the aim of creating BHF animal models that faithfully reproduce the human condition.

Only one study has thoroughly addressed the pathology of MACV infection in humans. In 1967, Child et al. described the autopsies of eight patients that died at Centro de Salud Hospital in San Joaquín, Bolivia, in 1963. The autopsy results were largely consistent among all eight cases, which were children of both sexes and male adults. Grossly, all organs were acutely congested, and generalized lymphadenopathy was a hallmark of the disease. Hemorrhages, the hallmark clinical sign of BHF, were detected infrequently in the form of dermal, renal, and hepatic petechiae but were widespread through the gastrointestinal walls and lumina of all patients and absent in the lungs. Some patients had focal ecchymoses in mesenteric lymph nodes and heart muscle tissue. Intracranial hemorrhages were found in four patients, both intracerebrally or subarachnoidally. On the histopathologic level, the reticuloendothelial system was characterized by slight lymphoid hyperplasia in spleens, lymph nodes, and GALT. Splenic and lymphadenic macrophages were sometimes found laden with debris, including remnants of erythrocytes. The bone marrow of all patients was slightly affected, with mild hyperplasia of all cells in three and mild hypoplasia of all elements, in two patients. Slight to moderate interstitial pneumonia was detected in all cases, sometimes with hyaline membrane formation. The kidneys were minimally affected, but collecting tubules contained clear hyaline casts associated with focal necrosis and regeneration of epithelial cells. Brains showed a characteristic widening of the Virchow–Robin spaces, and increased numbers of elongated microglia were typical. The most prominent and possibly most relevant pathologic changes were observed in hepatic tissues. Hepatic lobules were characterized by normal architecture with concomitant slight central congestion and increased eosinophilia of hepatocytes. Bile ducts appeared unaffected. However, while focal areas of hepatic necroses were found in only two cases, Kupffer cell numbers were clearly increased, and Kupffer cells were usually found to be increased in size due to erythrophagocytosis in every case. Mononuclear cells were typically also found in hepatic sinusoids that also contained unidentified acidophilic bodies 10–20 µm in diameter. These bodies were also found within Kupffer cells in several patients (Child et al. 1967).

The pathology of animal MAVC infection has been described for rhesus monkeys (*Macaca mulatta*), grivets (*Chlorocebus aethiops*), and suckling golden (Syrian) hamsters (*Mesocricetus auratus*).

Rhesus monkeys, subcutaneously injected with a brain suspension derived from suckling hamsters infected intracranially with MACV, uniformly die if this suspension contains equal to or more than 10^3 pfu (plaque-forming units) of the virus (and death occurs faster with higher doses). The gross pathology of these animals is defined by a petechial skin rash of the face, thorax, abdomen, and forelimbs; lymphadenopathy; splenomegaly; yellow hepatomegaly; meningeal edema; and hydropericardium. Hemorrhages (skin, myocardia, brain, and nose) are infrequently observed. Microscopically, these animals developed acute coagulation necroses of hepatocytes in the presence of spherical acidophilic hyaline bodies, necrotizing enteritis, epithelial necroses (skin and oral mucosa), and adrenal cortical necroses (zona fasciculata). Nonsuppurative meningoencephalitis and lymphoid necroses occur sometimes in animals with a protracted disease course. Myocardial degeneration, lymphoid and reticuloendothelial cell hyperplasia, or lymphoid depletion also develops sometimes. Overall, MACV infection of rhesus monkeys resembles BHF, with the notable exception of the occurrence of necroses in epithelia, gastrointestinal tract, and adrenal cortices (not observed in humans), hyaline tubular casts (not observed in monkeys), and lymphoid depletion or necroses (not recorded in humans) (Terrell et al. 1973a).

Grivets infected subcutaneously with a hamster brain suspension containing 10^3 pfu of MACV usually die. Gross lesions include yellow hepatic necrosis, necrotizing enteritis, acute suppurative bronchopneumonia, and hemorrhages (skin, lungs, intestine, liver, and lymph nodes). The microscopic pathology is characterized by necroses in liver, intestine, skin, oral cavities, and adrenal cortices. Acute thrombosis, possibly due to frequent bacterial superinfection, is a typical finding. Lymphoid depletion occurs roughly in every other grivet, and pancreatitis may be observed (McLeod et al. 1978). Together, these findings suggest that grivets are a less optimal animal model of BHF than rhesus monkeys.

Intracerebral infection of suckling golden hamsters with hamster-passaged MACV results in uniformly lethal infection that usually lasts 7–9 days. Importantly, there are no grossly visible lesions in deceased hamsters. Typical microscopic lesions include progressive splenic lymphoid depletion of the white pulp and centrilobular hepatic fatty metamorphosis, which begin around day 7 postinfection. Lymphoid depletion continues over the course of disease, and focal lymphoid necroses can be observed starting on day 9. The bone marrow is usually only affected by slight hypoplasia and depletion. Diffuse pulmonary congestion is the only pathological sign found in lungs. The CNS seems to be inconsistently involved (Terrell et al. 1973b). The lack of specific reagents for hamsters, and the fact that these animals only become sick after intracerebral infection, raises doubts about the suitability of these animals as BHF animal models.

The MACV literature also states that mature Geoffroy's tamarins (*Saguinus geoffroyi*), mature crab-eating macaques (*Macaca fascicularis*), mature strain 13 guinea pigs (*Cavia porcellus*), and suckling laboratory mice can be lethally infected with MACV, whereas adult common marmosets (*Callithrix jacchus*), hamsters, and laboratory mice only develop MACV-specific antibodies.

Descriptive pathologies have unfortunately not been published for either of the lethally infected animals (Johnson et al. 1965a; Eddy et al. 1975a; Webb et al. 1975).

A recently described alternative to suckling hamsters are adult signal transducer and activator of transcription 1 (STAT-1) knockout mice, which are defective in type I, II, and III interferon signaling. In these animals, MACV has an early tropism for the spleen and kidney (day 3 postinfection) and by day 5 can be found in all other organs. Viral titers remain high until the time of death (day 7). The most significant histopathologic findings are mild to moderate hepatic necroses; moderate to marked lymphoid depletion in lymph nodes, spleen, and thymus; and pancreatitis. Importantly, ribavirin, a drug partially effective against MACV in humans, protected these mice against MACV disease, thereby supporting the potential usefulness of these animals for drug efficacy studies (Bradfute et al. 2011).

19.6 TREATMENT AND VACCINES

Currently, there are no MACV-specific treatments approved for use against BHF. Treatment therefore consists primarily of supportive care.

Vaccines: Despite the biodefense concern and public health risks associated with BHF, FDA-licensed vaccines are not yet available. At present, vaccines for the prevention of human arenavirus disease are limited to a single, safe, efficacious, live-attenuated vaccine, called Candid 1, for the prevention of JUNV infection (Maiztegui et al. 1987, 1998; Barrera Oro and McKee 1991; Enria and Maiztegui 1994). JUNV Candid 1, which has been classified as an Investigational New Drug (IND) in the United States, was derived from the wild-type JUNV strain XJ13 through serial passage both in vivo and in vitro (Goni et al. 2006). Candid 1 has been evaluated in controlled trials among at-risk populations of agricultural workers in Argentina, where it showed a protective efficacy greater or equal to 84% and has led to a consistent reduction in AHF cases with no serious side effects (Maiztegui et al. 1998; Enria et al. 2008, 2010). The vaccine also cross protects experimental animals against MACV challenge (Barrera-Oro and Lupton 1988), which suggests that Candid 1 could be used during a BHF outbreak as an emergency measure to contain it.

19.6.1 PASSIVE ANTIBODY THERAPY

Passive antibody therapy might be useful for the treatment of BHF (Eddy et al. 1975b), but it has not been thoroughly evaluated in a clinical setting. Promisingly, the transfusion of immune convalescent plasma with defined doses of JUNV-neutralizing antibodies is the currently indicated therapeutic intervention against Junín/AHF. Immune serum treatment is effective in attenuating disease and reducing lethality to less than 1% if administered within the first 8 days of disease to provide an adequate dose of neutralizing antibodies (Maiztegui et al. 1979; Enria et al. 1984, 1985; Enria and Maiztegui 1994). However, about 10% of treated patients develop a transient cerebellar–cranial nerve syndrome 3–6 weeks later (Maiztegui et al. 1979; Enria et al. 1984, 1985; Enria and Maiztegui 1994). Even if a similar plasma therapy could be developed for BHF, maintaining adequate plasma stocks will be a challenge due to the limited number of BHF cases and the absence of a program for convalescent serum collection. The additional risk of transfusion-borne diseases emphasizes that alternative treatments ought to be developed (García et al. 2011).

19.6.2 ANTIVIRALS

Current anti-arenaviral therapy is limited to an off-label use of the nonimmunosuppressive guanosine analog, ribavirin (1-β-D-ribofuranosyl-1-*H*-1,2,4-triazole-3-carboxamide), an IMP dehydrogenase inhibitor. Recent studies suggest that the antiviral activity of ribavirin on arenaviruses might be exerted, at least partially, by lethal mutagenesis (Moreno et al. 2011). Unfortunately, ribavirin is only partially efficacious against some arenavirus infections and is associated with significant toxicity in humans (Petkevich et al. 1981; Kenyon et al. 1986; McCormick et al. 1986; Rodriguez et al. 1986; Weissenbacher et al. 1986; Huggins 1989; McKee et al. 1988; Enria and Maiztegui 1994; Barry et al. 1995; Kilgore et al. 1997; Snell 2001; Enria et al. 2008; Khan et al. 2008). Ribavirin can lead to adverse side effects such as thrombocytosis, severe anemia, and birth defects (Enria et al. 1987; McKee et al. 1988).

Recent independent small molecule high-throughput screens have identified additional promising antivirals. These efforts have to date identified six chemically distinct classes of small-molecule compounds that specifically inhibit GP-mediated membrane fusion with differing selectivities against New World and/or Old World arenaviruses (Bolken et al. 2006; Larson et al. 2008; Lee et al. 2008; York et al. 2008). One highly active and specific small-molecule inhibitor, ST-294, was found to inhibit MACV in vitro, as well as other New World pathogenic viruses (JUNV, GTOV, SABV) at concentrations in the nanomolar range. This molecule also demonstrated favorable pharmacodynamic properties (metabolically stable and orally bioavailable), as well as in vivo anti-arenaviral activity in a newborn mouse model (Bolken et al. 2006). Mechanism-of-action studies suggest that this compound targets GP2 and is a viral entry inhibitor (Bolken et al. 2006). Another inhibitor, ST-193, a benzimidazole derivative, was found to inhibit cell entry of various arenaviruses, including MACV, at least in vitro (Larson et al. 2008). Finally, two lead compounds, 16G8 and 17C8, are highly active against MACV and other South American hemorrhagic fever arenaviruses, as well as the more distantly related Lassa virus. These compounds act at the level of GP-mediated membrane fusion with an $IC_{50} \approx 200$–350 nm (Lee et al. 2008). Despite chemical differences, evidence suggests that these diverse inhibitors act through the pH-sensitive interface of the signal peptide and GP2 subunits in the GP spike complex and prevent virus entry by stabilizing the prefusion spike complex against pH-induced activation in the endosome (Bolken et al. 2006; Larson et al. 2008; York et al. 2008).

Other types of inhibitors that target viral RNA synthesis have also been reported. T-705, a pyrazine derivative with broad antiviral activity against RNA viruses, including orthomyxoviruses (Kiso et al. 2010), flaviviruses (Morrey et al. 2008; Julander et al. 2009), bunyaviruses, and several nonpathogenic arenaviruses (Gowen et al. 2007, 2008, 2010) was also found to be active in vitro against MACV, JUNV, and GTOV. T-705 most likely acts as a purine nucleoside analog specifically targeting L (Mendenhall et al. 2011). Studies employing the Pichindé virus hamster model of acute arenaviral disease demonstrated that T-705 could protect against viral infection with efficacy in late-stage infection (Gowen et al. 2008).

19.7   MAMMALIAN RESERVOIRS OF MACV

There is a remarkable rodent specificity seen among arenaviruses in nature. Field studies strongly support the concept of only a single major reservoir rodent host for each virus (Salazar-Bravo et al. 2002b). Non-reservoir rodents might at times develop chronic infection and viruria, such as is observed following MACV infection of golden hamsters (*Mesocricetus auratus*) (Johnson et al. 1965a). Rodents of the superfamily Muroidea are the natural hosts of arenaviruses (with the possible exception of Tacaribe virus, which might be transmitted by bats, and Chapare, Lujo, and Sabiá viruses, for which no reservoirs have yet been identified). The geographical distribution of each arenavirus is determined by the range of its corresponding rodent host. New World arenaviruses are found in rodents of the family Cricetidae, subfamily Sigmodontinae, in specialized ecological niches in South and North America (Clegg 2002; Salazar-Bravo et al. 2002b). Current evidence suggests a long-term "diffuse coevolution" between the arenaviruses and their rodent hosts. According to this model, a parallel phylogeny between the viruses and their corresponding rodent host(s) allows for host switches between rodents of closely related taxa (Hugot et al. 2001; Salazar-Bravo et al. 2002b). For example, the closely related MACV and JUNV (Clade B) are found in rodents of species that belong to the same genus, *Calomys*.

Arenaviruses establish chronic infections in their respective reservoirs accompanied by chronic viremia or viruria without clinical signs of disease (Johnson et al. 1965a; Sabattini et al. 1977; Fulhorst et al. 1999). The chronic carrier state in rodents usually results from exposure to infectious virus early in ontogeny (at or near birth) or later in life through aggressive or venereal behavior (Webb et al. 1975; Mills et al. 1992).

The big laucha (*Calomys callosus*) is the principal host for MACV. The virus was recovered repeatedly from this small pastoral and peridomestic mouse (Johnson et al. 1966). MACV induces a viremic immunotolerant infection in suckling lauchas and a split response in animals more than 9 days of age (Justines and Johnson 1969; Webb et al. 1975). The "immunocompetent" response of 50% of the mice is characterized by clearance of viremia, minimal or absent viruria, and the presence of circulating neutralizing antibodies. The other "immunotolerant" mice develop persistent viremia, viruria, little or no neutralizing antibodies, anemia, and splenomegaly. MACV antigen can be detected in most tissues of these animals, including the reproductive organs (Justines and Johnson 1969; Webb et al. 1975). Virus can be isolated from blood, spleen, and kidneys (Johnson et al. 1965a; Kuns 1965). Long-term effects of tolerant infection include mild runting, reduced survival rate, and almost total sterility among females, largely caused by virus infection of embryos. Selective breeding experiments, in laucha rodents, demonstrated that a complex polygenic inheritance accounts for the split response following MACV infection, suggesting a host genetic component as a determinant (Justines and Johnson 1969; Webb et al. 1975). In these experiments, the infection of newly born laucha rodents could occur neonatally through the milk of adult mice that were infected through sexual transmission of MACV, suggesting that horizontal transmission through venereal encounters might be an important natural mechanism for virus maintenance (Webb et al. 1975). These studies also predict a model for MACV maintenance in its reservoir: virus infection would be more common in larger wild colonies of lauchas where increasing venereal transmission occurs, and infected colonies would eventually pass through a phase of reduced population with near complete, tolerant infection as young infected females are rendered sterile. There are some differences in the maintenance mechanism of JUNV and MACV. First, horizontal venereal transmission does not seem to be predominant in JUNV transmission as it would not account for the greater prevalence of infection in male drylands lauchas (*Calomys musculinus*). Second, while viral infection is hypothesized to be an important driving force in reservoir population dynamics in the MACV-big laucha model, JUNV infection in drylands lauchas should have a much less severe effect. This is because in contrast to MACV-infected female mice, which abort (Webb et al. 1975), chronically JUNV-infected rodent females, when infected as adults, have a normal number of pups (Vitullo and Merani 1990). Finally, all pups born to MACV-viremic female mice in the laboratory are infected neonatally through the milk (Webb et al. 1975), whereas JUNV-infected female rodents transmit the virus to only half of their pups (Vitullo et al. 1987; Vitullo and Merani 1990).

BHF is an example of natural nidality: zoonoses associated with the hosts occur focally and have an incomplete pattern of overlap with the geographical host range. Big lauchas are found over a larger area than the currently defined endemic zone of BHF. Furthermore, the population of big lauchas responsible for the maintenance and transmission of MACV represents an independent monophyletic lineage, different from that in other areas of South America (Salazar-Bravo et al. 2002a). An independent evolutionary history of such a lineage could explain the phenomenon of natural nidality for BHF.

19.8 DIAGNOSTICS

The detection of a viral antigen and/or the viral genome is crucial for rapid diagnosis of patients with hemorrhagic fever caused by South American arenaviruses, especially for patients in the acute phase. In general, reverse transcriptase PCR (RT-PCR) is more sensitive in detecting viruses in patients' specimens than antigen-capture ELISA. The application of RT-PCR and TaqMan™ PCR for detection of MACV, JUNV, and GTOV genomes has been reported, but it has yet to be applied to clinical specimens. Therefore, the possibility that it does not detect novel arenavirus strains or arenaviruses cannot be ruled out (Vieth et al. 2005). Recently, TaqMan-based PCR assays for specific and absolute quantitative detection of multiple hemorrhagic fever viruses (filoviruses, arenaviruses, and bunyaviruses), including MACV, have been developed. The MACV target genes were GPC and L, and the limit of detection for the assays was 0.001 pfu/PCR. These assays are not only qualitative (presence of target) but also quantitative [measure a single DNA/RNA target sequence in an unknown sample and express the final results as an absolute value (e.g., viral load, pfus, or copies/ml)] on the basis of concentration of standard samples and can be used in viral load, vaccine, and antiviral drug studies. These assays also provide a repertoire of diagnostic tools that can serve as a foundation for the simultaneous identification and analysis of potential biothreat agents when multiple, rapid cycling, real-time PCR platforms are used (Trombley et al. 2010).

Recently, an antigen-capture ELISA has been developed using monoclonal antibodies against the JUNV NP. The antigens of all human pathogenic South American arenaviruses, including MACV, could be detected using monoclonal antibodies C11-12 or E-4-2. Therefore, the established antigen-capture ELISAs recognized highly conserved epitopes, suggesting that they may be applicable not only for the diagnosis of Junín/Argentinian hemorrhagic fever but also for the diagnosis of "VHF" and the other Southern American hemorrhagic fevers (Nakauchi et al. 2009).

DISCLAIMER

The content of this publication does not necessarily reflect the views or policies of the U.S. Department of Defense, the U.S. Department of the Army, and the U.S. Department of Health and Human Services or of the institutions and companies affiliated with the authors. JHK performed this work as an employee of Tunnell Consulting, Inc., a subcontractor to Battelle Memorial Institute, and FKM as an employee of Battelle Memorial Institute under its prime contract with NIAID, under Contract No. HHSN272200200016I.

REFERENCES

Abraham, J., J. A. Kwong, C. G. Albarino, J. J. G. Lu, S. R. Radoshitzky, J. Salazar-Bravo, M. Farzan, C. F. Spiropoulou, and H. Choe. 2009. Host-species transferrin receptor 1 orthologs are cellular receptors for nonpathogenic new world clade B Arenaviruses. *Plos Pathogens* 5 (4):e1000358.

Agnihothram, S. S., J. York, and J. H. Nunberg. 2006. Role of the stable signal peptide and cytoplasmic domain of G2 in regulating intracellular transport of the Junin virus envelope glycoprotein complex. *J Virol* 80 (11):5189–5198.

Aguilar, P. V., W. Camargo, J. Vargas, C. Guevara, Y. Roca, V. Felices, V. A. Laguna-Torres, R. Tesh, T. G. Ksiazek, and T. J. Kochel. 2009. Reemergence of Bolivian hemorrhagic fever, 2007–2008. *Emerg Infect Dis* 15 (9):1526–1528.

Albarino, C. G., D. M. Posik, P. D. Ghiringhelli, M. E. Lozano, and V. Romanowski. 1998. Arenavirus phylogeny: a new insight. *Virus Genes* 16 (1):39–46.

Arribalzaga, R. A. 1955. New epidemic disease due to unidentified germ: nephrotoxic, leukopenic and enanthematous hyperthermia. *Dia Med* 27 (40):1204–1210.

Auperin, D. D., R. W. Compans, and D. H. Bishop. 1982a. Nucleotide sequence conservation at the 3′ termini of the virion RNA species of New World and Old World arenaviruses. *Virology* 121 (1):200–203.

Auperin, D., K. Dimock, P. Cash, W. E. Rawls, W. C. Leung, and D. H. Bishop. 1982b. Analyses of the genomes of prototype pichinde arenavirus and a virulent derivative of Pichinde Munchique: evidence for sequence conservation at the 3′ termini of their viral RNA species. *Virology* 116 (1):363–367.

Auperin, D. D., M. Galinski, and D. H. Bishop. 1984. The sequences of the N protein gene and intergenic region of the S RNA of Pichinde arenavirus. *Virology* 134 (1):208–219.

Barrera-Oro, J. G. and H. W. Lupton. 1988. Cross-protection against Machupo virus with Candid # 1 live-attenuated Junin virus vaccine. I. The postvaccination prechallenge immune response. In *Second International Conference on the Impact of Viral Diseases on the Development of Latin American Countries and the Caribbean Region*. Buenos Aires, Argentina.

Barrera Oro, J. G. and K. T. McKee, Jr. 1991. Toward a vaccine against Argentine hemorrhagic fever. *Bull Pan Am Health Organ* 25 (2):118–126.

Barry, M., M. Russi, L. Armstrong, D. Geller, R. Tesh, L. Dembry, J. P. Gonzalez, A. S. Khan, and C. J. Peters. 1995. Brief report: treatment of a laboratory-acquired Sabia virus infection. *N Engl J Med* 333 (5):294–296.

Beyer, W. R., D. Popplau, W. Garten, D. von Laer, and O. Lenz. 2003. Endoproteolytic processing of the lymphocytic choriomeningitis virus glycoprotein by the subtilase SKI-1/S1P. *J Virol* 77 (5):2866–2872.

Bolken, T. C., S. Laquerre, Y. Zhang, T. R. Bailey, D. C. Pevear, S. S. Kickner, L. E. Sperzel, K. F. Jones, T. K. Warren, S. Amanda Lund, D. L. Kirkwood-Watts, D. S. King, A. C. Shurtleff, M. C. Guttieri, Y. Deng, M. Bleam, and D. E. Hruby. 2006. Identification and characterization of potent small molecule inhibitor of hemorrhagic fever New World arenaviruses. *Antiviral Res* 69 (2):86–97.

Bowen, M. D., C. J. Peters, and S. T. Nichol. 1996. The phylogeny of New World (Tacaribe complex) arenaviruses. *Virology* 219 (1):285–290.

Bowen, M. D., C. J. Peters, and S. T. Nichol. 1997. Phylogenetic analysis of the *Arenaviridae*: patterns of virus evolution and evidence for cospeciation between arenaviruses and their rodent hosts. *Mol Phylogenet Evol* 8 (3):301–316.

Bowen, M. D., P. E. Rollin, T. G. Ksiazek, H. L. Hustad, D. G. Bausch, A. H. Demby, M. D. Bajani, C. J. Peters, and S. T. Nichol. 2000. Genetic diversity among Lassa virus strains. *J Virol* 74 (15):6992–7004.

Bradfute, S. B., K. S. Stuthman, A. C. Shurtleff, and S. Bavari. 2011. A STAT-1 knockout mouse model for Machupo virus pathogenesis. *Virol J* 8:300.

Buchmeier, M. J., J. C. de La Torre, and C. J. Peters. 2006. *Arenaviridae*: the viruses and their replication. In *Fields Virology*, eds. D. M. Knipe and P. M. Howley. Philadelphia, PA: Lippincott Williams & Wilkins.

Buchmeier, M. J., P. J. Southern, B. S. Parekh, M. K. Wooddell, and M. B. Oldstone. 1987. Site-specific antibodies define a cleavage site conserved among arenavirus GP-C glycoproteins. *J Virol* 61 (4):982–985.

Cajimat, M. N. and C. F. Fulhorst. 2004. Phylogeny of the Venezuelan arenaviruses. *Virus Res* 102 (2):199–206.

Cajimat, M. N. B., M. L. Milazzo, R. D. Bradley, and C. F. Fulhorst. 2012. Ocozocoautla de Espinosa virus and hemorrhagic fever, Mexico. *Emerg Infect Dis* 8 (3):401–405.

Capul, A. A., M. Perez, E. Burke, S. Kunz, M. J. Buchmeier, and J. C. de la Torre. 2007. Arenavirus Z-glycoprotein association requires Z myristoylation but not functional RING or late domains. *J Virol* 81 (17):9451–9460.

Castilla, V., S. E. Mersich, N. A. Candurra, and E. B. Damonte. 1994. The entry of Junin virus into Vero cells. *Arch Virol* 136 (3–4):363–374.

Charrel, R. N. and X. de Lamballerie. 2003. Arenaviruses other than Lassa virus. *Antiviral Res* 57 (1–2):89–100.

Charrel, R. N., H. Feldmann, C. F. Fulhorst, R. Khelifa, R. de Chesse, and X. de Lamballerie. 2002. Phylogeny of New World arenaviruses based on the complete coding sequences of the small genomic segment identified an evolutionary lineage produced by intrasegmental recombination. *Biochem Biophys Res Commun* 296 (5):1118–1124.

Charrel, R. N., J. J. Lemasson, M. Garbutt, R. Khelifa, P. De Micco, H. Feldmann, and X. de Lamballerie. 2003. New insights into the evolutionary relationships between arenaviruses provided by comparative analysis of small and large segment sequences. *Virology* 317 (2):191–196.

Child, P. L., R. B. MacKenzie, L. R. Valverde, and K. M. Johnson. 1967. Bolivian hemorrhagic fever. A pathologic description. *Arch Pathol* 83 (5):434–445.

Clegg, J. C. 2002. Molecular phylogeny of the arenaviruses. *Curr Top Microbiol Immunol* 262:1–24.

Cornu, T. I. and J. C. de la Torre. 2001. RING finger Z protein of lymphocytic choriomeningitis virus (LCMV) inhibits transcription and RNA replication of an LCMV S-segment minigenome. *J Virol* 75 (19):9415–9426.

Cornu, T. I. and J. C. de la Torre. 2002. Characterization of the arenavirus RING finger Z protein regions required for Z-mediated inhibition of viral RNA synthesis. *J Virol* 76 (13):6678–6688.

Cornu, T. I., H. Feldmann, and J. C. de la Torre. 2004. Cells expressing the RING finger Z protein are resistant to arenavirus infection. *J Virol* 78 (6):2979–2983.

Daelli, M. G. and C. E. Coto. 1982. Inhibition of the production of infectious particles in cells infected with Junin virus in the presence of tunicamycin. *Rev Argent Microbiol* 14 (3):171–176.

de Manzione, N., R. A. Salas, H. Paredes, O. Godoy, L. Rojas, F. Araoz, C. F. Fulhorst, T. G. Ksiazek, J. N. Mills, B. A. Ellis, C. J. Peters, and R. B. Tesh. 1998. Venezuelan hemorrhagic fever: clinical and epidemiological studies of 165 cases. *Clin Infect Dis* 26 (2):308–313.

Delgado, S., B. R. Erickson, R. Agudo, P. J. Blair, E. Vallejo, C. G. Albarino, J. Vargas, A. J. Comer, P. E. Rollin, T. G. Ksiazek, J. G. Olson, and S. T. Nichol. 2008. Chapare virus, a newly discovered arenavirus isolated from a fatal hemorrhagic fever case in Bolivia. *PLoS Path* 18(4):e1000047.

Di Simone, C. and M. J. Buchmeier. 1995. Kinetics and pH dependence of acid-induced structural changes in the lymphocytic choriomeningitis virus glycoprotein complex. *Virology* 209 (1):3–9.

Di Simone, C., M. A. Zandonatti, and M. J. Buchmeier. 1994. Acidic pH triggers LCMV membrane fusion activity and conformational change in the glycoprotein spike. *Virology* 198 (2):455–465.

Eddy, G. A., S. K. Scott, F. S. Wagner, and O. M. Brand. 1975a. Pathogenesis of Machupo virus infection in primates. *Bull World Health Organ* 52 (4–6):517–521.

Eddy, G. A., F. S. Wagner, S. K. Scott, and B. J. Mahlandt. 1975b. Protection of monkeys against Machupo virus by the passive administration of Bolivian haemorrhagic fever immunoglobulin (human origin). *Bull World Health Organ* 52 (4–6):723–727.

Eichler, R., O. Lenz, W. Garten, and T. Strecker. 2006. The role of single N-glycans in proteolytic processing and cell surface transport of the Lassa virus glycoprotein GP-C. *Virol J* 3:41.

Eichler, R., O. Lenz, T. Strecker, M. Eickmann, H. D. Klenk, and W. Garten. 2003a. Identification of Lassa virus glycoprotein signal peptide as a trans-acting maturation factor. *EMBO Rep* 4 (11):1084–1088.

Eichler, R., O. Lenz, T. Strecker, and W. Garten. 2003b. Signal peptide of Lassa virus glycoprotein GP-C exhibits an unusual length. *FEBS Lett* 538 (1–3):203–206.

Eichler, R., T. Strecker, L. Kolesnikova, J. ter Meulen, W. Weissenhorn, S. Becker, H. D. Klenk, W. Garten, and O. Lenz. 2004. Characterization of the Lassa virus matrix protein Z: electron microscopic study of virus-like particles and interaction with the nucleoprotein (NP). *Virus Res* 100 (2):249–255.

Enria, D. A., A. M. Ambrosio, A. M. Briggiler, M. R. Feuillade, and E. Crivelli. 2010. Candid#1 vaccine against Argentine hemorrhagic fever produced in Argentina. Immunogenicity and safety. *Medicina (B Aires)* 70 (3):215–222.

Enria, D. A., A. M. Briggiler, N. J. Fernandez, S. C. Levis, and J. I. Maiztegui. 1984. Importance of dose of neutralising antibodies in treatment of Argentine haemorrhagic fever with immune plasma. *Lancet* 2 (8397):255–256.

Enria, D. A., A. M. Briggiler, S. Levis, D. Vallejos, J. I. Maiztegui, and P. G. Canonico. 1987. Tolerance and antiviral effect of ribavirin in patients with Argentine hemorrhagic fever. *Antiviral Res* 7 (6):353–359.

Enria, D. A., A. M. Briggiler, and Z. Sanchez. 2008. Treatment of Argentine hemorrhagic fever. *Antiviral Res* 78 (1):132–139.

Enria, D. A., A. J. de Damilano, A. M. Briggiler, A. M. Ambrosio, N. J. Fernandez, M. R. Feuillade, and J. I. Maiztegui. 1985. Late neurologic syndrome in patients with Argentinian hemorrhagic fever treated with immune plasma. *Medicina (B Aires)* 45 (6):615–620.

Enria, D. A. and J. I. Maiztegui. 1994. Antiviral treatment of Argentine hemorrhagic fever. *Antiviral Res* 23 (1):23–31.

Eschli, B., K. Quirin, A. Wepf, J. Weber, R. Zinkernagel, and H. Hengartner. 2006. Identification of an N-terminal trimeric coiled-coil core within arenavirus glycoprotein 2 permits assignment to class I viral fusion proteins. *J Virol* 80 (12):5897–5907.

Fan, L., T. Briese, and W. I. Lipkin. 2010. Z proteins of New World arenaviruses bind RIG-I and interfere with type I interferon induction. *J Virol* 84 (4):1785–1791.

Froeschke, M., M. Basler, M. Groettrup, and B. Dobberstein. 2003. Long-lived signal peptide of lymphocytic choriomeningitis virus glycoprotein pGP-C. *J Biol Chem* 278 (43):41914–41920.

Fulhorst, C. F., T. G. Ksiazek, C. J. Peters, and R. B. Tesh. 1999. Experimental infection of the cane mouse *Zygodontomys brevicauda* (family Muridae) with Guanarito virus (*Arenaviridae*), the etiologic agent of Venezuelan hemorrhagic fever. *J Infect Dis* 180 (4):966–969.

Gangemi, J. D., R. R. Rosato, E. V. Connell, E. M. Johnson, and G. A. Eddy. 1978. Structural polypeptides of Machupo virus. *J Gen Virol* 41 (1):183–188.

García, C. C., C. S. Sepúlveda, and E. B. Damonte. 2011. Novel therapeutic targets for arenavirus hemorrhagic fevers. *Future Virology* 6 (1):27–44.

Garcin, D. and D. Kolakofsky. 1990. A novel mechanism for the initiation of Tacaribe arenavirus genome replication. *J Virol* 64 (12):6196–6203.

Garcin, D. and D. Kolakofsky. 1992. Tacaribe arenavirus RNA synthesis in vitro is primer dependent and suggests an unusual model for the initiation of genome replication. *J Virol* 66 (3):1370–1376.

Glushakova, S. E., A. I. Iakuba, A. D. Vasiuchkov, R. F. Mar'iankova, T. M. Kukareko, T. A. Stel'makh, T. P. Kurash, and I. S. Lukashevich. 1990a. Lysosomotropic agents inhibit the penetration of arenaviruses into a culture of BHK-21 and Vero cells. *Vopr Virusol* 35 (2):146–150.

Glushakova, S. E. and I. S. Lukashevich. 1989. Early events in arenavirus replication are sensitive to lysosomotropic compounds. *Arch Virol* 104 (1–2):157–161.

Glushakova, S. E., I. S. Lukashevich, and L. A. Baratova. 1990b. Prediction of arenavirus fusion peptides on the basis of computer analysis of envelope protein sequences. *FEBS Lett* 269 (1):145–147.

Glushakova, S. E., V. G. Omelyanenko, I. S. Lukashevitch, A. A. Bogdanov, Jr., A. B. Moshnikova, A. T. Kozytch, and V. P. Torchilin. 1992. The fusion of artificial lipid membranes induced by the synthetic arenavirus 'fusion peptide'. *Biochim Biophys Acta* 1110 (2):202–208.

Goni, S. E., J. A. Iserte, A. M. Ambrosio, V. Romanowski, P. D. Ghiringhelli, and M. E. Lozano. 2006. Genomic features of attenuated Junin virus vaccine strain candidate. *Virus Genes* 32 (1):37–41.

Gonzalez, J. P., A. Sanchez, and R. Rico-Hesse. 1995. Molecular phylogeny of Guanarito virus, an emerging arenavirus affecting humans. *Am J Trop Med Hyg* 53 (1):1–6.

Gowen, B. B., D. F. Smee, M. H. Wong, J. O. Hall, K. H. Jung, K. W. Bailey, J. R. Stevens, Y. Furuta, and J. D. Morrey. 2008. Treatment of late stage disease in a model of arenaviral hemorrhagic fever: T-705 efficacy and reduced toxicity suggests an alternative to ribavirin. *PLoS One* 3 (11):e3725.

Gowen, B. B., M. H. Wong, K. H. Jung, A. B. Sanders, M. Mendenhall, K. W. Bailey, Y. Furuta, and R. W. Sidwell. 2007. In vitro and in vivo activities of T-705 against arenavirus and bunyavirus infections. *Antimicrob Agents Chemother* 51 (9):3168–3176.

Gowen, B. B., M. H. Wong, K. H. Jung, D. F. Smee, J. D. Morrey, and Y. Furuta. 2010. Efficacy of favipiravir (T-705) and T-1106 pyrazine derivatives in phlebovirus disease models. *Antiviral Res* 86 (2):121–127.

Harnish, D. G., S. J. Polyak, and W. E. Rawls. 1993. Arenavirus replication: molecular dissection of the role of protein and RNA. In *The Arenaviridae*, ed M. S. Salvato. New York: Plenum Press.

Harrison, L. H., N. A. Halsey, K. T. McKee, Jr., C. J. Peters, J. G. Barrera Oro, A. M. Briggiler, M. R. Feuillade, and J. I. Maiztegui. 1999. Clinical case definitions for Argentine hemorrhagic fever. *Clin Infect Dis* 28 (5):1091–1094.

Hass, M., U. Golnitz, S. Muller, B. Becker-Ziaja, and S. Gunther. 2004. Replicon system for Lassa virus. *J Virol* 78 (24):13793–13803.

Hass, M., M. Westerkofsky, S. Muller, B. Becker-Ziaja, C. Busch, and S. Gunther. 2006. Mutational analysis of the Lassa virus promoter. *J Virol* 80 (24):12414–12419.

Helguera, G., S. Jemielity, J. Abraham, S. M. Cordo, M. G. Martinez, J. A. Rodriguez, C. Bregni, J. J. Wang, M. Farzan, M. L. Penichet, N. A. Candurra, and H. Choe. 2012. An antibody recognizing the apical domain of human transferrin receptor 1 efficiently inhibits the entry of all New World hemorrhagic fever arenaviruses. *J Virol* 86 (7):4024–4028.

Huggins, J. W. 1989. Prospects for treatment of viral hemorrhagic fevers with ribavirin, a broad-spectrum antiviral drug. *Rev Infect Dis* 11 (Suppl 4):S750–S761.

Hugot, J. P., J. P. Gonzalez, and C. Denys. 2001. Evolution of the Old World *Arenaviridae* and their rodent hosts: generalized host-transfer or association by descent? *Infect Genet Evol* 1 (1):13–20.

Jacamo, R., N. Lopez, M. Wilda, and M. T. Franze-Fernandez. 2003. Tacaribe virus Z protein interacts with the L polymerase protein to inhibit viral RNA synthesis. *J Virol* 77 (19):10383–10393.

Jay, M. T., C. Glaser, and C. F. Fulhorst. 2005. The arenaviruses. *J Am Vet Med Assoc* 227 (6):904–915.

Johnson, K. M., M. L. Kuns, R. B. Mackenzie, P. A. Webb, and C. E. Yunker. 1966. Isolation of Machupo virus from wild rodent *Calomys callosus*. *Am J Trop Med Hyg* 15 (1):103–106.

Johnson, K. M., R. B. Mackenzie, P. A. Webb, and M. L. Kuns. 1965a. Chronic infection of rodents by Machupo virus. *Science* 150 (703):1618–1619.

Johnson, K. M., N. H. Wiebenga, R. B. Mackenzie, M. L. Kuns, N. M. Tauraso, A. Shelokov, P. A. Webb, G. Justines, and H. K. Beye. 1965b. Virus isolations from human cases of hemorrhagic fever in Bolivia. *Proc Soc Exp Biol Med* 118:113–118.

Julander, J. G., K. Shafer, D. F. Smee, J. D. Morrey, and Y. Furuta. 2009. Activity of T-705 in a hamster model of yellow fever virus infection in comparison with that of a chemically related compound, T-1106. *Antimicrob Agents Chemother* 53 (1):202–209.

Justines, G. and K. M. Johnson. 1969. Immune tolerance in *Calomys callosus* infected with Machupo virus. *Nature* 222 (5198):1090–1091.

Kenyon, R. H., P. G. Canonico, D. E. Green, and C. J. Peters. 1986. Effect of ribavirin and tributylribavirin on Argentine hemorrhagic fever (Junin virus) in guinea pigs. *Antimicrob Agents Chemother* 29 (3):521–523.

Khan, S. H., A. Goba, M. Chu, C. Roth, T. Healing, A. Marx, J. Fair, M. C. Guttieri, P. Ferro, T. Imes, C. Monagin, R. F. Garry, D. G. Bausch, and Network Mano River Union Lassa Fever. 2008. New opportunities for field research on the pathogenesis and treatment of Lassa fever. *Antiviral Res* 78 (1):103–115.

Kilgore, P. E., T. G. Ksiazek, P. E. Rollin, J. N. Mills, M. R. Villagra, M. J. Montenegro, M. A. Costales, L. C. Paredes, and C. J. Peters. 1997. Treatment of Bolivian hemorrhagic fever with intravenous ribavirin. *Clin Infect Dis* 24 (4):718–722.

Kiso, M., K. Takahashi, Y. Sakai-Tagawa, K. Shinya, S. Sakabe, Q. M. Le, M. Ozawa, Y. Furuta, and Y. Kawaoka. 2010. T-705 (favipiravir) activity against lethal H5N1 influenza A viruses. *Proc Natl Acad Sci USA* 107 (2):882–887.

Kranzusch, P. J., A. D. Schenk, A. A. Rahmeh, S. R. Radoshitzky, S. Bavari, T. Walz, and S. P. Whelan. 2011. Assembly of a functional Machupo virus polymerase complex. *Proc Natl Acad Sci USA* 107 (46):20069–20074.

Kranzusch, P. J. and S. P. Whelan. 2011. Arenavirus Z protein controls viral RNA synthesis by locking a polymerase-promoter complex. *Proc Natl Acad Sci USA* 108 (49):19743–19748.

Kuns, M. L. 1965. Epidemiology of Machupo virus infection. II. Ecological and control studies of hemorrhagic fever. *Am J Trop Med Hyg* 14 (5):813–816.

Kunz, S., P. Borrow, and M. B. Oldstone. 2002. Receptor structure, binding, and cell entry of arenaviruses. *Curr Top Microbiol Immunol* 262:111–137.

Kunz, S., K. H. Edelmann, J. C. de la Torre, R. Gorney, and M. B. Oldstone. 2003. Mechanisms for lymphocytic choriomeningitis virus glycoprotein cleavage, transport, and incorporation into virions. *Virology* 314 (1):168–178.

Larson, R. A., D. Dai, V. T. Hosack, Y. Tan, T. C. Bolken, D. E. Hruby, and S. M. Amberg. 2008. Identification of a broad-spectrum arenavirus entry inhibitor. *J Virol* 82 (21):10768–10775.

Lecompte, E., J. Ter Meulen, S. Emonet, S. Daffis, and R. N. Charrel. 2007. Genetic identification of Kodoko virus, a novel arenavirus of the African pigmy mouse (*Mus Nannomys minutoides*) in West Africa. *Virology* 364 (1):178–183.

Lee, A. M., J. M. Rojek, C. F. Spiropoulou, A. T. Gundersen, W. Jin, A. Shaginian, J. York, J. H. Nunberg, D. L. Boger, M. B. Oldstone, and S. Kunz. 2008. Unique small molecule entry inhibitors of hemorrhagic fever arenaviruses. *J Biol Chem* 283 (27):18734–18742.

Lee, K. J., I. S. Novella, M. N. Teng, M. B. Oldstone, and J. C. de La Torre. 2000. NP and L proteins of lymphocytic choriomeningitis virus (LCMV) are sufficient for efficient transcription and replication of LCMV genomic RNA analogs. *J Virol* 74 (8):3470–3477.

Lenz, O., J. ter Meulen, H. D. Klenk, N. G. Seidah, and W. Garten. 2001. The Lassa virus glycoprotein precursor GP-C is proteolytically processed by subtilase SKI-1/S1P. *Proc Natl Acad Sci USA* 98 (22):12701–12705.

Leung, W. C., H. P. Ghosh, and W. E. Rawls. 1977. Strandedness of Pichinde virus RNA. *J Virol* 22 (1):235–237.

Linero, F. N. and L. A. Scolaro. 2009. Participation of the phosphatidylinositol 3-kinase/Akt pathway in Junin virus replication in vitro. *Virus Res* 145 (1):166–170.

Lopez, N., R. Jacamo, and M. T. Franze-Fernandez. 2001. Transcription and RNA replication of Tacaribe virus genome and antigenome analogs require N and L proteins: Z protein is an inhibitor of these processes. *J Virol* 75 (24):12241–12251.

Mackenzie, R. B., H. K. Beye, L. Valverde, and H. Garron. 1964. Epidemic hemorrhagic fever in Bolivia. I. a preliminary report of the epidemiologic and clinical findings in a new epidemic area in South America. *Am J Trop Med Hyg* 13:620–625.

Maiztegui, J., F. Feinsod, A. Briggiler, C. J. Peters, D. Enria, H. Lupton, A. Ambrosio, C. Macdonald, E. Tiano, M. R. Feuillade, G. Gamboa, O. Conti, D. Vallejos, and J. B. Oro. 1987. Inoculation of the 1st Argentinean volunteers with attenuated candid-1 strain Junin virus. *Medicina-Buenos Aires* 47 (6):565–565.

Maiztegui, J. I., N. J. Fernandez, and A. J. de Damilano. 1979. Efficacy of immune plasma in treatment of Argentine haemorrhagic fever and association between treatment and a late neurological syndrome. *Lancet* 2 (8154):1216–1217.

Maiztegui, J. I., K. T. McKee, Jr., J. G. Barrera Oro, L. H. Harrison, P. H. Gibbs, M. R. Feuillade, D. A. Enria, A. M. Briggiler, S. C. Levis, A. M. Ambrosio, N. A. Halsey, and C. J. Peters. 1998. Protective efficacy of a live attenuated vaccine against Argentine hemorrhagic fever. AHF Study Group. *J Infect Dis* 177 (2):277–283.

Martinez, M. G., S. M. Cordo, and N. A. Candurra. 2007. Characterization of Junin arenavirus cell entry. *J Gen Virol* 88 (Pt 6):1776–1784.

Martinez, M. G., S. M. Cordo, and N. A. Candurra. 2008. Involvement of cytoskeleton in Junin virus entry. *Virus Res* 138 (1–2):17–25.

Martinez, M. G., M. B. Forlenza, and N. A. Candurra. 2009. Involvement of cellular proteins in Junin arenavirus entry. *Biotechnol J* 4 (6):866–870.

Martinez-Sobrido, L., P. Giannakas, B. Cubitt, A. Garcia-Sastre, and J. C. de la Torre. 2007. Differential inhibition of type I interferon induction by arenavirus nucleoproteins. *J Virol* 81 (22):12696–12703.

McCormick, J. B., I. J. King, P. A. Webb, C. L. Scribner, R. B. Craven, K. M. Johnson, L. H. Elliott, and R. Belmont-Williams. 1986. Lassa fever. Effective therapy with ribavirin. *N Engl J Med* 314 (1):20–26.

McKee, K. T., Jr., J. W. Huggins, C. J. Trahan, and B. G. Mahlandt. 1988. Ribavirin prophylaxis and therapy for experimental Argentine hemorrhagic fever. *Antimicrob Agents Chemother* 32 (9):1304–1309.

McLeod, C. G., Jr., J. L. Stookey, J. D. White, G. A. Eddy, and G. A. Fry. 1978. Pathology of Bolivian Hemorrhagic fever in the African green monkey. *Am J Trop Med Hyg* 27 (4):822–826.

Mendenhall, M., A. Russell, T. Juelich, E. L. Messina, D. F. Smee, A. N. Freiberg, M. R. Holbrook, Y. Furuta, J. C. de la Torre, J. H. Nunberg, and B. B. Gowen. 2011. T-705 (favipiravir) inhibition of arenavirus replication in cell culture. *Antimicrob Agents Chemother* 55 (2):782–787.

Meyer, B. J., J. C. de la Torre, and P. J. Southern. 2002. Arenaviruses: genomic RNAs, transcription, and replication. *Curr Top Microbiol Immunol* 262:139–157.

Meyer, B. J. and P. J. Southern. 1993. Concurrent sequence analysis of 5' and 3' RNA termini by intramolecular circularization reveals 5' nontemplated bases and 3' terminal heterogeneity for lymphocytic choriomeningitis virus mRNAs. *J Virol* 67 (5):2621–2627.

Meyer, B. J. and P. J. Southern. 1994. Sequence heterogeneity in the termini of lymphocytic choriomeningitis virus genomic and antigenomic RNAs. *J Virol* 68 (11):7659–7664.

Mills, J. N., B. A. Ellis, K. T. McKee, Jr., G. E. Calderon, J. I. Maiztegui, G. O. Nelson, T. G. Ksiazek, C. J. Peters, and J. E. Childs. 1992. A longitudinal study of Junin virus activity in the rodent reservoir of Argentine hemorrhagic fever. *Am J Trop Med Hyg* 47 (6):749–763.

Molteni, H. D., H. C. Guarinos, C. O. Petrillo, and F. Jaschek. 1961. Clinico-statistical study of 338 patients with epidemic hemorrhagic fever in the northwest of the province of Buenos Aires. *Sem Med* 118:839–855.

Moreno, H., I. Gallego, N. Sevilla, J. C. de la Torre, E. Domingo, and V. Martin. 2011. Ribavirin can be mutagenic for arenaviruses. *J Virol* 85 (14):7246–7255.

Morrey, J. D., B. S. Taro, V. Siddharthan, H. Wang, D. F. Smee, A. J. Christensen, and Y. Furuta. 2008. Efficacy of orally administered T-705 pyrazine analog on lethal West Nile virus infection in rodents. *Antiviral Res* 80 (3):377–379.

Murphy, F. A., P. A. Webb, K. M. Johnson, and S. G. Whitfield. 1969. Morphological comparison of Machupo with lymphocytic choriomeningitis virus: basis for a new taxonomic group. *J Virol* 4 (4):535–541.

Murphy, F. A., P. A. Webb, K. M. Johnson, S. G. Whitfield, and W. A. Chappell. 1970. Arenoviruses in Vero cells: ultrastructural studies. *J Virol* 6 (4):507–518.

Nakauchi, M., S. Fukushi, M. Saijo, T. Mizutani, A. E. Ure, V. Romanowski, I. Kurane, and S. Morikawa. 2009. Characterization of monoclonal antibodies to Junin virus nucleocapsid protein and application to the diagnosis of hemorrhagic fever caused by South American arenaviruses. *Clin Vaccine Immunol* 16 (8):1132–1138.

Oldenburg, J., T. Reignier, M. L. Flanagan, G. A. Hamilton, and P. M. Cannon. 2007. Differences in tropism and pH dependence for glycoproteins from the Clade B1 arenaviruses: implications for receptor usage and pathogenicity. *Virology* 364 (1):132–139.

Oldstone, M. B. 2002. Arenaviruses. I. The epidemiology molecular and cell biology of arenaviruses. Introduction. *Curr Top Microbiol Immunol* 262:V–XII.

Padula, P. J. and Z. M. de Martinez Segovia. 1984. Replication of Junin virus in the presence of tunicamycin. *Intervirology* 22 (4):227–231.

Perez, M., R. C. Craven, and J. C. de la Torre. 2003. The small RING finger protein Z drives arenavirus budding: implications for antiviral strategies. *Proc Natl Acad Sci USA* 100 (22):12978–12983.

Perez, M. and J. C. de la Torre. 2003. Characterization of the genomic promoter of the prototypic arenavirus lymphocytic choriomeningitis virus. *J Virol* 77 (2):1184–1194.

Perez, M., D. L. Greenwald, and J. C. de la Torre. 2004. Myristoylation of the RING finger Z protein is essential for arenavirus budding. *J Virol* 78 (20):11443–11448.

Petkevich, A. S., V. M. Sabynin, I. S. Lukashevich, G. A. Galegov, and V. I. Votiakov. 1981. Effect of riboivirin (virazole) on arenavirus reproduction in cell cultures. *Vopr Virusol* (2):244–245.

Pinschewer, D. D., M. Perez, and J. C. de la Torre. 2005. Dual role of the lymphocytic choriomeningitis virus intergenic region in transcription termination and virus propagation. *J Virol* 79 (7):4519–4526.

Radoshitzky, S. R., J. Abraham, C. F. Spiropoulou, J. H. Kuhn, D. Nguyen, W. Li, J. Nagel, P. J. Schmidt, J. H. Nunberg, N. C. Andrews, M. Farzan, and H. Choe. 2007. Transferrin receptor 1 is a cellular receptor for New World haemorrhagic fever arenaviruses. *Nature* 446 (7131):92–96.

Radoshitzky, S. R., J. H. Kuhn, C. F. Spiropoulou, C. G. Albarino, D. P. Nguyen, J. Salazar-Bravo, T. Dorfman, A. S. Lee, E. Wang, S. R. Ross, H. Choe, and M. Farzan. 2008. Receptor determinants of zoonotic transmission of New World hemorrhagic fever arenaviruses. *Proc Natl Acad Sci USA* 105 (7):2664–2669.

Raju, R., L. Raju, D. Hacker, D. Garcin, R. Compans, and D. Kolakofsky. 1990. Nontemplated bases at the 5′ ends of Tacaribe virus mRNAs. *Virology* 174 (1):53–59.

Rodriguez, M., J. B. McCormick, and M. C. Weissenbacher. 1986. Antiviral effect of ribavirin on Junin virus replication in vitro. *Rev Argent Microbiol* 18 (2):69–74.

Rojek, J. M., A. M. Lee, N. Nguyen, C. F. Spiropoulou, and S. Kunz. 2008. Site 1 protease is required for proteolytic processing of the glycoproteins of the South American hemorrhagic fever viruses Junin, Machupo, and Guanarito. *J Virol* 82 (12):6045–6051.

Rowe, W. P., W. E. Pugh, P. A. Webb, and C. J. Peters. 1970. Serological relationship of the Tacaribe complex of viruses to lymphocytic choriomeningitis virus. *J Virol* 5 (3):289–292.

Rugiero, H. R., H. Ruggiero, C. Gonzalezcambaceres, F. A. Cintora, F. Maglio, C. Magnoni, L. Astarloa, G. Squassi, A. Giacosa, and D. Fernandez. 1964. Argentine hemorrhagic fever. II. Descriptive clinical study. *Rev Asoc Med Argent* 78:281–294.

Sabattini, M. S., L. Gonzles del Rio, G. Diaz, and V. R. Vega. 1977. Infeccion natural y experimental de roedores con virus Junin. *Medicina (B Aires)* 37 (Suppl 3):149–161.

Salazar-Bravo, J., J. W. Dragoo, M. D. Bowen, C. J. Peters, T. G. Ksiazek, and T. L. Yates. 2002a. Natural nidality in Bolivian hemorrhagic fever and the systematics of the reservoir species. *Infect Genet Evol* 1 (3):191–199.

Salazar-Bravo, J., L. A. Ruedas, and T. L. Yates. 2002b. Mammalian reservoirs of arenaviruses. *Curr Top Microbiol Immunol* 262:25–63.

Salvato, M. S. 1993. Molecular biology of the prototype arenavirus, lymphocytic choriomeningitis virus. In *The Arenaviridae*, ed. M. S. Salvato. New York: Plenum Press.

Salvato, M. S., J. C. S. Clegg, M. J. Buchmeier, R. N. Charrel, J. P. Gonzalez, I. S. Lukashevich, C. J. Peters, and V. Romanowski. 2011. Family *Arenaviridae*. In *Virus Taxonomy—Ninth Report of the International Committee on Taxonomy of Viruses*, eds. A. M. Q. King, M. J. Adams, E. B. Carstens and E. J. Lefkowitz. London, U.K.: Elsevier/Academic Press.

Salvato, M. S., K. J. Schweighofer, J. Burns, and E. M. Shimomaye. 1992. Biochemical and immunological evidence that the 11 kDa zinc-binding protein of lymphocytic choriomeningitis virus is a structural component of the virus. *Virus Res* 22 (3):185–198.

Salvato, M., E. Shimomaye, and M. B. Oldstone. 1989. The primary structure of the lymphocytic choriomeningitis virus L gene encodes a putative RNA polymerase. *Virology* 169 (2):377–384.

Singh, M. K., F. V. Fuller-Pace, M. J. Buchmeier, and P. J. Southern. 1987. Analysis of the genomic L RNA segment from lymphocytic choriomeningitis virus. *Virology* 161 (2):448–456.

Snell, N. J. 2001. Ribavirin—Current status of a broad spectrum antiviral agent. *Expert Opin Pharmacother* 2 (8):1317–1324.

Southern, P. J., M. K. Singh, Y. Riviere, D. R. Jacoby, M. J. Buchmeier, and M. B. Oldstone. 1987. Molecular characterization of the genomic S RNA segment from lymphocytic choriomeningitis virus. *Virology* 157 (1):145–155.

Stinebaugh, B. J., F. X. Schloeder, K. M. Johnson, R. B. Mackenzie, G. Entwisle, and E. De Alba. 1966. Bolivian hemorrhagic fever. A report of four cases. *Am J Med* 40 (2):217–230.

Strecker, T., R. Eichler, J. Meulen, W. Weissenhorn, H. Dieter Klenk, W. Garten, and O. Lenz. 2003. Lassa virus Z protein is a matrix protein and sufficient for the release of virus-like particles [corrected]. *J Virol* 77 (19):10700–10705.

Strecker, T., A. Maisa, S. Daffis, R. Eichler, O. Lenz, and W. Garten. 2006. The role of myristoylation in the membrane association of the Lassa virus matrix protein Z. *Virol J* 3:93.

Terrell, T. G., J. L. Stookey, G. A. Eddy, and M. D. Kastello. 1973a. Pathology of Bolivian hemorrhagic fever in the rhesus monkey. *Am J Pathol* 73 (2):477–494.

Terrell, T. G., J. L. Stookey, R. O. Spertzel, and R. W. Kuehne. 1973b. Comparative histopathology of two strains of Bolivian hemorrhagic fever virus infections in suckling hamsters. *Am J Trop Med Hyg* 22 (6):814–818.

Tortorici, M. A., C. G. Albarino, D. M. Posik, P. D. Ghiringhelli, M. E. Lozano, R. Rivera Pomar, and V. Romanowski. 2001. Arenavirus nucleocapsid protein displays a transcriptional antitermination activity in vivo. *Virus Res* 73 (1):41–55.

Trombley, A. R., L. Wachter, J. Garrison, V. A. Buckley-Beason, J. Jahrling, L. E. Hensley, R. J. Schoepp, D. A. Norwood, A. Goba, J. N. Fair, and D. A. Kulesh. 2010. Comprehensive panel of real-time TaqMan polymerase chain reaction assays for detection and absolute quantification of filoviruses, arenaviruses, and New World hantaviruses. *Am J Trop Med Hyg* 82 (5):954–960.

Urata, S., T. Noda, Y. Kawaoka, H. Yokosawa, and J. Yasuda. 2006. Cellular factors required for Lassa virus budding. *J Virol* 80 (8):4191–4195.

Vela, E. M., L. Zhang, T. M. Colpitts, R. A. Davey, and J. F. Aronson. 2007. Arenavirus entry occurs through a cholesterol-dependent, non-caveolar, clathrin-mediated endocytic mechanism. *Virology* 369 (1):1–11.

Vieth, S., C. Drosten, R. Charrel, H. Feldmann, and S. Gunther. 2005. Establishment of conventional and fluorescence resonance energy transfer-based real-time PCR assays for detection of pathogenic New World arenaviruses. *J Clin Virol* 32 (3):229–235.

Vieth, S., A. E. Torda, M. Asper, H. Schmitz, and S. Gunther. 2004. Sequence analysis of L RNA of Lassa virus. *Virology* 318 (1):153–168.

Vitullo, A. D., V. L. Hodara, and M. S. Merani. 1987. Effect of persistent infection with Junin virus on growth and reproduction of its natural reservoir, *Calomys musculinus*. *Am J Trop Med Hyg* 37 (3):663–669.

Vitullo, A. D. and M. S. Merani. 1990. Vertical transmission of Junin virus in experimentally infected adult *Calomys musculinus*. *Intervirology* 31 (6):339–344.

Webb, P. A., G. Justines, and K. M. Johnson. 1975. Infection of wild and laboratory animals with Machupo and Latino viruses. *Bull World Health Organ* 52 (4–6):493–499.

Weissenbacher, M. C., M. A. Calello, M. S. Merani, J. B. McCormick, and M. Rodriguez. 1986. Therapeutic effect of the antiviral agent ribavirin in Junin virus infection of primates. *J Med Virol* 20 (3):261–267.

Wilson, S. M. and J. C. Clegg. 1991. Sequence analysis of the S RNA of the African arenavirus Mopeia: an unusual secondary structure feature in the intergenic region. *Virology* 180 (2):543–552.

Wright, K. E., R. C. Spiro, J. W. Burns, and M. J. Buchmeier. 1990. Post-translational processing of the glycoproteins of lymphocytic choriomeningitis virus. *Virology* 177 (1):175–183.

York, J., D. Dai, S. M. Amberg, and J. H. Nunberg. 2008. pH-induced activation of arenavirus membrane fusion is antagonized by small-molecule inhibitors. *J Virol* 82 (21):10932–10939.

York, J. and J. H. Nunberg. 2006. Role of the stable signal peptide of Junin arenavirus envelope glycoprotein in pH-dependent membrane fusion. *J Virol* 80 (15):7775–7780.

York, J. and J. H. Nunberg. 2009. Intersubunit interactions modulate pH-induced activation of membrane fusion by the Junin virus envelope glycoprotein GPC. *J Virol* 83 (9):4121–4126.

York, J., V. Romanowski, M. Lu, and J. H. Nunberg. 2004. The signal peptide of the Junin arenavirus envelope glycoprotein is myristoylated and forms an essential subunit of the mature G1-G2 complex. *J Virol* 78 (19):10783–10792.

Young, P. R. and C. R. Howard. 1983. Fine structure analysis of Pichinde virus nucleocapsids. *J Gen Virol* 64 (Pt 4):833–842.

20 "Venezuelan" Hemorrhagic Fever

Sheli R. Radoshitzky, Fabian de Kok-Mercado,
Peter B. Jahrling, Sina Bavari, and Jens H. Kuhn

CONTENTS

20.1 HISTORY OF "VENEZUELAN" HEMORRHAGIC FEVER

Guanarito virus (GTOV) emerged in September 1989 as the cause of a yet officially unnamed disease that is often referred to as "Venezuelan" hemorrhagic fever ("VHF"). This severe hemorrhagic illness was first recognized by physicians in the municipalities of Guanarito and Guanare in the state of Portuguesa in central Venezuela. The outbreak was initially thought to be dengue hemorrhagic fever (severe dengue). Between 1990 and 1991, a total of 104 cases were reported with a ≈25% case-fatality rate (Salas et al. 1991). The virus (GTOV prototype strain INH-95551) was later isolated from the spleen of a 20-year-old male farm worker during autopsy (Tesh et al. 1994). Humans become infected through contact with infected rodent reservoirs (predominantly short-tailed zygodonts) or inhalation of aerosolized virus from contaminated rodent excreta or secreta

(Charrel and de Lamballerie 2003). The disease has appeared in subsequent years mainly in the harvest season, between November and January, occurring predominantly among male agricultural workers. After a seemingly spontaneous drop in human cases between 1989 and 1992, a new outbreak occurred in 2002 with 18 reported cases (Charrel et al. 2003). By 2006, approximately 600 cases of "VHF" have been reported (Figure 20.1) (Fulhorst et al. 2008; de Manzione et al. 1998).

20.2 CLINICAL COURSE OF "VENEZUELAN" HEMORRHAGIC FEVER

"VHF" is clinically similar to other South American hemorrhagic fevers (Junín/Argentinian hemorrhagic fever and Machupo/Bolivian hemorrhagic fever) (Arribalzaga 1955; Harrison et al. 1999; de Manzione et al. 1998; Molteni et al. 1961; Rugiero et al. 1964; Salas et al. 1991; Stinebaugh et al. 1966). Most patients report an insidious progressive onset of symptoms. The disease usually begins after an incubation period of 1–2 weeks with a mild, nonspecific febrile illness that progresses in severity over the next 5–7 days. The most common presenting symptoms are malaise, weakness, nausea, myalgia, prostration, headache, arthralgia, and sore throat. Characteristic clinical signs include fever and cough, vomiting and diarrhea leading to dehydration, pharyngitis/tonsillitis, cervical lymphadenopathy, scattered pulmonary crackles, as well as signs of vascular damage, such as pronounced conjunctival injection, facial edema, and petechiae. The majority of patients also have thrombocytopenia and leukopenia upon admission. A variety of hemorrhagic manifestations include epistaxis, bleeding gums, menorrhagia, and melena. Hematemesis and hematuria might also be observed in some patients. Neurological signs may be prominent and include tremor of the hands, vertigo, and convulsions. Death usually occurs within 1–6 days of hospital admission. Although the initial platelet count appears to have no relation to the outcome of the case, a history of recent convulsions has been associated with poor prognosis (de Manzione et al. 1998; Salas et al. 1991). CNS manifestations, including encephalitis and convulsions, also have been reported to occur in cases of Lassa fever and Junín/Argentinian hemorrhagic fever where they mean a poor disease prognosis as well (Cummins et al. 1992; Peters 2002).

"VHF" patients who survive begin to improve during the 2nd week of illness, as the appearance of neutralizing antibodies signals the onset of an immune response. Convalescence is often lengthy and lasts several weeks with possible fatigue, hair loss, and hearing loss (de Manzione et al. 1998; Peters 2002; Salas et al. 1991; Vainrub and Salas 1994).

The gross and histopathological necropsy findings generally show the following features: diffuse pulmonary edema and congestion with intraparenchymal and subpleural hemorrhages, liver congestion with focal hemorrhages and yellow discoloration, cardiomegaly with epicardial hemorrhages, splenomegaly and congestion, renal edema with loss of corticomedullary border, and gastrointestinal hemorrhages and presence of blood in the bladder and uterus (Salas et al. 1991).

20.3 TAXONOMY AND PHYLOGENETIC RELATIONSHIPS OF GUANARITO VIRUS

The family *Arenaviridae* includes a single genus, *Arenavirus*, currently comprising 25 recognized species (Clegg 2002; Charrel et al. 2003; Delgado et al. 2008; Jay et al. 2005; Lecompte et al. 2007; Oldstone 2002; Salvato et al. 2011). Based on their antigenic properties, arenaviruses have been divided into two distinct groups: the Old World arenaviruses (Lassa–lymphocytic choriomeningitis serocomplex) that include viruses indigenous to Africa and the ubiquitous lymphocytic choriomeningitis virus (LCMV) and the New World arenaviruses (Tacaribe serocomplex) that include viruses indigenous to the Americas (Bowen et al. 1996; Cajimat and Fulhorst 2004; Clegg 2002; Rowe et al. 1970).

Sequences of the *NP* gene of all known arenaviruses have provided the basis for their phylogenetic analysis. This analysis supports the previously defined antigenic groupings and further defines virus relationships within them. Sequence data derived from other regions of the genome, where

Approximate locality of the residence of patients reported with
Guanarito virus (GTOV) in 1996

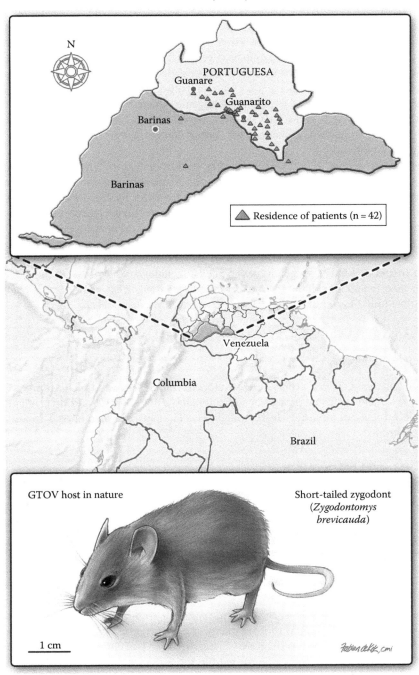

FIGURE 20.1 **(See color insert.)** Geography and ecology of Guanarito virus.

available, are largely consistent with this analysis. The 25 member viruses of the 25 species represent four phylogenetic groups. The Old World complex forms one monophyletic group that is deeply rooted to the three New World lineages (Albarino et al. 1998; Bowen et al. 1997, 2000; Vieth et al. 2004). New World arenaviruses are subdivided into three groups (A, B, and C). GTOV is a member of group B (Bowen et al. 1996; Gonzalez et al. 1995), which also contains the human pathogenic viruses Machupo virus (MACV), Junín virus (JUNV), Sabiá virus (SABV), and Chapare virus (CHPV) and the nonpathogenic Tacaribe virus, Amaparí virus, and Cupixi virus (Bowen et al. 1996; Charrel et al. 2002, 2003). This group has been further subdivided into B1, B2, and B3 sublineages. Interestingly, the four hemorrhagic fever-causing viruses do not group together but rather are distributed in the three sublineages together with nonpathogenic viruses. GTOV belongs to sublineage B2 and is most closely related to the nonpathogenic Amaparí virus. MACV and JUNV belong to sublineage B1, and SABV and CHPV to B3 (Charrel and de Lamballerie 2003). It is therefore apparent that during arenavirus evolution, the trait of human pathogenicity has arisen on at least two independent occasions.

In a study by Weaver et al., 29 GTOV isolates from rodents and humans from diverse localities in western Venezuela were analyzed by partial sequencing of the *NP* gene. Phylogenetic analysis revealed nine distinct lineages or genotypes that differ by 4%–10% in nucleotide and up to 9% in amino acid sequences. With the exception of one genotype (genotype 9), all appeared to be restricted to discrete, small geographic regions in western Venezuela. Genotype 9, however, occurred in several states and locations (variants were isolated from Portuguesa, Cojedes, Barinas, and Apure States). In addition, each of the nine genotypes included at least one variant recovered from a rodent, but only two genotypes (6 and 9) included isolates from human "VHF" cases. Genotype 6 also included isolates from short-tailed zygodonts and genotype 9 included isolates from both short-tailed zygodonts and Alston's cotton rats (Weaver et al. 2000).

20.4 GUANARITO VIRUS STRUCTURE AND LIFE CYCLE

20.4.1 VIRUS PROPERTIES

Like all arenaviruses, GTOV particles are enveloped, round, oval, and pleomorphic (70–280 nm in diameter) with short surface projections (Buchmeier 2002; Charrel and de Lamballerie 2003; Jay et al. 2005; Meyer et al. 2002; Tesh et al. 1994). The particles also appear sandy in electron microscopy sections.

20.4.2 GENOMIC ORGANIZATION

The arenavirus genome consists of two single-stranded RNA molecules, designated L (large) and S (small) (Figure 20.2). Each segment encodes two different proteins in two nonoverlapping reading frames of opposite polarities. The L segment ($\approx$7200 nt) encodes the viral RNA-dependent RNA polymerase (L) and a matrix protein with a zinc-binding really interesting new gene (RING) motif (Z). The S segment ($\approx$3500 nt) encodes the structural viral proteins: the nucleoprotein (NP), which associates with viral RNA in the form of bead-like structures to form nucleocapsids, and the envelope glycoprotein precursor (GPC). Extracted virion RNA is not infectious, and therefore, arenaviruses are considered by some as negative-sense RNA viruses.

The two open reading frames on each genomic segment are separated by an intergenic noncoding region (IGR) with the potential to form one or more energetically stable stem–loop (hairpin) structures (Auperin et al. 1984; Wilson and Clegg 1991). The IGR serves individual functions in structure-dependent transcription termination for enhanced gene expression (Meyer and Southern 1993, 1994; Tortorici et al. 2001) and in the virus assembly and/or budding (Pinschewer et al. 2005).

The 5′ and 3′ ends of the L and S RNA segments are noncoding untranslated regions (UTRs) and contain a conserved reverse complementary sequence of 19 or 20 nucleotides at each

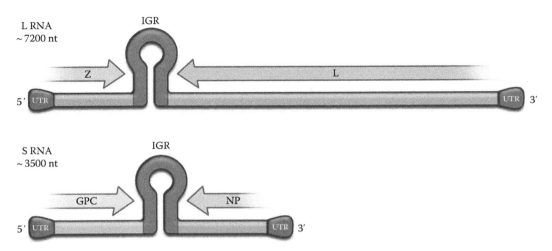

FIGURE 20.2 **(See color insert.)** Schematic of the Guanarito virus genome. GPC, glycoprotein precursor gene; IGR, intergenic region; L, RNA-dependent RNA polymerase; NP, nucleoprotein; UTR, untranslated region; Z, matrix protein.

extremity (Auperin et al. 1982a,b). These termini are thought to base-pair and form panhandle structures (Harnish et al. 1993; Salvato et al. 1989), a prediction that is supported by electron microscopic studies of arenaviral nucleocapsids (Young and Howard 1983). The 3' UTR of each segment contains the arenaviral promoter, which functions as a duplex and is composed of two elements: (1) a sequence-specific region at the 3' end of the promoter and (2) a variable complementary region at the 5' end of the promoter where solely base-pairing between the 3' and 5' ends of the segment is important (Hass et al. 2006; Perez and de la Torre 2003). In the case of MACV, RNA template recognition by the L polymerase requires a sequence-specific motif, 3' N_{1-2}HKUG 5' ($C_2G_3U_4G_5$), located at positions 2–5 in the 3' terminus of the viral genome. This motif is conserved in all experimentally determined arenavirus termini, including GTOV. Furthermore, L-RNA complex formation depends on a single-stranded nature of the RNA, indicating that intertermini dsRNA interactions must be partially broken for complex assembly to occur (Kranzusch et al. 2011).

20.4.3 VIRUS LIFE CYCLE

20.4.3.1 Virus Entry

Arenaviral cell entry is mediated by the GP spike complex on the virion surface. The arenaviral GP shares many characteristics with class I fusion proteins. GP is a trimer of heterodimers (Eschli et al. 2006; Gallaher et al. 2001) comprising the integral membrane protein GP2, which is noncovalently attached to the highly glycosylated peripheral protein GP1 (Daelli and Coto 1982; Eichler et al. 2006; Gangemi et al. 1978; Padula and de Martinez Segovia 1984; Wright et al. 1990). In accordance with the general class I architecture, the arenaviral GP1 subunit has been shown to mediate receptor binding. In the case of GTOV, as well as other pathogenic New World arenaviruses, GP1 binds the cellular receptor transferrin receptor 1 (TfR1) to initiate the entry process (Abraham et al. 2009; Radoshitzky et al. 2007, 2008). Following TfR1 binding, the virion is most likely endocytosed through clathrin-coated vesicles (Martinez et al. 2007; Vela et al. 2007) into intracellular endosomal compartments, where GP2-dependent acid-induced membrane fusion occurs (Castilla et al. 1994; Glushakova and Lukashevich 1989; Glushakova et al. 1990; Kunz et al. 2002; Martinez et al. 2007; Oldenburg et al. 2007). This internalization process is also dependent on the cellular clathrin coat–associated protein Eps15 and dynamin (Martinez et al. 2007). Exposure to low pH in late endosomes destabilizes the high-energy prefusion state of the GP spike complex resulting in irreversible conformational changes, in which the GP1 subunit is released from the GP spike (Di Simone and Buchmeier 1995; Di Simone et al. 1994). The hydrophobic fusion peptide at the N-terminus of GP2 (Glushakova et al. 1990, 1992) becomes exposed and inserts into the target membrane. This results in a series of conformational rearrangements in the GP2 subunit leading to the postfusion six α-helix core structure. This low-energy state is formed by C-terminal α-helices packing in an antiparallel manner into the hydrophobic grooves of an N-terminal trimeric coiled-coil core (Eschli et al. 2006). Thus, the N-terminus and the C-terminus of GP2 are drawn together and the viral envelope and the target membrane are pulled together to give rise to the fusion pore.

The signal peptide of the arenavirus GPC differs from conventional signal peptides by its length (58 amino acids) and by its stable association with the GP spike on the mature arenaviral virion (Froeschke et al. 2003; Lenz et al. 2001; York et al. 2004). The signal peptide associates noncovalently with the cytoplasmic domain of GP2 (Agnihothram et al. 2006), specifically with a highly conserved series of six cysteine and histidine residues that form two zinc-binding sites. An absolutely conserved cysteine residue (Cys-57) in the short cytoplasmic C-terminal tail of the signal peptide interacts with zinc cluster II of GP2 and is required for its retention in the GP complex (Briknarova et al. 2011; York and Nunberg 2007). In this context, the signal peptide modulates the pH-dependent membrane-fusion activity of the mature GP spike (Eichler et al. 2003b; York and Nunberg 2006). Both myristoylation of the signal peptide at its N-terminal glycine residue and a positively charged amino acid, lysine 33, are specifically required for this function (York and Nunberg 2006; York et al. 2004). K33 is located in the putative membrane proximal region of the signal-peptide ectodomain and may interact with GP2 to regulate membrane fusion (Agnihothram et al. 2006; Eichler et al. 2003b; York and Nunberg 2009; York et al. 2004). The current hypothesis is that the putative signal peptide–GP2 interface stabilizes the prefusion state of the GP complex at neutral pH and triggers the conformational changes leading to membrane fusion at acidic pH (York and Nunberg 2006, 2009). Following pH-dependent fusion of the virions with cellular membranes, viral nucleocapsids are delivered into the cell cytoplasm, where arenavirus transcription and replication exclusively occur (Meyer et al. 2002).

20.4.3.2 Virus Transcription and Protein Expression

Arenavirus RNA synthesis is initiated after delivery of the two encapsidated S and L segments, each associated with L, into the cytosol. The L polymerase initiates from the 3′ end of the genomic template and produces subgenomic, genome-complementary, NP and L mRNAs terminating at nonspecific sites at the IGR. These mRNAs are capped and not polyadenylated (Meyer and Southern 1993; Singh et al. 1987; Southern et al. 1987). NP, L, and the viral RNA form the transcriptionally active unit, the ribonucleoprotein (RNP) complex. Both proteins are the minimal *trans*-acting factors required for arenaviral RNA replication and transcription. Minimal *cis*-acting elements are the promoter sequences residing within the 3′ UTR at the end of each RNA segment, as well as the IGRs (Hass et al. 2004; Lee et al. 2000; Lopez et al. 2001; Meyer and Southern 1993).

The 5′ ends of viral mRNAs contain, in addition to a cap structure, several extra random non-templated bases, resembling the mRNAs of influenza A viruses and bunyaviruses (Garcin and Kolakofsky 1990; Meyer and Southern 1993; Raju et al. 1990). Therefore, it has been inferred that the arenaviral L protein uses a transcription-initiation mechanism involving an oligonucleotide-cap primer of host origin, reminiscent of the cap-snatching strategy of orthomyxoviruses.

Transcription of GPC and Z mRNAs occurs only after one round of arenavirus replication, in which S and L antigenomes are synthesized. Consequently, NP mRNA and NP protein accumulate earlier than GPC mRNA and the GP1 and GP2 glycoproteins. Furthermore, NP is much more abundant in cells than GP1 and GP2 (Buchmeier et al. 2006; Meyer et al. 2002). These observations illustrate how the ambisense coding arrangement of arenaviruses allows for independent regulation of viral gene expression separate from virion production.

GP is translated as a single precursor protein (GPC) into the lumen of the endoplasmic reticulum (ER), where the signal peptide is cleaved off (Eichler et al. 2003a). GPC then undergoes extensive *N*-linked glycosylation (Padula and de Martinez Segovia 1984; Wright et al. 1990) and is thought to oligomerize prior to being proteolytically processed by the subtilase SKI-1/S1P to generate the two mature glycoproteins GP1 and GP2 (Beyer et al. 2003; Buchmeier et al. 1987; Kunz et al. 2003; Lenz et al. 2001; Rojek et al. 2008). In addition to its role as a signal sequence for ER import, the signal peptide of arenaviruses is also specifically required for intracellular trafficking of GPC from the ER to the cell surface (Agnihothram et al. 2006; Eichler et al. 2003a), as well as for the proteolytic maturation of GPC (Eichler et al. 2003b; York and Nunberg 2006). The signal peptide associates with and masks endogenous dibasic ER retention/retrieval signals present in the cytoplasmic tail of GP2 and thereby ensures that only the fully assembled GP spike is transported to the plasma membrane (Agnihothram et al. 2006).

20.4.3.3 Virus Replication

During replication, the L polymerase reads through the IGR transcription-termination signal and generates uncapped antigenomic and genomic RNA species (Leung et al. 1977) that contain a single nontemplated G on the 5′ end (Garcin and Kolakofsky 1990; Raju et al. 1990). A model for this extra nontemplated G contends that a dinucleotide primer (pppGpC) used in replication initiation first base-pairs with positions 2 and 3 of the template RNA and then produces a single G overhang when the nascent RNA slips two bases "backward" on the template (Garcin and Kolakofsky 1992).

20.4.3.4 Virus Assembly and Budding

The newly synthesized full-length antigenomic and genomic RNA species are encapsidated by NP to generate RNP complexes for further mRNA transcription and for production of virus progeny (Raju et al. 1990). It is thought that arenaviral Z plays an important role in regulating the dynamics of arenaviral infection by a direct feedback loop where increased viral gene expression and higher levels of Z act to suppress RNA synthesis and balance the infection process (Cornu and de la Torre 2001, 2002; Cornu et al. 2004; Lopez et al. 2001). Z inhibits RNA synthesis by locking the L polymerase in a catalytically inactive promoter-bound state (Jacamo et al. 2003; Kranzusch and Whelan 2011). It is plausible that the resulting Z–L–RNA complex may serve as a critical intermediate to ensure genomic RNA is packaged along with the L polymerase poised to reinitiate viral RNA synthesis in the newly infected cell (Kranzusch and Whelan 2011). The integrity of the RING domain within Z is essential for Z–L interaction and therefore for Z inhibitory activity (Cornu and de la Torre 2002).

GTOV buds from the plasma membrane (Dalton et al. 1968; Tesh et al. 1994), where the virus RNP core associates with host-derived membrane that is highly enriched with the viral GP spike complex. In addition to its role as a potent inhibitor of virus RNA synthesis, Z is also essential for arenavirus assembly and budding (Eichler et al. 2004; Perez et al. 2003; Strecker et al. 2003). Z is strongly membrane-associated and accumulates near the inner surface of the plasma membrane where budding takes place (Strecker et al. 2003). Myristoylation of Z is critical for its binding to the viral GP complex, specifically with the signal-peptide component, and for its association with lipid membranes and, thus, for effective virus budding (Capul et al. 2007; Perez et al. 2004; Strecker et al. 2006). Interaction of NP and Z protein was also reported, suggesting that Z might recruit NP to cellular membranes where virus assembly takes place (Salvato 1993; Salvato et al. 1992).

Consistent with its features as a bona fide budding protein, GTOV Z contains the canonical proline-rich late (L) domain P(T/S)AP motif. L domains are highly conserved and have been shown to mediate interaction with host cell proteins required for virus budding. For example, the cellular multivesicular body pathway proteins Vps4A, Vps4B, and Tsg101 are involved in Lassa virus budding (Urata et al. 2006) and Z was shown to recruit Tsg101 to the plasma membrane (Perez et al. 2003).

20.5 PATHOLOGY AND PATHOGENESIS OF "VENEZUELAN" HEMORRHAGIC FEVER

20.5.1 HUMANS

Autopsies revealed diffuse pulmonary edema and congestion, liver congestion with focal hemorrhages and yellow discoloration, cardiomegaly, splenic enlargement and congestion, renal edema, and the presence of blood in the gastrointestinal tract and other hollow viscera (Table 20.1) (Salas et al. 1991).

20.5.2 NONHUMAN PRIMATES

Three rhesus monkeys (species *Macaca mulatta*) inoculated subcutaneously with GTOV became clinically ill, yet survived infection. The animals showed signs of lethargy and reduced food intake and were febrile and viremic. GTOV-infected monkeys had high levels of specific neutralizing antibodies in their convalescent sera at day 54 postinfection (Table 20.1) (Tesh et al. 1994).

20.5.3 MICE

GTOV (INH-95551) is pathogenic for baby mice on initial passage. Newborn mice (CD-1 strain suckling mice) inoculated intracerebrally with the virus exhibited first sign of illness

TABLE 20.1
Animal Models Used to Understand the Pathology and Pathogenesis of "Venezuelan" Hemorrhagic Fever

Disease	Animal Model	Symptoms	Necrosis	Hemorrhages	Neurological Syndrome	Reference
"VHF"	Humans	Cardiomegaly, splenomegaly, leukopenia, thrombocytopenia, conjunctivitis, facial edema		Liver, lung (edema and congestion), epicardium, GI tract, kidney (edema), spleen (congestion), petechiae, hematemesis, epistaxis, bleeding gums, melena, menorrhagia	Hand tremor, convulsions	(Salas et al. 1991)
Strain 95551	Rhesus monkeys	Lethargy, febrile, all recovered				(Tesh et al. 1994)
Strain 95551	Guinea pigs	Interstitial pneumonia, lymphoid depletion, occasional platelet thrombi in blood vessels associated with hemorrhage	Epithelium of the GI tract and esophagus, adrenal cortex, hepatocellular, lymphoid, and hematopoietic cells in the spleen, lymph nodes, intestine, and lung	Lung, stomach, spleen (congestion), adrenal cortex		(Hall et al. 1996)

approximately 10 days later. Sick animals showed signs of lethargy, runting, ataxia, and hind limb paralysis. Scattered death occurred in the inoculated mice beginning on the 12th day, but a few of the animals survived infection. In contrast to newborn mice, adult mice inoculated intraperitoneally with a 10% suspension of GTOV-infected mice brain showed no signs of illness (Tesh et al. 1994).

20.5.4 GUINEA PIGS

GTOV (INH-95551) is also fatal to adult strain 13 guinea pigs. Animals die between day 11 and 14 post inoculation. Adult strain 13 guinea pigs die at day 12–23 following subcutaneous inoculation of four additional GTOV isolates (Tesh et al. 1994). Similar lesions were observed between strain 13 and Hartley guinea pigs. Lesions were characterized by single-cell necrosis of epithelium of the gastrointestinal tract, interstitial pneumonia, lymphoid and hematopoietic cell necrosis, and the presence of platelet thrombi in occasional blood vessels associated with hemorrhage. Viral antigen was demonstrated in lymphoid tissues and macrophages, endothelial cells of multiple organs, pulmonary epithelium, epithelium of the gastrointestinal tract, and in miscellaneous other tissues and cells. GTOV titers in the spleen, liver, and kidney samples were higher than those in blood and brain, indicating viral replication in these organs. Intact virions and typical arenavirus inclusions were demonstrated by immunoelectron microscopy in macrophages of the spleen, epithelia of the lung, intestinal tract and macrophages, liver, and kidney, but not brain tissues (Table 20.1) (Hall et al. 1996).

With minor variation, GTOV infection is similar to that of other arenaviruses. JUNV, MACV, and Lassa virus show a similar course in the guinea pig and primate models. Characteristic lesions are lymphoid necrosis, bone marrow depletion, and interstitial pneumonia. Moreover, macrophages appear to be the predominant cells involved in all three diseases. One difference is the minimal liver necrosis in GTOV compared with Lassa virus, in which it is more extensive. GTOV more nearly resembles viscerotropic strains of JUNV, and viral antigen is recovered from the same tissues in JUNV-infected animals as observed in the GTOV-infected guinea pigs.

Because GTOV infection did not produce hemorrhagic disease or serious illness in rhesus monkeys, guinea pigs might be a more valid animal model of the human disease and therefore preferable for pathogenesis studies, as well as for treatment and drug evaluations (Table 20.1).

20.6 TREATMENT AND VACCINES

Currently, there are no virus-specific treatments approved for use against "VHF" and treatment consists primarily of supportive care.

20.6.1 VACCINES

Despite the biodefense and public health risks associated with "VHF," there are no FDA-licensed vaccines. At present, the only vaccines for the prevention of human arenavirus disease are limited to a single, safe, efficacious, live-attenuated vaccine, called Candid 1, for the prevention of JUNV infection (Barrera Oro and McKee 1991; Enria and Maiztegui 1994; Maiztegui et al. 1987, 1998). JUNV Candid 1, which is classified as an Investigational New Drug in the United States, was derived from the wild-type (WT) JUNV strain XJ13 through serial passage both in vivo and in vitro (Goni et al. 2006). Candid 1 has been evaluated in controlled trials among at-risk population of agricultural workers in Argentina, where it showed a protective efficacy greater or equal to 84% and has led to a consistent reduction in Junín/Argentinian hemorrhagic fever cases with no serious side effects (Enria et al. 2008, 2010; Maiztegui et al. 1998). It is not known if the Candid 1 vaccine would be useful against other arenavirus hemorrhagic fevers.

20.6.2 Passive Antibody Therapy

Studies to assess the effect of immune plasma on the course and outcome of "VHF" have not been reported to date. However, the transfusion of immune convalescent plasma with defined doses of JUNV-neutralizing antibodies is the present therapeutic intervention against the related Junín/Argentinian hemorrhagic fever. Immune serum treatment has been shown to be effective in attenuating disease and reducing lethality to less than 1% if administered within the first 8 days of disease to provide an adequate dose of neutralizing antibodies (Enria et al. 1984, 1985; Enria and Maiztegui 1994; Maiztegui et al. 1979). A few studies in animal models suggest that passive antibody therapy might be useful for the treatment of Machupo/Bolivian hemorrhagic fever as well (Eddy et al. 1975). An in vitro study with Vero E6 cells showed that convalescent sera from six of seven putative "VHF" cases neutralized the infectivity of GTOV strain INH-95551 and the neutralizing titers in the positive sera ranged from 160 to 640 (Fulhorst et al. 2008). These results suggest that passive antibody therapy may prove beneficial in the treatment of "VHF" caused by strains of GTOV that are enzootic in the region in which "VHF" is prevalent.

However, many considerations argue for the need for alternative treatments. First, plasma therapy is not as efficient in preventing Junín/Argentinian hemorrhagic fever when initiated after 8 days of illness; second, a late neurological syndrome is observed in 10% of plasma-treated patients (Enria et al. 1985); third, maintaining adequate stocks of plasma is difficult due to the limited number of "VHF" cases and the absence of a program for convalescent serum collection; fourth, there is the risk of transfusion-borne diseases (García et al. 2011).

20.6.3 Antivirals

Current antiviral therapy is limited to an off-label use of the nonimmunosuppressive guanosine analog, ribavirin (1-β-D-ribofuranosyl-1-H-1,2,4-triazole-3-carboxamide), an IMP dehydrogenase inhibitor, which has had only partial efficacy against some arenavirus infections and is associated with significant toxicity in humans (Barry et al. 1995; Enria and Maiztegui 1994; Enria et al. 2008; Huggins 1989; Kenyon et al. 1986; Khan et al. 2008; Kilgore et al. 1997; McCormick et al. 1986; McKee et al. 1988; Petkevich et al. 1981; Rodriguez et al. 1986; Snell 2001; Weissenbacher et al. 1986). Ribavirin can lead to adverse side effects such as thrombocytosis, severe anemia, and birth defects (Enria et al. 1987; McKee et al. 1988). Recent studies suggest that the antiviral activity of ribavirin on arenaviruses might be exerted, at least partially, by lethal mutagenesis (Moreno et al. 2011).

Recent independent small-molecule high-throughput screens have also identified promising antivirals. These efforts have to date identified six chemically distinct classes of small-molecule compounds that specifically inhibit GP-mediated membrane fusion with differing selectivities against New World and/or Old World arenaviruses (Bolken et al. 2006; Larson et al. 2008; Lee et al. 2008; York et al. 2008). One highly active and specific small-molecule inhibitor, ST-294, was found to inhibit GTOV in vitro, as well as other New World pathogenic viruses (JUNV, MACV, SABV) at concentrations in the nanomolar range. This molecule also demonstrated favorable pharmacodynamic properties (metabolically stable and orally bioavailable), as well as in vivo antiarenaviral activity in a newborn mouse model (Bolken et al. 2006). Mechanism of action studies suggests that this compound targets GP2 and is a viral entry inhibitor (Bolken et al. 2006). Another inhibitor, ST-193, a benzimidazole derivative, was found to have a potent in vitro activity against arenaviral entry mediated by diverse arenavirus envelope proteins, including GTOV and other New World pathogenic arenaviruses (Larson et al. 2008). Finally, two lead compounds, 16G8 and 17C8, showed high activity against GTOV and other South American hemorrhagic fever arenaviruses, as well as the more distantly related Lassa virus. These compounds act at the level of GP-mediated membrane fusion with an $IC_{50} \approx 200$–350 nm (Lee et al. 2008). Despite chemical differences, evidence suggests that these diverse inhibitors act through the pH-sensitive interface of the signal peptide and

GP2 subunits in the GP spike complex and prevent virus entry by stabilizing the prefusion spike complex against pH-induced activation in the endosome (Bolken et al. 2006; Larson et al. 2008; York et al. 2008).

T-705, a pyrazine derivative with broad antiviral activity against RNA viruses, including influenza viruses (Kiso et al. 2010), flaviviruses (Julander et al. 2009; Morrey et al. 2008), bunyaviruses, and several nonpathogenic arenaviruses (Gowen et al. 2007, 2008, 2010), was also found to be active against GTOV and the other pathogenic New World arenaviruses Machupo and Junín. T-705 may act as a purine nucleoside analog specifically targeting the viral RNA-dependent RNA polymerase (Mendenhall et al. 2011).

20.7 MAMMALIAN RESERVOIRS OF GUANARITO VIRUS

Rodents of the superfamily Muroidea are the natural hosts of the currently classified arenaviruses with the possible exception of Tacaribe virus (bats?), Lujo virus and SABV (unknown). New World arenaviruses are found in rodents of the family Cricetidae, subfamily Sigmodontinae. The geographic distribution of each arenavirus is determined by the range of its corresponding rodent host(s). Current evidence suggests a long-term "diffuse coevolution" between the arenaviruses and their rodent hosts. According to this model, a parallel phylogeny between the viruses and their corresponding rodent host(s) is the most common transmission mechanism and allows for host switches between closely related rodents (Hugot et al. 2001; Salazar-Bravo et al. 2002). There is a remarkable rodent specificity seen among arenaviruses in nature. Field studies strongly support the concept of only a single major reservoir host for each virus (Salazar-Bravo et al. 2002).

Experimental work by Fulhorst and colleagues identified the nocturnal short-tailed zygodonts (species *Zygodontomys brevicauda*) as the reservoirs of GTOV. Alston's cotton rats (species *Sigmodon alstoni*) were indicated as a secondary host and fulvous colilargos, guaira spiny-rats, and roof rats (species *Oligoryzomys fulvescens*, *Proechimys guairae*, and *Rattus rattus*, respectively) were found to be seropositive (Tesh et al. 1993). Virus could be isolated from throat swabs, urine, spleens, lungs, or kidneys of infected animals, and antibodies were found in the sera (Milazzo et al. 2011).

Short-tailed zygodonts are native to the plains of western Venezuela and can reach high densities in tall grassy (weedy) areas found in pastoral and agricultural areas along roadsides and fence lines and in the naturally occurring savannah that dominates the landscape of the "VHF"-endemic region (Fulhorst et al. 1997, 1999; Salazar-Bravo et al. 2002; Tesh et al. 1993). The presence of GTOV-infected short-tailed zygodonts in Apure, Barinas, Cojedes, and Guárico indicates that GTOV was enzootic in Portuguesa long before 1989. As such, the emergence of "VHF" was likely a consequence of demographic and/or ecological changes in rural areas of Portuguesa that eventually resulted in a significant increase in the frequency of contact between humans and GTOV-infected rodents (Fulhorst et al. 2008). However, during 4 years of rodent trapping in the region of "VHF" endemicity, short-tailed zygodonts or Alston's cotton rats were never found within houses or farm buildings (de Manzione et al. 1998). Presumably, human infection therefore occurs outdoors. Thus, one might expect persons having frequent contact with rodent-infested grassland habitats to be at higher risk of contracting "VHF."

Laboratory infection of short-tailed zygodonts with GTOV produces chronic viremia characterized by persistent shedding of infectious virus in oral and respiratory secretions and urine without clinical signs of disease through day 208 post inoculation (Fulhorst et al. 1999). Analyses of field and laboratory data suggest that horizontal transmission is the dominant mode of GTOV transmission in short-tailed zygodonts as most GTOV infections in these mice are acquired in an age-dependent manner. Therefore, the chronic carrier state in short-tailed zygodonts most likely results from exposure to infectious virus later in life through aggressive or venereal behavior like allogrooming, mating, intraspecies aggression, and other activities that entail close physical contact. Evidence also suggests that male and female mice contribute equally to GTOV transmission (Milazzo et al. 2011).

"VHF" is an example of natural nidality; zoonoses within the host reservoir occur focally and have an incomplete pattern of overlap with the host species range. Short-tailed zygodonts are found in larger distribution area than the endemic areas of "VHF" (Fulhorst et al. 1997; Weaver et al. 2000). Indeed, some of the GTOV variants were isolated from locations outside of the endemic/epidemic region (outlying locations in Cojedes, Barinas, and Apure States) and yet were found to belong to genotypes that included variants isolated from human cases of "VHF" from areas surrounding Guanarito. This suggests that pathogenic GTOV variants occur in these outlying areas, but do not frequently infect people and/or cause inapparent disease there. Furthermore, "VHF" does not appear to be associated with a specific GTOV genotype that is restricted to a particular rodent host. With the exception of genotype 1, isolated only from fulvous colilargos, all of the GTOV genotypes included variants that were recovered from short-tailed zygodonts. Variants belonging to genotypes 7, 8, and 9 were also recovered from Alston's cotton rats. Quasispecies diversity analysis suggests that rodent isolates have higher sequence variation than human isolates. One rodent isolate included a mixture of two phylogenetically distinct genotypes, suggesting a dual infection (Weaver et al. 2000).

20.8 DIAGNOSTICS

Detection of a viral antigen and/or the viral genome is crucial for rapid diagnosis of patients with hemorrhagic fever caused by South American arenaviruses, especially for patients in the acute phase. In general, reverse transcriptase PCR (RT-PCR) is more sensitive in detecting viruses in patients' specimens than antigen-capture ELISA. The application of RT-PCR and TaqMan PCR for detection of the GTOV, JUNV, and MACV genomes has been reported, but it has yet to be applied to clinical specimens. Therefore, the possibility that it does not detect novel arenavirus strains or arenaviruses cannot be ruled out (Vieth et al. 2005). Recently, TaqMan™-based PCR assays for specific and absolute quantitative detection of multiple hemorrhagic fever viruses (filoviruses, arenaviruses, and bunyaviruses), including GTOV, have been developed. The GTOV target genes were GP1 and the L polymerase and the limit of detection for the assays ranged from 10 to 0.001 plaque-forming units (PFU)/PCR (0.1–0.001 for GTOV). These assays are not only qualitative (presence of target) but also quantitative [measure a single DNA/RNA target sequence in an unknown sample and express the final results as an absolute value (e.g., viral load, PFUs, or copies/ml)] on the basis of concentration of standard samples and can be used in viral load, vaccine, and antiviral drug studies. These assays also provide a repertoire of diagnostic tools that can serve as a foundation for the simultaneous identification and analysis of potential biothreat agents when multiple, rapid cycling, real-time PCR platforms are used (Trombley et al. 2010).

Recently, an antigen-capture ELISA has been developed using monoclonal antibodies against the JUNV NP. The antigens of all human pathogenic South American arenaviruses, including GTOV, could be detected using monoclonal antibodies C11–12 or E-4-2. Therefore, the established antigen-capture ELISAs recognized highly conserved epitopes, suggesting that they may be applicable not only for the diagnosis of Junín/Argentinian hemorrhagic fever but also for the diagnosis of "VHF" and the other Southern American hemorrhagic fevers (Nakauchi et al. 2009).

DISCLAIMER

The content of this publication does not necessarily reflect the views or policies of the U.S. Department of Defense, the U.S. Department of the Army, the U.S. Department of Health and Human Services, or the institutions and companies affiliated with the authors. JHK performed this work as an employee of Tunnell Consulting, Inc., a subcontractor to Battelle Memorial Institute, and

FKM as an employee of Battelle Memorial Institute under its prime contract with NIAID, under Contract No. HHSN272200200016I.

REFERENCES

Abraham, J., J. A. Kwong, C. G. Albarino, J. J. G. Lu, S. R. Radoshitzky, J. Salazar-Bravo, M. Farzan, C. F. Spiropoulou, and H. Choe. 2009. Host-species transferrin receptor 1 orthologs are cellular receptors for nonpathogenic new world clade B Arenaviruses. *Plos Pathogens* **5**(4):e1000358.

Agnihothram, S. S., J. York, and J. H. Nunberg. 2006. Role of the stable signal peptide and cytoplasmic domain of G2 in regulating intracellular transport of the Junin virus envelope glycoprotein complex. *J Virol* **80**(11):5189–5198.

Albarino, C. G., D. M. Posik, P. D. Ghiringhelli, M. E. Lozano, and V. Romanowski. 1998. Arenavirus phylogeny: A new insight. *Virus Genes* **16**(1):39–46.

Arribalzaga, R. A. 1955. New epidemic disease due to unidentified germ: Nephrotoxic, leukopenic and enanthematous hyperthermia. *Dia Med* **27**(40):1204–1210.

Auperin, D. D., R. W. Compans, and D. H. Bishop. 1982a. Nucleotide sequence conservation at the 3′ termini of the virion RNA species of New World and Old World Arenaviruses. *Virology* **121**(1):200–203.

Auperin, D., K. Dimock, P. Cash, W. E. Rawls, W. C. Leung, and D. H. Bishop. 1982b. Analyses of the genomes of prototype pichinde arenavirus and a virulent derivative of Pichinde Munchique: Evidence for sequence conservation at the 3′ termini of their viral RNA species. *Virology* **116**(1):363–367.

Auperin, D. D., M. Galinski, and D. H. Bishop. 1984. The sequences of the N protein gene and intergenic region of the S RNA of pichinde arenavirus. *Virology* **134**(1):208–219.

Barrera Oro, J. G. and K. T. McKee, Jr. 1991. Toward a vaccine against Argentine hemorrhagic fever. *Bull Pan Am Health Organ* **25**(2):118–126.

Barry, M., M. Russi, L. Armstrong, D. Geller, R. Tesh, L. Dembry, J. P. Gonzalez, A. S. Khan, and C. J. Peters. 1995. Brief report: Treatment of a laboratory-acquired Sabia virus infection. *N Engl J Med* **333**(5):294–296.

Beyer, W. R., D. Popplau, W. Garten, D. von Laer, and O. Lenz. 2003. Endoproteolytic processing of the lymphocytic choriomeningitis virus glycoprotein by the subtilase SKI-1/S1P. *J Virol* **77**(5):2866–2872.

Bolken, T. C., S. Laquerre, Y. Zhang, T. R. Bailey, D. C. Pevear, S. S. Kickner, L. E. Sperzel, K. F. Jones, T. K. Warren, S. Amanda Lund, D. L. Kirkwood-Watts, D. S. King, A. C. Shurtleff, M. C. Guttieri, Y. Deng, M. Bleam, and D. E. Hruby. 2006. Identification and characterization of potent small molecule inhibitor of hemorrhagic fever New World Arenaviruses. *Antiviral Res* **69**(2):86–97.

Bowen, M. D., C. J. Peters, and S. T. Nichol. 1996. The phylogeny of New World (Tacaribe complex) Arenaviruses. *Virology* **219**(1):285–290.

Bowen, M. D., C. J. Peters, and S. T. Nichol. 1997. Phylogenetic analysis of the Arenaviridae: Patterns of virus evolution and evidence for cospeciation between arenaviruses and their rodent hosts. *Mol Phylogenet Evol* **8**(3):301–316.

Bowen, M. D., P. E. Rollin, T. G. Ksiazek, H. L. Hustad, D. G. Bausch, A. H. Demby, M. D. Bajani, C. J. Peters, and S. T. Nichol. 2000. Genetic diversity among Lassa virus strains. *J Virol* **74**(15):6992–7004.

Briknarova, K., C. J. Thomas, J. York, and J. H. Nunberg. 2011. Structure of a zinc-binding domain in the Junin virus envelope glycoprotein. *J Biol Chem* **286**(2):1528–1536.

Buchmeier, M. J. 2002. Arenaviruses: Protein structure and function. *Curr Top Microbiol Immunol* **262**:159–173.

Buchmeier, M. J., J. C. de La Torre, and C. J. Peters. 2006. *Arenaviridae*: The viruses and their replication. In *Fields Virology*, eds. D. M. Knipe and P. M. Howley. Philadelphia, PA: Lippincott Williams & Wilkins.

Buchmeier, M. J., P. J. Southern, B. S. Parekh, M. K. Wooddell, and M. B. Oldstone. 1987. Site-specific antibodies define a cleavage site conserved among arenavirus GP-C glycoproteins. *J Virol* **61**(4):982–985.

Cajimat, M. N. and C. F. Fulhorst. 2004. Phylogeny of the Venezuelan arenaviruses. *Virus Res* **102**(2):199–206.

Capul, A. A., M. Perez, E. Burke, S. Kunz, M. J. Buchmeier, and J. C. de la Torre. 2007. Arenavirus Z-glycoprotein association requires Z myristoylation but not functional RING or late domains. *J Virol* **81**(17):9451–9460.

Castilla, V., S. E. Mersich, N. A. Candurra, and E. B. Damonte. 1994. The entry of Junin virus into Vero cells. *Arch Virol* **136**(3–4):363–374.

Charrel, R. N., H. Feldmann, C. F. Fulhorst, R. Khelifa, R. de Chesse, and X. de Lamballerie. 2002. Phylogeny of New World arenaviruses based on the complete coding sequences of the small genomic segment identified an evolutionary lineage produced by intrasegmental recombination. *Biochem Biophys Res Commun* **296**(5):1118–1124.

Charrel, R. N. and X. de Lamballerie. 2003. Arenaviruses other than Lassa virus. *Antiviral Res* **57**(1–2): 89–100.

Charrel, R. N., J. J. Lemasson, M. Garbutt, R. Khelifa, P. De Micco, H. Feldmann, and X. de Lamballerie. 2003. New insights into the evolutionary relationships between arenaviruses provided by comparative analysis of small and large segment sequences. *Virology* **317**(2):191–196.

Clegg, J. C. 2002. Molecular phylogeny of the arenaviruses. *Curr Top Microbiol Immunol* **262**:1–24.

Cornu, T. I., H. Feldmann, and J. C. de la Torre. 2004. Cells expressing the RING finger Z protein are resistant to arenavirus infection. *J Virol* **78**(6):2979–2983.

Cornu, T. I. and J. C. de la Torre. 2001. RING finger Z protein of lymphocytic choriomeningitis virus (LCMV) inhibits transcription and RNA replication of an LCMV S-segment minigenome. *J Virol* **75**(19):9415–9426.

Cornu, T. I. and J. C. de la Torre. 2002. Characterization of the arenavirus RING finger Z protein regions required for Z-mediated inhibition of viral RNA synthesis. *J Virol* **76**(13):6678–6688.

Cummins, D., D. Bennett, S. P. Fisher-Hoch, B. Farrar, S. J. Machin, and J. B. McCormick. 1992. Lassa fever encephalopathy: Clinical and laboratory findings. *J Trop Med Hyg* **95**(3):197–201.

Daelli, M. G. and C. E. Coto. 1982. Inhibition of the production of infectious particles in cells infected with Junin virus in the presence of tunicamycin. *Rev Argent Microbiol* **14**(3):171–176.

Dalton, A. J., W. P. Rowe, G. H. Smith, R. E. Wilsnack, and W. E. Pugh. 1968. Morphological and cytochemical studies on lymphocytic choriomeningitis virus. *J Virol* **2**(12):1465–1478.

Delgado, S., B. R. Erickson, R. Agudo, P. J. Blair, E. Vallejo, C. G. Albarino, J. Vargas, A. J. Comer, P. E. Rollin, T. G. Ksiazek, J. G. Olson, and S. T. Nichol. 2008. Chapare virus, a newly discovered arenavirus isolated from a fatal hemorrhagic fever case in Bolivia. *PLoS Path* **18**(4):e1000047.

Di Simone, C. and M. J. Buchmeier. 1995. Kinetics and pH dependence of acid-induced structural changes in the lymphocytic choriomeningitis virus glycoprotein complex. *Virology* **209**(1):3–9.

Di Simone, C., M. A. Zandonatti, and M. J. Buchmeier. 1994. Acidic pH triggers LCMV membrane fusion activity and conformational change in the glycoprotein spike. *Virology* **198**(2):455–465.

Eddy, G. A., F. S. Wagner, S. K. Scott, and B. J. Mahlandt. 1975. Protection of monkeys against Machupo virus by the passive administration of Bolivian haemorrhagic fever immunoglobulin (human origin). *Bull World Health Organ* **52**(4–6):723–727.

Eichler, R., O. Lenz, W. Garten, and T. Strecker. 2006. The role of single N-glycans in proteolytic processing and cell surface transport of the Lassa virus glycoprotein GP-C. *Virol J* **3**:41.

Eichler, R., O. Lenz, T. Strecker, M. Eickmann, H. D. Klenk, and W. Garten. 2003a. Identification of Lassa virus glycoprotein signal peptide as a trans-acting maturation factor. *EMBO Rep* **4**(11):1084–1088.

Eichler, R., O. Lenz, T. Strecker, and W. Garten. 2003b. Signal peptide of Lassa virus glycoprotein GP-C exhibits an unusual length. *FEBS Lett* **538**(1–3):203–206.

Eichler, R., T. Strecker, L. Kolesnikova, J. Ter Meulen, W. Weissenhorn, S. Becker, H. D. Klenk, W. Garten, and O. Lenz. 2004. Characterization of the Lassa virus matrix protein Z: Electron microscopic study of virus-like particles and interaction with the nucleoprotein (NP). *Virus Res* **100**(2):249–255.

Enria, D. A., A. M. Ambrosio, A. M. Briggiler, M. R. Feuillade, and E. Crivelli. 2010. Candid#1 vaccine against Argentine hemorrhagic fever produced in Argentina. Immunogenicity and safety. *Medicina (B Aires)* **70**(3):215–222.

Enria, D. A., A. M. Briggiler, N. J. Fernandez, S. C. Levis, and J. I. Maiztegui. 1984. Importance of dose of neutralising antibodies in treatment of Argentine haemorrhagic fever with immune plasma. *Lancet* **2**(8397):255–256.

Enria, D. A., A. M. Briggiler, S. Levis, D. Vallejos, J. I. Maiztegui, and P. G. Canonico. 1987. Tolerance and antiviral effect of ribavirin in patients with Argentine hemorrhagic fever. *Antiviral Res* **7**(6):353–359.

Enria, D. A., A. M. Briggiler, and Z. Sanchez. 2008. Treatment of Argentine hemorrhagic fever. *Antiviral Res* **78**(1):132–139.

Enria, D. A., A. J. de Damilano, A. M. Briggiler, A. M. Ambrosio, N. J. Fernandez, M. R. Feuillade, and J. I. Maiztegui. 1985. Late neurologic syndrome in patients with Argentinian hemorrhagic fever treated with immune plasma. *Medicina (B Aires)* **45**(6):615–620.

Enria, D. A. and J. I. Maiztegui. 1994. Antiviral treatment of Argentine hemorrhagic fever. *Antiviral Res* **23**(1):23–31.

Eschli, B., K. Quirin, A. Wepf, J. Weber, R. Zinkernagel, and H. Hengartner. 2006. Identification of an N-terminal trimeric coiled-coil core within arenavirus glycoprotein 2 permits assignment to class I viral fusion proteins. *J Virol* **80**(12):5897–5907.

Froeschke, M., M. Basler, M. Groettrup, and B. Dobberstein. 2003. Long-lived signal peptide of lymphocytic choriomeningitis virus glycoprotein pGP-C. *J Biol Chem* **278**(43):41914–41920.

Fulhorst, C. E., M. D. Bowen, R. A. Salas, N. M. de Manzione, G. Duno, A. Utrera, T. G. Ksiazek, C. et al. 1997. Isolation and characterization of pirital virus, a newly discovered South American arenavirus. *Am J Trop Med Hyg* **56**(5):548–553.

Fulhorst, C. F., M. N. Cajimat, M. L. Milazzo, H. Paredes, N. M. de Manzione, R. A. Salas, P. E. Rollin, and T. G. Ksiazek. 2008. Genetic diversity between and within the arenavirus species indigenous to western Venezuela. *Virology (New York)* **378**(2):205–213.

Fulhorst, C. F., T. G. Ksiazek, C. J. Peters, and R. B. Tesh. 1999. Experimental infection of the cane mouse *Zygodontomys brevicauda* (family Muridae) with guanarito virus (Arenaviridae), the etiologic agent of Venezuelan hemorrhagic fever. *J Infect Dis* **180**(4):966–969.

Gallaher, W. R., C. DiSimone, and M. J. Buchmeier. 2001. The viral transmembrane superfamily: Possible divergence of Arenavirus and Filovirus glycoproteins from a common RNA virus ancestor. *BMC Microbiol* **1**:1.

Gangemi, J. D., R. R. Rosato, E. V. Connell, E. M. Johnson, and G. A. Eddy. 1978. Structural polypeptides of Machupo virus. *J Gen Virol* **41**(1):183–188.

Garcin, D. and D. Kolakofsky. 1990. A novel mechanism for the initiation of Tacaribe arenavirus genome replication. *J Virol* **64**(12):6196–6203.

Garcin, D. and D. Kolakofsky. 1992. Tacaribe arenavirus RNA synthesis in vitro is primer dependent and suggests an unusual model for the initiation of genome replication. *J Virol* **66**(3):1370–1376.

García, C. C., C. S. Sepúlveda, and E. B. Damonte. 2011. Novel therapeutic targets for arenavirus hemorrhagic fevers. *Future Virology* **6**(1):27–44.

Glushakova, S. E., A. I. Iakuba, A. D. Vasiuchkov, R. F. Mar'iankova, T. M. Kukareko, T. A. Stel'makh, T. P. Kurash, and I. S. Lukashevich. 1990. Lysosomotropic agents inhibit the penetration of arenaviruses into a culture of BHK-21 and Vero cells. *Vopr Virusol* **35**(2):146–150.

Glushakova, S. E. and I. S. Lukashevich. 1989. Early events in arenavirus replication are sensitive to lysosomotropic compounds. *Arch Virol* **104**(1–2):157–161.

Glushakova, S. E., I. S. Lukashevich, and L. A. Baratova. 1990. Prediction of arenavirus fusion peptides on the basis of computer analysis of envelope protein sequences. *FEBS Lett* **269**(1):145–147.

Glushakova, S. E., V. G. Omelyanenko, I. S. Lukashevitch, A. A. Bogdanov, Jr., A. B. Moshnikova, A. T. Kozytch, and V. P. Torchilin. 1992. The fusion of artificial lipid membranes induced by the synthetic arenavirus 'fusion peptide'. *Biochim Biophys Acta* **1110**(2):202–208.

Goni, S. E., J. A. Iserte, A. M. Ambrosio, V. Romanowski, P. D. Ghiringhelli, and M. E. Lozano. 2006. Genomic features of attenuated Junin virus vaccine strain candidate. *Virus Genes* **32**(1):37–41.

Gonzalez, J. P., A. Sanchez, and R. Rico-Hesse. 1995. Molecular phylogeny of Guanarito virus, an emerging arenavirus affecting humans. *Am J Trop Med Hyg* **53**(1):1–6.

Gowen, B. B., D. F. Smee, M. H. Wong, J. O. Hall, K. H. Jung, K. W. Bailey, J. R. Stevens, Y. Furuta, and J. D. Morrey. 2008. Treatment of late stage disease in a model of arenaviral hemorrhagic fever: T-705 efficacy and reduced toxicity suggests an alternative to ribavirin. *PLoS One* **3**(11):e3725.

Gowen, B. B., M. H. Wong, K. H. Jung, A. B. Sanders, M. Mendenhall, K. W. Bailey, Y. Furuta, and R. W. Sidwell. 2007. In vitro and in vivo activities of T-705 against arenavirus and bunyavirus infections. *Antimicrob Agents Chemother* **51**(9):3168–3176.

Gowen, B. B., M. H. Wong, K. H. Jung, D. F. Smee, J. D. Morrey, and Y. Furuta. 2010. Efficacy of favipiravir (T-705) and T-1106 pyrazine derivatives in phlebovirus disease models. *Antiviral Res* **86**(2):121–127.

Hall, W. C., T. W. Geisbert, J. W. Huggins, and P. B. Jahrling. 1996. Experimental infection of guinea pigs with Venezuelan hemorrhagic fever virus (Guanarito): A model of human disease. *Am J Trop Med Hyg* **55**(1):81–88.

Harnish, D. G., S. J. Polyak, and W. E. Rawls. 1993. Arenavirus replication: Molecular dissection of the role of protein and RNA. In *The Arenaviridae*, ed. M. S. Salvato. New York: Plenum Press.

Harrison, L. H., N. A. Halsey, K. T. McKee, Jr., C. J. Peters, J. G. Barrera Oro, A. M. Briggiler, M. R. Feuillade, and J. I. Maiztegui. 1999. Clinical case definitions for Argentine hemorrhagic fever. *Clin Infect Dis* **28**(5):1091–1094.

Hass, M., U. Golnitz, S. Muller, B. Becker-Ziaja, and S. Gunther. 2004. Replicon system for Lassa virus. *J Virol* **78**(24):13793–13803.

Hass, M., M. Westerkofsky, S. Muller, B. Becker-Ziaja, C. Busch, and S. Gunther. 2006. Mutational analysis of the Lassa virus promoter. *J Virol* **80**(24):12414–12419.

Huggins, J. W. 1989. Prospects for treatment of viral hemorrhagic fevers with ribavirin, a broad-spectrum antiviral drug. *Rev Infect Dis* **11**(Suppl 4):S750–S761.

Hugot, J. P., J. P. Gonzalez, and C. Denys. 2001. Evolution of the Old World Arenaviridae and their rodent hosts: Generalized host-transfer or association by descent? *Infect Genet Evol* **1**(1):13–20.

Jacamo, R., N. Lopez, M. Wilda, and M. T. Franze-Fernandez. 2003. Tacaribe virus Z protein interacts with the L polymerase protein to inhibit viral RNA synthesis. *J Virol* **77**(19):10383–10393.

Jay, M. T., C. Glaser, and C. F. Fulhorst. 2005. The arenaviruses. *J Am Vet Med Assoc* **227**(6):904–915.

Julander, J. G., K. Shafer, D. F. Smee, J. D. Morrey, and Y. Furuta. 2009. Activity of T-705 in a hamster model of yellow fever virus infection in comparison with that of a chemically related compound, T-1106. *Antimicrob Agents Chemother* **53**(1):202–209.

Kenyon, R. H., P. G. Canonico, D. E. Green, and C. J. Peters. 1986. Effect of ribavirin and tributylribavirin on argentine hemorrhagic fever (Junin virus) in guinea pigs. *Antimicrob Agents Chemother* **29**(3):521–523.

Khan, S. H., A. Goba, M. Chu, C. Roth, T. Healing, A. Marx, J. Fair et al. and Network Mano River Union Lassa Fever. 2008. New opportunities for field research on the pathogenesis and treatment of Lassa fever. *Antiviral Res* **78**(1):103–115.

Kilgore, P. E., T. G. Ksiazek, P. E. Rollin, J. N. Mills, M. R. Villagra, M. J. Montenegro, M. A. Costales, L. C. Paredes, and C. J. Peters. 1997. Treatment of Bolivian hemorrhagic fever with intravenous ribavirin. *Clin Infect Dis* **24**(4):718–722.

Kiso, M., K. Takahashi, Y. Sakai-Tagawa, K. Shinya, S. Sakabe, Q. M. Le, M. Ozawa, Y. Furuta, and Y. Kawaoka. 2010. T-705 (favipiravir) activity against lethal H5N1 influenza A viruses. *Proc Natl Acad Sci USA* **107**(2):882–887.

Kranzusch, P. J., A. D. Schenk, A. A. Rahmeh, S. R. Radoshitzky, S. Bavari, T. Walz, and S. P. Whelan. 2011. Assembly of a functional Machupo virus polymerase complex. *Proc Natl Acad Sci USA* **107**(46):20069–20074.

Kranzusch, P. J. and S. P. Whelan. 2011. Arenavirus Z protein controls viral RNA synthesis by locking a polymerase-promoter complex. *Proc Natl Acad Sci USA* **108**(49):19743–19748.

Kunz, S., P. Borrow, and M. B. Oldstone. 2002. Receptor structure, binding, and cell entry of arenaviruses. *Curr Top Microbiol Immunol* **262**:111–137.

Kunz, S., K. H. Edelmann, J. C. de la Torre, R. Gorney, and M. B. Oldstone. 2003. Mechanisms for lymphocytic choriomeningitis virus glycoprotein cleavage, transport, and incorporation into virions. *Virology* **314**(1):168–178.

Larson, R. A., D. Dai, V. T. Hosack, Y. Tan, T. C. Bolken, D. E. Hruby, and S. M. Amberg. 2008. Identification of a broad-spectrum arenavirus entry inhibitor. *J Virol* **82**(21):10768–10775.

Lecompte, E., J. Ter Meulen, S. Emonet, S. Daffis, and R. N. Charrel. 2007. Genetic identification of Kodoko virus, a novel arenavirus of the African pigmy mouse (Mus Nannomys minutoides) in West Africa. *Virology* **364**(1):178–183.

Lee, A. M., J. M. Rojek, C. F. Spiropoulou, A. T. Gundersen, W. Jin, A. Shaginian, J. York et al. 2008. Unique small molecule entry inhibitors of hemorrhagic fever arenaviruses. *J Biol Chem* **283**(27):18734–18742.

Lee, K. J., I. S. Novella, M. N. Teng, M. B. Oldstone, and J. C. de La Torre. 2000. NP and L proteins of lymphocytic choriomeningitis virus (LCMV) are sufficient for efficient transcription and replication of LCMV genomic RNA analogs. *J Virol* **74**(8):3470–3477.

Lenz, O., J. ter Meulen, H. D. Klenk, N. G. Seidah, and W. Garten. 2001. The Lassa virus glycoprotein precursor GP-C is proteolytically processed by subtilase SKI-1/S1P. *Proc Natl Acad Sci USA* **98**(22):12701–13705.

Leung, W. C., H. P. Ghosh, and W. E. Rawls. 1977. Strandedness of Pichinde virus RNA. *J Virol* **22**(1):235–237.

Lopez, N., R. Jacamo, and M. T. Franze-Fernandez. 2001. Transcription and RNA replication of Tacaribe virus genome and antigenome analogs require N and L proteins: Z protein is an inhibitor of these processes. *J Virol* **75**(24):12241–12251.

Maiztegui, J., F. Feinsod, A. Briggiler, C. J. Peters, D. Enria, H. Lupton, A. Ambrosio et al. 1987. Inoculation of the 1st Argentinean Volunteers with Attenuated CANDID-1 Strain Junin Virus. *Medicina-Buenos Aires* **47**(6):565–565.

Maiztegui, J. I., N. J. Fernandez, and A. J. de Damilano. 1979. Efficacy of immune plasma in treatment of Argentine haemorrhagic fever and association between treatment and a late neurological syndrome. *Lancet* **2**(8154):1216–1217.

Maiztegui, J. I., K. T. McKee, Jr., J. G. Barrera Oro, L. H. Harrison, P. H. Gibbs, M. R. Feuillade, D. A. Enria et al. 1998. Protective efficacy of a live attenuated vaccine against Argentine hemorrhagic fever. AHF Study Group. *J Infect Dis* **177**(2):277–283.

de Manzione, N., R. A. Salas, H. Paredes, O. Godoy, L. Rojas, F. Araoz, C. F. Fulhorst et al. 1998. Venezuelan hemorrhagic fever: Clinical and epidemiological studies of 165 cases. *Clin Infect Dis* **26**(2):308–313.

Martinez, M. G., S. M. Cordo, and N. A. Candurra. 2007. Characterization of Junin arenavirus cell entry. *J Gen Virol* **88**(Pt 6):1776–1784.

McCormick, J. B., I. J. King, P. A. Webb, C. L. Scribner, R. B. Craven, K. M. Johnson, L. H. Elliott, and R. Belmont-Williams. 1986. Lassa fever. Effective therapy with ribavirin. *N Engl J Med* **314**(1):20–26.

McKee, K. T., Jr., J. W. Huggins, C. J. Trahan, and B. G. Mahlandt. 1988. Ribavirin prophylaxis and therapy for experimental Argentine hemorrhagic fever. *Antimicrob Agents Chemother* **32**(9):1304–1309.

Mendenhall, M., A. Russell, T. Juelich, E. L. Messina, D. F. Smee, A. N. Freiberg, M. R. Holbrook et al. 2011. T-705 (favipiravir) inhibition of arenavirus replication in cell culture. *Antimicrob Agents Chemother* **55**(2):782–787.

Meyer, B. J., J. C. de la Torre, and P. J. Southern. 2002. Arenaviruses: Genomic RNAs, transcription, and replication. *Curr Top Microbiol Immunol* **262**:139–157.

Meyer, B. J. and P. J. Southern. 1993. Concurrent sequence analysis of 5' and 3' RNA termini by intramolecular circularization reveals 5' nontemplated bases and 3' terminal heterogeneity for lymphocytic choriomeningitis virus mRNAs. *J Virol* **67**(5):2621–2627.

Meyer, B. J. and P. J. Southern. 1994. Sequence heterogeneity in the termini of lymphocytic choriomeningitis virus genomic and antigenomic RNAs. *J Virol* **68**(11):7659–7664.

Milazzo, M. L., M. N. Cajimat, G. Duno, F. Duno, A. Utrera, and C. F. Fulhorst. 2011. Transmission of Guanarito and Pirital Viruses among Wild Rodents, Venezuela. *Emerg Infect Dis* **17**(12):2209–2215.

Molteni, H. D., H. C. Guarinos, C. O. Petrillo, and F. Jaschek. 1961. Clinico-statistical study of 338 patients with epidemic hemorrhagic fever in the northwest of the province of Buenos Aires. *Sem Med* **118**:839–855.

Moreno, H., I. Gallego, N. Sevilla, J. C. de la Torre, E. Domingo, and V. Martin. 2011. Ribavirin can be mutagenic for arenaviruses. *J Virol* **85**(14):7246–7255.

Morrey, J. D., B. S. Taro, V. Siddharthan, H. Wang, D. F. Smee, A. J. Christensen, and Y. Furuta. 2008. Efficacy of orally administered T-705 pyrazine analog on lethal West Nile virus infection in rodents. *Antiviral Res* **80**(3):377–379.

Nakauchi, M., S. Fukushi, M. Saijo, T. Mizutani, A. E. Ure, V. Romanowski, I. Kurane, and S. Morikawa. 2009. Characterization of monoclonal antibodies to Junin virus nucleocapsid protein and application to the diagnosis of hemorrhagic fever caused by South American arenaviruses. *Clin Vaccine Immunol* **16**(8):1132–1138.

Oldenburg, J., T. Reignier, M. L. Flanagan, G. A. Hamilton, and P. M. Cannon. 2007. Differences in tropism and pH dependence for glycoproteins from the Clade B1 arenaviruses: Implications for receptor usage and pathogenicity. *Virology* **364**(1):132–139.

Oldstone, M. B. 2002. Arenaviruses. I. The epidemiology molecular and cell biology of arenaviruses. Introduction. *Curr Top Microbiol Immunol* **262**:V–XII.

Padula, P. J. and Z. M. de Martinez Segovia. 1984. Replication of Junin virus in the presence of tunicamycin. *Intervirology* **22**(4):227–231.

Perez, M., R. C. Craven, and J. C. de la Torre. 2003. The small RING finger protein Z drives arenavirus budding: Implications for antiviral strategies. *Proc Natl Acad Sci USA* **100**(22):12978–12983.

Perez, M. and J. C. de la Torre. 2003. Characterization of the genomic promoter of the prototypic arenavirus lymphocytic choriomeningitis virus. *J Virol* **77**(2):1184–1194.

Perez, M., D. L. Greenwald, and J. C. de la Torre. 2004. Myristoylation of the RING finger Z protein is essential for arenavirus budding. *J Virol* **78**(20):11443–11448.

Peters, C. J. 2002. Human infection with arenaviruses in the Americas. *Curr Top Microbiol Immunol* **262**:65–74.

Petkevich, A. S., V. M. Sabynin, I. S. Lukashevich, G. A. Galegov, and V. I. Votiakov. 1981. Effect of ribavirin (virazole) on arenavirus reproduction in cell cultures. *Vopr Virusol* 1981 Mar–Apr (2):244–245.

Pinschewer, D. D., M. Perez, and J. C. de la Torre. 2005. Dual role of the lymphocytic choriomeningitis virus intergenic region in transcription termination and virus propagation. *J Virol* **79**(7):4519–4526.

Radoshitzky, S. R., J. Abraham, C. F. Spiropoulou, J. H. Kuhn, D. Nguyen, W. Li, J. Nagel et al. 2007. Transferrin receptor 1 is a cellular receptor for New World haemorrhagic fever arenaviruses. *Nature* **446**(7131):92–96.

Radoshitzky, S. R., J. H. Kuhn, C. F. Spiropoulou, C. G. Albarino, D. P. Nguyen, J. Salazar-Bravo, T. Dorfman et al. 2008. Receptor determinants of zoonotic transmission of New World hemorrhagic fever arenaviruses. *Proc Natl Acad Sci USA* **105**(7):2664–2669.

Raju, R., L. Raju, D. Hacker, D. Garcin, R. Compans, and D. Kolakofsky. 1990. Nontemplated bases at the 5' ends of Tacaribe virus mRNAs. *Virology* **174**(1):53–59.

Rodriguez, M., J. B. McCormick, and M. C. Weissenbacher. 1986. Antiviral effect of ribavirin on Junin virus replication in vitro. *Rev Argent Microbiol* **18**(2):69–74.

Rojek, J. M., A. M. Lee, N. Nguyen, C. F. Spiropoulou, and S. Kunz. 2008. Site 1 protease is required for proteolytic processing of the glycoproteins of the South American hemorrhagic fever viruses Junin, Machupo, and Guanarito. *J Virol* **82**(12):6045–6051.

Rowe, W. P., W. E. Pugh, P. A. Webb, and C. J. Peters. 1970. Serological relationship of the Tacaribe complex of viruses to lymphocytic choriomeningitis virus. *J Virol* **5**(3):289–292.

Rugiero, H. R., H. Ruggiero, C. Gonzalezcambaceres, F. A. Cintora, F. Maglio, C. Magnoni, L. Astarloa et al. 1964. Argentine hemorrhagic fever. II. Descriptive clinical study. *Rev Assoc Med Argent* **78**:281–294.

Salas, R., N. de Manzione, R. B. Tesh, R. Rico-Hesse, R. E. Shope, A. Betancourt, O. Godoy et al. 1991. Venezuelan haemorrhagic fever. *Lancet* **338**(8774):1033–1036.

Salazar-Bravo, J., L. A. Ruedas, and T. L. Yates. 2002. Mammalian reservoirs of arenaviruses. *Curr Top Microbiol Immunol* **262**:25–63.

Salvato, M. S. 1993. Molecular biology of the prototype arenavirus, lymphocytic choriomeningitis virus. In *The Arenaviridae*, ed. M. S. Salvato. New York: Plenum Press.

Salvato, M. S., J. C. S. Clegg, M. J. Buchmeier, R. N. Charrel, J. P. Gonzalez, I. S. Lukashevich, C. J. Peters, and V. Romanowski. 2011. Family *Arenaviridae*. In *Virus Taxonomy—Ninth Report of the International Committee on Taxonomy of Viruses*, eds. A. M. Q. King, M. J. Adams, E. B. Carstens, and E. J. Lefkowitz. London, U.K.: Elsevier/Academic Press.

Salvato, M. S., K. J. Schweighofer, J. Burns, and E. M. Shimomaye. 1992. Biochemical and immunological evidence that the 11 kDa zinc-binding protein of lymphocytic choriomeningitis virus is a structural component of the virus. *Virus Res* **22**(3):185–198.

Salvato, M., E. Shimomaye, and M. B. Oldstone. 1989. The primary structure of the lymphocytic choriomeningitis virus L gene encodes a putative RNA polymerase. *Virology* **169**(2):377–384.

Singh, M. K., F. V. Fuller-Pace, M. J. Buchmeier, and P. J. Southern. 1987. Analysis of the genomic L RNA segment from lymphocytic choriomeningitis virus. *Virology* **161**(2):448–456.

Snell, N. J. 2001. Ribavirin—Current status of a broad spectrum antiviral agent. *Expert Opin Pharmacother* **2**(8):1317–1324.

Southern, P. J., M. K. Singh, Y. Riviere, D. R. Jacoby, M. J. Buchmeier, and M. B. Oldstone. 1987. Molecular characterization of the genomic S RNA segment from lymphocytic choriomeningitis virus. *Virology* **157**(1):145–155.

Stinebaugh, B. J., F. X. Schloeder, K. M. Johnson, R. B. Mackenzie, G. Entwisle, and E. De Alba. 1966. Bolivian hemorrhagic fever. A report of four cases. *Am J Med* **40**(2):217–230.

Strecker, T., R. Eichler, J. Meulen, W. Weissenhorn, H. Dieter Klenk, W. Garten, and O. Lenz. 2003. Lassa virus Z protein is a matrix protein and sufficient for the release of virus-like particles [corrected]. *J Virol* **77**(19):10700–10705.

Strecker, T., A. Maisa, S. Daffis, R. Eichler, O. Lenz, and W. Garten. 2006. The role of myristoylation in the membrane association of the Lassa virus matrix protein Z. *Virol J* **3**:93.

Tesh, R. B., P. B. Jahrling, R. Salas, and R. E. Shope. 1994. Description of Guanarito virus (Arenaviridae: Arenavirus), the etiologic agent of Venezuelan hemorrhagic fever. *Am J Trop Med Hyg* **50**(4):452–459.

Tesh, R. B., M. L. Wilson, R. Salas, N. M. De Manzione, D. Tovar, T. G. Ksiazek, and C. J. Peters. 1993. Field studies on the epidemiology of Venezuelan hemorrhagic fever: Implication of the cotton rat *Sigmodon alstoni* as the probable rodent reservoir. *Am J Trop Med Hyg* **49**(2):227–235.

Tortorici, M. A., C. G. Albarino, D. M. Posik, P. D. Ghiringhelli, M. E. Lozano, R. Rivera Pomar, and V. Romanowski. 2001. Arenavirus nucleocapsid protein displays a transcriptional antitermination activity in vivo. *Virus Res* **73**(1):41–55.

Trombley, A. R., L. Wachter, J. Garrison, V. A. Buckley-Beason, J. Jahrling, L. E. Hensley, R. J. Schoepp et al. 2010. Comprehensive panel of real-time TaqMan polymerase chain reaction assays for detection and absolute quantification of filoviruses, arenaviruses, and New World hantaviruses. *Am J Trop Med Hyg* **82**(5):954–960.

Urata, S., T. Noda, Y. Kawaoka, H. Yokosawa, and J. Yasuda. 2006. Cellular factors required for Lassa virus budding. *J Virol* **80**(8):4191–4195.

Vainrub, B. and R. Salas. 1994. Latin American hemorrhagic fever. *Infect Dis Clin North Am* **8**(1):47–59.

Vela, E. M., L. Zhang, T. M. Colpitts, R. A. Davey, and J. F. Aronson. 2007. Arenavirus entry occurs through a cholesterol-dependent, non-caveolar, clathrin-mediated endocytic mechanism. *Virology* **369**(1):1–11.

Vieth, S., C. Drosten, R. Charrel, H. Feldmann, and S. Gunther. 2005. Establishment of conventional and fluorescence resonance energy transfer-based real-time PCR assays for detection of pathogenic New World arenaviruses. *J Clin Virol* **32**(3):229–235.

Vieth, S., A. E. Torda, M. Asper, H. Schmitz, and S. Gunther. 2004. Sequence analysis of L RNA of Lassa virus. *Virology* **318**(1):153–168.

Weaver, S. C., R. A. Salas, N. de Manzione, C. F. Fulhorst, G. Duno, A. Utrera, J. N. Mills, T. G. Ksiazek, D. Tovar, and R. B. Tesh. 2000. Guanarito virus (*Arenaviridae*) isolates from endemic and outlying localities in Venezuela: Sequence comparisons among and within strains isolated from Venezuelan hemorrhagic fever patients and rodents. *Virology* **266**(1):189–195.

Weissenbacher, M. C., M. A. Calello, M. S. Merani, J. B. McCormick, and M. Rodriguez. 1986. Therapeutic effect of the antiviral agent ribavirin in Junin virus infection of primates. *J Med Virol* **20**(3):261–267.

Wilson, S. M. and J. C. Clegg. 1991. Sequence analysis of the S RNA of the African arenavirus Mopeia: An unusual secondary structure feature in the intergenic region. *Virology* **180**(2):543–552.

Wright, K. E., R. C. Spiro, J. W. Burns, and M. J. Buchmeier. 1990. Post-translational processing of the glycoproteins of lymphocytic choriomeningitis virus. *Virology* **177**(1):175–183.

York, J., D. Dai, S. M. Amberg, and J. H. Nunberg. 2008. pH-induced activation of arenavirus membrane fusion is antagonized by small-molecule inhibitors. *J Virol* **82**(21):10932–10939.

York, J. and J. H. Nunberg. 2006. Role of the stable signal peptide of Junin arenavirus envelope glycoprotein in pH-dependent membrane fusion. *J Virol* **80**(15):7775–7780.

York, J. and J. H. Nunberg. 2007. A novel zinc-binding domain is essential for formation of the functional Junin virus envelope glycoprotein complex. *J Virol* **81**(24):13385–13391.

York, J. and J. H. Nunberg. 2009. Intersubunit interactions modulate pH-induced activation of membrane fusion by the Junin virus envelope glycoprotein GPC. *J Virol* **83**(9):4121–4126.

York, J., V. Romanowski, M. Lu, and J. H. Nunberg. 2004. The signal peptide of the Junin arenavirus envelope glycoprotein is myristoylated and forms an essential subunit of the mature G1-G2 complex. *J Virol* **78**(19):10783–10792.

Young, P. R. and C. R. Howard. 1983. Fine structure analysis of Pichinde virus nucleocapsids. *J Gen Virol* **64**(Pt 4):833–842.

21 Rift Valley Fever Virus and Hemorrhagic Fever

Alexander N. Freiberg and Ramon Flick

CONTENTS

21.1 INTRODUCTION

Rift Valley fever virus (RVFV; *Bunyaviridae: Phlebovirus*) is a zoonotic arthropod-borne pathogen that often results in severe morbidity and mortality in both humans and livestock (Bird et al., 2009; Bouloy and Flick, 2009; Bouloy and Weber, 2010; Ikegami and Makino, 2009; LaBeaud et al., 2010; Pepin et al., 2010) and was first identified in 1930 during an outbreak in sheep in the greater Rift Valley in Kenya (Daubney and Hudson, 1932; Daubney et al., 1931). Since its geographic range continues to spread and now includes most countries in Africa, Madagascar, and the Arabian Peninsula, it presents a real threat to naïve populations around the world by accidental introduction (i.e., the result of increased world travel) or a bioterror/agroterror event. RVFV infection can cause a wide range of disease manifestations in humans, varying from an asymptomatic undifferentiated febrile illness to hemorrhagic fever, encephalitis, retinitis, and death. Infections in pregnant ruminants can result in the so-called abortion storms and result in disastrous economic consequences. There are currently no Food and Drug Administration (FDA)- or U.S. Department of Agriculture (USDA)-licensed vaccines, and therefore, the development of effective countermeasures and implementing surveillance and diagnostic capabilities are critical. Overall, the lack of prophylactic and therapeutic measures, the potential for human-to-human transmission, and the significant threat to human and livestock associated with RVFV make this pathogen a serious public health concern.

21.2   BIOLOGICAL PROPERTIES

RVFV is a member of the *Phlebovirus* genus within the family *Bunyaviridae*. As typical for bunyaviruses, RVFV has a trisegmented single-stranded RNA genome of negative or ambisense polarity (Figure 21.1a). The three genome segments encode four structural proteins: the viral nucleoprotein (N) on the small (S; 1690 nt) segment, the two surface glycoproteins (Gn and Gc) on the medium (M; 3885 nt) segment, and the viral RNA-dependent RNA polymerase (L) on the large (L; 6404 nt) segment (Struthers et al., 1984). In addition, RVFV encodes one nonstructural protein on the S segment (NSs) and two nonstructural proteins on the M segment (NSm and 78 kDa protein). While these proteins are dispensable for viral multiplication in cell culture, they are important for pathogenesis in vivo (Bird et al., 2007a; Gerrard et al., 2007; Vialat et al., 2000; Won et al., 2006).

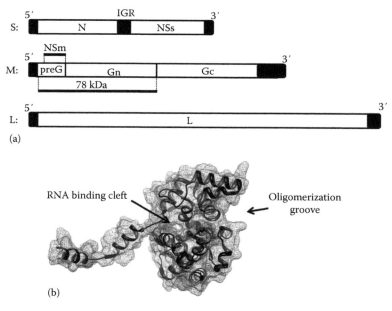

FIGURE 21.1   *RVFV genome organization.* (a) Schematic representation of the antigenomic sense S, M, and L segments and the encoded ORF are represented. The UTRs are indicated as black boxes at the 5′ and 3′ ends. The intergenic region (IGR) present on the S segment is indicated as well. (b) A 3D structure of the monomeric nucleoprotein is shown (Ferron et al., 2011). The secondary structural elements are indicated as ribbons surrounded by the surface area represented as a mesh. The oligomerization groove and RNA-binding cleft are indicated by the arrows.

21.2.1 VIRUS STRUCTURE

Like all bunyaviruses, RVFV is a lipid-enveloped virus. In comparison with bunyaviruses from the other genera, the phleboviruses seem to have a more regular size distribution and spherical, rather than pleomorphic, shape (Ellis et al., 1979; Freiberg et al., 2008). Early studies of bunyaviruses by negative-stain electron microscopy (EM) reported the presence of hollow cylinders arranged on a regular pattern on the virus surface (von Bonsdorff and Pettersson, 1975; Ellis et al., 1979; Pettersson, 1975). Recent studies utilizing cryo-electron tomography and microscopy (cryo-EM) revealed that RVFV particles are actually highly ordered and assemble into an icosahedral symmetry (Figure 21.2a) (Freiberg et al., 2008; Huiskonen et al., 2009; Sherman et al., 2009). The cryo-EM data indicate that the surface glycoproteins, Gn and Gc, assemble in capsomer-like structures, which are arranged on an icosahedral lattice with a triangulation number (T) of 12. This T number has so far only been described for one other virus, Uukuniemi virus (UUKV), which belongs like RVFV to the *Phlebovirus* genus (Overby et al., 2008). The 3D reconstructions and calculations of the capsomer and glycoprotein volumes estimated 122 capsomers present on the virus surface formed by 720 Gn–Gc heterodimers (Freiberg et al., 2008; Huiskonen et al., 2009). There are 12 pentameric (pentons) and 110 hexameric (hexons) capsomers and the protrusions extend approximately 96 Å above the lipid envelope (Figure 21.2c). The cryo-EM 3D structures also showed the presence of a regular network of ridges connecting neighboring capsomers, suggestive of their potential importance for virus

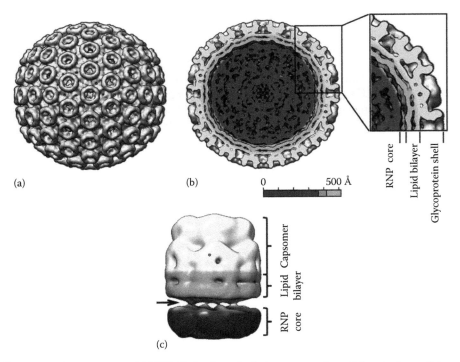

FIGURE 21.2 *A 3D structure of RVFV.* (a) Surface-shaded representation of the single-particle cryo-EM map of RVFV vaccine strain MP-12. (Adapted from Sherman, M.B., *Virology*, 387, 11, 2009.) The cryo-EM 3D map represents the T=12 icosahedral symmetry. The two glycoproteins, Gn and Gc, are arranged in 122 capsomers on the virus surface. (b) Representation as in panel A but with the front half of the map removed. An enlarged area of the virus particle indicates the outer glycoprotein shell, lipid bilayer, and RNP core. The RNP core has no specific symmetry. (c) Extracted capsomer indicating one of the side channels inside ridges connecting neighboring capsomers (asterisk). The arrow indicates densities spanning the gap separating the RNP core from the lipid envelope. These densities most likely represent the glycoprotein cytoplasmic tails.

assembly. Inside the virion, no symmetry could be described, indicative of a random arrangement of the ribonucleoprotein (RNP) segments within the virion (Figure 21.2b).

The RNP core is separated from the inner surface of the lipid envelope by an approximately 20 Å gap. In contrast to other negative-sense RNA strand viruses, bunyaviruses do not contain a matrix protein beneath the lipid envelope. In the 3D reconstructions, a layer of RNP seems to be located proximal to the inner leaflet of the membrane and densities spanning the space between the RNP core and the lipid envelope are visible (Huiskonen et al., 2009; Sherman et al., 2009). These densities are most likely representing the glycoprotein cytoplasmic tails interacting with the RNP complexes (Figure 21.2c). This interaction would compensate for the absence of a matrix protein and the importance of the cytoplasmic tail domains has been described by the use of minigenome systems for members of different bunyavirus genera (Overby et al., 2007; Shi et al., 2007; Snippe et al., 2007).

21.2.2 GENOME ORGANIZATION

The three RNA segments contain untranslated regions (UTRs) at the 3′ and 5′ ends. These UTRs are segment specific and serve as promoters for transcription and contain *cis*-acting RNA replication signals of approximately 30 nt at their termini (Schmaljohn and Nichol, 2007). The 3′ and 5′ sequences of the UTRs of each segment possess the sequence 5′-ACACAAAG … CUUUGUGU-3′, and the first eight nucleotides are strictly conserved within the S, M, and L segments of all phleboviruses. The UTRs of all three segments are complementary to each other (Kolakofsky and Hacker, 1991; Pardigon et al., 1982; Patterson et al., 1983), which allows for the formation of a panhandle structure, explaining why the RNPs have a circular appearance when observed by EM (Flick et al., 2002; Obijeski et al., 1976; Pettersson and von Bonsdorff, 1975; Samso et al., 1975). The involvement of *cis*-acting elements within the flanking UTRs in regulation of RNA synthesis had previously been demonstrated by Flick et al. (2004) for the closely related phlebovirus UUKV. The M and L segments are of negative polarity, while the phlebovirus S segment utilizes an ambisense coding strategy for the N and NSs proteins (Giorgi et al., 1991). The coding capacity of the genome is depicted in Figure 21.1a. In purified virus particles, not only the viral negative-sense genomes (vRNA) were detected but also a small fraction of the so-called complementary RNA (cRNA) (i.e., replicative intermediates) (Ikegami et al., 2005). This finding confirms past studies, which showed that the genome and cRNA polarity of the S segment could be detected in purified UUK virions (Simons et al., 1990). The presence of the S cRNA together with the genome in the virus assures that both the N and NSs proteins are expressed at early stages of the virus life cycle.

21.3 PROTEIN FUNCTIONS

21.3.1 NUCLEOPROTEIN

The N protein is encoded on the S segment in antisense orientation and is the most abundant viral protein in the phlebovirus virion. N plays an important role for encapsidation of the viral RNA, for multiple steps during the replication cycle, including transcription and replication, and packaging of the viral genomes into the virus particles (Schmaljohn and Nichol, 2007). Recently, the structure of RVFV N protein was solved (Figure 21.1b) and the structures revealed that the monomeric N protein has a novel all-helical fold unique among negative-sense RNA viruses (Ferron et al., 2011; Raymond et al., 2010). As an oligomer, RVFV N protein assembles into a ring-shaped hexamer (Ferron et al., 2011). Compared to the N oligomers from other negative-strand RNA viruses (e.g., vesicular stomatitis virus (VSV), respiratory syncytial virus (RSV), and influenza virus), the ring forms by only six RVFV N subunits and a diameter of 100 Å is the smallest (Ferron et al., 2011). The RVFV RNPs do not assemble into highly ordered structures but rather into flexible filamentous assemblies. A positively charged patch has been identified within the inner part on one side of the hexamer, and mutagenesis experiments supported the possibility of this patch as an RNA-binding cleft. Piper et al. (2011) reported that N interacts with the cytoplasmic tail domain of Gn and that the RNPs localize to the interior of the virion. While nucleoproteins are not known to be directly involved in pathogenesis, experiments using recombinant expressed N have shown it to elicit a partial immune protection (Lagerqvist et al., 2009; Lorenzo et al., 2010). However, although RVFV N protein is strongly immunogenic, it cannot elicit neutralizing antibodies.

21.3.2 Nonstructural Protein NSs

The phlebovirus NSs is the major virulence factor, primarily acting as an interferon antagonist by inhibiting IFN-β mRNA expression (Billecocq et al., 2004; Bouloy et al., 2001; Elliott and Weber, 2009). RVFV NSs is phosphorylated and accumulates in the nuclei of infected cells, where it forms fibrillar structures (Struthers and Swanepoel, 1982; Swanepoel and Blackburn, 1977). Mansuroglu et al. (2010) showed that cellular DNA is mainly excluded from the NSs filaments except for some clusters of a pericentromeric gamma-satellite sequence, which has been correlated with the induction of chromosome cohesion and segregation defects in RVFV-infected murine and ovine cells.

Ikegami et al. (2005) reported that virions can incorporate cRNA serving as templates for NSs mRNA, which then results in the early presence of NSs during infection without the need of a replication step of the S segment. The molecular inhibitory effect of NSs is mediated through multiple mechanisms and involves several cellular proteins. NSs is able to inhibit the formation of the transcription factor TFIIH by interacting with p44, a subunit of TFIIH (Le May et al., 2004). This interaction results in a competitive inhibition with XPD, another component of TFIIH. This transcriptional inhibition is resulting in the general suppression not only of the host cell transcription but also of the antiviral response (Le May et al., 2004). At an early stage of infection, the transcription of IFN-β is inhibited (3–4 h postinfection), while the general host cell transcription is blocked at later stages of the viral life cycle (8–9 h postinfection). Another mode of transcriptional suppression is the specific inhibition of the IFN-β expression through the interaction of NSs with SAP30 (Le May et al., 2008). Further, RVFV NSs is able to disrupt the activity of the double-stranded RNA-dependent protein kinase R (PKR), therefore interfering with the host cellular antiviral response (Habjan et al., 2009b; Ikegami et al., 2009a and b). This function of RVFV NSs represents an inhibition of the translational activity of the host cell. NSs promotes the posttranscriptional downregulation of PKR through the proteasomal pathway and prevents the phosphorylation of the eukaryotic initiation factor 2 alpha (eIF2α). RVFV strains, which are deficient in functional NSs, are less virulent than the wild-type strains and have been explored as potential attenuated vaccine candidates [reviewed in (Bird et al., 2009; Boshra et al., 2011; Bouloy and Flick, 2009; Ikegami and Makino, 2009; LaBeaud et al., 2008)].

21.3.3 Nonstructural Protein NSm and 78 kDa Protein

The M segment encodes NSm, 78 kDa, Gn and Gc proteins, and those are synthesized from a single open reading frame (ORF) of the M mRNA. Different AUGs are present at the 5′ region of the M mRNA, resulting in either the 78 kDa protein (1st AUG) or NSm (2nd AUG) (Figure 21.1a). While the C-terminus of the 78 kDa protein is identical to that of Gn, the C-terminus of NSm is generated by cleavage at the N-terminus of Gn. Both NSm and 78 kDa proteins are dispensable for viral replication in cell cultures (Gerrard et al., 2007; Ikegami et al., 2006; Muller et al., 1995; Won et al., 2006) and are not essential for virulence and lethality

(Bird et al., 2007a). However, a recombinant RVFV lacking the expression of both the 78 kDa protein and NSm indicated the development of more apoptosis compared to the parent RVFV strain in cell culture (Won et al., 2007). Expression of NSm is able to inhibit the cleavage of caspase-8 and caspase-9, demonstrating that NSm suppresses apoptosis. It is therefore possible that NSm affects viral pathogenicity in the infected host.

21.3.4 GLYCOPROTEINS

The RVFV glycoproteins, Gn and Gc, are important for attachment to the host cell, entry, and maturation of the virus. It has been suggested that Gn is required for virus budding, since recombinant expressed Gn localizes to the Golgi complex through its Golgi localization signal, while Gc retains in the ER if expressed alone (Gerrard and Nichol, 2002; Piper et al., 2011). Gc associates with Gn in the ER and moves to the Golgi apparatus through its physical interaction with Gn.

A recent study by Morrill and colleagues associated the virulence phenotype of the RVFV ZH501 strain with a single amino acid substitution in the Gn protein (at nucleotide 847) of the M segment (Morrill et al., 2010). Recombinant expressed Gn and Gc can form virus-like particles (VLPs) and are recognized by the immune system (de Boer et al., 2010; Liu et al., 2008). RVF VLPs induce the production of neutralizing antibodies and result in complete immune protection from challenge with wild-type RVFV in the mouse and rat model (de Boer et al., 2010; Mandell et al., 2010a,b; Naslund et al., 2009; Pichlmair et al., 2010).

21.3.5 RNA-DEPENDENT RNA POLYMERASE

The polymerase is expressed on the L segment, represents the largest viral protein expressed, and has two functions, transcription of viral mRNA and replication of viral genomes (Lopez et al., 1995). These two functions should require a switch between the transcriptase and replicase activities, but the mechanism is still unknown. Zamoto-Niikura et al. (2009) demonstrated that the conserved SDD motif (serine-aspartic acid-aspartic acid) in the L protein is mediating activity and that oligomerization of the L protein is important for polymerase activity. Recently, the crystal structure of the endonuclease domain of the orthobunyavirus La Crosse L protein has been solved, and alignments with other bunyaviral polymerase sequences revealed that this domain was conserved (Reguera et al., 2010).

21.3.6 ENTRY, REPLICATION CYCLE, AND ASSEMBLY

The two viral glycoproteins are essential for binding to the host cell and entry of the virus. A recent study by Lozach et al. (2011) showed that RVFV and several other phleboviruses use dendritic cell-specific intercellular adhesion molecule-3-grabbing non-integrin (DC-SIGN) to infect immature dendritic cells and other DC-SIGN-expressing cells. Phlebovirus glycoproteins have several N-linked oligosaccharides, which are necessary to bind DC-SIGN. However, bunyaviruses have the ability to infect a wide spectrum of cells and many of those do not express DC-SIGN (Schmaljohn and Nichol, 2007; Svajger et al., 2010). Therefore, the identity of the host cell receptor(s) has still not been solved. Bioinformatic analysis has predicted that the bunyavirus Gc protein resembles a class II fusion protein (Garry and Garry, 2004) and low pH following endocytosis of the virus particle is necessary to activate entry (Filone et al., 2006). Further, it has been shown that UUKV is endocytosed within 10 min after binding to the host cell and that this step involves noncoated vesicles (Lozach et al., 2010). Acidification of the endocytic vesicles is thought to promote a conformational change in the glycoproteins (presumably mainly in Gc) that facilitates fusion of the viral and cellular membrane. Upon penetration of the host cell from late endosomal compartments and uncoating the encapsidated genome, RNA-dependent RNA polymerase is released into the cytoplasm, where the primary transcription takes place to synthesize viral mRNA (Ikegami et al., 2005).

After uncoating of viral genomes (vRNA), primary transcription of negative-sense vRNA to mRNA is initiated by interaction of the virion-associated L protein and the three viral RNPs (Bouloy and Hannoun, 1976). All replication steps of the phlebovirus life cycle occur in the cytoplasm of the infected cell and start around 1–2 h postinfection (Ikegami et al., 2005). The replication process involves a replicative intermediate of each of the three genomic RNA molecules, representing an exact copy of the viral genome (cRNA). The RNA-dependent RNA polymerase acts both as a transcriptase and replicase (Jin and Elliott, 1993) but requires N for both activities (Lopez et al., 1995). The vRNA and cRNA are always encapsidated by the N proteins and associated with the L protein (Dunn et al., 1995; Lopez et al., 1995). For the S segment, the cRNA serves as the template for the transcription of NSs mRNA and the vRNA as a template for N mRNA transcription.

Synthesis of the mRNA is initiated through a cap-snatching mechanism, while the synthesis of the cRNA is initiated with 5′ nucleoside triphosphates (Gro et al., 1992; Reguera et al., 2010; Simons and Pettersson, 1991). In contrast to cellular and most other viral mRNAs, bunyavirus mRNA is not polyadenylated (Schmaljohn and Nichol, 2007). During mRNA synthesis, a premature termination of the mRNAs on their templates indicates that the L protein can recognize a transcription termination signal. For the M mRNA, termination of the transcription occurs approximately 100–150 nt before the end of the M template (Collett, 1986) and for the S mRNA within the IGR of the S segment (Gro et al., 1992; Ihara et al., 1985; Simons and Pettersson, 1991). Two types of transcription termination signals were suggested for different phleboviruses (Giorgi et al., 1991; Gro et al., 1992): (a) a C-rich sequence motif in the IGR found for RVFV and Toscana virus or (b) a hairpin structure in the IGR found for viruses like UUK and Punta Toro (Iapalucci et al., 1991; Lopez et al., 2007). In addition, the importance of a pentanucleotide sequence, 5′-GCUGC-3′, as a transcription terminator for N, NSs, and M mRNA has then been described (Albarino et al., 2007; Ikegami et al., 2007). Recently, Lara et al. (2011) could demonstrate that the pentanucleotide is also likely the signal for the termination of the 3′ end of the L mRNA.

The S and L mRNAs are translated on free ribosomes, while the M mRNAs are translated on membrane-bound ribosomes. The glycoproteins enter the secretory pathway as a precursor polyprotein, which is cleaved by signal peptidase to yield mature Gn and Gc (Gerrard et al., 2007). Following translocation and cleavage, both glycoproteins are glycosylated at asparagine residues. Gn and Gc are targeted to the Golgi and the transport signal has been localized to Gn and complexing of Gn and Gc in the ER is necessary for efficient transport of Gc to the Golgi. The phlebovirus Gc protein has a characteristic C-terminal ER retention motif, which evidently can be overcome when Gn and Gc oligomerize (Gerrard and Nichol, 2002). Localization of the glycoproteins in the Golgi complex is essential for proper maturation of new infectious particles, which occurs at smooth membranes of the Golgi and differentiates bunyaviruses from other enveloped viruses, which bud at the cell plasma membrane and/or intracellular membranes of the endoplasmic reticulum (such as paramyxoviruses or flaviviruses). The only exception is RVFV-infected hepatocytes, in which budding of the virus particles was observed to occur at the plasma membrane in addition to the Golgi vesicles (Anderson and Smith, 1987). For the production of infectious bunyavirus particles, the virus must carry all three RNA segments. A study by Terasaki et al. (2011) recently demonstrated that a direct or indirect intermolecular interaction between the S, M, and L segments is required to drive the co-packaging of the three viral RNA segments to produce infectious RVFV. A recent study by Piper et al. (2011) demonstrated that Gn is necessary and sufficient to package the RNP complexes. While the cytoplasmic tail of Gc is dispensable for recruitment and packaging of the genomes, N proteins, and polymerase, the first 30 amino acids of the Gn cytoplasmic tail are important for binding of N and the last 40 amino acids of the 70 amino acid cytoplasmic tail are required for binding of the polymerase. These studies indicate that interaction between multiple Gn's and encapsidated genomes results in a change of membrane curvature, which ultimately leads to maturation of the virions into the Golgi lumen (Piper et al., 2011). Late in infection, the Golgi complex undergoes morphological changes, and after budding into the Golgi cisternae, virus particles are transported to the cell surface within vesicles analogous to those in the secretory pathway. Fusion with the plasma membrane then allows for the release of virions into the extracellular medium.

21.4 CLINICAL PRESENTATION

Among the phleboviruses known to cause disease in humans, RVFV is the most severe and a pathogen of concern for public and animal health, which can result in disastrous economic consequences (Mandell and Flick, 2010). Infection can cause a wide range of disease in humans, varying from an asymptomatic (30%–60%), over a weeklong undifferentiated febrile illness, to hemorrhagic fever and death (the overall case fatality rate is estimated to be between 0.5% and 2%). The most frequent course of the disease is a febrile illness. However, during recent outbreaks, the case fatality of hospitalized cases was increased (Kahlon et al., 2010; LaBeaud et al., 2008; Madani et al., 2003). While less than 2% of human infections result in severe disease (hemorrhagic fever, encephalitis, or retinitis), this can actually add up to a substantial number of cases given the large scale of some epizootics. For example, an estimated 27,500 RVFV infections occurred during the 1997/1998 outbreak in the Garissa district of Kenya (Woods et al., 2002), and approximately 200,000 human cases with 600 deaths were estimated to have occurred during the outbreak in Egypt in 1977 and 1978 (Meegan, 1981). Human infections normally occur in groups having close contact with livestock, such as herdsmen, veterinarians, and abattoir workers.

Infected patients either experience no detectable symptoms or develop a mild form of the disease, which is characterized by a feverish syndrome with a sudden onset of flu-like fever, muscle pain, joint pain, and headache. Some patients also develop neck stiffness, sensitivity to light, loss of appetite, and vomiting. The symptoms of RVF usually last from 4 to 7 days. Then the immune response becomes detectable with the appearance of antibodies and the virus gradually disappears from the blood. However, a small percentage of patients develop a much more severe form of the disease. This usually appears as one or more of three distinct syndromes: ocular disease (0.5%–2%), meningoencephalitis (less than 1%), or hemorrhagic fever (less than 1%). Ophthalmologic complications remain important sequelae of human RVF disease and result in long-lasting visual disturbance in affected individuals (LaBeaud et al., 2008). In its most severe form, RVF can also lead to acute renal dysfunction (El Imam et al., 2009).

While the general signs and symptoms of a patient infected with RVFV mimic many other febrile tropical illnesses present in the geographic areas where RVF is endemic, no distinctive clinical syndrome for RVF has been described. A recent publication by Kahlon et al. (2010) assessed a small case series of six acutely ill RVF hemorrhagic fever patients during an outbreak in Kenya in 2007 and described the clinical phenotype of severe RVF. The illness was characterized by an early syndrome of fever, throbbing headaches, malaise, proximal large-joint arthralgias in knees, elbows, and shoulders, and anorexia. These signs of disease were then followed by vomiting, nausea, and vague midepigastric discomfort. While currently RVF in humans is distinguished from other febrile illnesses mainly by its association with concurrent illness in livestock, this more thorough documentation of the clinical syndromes of RVF will greatly help the clinician at the bedside to recognize the disease. However, more work is needed to describe specific manifestations of RVF in humans to recognize the disease in its early phase and to help limit its spread.

21.5 DIAGNOSIS

RVFV belongs to the group of RNA viruses, including Lassa, Ebola, Marburg, and Crimean–Congo hemorrhagic fever and yellow fever viruses, which have the potential to cause viral hemorrhagic fever (VHF). Infections by VHF viruses are clinically difficult to recognize, if no hemorrhagic or specific organ manifestations are present. Further, work with these viruses requires high-biocontainment facilities and currently only a limited number of reference laboratories worldwide are available.

In endemic areas, diagnosis of RVFV infection is often performed by combining the analysis of clinical signs of disease and available diagnostic testing. RVF may be suspected when there is an outbreak of febrile illness in humans associated with headache and myalgia and the occurrence of abortions in domestic ruminants and death of young animals. Due to the fact that RVF outbreaks normally occur in very remote locations, field deployable techniques are essential for the rapid detection of RVFV so that effective control of the outbreak can be initiated. A major obstacle in the management of VHF outbreaks in general is the inability to identify cases with poor prognosis early enough to allow for a more intense supportive therapy and possibly the administration of experimental chemotherapeutic drugs.

Diagnosis of RVF is achieved by a variety of techniques, such as virus isolation (Anderson et al., 1989), viral antigen detection (Meegan et al., 1989; Niklasson et al., 1983), detection of specific antibodies (Swanepoel et al., 1986a), and amplification of viral nucleic acids (Drosten et al., 2002; Garcia et al., 2001; Ibrahim et al., 1997). While isolation of the virus can be achieved from whole blood, serum, or tissue samples, it is a lengthy and expensive procedure and requires high-containment laboratories. Therefore, diagnostic assays based on the detection of nucleic acids are preferable to rapidly and reliably analyze samples for RVFV.

The development of multiplexed PCR, reverse transcription (RT)-PCR enzyme hybridization assays, and real-time detection PCR is very helpful, since detection can be performed simultaneously for many hemorrhagic fever viruses (Bird et al., 2007b; Drosten et al., 2002; Garcia et al., 2001; He et al., 2009; Ibrahim et al., 1997; Sall et al., 2002). During an RVF outbreak in Kenya from December 2006 through March 2007, individuals with high viremia could rapidly be identified by using field RT-PCR testing (Njenga et al., 2009). This study demonstrated for the first time a strong correlation between high viremia levels in RVFV cases and fatality and demonstrated that quantitative (q)RT-PCR test can be performed quickly in a field-based laboratory setting. Another practical technique for use in a field laboratory setting is the real-time RT-loop-mediated isothermal amplification (RT-LAMP) assay (LeRoux et al., 2009; Peyrefitte et al., 2008). The advantage of this assay is that it does not rely on thermocycler equipment; it only requires a one-step, single-tube reaction, has a comparable sensitivity and specificity to RT-PCR, and is faster than a traditional PCR reaction (<30 min).

Diagnostics should not only rely on PCR but rather should be combined with other tests, that is, detection of RVFV-specific antibodies. This prevents potentially false-negative results due to the rather short viremic phase during which RVFV genomes can be detected.

RVFV antigen can rapidly be detected in blood and tissue samples by using immunological techniques, such as immunohistochemistry staining of tissue samples or enzyme-linked immunosorbent assays (ELISA) (Meegan et al., 1989; Niklasson et al., 1983; Zaki et al., 2006). A highly specific, safe, and simple sandwich ELISA for the detection of RVFV N in inactivated specimens has recently been described and represents a valuable diagnostic tool for less-equipped laboratories in Africa (Jansen van Vuren and Paweska, 2009). Serological tests, such as detection of type-specific IgG and IgM antibodies, can also easily be performed shortly after exposure to the virus (Paweska et al., 2003a,b, 2005a,b). Other serodiagnostic tests available for detection of RVFV are hemagglutination-inhibition, indirect immunofluorescence, and virus neutralization test (VNT) (Paweska et al., 2005b, 2007; Swanepoel et al., 1986b). However, disadvantages of these techniques are high costs, long duration of the assay, and potential health risks for the laboratory personnel. Alternatively, the MP-12 vaccine strain can be used instead of the wild-type virus eliminating the need of a high-containment biosafety (BSL)-3+ or BSL-4 laboratory. The main issue with VNT as a diagnostic tool is that RVFV induces a long-lasting neutralizing immunity and therefore previously infected individuals will also give a positive result.

21.6 PATHOLOGY AND PATHOGENESIS

A majority of the patients infected with RVFV suffer from a self-limiting, mild to often subclinical febrile illness, with an incubation time of typically 4–6 days (Daubney et al., 1931; Findlay and Daubney, 1931; Joubert et al., 1951; Madani et al., 2003; Mundel and Gear, 1951; Strausbaugh et al., 1978). Symptoms for RVF include severe chills, weakness, malaise, throbbing headache, and dizziness (Findlay and Daubney, 1931; Kahlon et al., 2010). Further, patients can develop a sensation of fullness over the liver region. Following these symptoms, RVF patients have a fever, decreased blood pressure, proximal large-joint arthralgia (particularly shoulders, elbows, and knees), and anorexia followed by nausea and vomiting. Between the 3rd and 4th day after onset of symptoms, a lessening can be observed and the body temperature returns to a normal level. However, in some patients, the fever can last for as much as 10 days, or within 1–3 days after the recovery, a recurrence of high fever can be observed (Findlay and Daubney, 1931; Mundel and Gear, 1951). This biphasic febrile phase is normally combined with a severe headache. During the febrile period, virus can be detected in the blood, as well as neutralizing antibodies, which appear around day 4 after onset of symptoms (Findlay and Daubney, 1931; Sabin and Blumberg, 1947). Patients can also develop a massive coronary thrombosis after the febrile phase (Mundel and Gear, 1951; Schwentker and Rivers, 1934).

In some cases, RVF patients can develop a maculopathy or retinopathy (0.5%–2%), which can affect either one or both eyes (Deutman and Klomp, 1981; Salib and Sobhy, 1978; Siam et al., 1980). Patients described a blurry vision and noticed floating black spots within their visual fields or even the loss of central vision. This pathology of RVF can occur either very early after infection or months later. In many cases, no complete recovery of vision occurs and chorioretinal scarring can remain in macular and paramacular areas (Al-Hazmi et al., 2005; Ayoub et al., 1978; Deutman and Klomp, 1981; Freed, 1951; Salib and Sobhy, 1978; Siam and Meegan, 1980; Siam et al., 1978). However, some patients show a partial improvement in vision over time (Salib and Sobhy, 1978; Schrire, 1951; Siam and Meegan, 1980; Siam et al., 1978).

Human cases of RVF, which develop jaundice, hemorrhagic fever, or neurological disease, are at increased risk of fatality (Laughlin et al., 1979; Madani et al., 2003). Patients involving hemorrhagic manifestation typically have a sudden onset of illness, including fever, headache, lethargy, vomiting, nausea, midepigastric discomfort, and body pain. Further, these patients may present with a low blood pressure, hematemesis, diarrhea, jaundice, macular rash over the entire trunk, subconjunctival hemorrhage, and bleeding from the gums and/or gastrointestinal mucosal membrane (Kahlon et al., 2010; Swanepoel et al., 1979; Yassin, 1978). Typically, elevated aspartate aminotransferase (AST), lactate dehydrogenase (LDH), alanine aminotransferase (ALT), and reduction of platelet count and hemoglobin are observed (Al-Hazmi et al., 2003; Swanepoel et al., 1979). Death can occur within 3–17 days after onset of symptoms. Fatal RVF cases show diffuse hepatic and gastrointestinal necroses (Abdel-Wahab et al., 1978; Swanepoel et al., 1979).

RVF patients developing encephalitis may present with a sudden onset of fever, rigor, retroorbital and throbbing headache, mild arthralgias of the knees and elbows, and bilateral retinal hemorrhage (Alrajhi et al., 2004; Kahlon et al., 2010; Maar et al., 1979). Eventually, patients experience neck rigidity after onset of illness, confusion, hypersalivation, fatigue, malaise, stupor, and coma, and temporal or permanent vision loss has been described as well (Alrajhi et al., 2004; Maar et al., 1979; van Velden et al., 1977). Indicative of the possible development of meningitis or meningoencephalitis was the increased number of white blood cells (mainly consisting of lymphocytes) in the cerebrospinal fluid (CSF). The consciousness level can be decreased in these patients as well (Alrajhi et al., 2004). Histopathological lesions in the brains of these patients were characterized by focal necrosis associated with infiltration of lymphocytes and macrophages and perivascular cuffing (van Velden et al., 1977).

Multiple animal models (rodent, livestock, and primates) have been developed to study the pathogenesis caused by RVFV and to evaluate therapeutics, antivirals, and vaccines. Detailed descriptions of individual animal models have recently been reviewed in Ikegami and Makino (2011), Boshra et al. (2011), and Ross et al. (2011).

21.7   EPIDEMIOLOGY

The geographic distribution of RVFV covers much of Africa, ranging from Egypt to South Africa and from the Senegal to Madagascar. However, the first confirmed outbreak of RVFV outside Africa was reported in 2000 in Saudi Arabia and Yemen (Ahmad, 2000; Shoemaker et al., 2002). The evolution of RVFV has been influenced by changes in the environment during the past 150 years (Battles and Dalrymple, 1988; Bird et al., 2007c, 2008b; Sall et al., 1997, 1998, 1999; Shoemaker et al., 2002; Xu et al., 2007). No exclusive correlation between geographic location and virus genotype can be observed; however, representative genotypes from one area tend to cluster within a specific lineage (Bird et al., 2007c; Pepin et al., 2010). At least three separate introductions of RVFV have been recognized across significant natural geographic barriers. The large outbreak in Egypt during 1977–1979 was the first time when the virus was recognized north of the Sahara desert and was associated with the construction of the Aswan High Dam along the Nile River (Johnson et al., 1978). Later then in 1979, RVFV was isolated for the first time in Madagascar, outside of continental Africa, and is since then endemic (Morvan et al., 1991a,b, 1992a,b). As already mentioned, more recently, RVFV has been detected in Saudi Arabia and Yemen (Ahmad, 2000; Shoemaker et al., 2002).

RVFV has the potential to infect a remarkable number of different vectors, including mosquitoes, ticks, and flies. However, mosquitoes, especially floodwater *Aedes* mosquitoes, are the principal vector and play an important role in RVFV epizootics, which frequently occur at times of unusual high precipitation (Linthicum et al., 1999). In general, the vectors can be classified into two groups: (1) reservoir vectors, including certain *Aedes* species (spp.), and (2) amplifying vectors, consisting of *Culex* spp. (Fontenille et al., 1994, 1995; Gear et al., 1955). Damboes are thought to play a central role, because they flood during heavy rainfall, which can result in an explosive population growth in floodwater *Aedes* mosquitoes (Swanepoel, 1981). During interepidemic periods, RVFV was isolated from unfed mosquitoes, which demonstrated that the virus can be maintained between epidemics through transovarial transmission in mosquitoes (Linthicum et al., 1985). Following heavy rainfall and flooding, infected mosquitoes then hatch and feed on nearby livestock, which either can initiate an epizootic or the viremic livestock can act as an amplifying host and transmit the virus to other mosquitoes (Swanepoel, 1981). Once livestock is infected, the classical hallmark of RVF epizootics is the large number of abortions observed among pregnant ruminants. Usually 1–2 weeks after these abortions, initial human cases can be seen, normally involving farmers and others living in close proximity to livestock. RVFV infection can be acquired either through the bite from an infected mosquito or more importantly through direct contact with infected livestock. People involved in the birth or abortions of livestock, butchering animals, or abattoir workers are at high risk of infection during epizootics (Abu-Elyazeed et al., 1996; Chambers and Swanepoel, 1980). Aerosol exposure has been demonstrated to be another potential route of infection, especially in a laboratory setting (Francis and Magill, 1935; Smithburn et al., 1949).

21.8   TREATMENT

Different strategies are being employed to generate efficacious and safe vaccine candidates and therapeutics for animal and human use. However, the development and approval process has proven to be quite difficult (Bird et al., 2009; Boshra et al., 2011; Bouloy and Flick, 2009; Ikegami and Makino, 2009; Labeaud, 2010; Mandell and Flick, 2010; Pepin et al., 2010). The current focus is to generate safer live-attenuated RVFV strains by reverse genetics, the use of chimeric viral replicon systems, DNA vaccines, and VLPs (Bird et al., 2009; Boshra et al., 2011; Bouloy and Flick, 2009; Labeaud, 2010; LaBeaud et al., 2010; Mandell and Flick, 2010).

21.8.1 RVFV Vaccines for Veterinary Applications

A vaccine targeting RVFV infections in livestock is a promising approach to prevent epizootics. The currently used Smithburn vaccine, a modified live virus, is only partially attenuated and leads to high abortion rates or teratology (10%–25%) of pregnant animals (Coetzer and Barnard, 1977; Kamal, 2009; Swanepoel and Coetzer, 2004) and exhibits pathogenicity in European cattle (Botros et al., 2006). In addition, the risk of reversion to full virulence precludes its use in countries where RVFV is not known to be endemic (Swanepoel and Coetzer, 2004). Inactivated derivatives of the Smithburn vaccines have been also extensively tested in livestock (Barnard, 1979; Barnard and Botha, 1977; Swanepoel and Coetzer, 2004). However, formalin-inactivated vaccines are more expensive to produce and require multiple inoculations to protect animals. Therefore, they are not a vaccine of choice during outbreaks when a rapid onset of immunity is required.

21.8.2 RVFV Vaccines for Human Use

There is currently no licensed vaccine available for use in the United States or Europe. TSI-GSD-200, a formalin-inactivated vaccine, is the only RVFV vaccine presently available for use in humans (Pittman et al., 1999). TSI-GSD-200 is produced in rhesus lung cells and is available for military personnel, veterinarians working in endemic areas, high-containment laboratory workers, and others at high risk for contracting RVFV. However, this vaccine requires multiple inoculations, is expensive, and is in short supply (Frank-Peterside, 2000; Kark et al., 1982, 1985; Niklasson et al., 1985; Pittman et al., 1999), making its administration prohibitive for widespread use.

21.8.2.1 Live-Attenuated Viruses

A live-attenuated vaccine, MP-12, is efficacious in livestock. However, MP-12 may have similar safety concerns associated with the Smithburn strain, since MP-12 can be teratogenic in sheep: one study showed that pregnant sheep vaccinated with MP-12 within the first trimester resulted in abortion (4%) and teratogenic effects (14%) (Hunter et al., 2002). MP-12 was derived by mutagenesis (5-flourouracil) (Caplen et al., 1985) of an RVFV strain (ZH548) that was isolated during the 1977 outbreak in Egypt. MP-12 was recently tested in human clinical trials, to determine adverse effects in humans using a single injection dose escalation protocol, with promising results (Bettinger et al., 2009). A 95% seroconversion rate was reported with a high titer in a plaque reduction neutralization test (PRNT). In addition, a genetic analysis of MP-12 isolated from vaccinated individuals showed no reversions of vaccine virus in attenuated regions to wild-type RVFV. This was important since MP-12 is only based on one or more mutations in all three segments and a reversion to wild-type virus was a valid concern. The use of MP-12 as a livestock vaccine is limited based on the fact that no differentiation can be easily made between vaccinated and infected animals (differentiation of infected from vaccinated animals [DIVA] concept).

Several alternative live-attenuated RVFV vaccines are being developed: (i) the natural isolate Clone 13 (Muller et al., 1995), (ii) the reassortant R566 (Bouloy and Flick, 2009), and (iii) a ΔNSs-ΔNSm recombinant virus (Bird et al., 2008a). RVFV Clone 13 is characterized by a large deletion of the NSs gene from the S segment, R566 is a reassortant of the L and M segments of MP-12 and the S segment of Clone 13 that combines attenuation markers from both strains (Bouloy and Flick, 2009), and the ΔNSs-ΔNSm ZH501 strain is a reverse genetics-based vaccine candidate lacking both the NSs and NSm genes. Since all three vaccine candidates have NSs or both NSs/NSm gene deletions, which prevent the virus from hijacking the type 1 IFN pathway, make reversion almost impossible, and satisfy the DIVA concept, they all fulfill important features for a safe RVFV vaccine candidate.

All three vaccine candidates have promising results in livestock studies. Clone 13 has been tested in sheep and cattle (Hunter, 2001) and all vaccinated animals survived a challenge with virulent RVFV and no abortions were observed (Dungu et al., 2010; von Teichman et al., 2011).

R566 has also been used in immunological studies in sheep and pregnant ewes in Senegal. None of the vaccinated animals exhibited signs of illness and none of the pregnant ewes aborted (M. Bouloy and Y. Thionganne, personal communication). The ΔNSs-ΔNSm recombinant RVFV was also tested in pregnant sheep and protected all animals against RVFV challenge, and no abortions were reported (Bird et al., 2011). Because the deletion of NSs and/or NSm is responsible for attenuation, Clone 13, R566, and the recombinant ΔNSs-ΔNSm RVFV can be considered safe from reversion to a replication-competent, fully infectious virus. However, the interactions of vaccines, modeled on particularly virulent viruses, with wild virus need to be investigated before they are put into common use. The use of reverse genetics to generate a vaccine candidate with particular features (Bouloy and Flick, 2009), that is, deletion of genes, inserted reporter genes, e.g., GFP (Bird et al., 2008a), seems to be a promising approach and allows the generation and screening of different candidates within a short time frame.

21.8.2.2 Recombinant Vectors

Recombinant vectors are another approach to generate a safe and efficacious RVFV vaccine candidate (Bird et al., 2009; Bouloy and Flick, 2009; Labeaud, 2010). Main features of recombinant vectors are as follows: (i) the antigen of interest is not expressed through the disease-causing virus or attenuated virus variant, thereby eliminating the risks of reversion to a pathogenic phenotype, and (ii) the antigen of interest is expressed through viral infection, allowing the recombinant proteins to be processed similarly to that during an infection of the disease-causing virus (Boshra et al., 2011).

High expression of RVFV glycoproteins can be achieved using different recombinant systems: (i) recombinant lumpy skin disease virus (LSDV) (Wallace et al., 2006); (ii) adenovirus-based platform (Holman et al., 2009); (iii) Newcastle disease virus (de Boer et al., 2010; Kortekaas et al., 2010); (iv) alphavirus replicon vector (Bhardwaj et al., 2010; Gorchakov et al., 2007; Heise et al., 2009), that is, Sindbis or Venezuelan equine encephalitis viruses; (v) capripoxvirus KS1 vector (Soi et al., 2010); and (vi) modified vaccinia Ankara (MVA) vector (Boshra et al., 2011). All vaccine candidates showed promising protection in mouse lethal challenge studies, and some were tested in immunological studies (Kortekaas et al., 2010) and initial challenge studies (Soi et al., 2010) in livestock. However, efficacy and safety have to be confirmed in challenge studies in an appropriate animal model (Ross et al., 2011; Smith et al., 2011) for the development of a human vaccine or in an appropriate livestock, i.e., sheep and cattle, for the development of a veterinary vaccine to encourage further transition steps to a clinical product.

21.8.2.3 Virus-Like Particles

To avoid issues associated with formalin-inactivated and live-attenuated vaccines, recent RVFV vaccine developments focus on VLP-based candidates (de Boer et al., 2010; Habjan et al., 2009a; Mandell et al., 2010a,b; Naslund et al., 2009; Pichlmair et al., 2010). VLPs offer the advantage of expressing the immunogenic components of a virus of interest without the risks of using a functional replicating virus. Furthermore, VLPs may be more immunogenic than recombinant proteins as they maintain conformational epitopes that can induce neutralizing antibodies (Grgacic and Anderson, 2006). Structural similarity with the virus of interest combined with the lack of viral genetic material makes this technology ideal to generate safe vaccine candidates (Grgacic and Anderson, 2006; Ludwig and Wagner, 2007). VLPs can be generated in mammalian or insect cell cultures, the latter being the most promising system because of the potential for high yield VLP production using the recombinant baculovirus expression system and the ease of scale-up for manufacturing (Gheysen et al., 1989; Latham and Galarza, 2001; Loudon and Roy, 1991; Nermut et al., 1994; Rose et al., 1993; Yamshchikov et al., 1995; Zeng et al., 1996). However, the recently described mammalian cell approach seems to be, at least on a laboratory scale, more efficient and cost-effective (Mandell et al., 2010b) compared to the use of recombinant baculovirus infection of insect cells to produce RVF VLPs. Immunological data (i.e., high neutralizing antibody titers) and complete protection in rodent challenge studies (mice and rats) encourage the further development of VLP-based RVFV vaccine candidates (de Boer et al., 2010; Habjan et al., 2009a; Mandell et al., 2010a,b; Naslund et al., 2009). Overall, the successful FDA approval of a VLP-based vaccine for human papilloma virus types 16 and 18 (Cervarix) (2010; Monie et al., 2008) demonstrates that the VLP technology is a suitable vaccine approach to combat viral diseases.

21.8.2.4 DNA Vaccines

A virus-free approach using DNA-based vaccine technologies is an alternative to avoid any virus-related adverse effects. In addition, DNA vaccines do not require a cold chain for vaccine distribution and can be manufactured cost-effective. However, gene-gun immunizations with a cDNA encoding the RVFV M segment (glycoproteins and NSm) resulted in only partial protection in a mouse challenge study (Spik et al., 2006). Nevertheless, neutralizing antibodies were generated after three vaccinations with a similar construct expressing the RVFV glycoproteins but lacking the NSm protein, and 100% survival from a 100 LD_{50} lethal challenge was observed.

Promising immunological data are also reported after DNA vaccination of sheep using an RVFV glycoprotein-expressing construct (Lorenzo et al., 2008), and the immune response was further improved by combination with an N-expressing component. A booster with MP-12 resulted in a transient cellular response demonstrating that DNA vaccines can be successfully used in a prime-boost vaccination strategy.

An alternative DNA vaccine approach followed up on the question if RVFV N is an important vaccine component. Using a nonlethal challenge with RVFV ZH548, four of eight and five of eight mice vaccinated with DNA constructs encoding RVFV N or Gn/Gc, respectively, were aviremic (Lagerqvist et al., 2009). These findings suggest that N as a vaccine component can contribute to inducing protective immunity.

A more recent approach used the complement protein C3d fused onto the RVFV Gn protein to provide a self-adjuvant. Interestingly, the C3d-trimer-Gn vaccine candidate decreased morbidity in a mouse challenge study compared to only a Gn-based vaccine candidate (Bhardwaj et al., 2010).

21.8.2.5 Antivirals

While vaccines are the primary defense against viral diseases (Plotkin, 2005), the development of antiviral compounds with therapeutic efficacy against highly pathogenic RNA viruses is also important (Beigel and Bray, 2008; Bray, 2008). However, with the exception of ribavirin, few compounds are licensed for treatment of hemorrhagic fever viruses (Gowen and Holbrook, 2008).

MxA, a large dynamin-like GTPase, exhibits antiviral capabilities against a wide array of RNA viruses (Kochs et al., 2002; Pavlovic et al., 1992; Reichelt et al., 2004). Although RVFV is a strong antagonist of IFN production, it is very sensitive to the action of IFN or to IFN inducers (Bouloy et al., 2001). The infection cycle of bunyaviruses is significantly affected by human interferon-induced protein MxA by blocking the immediate early primary viral transcription step (Kochs et al., 2002; Reichelt et al., 2004).

Ribavirin and polyriboinosinic acid complexed with poly-L-lysine and carboxymethylcellulose (poly(ICLC)) prevents clinical symptoms of RVFV in mice and hamsters (Peters et al., 1986). Furthermore, RVFV-infected rhesus macaques treated with ribavirin showed no viremia after challenge (Peters et al., 1986). Unfortunately, ribavirin might have undesired effects. When used in Saudi Arabia to treat RVFV patients, ribavirin seemed active but did not prevent the development of severe meningoencephalic complications, including hallucinations, lethargy, and coma (M. Bouloy and B. Swanepoel, personal communication). These and other similar reports have limited ribavirin's utility because of significant adverse effects and lack of specificity (Kilgore et al., 1997; McCormick et al., 1986; Monath, 2008).

Using real-time RT-PCR for RVFV quantification, different antiviral compounds, that is, ribavirin, IFN-α, 6-azauridine, and glycyrrhizin (Garcia et al., 2001), were tested. At the lowest tested concentrations, 62.5 µg/mL ribavirin, 1 IU/mL interferon alpha (IFN-α), and 0.3 µg/mL 6-azauridine were still inhibitory, while glycyrrhizin showed no antiviral activity with viral replication at the lowest tested concentration (156 µg/µL), with inhibition only at high concentrations (1250 and 2500 µg/mL). IFN-α and ribavirin showed dose-dependent reduction in viral replication.

Favipiravir (compound T-705), a pyrazinecarboxamide, inhibits RNA-dependent RNA polymerase activity, and host cell kinases convert T-705 into T-705 ribofuranosyl phosphate (T-705RTP), a form that inhibits virus polymerase without affecting host cellular RNA or DNA synthesis. T-705 and the related pyrazine derivative T-1106 (Gowen et al., 2010) protect mice and hamsters against a lethal challenge with Punta Toro virus (PTV), a RVFV-related phlebovirus. In fact, favipiravir and T-1106 were shown to be efficacious for several viruses in the bunyavirus family including La Crosse, Punta Toro, RVF, and sandfly fever Sicilian viruses and in the arenavirus family including Junín, Pichinde, and Tacaribe viruses (Gowen et al., 2007, 2008; Mendenhall et al., 2011); influenza virus (Kiso et al., 2010; Smee et al., 2010); West Nile virus (Morrey et al., 2008); yellow fever virus (Julander et al., 2009a); and Western equine encephalitis virus (Julander et al., 2009b). Favipiravir was also shown to be less toxic compared to ribavirin (Gowen et al., 2008). Antiviral therapeutics such as T-705/T-1106 may therefore present a more viable choice than ribavirin for postexposure treatment of RVFV infections (Furuta et al., 2009; Julander et al., 2009a).

LJ001, an aryl-methyldiene rhodanine derivative, intercalates in viral membranes and prevents virus–cell fusion. Its broad-spectrum activity against enveloped viruses, including RVFV, might encourage further testing and development (Wolf et al., 2010).

Another antiviral strategy is based on a chimeric antibody, bavituximab, which targets anionic phospholipids on the cell membranes of enveloped viruses (e.g., Pichinde virus) and virus-infected cells (Soares et al., 2008). However, more testing has to be done to evaluate the broad-spectrum antiviral activity as well as the specific activity against RVFV.

Overall, many promising RVFV vaccine and therapeutic concepts are currently employed to generate an urgently needed safe and efficacious countermeasure against RVFV. It will be important to raise awareness of this potential bioterror threat to ensure continuous funding for further development of RVFV countermeasures.

21.9 CONCLUSIONS

RVFV continues to circulate among animals and humans in sub-Saharan Africa and parts of the Middle East and is a significant health and economic burden in many of those areas. Furthermore, it remains a serious threat to other parts of the world. Bioterrorism/agroterrorism, trade, world travel, and the presence of mosquito species capable of transmitting the virus make RVFV a major threat to Western countries (Mandell and Flick, 2010). Coupled with the fact that there is currently no FDA- or USDA-approved RVFV vaccine for human or veterinary use, there is a clear need for more RVFV vaccine and antiviral research (Bird et al., 2009; Bouloy and Flick, 2009; Ikegami and Makino, 2009). Development of more effective methods for RVFV outbreak prevention and control remains not only a priority for the African countries but also a global health priority. It has been proposed that a single infected person or animal, which is able to enter Europe or the United States, would be sufficient to initiate a major outbreak before RVFV would be detected. This would quickly lead to the spread of RVFV and cause a severe strain on health-care systems as human infections become more common. Widespread public panic might also ensue because of the knowledge that a hemorrhagic fever is circulating in the population. A severe economic impact is inevitable and will be felt in almost all economic sectors, from agriculture to health care and travel, likely for years post-identification.

Our understanding of the molecular biology of RVFV and its interaction with the host cell has greatly increased over the last decade, specifically through the availability of reverse genetics and recombinant vectors. However, little is still known about the detailed mechanisms of how the virus spreads inside an infected host and the development of neuropathogenesis and retinitis. Further, only very limited information is available about the molecular mechanisms on how arthropods limit virus spread, and this knowledge could be used for developing new antiviral strategies against this important arboviral pathogen.

ACKNOWLEDGMENT

The authors would like to thank Dr. John Morrill for his comments on the manuscript.

REFERENCES

(2010). FDA licensure of bivalent human papillomavirus vaccine (HPV2, Cervarix) for use in females and updated HPV vaccination recommendations from the Advisory Committee on Immunization Practices (ACIP). *MMWR Morb Mortal Wkly Rep* **59**(20), 626–629. Also available at: http://www.cdc.gov/mmwr/pdf/wk/mm5920.pdf

Abdel-Wahab, K. S., El Baz, L. M., El-Tayeb, E. M., Omar, H., Ossman, M. A., and Yasin, W. (1978). Rift Valley fever virus infections in Egypt: Pathological and virological findings in man. *Trans R Soc Trop Med Hyg* **72**(4), 392–396.

Abu-Elyazeed, R., el-Sharkawy, S., Olson, J., Botros, B., Soliman, A., Salib, A., Cummings, C., and Arthur, R. (1996). Prevalence of anti-Rift-Valley-fever IgM antibody in abattoir workers in the Nile delta during the 1993 outbreak in Egypt. *Bull World Health Organ* **74**(2), 155–158.

Ahmad, K. (2000). More deaths from Rift Valley fever in Saudi Arabia and Yemen. *Lancet* **356**(9239), 1422.

Albarino, C. G., Bird, B. H., and Nichol, S. T. (2007). A shared transcription termination signal on negative and ambisense RNA genome segments of Rift Valley fever, sandfly fever Sicilian, and Toscana viruses. *J Virol* **81**(10), 5246–5256.

Al-Hazmi, A., Al-Rajhi, A. A., Abboud, E. B., Ayoola, E. A., Al-Hazmi, M., Saadi, R., and Ahmed, N. (2005). Ocular complications of Rift Valley fever outbreak in Saudi Arabia. *Ophthalmology* **112**(2), 313–318.

Al-Hazmi, M., Ayoola, E. A., Abdurahman, M., Banzal, S., Ashraf, J., El-Bushra, A., Hazmi, A., Abdullah, M., Abbo, H., Elamin, A., Al-Sammani el, T., Gadour, M., Menon, C., Hamza, M., Rahim, I., Hafez, M., Jambavalikar, M., Arishi, H., and Aqeel, A. (2003). Epidemic Rift Valley fever in Saudi Arabia: A clinical study of severe illness in humans. *Clin Infect Dis* **36**(3), 245–252.

Alrajhi, A. A., Al-Semari, A., and Al-Watban, J. (2004). Rift Valley fever encephalitis. *Emerg Infect Dis* **10**(3), 554–555.

Anderson, G. W., Jr., Saluzzo, J. F., Ksiazek, T. G., Smith, J. F., Ennis, W., Thureen, D., Peters, C. J., and Digoutte, J. P. (1989). Comparison of in vitro and in vivo systems for propagation of Rift Valley fever virus from clinical specimens. *Res Virol* **140**(2), 129–138.

Anderson, G. W., Jr. and Smith, J. F. (1987). Immunoelectron microscopy of Rift Valley fever viral morphogenesis in primary rat hepatocytes. *Virology* **161**(1), 91–100.

Ayoub, M., Barhoma, G., and Zaghlol, I. (1978). Ocular manifestations of Rift Valley fever. *Bull Ophthalmol Soc Egypt* **71**(75), 125–133.

Barnard, B. J. (1979). Rift Valley fever vaccine—antibody and immune response in cattle to a live and an inactivated vaccine. *J S Afr Vet Assoc* **50**(3), 155–157.

Barnard, B. J. and Botha, M. J. (1977). An inactivated rift valley fever vaccine. *J S Afr Vet Assoc* **48**(1), 45–48.

Battles, J. K. and Dalrymple, J. M. (1988). Genetic variation among geographic isolates of Rift Valley fever virus. *Am J Trop Med Hyg* **39**(6), 617–631.

Beigel, J. and Bray, M. (2008). Current and future antiviral therapy of severe seasonal and avian influenza. *Antiviral Res* **78**(1), 91–102.

Bettinger, G. E., Peters, C. J., Pittman, P., Morrill, J. C., Ranadive, M., Kormann, R. N., and Lokukamage, N. (2009). *Rift Valley Fever Workshop, Dokki, Giza, Cairo, Egypt.*

Bhardwaj, N., Heise, M. T., and Ross, T. M. (2010). Vaccination with DNA plasmids expressing Gn coupled to C3d or alphavirus replicons expressing gn protects mice against Rift Valley fever virus. *PLoS Negl Trop Dis* **4**(6), e725.

Billecocq, A., Spiegel, M., Vialat, P., Kohl, A., Weber, F., Bouloy, M., and Haller, O. (2004). NSs protein of Rift Valley fever virus blocks interferon production by inhibiting host gene transcription. *J Virol* **78**(18), 9798–9806.

Bird, B. H., Albarino, C. G., Hartman, A. L., Erickson, B. R., Ksiazek, T. G., and Nichol, S. T. (2008a). Rift valley fever virus lacking the NSs and NSm genes is highly attenuated, confers protective immunity from virulent virus challenge, and allows for differential identification of infected and vaccinated animals. *J Virol* **82**(6), 2681–2691.

Bird, B. H., Albarino, C. G., and Nichol, S. T. (2007a). Rift Valley fever virus lacking NSm proteins retains high virulence in vivo and may provide a model of human delayed onset neurologic disease. *Virology* **362**(1), 10–15.

Bird, B. H., Bawiec, D. A., Ksiazek, T. G., Shoemaker, T. R., and Nichol, S. T. (2007b). Highly sensitive and broadly reactive quantitative reverse transcription-PCR assay for high-throughput detection of Rift Valley fever virus. *J Clin Microbiol* **45**(11), 3506–3513.

Bird, B. H., Githinji, J. W., Macharia, J. M., Kasiiti, J. L., Muriithi, R. M., Gacheru, S. G., Musaa, J. O., Towner, J. S., Reeder, S. A., Oliver, J. B., Stevens, T. L., Erickson, B. R., Morgan, L. T., Khristova, M. L., Hartman, A. L., Comer, J. A., Rollin, P. E., Ksiazek, T. G., and Nichol, S. T. (2008b). Multiple virus lineages sharing recent common ancestry were associated with a Large Rift Valley fever outbreak among livestock in Kenya during 2006–2007. *J Virol* **82**(22), 11152–11166.

Bird, B. H., Khristova, M. L., Rollin, P. E., Ksiazek, T. G., and Nichol, S. T. (2007c). Complete genome analysis of 33 ecologically and biologically diverse Rift Valley fever virus strains reveals widespread virus movement and low genetic diversity due to recent common ancestry. *J Virol* **81**(6), 2805–2816.

Bird, B. H., Ksiazek, T. G., Nichol, S. T., and MacLachlan, N. J. (2009). Rift Valley fever virus. *J Am Vet Med Assoc* **234**(7), 883–893.

Bird, B. H., Maartens, L. H., Campbell, S., Erasmus, B. J., Erickson, B. R., Dodd, K. A., Spiropoulou, C. F., Cannon, D., Drew, C. P., Knust, B., McElroy, A. K., Khristova, M. L., Albarino, C. G., and Nichol, S. T. (2011). Rift Valley fever virus vaccine lacking the NSs and NSm genes is safe, nonteratogenic, and confers protection from viremia, pyrexia, and abortion following challenge in adult and pregnant sheep. *J Virol* **85**(24), 12901–12909.

de Boer, S. M., Kortekaas, J., Antonis, A. F., Kant, J., van Oploo, J. L., Rottier, P. J., Moormann, R. J., and Bosch, B. J. (2010). Rift Valley fever virus subunit vaccines confer complete protection against a lethal virus challenge. *Vaccine* **28**(11), 2330–2339.

von Bonsdorff, C. H. and Pettersson, R. (1975). Surface structure of Uukuniemi virus. *J Virol* **16**(5), 1296–1307.

Boshra, H., Lorenzo, G., Busquets, N., and Brun, A. (2011). Rift valley fever: Recent insights into pathogenesis and prevention. *J Virol* **85**(13), 6098–6105.

Botros, B., Omar, A., Elian, K., Mohamed, G., Soliman, A., Salib, A., Salman, D., Saad, M., and Earhart, K. (2006). Adverse response of non-indigenous cattle of European breeds to live attenuated Smithburn Rift Valley fever vaccine. *J Med Virol* **78**(6), 787–791.

Bouloy, M. and Flick, R. (2009). Reverse genetics technology for Rift Valley fever virus: Current and future applications for the development of therapeutics and vaccines. *Antiviral Res* **84**(2), 101–118.

Bouloy, M. and Hannoun, C. (1976). Studies on lumbo virus replication. I. RNA-dependent RNA polymerase associated with virions. *Virology* **69**(1), 258–264.

Bouloy, M., Janzen, C., Vialat, P., Khun, H., Pavlovic, J., Huerre, M., and Haller, O. (2001). Genetic evidence for an interferon-antagonistic function of rift valley fever virus nonstructural protein NSs. *J Virol* **75**(3), 1371–1377.

Bouloy, M. and Weber, F. (2010). Molecular biology of rift valley fever virus. *Open Virol J* **4**, 8–14.

Bray, M. (2008). Highly pathogenic RNA viral infections: Challenges for antiviral research. *Antiviral Res* **78**(1), 1–8.

Caplen, H., Peters, C. J., and Bishop, D. H. (1985). Mutagen-directed attenuation of Rift Valley fever virus as a method for vaccine development. *J Gen Virol* **66**(Pt 10), 2271–2277.

Chambers, P. G. and Swanepoel, R. (1980). Rift valley fever in abattoir workers. *Cent Afr J Med* **26**(6), 122–126.

Coetzer, J. A. and Barnard, B. J. (1977). Hydrops amnii in sheep associated with hydranencephaly and arthrogryposis with wesselsbron disease and rift valley fever viruses as aetiological agents. *Onderstepoort J Vet Res* **44**(2), 119–126.

Collett, M. S. (1986). Messenger RNA of the M segment RNA of Rift Valley fever virus. *Virology* **151**(1), 151–156.

Daubney, R. and Hudson, J. R. (1932). Rift Valley fever. *Lancet* **1**, 611–612.

Daubney, R., Hudson, J. R., and Garnham, P. C. (1931). Enzootic hepatitis of Rift Valley fever. An undescribed virus disease of sheep, cattle and man from East Africa. *J. Pathol. Bacteriol* **34**, 545–579.

Deutman, A. F. and Klomp, H. J. (1981). Rift Valley fever retinitis. *Am J Ophthalmol* **92**(1), 38–42.

Drosten, C., Gottig, S., Schilling, S., Asper, M., Panning, M., Schmitz, H., and Gunther, S. (2002). Rapid detection and quantification of RNA of Ebola and Marburg viruses, Lassa virus, Crimean-Congo hemorrhagic fever virus, Rift Valley fever virus, dengue virus, and yellow fever virus by real-time reverse transcription-PCR. *J Clin Microbiol* **40**(7), 2323–2330.

Dungu, B., Louw, I., Lubisi, A., Hunter, P., von Teichman, B. F., and Bouloy, M. (2010). Evaluation of the efficacy and safety of the Rift Valley Fever Clone 13 vaccine in sheep. *Vaccine* **28**(29), 4581–4587.

Dunn, E. F., Pritlove, D. C., Jin, H., and Elliott, R. M. (1995). Transcription of a recombinant bunyavirus RNA template by transiently expressed bunyavirus proteins. *Virology* **211**(1), 133–143.

El Imam, M., El Sabiq, M., Omran, M., Abdalkareem, A., El Gaili Mohamed, M. A., Elbashir, A., and Khalafala, O. (2009). Acute renal failure associated with the Rift Valley fever: A single center study. *Saudi J Kidney Dis Transpl* **20**(6), 1047–1052.

Elliott, R. M. and Weber, F. (2009). Bunyaviruses and the type I interferon system. *Viruses* **1**(3), 1003–1021.

Ellis, D. S., Simpson, D. I., Stamford, S., and Abdel Wahab, K. S. (1979). Rift Valley fever virus: Some ultrastructural observations on material from the outbreak in Egypt 1977. *J Gen Virol* **42**(2), 329–337.

Ferron, F., Li, Z., Danek, E. I., Luo, D., Wong, Y., Coutard, B., Lantez, V., Charrel, R., Canard, B., Walz, T., and Lescar, J. (2011). The hexamer structure of Rift Valley fever virus nucleoprotein suggests a mechanism for its assembly into ribonucleoprotein complexes. *PLoS Pathog* **7**(5), e1002030.

Filone, C. M., Heise, M., Doms, R. W., and Bertolotti-Ciarlet, A. (2006). Development and characterization of a Rift Valley fever virus cell–cell fusion assay using alphavirus replicon vectors. *Virology* **356**(1–2), 155–164.

Findlay, G. M. and Daubney, R. (1931). The virus of rift valley fever or enzootic hepatitis. *Lancet* **221**, 1350–1351.

Flick, K., Katz, A., Overby, A., Feldmann, H., Pettersson, R. F., and Flick, R. (2004). Functional analysis of the noncoding regions of the Uukuniemi virus (Bunyaviridae) RNA segments. *J Virol* **78**(21), 11726–11738.

Flick, R., Elgh, F., and Pettersson, R. F. (2002). Mutational analysis of the Uukuniemi virus (Bunyaviridae family) promoter reveals two elements of functional importance. *J Virol* **76**(21), 10849–10860.

Fontenille, D., Traore-Lamizana, M., Trouillet, J., Leclerc, A., Mondo, M., Ba, Y., Digoutte, J. P., and Zeller, H. G. (1994). First isolations of arboviruses from phlebotomine sand flies in West Africa. *Am J Trop Med Hyg* **50**(5), 570–574.

Fontenille, D., Traore-Lamizana, M., Zeller, H., Mondo, M., Diallo, M., and Digoutte, J. P. (1995). Short report: Rift Valley fever in western Africa: Isolations from *Aedes* mosquitoes during an interepizootic period. *Am J Trop Med Hyg* **52**(5), 403–404.

Francis, T. and Magill, T. P. (1935). Rift Valley Fever: A report of three cases of laboratory infection and the experimental transmission of the disease to ferrets. *J Exp Med* **62**(3), 433–448.

Frank-Peterside, N. (2000). Response of laboratory staff to vaccination with an inactivated Rift Valley fever vaccine—TSI-GSD 200. *Afr J Med Med Sci* **29**(2), 89–92.

Freed, I. (1951). Rift valley fever in man, complicated by retinal changes and loss of vision. *S Afr Med J* **25**(50), 930–932.

Freiberg, A. N., Sherman, M. B., Morais, M. C., Holbrook, M. R., and Watowich, S. J. (2008). Three-dimensional organization of Rift Valley fever virus revealed by cryoelectron tomography. *J Virol* **82**(21), 10341–10348.

Furuta, Y., Takahashi, K., Shiraki, K., Sakamoto, K., Smee, D. F., Barnard, D. L., Gowen, B. B., Julander, J. G., and Morrey, J. D. (2009). T-705 (favipiravir) and related compounds: Novel broad-spectrum inhibitors of RNA viral infections. *Antiviral Res* **82**(3), 95–102.

Garcia, S., Crance, J. M., Billecocq, A., Peinnequin, A., Jouan, A., Bouloy, M., and Garin, D. (2001). Quantitative real-time PCR detection of Rift Valley fever virus and its application to evaluation of antiviral compounds. *J Clin Microbiol* **39**(12), 4456–4461.

Garry, C. E. and Garry, R. F. (2004). Proteomics computational analyses suggest that the carboxyl terminal glycoproteins of Bunyaviruses are class II viral fusion protein (beta-penetrenes). *Theor Biol Med Model* **1**, 10.

Gear, J., De Meillon, B., Le Roux, A. F., Kofsky, R., Innes, R. R., Steyn, J. J., Oliff, W. D., and Schulz, K. H. (1955). Rift valley fever in South Africa; A study of the 1953 outbreak in the Orange Free State, with special reference to the vectors and possible reservoir hosts. *S Afr Med J* **29**(22), 514–518.

Gerrard, S. R., Bird, B. H., Albarino, C. G., and Nichol, S. T. (2007). The NSm proteins of Rift Valley fever virus are dispensable for maturation, replication and infection. *Virology* **359**(2), 459–465.

Gerrard, S. R. and Nichol, S. T. (2002). Characterization of the Golgi retention motif of Rift Valley fever virus G(N) glycoprotein. *J Virol* **76**(23), 12200–12210.

Gheysen, D., Jacobs, E., de Foresta, F., Thiriart, C., Francotte, M., Thines, D., and De Wilde, M. (1989). Assembly and release of HIV-1 precursor Pr55gag virus-like particles from recombinant baculovirus-infected insect cells. *Cell* **59**(1), 103–112.

Giorgi, C., Accardi, L., Nicoletti, L., Gro, M. C., Takehara, K., Hilditch, C., Morikawa, S., and Bishop, D. H. (1991). Sequences and coding strategies of the S RNAs of Toscana and Rift Valley fever viruses compared to those of Punta Toro, Sicilian Sandfly fever, and Uukuniemi viruses. *Virology* **180**(2), 738–753.

Gorchakov, R., Volkova, E., Yun, N., Petrakova, O., Linde, N. S., Paessler, S., Frolova, E., and Frolov, I. (2007). Comparative analysis of the alphavirus-based vectors expressing Rift Valley fever virus glycoproteins. *Virology* **366**(1), 212–225.

Gowen, B. B. and Holbrook, M. R. (2008). Animal models of highly pathogenic RNA viral infections: Hemorrhagic fever viruses. *Antiviral Res* **78**(1), 79–90.

Gowen, B. B., Smee, D. F., Wong, M. H., Hall, J. O., Jung, K. H., Bailey, K. W., Stevens, J. R., Furuta, Y., and Morrey, J. D. (2008). Treatment of late stage disease in a model of arenaviral hemorrhagic fever: T-705 efficacy and reduced toxicity suggests an alternative to ribavirin. *PLoS One* **3**(11), e3725.

Gowen, B. B., Wong, M. H., Jung, K. H., Sanders, A. B., Mendenhall, M., Bailey, K. W., Furuta, Y., and Sidwell, R. W. (2007). In vitro and in vivo activities of T-705 against arenavirus and bunyavirus infections. *Antimicrob Agents Chemother* **51**(9), 3168–3176.

Gowen, B. B., Wong, M. H., Jung, K. H., Smee, D. F., Morrey, J. D., and Furuta, Y. (2010). Efficacy of favipiravir (T-705) and T-1106 pyrazine derivatives in phlebovirus disease models. *Antiviral Res* **86**(2), 121–127.

Grgacic, E. V. and Anderson, D. A. (2006). Virus-like particles: Passport to immune recognition. *Methods* **40**(1), 60–65.

Gro, M. C., Di Bonito, P., Accardi, L., and Giorgi, C. (1992). Analysis of 3′ and 5′ ends of N and NSs messenger RNAs of Toscana Phlebovirus. *Virology* **191**(1), 435–438.

Habjan, M., Penski, N., Wagner, V., Spiegel, M., Overby, A. K., Kochs, G., Huiskonen, J. T., and Weber, F. (2009a). Efficient production of Rift Valley fever virus-like particles: The antiviral protein MxA can inhibit primary transcription of bunyaviruses. *Virology* **385**(2), 400–408.

Habjan, M., Pichlmair, A., Elliott, R. M., Overby, A. K., Glatter, T., Gstaiger, M., Superti-Furga, G., Unger, H., and Weber, F. (2009b). NSs protein of Rift Valley fever virus induces the specific degradation of the double-stranded RNA-dependent protein kinase. *J Virol* **83**(9), 4365–4375.

He, J., Kraft, A. J., Fan, J., Van Dyke, M., Wang, L., Bose, M. E., Khanna, M., Metallo, J. A., and Henrickson, K. J. (2009). Simultaneous detection of CDC category "A" DNA and RNA bioterrorism agents by use of multiplex PCR & RT-PCR enzyme hybridization assays. *Viruses* **1**(3), 441–459.

Heise, M. T., Whitmore, A., Thompson, J., Parsons, M., Grobbelaar, A. A., Kemp, A., Paweska, J. T., Madric, K., White, L. J., Swanepoel, R., and Burt, F. J. (2009). An alphavirus replicon-derived candidate vaccine against Rift Valley fever virus. *Epidemiol Infect*, 1–10.

Holman, D. H., Penn-Nicholson, A., Wang, D., Woraratanadharm, J., Harr, M. K., Luo, M., Maher, E. M., Holbrook, M. R., and Dong, J. Y. (2009). A complex adenovirus-vectored vaccine against Rift Valley fever virus protects mice against lethal infection in the presence of preexisting vector immunity. *Clin Vaccine Immunol* **16**(11), 1624–1632.

Huiskonen, J. T., Overby, A. K., Weber, F., and Grunewald, K. (2009). Electron cryo-microscopy and single-particle averaging of Rift Valley fever virus: Evidence for GN–GC glycoprotein heterodimers. *J Virol* **83**(8), 3762–3769.

Hunter, B. (2001). *Proceedings of 5th International Sheep Veterinary Congress,* Stellenbosch, South Africa. University of Pretoria, South Africa.

Hunter, P., Erasmus, B. J., and Vorster, J. H. (2002). Teratogenicity of a mutagenised Rift Valley fever virus (MVP 12) in sheep. *Onderstepoort J Vet Res* **69**(1), 95–98.

Iapalucci, S., Lopez, N., and Franze-Fernandez, M. T. (1991). The 3′ end termini of the Tacaribe arenavirus subgenomic RNAs. *Virology* **182**(1), 269–278.

Ibrahim, M. S., Turell, M. J., Knauert, F. K., and Lofts, R. S. (1997). Detection of Rift Valley fever virus in mosquitoes by RT-PCR. *Mol Cell Probes* **11**(1), 49–53.

Ihara, T., Matsuura, Y., and Bishop, D. H. (1985). Analyses of the mRNA transcription processes of Punta Toro phlebovirus (Bunyaviridae). *Virology* **147**(2), 317–325.

Ikegami, T. and Makino, S. (2009). Rift valley fever vaccines. *Vaccine* **27**(Suppl 4), D69–D72.

Ikegami, T. and Makino, S. (2011). The pathogenesis of Rift Valley fever. *Viruses* **3**(5), 493–519.

Ikegami, T., Narayanan, K., Won, S., Kamitani, W., Peters, C. J., and Makino, S. (2009a). Dual functions of Rift Valley fever virus NSs protein: Inhibition of host mRNA transcription and post-transcriptional downregulation of protein kinase PKR. *Ann N Y Acad Sci* **1171**(Suppl 1), E75–E85.

Ikegami, T., Narayanan, K., Won, S., Kamitani, W., Peters, C. J., and Makino, S. (2009b). Rift Valley fever virus NSs protein promotes post-transcriptional downregulation of protein kinase PKR and inhibits eIF2alpha phosphorylation. *PLoS Pathog* **5**(2), e1000287.

Ikegami, T., Won, S., Peters, C. J., and Makino, S. (2005). Rift Valley fever virus NSs mRNA is transcribed from an incoming anti-viral-sense S RNA segment. *J Virol* **79**(18), 12106–12111.

Ikegami, T., Won, S., Peters, C. J., and Makino, S. (2006). Rescue of infectious rift valley fever virus entirely from cDNA, analysis of virus lacking the NSs gene, and expression of a foreign gene. *J Virol* **80**(6), 2933–2940.

Ikegami, T., Won, S., Peters, C. J., and Makino, S. (2007). Characterization of Rift Valley fever virus transcriptional terminations. *J Virol* **81**(16), 8421–8438.

Jansen van Vuren, P. and Paweska, J. T. (2009). Laboratory safe detection of nucleocapsid protein of Rift Valley fever virus in human and animal specimens by a sandwich ELISA. *J Virol Methods* **157**(1), 15–24.

Jin, H. and Elliott, R. M. (1993). Characterization of Bunyamwera virus S RNA that is transcribed and replicated by the L protein expressed from recombinant vaccinia virus. *J Virol* **67**(3), 1396–1404.

Johnson, B. K., Chanas, A. C., el-Tayeb, E., Abdel-Wahab, K. S., Sheheta, F. A., and Mohamed, A. e.-D. (1978). Rift Valley fever in Egypt, 1978. *Lancet* **2**(8092 Pt 1), 745.

Joubert, J. D., Ferguson, A. L., and Gear, J. (1951). Rift Valley fever in South Africa: 2. The occurrence of human cases in the Orange Free State, the north-western Cape province, the western and southern Transvaal. A Epidemiological and clinical findings. *S Afr Med J* **25**(48), 890–891.

Julander, J. G., Shafer, K., Smee, D. F., Morrey, J. D., and Furuta, Y. (2009a). Activity of T-705 in a hamster model of yellow fever virus infection in comparison with that of a chemically related compound, T-1106. *Antimicrob Agents Chemother* **53**(1), 202–209.

Julander, J. G., Smee, D. F., Morrey, J. D., and Furuta, Y. (2009b). Effect of T-705 treatment on western equine encephalitis in a mouse model. *Antiviral Res* **82**(3), 169–171.

Kahlon, S. S., Peters, C. J., Leduc, J., Muchiri, E. M., Muiruri, S., Njenga, M. K., Breiman, R. F., White, A. C., Jr., and King, C. H. (2010). Severe Rift Valley fever may present with a characteristic clinical syndrome. *Am J Trop Med Hyg* **82**(3), 371–375.

Kamal, S. A. (2009). Pathological studies on postvaccinal reactions of Rift Valley fever in goats. *Virol J* **6**, 94.

Kark, J. D., Aynor, Y., and Peters, C. J. (1982). A rift Valley fever vaccine trial. I. Side effects and serologic response over a six-month follow-up. *Am J Epidemiol* **116**(5), 808–820.

Kark, J. D., Aynor, Y., and Peters, C. J. (1985). A Rift Valley fever vaccine trial: 2. Serological response to booster doses with a comparison of intradermal versus subcutaneous injection. *Vaccine* **3**(2), 117–122.

Kilgore, P. E., Ksiazek, T. G., Rollin, P. E., Mills, J. N., Villagra, M. R., Montenegro, M. J., Costales, M. A., Paredes, L. C., and Peters, C. J. (1997). Treatment of Bolivian hemorrhagic fever with intravenous ribavirin. *Clin Infect Dis* **24**(4), 718–722.

Kiso, M., Takahashi, K., Sakai-Tagawa, Y., Shinya, K., Sakabe, S., Le, Q. M., Ozawa, M., Furuta, Y., and Kawaoka, Y. (2010). T-705 (favipiravir) activity against lethal H5N1 influenza A viruses. *Proc Natl Acad Sci USA* **107**(2), 882–887.

Kochs, G., Janzen, C., Hohenberg, H., and Haller, O. (2002). Antivirally active MxA protein sequesters La Crosse virus nucleocapsid protein into perinuclear complexes. *Proc Natl Acad Sci USA* **99**(5), 3153–3158.

Kolakofsky, D. and Hacker, D. (1991). Bunyavirus RNA synthesis: Genome transcription and replication. *Curr Top Microbiol Immunol* **169**, 143–159.

Kortekaas, J., de Boer, S. M., Kant, J., Vloet, R. P., Antonis, A. F., and Moormann, R. J. (2010). Rift Valley fever virus immunity provided by a paramyxovirus vaccine vector. *Vaccine* **28**(27), 4394–4401.

LaBeaud, A. D., Kazura, J. W., and King, C. H. (2010). Advances in Rift Valley fever research: Insights for disease prevention. *Curr Opin Infect Dis* **23**(5), 403–408.

LaBeaud, A. D., Muchiri, E. M., Ndzovu, M., Mwanje, M. T., Muiruri, S., Peters, C. J., and King, C. H. (2008). Interepidemic Rift Valley fever virus seropositivity, northeastern Kenya. *Emerg Infect Dis* **14**(8), 1240–1246.

Labeaud, D. (2010). Towards a safe, effective vaccine for Rift Valley fever virus. *Future Virol* **5**(6), 675–678.

Lagerqvist, N., Naslund, J., Lundkvist, A., Bouloy, M., Ahlm, C., and Bucht, G. (2009). Characterisation of immune responses and protective efficacy in mice after immunisation with Rift Valley Fever virus cDNA constructs. *Virol J* **6**, 6.

Lara, E., Billecocq, A., Leger, P., and Bouloy, M. (2011). Characterization of wild-type and alternate transcription termination signals in the Rift Valley fever virus genome. *J Virol* **85**(23), 12134–12145.

Latham, T. and Galarza, J. M. (2001). Formation of wild-type and chimeric influenza virus-like particles following simultaneous expression of only four structural proteins. *J Virol* **75**(13), 6154–6165.

Laughlin, L. W., Meegan, J. M., Strausbaugh, L. J., Morens, D. M., and Watten, R. H. (1979). Epidemic Rift Valley fever in Egypt: Observations of the spectrum of human illness. *Trans R Soc Trop Med Hyg* **73**(6), 630–633.

Le May, N., Dubaele, S., Proietti De Santis, L., Billecocq, A., Bouloy, M., and Egly, J. M. (2004). TFIIH transcription factor, a target for the Rift Valley hemorrhagic fever virus. *Cell* **116**(4), 541–550.

Le May, N., Mansuroglu, Z., Leger, P., Josse, T., Blot, G., Billecocq, A., Flick, R., Jacob, Y., Bonnefoy, E., and Bouloy, M. (2008). A SAP30 complex inhibits IFN-beta expression in Rift Valley fever virus infected cells. *PLoS Pathog* **4**(1), e13.

Le Roux, C. A., Kubo, T., Grobbelaar, A. A., van Vuren, P. J., Weyer, J., Nel, L. H., Swanepoel, R., Morita, K., and Paweska, J. T. (2009). Development and evaluation of a real-time reverse transcription-loop-mediated isothermal amplification assay for rapid detection of Rift Valley fever virus in clinical specimens. *J Clin Microbiol* **47**(3), 645–651.

Linthicum, K. J., Anyamba, A., Tucker, C. J., Kelley, P. W., Myers, M. F., and Peters, C. J. (1999). Climate and satellite indicators to forecast Rift Valley fever epidemics in Kenya. *Science* **285**(5426), 397–400.

Linthicum, K. J., Davies, F. G., Kairo, A., and Bailey, C. L. (1985). Rift Valley fever virus (family Bunyaviridae, genus *Phlebovirus*). Isolations from Diptera collected during an inter-epizootic period in Kenya. *J Hyg (Lond)* **95**(1), 197–209.

Liu, L., Celma, C. C., and Roy, P. (2008). Rift Valley fever virus structural proteins: Expression, characterization and assembly of recombinant proteins. *Virol J* **5**, 82.

Lopez, N. and Franze-Fernandez, M. T. (2007). A single stem-loop structure in Tacaribe arenavirus intergenic region is essential for transcription termination but is not required for a correct initiation of transcription and replication. *Virus Res* **124**(1–2), 237–244.

Lopez, N., Muller, R., Prehaud, C., and Bouloy, M. (1995). The L protein of Rift Valley fever virus can rescue viral ribonucleoproteins and transcribe synthetic genome-like RNA molecules. *J Virol* **69**(7), 3972–3979.

Lorenzo, G., Martin-Folgar, R., Hevia, E., Boshra, H., and Brun, A. (2010). Protection against lethal Rift Valley fever virus (RVFV) infection in transgenic IFNAR(-/-) mice induced by different DNA vaccination regimens. *Vaccine* **28**(17), 2937–2944.

Lorenzo, G., Martin-Folgar, R., Rodriguez, F., and Brun, A. (2008). Priming with DNA plasmids encoding the nucleocapsid protein and glycoprotein precursors from Rift Valley fever virus accelerates the immune responses induced by an attenuated vaccine in sheep. *Vaccine* **26**(41), 5255–5262.

Loudon, P. T. and Roy, P. (1991). Assembly of five bluetongue virus proteins expressed by recombinant baculoviruses: Inclusion of the largest protein VP1 in the core and virus-like proteins. *Virology* **180**(2), 798–802.

Lozach, P. Y., Kuhbacher, A., Meier, R., Mancini, R., Bitto, D., Bouloy, M., and Helenius, A. (2011). DC-SIGN as a receptor for phleboviruses. *Cell Host Microbe* **10**(1), 75–88.

Lozach, P. Y., Mancini, R., Bitto, D., Meier, R., Oestereich, L., Overby, A. K., Pettersson, R. F., and Helenius, A. (2010). Entry of bunyaviruses into mammalian cells. *Cell Host Microbe* **7**(6), 488–499.

Ludwig, C. and Wagner, R. (2007). Virus-like particles-universal molecular toolboxes. *Curr Opin Biotechnol* **18**(6), 537–545.

Maar, S. A., Swanepoel, R., and Gelfand, M. (1979). Rift Valley fever encephalitis. A description of a case. *Cent Afr J Med* **25**(1), 8–11.

Madani, T. A., Al-Mazrou, Y. Y., Al-Jeffri, M. H., Mishkhas, A. A., Al-Rabeah, A. M., Turkistani, A. M., Al-Sayed, M. O., Abodahish, A. A., Khan, A. S., Ksiazek, T. G., and Shobokshi, O. (2003). Rift Valley fever epidemic in Saudi Arabia: Epidemiological, clinical, and laboratory characteristics. *Clin Infect Dis* **37**(8), 1084–1092.

Mandell, R. B. and Flick, R. (2010). Rift Valley fever virus: An unrecognized emerging threat? *Hum Vaccin* **6**(7), 597–601.

Mandell, R. B., Koukuntla, R., Mogler, L. J., Carzoli, A. K., Freiberg, A. N., Holbrook, M. R., Martin, B. K., Staplin, W. R., Vahanian, N. N., Link, C. J., and Flick, R. (2010a). A replication-incompetent Rift Valley fever vaccine: Chimeric virus-like particles protect mice and rats against lethal challenge. *Virology* **397**(1), 187–198.

Mandell, R. B., Koukuntla, R., Mogler, L. J., Carzoli, A. K., Holbrook, M. R., Martin, B. K., Vahanian, N., Link, C. J., and Flick, R. (2010b). Novel suspension cell-based vaccine production systems for Rift Valley fever virus-like particles. *J Virol Methods* **169**(2), 259–268.

Mansuroglu, Z., Josse, T., Gilleron, J., Billecocq, A., Leger, P., Bouloy, M., and Bonnefoy, E. (2010). Nonstructural NSs protein of rift valley fever virus interacts with pericentromeric DNA sequences of the host cell, inducing chromosome cohesion and segregation defects. *J Virol* **84**(2), 928–939.

McCormick, J. B., King, I. J., Webb, P. A., Scribner, C. L., Craven, R. B., Johnson, K. M., Elliott, L. H., and Belmont-Williams, R. (1986). Lassa fever. Effective therapy with ribavirin. *N Engl J Med* **314**(1), 20–26.

Meegan, J. M. (1981). Rift valley fever in Egypt: An overview of the epizootics in 1977 and 1978. In: T. A. Swartz, M. A. Klinberg, N. Goldblum, and C. M. Papier (Eds.), *Contributions to epidemiology and biostatistics: Rift Valley fever*, pp. 100–113. Karger AG, Basel.

Meegan, J., Le Guenno, B., Ksiazek, T., Jouan, A., Knauert, F., Digoutte, J. P., and Peters, C. J. (1989). Rapid diagnosis of Rift Valley fever: A comparison of methods for the direct detection of viral antigen in human sera. *Res Virol* **140**(1), 59–65.

Mendenhall, M., Russell, A., Smee, D. F., Hall, J. O., Skirpstunas, R., Furuta, Y., and Gowen, B. B. (2011). Effective oral favipiravir (T-705) therapy initiated after the onset of clinical disease in a model of arenavirus hemorrhagic Fever. *PLoS Negl Trop Dis* **5**(10), e1342.

Monath, T. P. (2008). Treatment of yellow fever. *Antiviral Res* **78**(1), 116–124.

Monie, A., Hung, C. F., Roden, R., and Wu, T. C. (2008). Cervarix: A vaccine for the prevention of HPV 16, 18-associated cervical cancer. *Biologics* **2**(1), 97–105.

Morrey, J. D., Taro, B. S., Siddharthan, V., Wang, H., Smee, D. F., Christensen, A. J., and Furuta, Y. (2008). Efficacy of orally administered T-705 pyrazine analog on lethal West Nile virus infection in rodents. *Antiviral Res* **80**(3), 377–379.

Morrill, J. C., Ikegami, T., Yoshikawa-Iwata, N., Lokugamage, N., Won, S., Terasaki, K., Zamoto-Niikura, A., Peters, C. J., and Makino, S. (2010). Rapid accumulation of virulent rift valley fever virus in mice from an attenuated virus carrying a single nucleotide substitution in the m RNA. *PLoS One* **5**(4), e9986.

Morvan, J., Fontenille, D., Saluzzo, J. F., and Coulanges, P. (1991a). Possible Rift Valley fever outbreak in man and cattle in Madagascar. *Trans R Soc Trop Med Hyg* **85**(1), 108.

Morvan, J., Lesbordes, J. L., Rollin, P. E., Mouden, J. C., and Roux, J. (1992a). First fatal human case of Rift Valley fever in Madagascar. *Trans R Soc Trop Med Hyg* **86**(3), 320.

Morvan, J., Rollin, P. E., Laventure, S., Rakotoarivony, I., and Roux, J. (1992b). Rift Valley fever epizootic in the central highlands of Madagascar. *Res Virol* **143**(6), 407–415.

Morvan, J., Saluzzo, J. F., Fontenille, D., Rollin, P. E., and Coulanges, P. (1991b). Rift Valley fever on the east coast of Madagascar. *Res Virol* **142**(6), 475–482.

Muller, R., Saluzzo, J. F., Lopez, N., Dreier, T., Turell, M., Smith, J., and Bouloy, M. (1995). Characterization of clone 13, a naturally attenuated avirulent isolate of Rift Valley fever virus, which is altered in the small segment. *Am J Trop Med Hyg* **53**(4), 405–411.

Mundel, B. and Gear, J. (1951). Rift valley fever; I. The occurrence of human cases in Johannesburg. *S Afr Med J* **25**(44), 797–800.

Naslund, J., Lagerqvist, N., Habjan, M., Lundkvist, A., Evander, M., Ahlm, C., Weber, F., and Bucht, G. (2009). Vaccination with virus-like particles protects mice from lethal infection of Rift Valley Fever Virus. *Virology* **385**(2), 409–415.

Nermut, M. V., Hockley, D. J., Jowett, J. B., Jones, I. M., Garreau, M., and Thomas, D. (1994). Fullerene-like organization of HIV gag-protein shell in virus-like particles produced by recombinant baculovirus. *Virology* **198**(1), 288–296.

Niklasson, B., Grandien, M., Peters, C. J., and Gargan, T. P., 2nd (1983). Detection of Rift Valley fever virus antigen by enzyme-linked immunosorbent assay. *J Clin Microbiol* **17**(6), 1026–1031.

Niklasson, B., Peters, C. J., Bengtsson, E., and Norrby, E. (1985). Rift Valley fever virus vaccine trial: Study of neutralizing antibody response in humans. *Vaccine* **3**(2), 123–127.

Njenga, M. K., Paweska, J., Wanjala, R., Rao, C. Y., Weiner, M., Omballa, V., Luman, E. T., Mutonga, D., Sharif, S., Panning, M., Drosten, C., Feikin, D. R., and Breiman, R. F. (2009). Using a field quantitative real-time PCR test to rapidly identify highly viremic rift valley fever cases. *J Clin Microbiol* **47**(4), 1166–1171.

Obijeski, J. F., Bishop, D. H., Palmer, E. L., and Murphy, F. A. (1976). Segmented genome and nucleocapsid of La Crosse virus. *J Virol* **20**(3), 664–675.

Overby, A. K., Pettersson, R. F., Grunewald, K., and Huiskonen, J. T. (2008). Insights into bunyavirus architecture from electron cryotomography of Uukuniemi virus. *Proc Natl Acad Sci USA* **105**(7), 2375–2379.

Overby, A. K., Pettersson, R. F., and Neve, E. P. (2007). The glycoprotein cytoplasmic tail of Uukuniemi virus (Bunyaviridae) interacts with ribonucleoproteins and is critical for genome packaging. *J Virol* **81**(7), 3198–3205.

Pardigon, N., Vialat, P., Girard, M., and Bouloy, M. (1982). Panhandles and hairpin structures at the termini of germiston virus RNAs (Bunyavirus). *Virology* **122**(1), 191–197.

Patterson, J. L., Kolakofsky, D., Holloway, B. P., and Obijeski, J. F. (1983). Isolation of the ends of La Crosse virus small RNA as a double-stranded structure. *J Virol* **45**(2), 882–884.

Pavlovic, J., Haller, O., and Staeheli, P. (1992). Human and mouse Mx proteins inhibit different steps of the influenza virus multiplication cycle. *J Virol* **66**(4), 2564–2569.

Paweska, J. T., Burt, F. J., Anthony, F., Smith, S. J., Grobbelaar, A. A., Croft, J. E., Ksiazek, T. G., and Swanepoel, R. (2003a). IgG-sandwich and IgM-capture enzyme-linked immunosorbent assay for the detection of antibody to Rift Valley fever virus in domestic ruminants. *J Virol Methods* **113**(2), 103–112.

Paweska, J. T., Burt, F. J., and Swanepoel, R. (2005a). Validation of IgG-sandwich and IgM-capture ELISA for the detection of antibody to Rift Valley fever virus in humans. *J Virol Methods* **124**(1–2), 173–181.

Paweska, J. T., Jansen van Vuren, P., and Swanepoel, R. (2007). Validation of an indirect ELISA based on a recombinant nucleocapsid protein of Rift Valley fever virus for the detection of IgG antibody in humans. *J Virol Methods* **146**(1–2), 119–124.

Paweska, J. T., Mortimer, E., Leman, P. A., and Swanepoel, R. (2005b). An inhibition enzyme-linked immunosorbent assay for the detection of antibody to Rift Valley fever virus in humans, domestic and wild ruminants. *J Virol Methods* **127**(1), 10–18.

Paweska, J. T., Smith, S. J., Wright, I. M., Williams, R., Cohen, A. S., Van Dijk, A. A., Grobbelaar, A. A., Croft, J. E., Swanepoel, R., and Gerdes, G. H. (2003b). Indirect enzyme-linked immunosorbent assay for the detection of antibody against Rift Valley fever virus in domestic and wild ruminant sera. *Onderstepoort J Vet Res* **70**(1), 49–64.

Pepin, M., Bouloy, M., Bird, B. H., Kemp, A., and Paweska, J. (2010). Rift Valley fever virus (Bunyaviridae: Phlebovirus): An update on pathogenesis, molecular epidemiology, vectors, diagnostics and prevention. *Vet Res* **41**(6), 61.

Peters, C. J., Reynolds, J. A., Slone, T. W., Jones, D. E., and Stephen, E. L. (1986). Prophylaxis of Rift Valley fever with antiviral drugs, immune serum, an interferon inducer, and a macrophage activator. *Antiviral Res* **6**(5), 285–297.

Pettersson, R. F. (1975). The structure of Uukuniemi virus, a proposed member of the bunyaviruses. *Med Biol* **53**(5), 418–424.

Pettersson, R. F. and von Bonsdorff, C. H. (1975). Ribonucleoproteins of Uukuniemi virus are circular. *J Virol* **15**(2), 386–392.

Peyrefitte, C. N., Boubis, L., Coudrier, D., Bouloy, M., Grandadam, M., Tolou, H. J., and Plumet, S. (2008). Real-time reverse-transcription loop-mediated isothermal amplification for rapid detection of rift valley Fever virus. *J Clin Microbiol* **46**(11), 3653–3659.

Pichlmair, A., Habjan, M., Unger, H., and Weber, F. (2010). Virus-like particles expressing the nucleocapsid gene as an efficient vaccine against Rift Valley fever virus. *Vector Borne Zoonotic Dis* **10**(7), 701–703.

Piper, M. E., Sorenson, D. R., and Gerrard, S. R. (2011). Efficient cellular release of Rift Valley fever virus requires genomic RNA. *PLoS One* **6**(3), e18070.

Pittman, P. R., Liu, C. T., Cannon, T. L., Makuch, R. S., Mangiafico, J. A., Gibbs, P. H., and Peters, C. J. (1999). Immunogenicity of an inactivated Rift Valley fever vaccine in humans: A 12-year experience. *Vaccine* **18**(1–2), 181–189.

Plotkin, S. A. (2005). Vaccines: Past, present and future. *Nat Med* **11**(4 Suppl), S5–S11.

Raymond, D. D., Piper, M. E., Gerrard, S. R., and Smith, J. L. (2010). Structure of the Rift Valley fever virus nucleocapsid protein reveals another architecture for RNA encapsidation. *Proc Natl Acad Sci USA* **107**(26), 11769–11774.

Reguera, J., Weber, F., and Cusack, S. (2010). Bunyaviridae RNA polymerases (L-protein) have an N-terminal, influenza-like endonuclease domain, essential for viral cap-dependent transcription. *PLoS Pathog* **6**(9), e1001101.

Reichelt, M., Stertz, S., Krijnse-Locker, J., Haller, O., and Kochs, G. (2004). Missorting of LaCrosse virus nucleocapsid protein by the interferon-induced MxA GTPase involves smooth ER membranes. *Traffic* **5**(10), 772–784.

Rose, R. C., Bonnez, W., Reichman, R. C., and Garcea, R. L. (1993). Expression of human papillomavirus type 11 L1 protein in insect cells: in vivo and in vitro assembly of viruslike particles. *J Virol* **67**(4), 1936–1944.

Ross, T. M., Bhardwaj, N., Bissel, S. J., Hartman, A. L., and Smith, D. R. (2012). Animal models of Rift Valley fever virus infection. *Virus Res* **163**(2):417–423.

Sabin, A. B. and Blumberg, R. W. (1947). Human infection with Rift Valley fever virus and immunity twelve years after single attack. *Proc Soc Exp Biol Med* **64**(4), 385–389.

Salib, M. and Sobhy, M. I. (1978). Epidemic maculopathy. *Bull Ophthalmol Soc Egypt* **71**(75), 103–106.

Sall, A. A., de, A. Z. P. M., Vialat, P., Sene, O. K., and Bouloy, M. (1998). Origin of 1997–1998 Rift Valley fever outbreak in East Africa. *Lancet* **352**(9140), 1596–1597.

Sall, A. A., de, A. Z. P. M., Zeller, H. G., Digoutte, J. P., Thiongane, Y., and Bouloy, M. (1997). Variability of the NS(S) protein among Rift Valley fever virus isolates. *J Gen Virol* **78** (Pt 11), 2853–2858.

Sall, A. A., Macondo, E. A., Sene, O. K., Diagne, M., Sylla, R., Mondo, M., Girault, L., Marrama, L., Spiegel, A., Diallo, M., Bouloy, M., and Mathiot, C. (2002). Use of reverse transcriptase PCR in early diagnosis of Rift Valley fever. *Clin Diagn Lab Immunol* **9**(3), 713–715.

Sall, A. A., Zanotto, P. M., Sene, O. K., Zeller, H. G., Digoutte, J. P., Thiongane, Y., and Bouloy, M. (1999). Genetic reassortment of Rift Valley fever virus in nature. *J Virol* **73**(10), 8196–8200.

Samso, A., Bouloy, M., and Hannoun, C. (1975). [Circular ribonucleoproteins in the virus Lumbo (Bunyavirus)]. *C R Acad Sci Hebd Seances Acad Sci D* **280**(6), 779–782.

Schmaljohn, C. and Nichol, S. T. (2007). Bunyaviridae. 5th edn. In: D. M. Knipe, P. M. Howley, D. E. Griffin, R. A. Lamb, M. A. Martin, B. Roizman, and S. E. Straus (Eds.), *Fields Virology*, pp. 1741–1789. Lippincott, Williams and Wilkins, Philadelphia, PA, USA.

Schrire, L. (1951). Macular changes in Rift Valley fever. *S Afr Med J* **25**(50), 926–930.

Schwentker, F. F. and Rivers, T. M. (1934). Rift Valley fever in man: Report of a fatal laboratory infection complicated by thrombophlebitis. *J Exp Med* **59**(3), 305–313.

Sherman, M. B., Freiberg, A. N., Holbrook, M. R., and Watowich, S. J. (2009). Single-particle cryo-electron microscopy of Rift Valley fever virus. *Virology* **387**(1), 11–15.

Shi, X., Kohl, A., Li, P., and Elliott, R. M. (2007). Role of the cytoplasmic tail domains of Bunyamwera orthobunyavirus glycoproteins Gn and Gc in virus assembly and morphogenesis. *J Virol* **81**(18), 10151–10160.

Shoemaker, T., Boulianne, C., Vincent, M. J., Pezzanite, L., Al-Qahtani, M. M., Al-Mazrou, Y., Khan, A. S., Rollin, P. E., Swanepoel, R., Ksiazek, T. G., and Nichol, S. T. (2002). Genetic analysis of viruses associated with emergence of Rift Valley fever in Saudi Arabia and Yemen, 2000–2001. *Emerg Infect Dis* **8**(12), 1415–1420.

Siam, A. L., Gharbawi, K. F., and Meegan, J. M. (1978). Ocular complications of Rift Valley fever. *J Egypt Public Health Assoc* **53**(3–4), 185–186.

Siam, A. L. and Meegan, J. M. (1980). Ocular disease resulting from infection with Rift Valley fever virus. *Trans R Soc Trop Med Hyg* **74**(4), 539–541.

Siam, A. L., Meegan, J. M., and Gharbawi, K. F. (1980). Rift Valley fever ocular manifestations: Observations during the 1977 epidemic in Egypt. *Br J Ophthalmol* **64**(5), 366–374.

Simons, J. F., Hellman, U., and Pettersson, R. F. (1990). Uukuniemi virus S RNA segment: ambisense coding strategy, packaging of complementary strands into virions, and homology to members of the genus Phlebovirus. *J Virol* **64**(1), 247–255.

Simons, J. F. and Pettersson, R. F. (1991). Host-derived 5′ ends and overlapping complementary 3′ ends of the two mRNAs transcribed from the ambisense S segment of Uukuniemi virus. *J Virol* **65**(9), 4741–4748.

Smee, D. F., Hurst, B. L., Wong, M. H., Bailey, K. W., Tarbet, E. B., Morrey, J. D., and Furuta, Y. (2010). Effects of the combination of favipiravir (T-705) and oseltamivir on influenza A virus infections in mice. *Antimicrob Agents Chemother* **54**(1), 126–133.

Smith, D. R., Bird, B. H., Lewis, B., Johnston, S. C., McCarthy, S., Keeney, A., Botto, M., Donnelly, G., Shamblin, J., Albarino, C. G., Nichol, S. T., and Hensley, L. E. (2012). Development of a novel non-human primate model for Rift Valley fever. *J Virol* **86**(4):2109–2120.

Smithburn, K. C., Mahaffy, A. F. et al. (1949). Rift Valley fever; accidental infections among laboratory workers. *J Immunol* **62**(2), 213–227.

Snippe, M., Willem Borst, J., Goldbach, R., and Kormelink, R. (2007). Tomato spotted wilt virus Gc and N proteins interact in vivo. *Virology* **357**(2), 115–123.

Soares, M. M., King, S. W., and Thorpe, P. E. (2008). Targeting inside-out phosphatidylserine as a therapeutic strategy for viral diseases. *Nat Med* **14**(12), 1357–1362.

Soi, R. K., Rurangirwa, F. R., McGuire, T. C., Rwambo, P. M., DeMartini, J. C., and Crawford, T. B. (2010). Protection of sheep against Rift Valley fever virus and sheep poxvirus with a recombinant capripoxvirus vaccine. *Clin Vaccine Immunol* **17**(12), 1842–1849.

Spik, K., Shurtleff, A., McElroy, A. K., Guttieri, M. C., Hooper, J. W., and SchmalJohn, C. (2006). Immunogenicity of combination DNA vaccines for Rift Valley fever virus, tick-borne encephalitis virus, Hantaan virus, and Crimean Congo hemorrhagic fever virus. *Vaccine* **24**(21), 4657–4666.

Strausbaugh, L. J., Laughlin, L. W., Meegan, J. M., and Watten, R. H. (1978). Clinical studies on Rift Valley fever, Part I: Acute febrile and hemorrhagic-like diseases. *J Egypt Public Health Assoc* **53**(3–4), 181–182.

Struthers, J. K. and Swanepoel, R. (1982). Identification of a major non-structural protein in the nuclei of Rift Valley fever virus-infected cells. *J Gen Virol* **60**(Pt 2), 381–384.

Struthers, J. K., Swanepoel, R., and Shepherd, S. P. (1984). Protein synthesis in Rift Valley fever virus-infected cells. *Virology* **134**(1), 118–124.

Svajger, U., Anderluh, M., Jeras, M., and Obermajer, N. (2010). C-type lectin DC-SIGN: An adhesion, signalling and antigen-uptake molecule that guides dendritic cells in immunity. *Cell Signal* **22**(10), 1397–1405.

Swanepoel, R. (1981). Observations on Rift Valley fever in Zimbabwe. *Contrib Epidemiol Biostat* **3**, 83–91.

Swanepoel, R. and Blackburn, N. K. (1977). Demonstration of nuclear immunofluorescence in Rift Valley fever infected cells. *J Gen Virol* **34**(3), 557–561.

Swanepoel, R. and Coetzer, J. A. (2004). Rift Valley fever. In: J. A. Coetzer and R. C. Tustin (Eds.), *Infectious Diseases of Livestock*, pp. 1037–1070, Oxford University Press Southern Africa, Cape Town.

Swanepoel, R., Manning, B., and Watt, J. A. (1979). Fatal Rift Valley fever of man in Rhodesia. *Cent Afr J Med* **25**(1), 1–8.

Swanepoel, R., Struthers, J. K., Erasmus, M. J., Shepherd, S. P., McGillivray, G. M., Erasmus, B. J., and Barnard, B. J. (1986a). Comparison of techniques for demonstrating antibodies to Rift Valley fever virus. *J Hyg (Lond)* **97**(2), 317–329.

Swanepoel, R., Struthers, J. K., Erasmus, M. J., Shepherd, S. P., McGillivray, G. M., Shepherd, A. J., Hummitzsch, D. E., Erasmus, B. J., and Barnard, B. J. (1986b). Comparative pathogenicity and antigenic cross-reactivity of Rift Valley fever and other African phleboviruses in sheep. *J Hyg (Lond)* **97**(2), 331–346.

von Teichman, B., Engelbrecht, A., Zulu, G., Dungu, B., Pardini, A., and Bouloy, M. (2011). Safety and efficacy of Rift Valley fever Smithburn and Clone 13 vaccines in calves. *Vaccine* **29**(34), 5771–5777.

Terasaki, K., Murakami, S., Lokugamage, K. G., and Makino, S. (2011). Mechanism of tripartite RNA genome packaging in Rift Valley fever virus. *Proc Natl Acad Sci USA* **108**(2), 804–809.

van Velden, D. J., Meyer, J. D., Olivier, J., Gear, J. H., and McIntosh, B. (1977). Rift Valley fever affecting humans in South Africa: A clinicopathological study. *S Afr Med J* **51**(24), 867–871.

Vialat, P., Billecocq, A., Kohl, A., and Bouloy, M. (2000). The S segment of rift valley fever phlebovirus (Bunyaviridae) carries determinants for attenuation and virulence in mice. *J Virol* **74**(3), 1538–1543.

Wallace, D. B., Ellis, C. E., Espach, A., Smith, S. J., Greyling, R. R., and Viljoen, G. J. (2006). Protective immune responses induced by different recombinant vaccine regimes to Rift Valley fever. *Vaccine* **24**(49–50), 7181–7189.

Wolf, M. C., Freiberg, A. N., Zhang, T., Akyol-Ataman, Z., Grock, A., Hong, P. W., Li, J., Watson, N. F., Fang, A. Q., Aguilar, H. C., Porotto, M., Honko, A. N., Damoiseaux, R., Miller, J. P., Woodson, S. E., Chantasirivisal, S., Fontanes, V., Negrete, O. A., Krogstad, P., Dasgupta, A., Moscona, A., Hensley, L. E., Whelan, S. P., Faull, K. F., Holbrook, M. R., Jung, M. E., and Lee, B. (2010). A broad-spectrum antiviral targeting entry of enveloped viruses. *Proc Natl Acad Sci USA* **107**(7), 3157–3162.

Won, S., Ikegami, T., Peters, C. J., and Makino, S. (2006). NSm and 78-kilodalton proteins of Rift Valley fever virus are nonessential for viral replication in cell culture. *J Virol* **80**(16), 8274–8278.

Won, S., Ikegami, T., Peters, C. J., and Makino, S. (2007). NSm protein of Rift Valley fever virus suppresses virus-induced apoptosis. *J Virol* **81**(24), 13335–13345.

Woods, C. W., Karpati, A. M., Grein, T., McCarthy, N., Gaturuku, P., Muchiri, E., Dunster, L., Henderson, A., Khan, A. S., Swanepoel, R., Bonmarin, I., Martin, L., Mann, P., Smoak, B. L., Ryan, M., Ksiazek, T. G., Arthur, R. R., Ndikuyeze, A., Agata, N. N., and Peters, C. J. (2002). An outbreak of Rift Valley fever in Northeastern Kenya, 1997–1998. *Emerg Infect Dis* **8**(2), 138–144.

Xu, F., Liu, D., Nunes, M. R., AP, D. A. R., Tesh, R. B., and Xiao, S. Y. (2007). Antigenic and genetic relationships among Rift Valley fever virus and other selected members of the genus Phlebovirus (Bunyaviridae). *Am J Trop Med Hyg* **76**(6), 1194–1200.

Yamshchikov, G. V., Ritter, G. D., Vey, M., and Compans, R. W. (1995). Assembly of SIV virus-like particles containing envelope proteins using a baculovirus expression system. *Virology* **214**(1), 50–58.

Yassin, W. (1978). Clinico-pathological picture in five human cases died with Rift Valley fever. *J Egypt Public Health Assoc* **53**(3–4), 191–193.

Zaki, A., Coudrier, D., Yousef, A. I., Fakeeh, M., Bouloy, M., and Billecocq, A. (2006). Production of monoclonal antibodies against Rift Valley fever virus Application for rapid diagnosis tests (virus detection and ELISA) in human sera. *J Virol Methods* **131**(1), 34–40.

Zamoto-Niikura, A., Terasaki, K., Ikegami, T., Peters, C. J., and Makino, S. (2009). Rift valley fever virus L protein forms a biologically active oligomer. *J Virol* **83**(24), 12779–12789.

Zeng, C. Q., Wentz, M. J., Cohen, J., Estes, M. K., and Ramig, R. F. (1996). Characterization and replicase activity of double-layered and single-layered rotavirus-like particles expressed from baculovirus recombinants. *J Virol* **70**(5), 2736–2742.

22 Crimean–Congo Hemorrhagic Fever Virus

Onder Ergonul

CONTENTS

22.1 INTRODUCTION

Crimean–Congo hemorrhagic fever (CCHF) is a fatal viral infection described in Asia, Africa, and Europe. Humans become infected through the bites of ticks, by contact with a patient with CCHF during the acute phase of infection, or by contact with blood or tissues from viremic livestock. The occurrence of CCHF closely approximates the known world distribution of *Hyalomma* spp. ticks. The novel studies of phylogenetic analyses reveal the interesting relations between the strains from distant outbreaks. CCHF is a fatal viral infection described in parts of Africa, Asia, Eastern Europe, and Middle East (Hoogstraal, 1979; Watts et al., 1988). The virus belongs to the genus *Nairovirus* in the *Bunyaviridae* family and causes severe diseases in humans, with the reported mortality rate of 3%–30% (Ergonul, 2006; Watts et al., 1988). The geographic range of CCHFV is known to be the most extensive one among the tickborne viruses related to human health and the second most widespread of all medically important arboviruses after dengue viruses (Ergonul, 2006). Humans become infected through the bites of ticks, by contact with a patient with CCHF during the acute phase of infection, or by contact with blood or tissues from viremic livestock (Ergonul, 2006; Hoogstraal, 1979; Watts et al., 1988). The health care workers (HCWs) are under serious risk of transmission of the infection, particularly during the follow-up of the patient, with hemorrhages from the nose, mouth, gums, vagina, and injection sites.

22.2 EPIDEMIOLOGY

Crimean hemorrhagic fever (CHF) was firstly described as a clinical entity in 1944–1945, when about 200 Soviet military personnel were infected while assisting peasants in devastated Crimea after Nazi invasion (Butenko and Karganova, 2007; Hoogstraal, 1979). Almost a decade later, Congo hemorrhagic fever was described in Congo by American scientists, and, more than a decade later, the virus was named as Crimean–Congo Hemorrhagic Fever (Woodall, 2007). After the Second World War, epidemics were reported from Asian, African, and Southeast European countries (Ergonul, 2006). In the first decade of the twenty-first century, new outbreaks have been reported from Pakistan (Athar et al., 2002), Iran (Mardani et al., 2003), Senegal (Nabeth et al., 2004), Albania (Papa et al., 2002), Kosovo (Drosten et al., 2002), Bulgaria (Papa et al., 2004), Turkey (Ergonul et al., 2004), Greece (Maltezou et al., 2009), Kenya (Dunster et al., 2002), Mauritania (Nabeth et al., 2004), Kazakhstan (Knust et al., 2012), Tajikistan (Tishkova et al., 2012), and India (Mishra et al., 2011; Patel et al., 2011). The serologic evidence for CCHFV was documented from Egypt, Portugal, Hungary, France, and Benin, although no human case was reported yet (Ergonul and Whitehouse, 2007). The cases were distributed among the actively working ages (Hoogstraal, 1979) that were eventually exposed to the tick population. The great majority of the affected cases deal with agriculture and/or husbandry. Almost 90% of the cases in the recent outbreak in Turkey were farmers (Ergonul and Whitehouse, 2007; Yilmaz et al., 2008). The HCWs are the second most affected groups in the literature. The gender distribution differs between countries, according to the participation of the women to the agricultural work. In Turkey, male to female ratio was reported as one to one (Yilmaz et al., 2009).

There are eight genetically distinct clades based on S segment of the genome, and only one of these clades, AP92, was previously known as causing no symptomatic disease. A study from Turkey reported a mild symptomatic child infected by a strain very closely related to AP92 (Midilli et al., 2009). Multiple strains of CCHFV within the same countries were reported from Turkey and Sudan (Aradaib et al., 2011; Gargili et al., 2011).

CCHFV circulates in the nature in an enzootic tick–vertebrate–tick cycle. Humans have been infected with CCHFV after contact with livestock and other animals, and there is evidence that the virus causes to disease among animals (Ergonul, 2006; Shepherd et al., 1987). Antibody surveys among livestock in endemic areas have shown high prevalence among both cattle and sheep. CCHF viral infection has been demonstrated more commonly among smaller wildlife species such as hares and hedgehogs that act as hosts for the immature stages of the tick vectors (Hoogstraal, 1979). The birds may play a role in the transportation of CCHFV-infected ticks between the countries. The spatial distribution of CCHF well was described in Turkey since the onset of the CCHF epidemic in 2003. Case notifications and spatial analysis of CCHF disease patterns were studied in detail, and climate data (temperature and normalized difference vegetation index) were used to develop a predictive model of the habitat suitability (HS) for *H. marginatum* in Turkey (Vatansever et al., 2007). Approximately 62% of CCHF cases resided in areas of HS above 50, indicating a strong spatial correlation between favorable environmental features for tick populations and CCHF cases. Interestingly, positive HS was predicted to exist outside of the main foci of disease, underlining the existence of additional factors involved in the maintenance of disease foci.

22.3 PATHOGENESIS

Microvascular instability and impaired hemostasis are the hallmarks of CCHFV infection. Interpretation of data derived from animal studies may be confounded by a series of factors, such as the species of the animal, the route of inoculation, and the virus dose. Recently, Bente et al. described a new mouse model, which was reported as exhibiting key features of fatal human CCHF (Bente et al., 2010). This model could be useful for the testing of therapeutic strategies and can be used to study virus attenuation.

After inoculation, virus first replicates in dendritic cells and other local tissues, with subsequent migration to regional lymph nodes and then dissemination through the lymph and blood monocytes to a broad range of tissues and organs, including the liver, spleen, and lymph nodes. Migration of tissue macrophages results in secondary infection of permissive parenchymal cells. Although lymphocytes remain free of infection, they may be destroyed in massive numbers over the course of illness through apoptosis, as seen in other forms of septic shock. The synthesis of cell surface tissue factor triggers the extrinsic coagulation pathway. Impaired hemostasis may entail endothelial cell, platelet, and/or coagulation factor dysfunction. Disseminated intravascular coagulopathy (DIC) is frequently noted in CCHF virus infections. Reduced levels of coagulation factors may be secondary to hepatic dysfunction and/or disseminated intravascular coagulation. In addition, CCHFV may lead to a hemorrhagic diathesis through direct damage of platelets and endothelial cells and/or indirectly through immunological and inflammatory pathways (Chen and Cosgriff, 2000; Geisbert and Jahrling, 2004; Peters and Zaki, 2002). These changes appear to be largely the consequence of the release of cytokines, chemokines, and other proinflammatory mediators from virus-infected monocytes and macrophages (Bray, 2007; Ergonul et al., 2006). Tissue damage may be mediated through direct necrosis of infected cells or indirectly through apoptosis of immune cells. The hepatocytes were reported to be particularly effected in CCHFV infection (Rodrigues et al., 2012).

22.4 CLINICAL MANIFESTATIONS

The incubation period for CCHFV ranges from 1 to 9 days. Thirty days of incubation period was reported from Turkey (Meric Koc and Willke, 2012). Patients initially exhibit a nonspecific prodrome, which typically lasts less than 1 week. Symptoms typically include high fever, headache, malaise, arthralgias, myalgias, nausea, abdominal pain, and rarely diarrhea (Bray, 2007). Early signs typically include fever, hypotension, conjunctivitis, and cutaneous flushing or a skin rash. Later, patients may develop signs of progressive hemorrhagic diathesis, such as petechiae, mucous membrane and conjunctival hemorrhage; hematuria; hematemesis; and melena. Disseminated intravascular coagulation and circulatory shock may ensue. Death is typically preceded by hemorrhagic diathesis, shock, and multi-organ system failure 1–2 weeks following the onset of symptoms. The disease was reported to be milder among the children (Tezer et al., 2010).

Laboratory abnormalities include usually leukopenia, thrombocytopenia, and elevated liver enzymes. Anemia is not usually seen at the early phase of the disease, but may develop late in the disease course. Coagulation abnormalities may include prolonged bleeding time, prothrombin time, and activated partial thromboplastin time; elevated fibrin degradation products; and decreased fibrinogen (Table 22.1).

TABLE 22.1
Laboratory Characteristics of CCHFV Infection

Tests	Findings and Comments
Complete blood count	
White blood cell count	Moderate or severe leukopenia, sometimes leukocytosis
Platelet count	Mild to severe decrease
Hemoglobin and/or hematocrit	Could be decreased later in disease course
Liver enzymes	Increased, usually AST > ALT
Coagulation studies (INR, PT, PTT, fibrinogen, fibrin split products, platelets, D-dimer)	Hemophagocytosis and DIC are common
Lactate dehydrogenase	A level greater than 4 mmol/L (36 mg/dL) may indicate persistent hypo-perfusion and sepsis
Creatinine phosphokinase	Elevated
Blood urea nitrogen and creatinine	Renal failure may occur late in disease course, but proteinuria may occur

22.5 DIAGNOSIS

The diagnosis is performed by the detection of the viral RNA genome and/or the antigen and the detection of specific IgM antibodies in human serum or blood (Table 22.2). Antigen detection (by ELISA) and RT-PCR are the most useful diagnostic techniques in the acute clinical setting. Viral isolation is of limited value because it requires a biosafety level 4 (BSL-4) laboratory. Leukopenia, particularly neutropenia, thrombocytopenia, high levels of liver enzymes alanine aminotransferase (ASL) and aspartate aminotransferase (AST), and lactate dehydrogenase (LDH) are regularly reported in patients.

TABLE 22.2
Case Management for CCHFV Infection

Evaluation of the cases

Clinical symptoms

Early symptoms (first days at the end of incubation): myalgia, fever, diarrhea

Late symptoms (3–10 days after incubation): bleeding from various sites

Patient history

1. Referral from endemic area

2. Outdoor activities (picnic, tracking, etc.) in endemic area

3. History of tick bite or contact with body fluids of the infected people

4. Dealing with husbandry in endemic area

Laboratory tests: Findings compatible with hemophagocytosis (low thrombocyte and white blood cell count, elevated AST, ALT, LDH, CPK)

Preventive measures

a. Isolate the patient

b. Inform and educate HCWs and caregivers of the patients

c. Assess the risk for transmission and use the barrier precautions accordingly

Investigations for confirmation

Serum for PCR and ELISA

a. IgM positivity or PCR positive confirms diagnosis, IgG positivity cannot

b. Sera for differential diagnosis

Decision making for therapy

1. Start ribavirin for early cases

2. Do not neglect other causes of clinical picture. Doxycycline or equivalent should be considered for the diseases in differential diagnosis list

3. Hematological support

a. Fresh frozen plasma to improve the homeostasis

b. Thrombocyte solutions

4. Respiratory support

Follow-up

No recurrence was reported. Therefore, there is no definition of follow-up

Postexposure prophylaxis

HCWs or other individuals who are exposed to the virus should be assessed for the level of the risk. Individuals in high-risk group should receive ribavirin, whereas individuals in low-risk group should be followed up with complete blood counts and biochemical tests for 14 days

---

22.5.1 Viral Isolation

The most definitive test is viral culture; however, the time to diagnosis, which is 2–10 days for the virus to grow, is too long for the management of acute cases. Moreover, the need for high containment facilities renders this technique more of a confirmatory test and research tool. CCHFV has been isolated most frequently by intracranial inoculation of newborn suckling mice (Zeller, 2007).

22.5.2 Molecular Detection: RT-PCR

Reverse transcriptase polymerase chain reaction (RT-PCR) is a very sensitive rapid test, although careful attention must be paid to the potential for false-positive results. One-step real-time RT-PCR assays using primers to the same nucleoprotein gene have been developed, but the development of these assays has been hampered by the high diversity of genome sequence (Atkinson et al., 2012; Drosten et al., 2002, 2003). One of the recently developed real-time RT-PCR assays designed to amplify a conserved region of the CCHF virus S segment is capable of detecting strains from all seven groups of CCHF, including the AP92 strain, which until recently represented a lineage of strains that were not associated with human disease (Atkinson et al., 2012).

22.5.3 Indirect Serological Diagnosis

The serological diagnosis of VHF infection is based on the detection of specific IgM and IgG antibodies induced by the immune response principally to the nucleoprotein, which is recognized as the predominant antigen (Tezer et al., 2010). Seroconversion with detection of CCHFV IgM antibodies or a $\geq$ fourfold increase in antibody titer between two successive blood samples is the evidence of a recent infection (Paragas et al., 2004; Watts et al., 1999). The serological diagnosis is valid after several days post onset of the disease; nevertheless, the antibody response rarely is observed in fatal cases (Ergonul et al., 2006). The ELISA is the most common technique for CCHFV antibody detection with the sensitivity of more than 90%. The ELISA was reported to be more sensitive than IFA (Burt et al., 1993). All native antigens have to be produced in a BSL-4 laboratory and irradiated prior to use. Usually IgM and IgG antibodies are detected 4–5 days post onset of symptoms. The IgM titer is maximal 2–3 weeks after onset of the disease, and the IgM antibodies generally disappear within 4 months. The IgG antibodies remain detectable for several years (Zeller, 2007).

22.5.4 Differential Diagnosis

The combined approach of viral genome detection and IgM detection is recommended for the diagnosis of CCHF in humans. CCHFV has an extended geographical distribution including Africa, southern Europe, Middle East, Russia, India, and China. Other viral etiologies have to be considered according to the origin of the patient and the risks of potential exposure. These would include Alkhurma and Rift Valley fever in the Middle East; Omsk hemorrhagic fever in Russia; Kyasanur Forest disease in India; hantaviruses in Europe and Asia; Lassa, Ebola, Marburg, Rift Valley fever, yellow fever in Africa; and dengue in various locations (Ergonul, 2007a). In tropical and subtropical countries, malaria is the most important alternative diagnosis to be excluded in cases of suspected VHF. The differential diagnosis list should include hepatitis viruses, influenza, *Neisseria meningitidis*, leptospirosis, borreliosis, typhoid, rickettsiosis, and Q fever (*Coxiella burnetii*) staphylococcal or gram-negative sepsis, toxic shock syndrome, salmonellosis and shigellosis, psittacosis, trypanosomiasis, septicemic plague, rubella, measles, and hemorrhagic smallpox (Borio et al., 2002; Ergonul, 2006; Zeller, 2007) (Table 22.3).

TABLE 22.3

Differential Diagnosis List for Viral Hemorrhagic Fevers

Disease Categories	Differentials
Infections	
Brucellosis	Pancytopenia, Wright agglutination
Q fever	Serology (ELISA or IFAT)
Rickettsia	Weil–Felix test
Ehrlichiosis	Serology (ELISA)
Hanta	Pulmonary or renal involvement, serology, PCR
Other viral hemorrhagic infections	
Ebola	Geographic location
Marburg	Geographic location
Leptospira	Agglutination
Salmonella	Widal test
Noninfectious reasons	
Vitamin B_{12} deficiency	Pancytopenia, and B_{12} level in serum
Febrile neutropenia	Underlying disease
HELLP syndrome	Geographic location
Drug side effects	
Metamizole	History

Noninfectious processes associated with bleeding diathesis that should be included in the differential diagnosis include idiopathic or thrombotic thrombocytopenic purpura, HELP syndrome among pregnant women, hemolytic uremic syndrome, acute leukemia, vitamin B_{12} deficiency, and collagen-vascular diseases (Ergonul, 2006; Ergonul et al., 2010).

22.6 THERAPY

Supportive therapy makes the essential part of the case management. Thrombocyte solutions and fresh frozen plasma are given by monitoring the bleeding status of the patients. Ribavirin is the only antiviral drug that has been used to treat viral hemorrhagic fever syndromes, including CCHF and Lassa fever (Bausch et al., 2010; Ergonul, 2007b; McCormick et al., 1986). Viruses in the *Bunyaviridae* family are generally sensitive to ribavirin (Sidwell and Smee, 2003). Ribavirin was earlier shown to be effective against CCHFV *in vitro* (Paragas et al., 2004; Tignor and Hanham, 1993; Watts et al., 1989). In suckling mice, ribavirin treatment reduced CCHF virus growth in the liver; significantly decreased, but did not prevent viremia; and significantly reduced mortality and extended the geometric mean time to death (Tignor and Hanham, 1993). In clinical practice, the ribavirin was found to be beneficial, especially at the earlier phase of the infection (Ergonul, 2008; Izadi and Salehi, 2009; Ozbey, 2010; Tasdelen Fisgin et al., 2009). Ribavirin is placed on the WHO essential medicines list (15th Model List of Essential Medicines, March 2007) to be used against CCHFV infection. In a recent review, ethical concerns about conducting a randomized controlled trial of ribavirin in the treatment of CCHF were detailed (Arda et al., 2012). Despite the need for more evidence about the impact of ribavirin in treatment, the authors described why it was not ethical to conduct an RCT in such a fatal disease with only one antiviral alternative (Arda et al., 2012; Ergonul, 2012).

Bulgarian investigators suggested that immunotherapy treatment of seven patients with severe CCHF via passive simultaneous transfer of two different specific immunoglobulin preparations, CCHF-bulin (for intramuscular use) and CCHF-venin (for intravenous use), prepared from the plasma of CCHF survivor donors boosted with one dose of CCHF vaccine, resulted in quick recovery of all patients (Vassilenko et al., 1990), and they suggested that the intravenous preparation be used for the treatment of all cases of CCHF (Dimitrov, 2007). In a recent study that included 22 severe patients from Turkey, prompt administration of CCHFV hyperimmunoglobulin was suggested as an alternative treatment approach, especially for high-risk individuals (Kubar et al., 2011). Further studies with larger sample size and more detailed design are necessary.

22.6.1 Infection Control Measures

Bunyaviruses are highly infectious after direct contact with infected blood and bodily secretions. A suspected case of CCHF must be immediately reported to the infection control professional and to the local or state health department. Infection control professional should notify the clinical laboratory as well as other clinicians and public health authorities.

22.6.2 Isolation Precautions

Direct contact with infected blood and bodily fluids has accounted for the majority of person-to-person transmission. Therefore, it was recommend that in the case of any patient with suspected or documented CCHFV infection, specific barrier precautions should be implemented immediately. Airborne transmission was suspected in one Lassa fever outbreak in 1969 (Carey et al., 1972), but there has never been any documented case of airborne transmission of that virus to humans. A prospective serological study from Sierra Leone suggested that the hospital staff who cared for Lassa fever patients using simple barrier nursing methods have no higher risk of infection than the local population (Helmick et al., 1986). A similar study from Turkey reported the lack of airborne transmission of CCHF to the HCWs after CCHF epidemic (Ergonul et al., 2007). Usually the standard precautions including health hygiene, using gloves, gowns, face shields, and masks, are sufficient to be protected. However, in case of procedure that may generate an aerosol, HCW should consider wearing an N95 or FFP2 respirator (European Norm (EN) 61010-1) (Tarantola et al., 2007).

An integrated strategy for the control of accidental exposure to blood and body fluids is the critical step while providing the protection among the HCWs (Tarantola et al., 2007). Sharp containers are the foremost safety equipment that should be available at all times to all units. The use of safety-engineered devices should also be considered in order to decrease the risk of needlestick injuries (Tarantola et al., 2007).

22.6.3 Postexposure Management

Postexposure management systems are an integral part of an effort to enhance HCW safety. In CCHF (Ergonul, 2008) and Lassa fever (Bausch et al., 2010), use of oral ribavirin as postexposure prophylaxis was well described as effective and beneficial drug. Ribavirin prophylaxis is generally well tolerated, potentially useful, and should therefore be recommended for HCWs who are at high risk of exposures such as percutaneous injuries or splash of contaminated blood or body fluid to the face or mucosal surfaces of the HCWs (Tarantola et al., 2007). In conclusion, many recent developments were noted in epidemiology, pathogenesis, and early detection; however, there is neither alternative antiviral therapy nor vaccine on the pipeline yet.

REFERENCES

Aradaib, I. E., Erickson, B. R., Karsany, M. S. et al. 2011. Multiple Crimean-Congo hemorrhagic fever virus strains are associated with disease outbreaks in Sudan, 2008–2009. *PLoS Negl Trop Dis* **5**(5):e1159.

Arda, B., Aciduman, A., and J. C. Johnston. 2012. A randomised controlled trial of ribavirin in Crimean Congo haemorrhagic fever: Ethical considerations. *J Med Ethics* **38**(2):117–120.

Athar, M. N., Baqai, H. Z., Ahmad, M. et al. 2002. Short report: Crimean-Congo hemorrhagic fever outbreak in Rawalpindi, Pakistan, February 2002. *Am J Trop Med Hyg* **69**(3):284–277.

Atkinson, B., Chamberlain, J., Logue, C. H. et al. 2012. Development of a real-time RT-PCR assay for the detection of Crimean-Congo hemorrhagic fever virus. *Vector Borne Zoonotic Dis* **12**(9):786–793.

Bausch, D. G., Hadi, C. M., Khan, S. H., and J. J. Lertora. 2010. Review of the literature and proposed guidelines for the use of oral ribavirin as postexposure prophylaxis for Lassa fever. *Clin Infect Dis* **51**(12):1435–1441.

Bente, D. A., Alimonti, J. B., Shieh, W. J. et al. 2010. Pathogenesis and immune response of Crimean-Congo hemorrhagic fever virus in a STAT-1 knockout mouse model. *J Virol* **84**(21):11089–11100.

Borio, L., Inglesby, T., Peters, C. J. et al. 2002. Hemorrhagic fever viruses as biological weapons: Medical and public health management. *JAMA* **287**(18):2391–2405.

Bray, M. 2007. Comparative pathogenesis of Crimean Congo hemorrhagic fever and Ebola hemorrhagic fever. In: Ergonul, O. and Whitehouse, C. A., eds. *Crimean Congo Hemorrhagic Fever: A Global Perspective.* Dordrecht, the Netherlands: Springer, pp. 221–231.

Burt, F. J., Swanepoel, R., and L. E. Braack. 1993. Enzyme-linked immunosorbent assays for the detection of antibody to Crimean-Congo haemorrhagic fever virus in the sera of livestock and wild vertebrates. *Epidemiol Infect* **111**(3):547–557.

Butenko, A. M. and G. Karganova. 2007. Crimean Congo hemorrhagic fever in Russia and other countries of the Former Soviet Union. In: Ergonul, O. and Whitehouse, C. A., eds. *Congo Hemorrhagic Fever: A Global Perspective.* Dordrecht, the Netherlands: Springer, pp. 99–115.

Carey, D. E., Kemp, G. E., White, H. A. et al. 1972. Lassa fever. Epidemiological aspects of the 1970 epidemic, Jos, Nigeria. *Trans R Soc Trop Med Hyg* **66**(3):402–408.

Chen, J. P. and T.M. Cosgriff. 2000. Hemorrhagic fever virus-induced changes in hemostasis and vascular biology. *Blood Coagul Fibrinolysis* **11**(5):461–483.

Dimitrov, D. S. 2007. Antibodies to CCHFV for prophylaxis and treatment. In: Ergonul, O. and Whitehouse, C. A., eds. *Crimean-Congo Hemorrhagic Fever: A Global Perspective.* Dordrecht, the Netherlands: Springer, pp. 261–269.

Drosten, C., Gottig, S., Schilling, S. et al. 2002. Rapid detection and quantification of RNA of Ebola and Marburg viruses, Lassa virus, Crimean-Congo hemorrhagic fever virus, Rift Valley fever virus, dengue virus, and yellow fever virus by real-time reverse transcription-PCR. *J Clin Microbiol* **40**(7):2323–2330.

Drosten, C., Kummerer, B.M., Schmitz, H., and S. Gunther. 2003. Molecular diagnostics of viral hemorrhagic fevers. *Antiviral Res* **57**(1–2):61–87.

Drosten, C., Minnak, D., Emmerich, P., Schmitz, H., and T. Reinicke. 2002. Crimean-Congo hemorrhagic fever in Kosovo. *J Clin Microbiol* **40**(3):1122–1123.

Dunster, L., Dunster, M., Ofula, V. et al. 2002. First documentation of human Crimean-Congo hemorrhagic fever, Kenya. *Emerg Infect Dis* **8**(9):1005–1006.

Ergonul, O. 2006. Crimean-Congo haemorrhagic fever. *Lancet Infect Dis* **6**(4):203–214.

Ergonul, O. 2007a. Clinical and pathologic features of Crimean Congo hemorrhagic fever. In: Ergonul, O. and Whitehouse, C. A., eds. *Crimean Congo Hemorrhagic Fever: A Global Perspective.* Dordrecht, the Netherlands: Springer, pp. 207–220.

Ergonul, O. 2007b. Treatment of Crimean Congo hemorrhagic fever. In: Ergonul, O. and Whitehouse, C. A., eds. *Crimean Congo Hemorrhagic Fever: A Global Perspective.* Dordrecht, the Netherlands: Springer, pp. 245–260.

Ergonul, O. 2008. Treatment of Crimean-Congo hemorrhagic fever. *Antiviral Res* **78**(1):125–131.

Ergonul, O. 2012. Crimean-Congo hemorrhagic fever virus: New outbreaks, new discoveries. *Curr Opin Virol* **2**(2):215–220.

Ergonul, O., Celikbas, A., Baykam, N., Eren, S., and B. Dokuzoguz. 2006. Analysis of risk-factors among patients with Crimean-Congo haemorrhagic fever virus infection: Severity criteria revisited. *Clin Microbiol Infect* **12**(6):551–554.

Ergonul, O., Celikbas, A., Dokuzoguz, B., Eren, S., Baykam, N., and H. Esener. 2004. Characteristics of patients with Crimean-Congo hemorrhagic fever in a recent outbreak in Turkey and impact of oral ribavirin therapy. *Clin Infect Dis* **39**(2):284–287.

Ergonul, O., Celikbas, A., Yildirim, U. et al. 2010. Pregnancy and Crimean-Congo haemorrhagic fever. *Clin Microbiol Infect* **16**(6):647–650.

Ergonul, O., Tuncbilek, S., Baykam, N., Celikbas, A., and B. Dokuzoguz. 2006. Evaluation of serum levels of interleukin (IL)-6, IL-10, and tumor necrosis factor-alpha in patients with Crimean-Congo hemorrhagic fever. *J Infect Dis* **193**(7):941–944.

Ergonul, O. and C. A. Whitehouse. 2007. Introduction. In: Ergonul, O. and Whitehouse, C. A., eds. *Crimean Congo Hemorrhagic Fever: A Global Perspective*. Dordrecht, the Netherlands: Springer, pp. 3–11.

Ergonul, O., Zeller, H. G., Celikbas, A., and B. Dokuzoguz. 2007. The lack of Crimean-Congo hemorrhagic fever virus antibodies in healthcare workers in an endemic region. *Int J Infect Dis* 11:48–51.

Gargili, A., Midilli, K., Ergonul, O. et al. 2011. Crimean-Congo hemorrhagic fever in European part of Turkey: Genetic analysis of the virus strains from ticks and a seroepidemiological study in humans. *Vector Borne Zoonotic Dis* 11(6):747–752.

Geisbert, T.W. and P.B. Jahrling. 2004. Exotic emerging viral diseases: Progress and challenges. *Nat Med* 10(12 Suppl):S110–S121.

Helmick, C. G., Webb, P. A., Scribner, C. L, Krebs, J. W., and J. B. McCormick. 1986. No evidence for increased risk of Lassa fever infection in hospital staff. *Lancet* 2(8517):1202–1205.

Hoogstraal, H. 1979. The epidemiology of tick-borne Crimean-Congo hemorrhagic fever in Asia, Europe, and Africa. *J Med Entomol* 15(4):307–417.

Izadi, S. and M. Salehi. 2009. Evaluation of the efficacy of ribavirin therapy on survival of Crimean-Congo hemorrhagic fever patients: A case-control study. *Jpn J Infect Dis* 62(1):11–15.

Knust, B., Medetov, Z. B., Kyraubayev, K. B. et al. 2012. Crimean-Congo hemorrhagic fever, Kazakhstan, 2009–2010. *Emerg Infect Dis* 18(4):643–645.

Kubar, A., Haciomeroglu, M., Ozkul, A. et al. 2011. Prompt administration of Crimean-Congo hemorrhagic fever (CCHF) virus hyperimmunoglobulin in patients diagnosed with CCHF and viral load monitorization by reverse transcriptase-PCR. *Jpn J Infect Dis* 64(5):439–443.

Maltezou, H. C., Papa, A., Tsiodras, S., Dalla, V., Maltezos, E., and A. Antoniadis. 2009. Crimean-Congo hemorrhagic fever in Greece: A public health perspective. *Int J Infect Dis* 13(6):713–716.

Mardani, M., Jahromi, M. K., Naieni, K. H., and M. Zeinali. 2003. The efficacy of oral ribavirin in the treatment of crimean-congo hemorrhagic fever in Iran. *Clin Infect Dis* 36(12):1613–1618.

McCormick, J. B., King, I. J., Webb, P. A. et al. 1986. Lassa fever. Effective therapy with ribavirin. *N Engl J Med* 314(1):20–26.

Meric Koc, M. and A. Willke. 2012. A case of Crimean-Congo hemorrhagic fever with long incubation period in Kocaeli, Turkey. *Mikrobiyol Bul* 46(1):129–133.

Midilli, K., Gargili, A., Ergonul, O. et al. 2009. The first clinical case due to AP92 like strain of Crimean-Congo hemorrhagic fever virus and a field survey. *BMC Infect Dis* 9:90.

Mishra, A. C., Mehta, M., Mourya, D. T., and S. Gandhi. 2011. Crimean-Congo haemorrhagic fever in India. *Lancet* 378(9788):372.

Nabeth, P., Cheikh, D. O., Lo, B. et al. 2004. Crimean-Congo hemorrhagic fever, Mauritania. *Emerg Infect Dis* 10(12):2143–2149.

Nabeth, P., Thior, M., Faye, O., and F. Simon. 2004. Human Crimean-Congo hemorrhagic fever, Senegal. *Emerg Infect Dis* 10(10):1881–1882.

Ozbey, S.B. 2010. Impact of early Ribavirin use on fatality of CCHF. *Klimik J* 23:6–10.

Papa, A., Bino, S., Llagami, A. et al. 2002. Crimean-Congo hemorrhagic fever in Albania, 2001. *Eur J Clin Microbiol Infect Dis* 21(8):603–606.

Papa, A., Christova, I., Papadimitriou, E., and A. Antoniadis. 2004. Crimean-Congo hemorrhagic fever in Bulgaria. *Emerg Infect Dis* 10(8):1465–1467.

Paragas, J., Whitehouse, C. A., Endy, T. P., and M. Bray. 2004. A simple assay for determining antiviral activity against Crimean-Congo hemorrhagic fever virus. *Antiviral Res* 62(1):21–25.

Patel, A. K., Patel, K. K., Mehta, M., Parikh, T. M., Toshniwal, H., and K. Patel. 2011. First Crimean-Congo hemorrhagic fever outbreak in India. *J Assoc Physicians India* 59:585–589.

Peters, C. J. and S. R. Zaki. 2002. Role of the endothelium in viral hemorrhagic fevers. *Crit Care Med* 30(5 Suppl):S268–S273.

Rodrigues, R., Paranhos-Baccala, G., Vernet, G., and C. N. Peyrefitte. 2012. Crimean-Congo hemorrhagic fever virus-infected hepatocytes induce ER-stress and apoptosis crosstalk. *PLoS One* 7(1):e29712.

Shepherd, A. J., Swanepoel, R., Leman, P. A., and S. P. Shepherd. 1987. Field and laboratory investigation of Crimean-Congo haemorrhagic fever virus (Nairovirus, family Bunyaviridae) infection in birds. *Trans R Soc Trop Med Hyg* 81(6):1004–1007.

Sidwell, R. W. and D. F. Smee. 2003. Viruses of the Bunya- and Togaviridae families: Potential as bioterrorism agents and means of control. *Antiviral Res* 57(1–2):101–111.

Tarantola, A., Ergonul, O., and P. Tattevin. 2007. Estimates and prevention of Crimean Congo hemorrhagic fever risks for health care workers. In: Ergonul, O. and Whitehouse, C. A., eds. *Crimean-Congo Hemorrhagic Fever: A Global Perspective*. Dordrecht, the Netherlands: Springer, pp. 281–294.

Tasdelen Fisgin, N., Ergonul, O., Doganci, L., and N. Tulek. 2009. The role of ribavirin in the therapy of Crimean-Congo hemorrhagic fever: Early use is promising. *Eur J Clin Microbiol Infect Dis* **28**(8):929–933.

Tezer, H., Sucakli, I.A., Sayli, T. R. et al. 2010. Crimean-Congo hemorrhagic fever in children. *J Clin Virol* **48**(3):184–186.

Tignor, G. H. and C. A. Hanham. 1993. Ribavirin efficacy in an in vivo model of Crimean-Congo hemorrhagic fever virus (CCHF) infection. *Antiviral Res* **22**(4):309–325.

Tishkova, F. H., Belobrova, E. A., Valikhodzhaeva, M., Atkinson, B., Hewson, R., and M. Mullojonova. 2012. Crimean-Congo hemorrhagic fever in Tajikistan. *Vector Borne Zoonotic Dis* **12**(9):722–726.

Vassilenko, S. M., Vassilev, T. L., Bozadjiev, L. G., Bineva, I. L., and G. Z. Kazarov. 1990. Specific intravenous immunoglobulin for Crimean-Congo haemorrhagic fever. *Lancet* **335**(8692):791–792.

Vatansever, Z., Uzun, R., Estrada-Pena, A., and O. Ergonul. 2007. Crimean Congo hemorrhagic fever in Turkey. In: Ergonul, O. and Whitehouse, C. A., eds. *Crimean Congo Hemorrhagic Fever: A Global Perspective.* Dordrecht, the Netherlands: Springer, pp. 59–74.

Watts, D. M., Ksiasek, T. G., Linthicum, K. J., and H. Hoogstraal. 1988. Crimean-Congo hemorrhagic fever. In Monath, T. P., ed. *The Arboviruses: Epidemiology and Ecology.* Boca Raton, FL: CRC.

Watts, D. M, Ussery, M. A., Nash, D., and C. J. Peters. 1989. Inhibition of Crimean-Congo hemorrhagic fever viral infectivity yields in vitro by ribavirin. *Am J Trop Med Hyg* **41**(5):581–585.

Woodall, J. P. 2007. Personal reflections. In: Ergonul, O. and Whitehouse, C. A., eds. *Crimean Congo Hemorrhagic Fever: A Global Perspective.* Dordrecht, the Netherlands: Springer, pp. 23–32.

Yilmaz, G. R., Buzgan, T., Irmak, H. et al. 2008. The epidemiology of Crimean-Congo hemorrhagic fever in Turkey, 2002–2007. *Int J Infect Dis* **13**(3):380–386.

Yilmaz, G. R., Buzgan, T., Irmak, H. et al. 2009. The epidemiology of Crimean-Congo hemorrhagic fever in Turkey, 2002–2007. *Int J Infect Dis* **13**(3):380–386.

Zeller, H. 2007. Laboratory diagnosis of Crimean Congo hemorrhagic fever. In: Ergonul, O. and Whitehouse, C. A., eds. *Crimean Congo Hemorrhagic Fever: A Global Perspective.* Dordrecht, the Netherlands: Springer, pp. 233–243.

23 Hemorrhagic Fever with Renal Syndrome

*Paul Heyman, Tatjana Avšič-Županc, Christel Cochez,
Ana Saksida, and Ake Lundkvist*

CONTENTS

23.1 INTRODUCTION

In February 1982, on a meeting of the Working Group on Hemorrhagic Fever with Renal Syndrome, it was decided that infections characterized by fever, hemorrhage, and renal involvement should in the future be referred to as "hemorrhagic fever with renal syndrome" (HFRS) (Anon 1983). HFRS is prevalent in large parts of Europe and Asia. The viruses that cause the earlier described conditions belong to the genus Hantavirus, a member of the family *Bunyaviridae* (Plyusnin et al. 1996; Vapalahti et al. 2003). Rodents and insectivores act as carriers [for the current taxonomy see, Wilson and Reed (2005)]. In rodents, hantaviruses are found in the *Cricetidae* family (subfamilies *Arvicolinae*, *Neotominae*, and *Sigmodontinae*) and in the *Muridae* family (Henttonen et al. 2008). Each hantavirus type is—as a rule—carried by a specific rodent host species. Recently, new hantaviruses have been discovered in the *Soricidae* and *Talpidae* families (subfamilies *Crocidurinae* and *Scalopinae* respectively) in which the very first hantavirus, Thottapalayam virus (TPMV), (Carey et al. 1971) was discovered in *Suncus murinus,* the Asian House shrew (family *Soricidae*, subfamily *Crocidurinae*) as such preceding Hantaan virus (HTNV) (Lee et al. 1978). In nature, their zoonotic cycle is maintained by transmission from rodent to rodent (and probably from insectivore to insectivore) via aerosolized contaminated excreta (Olsson et al. 2010).

In Europe, HFRS can be caused by several hantaviruses based on the distribution range of their hosts: Puumala virus (PUUV), carried by *Myodes glareolus* (the bank vole) (Brummer-Korvenkontio et al. 1982), and Dobrava virus (DOBV, carried by *Apodemus flavicollis* (the yellow-necked mouse)) (Avsic-Zupanc et al. 1992). *A. agrarius* (the striped field mouse) is the carrier of Saaremaa virus (SAAV) (Sjölander et al. 2002) or DOBV-A.a., in the Baltic States and Kurkino virus in Denmark and Northwest Russia (Klempa et al. 2003). In the Krasnodar region, in the southwest of Russia, infection with Sochi virus carried by *A. ponticus* (Caucasus field mouse)—a close relative of *A. flavicollis* with a distribution range that is limited to S-Russia, Georgia, Armenia, and Azerbaijan—can also produce HFRS (Klempa et al. 2008). The HFRS-inducing hantaviruses in Europe are Puumala (PUUV) carried by bank voles and two interrelated viruses carried by Apodemus mice, Dobrava virus (DOBV), and Saaremaa virus (SAAV). These are the species listed by the International Committee on Taxonomy of Viruses, but the nomenclature of the European Apodemus-derived hantaviruses has been, and still is, under debate and revision: in the literature DOBV-variants in *A. flavicollis* are also referred to as DOBV-Af, and variants in *A. ponticus* as DOBV-Ap. Some strains recovered from *A. agrarius* are described as a genotype DOBV-Aa. Infection with Seoul virus (SEOV) carried by *Rattus rattus* and *R. norvegicus* (black and brown rat respectively) could also cause HFRS but—in Europe—only one confirmed case (France) is known (Heyman et al., 2009; Lundkvist and Zeller, unpublished)

In Asia, HTNV, SEOV, PUUV, and Amur virus (AMRV, carried by *A. peninsulae*) (Lokugamage et al. 2004) are responsible for about 100,000 HFRS cases each year. Of Soochong (*A. peninsulae*) (Baek et al. 2006), Khabarovsk virus (KHAV, *Microtus fortis*) (Hörling et al. 1996), Muju virus (MUJV, *Myodes regulus*) (Song et al. 2007), and Topografov virus (TOPV, *Lemmus sibericus*) (Vapalahti et al. 1999), it is currently unknown whether these viruses are pathogenic to humans.

23.2 HISTORY

It is remarkable that HFRS "emerged" relatively recently although the viruses along with their carrier rodents are present since millions of years, significant human disturbances of the environment (e.g., war), the human population explosion—from approximately 1.7 billion in 1900 to 6.1 billion in 2000 and 7 billion in 2011 probably triggered the emergence. Subsequent invasions of previously undisturbed habitats for settlement, deforestation, expanding agricultural land use, and climate changes since the last glacial maximum (Berglund 2003) paved the way for HFRS and other zoonotic diseases. The twentieth century "consumer revolution" (Allen 2001) seems to coincide with the emergence of HFRS in Eurasia.

Clinical pictures possibly caused by hantaviruses were noted in China backdating to the first millennium (Lee et al. 1982). Hantavirus infection—caused by a still unknown and apparently disappeared hantavirus—has also been suggested as the cause of the multiple outbreaks (1485, 1502, 1508, 1517, 1528, 1551) of English sweating sickness (Sudor Anglicus) (Caius 1552; Vergil 1838; Hecker 1859; Wylie and Collier 1981; Dyer 1997; Taviner et al. 1998). Although commonly thought to be restricted to England—Wales and Scotland were never affected, which is on itself a peculiar epidemiological feature—the 1528 outbreak swept over Middle and Northern Europe (Roberts 1945). A more benign variant known as Picardy sweat or "suette miliaire"—the term "miliary fever" was later, during WW I, again used—caused outbreaks from 1718 to 1874 in France but was also observed in Italy—where the condition was known as "febbre miliare"—and South Germany, where it was called "Schweiss-friesel." The disease however never reached the United Kingdom (Hirsch 1883; MacNalty 1947). Hantavirus disease was also suggested as a possible cause for—in at least some percentage of the cases—the 1862–1863 "trench nephritis," "epidemic nephritis," or "war nephritis" epidemic during the American Civil War, in which around 14,000 individuals apparently developed an nephropathia epidemica (NE)-like condition (Langdon-Brown 1916; Lee 1996). Historical sources suggested that an epidemic of trench nephritis during World War I also may have been—at least in some cases—hantavirus infection (Rutherford 1916; Byam and Lloyd 1920). One argument in favor of this statement might be the already suggested role of voles in the transmission cycle in 1916; Rutherford—linking trench fever to voles, to the medieval sweating sickness in England and to the Picardy sweat in France—wrote: *In the trenches field voles have been observed (as is only natural) lying dead in considerable numbers as the result of a gas attack ... Field mice have been blamed as the source of a disease now extinct, the sweating sickness or sudor anglicus of the Middle Ages, of which a small epidemic, however, is said to have broken out in the south-west of France in the early years of the present century; it seems reasonable to suggest, in view of what has gone before, that they may perhaps be the source from which the infection of this new "trench fever" is derived as well* (Rutherford 1916). Raw adds to the argument with a description of five cases that closely match nowadays NE infections (Raw 1915). From the literature between 1915 and 1925, it however becomes clear that the terms "trench fever" (probably mostly *Bartonella quintana* infections, also called quintana fever, 5-day fever, or Shinbone fever), Werner–His disease, Volhynia fever, epidemic nephritis, or war nephritis stand for a mixture of medical conditions acquired in the trenches during WW I, and "confusion of tongues" in reports was never far away. The Canadian doctor R. D. Rudolf wrote in 1917: *The term "trench" has been used in connexion with nearly every disease that has occurred in this campaign* (WWI; Authors), *and usually with very insufficient reason. Probably the only good use of the word is in the disease called "trench foot"* (Rudolf 1917). The term "trench fever" was for instance also used to describe an endemic disease—probably caused by *B. quintana* or *B. vinsonii*—in the city of Belfast (N-Ireland) in the first half of the nineteenth century (Logan 1989). The horrible conditions in the trenches during WW I also caused trench mouth (gingivitis, also called Vincent's angina), which was common among soldiers in the trenches, trench foot (frostbite caused by prolonged exposure to cold and dampness), and trench back (pain and stiffness in the dorsal and lumbar regions as a result of the continuous search for cover). Among this multitude of medical problems related to WW I conditions in the trenches, there were however a number of clinical descriptions that closely match hantavirus infection due to PUUV (Raw 1915; Ameuille 1916; Langdon-Brown 1916; Rutherford 1916).

Around 1934, a condition called "nephropathia epidemica" was described in Sweden (Myhrman 1934), which in the 1980s was proven to be a mild form of hantavirus disease caused by an infection with PUUV (Brummer-Korvenkontio et al. 1982). During World War II, a rodent-borne disease ("Kriegs- or Feldnephritis") struck thousands of German troops in Finnish Lapland (1942) (Stuhlfauth 1943) and on the Eastern front (Sarre 1959). And in their search for the cause of death of Mozart, Zegers and collaborators may even have described the first documented outbreak of Feldnephritis dating from 1791 (Zegers et al. 2009) and commentaries on this paper.

In Korea, Manchuria, and the Far-Eastern part of Russia, hantavirus disease was probably endemic for centuries although it was first mentioned in the records of the Vladivostok regional hospital in 1913–1914 (Gavriluyk 1968). Korean hemorrhagic fever, which later was found to be caused by a hantavirus, caught the eye of Western medicine when during the Korean war (1951–1953), more than 3000 American and Korean soldiers fell severely ill with an infectious disease characterized by renal failure, generalized hemorrhage, and shock with a mortality rate of more than 10% (Cohen 1982; Bennett and Hart 1994).

Meanwhile in Russia, Gavrilyuk and Smorodintsev isolated the "hemorrhagic nephrosonephritis virus" from a patient's blood (Gavrilyuk and Smorodintsev 1971). In the mid-1960s, the shrew-borne TPMV was isolated from Asian house shrew or musk shrew (*Suncus murinus*) (Carey et al. 1971), and in 1976, HTNV was isolated from *A. agrarius coreae* (Lee et al. 1978). A list of denominations for hantavirus disease dating from before 1983 can be found in Table 23.1.

In 1992, a hantavirus with a mortality rate of up to 20% was characterized as a new hantavirus serotype and was named Dobrava (DOBV) after isolation from its rodent vector, *A. flavicollis* (yellow-necked field mouse) captured near Dobrava village, in Slovenia (Avsic-Zupanc et al. 1992). During the military conflict in Bosnia (1995), more than 300 individuals, most of them soldiers, were hospitalized in the Tuzla region (North-East Bosnia) with acute hantavirus symptoms (Lundkvist et al., 1997; Hukic et al., 2010), but several important outbreaks had already taken place in the Balkans regions prior to this epidemic. Hantaviruses again drew worldwide attention when, in May 1993, a sudden outbreak of a mysterious influenza-like prodrome of fever and myalgia, which rapidly evolved into an often fatal syndrome of shock and edema, occurred in the Four Corners region of the United States. Sin Nombre virus (SNV) was coined as the causative agent of a syndrome now known as hantavirus cardiopulmonary syndrome (HCPS) (Nichol et al. 1993). Unlike with HFRS, where the kidneys are the target organs, in HCPS, the lungs are affected (Rasmuson et al. 2011). The deer mouse

TABLE 23.1

Denominations for HFRS Encountered in the Literature and in Use before the WHO Memorandum of 1983

Trench nephritis

Wolhynia fever

His fever (Werner–His fever)

Far-Eastern hemorrhagic fever

Epidemic hemorrhagic fever

Hemorrhagic fever

Korean hemorrhagic fever

Songo fever

Hantaan fever

Epidemic hemorrhagic fever (Ryukosei syukketu netu)

Erhtaokiang disease (Nikodo disease—Nikodo fever)

Heiho fever

Hulin fever (Korin fever)

Kokko fever (Kokuko fever)

Manchurian epidemic hemorrhagic fever

Manchurian gastritis

Purpura hemorrhagica epidemica (epidemic hemorrhagic purpura)

Songo fever (Sunwu fever)

Tayinshan fever

Churilov's fever

Far-Eastern hemorrhagic fever

Far-Eastern hemorrhagic nephrosonephritis

Far-Eastern nephrosonephritis

Endemic hemorrhagic nephrosonephritis

Endemic nephrosonephritis

Epidemic hemorrhagic nephrosonephritis

Hemorrhagic nephrosonephritis

Infectious hemorrhagic nephrosonephritis

Infectious nephrosonephritis

Kalinin hemorrhagic fever

Transcarpatian hemorrhagic fever

Tula hemorrhagic fever

Ural hemorrhagic fever

Volga hemorrhagic fever

Yaroslav hemorrhagic fever

Inner Mongolian hemorrhagic fever

Bulgarian epidemic hemorrhagic fever

Hungarian epidemic hemorrhagic fever

Czechoslowakian epidemic hemorrhagic fever

Yugoslavian acute epidemic glomerulonephritis

Nephropathia epidemica

Puumala fever (Finland)

Lemming fever (Tularemia?)

Source: Anon, *Bull. World Health Organ.*, 61, 269, 1983.

(*Peromyscus maniculatus*) was shown to carry the virus; several so-called New World hantaviruses were soon discovered in other North and South-American rodent species (Mills et al. 2010).

Meanwhile, HFRS is recognized and considered endemic in most countries on the Eurasian continent. Through increased awareness and improved diagnostic tools, it is no longer problematic to detect the infection, and yearly reported case numbers are increasing in almost all countries. Whether forenamed factors are the only reason for this observation remains unclear.

23.3 BIOLOGICAL PROPERTIES (VIRUS STRUCTURE, GENOME ORGANIZATION, PROTEIN FUNCTIONS, REPLICATION CYCLE)

Hantaviruses are negative-stranded RNA viruses with a tripartite genome. The small segment (S) encodes for the nucleocapsid protein (N) and is approximately 1800 nucleotides (nt) long, the medium segment (M) has approx. 3600 nt and encodes the two surface glycoproteins, G1 (Gn) and G2 (Gc), and the surface of the virion is covered with Gn and Gc proteins. Finally, the large segment (L) is about 6500 nt long and encodes the RNA polymerase (Plyusnin et al. 1996). Some (PUUV and TULV) but not all hantaviruses possess a nonstructural protein (NS) in the S-segment open reading frame (Virtanen et al. 2010).

Morphologically, the virion may show some variations; in general, the particle is spherical in shape and about 120 nm in diameter, but a diameter of 60 nm and tubular particles of 350 nm long and 80 nm in diameter have also been reported (Huiskonen et al. 2010). Pathogenic and nonpathogenic hantaviruses use endothelial cells to replicate in, but with little or no damage to the infected endothelial cell (Hussein et al. 2011). The viruses in general (e.g., HTNV, PUUV) enter the cell only through the apical surface (Krautkrämer and Zeier 2008); the South-American Andes hantavirus (ANDV) enters via both the apical and the basolateral epithelial membrane (Rowe and Pekosz 2006; Rowe et al. 2008). After attachment to the epithelial the virus initiates interaction with the host cell receptor and especially integrins act as host cell receptors. Gavrilovskaya and coworkers found that pathogenic hantaviruses, e.g., PUUV, HTNV, SEOV, but also the North-American SNV, use β3 integrin whereas nonpathogenic hantaviruses such as PHV use β1 integrin (Gavrilovskaya et al. 2002, 2010). Following attachment to the cell, a process described as receptor-mediated endocytosis takes place (Jin et al. 2002). Once in the cell, the virus begins to reproduce itself. This process goes through several steps like transcription (mRNA synthesis), translation (virus protein synthesis), and replication (vRNA synthesis) and is exclusively cytoplasmic (Mir et al. 2008; Hussein et al. 2011). After transcription and replication of the viral genome, the viral mRNA and that of S, M, and L segment vRNAs are present in the cytoplasm of the affected cell. The N protein encapsidates the vRNA, and the newly formed virion can reenter the system.

23.4 CLINICAL PRESENTATION

Hantavirus infection in humans can result in two clinical syndromes: HFRS or HCPS caused by Old World or New World hantaviruses, respectively. However, the initial symptoms of all hantavirus infections are very similar, including an abrupt onset of high fever, malaise, myalgia, and other flu-like symptoms. Common factors of HFRS and HCPS are also increased vascular permeability leading to hypotension, thrombocytopenia, and leukocytosis with a left shift (Schmaljohn and Hjelle 1997; Khaiboullina et al. 2005; Schonrich et al. 2008).

The clinical presentation of HFRS varies from subclinical, mild, and moderate to severe, depending in part on the causative agent of the disease. In general, HFRS caused by HTNV, AMRV, and DOBV is more severe with mortality rates from 5% to 15%, while SEOV causes moderate and PUUV and SAAV cause mild form of disease with mortality rates below 1%. Nevertheless, individual PUUV case may be severe, an individual HTNV infection may be mild, or cases may even present with subclinical seroconversion (Linderholm and Elgh 2001; Jonsson et al. 2010).

A typical course of HFRS can be divided into five distinct phases: febrile, hypotensive, oliguric, polyuric, and convalescent. These phases are better distinguishable in severe forms of disease caused by HTNV and DOBV. After an incubation period between 2 and 4 weeks, the disease starts abruptly with high fever, chills, headache, backache, abdominal pains, nausea, and vomiting. Somnolence and visual disturbances (blurred vision) are frequently reported. This febrile phase usually lasts for 3–7 days. Toward the end of this phase, conjunctival hemorrhages and fine petechiae occur initially on the palate. The hypotensive stage can last from several hours to 2 days. In severe cases, hypotension, even shock, may develop rapidly, and one-third of HFRS deaths are associated with fulminant irreversible shock at this stage. Thrombocytopenia and leukocytosis are characteristic for this phase, and if severe hemorrhagic disease occurs, its onset is at this stage. Hemorrhagic manifestations can include petechiae on the skin and mucosa, ecchymoses, conjunctival suffusion, hematemesis, epistaxis, hematuria, melena, and fatal intracranial hemorrhages. In the oliguric phase, which lasts for 3–7 days, blood pressure becomes normalized while kidney function is transiently decreased, leading to oliguria or even anuria, proteinuria, abnormal urinary sediment, including microscopic hematuria, and azotemia. During oliguric phase, which is usually accompanied by abdominal or back pain, patients with severe symptoms have to be treated by hemodialysis. One half of fatalities occur during this phase. Typical laboratory findings are elevated levels of serum creatinine and urea. In the polyuric phase, renal function starts to recover and urinary output increases. The onset of the diuretic phase is a positive prognostic sign for the patient. It can last for days or weeks with patient passing several liters of urine per day. Convalescence, characterized by recovery of clinical and laboratory markers, is usually prolonged and can last up to 6 months. Full recovery is usually reached, and longer-lasting complications are rare but can include chronic renal failure and hypertension (McCaughey and Hart 2000; Linderholm and Elgh 2001; Heyman et al. 2009).

The clinical picture of PUUV infection (nephropathia epidemica, NE) is basically similar to DOBV, but shows a limited spectrum of symptoms resulting in lower mortality rates (0%, 1%). In NE, severe hemorrhagic manifestations and shock usually do not occur, but mild hemorrhagic symptom such as petechiae are seen in about one-third of patients. Instead of full-blown shock syndrome, hypotension is observed. Although most patients have signs of kidney function failure, it is generally less prominent than that in HFRS caused by more virulent hantaviruses, with oliguria or anuria manifesting in less than half of patients. Altogether, since the clinical course of NE is often uncharacteristic and resembles more a febrile disease with abdominal pain, it is often not diagnosed as NE (Kanerva et al. 1998; Settergren 2000).

SEOV infections cause a moderate form of HFRS with clinical presentation and course very similar to HFRS caused by HTNV. However, SEOV infections are often associated with the presence of hepatitis, which is generally not present in other hantavirus infections (Kim et al. 1995).

23.5 DIAGNOSIS

Almost all acute HFRS cases have IgM, and many have also already IgG, reactive with the viral N protein at the onset of the clinical symptoms. Although a number of highly sensitive PCRs, and real-time PCRs, have been developed, the number of non-viremic acute-phase patients is significant. Based on this, serological tests that detect serum IgM and/or IgG to hantaviral N are the most common methods for laboratory diagnosis of suspected HFRS cases. The first serological test that was developed for diagnoses of HFRS in Europe and Asia was the indirect immunofluorescence assay, which uses hantavirus-infected cells fixed on microscope slides. Today, most hantavirus antigens used for serological tests are recombinant-expressed antigens. The hantavirus N proteins have been expressed and purified from various recombinant expression systems, including bacterial (Kallio-Kokko et al. 2000), baculovirus (Schmaljohn et al. 1988; Kallio-Kokko et al. 2000), insect (Vapalahti et al. 1996), *Saccharomyces* spp. (Razanskiene et al. 2004; Schmidt et al. 2005), and plant (Kehm et al. 2001). All three viral structural proteins (Gn, Gc, and N) may induce high levels of IgM detectable at the onset of symptoms, but the responses to the two glycoproteins are often delayed (Lundkvist et al. 1993; Vapalahti et al. 1995).

The viral N protein induces a strong humoral immune response in humans and rodents and includes three major epitopes of cross-reactive antigens for hantaviruses. A number of studies have demonstrated that these epitopes are all located in the amino-terminal region of the N protein (Lundkvist et al. 1996; Hörling and Lundkvist 1997; Lindkvist et al. 2007).

The N protein is suitable for use as an antigen in enzyme-linked immunosorbent assays (ELISAs) for the diagnosis of hantavirus infection (Vapalahti et al. 1996; Hjelle et al. 1997; Figueiredo et al. 2008), as well as strip immunoblot tests (Hjelle et al. 1997). However, the most common serological tests for hantaviruses are indirect IgG and IgM (ELISAs) as well as IgM-capture ELISAs (Vapalahti et al. 1996; Sjölander et al. 2000).

These tests take about 4–6 h to perform. Virus-infected lysates or purified N protein may also be efficiently used as antigens in ELISAs (Vapalahti et al. 1996; Lundkvist et al. 1997; Sjölander et al. 2000). It is widely recognized that a neutralization test (focus reduction neutralization test—FRNT or plaque reduction neutralization test) is the only definitive method to differentiate the antibody-responses to various hantaviruses (Schmaljohn et al. 1985; Chu et al. 1994; Lundkvist et al. 1997). It is a highly specific test that detects and measures the levels of neutralizing antibodies. Cross-FRNT permits serotypic classification of hantavirus infections in humans and rodents (Lundkvist et al. 1995). Although the assay is highly specific and is capable of distinguishing also closely related hantaviruses, it has been shown to be less specific when human acute-phase sera from HFRS patients were used (Lundkvist et al. 1995, 1997). Unfortunately, neutralization tests are most laborious and require BSL-3 laboratories, and only very few laboratories are running them regularly.

As mentioned earlier, the diagnostic value of PCRs is limited. However, highly sensitive diagnostic tests have been developed based on the detection of the viral RNA genome. The hantavirus genome can be detected by reverse transcription PCR (RT-PCR) in clinical samples such as blood, serum, or organ fragments, at the onset of illness. The detection of viral genomes in patients already before the 1st day of symptoms has also been reported (Ferres et al. 2007; Padula et al. 2007). However, the low levels of viral RNA present in human and rodent tissue samples can require nested-RT-PCR techniques using primers selected for regions with high homology. Nested RT-PCR tests have been developed, e.g., for HTNV (Li et al. 2009) and PUUV (Evander et al. 2007).

23.6 PATHOLOGY AND PATHOGENESIS

In both animals and humans, hantavirus infections mainly occur in pulmonary or renal endothelial cells and macrophages; therefore, the viral antigen is present also in many different organs (Hughes et al. 1993; Schmaljohn 1996). In contrast to the humans, animals are persistently infected throughout their entire life. Thus, the lack of apparent disease in natural host and the lack of proper animal models have limited our understanding of hantavirus pathogenesis (Mackow and Gavrilovskaya 2001). Until recently, there was no animal model for HFRS. However, cynomolgus monkeys (*Macaca fascicularis*) infected with wild-type PUUV strains produce disease symptoms that resemble an NE clinical pathology (Klingström et al. 2002; Sironen et al. 2008).

The central phenomena behind the pathogenesis of HFRS are increased vascular permeability and acute thrombocytopenia with marked permeability of microvascular beds (Vapalahti et al. 2001; Mackow and Gavrilovskaya 2009). Hantavirus replication occurs in the vascular endothelium but does not seem to cause direct cytopathic effects (Zaki et al. 1995; Terajima et al. 2007). The hantavirus replication cycle is rather slow, resulting in late viremia on days 5–10 post infection, which would suggest a virus persistence rather than acute lytic progression seen in other viral hemorrhagic fevers (Mackow and Gavrilovskaya 2009). In human kidney tissues of NE patients, the viral antigen was detected along with inflammatory cell infiltrations and tubular damage, suggesting that viral replication together with the immune response is involved in tissue injury. The peritubular area of the distal nephron is the main site where an increased expression of several cytokines and endothelial adhesion molecules is seen (Temonen et al. 1996). The renal involvement in acute NE is characterized by markedly decreased glomerular filtration rate and renal plasma flow. Increased glomerular permeability leads to massive proteinuria and as a sign of tubular dysfunction (Ala-Houhala et al. 2002).

It is not yet completely understood how hantaviruses disseminate in the human body: after inhalation, the infection begins with an interaction of Gn and Gc surface proteins with β-integrin receptors at the target cell membrane. It has been shown that both pathogenic (HTNV, SEOV, PUUV, SNV) and nonpathogenic (TULV, PHV) hantaviruses infect human endothelial cells, but they use different integrin receptor ($\alpha_v\beta_1$ vs. $\alpha_5\beta_3$) (Gavrilovskaya et al. 1999, 2002). Probably immature dendritic cells play a pivot role in hantavirus dissemination, as they express β_3 integrin receptors, and are located in the vicinity of epithelial cells. They can also serve as vehicles for the transport of the virions through the lymphatic vessels to the regional lymph nodes, where after further replication, virions can reach endothelial cells. These cells allow virus replication, which induces immune activation. It has been shown that type I interferon response has been delayed in cells infected with pathogenic hantaviruses, resulting in higher viral titters (Schonrich et al. 2008; Jonsson et al. 2010). Inflammatory cytokines and chemokines produced by antiviral innate immune response can act as double-edged sword. In NE patients, cytotoxic T-cells may contribute to the capillary damage via immunopathology, and also by increased concentrations of nitric oxide and tumor necrosis factor (TNF-α) (Groeneveld et al. 1995; Linderholm et al. 1996). In contrast with other hemorrhagic fever viruses, which inhibit maturation of infected dendritic cells, hantaviruses induce their maturation and thus elicit a vigorous T-cell response during acute infection (Kilpatrick et al. 2004). In NE patients, the cytotoxic T lymphocyte response enhanced a number of activated CD8[+] T cells and reversed CD4[+] vs. CD8[+] T cell ratio, which has coincided with the onset of clinical disease (Tuuminen et al. 2007). A mixed pattern of Th1 and Th2 immune response patterns, high levels of proinflammatory cytokines, and their insufficient suppression by regulatory cytokines leads to the harmful effect of immune response in HFRS-infected patients (Schonrich et al. 2008). It has been further shown that an imbalance in cytokine production might contribute to a more severe clinical course of HFRS (Saksida et al. 2011). Hantavirus pathogenesis is likely to be a complex multifactorial process that includes contributions from immune responses, platelet dysfunction, and the dysregulation of endothelial cell barrier functions (Mackow and Gavrilovskaya 2009). Above that, a genetic predisposition toward severe PUUV infection was shown to be related to HLA type, especially HLA-B8, HLA-DRB1*03, and HLA-DRB1*13 (Mustonen et al. 1996; Mäkelä et al. 2002; Korva et al. 2011), whereas HLA-B*35 might be a genetic risk factor for DOBV infection (Korva et al. 2011).

23.7 EPIDEMIOLOGY

With yearly 20,000–50,000 cases of HFRS—or about 90% of the total number of cases worldwide—the People's Republic of China is the most heavily affected country in the world (Zhang et al. 2010). HFRS indeed is a serious public health problem in the People's Republic of China, where the condition has been reported in 29 of 31 provinces. Primarily, HTNV and SEOV caused disease in humans, and Gou virus (GOUV) was isolated from *R. rattus* in China (Wang et al. 2000). Between 1950 and 2007, a total of 1,557,622 cases of HFRS occurred, and overall mortality was 3% (46,427 fatalities). The most important epidemic took place in 1986 with more than 115,000 cases (incidence $11.1/10^5$). After implementation of preventive measures such as rodent control, education of public health workers, and vaccination of at-risk groups, the incidence of HFRS has significantly decreased. Mortality rates also declined from 14.2% in 1969 to 1% now. Before 1982, HFRS cases were defined by a national standard of clinical criteria. After 1982, a combination of clinical criteria and serology was applied (Zhang et al. 2010; Liu et al. 2011). Only the Xinjiang and Tibet provinces seem relatively HFRS-free; all other provinces reported cases from 1986 onward. Especially the Shandong province—that historically bears 30% of the national cumulative number of cases—is heavily affected. Interestingly, seasonal shifts—i.e., the majority of the cases taking place from February to June (spring) or from September to January (autumn–winter)—were observed in this province. This might be explained by the fact the *A. agrarius* populations (transmitting HTNV) typically peak in winter and *R. norvegicus* populations (transmitting SEOV) in spring. Mongolia (Inner Mongolia Autonomous Region China) experiences HFRS outbreaks, caused by HTNV and SEOV, since at least 40 years (Zhang et al. 2009).

In Taiwan, *R. norvegicus, R. flavipectus* (the buff-breasted rat, a sub-species of *R. tanezumi*), and *Mus caroli* (Ryukyu mouse) figure as SEOV reservoirs in Taiwan, where also a Borneo pet orangutan showed signs of clinical SEOV infection (Chen et al. 2011).

In the Republic of Korea—the country where the first rodent-borne hantavirus (HTNV) was identified—HFRS (with HTNV and SEOV as causal viruses) is also endemic (Seo et al. 2007; Ryou et al. 2011) and widespread since decades. Again, due to vector characteristics, a spring and winter HFRS peak takes place each year (Song et al. 2006). As in the People's Republic of China, a significant decrease in the yearly number of cases was however achieved by implementing the Hantavax vaccine (Cho et al. 2002; Park et al. 2004).

In the remainder of the far-eastern region, hantavirus infections are less predominant but nevertheless present. Blasdell and coworkers published an excellent review of rodent-borne zoonotic viruses in Southeast Asia (Blasdell et al. 2009).

In Vietnam, the HFRS problem is mainly located in the north of the country and caused by SEOV (Huong et al. 2010). From Myanmar, only scarce information is available, but yet unidentified hantavirus(es) was reported. Whether there are human cases is unknown (Lee et al. 1999). In Cambodia, only hantaviral presence (SEOV) was confirmed in rodents (Reynes et al. 2003; Henttonen et al. 2008).

In India, a seroprevalence of 4% was measured in the general population, but endemic regions were not yet identified. Indications that TMPV could induce HFRS were also found (Chandy et al. 2008, 2009). A similar TMPV-related case also occurred in Thailand in 2006 (Okumura et al. 2007). In Thailand, HFRS is mainly caused by SEOV (Suputthamongkol et al. 2005; Pattamadilok et al. 2006). TMPV was also found in *S. murinus* in Nepal, but no human cases were reported (Kang et al. 2011). Serang virus (SERV) was isolated from the lung tissue of *R. tanezumi* captured in Indonesia (Plyusnina et al. 2009). Interesting is that SERV seems phylogenetically more closely related to Thailand virus (THAIV, carried by *Bandicota indica*), than to SEOV or GOUV, thus suggesting the possibility of a host-switching event between ancestors of the bandicoot rat (*Bandicota indica*) to the black and/or Norway rat (*R. rattus and R. norvegicus* respectively). In the Philippines, SEOV is also present, and a seroprevalence varying between human population groups of 5%–7% was established (Quelapio et al. 2000). Serological evidence of human hantavirus infections have been reported in Indonesia (Groen et al. 2002). Even the presence of novel Hantaviruses have been reported in Asian House rat (*Rattus tanezumi*) in Indonesia (Plyusnina et al. 2009). In Malaysia, HTNV seems to be the most prevalent hantavirus (Lam et al. 2001); from Sri Lanka, at least 16 human cases and a seroprevalence of 5.6% was reported, but no serotype was coined (Vitarana et al. 1988; Bi et al. 2008). Japan has not reported HFRS outbreaks or cases in the last 20 years. Several rodent species including *R. norvegicus* have been identified as seropositive for hantaviruses in different locations throughout the country (Kariwa et al. 2007). No reports are available about the presence of hantaviruses in rodents or humans from Bangladesh, Bhutan, East Timor, Bhutan, or Brunei. Australia also seems to be HFRS-free, although a number of rodents from Queensland, Victoria, the Northern Territory, South Australia, and New South Wales were found seropositive for hantavirus antibodies; only from Western Australia, all rodents were seronegative (LeDuc et al. 1986). There are however no HFRS cases known from Australia (Anon 2008). The Arabian Peninsula (Saudi Arabia, Kuwait, Qatar, Yemen, and United Arab Emirates) is apparently not threatened by hantaviruses. Only from Kuwait, seropositive rodents and humans were reported (Pacsa et al. 2002).

From Israel, one report mentioned a seropositivity of 12.3% in hemodialyzed patients and 9% in renal failure patients, versus 2% in the normal population (George et al. 1998). From the neighboring countries (Jordan, Syria, Lebanon) as well as from Uzbekistan, Turkmenistan, Armenia, Pakistan, Afghanistan, or Iraq, no information is available.

Turkey was first confronted with a hantavirus epidemic in the year 2009. From 2009 on, an increasing number of provinces reported hantavirus infections (Kaya et al. 2010; Heyman et al. 2011). Only the presence of TULV was described from Kazakhstan (Ketai et al. 1994; Plyusnina et al. 2008). In Siberia (Asian part of Russia), hantavirus infections occur, but the disease is rather rare (Garanina et al. 2009) and Georgia (Kuchuloria et al. 2009). For comparison, Russia accounts for an average yearly number of 10,000 hantavirus cases: 95% occur in European Russia and the remaining 5% in Asian part of the country.

Until fairly recently, no detailed data on hantavirus infections in Africa were available, despite the fact that members of the order *Rodentia*, i.e., 19 families, 98 genera with 345 species (http://project.biodiversity.be/africanrodentia), are readily present on the continent. The *Muridae* family is not only important as reservoir and vector for several human diseases but is of particular importance for its detrimental effect on crops and food stocks. Gonzalez and coworkers published in 1984 a report regarding the serological evidence of hantaviral infections in human and rodents in Benin, Burkino Faso, the Central African Republic, and Gabon (Gonzalez et al. 1984). Soon after this first report, human hantavirus infections have also been demonstrated in Senegal (Saluzzo et al. 1985), Nigeria (Tomori et al. 1986), Djibouti (Rodier et al. 1993), and Egypt (Baddour et al. 1996; Botros et al. 2004) through serological surveys. The first African hantavirus, Sangassou virus (SANGV), was identified in Guinea (Klempa et al. 2006). There probably exists a potential for pathogenicity of SANGV and other African hantaviruses as witnessed by recent work by Klempa et al. (2010).

In Europe, hantavirus infections are known since decades: from the 1980s on, multiple reports from the majority of European countries are available. As already mentioned in the historical section of this chapter, the mild form of hantavirus infection due to PUUV infection— NE— was already described in the 1930s (Myhrman 1934). From the early 1980s on, serological tests for hantavirus infections were applied in Scandinavia, Western Europe, and Russia. Hantavirus infections in the European Union cause yearly on average around 10,000 hospitalizations. Most countries report a seroprevalence in the healthy population between 1% and 2% (Heyman and Vaheri 2008). Only the Southern European countries (Spain, Italy) seem to be free of HFRS cases, although there is a seroprevalence detected in the general population (Kallio-Kokko et al. 2006; Sanfeliu et al. 2011).

For reasons so far unknown, there is a tendency of increase in the yearly recorded number of cases in Europe (Heyman et al. 2011). The milder climatic conditions of the past 10–15 years could be one explanation. The circulation of hantaviruses in rodent populations corresponds linearly to the rodent population density and to the number of human cases (Heyman et al. 2001). The hot topic however is which concepts may drive rodent ecology, i.e. climate, food, reproduction opportunities, predation, disease, etc. Favorable climatic conditions in terms of a mild winter and spring certainly favor the winter survival rate of rodents, and a relationship between climatic conditions and food availability was confirmed (Tersago et al. 2009). As far as food supply is concerned, both quantity and quality of the available food supplies influence rodent population density. Food availability also limits the average density of populations, and peaks of rodent population densities are caused by changes in their food supplies. It is therefore important to know what rodents eat as this can vary according to habitat features. Most hantavirus carrying rodent species are generalists and eat a variety of nutrients, but will preferably forage on what is most abundant and easiest to obtain, i.e., insects, hard or soft fruits, sprouts, leaves, etc. Their diet may—and must—therefore change from year to year and especially from season to season. Once a population has reached its peak density, social interactions like territoriality, infanticide, and increased mortality due to habitat saturation and diseases can begin to downsize populations. Predation has been proven to regulate rodent populations. Particularly, the generalist predators follow the same principles in procuring food as their prey, i.e., focusing on peaking— and thus abundant—rodent populations. Predation can decrease rodent population density below those limits dictated by food availability and contribute to population limitation and thus work as control, even pest control, instrument. Care should therefore be taken when applying predator control programs like—for instance—elimination of foxes or uncontrolled removal of old buildings that serve as shelters for birds of prey, i.e., owls. In Belgium, the hunting lobby promoted and proposed in 2010 an almost unlimited "stamp-out" action against foxes. Such ill-advised actions could result in rodent populations—and not only rodents—definitively going out of control with all detrimental consequences for public health as result.

23.8   PROGNOSIS AND TREATMENT

The long-term prognosis, especially after PUUV infection, is favorable although some reports mention persisting hypertension, glomerular hyperfiltration, and hormonal deficiencies (Miettinen et al. 2006; Mäkelä et al. 2010).

There is no specific therapy available for HFRS; therefore, the treatment consists mainly of supportive measures. Timely admission to an intensive care unit (if necessary) for taking all appropriate measures is necessary (Settergren 2000; Nguyên et al. 2001). Monitoring the patient's fluid balance and kidney function is important, and hemodialysis treatment may be required (Settergren 2000). In severe HFRS cases, corticosteroids can be used in the treatment of hantavirus infection (Seitsonen et al. 2006). Ribavirin was demonstrated to have beneficial effect both *in vitro* and *in vivo* because it inhibits hantavirus replication (Mertz et al. 2004). Ribavirin was more particularly used for treating HFRS in China, where it has been shown that it can significantly reduce mortality (Huggins et al. 1991), and the antiviral effectiveness of the combination of ribavirin and amixine, an interferon inducer, was demonstrated in suckling mice (Loginova et al. 2005). Reducing human exposure to infected rodents and their excreta is thus still the most effective way of preventing hantavirus infections.

Four types of vaccines based on Gold hamster (*Mesocricetus auratus*) or Mongolian gerbil (*Meriones unguiculatus*) kidney cells are in use in China (Park et al. 2004; Dong et al. 2005), while only the vaccine based on sucking mouse brain—Hantavax that was approved in 1990—is in use in Korea (Cho et al. 2002). A Vero cell culture-derived inactivated vaccine against HTNV with enhanced protective qualities compared to Hantavax was recently developed in Korea (Choi et al. 2003). More recently, efforts have been made to develop recombinant vaccines based on vaccinia viruses or recombinant bacteria (Schmaljohn 2009). A vaccine against PUUV based on chimeric hepatitis virus particles was developed in Europe (Koletzki et al. 1999). DNA vaccine is based on the introduction of an antigen-encoding plasmid into the organism directing the *de novo* synthesis *in vivo*. The next logical step might be DNA vaccines, for which experiments have shown encouraging results (Hooper et al. 2001; Spik et al. 2008; Badger et al. 2011).

23.9   CONCLUSIONS AND FUTURE PERSPECTIVES

Hantaviruses and HFRS are a major public health issue since decades, probably centuries. Fairly recently however, the problem has grown significantly, and we are as yet not entirely sure why this has been happening. There exists now sufficient proof that climate change, trade, human behavior, and alteration of the landscape by agriculture and the increasing need of housing facilities have affected and promoted the transmission of hantaviruses. These changes each appear to be influenced by a number of factors including the hantavirus species, the geographic region, type of alteration, although the disturbance is largely—if not entirely—due to human interference. Also have the hantavirus diagnostic tools in rodents and humans have been available only for merely two decades; longitudinal ecological studies to evaluate the long-term effects are thus limited. With on the one hand an unforeseeable evolution of human–habitat interaction in many regions of the world and on the other hand the fact that the underlying mechanisms that mediate hantavirus transmission are largely unknown, predictions on how the transmission dynamics of hantaviruses will change globally or even locally may be speculative. The continuation and establishment of long-term ecological efforts on hantaviruses should be valued and encouraged as these studies will be critical in predicting future human risk. Establishment of habitat, reservoir, and host surveillance in regions where large numbers of human cases (i.e. Scandinavia, Europe, Russia, China) persistently occur is absolutely necessary for managing the risk for humans on a larger scale.

Currently, there are no treatments or generally (e.g., EC of FDA accepted) accepted vaccines available. Even if such a vaccine was experimentally developed, the small target population (at-risk groups) would make production costly and perhaps not interesting for the pharmaceutical industry. Reducing the pass of climate change and minimizing changes in habitats, ecosystems, and landscape could have a stabilizing effect, but prevention and simultaneously host–pathogen ecology surveillance are in practice the best—and so far only—ways to safeguard human health from the increasing threat of HFRS.

ACKNOWLEDGMENTS

The authors are indebted to the ENIVD-CLRN hantavirus workgroup and the ECDC for their continued support. This study was partially funded by EU grant FP7-261504 EDENext and is catalogued by the EDENext Steering Committee as EDENext000 (http://www.edenext.eu). The contents of this publication are the sole responsibility of the authors and don't necessarily reflect the views of the European Commission.

REFERENCES

Ala-Houhala, I, M Koskinen, T Ahola, A Harmoinen, T Kouri, K Laurila, J Mustonen, and A Pasternack. 2002. Increased glomerular permeability in patients with nephropathia epidemica caused by Puumala hantavirus. *Nephrology, Dialysis, Transplantation: Official Publication of the European Dialysis and Transplant Association—European Renal Association* 17(2): 246–252.

Allen, R C. 2001. The great divergence in European wages and prices from the middle ages to the First World War. *Explorations in Economic History* 38: 411–447. doi:10.1006/exeh.2001.0775.

Ameuille, P. 1916. Du rôle de l'infection dans les néphrits de guerre. *Ann de Med* 3: 298–323.

Anon, 1983. Haemorrhagic fever with renal syndrome: Memorandum from a WHO meeting. *Bulletin of the World Health Organization* 61(2): 269–275.

Anon, 2008. Biosecurity Australia Device 2008/10. Australian Government, March 26.

Avsic-Zupanc, T, S Y Xiao, R Stojanovic, A Gligic, G van der Groen, and J W LeDuc. 1992. Characterization of Dobrava virus: A Hantavirus from Slovenia, Yugoslavia. *Journal of Medical Virology* 38(2): 132–137.

Baddour, M M, A S Mourad, A M Awwad, and J R Murphy. 1996. Hantaan virus infection among human and rat populations in Alexandria. *The Journal of the Egyptian Public Health Association* 71(3–4): 213–228.

Badger, C V, J D Richardson, R L Dasilva, M J Richards, M D Josleyn, L C Dupuy, J W Hooper, and C S Schmaljohn. 2011. Development and application of a flow cytometric potency assay for DNA vaccines. *Vaccine* 29(39): 6728–6735. doi:10.1016/j.vaccine.2010.12.053. http://www.ncbi.nlm.nih.gov/pubmed/21219978.

Baek, L J, H Kariwa, K Lokugamage, K Yoshimatsu, J Arikawa, I Takashima, J-I Kang et al. 2006. Soochong virus: An antigenically and genetically distinct hantavirus isolated from Apodemus peninsulae in Korea. *Journal of Medical Virology* 78(2): 290–297. doi:10.1002/jmv.20538.

Bennett, M and C A Hart. 1994. Hantavirus infection. *Journal of Medical Microbiology* 41(2): 71–73.

Berglund, B E. 2003. Human impact and climate change—Synchronous events and a causal link? *Quaternary International* 105: 7–12.

Bi, Z, P B H Formenty, and C E Roth. 2008. Hantavirus infection: A review and global update. *Journal of Infection in Developing Countries* 2(1): 3–23.

Blasdell, K, V Herbreteau, H Henttonen, D Phonekeo, J P Hugot, and Ph Buchy. 2009. Rodent-borne Zoonotic viruses in Southeast Asia. *Kasetart Journal (Natural Science)* 43: 94–105.

Botros, B A, M Sobh, T Wierzba, R R Arthur, E W Mohareb, R Frenck, A El Refaie, I Mahmoud, G D Chapman, and R R Graham. 2004. Prevalence of hantavirus antibody in patients with chronic renal disease in Egypt. *Transactions of the Royal Society of Tropical Medicine and Hygiene* 98(6): 331–336. doi:10.1016/S0035-9203(03)00063-4.

Brummer-Korvenkontio, M, H Henttonen, and A Vaheri. 1982. Hemorrhagic fever with renal syndrome in Finland: Ecology and virology of nephropathia epidemica. *Scandinavian Journal of Infectious Diseases. Supplementum* 36: 88–91.

Brus Sjölander, K, I Golovljova, A Plyusnin, and A Lundkvist. 2000. Diagnostic potential of puumala virus nucleocapsid protein expressed in Drosophila melanogaster cells. *Journal of Clinical Microbiology* 38(6): 2324–2329.

Byam, W and L Lloyd. 1920. Trench fever: Its epidemiology and endemiology. *Proceedings of the Royal Society of Medicine* 13(*Sect Epidemiol State Med*): 1–27.

Caius, J. 1552. *Sweatyng Sicknesse*. Appendix to J F C Hecker. *The Epidemics of The Middle Ages*, Translated by B G Babington, London, G. Woodfall and Son, 1844, pp. 353–380.

Carey, D E, R Reuben, K N Panicker, R E Shope, and R M Myers. 1971. Thottapalayam virus: A presumptive arbovirus isolated from a shrew in India. *The Indian Journal of Medical Research* 59(11): 1758–1760.

Chandy, S, S Abraham, and G Sridharan. 2008. Hantaviruses: An emerging public health threat in India? A review. *Journal of Biosciences* 33(4): 495–504.

Chandy, S, H Boorugu, A Chrispal, K Thomas, P Abraham, and G Sridharan. 2009. Hantavirus infection: A case report from India. *Indian Journal of Medical Microbiology* 27(3): 267–270. doi:10.4103/0255-0857.53215.

Chen, C-C, K J-C Pei, C-M Yang, M-D Kuo, S-T Wong, S-C Kuo, and F-G Lin. 2011. A possible case of hantavirus infection in a Borneo orangutan and its conservation implication. *Journal of Medical Primatology* 40(1): 2–5. doi:10.1111/j.1600-0684.2010.00442.x.

Cho, H-W, C R Howard, and H-W Lee. 2002. Review of an inactivated vaccine against hantaviruses. *Intervirology* 45(4–6): 328–333.

Choi, Y, C-J Ahn, K-M Seong, M-Y Jung, and B-Y Ahn. 2003. Inactivated Hantaan virus vaccine derived from suspension culture of Vero cells. *Vaccine* 21(17–18): 1867–1873.

Chu, Y K, C Rossi, J W Leduc, H W Lee, C S Schmaljohn, and J M Dalrymple. 1994. Serological relationships among viruses in the Hantavirus genus, family Bunyaviridae. *Virology* 198(1): 196–204. doi:10.1006/viro.1994.1022.

Cohen, M S. 1982. Epidemic hemorrhagic fever revisited. *Reviews of Infectious Diseases* 4(5): 992–995.

Dong, G-M, L Han, Q An, W-X Liu, Y Kong, and L-H Yang. 2005. Immunization effect of purified bivalent vaccine to haemorrhagic fever with renal syndrome manufactured from primary cultured hamster kidney cells. *Chinese Medical Journal* 118(9): 766–768.

Dyer, A. 1997. The English sweating sickness of 1551: An epidemic anatomized. *Medical History* 41(3): 362–384.

Evander, M, I Eriksson, L Pettersson, P Juto, C Ahlm, G E Olsson, G Bucht, and A Allard. 2007. Puumala hantavirus viremia diagnosed by real-time reverse transcriptase PCR using samples from patients with hemorrhagic fever and renal syndrome. *Journal of Clinical Microbiology* 45(8): 2491–2497. doi:10.1128/JCM.01902-06.

Ferres, M, P Vial, C Marco, L Yanez, P Godoy, C Castillo, B Hjelle, I Delgado, S-J Lee, and G J Mertz. 2007. Prospective evaluation of household contacts of persons with hantavirus cardiopulmonary syndrome in chile. *The Journal of Infectious Diseases* 195(11): 1563–1571. doi:10.1086/516786.

Figueiredo, L T, M L Moreli, A A Borges, G G Figueiredo, R L Souza, and V H Aquino. 2008. Expression of a hantavirus N protein and its efficacy as antigen in immune assays. *Brazilian Journal of Medical and Biological Research* 41(7): 596–599.

Garanina, S B, A E Platonov, V I Zhuravlev, A N Murashkina, V V Yakimenko, A G Korneev, and G A Shipulin. 2009. Genetic diversity and geographic distribution of hantaviruses in Russia. *Zoonoses and Public Health* 56(6–7): 297–309. doi:10.1111/j.1863-2378.2008.01210.x.

Gavrilovskaya, I N, E J Brown, M H Ginsberg, and E R Mackow. 1999. Cellular entry of hantaviruses which cause hemorrhagic fever with renal syndrome is mediated by beta3 integrins. *Journal of Virology* 73(5): 3951–3959.

Gavrilovskaya, I N, E E Gorbunova, and E R Mackow. 2010. Pathogenic hantaviruses direct the adherence of quiescent platelets to infected endothelial cells. *Journal of Virology* 84(9): 4832–4839. doi:10.1128/JVI.02405-09.

Gavrilovskaya, I N, T Peresleni, E Geimonen, and E R Mackow. 2002. Pathogenic hantaviruses selectively inhibit beta3 integrin directed endothelial cell migration. *Archives of Virology* 147(10): 1913–1931. doi:10.1007/s00705-002-0852-0.

Gavriluyk, B K. 1968. Hemorrhagic nephroso-nephritis in the Primo'ye region. *Autoref Diss Sosh Uchen Step Dokt Med Nauk* 44.

Gavrilyuk, B K and V V. Smorodintsev. 1971. Isolation of hemorrhagic nephrosonephritis virusin cell cultures. *Arch Gesamte Virusforschun* 34: 171–179.

George, J, M Patnaik, E Bakshi, Y Levy, A Ben-David, A Ahmed, J B Peter, and Y Shoenfeld. 1998. Hantavirus seropositivity in Israeli patients with renal failure. *Viral Immunology* 11(2): 103–108.

Gonzalez, J P, J B McCormick, D Baudon, J P Gautun, D Y Meunier, E Dournon, and A J Georges. 1984. Serological evidence for Hantaan-related virus in Africa. *Lancet* 2(8410) (November 3): 1036–1037.

Groen, J, C Suharti, P Koraka, E C M van Gorp, J Sutaryo, A Lundkvist, and A D M E Osterhaus. 2002. Serological evidence of human hantavirus infections in Indonesia. *Infection* 30(5): 326–327. doi:10.1007/s15010-002-2194-y.

Groeneveld, P H, P Colson, K M Kwappenberg, and J Clement. 1995. Increased production of nitric oxide in patients infected with the European variant of hantavirus. *Scandinavian Journal of Infectious Diseases* 27(5): 453–456.

Hecker, J F C. 1859. *The Epidemics of the Middle Ages*, 3rd edn. London, U.K.: Trübner&Co.

Henttonen, H, Ph Buchy, Y Suputtamongkol, S Jittapalapong, V Herbreteau, J Laakkonen, Y Chaval et al. 2008. Recent discoveries of new hantaviruses widen their range and question their origins. *Annals of the New York Academy of Sciences* 1149: 84–89. doi:10.1196/annals.1428.064.

Heyman, P, C Ceianu, I Christova, N Tordo, M Beersma, M Joao Alves, A Lundkvist et al. 2011. A five-year perspective on the situation of haemorrhagic fever with renal syndrome and status of the hantavirus reservoirs in Europe, 2005–2010. *Euro Surveillance* 16(36): 19961. http://www.ncbi.nlm.nih.gov/pubmed/21924118.

Heyman, P, C Cochez, G Korukluoglu, A Gözalan, Y Uyar, and Å Lundkvist. 2011. Bridging continents; Hantaviruses of Europe and Asia Minor. *Türk Hijyen ve Deneysel Biyoloji Dergisi* 68(1): 41–48.

Heyman, P and A Vaheri. 2008. Situation of hantavirus infections and haemorrhagic fever with renal syndrome in European countries as of December 2006. *Euro Surveillance* 13(28): 18925. http://www.ncbi.nlm.nih.gov/pubmed/18761927.

Heyman, P, A Vaheri, A Lundkvist, and T Avsic-Zupanc. 2009. Hantavirus infections in Europe: From virus carriers to a major public-health problem. *Expert Review of Anti-Infective Therapy* 7(2): 205–217. doi:10.1586/14787210.7.2.205.

Heyman, P, T Vervoort, S Escutenaire, E Degrave, J Konings, C Vandenvelde, and R Verhagen. 2001. Incidence of hantavirus infections in Belgium. *Virus Research* 77(1): 71–80.

Hirsch, A. 1883. *Handbook of Geographical and Historical Pathology 1*. Vol. 1. London, U.K.: The New Sydenham Society.

Hjelle, B, S Jenison, N Torrez-Martinez, B Herring, S Quan, A Polito, S Pichuantes et al. 1997. Rapid and specific detection of Sin Nombre virus antibodies in patients with hantavirus pulmonary syndrome by a strip immunoblot assay suitable for field diagnosis. *Journal of Clinical Microbiology* 35(3): 600–608.

Hooper, J W, D M Custer, E Thompson, and C S Schmaljohn. 2001. DNA vaccination with the Hantaan virus M gene protects Hamsters against three of four HFRS hantaviruses and elicits a high-titer neutralizing antibody response in Rhesus monkeys. *Journal of Virology* 75(18): 8469–8477.

Hörling, J, V Chizhikov, A Lundkvist, M Jonsson, L Ivanov, A Dekonenko, B Niklasson et al. 1996. Khabarovsk virus: a phylogenetically and serologically distinct hantavirus isolated from Microtus fortis trapped in far-east Russia. *The Journal of General Virology* 77(Pt 4): 687–694.

Hörling, J and A Lundkvist. 1997. Single amino acid substitutions in Puumala virus envelope glycoproteins G1 and G2 eliminate important neutralization epitopes. *Virus Research* 48(1): 89–100.

Huggins, J W, C M Hsiang, T M Cosgriff, M Y Guang, J I Smith, Z O Wu, J W LeDuc, Z M Zheng, J M Meegan, and Q N Wang. 1991. Prospective, double-blind, concurrent, placebo-controlled clinical trial of intravenous ribavirin therapy of hemorrhagic fever with renal syndrome. *The Journal of Infectious Diseases* 164(6): 1119–1127.

Hughes, J M, C J Peters, M L Cohen, and B W Mahy. 1993. Hantavirus pulmonary syndrome: An emerging infectious disease. *Science (New York)* 262(5135): 850–851.

Huiskonen, J T, J Hepojoki, P Laurinmäki, A Vaheri, H Lankinen, S J Butcher, and K Grünewald. 2010. Electron cryotomography of Tula hantavirus suggests a unique assembly paradigm for enveloped viruses. *Journal of Virology* 84(10): 4889–4897. doi:10.1128/JVI.00057-10.

Hukic, M, J Nikolic, A Valjevac, M Seremet, G Tesic, and A Markotic. 2010. A serosurvey reveals Bosnia and Herzegovina as a Europe's hotspot in hantavirus seroprevalence. *Epidemiology and Infection* 138(8): 1185–1193. doi:10.1017/S0950268809991348.

Huong, V T Q, K Yoshimatsu, V D Luan, L V Tuan, L Nhi, J Arikawa, and T M N Nguyen. 2010. Hemorrhagic fever with renal syndrome, Vietnam. *Emerging Infectious Diseases* 16(2): 363–365.

Hussein, I T M, A Haseeb, A Haque, and M A Mir. 2011. Recent advances in hantavirus molecular biology and disease. *Advances in Applied Microbiology* 74: 35–75. doi:10.1016/B978-0-12-387022-3.00006-9.

Jin, M, J Park, S Lee, B Park, J Shin, K-J Song, T-I Ahn, S-Y Hwang, B-Y Ahn, and K Ahn. 2002. Hantaan virus enters cells by clathrin-dependent receptor-mediated endocytosis. *Virology* 294(1): 60–69. doi:10.1006/viro.2001.1303.

Jonsson, C B, L T M Figueiredo, and O Vapalahti. 2010. A global perspective on hantavirus ecology, epidemiology, and disease. *Clinical Microbiology Reviews* 23(2): 412–441. doi:10.1128/CMR.00062-09.

Kallio-Kokko, H, J Laakkonen, A Rizzoli, V Tagliapietra, I Cattadori, S E Perkins, P J Hudson et al. 2006. Hantavirus and arenavirus antibody prevalence in rodents and humans in Trentino, Northern Italy. *Epidemiology and Infection* 134(4): 830–836. doi:10.1017/S0950268805005431.

Kallio-Kokko, H, A Lundkvist, A Plyusnin, T Avsic-Zupanc, A Vaheri, and O Vapalahti. 2000. Antigenic properties and diagnostic potential of recombinant dobrava virus nucleocapsid protein. *Journal of Medical Virology* 61(2): 266–274.

Kanerva H, J Mustonen, and A Vaheri. 1998. Pathogenesis of Puumala and other hantavirus infections. *Reviews in Medical Virology* 8(2): 67–86.

Kang, H J, M Y Kosoy, S K Shrestha, M P Shrestha, R V Gibbons, and R Yanagihara. 2011. Genetic diversity of Thottapalayam virus, a Hantavirus Harbored by the Asian House Shrew (*Suncus murinus*) in Nepal. *American Journal of Tropical Medicine and Hygiene* 85(3): 540–545.

Kariwa, H, K Lokugamage, N Lokugamage, H Miyamoto, K Yoshii, M Nakauchi, K Yoshimatsu et al. 2007. A comparative epidemiological study of hantavirus infection in Japan and Far East Russia. *The Japanese Journal of Veterinary Research* 54(4): 145–161.

Kaya, S, G Yılmaz, S Erensoy, D Yağçı Çağlayık, Y Uyar, and I Köksal. 2010. Hantavirus infection: Two case reports from a province in the Eastern Blacksea Region, Turkey. *Mikrobiyoloji Bulteni* 44(3): 479–487.

Kehm, R, N J Jakob, T M Welzel, E Tobiasch, O Viczian, S Jock, K Geider, S Süle, and G Darai. 2001. Expression of immunogenic Puumala virus nucleocapsid protein in transgenic tobacco and potato plants. *Virus Genes* 22(1): 73–83.

Ketai, L H, M R Williamson, R J Telepak, H Levy, F T Koster, K B Nolte, and S E Allen. 1994. Hantavirus pulmonary syndrome: Radiographic findings in 16 patients. *Radiology* 191(3): 665–668.

Khaiboullina, S F, S P Morzunov, and S C St Jeor. 2005. Hantaviruses: Molecular biology, evolution and pathogenesis. *Current Molecular Medicine* 5(8): 773–790.

Kilpatrick, E D, M Terajima, F T Koster, M D Catalina, J Cruz, and F A Ennis. 2004. Role of specific CD8+ T cells in the severity of a fulminant zoonotic viral hemorrhagic fever, hantavirus pulmonary syndrome. *Journal of Immunology (Baltimore, MD: 1950)* 172(5): 3297–3304.

Kim, Y S, C Ahn, J S Han, S Kim, J S Lee, and P W Lee. 1995. Hemorrhagic fever with renal syndrome caused by the Seoul virus. *Nephron* 71(4): 419–427.

Klempa, B, E Fichet-Calvet, E Lecompte, B Auste, V Aniskin, H Meisel, C Denys, L Koivogui, J ter Meulen, and D H Krüger. 2006. Hantavirus in African wood mouse, Guinea. *Emerging Infectious Diseases* 12(5): 838–840.

Klempa, B, L Koivogui, O Sylla, K Koulemou, B Auste, D H Krüger, and J ter Meulen. 2010. Serological evidence of human hantavirus infections in Guinea, West Africa. *The Journal of Infectious Diseases* 201(7) (April 1): 1031–1034. doi:10.1086/651169.

Klempa, B, H A Schmidt, R Ulrich, S Kaluz, M Labuda, H Meisel, B Hjelle, and D H Krüger. 2003. Genetic interaction between distinct Dobrava hantavirus subtypes in *Apodemus agrarius* and *A. flavicollis* in nature. *Journal of Virology* 77(1): 804–809.

Klempa, B, E A Tkachenko, T K Dzagurova, Y V Yunicheva, V G Morozov, N M Okulova, G P Slyusareva, A Smirnov, and D H Kruger. 2008. Hemorrhagic fever with renal syndrome caused by 2 lineages of Dobrava hantavirus, Russia. *Emerging Infectious Diseases* 14(4): 617–625.

Klingström, J, A Plyusnin, A Vaheri, and A Lundkvist. 2002. Wild-type Puumala hantavirus infection induces cytokines, C-reactive protein, creatinine, and nitric oxide in cynomolgus macaques. *Journal of Virology* 76(1): 444–449.

Koletzki, D, S S Biel, H Meisel, E Nugel, H R Gelderblom, D H Krüger, and R Ulrich. 1999. HBV core particles allow the insertion and surface exposure of the entire potentially protective region of Puumala hantavirus nucleocapsid protein. *Biological Chemistry* 380(3): 325–333. doi:10.1515/BC.1999.044.

Korva, M, A Saksida, S Kunilo, B Vidan Jeras, and T Avsic-Zupanc. 2011. HLA-associated hemorrhagic fever with renal syndrome disease progression in slovenian patients. *Clinical and Vaccine Immunology: CVI* 18(9): 1435–1440. doi:10.1128/CVI.05187–11.

Krautkrämer, E and M Zeier. 2008. Hantavirus causing hemorrhagic fever with renal syndrome enters from the apical surface and requires decay-accelerating factor (DAF/CD55). *Journal of Virology* 82(9): 4257–4264. doi:10.1128/JVI.02210–07.

Kuchuloria, T, D V Clark, M J Hepburn, T Tsertsvadze, G Pimentel, and P Imnadze. 2009. Hantavirus infection in the Republic of Georgia. *Emerging Infectious Diseases* 15(9): 1489–1491.

Lam, S K, K B Chua, T Myshrall, S Devi, D Zainal, S A Afifi, K Nerome, Y K Chu, and H W Lee. 2001. Serological evidence of hantavirus infections in Malaysia. *The Southeast Asian Journal of Tropical Medicine and Public Health* 32(4): 809–813.

Langdon-Brown, W. 1916. Trench nephritis. *Lancet* i: 391–395.

LeDuc, J W, G A Smith, J E Childs, F P Pinheiro, J I Maiztegul, B Niklasson, A Antoniades et al. 1986. Global survey of antibody to Hantaan-related viruses among peridomestic rodents. *Bulletin of the World Health Organization* 64(1): 139–144.

Lee, H W. 1982. Hemorrhagic fever with renal syndrome (HFRS). *Scandinavian Journal of Infectious Diseases* 36(Suppl.): 82–85.

Lee, H W. 1996. Hantaviruses: An emerging disease. *Philippine Journal of Microbiology and Infectious Diseases* 25(1): S19–S24.

Lee, H W, C Calisher, and C S Schmaljohn. 1999. Epidemiology and epizootiology. In *Manual of Hemorrhagic Fever with Renal Syndrome and Hantavirus Pulmonary Syndrome.* pp. 40–48. Seoul, South Korea: WHO Collaborating Cenbter for virus Reference and Research.

Lee, H W, P W Lee, and K M Johnson. 1978. Isolation of the etiologic agent of Korean hemorrhagic fever. *The Journal of Infectious Diseases* 137(3): 298–308.

Li, J, Y X Liu, and Z T Zhao. 2009. Genotyping of hantaviruses occurring in Linyi, China, by nested RT-PCR combined with single-strand conformation polymorphism analysis. *Acta Virologica* 53(2): 121–124.

Linderholm, M and F Elgh. 2001. Clinical characteristics of hantavirus infections on the Eurasian continent. *Current Topics in Microbiology and Immunology* 256: 135–151.

Linderholm, M, P H Groeneveld, and A Tärnvik. 1996. Increased production of nitric oxide in patients with hemorrhagic fever with renal syndrome—Relation to arterial hypotension and tumor necrosis factor. *Infection* 24(5): 337–340.

Lindkvist, M, K Lahti, B Lilliehöök, A Holmström, C Ahlm, and G Bucht. 2007. Cross-reactive immune responses in mice after genetic vaccination with cDNA encoding hantavirus nucleocapsid proteins. *Vaccine* 25(9) (February 19): 1690–1699. doi:10.1016/j.vaccine.2006.09.082.

Liu, X, B-G Jiang, P Bi, W Yang, and Q Liu. 2011. Prevalence of haemorrhagic fever with renal syndrome in mainland China: Analysis of National Surveillance Data, 2004–2009. *Epidemiology and Infection.* doi:CJO 2011 doi:10.1017/S0950268811001063.

Logan, J S. 1989. Trench fever in Belfast, and the nature of the 'relapsing fevers' in the United Kingdom in the nineteenth century. *The Ulster Medical Journal* 58(1): 83–88.

Loginova, S Ia, A V Koval'chuk, S V Borisevich, N K Kopylova, Iu I Pashchenko, R A Khamitov, V A Maksimov, and A M Shuster. 2005. Antiviral effectiveness of the combined use of amixine and virasole in experimental hemorrhagic fever with renal syndrome in sucking albino mice. *Voprosy Virusologii* 50(6): 30–32.

Lokugamage, K, H Kariwa, N Lokugamage, H Miyamoto, M Iwasa, T Hagiya, K Araki et al. 2004. Genetic and antigenic characterization of the Amur virus associated with hemorrhagic fever with renal syndrome. *Virus Research* 101(2): 127–134. doi:10.1016/j.virusres.2003.12.031.

Lundkvist, A et al. 1997. Puumala and Dobrava viruses cause hemorrhagic fever with renal syndrome in Bosnia-Herzegovina: Evidence of highly cross-neutralizing antibody responses in early patient sera. *Journal of Medical Virology* 53(1), 51–59.

Lundkvist, A, S Björsten, B Niklasson, and N Ahlborg. 1995. Mapping of B-cell determinants in the nucleocapsid protein of Puumala virus: Definition of epitopes specific for acute immunoglobulin G recognition in humans. *Clinical and Diagnostic Laboratory Immunology* 2(1): 82–86.

Lundkvist, A, J Hörling, S Björsten, and B Niklasson. 1995. Sensitive detection of hantaviruses by biotin-streptavidin enhanced immunoassays based on bank vole monoclonal antibodies. *Journal of Virological Methods* 52(1–2): 75–86.

Lundkvist, A, J Hörling, and B Niklasson. 1993. The humoral response to Puumala virus infection (nephropathia epidemica) investigated by viral protein specific immunoassays. *Archives of Virology* 130(1–2): 121–130.

Lundkvist, A, O Vapalahti, A Plyusnin, K B Sjölander, B Niklasson, and A Vaheri. 1996. Characterization of Tula virus antigenic determinants defined by monoclonal antibodies raised against baculovirus-expressed nucleocapsid protein. *Virus Research* 45(1): 29–44.

Mackow, E R and I N Gavrilovskaya. 2001. Cellular receptors and hantavirus pathogenesis. *Current Topics in Microbiology and Immunology* 256: 91–115.

Mackow, E R and I N Gavrilovskaya. 2009. Hantavirus regulation of endothelial cell functions. *Thrombosis and Haemostasis* 102(6): 1030–1041. doi:10.1160/TH09-09-0640.

MacNalty, A S. 1947. The renaissance and its influence on English medicine, surgery and public health. *Annals of the Royal College of Surgeons of England* 1(1). The Thomas Vicary Lecture given at the Royal College of Surgeons of England. 8–30.

Mäkelä, S, P Jaatinen, M Miettinen, J Salmi, I Ala-Houhala, H Huhtala, M Hurme, I Pörsti, A Vaheri, and J Mustonen. 2010. Hormonal deficiencies during and after Puumala hantavirus infection. *European Journal of Clinical Microbiology and Infectious Diseases: Official Publication of the European Society of Clinical Microbiology* 29(6): 705–713. doi:10.1007/s10096-010-0918-y.

Mäkelä, S, J Mustonen, I Ala-Houhala, M Hurme, J Partanen, O Vapalahti, A Vaheri, and A Pasternack. 2002. Human leukocyte antigen-B8-DR3 is a more important risk factor for severe Puumala hantavirus infection than the tumor necrosis factor-alpha(-308) G/A polymorphism. *The Journal of Infectious Diseases* 186(6) (September 15): 843–846. doi:10.1086/342413.

McCaughey, C and C A Hart. 2000. Hantaviruses. *Journal of Medical Microbiology* 49(7): 587–599.

Mertz, G J, L Miedzinski, D Goade, A T Pavia, B Hjelle, C O Hansbarger, H Levy et al. 2004. Placebo-controlled, double-blind trial of intravenous ribavirin for the treatment of hantavirus cardiopulmonary syndrome in North America. *Clinical Infectious Diseases: An Official Publication of the Infectious Diseases Society of America* 39(9) (November 1): 1307–1313. doi:10.1086/425007.

Miettinen, M H, S M Mäkelä, I O Ala-Houhala, H S A Huhtala, T Kööbi, A I Vaheri, A I Pasternack, I H Pörsti, and J T Mustonen. 2006. Ten-year prognosis of Puumala hantavirus-induced acute interstitial nephritis. *Kidney International* 69(11): 2043–2048. doi:10.1038/sj.ki.5000334.

Mills, J N, B R Amman, and G E Glass. 2010. Ecology of hantaviruses and their hosts in North America. *Vector Borne and Zoonotic Diseases (Larchmont, NY)* 10(6): 563–574. doi:10.1089/vbz.2009.0018.

Mir, M A, W A Duran, B L Hjelle, C Ye, and A T Panganiban. 2008. Storage of cellular 5′ mRNA caps in P bodies for viral cap-snatching. *Proceedings of the National Academy of Sciences of the United States of America* 105(49) (December 9): 19294–19299. doi:10.1073/pnas.0807211105.

Mustonen, J, J Partanen, M Kanerva, K Pietilä, O Vapalahti, A Pasternack, and A Vaheri. 1996. Genetic susceptibility to severe course of nephropathia epidemica caused by Puumala hantavirus. *Kidney International* 49(1): 217–221.

Myhrman, G. A. 1934. A renal disease with particular symptoms. *Nordisk Medicinsk Tidskrif* 7: 793–794.

Nguyên, A T, C Penalba, P Bernadac, S Jaafar, M Kessler, P Canton, and B Hoen. 2001. Respiratory manifestations of hemorrhagic fever with renal syndrome. Retrospective study of 129 cases in Champagne-Ardenne and Lorraine. *Presse Médicale (Paris, France: 1983)* 30(2) (January 20): 55–58.

Nichol, S T, C F Spiropoulou, S Morzunov, P E Rollin, T G Ksiazek, H Feldmann, A Sanchez, J Childs, S Zaki, and C J Peters. 1993. Genetic identification of a hantavirus associated with an outbreak of acute respiratory illness. *Science (New York)* 262(5135) (November 5): 914–917.

Okumura, M, K Yoshimatsu, S Kumperasart, I Nakamura, M Ogino, M Taruishi, A Sungdee et al. 2007. Development of serological assays for Thottapalayam virus, an insectivore-borne Hantavirus. *Clinical and Vaccine Immunology: CVI* 14(2): 173–181. doi:10.1128/CVI.00347-06.

Olsson, G E, H Leirs, and H Henttonen. 2010. Hantaviruses and their hosts in Europe: Reservoirs here and there, but not everywhere? *Vector Borne and Zoonotic Diseases (Larchmont, NY)* 10(6): 549–561. doi:10.1089/vbz.2009.0138.

Pacsa, A S, E A Elbishbishi, U C Chaturvedi, K Y Chu, and A S Mustafa. 2002. Hantavirus-specific antibodies in rodents and humans living in Kuwait. *FEMS Immunology and Medical Microbiology* 33(2) (June 3): 139–142.

Padula, P, V P Martinez, C Bellomo, S Maidana, J San Juan, P Tagliaferri, S Bargardi et al. 2007. Pathogenic hantaviruses, northeastern Argentina and eastern Paraguay. *Emerging Infectious Diseases* 13(8): 1211–1214.

Park, K, C S Kim, and K-T Moon. 2004. Protective effectiveness of hantavirus vaccine. *Emerging Infectious Diseases* 10(12): 2218–2220.

Pattamadilok, S, B-H Lee, S Kumperasart, K Yoshimatsu, M Okumura, I Nakamura, K Araki et al. 2006. Geographical distribution of hantaviruses in Thailand and potential human health significance of Thailand virus. *The American Journal of Tropical Medicine and Hygiene* 75(5): 994–1002.

Plyusnin, A, O Vapalahti, and A Vaheri. 1996. Hantaviruses: Genome structure, expression and evolution. *The Journal of General Virology* 77(Pt 11): 2677–2687.

Plyusnina, A, I-N Ibrahim, and A Plyusnin. 2009. A newly recognized hantavirus in the Asian house rat (*Rattus tanezumi*) in Indonesia. *The Journal of General Virology* 90(Pt 1): 205–209. doi:10.1099/vir.0.006155-0.

Plyusnina, A, J Laakkonen, J Niemimaa, H Henttonen, and A Plyusnin. 2008. New genetic lineage of Tula Hantavirus in *Microtus arvalis* obscurus in Eastern Kazakhstan. *The Open Virology Journal* 2: 32–36. doi:10.2174/1874357900802010032.

Quelapio, I D, L Villa, S M Clarin, M Bacosa, and T E Tupasi. 2000. Seroepidemiology of Hantavirus in the Philippines. *International Journal of Infectious Diseases: IJID: Official Publication of the International Society for Infectious Diseases* 4(2): 104–107.

Rasmuson, J, C Andersson, E Norrman, M Haney, M Evander, and C Ahlm. 2011. Time to revise the paradigm of hantavirus syndromes? Hantavirus pulmonary syndrome caused by European hantavirus. *European Journal of Clinical Microbiology and Infectious Diseases: Official Publication of the European Society of Clinical Microbiology* 30(5): 685–690. doi:10.1007/s10096-010-1141-6.

Raw, N. 1915. Trench nephritis: A record of five cases. *British Medical Journal* 2(2856) (September 25): 468.

Razanskiene, A, J Schmidt, A Geldmacher, A Ritzi, M Niedrig, A Lundkvist, D H Krüger, H Meisel, K Sasnauskas, and R Ulrich. 2004. High yields of stable and highly pure nucleocapsid proteins of different hantaviruses can be generated in the yeast *Saccharomyces cerevisiae*. *Journal of Biotechnology* 111(3) (August 5): 319–333. doi:10.1016/j.jbiotec.2004.04.010.

Reynes, J-M, J-L Soares, T Hüe, M Bouloy, S Sun, S L Kruy, F Flye Sainte Marie, and H Zeller. 2003. Evidence of the presence of Seoul virus in Cambodia. *Microbes and Infection/Institut Pasteur* 5(9): 769–773.

Roberts, L. 1945. Sweating sickness and picardy sweat. *British Medical Journal* 2(4144) (August 11): 195.

Rodier, G, A K Soliman, J Bouloumié, and D Kremer. 1993. Presence of antibodies to Hantavirus in rat and human populations of Djibouti. *Transactions of the Royal Society of Tropical Medicine and Hygiene* 87(2): 160–161.

Rowe, R K and A Pekosz. 2006. Bidirectional virus secretion and nonciliated cell tropism following Andes virus infection of primary airway epithelial cell cultures. *Journal of Virology* 80(3): 1087–1097. doi:10.1128/JVI.80.3.1087-1097.2006.

Rowe, R K, J W Suszko, and A Pekosz. 2008. Roles for the recycling endosome, Rab8, and Rab11 in hantavirus release from epithelial cells. *Virology* 382(2) (December 20): 239–249. doi:10.1016/j.virol.2008.09.021.

Rudolf, R D. 1917. War nephritis. *The Canadian Medical Association Journal* 7(4): 289–297.

Rutherford, W J. 1916. Trench fever: The field vole a possible origin. *British Medical Journal* 2(2907) (September 16): 386–387.

Ryou, J, H I Lee, Y J Yoo, Y T Noh, S-M Yun, S Y Kim, E-H Shin, M G Han, and Y R Ju. 2011. Prevalence of hantavirus infection in wild rodents from five provinces in Korea, 2007. *Journal of Wildlife Diseases* 47(2): 427–432.

Saksida, A, B Wraber, and T Avšič-Županc. 2011. Serum levels of inflammatory and regulatory cytokines in patients with hemorrhagic fever with renal syndrome. *BMC Infectious Diseases* 11(1): 142. doi:10.1186/1471-2334-11-142.

Saluzzo J F, J P Digoutte, F Adam, S P Bauer, and J B McCormick. 1985. Serological evidence for Hantaan-related virus infection in rodents and man in Senegal. *Transactions of the Royal Society of Tropical Medicine and Hygiene* 79: 874–875.

Sanfeliu, I, M M Nogueras, M I Gegúndez, F Segura, L Lledó, B Font, and J V Saz. 2011. Seroepidemiological survey of hantavirus infection in healthy people in vallès occidental, barcelona. *Vector Borne and Zoonotic Diseases (Larchmont, NY)* 11(6): 697–700. doi:10.1089/vbz.2010.0165.

Sarre, H. 1959. *Nierenkrankheiten*, 2nd edn. Stuttgart, Germany: Thieme.

Schmaljohn, C. 1996. Molecular biology of hantaviruses. In *The Bunyaviridae*. New York: Elliot RM, Plenum Press.

Schmaljohn, C. 2009. Vaccines for hantaviruses. *Vaccine* 27 (Suppl 4) (November 5): D61–D64. doi:10.1016/j.vaccine.2009.07.096.

Schmaljohn, C S, S E Hasty, J M Dalrymple, J W LeDuc, H W Lee, C H von Bonsdorff, M Brummer-Korvenkontio, A Vaheri, T F Tsai, and H L Regnery. 1985. Antigenic and genetic properties of viruses linked to hemorrhagic fever with renal syndrome. *Science (New York)* 227(4690) (March 1): 1041–1044.

Schmaljohn, C and B Hjelle. 1997. Hantaviruses: A global disease problem. *Emerging Infectious Diseases* 3(2): 95–104.

Schmaljohn, C S, K Sugiyama, A L Schmaljohn, and D H Bishop. 1988. Baculovirus expression of the small genome segment of Hantaan virus and potential use of the expressed nucleocapsid protein as a diagnostic antigen. *The Journal of General Virology* 69(Pt 4): 777–786.

Schmidt, J, B Jandrig, B Klempa, K Yoshimatsu, J Arikawa, H Meisel, M Niedrig, C Pitra, D H Krüger, and R Ulrich. 2005. Nucleocapsid protein of cell culture-adapted Seoul virus strain 80–39: Analysis of its encoding sequence, expression in yeast and immuno-reactivity. *Virus Genes* 30(1): 37–48. doi:10.1007/s11262-004-4580-2.

Schonrich, G, A Rang, N Lutteke, M J Raferty, N Charbonnel, and R G Ulrich. 2008. Hantavirus-induced immunity in rodent reservoirs and humans. *Immunological Reviews* 225: 163–189.

Seitsonen, E, M Hynninen, E Kolho, H Kallio-Kokko, and V Pettilä. 2006. Corticosteroids combined with continuous veno-venous hemodiafiltration for treatment of hantavirus pulmonary syndrome caused by Puumala virus infection. *European Journal of Clinical Microbiology and Infectious Diseases: Official Publication of the European Society of Clinical Microbiology* 25(4): 261–266. doi:10.1007/s10096-006-0117-z.

Seo, J-H, K-H Park, J-Y Lim, and H-S Youn. 2007. Hemorrhagic fever with renal syndrome (HFRS, Korean hemorrhagic fever). *Pediatric Nephrology (Berlin, Germany)* 22(1): 156–157. doi:10.1007/s00467-006-0234-z.

Settergren, B. 2000. Clinical aspects of nephropathia epidemica (Puumala virus infection) in Europe: A review. *Scandinavian Journal of Infectious Diseases* 32(2): 125–132.

Sironen, T, J Klingström, A Vaheri, L C Andersson, A Lundkvist, and A Plyusnin. 2008. Pathology of Puumala hantavirus infection in macaques. *PloS One* 3(8): e3035. doi:10.1371/journal.pone.0003035.

Sjölander, K B, I Golovljova, V Vasilenko, A Plyusnin, and A Lundkvist. 2002. Serological divergence of Dobrava and Saaremaa hantaviruses: Evidence for two distinct serotypes. *Epidemiology and Infection* 128(1): 99–103.

Song, J, B Chun, S Kim, L J Baek, S H Kim, J W Sohn, H J Cheong, W J Kim, S C Park, and M J Kim. 2006. Epidemiology of hemorrhagic fever with renal syndrome in endemic area in the Republic of Korea, 1995–1998. *Journal of Korean Medical Science* 21: 614–620.

Song, K-J, L J Baek, S Moon, S J Ha, S H Kim, K S Park, T A Klein et al. 2007. Muju virus, a novel hantavirus harboured by the arvicolid rodent Myodes regulus in Korea. *The Journal of General Virology* 88(Pt 11): 3121–3129. doi:10.1099/vir.0.83139-0.

Spik, K W, C Badger, I Mathiessen, T Tjelle, J W Hooper, and C Schmaljohn. 2008. Mixing of M segment DNA vaccines to Hantaan virus and Puumala virus reduces their immunogenicity in hamsters. *Vaccine* 26(40) (September 19): 5177–5181. doi:10.1016/j.vaccine.2008.03.097.

Stuhlfauth, K. 1943. Bericht über ein neues schlammfieberähnliches Krankheitsbild bei Deutscher Truppen in Lappland. *Deutsche Medizinische Wochenschrift* 439(1946): 474–477.

Suputthamongkol, Y, N Nitatpattana, M Chayakulkeeree, S Palabodeewat, S Yoksan, and J-P Gonzalez. 2005. Hantavirus infection in Thailand: First clinical case report. *The Southeast Asian Journal of Tropical Medicine and Public Health* 36(3): 700–703.

Taviner, M, G Thwaites, and V Gant. 1998. The English sweating sickness, 1485–1551: A viral pulmonary disease? *Medical History* 42(1): 96–98.

Temonen, M, J Mustonen, H Helin, A Pasternack, A Vaheri, and H Holthöfer. 1996. Cytokines, adhesion molecules, and cellular infiltration in nephropathia epidemica kidneys: An immunohistochemical study. *Clinical Immunology and Immunopathology* 78(1): 47–55.

Terajima, M, D Hayasaka, K Maeda, and F A Ennis. 2007. Immunopathogenesis of hantavirus pulmonary syndrome and hemorrhagic fever with renal syndrome: Do CD8+ T cells trigger capillary leakage in viral hemorrhagic fevers? *Immunology Letters* 113(2) (November 15): 117–120. doi:10.1016/j.imlet.2007.08.003.

Tersago, K, R Verhagen, A Servais, P Heyman, G Ducoffre, and H Leirs. 2009. Hantavirus disease (nephropathia epidemica) in Belgium: Effects of tree seed production and climate. *Epidemiology and Infection* 137(2): 250–256. doi:10.1017/S0950268808000940.

Tomori, O, S Morikawa, Y Matsuura, and T Kitamura. 1986. Antibody to Japanese strain of haemorrhagic fever with renal syndrome (HFRS) virus in Nigerian sera. *Transactions of the Royal Society of Tropical Medicine and Hygiene* 80(6): 1008–1009.

Tuuminen, T, E Kekäläinen, S Mäkelä, I Ala-Houhala, F A Ennis, K Hedman, J Mustonen, A Vaheri, and T Petteri Arstila. 2007. Human CD8+ T cell memory generation in Puumala hantavirus infection occurs after the acute phase and is associated with boosting of EBV-specific CD8+ memory T cells. *Journal of Immunology (Baltimore, MD: 1950)* 179(3) (August 1): 1988–1995.

Vapalahti, O, H Kallio-Kokko, A Närvänen, I Julkunen, A Lundkvist, A Plyusnin, H Lehväslaiho, M Brummer-Korvenkontio, A Vaheri, and H Lankinen. 1995. Human B-cell epitopes of Puumala virus nucleocapsid protein, the major antigen in early serological response. *Journal of Medical Virology* 46(4): 293–303.

Vapalahti, O, A Lundkvist, V Fedorov, C J Conroy, S Hirvonen, A Plyusnina, K Nemirov et al. 1999. Isolation and characterization of a hantavirus from *Lemmus sibiricus*: Evidence for host switch during hantavirus evolution. *Journal of Virology* 73(7): 5586–5592.

Vapalahti, O, A Lundkvist, H Kallio-Kokko, K Paukku, I Julkunen, H Lankinen, and A Vaheri. 1996. Antigenic properties and diagnostic potential of puumala virus nucleocapsid protein expressed in insect cells. *Journal of Clinical Microbiology* 34(1): 119–125.

Vapalahti, O, A Lundkvist, and A Vaheri. 2001. Human immune response, host genetics, and severity of disease. *Current Topics in Microbiology and Immunology* 256: 153–169.

Vapalahti, O, J Mustonen, A Lundkvist, H Henttonen, A Plyusnin, and A Vaheri. 2003. Hantavirus infections in Europe. *The Lancet Infectious Diseases* 3(10): 653–661.

Vergil, P. 1838. *Polydore Vergil's English History.* H. Ellis (Ed.). London: John Bowyer Nichols and Son.

Virtanen, J O, K M Jääskeläinen, J Djupsjöbacka, A Vaheri, and A Plyusnin. 2010. Tula hantavirus NSs protein accumulates in the perinuclear area in infected and transfected cells. *Archives of Virology* 155(1): 117–121. doi:10.1007/s00705-009-0546-y.

Vitarana, T, G Colombage, V Bandaranayake, and H W Lee. 1988. Hantavirus disease in Sri Lanka. *Lancet* 2(8622) (November 26): 1263.

Wang, H, K Yoshimatsu, H Ebihara, M Ogino, K Araki, H Kariwa, Z Wang et al. 2000. Genetic diversity of hantaviruses isolated in china and characterization of novel hantaviruses isolated from *Niviventer confucianus* and *Rattus rattus. Virology* 278(2) (December 20): 332–345. doi:10.1006/viro.2000.0630.

Wilson, D E and D A M Reed. 2005. *Mammal Species of the World. A Taxonomic and Geographic Reference.* 3rd edn. Baltimore, MD: John Hopkins University Press. http://www.bucknell.edu/msw3/.

Wylie, J A and L H Collier. 1981. The English sweating sickness (sudor Anglicus): A reappraisal. *Journal of the History of Medicine and Allied Sciences* 36(4): 425–445.

Zaki, S R, P W Greer, L M Coffield, C S Goldsmith, K B Nolte, K Foucar, R M Feddersen, R E Zumwalt, G L Miller, and A S Khan. 1995. Hantavirus pulmonary syndrome. Pathogenesis of an emerging infectious disease. *The American Journal of Pathology* 146(3): 552–579.

Zegers, R H C, A Weigl, and A Steptoe. 2009. The death of Wolfgang Amadeus Mozart: An epidemiologic perspective. *Annals of Internal Medicine* 151: 274–278.

Zhang, Y-Z, F-X Zhang, J-B Wang, Z-W Zhao, M-H Li, H-X Chen, Y Zou, and A Plyusnin. 2009. Hantaviruses in rodents and humans, Inner Mongolia Autonomous Region, China. *Emerging Infectious Diseases* 15(6): 885–891.

Zhang, Y-Z, Y Zou, Z F Fu, and A Plyusnin. 2010. Hantavirus infections in humans and animals, China. *Emerging Infectious Diseases* 16(8): 1195–1203.

24 *Ebola virus* Infection

*Anthony Griffiths, Andrew Hayhurst, Robert Davey,
Olena Shtanko, Ricardo Carrion Jr., and Jean L. Patterson*

CONTENTS

24.1 INTRODUCTION

The viral genus *Ebola virus* or Ebola is the causative agent of Ebola hemorrhagic fever. Ebola was originally recognized in 1976 near the Ebola River Valley in the Democratic Republic of the Congo (DRC). In March 2002, the International Committee on Taxonomy of Viruses (ICTV) changed the *Filovirus* genus to the *Filoviridae* family with two specific genera: *Ebola virus* and *Marburg virus* (ICTV 2002). Currently, there are five recognized *Ebola virus* species: *Zaire ebolavirus* (ZEBOV) was the first to be recognized when in 1976, a schoolteacher developed symptoms initializing resembling malaria that are now well-characterized symptoms of rapid onset of fever, malaise, muscle pain, headache, and later vomiting, diarrhea, and over a maculopapular rash with bleeding at need sites and bodily orifices (Isaacson et al. 1977).

Sudan ebolavirus (SEBOV) also emerged in 1976 and was originally thought to be similar to ZEBOV. In 1989, an outbreak in crab-eating macaques from the Hazleton Laboratories (now Covance), originally thought to be simian hemorrhagic fever virus (SHFV), became known as *Reston ebolavirus* (REBOV). The virus has since been found in other affected animals that had been imported from the Philippines (Center for Disease Control 2008).

Ebola from the Ivory Coast, *Cote d'Ivoire ebolavirus* (CIEBOV) or *Taï Forest ebolavirus*, was isolated from chimpanzees from the Taï Forest in Cote D'Ivoire, Africa, in 1994 (Waterman 1999). The most recent identification of an *Ebola virus* was confirmed in 2007; this virus was identified as *Bundibugyo ebolavirus* from the Bundibugyo district of Uganda (World Health Organization 2008a).

Due to the high mortality rate of *Ebola virus* infection, the lack of an approved treatment or vaccine and the ability to be transmitted in the laboratory and known human-to-human transmission, all filoviruses are classified as biosafety level 4 (BSL-4) pathogens by the World Health Organization (WHO 2008b) and the Centers for Disease Control and Prevention (CDC) and all work with filoviruses requires maximum containment (BMBL 5th edn, CDC). *Ebola virus*, like *Marburg virus*, is classified as CDC Category A agents (http://www.bt.cdc.gov/agent/agentlist-category.asp#a). This classification is part of an ever-changing system for prioritizing initial public health preparedness risk and grading of the potential of these pathogens as biological weapons. Filoviruses are classified as select agents and are federally regulated within the United States by the CDC Select Agent Program (http://www.cdc.gov/of/sap/).

24.2   VIRUS STRUCTURE

The electron micrographs of viruses appearing during the initial outbreak showed filamentous particles that exhibited a variable length of between 300 nm and greater than 1500 nm, a diameter of 80 nm, and a 20–30 nm central axis running through the entire particle (Murphy et al. 1978). These studies noted that the particles were indistinguishable from *Marburg virus* particles, first described in 1968 (Kissling et al. 1968). Filovirus particles are pleomorphic and the unusual structures they form have become a defining characteristic. They may be long filaments, "U" shaped, "6" shaped, or circular (Bowen et al. 1977; Johnson et al. 1977; Pattyn et al. 1977; Murphy et al. 1978). The particle consists of a lipid envelope derived from host membranes decorated with membrane-bound peplomers of GP, which protrude approximately 10 nm. The envelope surrounds the matrix (comprised of VP40 and VP24), which coats the 40–50 nm diameter nucleocapsid (comprised of VP30, VP35, NP, and L).

24.3   GENOME ORGANIZATION

The genetic material of the ebolaviruses is a negative-sense single-stranded RNA (Regnery et al. 1980; Sanchez et al. 1993). The genomes of all the *Ebola viruses* are organized similarly (Figure 24.1) and are similar to *Marburg virus* [although *Marburg virus* represents a distinct lineage (Suzuki and Gojobori 1997)]. Additionally, it has been observed that the filovirus genomes resemble those of rhabdoviruses and paramyxoviruses, although the complexity of the filovirus genome suggests a closer relationship to the family *Paramyxoviridae* (Sanchez et al. 1993). The filoviruses are the largest viruses in the order Mononegavirales, having a genome of approximately 19,000 nucleotides.

The order of the *Ebola virus* genes is 3'-NP-VP35-VP40-GP-VP30-VP24-L (Figure 24.1). There are conserved transcriptional start and polyadenylation sequences that delineate the genes. In ebolaviruses, several of the genes have overlaps where the stop site of an upstream gene overlaps the start of a downstream gene (Sanchez et al. 1993; Ikegami et al. 2001; Groseth et al. 2002; Sanchez and Rollin 2005; Towner et al. 2008). This contrasts to *Marburg virus*, which has only one gene overlap (Feldmann et al. 1992). The 3' leader sequences are short and of fairly consistent length, ranging from 50 to 70 nt, while the 5' leader is variable in length, ranging from 25 nt to greater than 600 nt (Sanchez et al. 1993; Ikegami et al. 2001; Groseth et al. 2002; Sanchez and Rollin 2005; Towner et al. 2008).

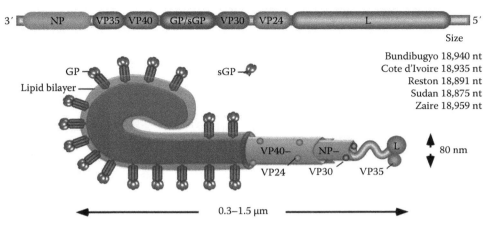

FIGURE 24.1 **(See color insert.)** Schematic representation of *Ebola virus* genome and structure. Upper panel. The arrangement and the relative sizes of *Ebola virus* genes are shown. Gene sizes are similar between all family members with some differences in the intragenic regions. The size of genomes for family members is indicated at right. Lower panel. The structure of the virion is shown. Particles are pleiomorphic and range in length from 300 to 1500 nm. The diameter of particles is relatively constant at 80 nm, except at ends where the virus filament can loop back (as shown at left) as well as form ring structures. The virus is enveloped, being coated with a lipid bilayer derived from the host cell membrane at time of budding. The homotrimeric GP is anchored in the bilayer by a transmembrane domain with the surface portion being exposed to the medium. Following an RNA-editing event, a truncated, secreted form of GP is generated (sGP). Underlying the lipid envelope is the matrix, which is composed mostly of VP40 and a lesser amount of VP24. The matrix tightly encases the nucleocapsid that is made mostly of NP with some VP30. The nucleocapsid coats the vRNA that interacts with the polymerase complex composed of L and the accessory protein VP35.

24.3.1 SPECIATION AND SEQUENCE DIVERSITY

The initial phylogenetic analyses of *Ebola virus* genomes compared coding regions (particularly the glycoprotein (GP) gene) (Sanchez et al. 1989, 1996; Georges-Courbot et al. 1997; Suzuki and Gojobori 1997; Volchkov et al. 1997) and more recently the full-length genome (Towner et al. 2008). These analyses showed that *Ebola virus* and *Marburg virus* have a distinct lineage and each of the five *Ebola virus* species represents a monophyletic group (Figure 24.2a). It has been noted that the close phylogenetic relationship of the Reston and Sudan species suggests that the *Reston virus*, associated with monkeys and pigs in the Philippines, may not be indigenous to Asia but may have been introduced to this region from Africa. However, there has yet to be an explanation for repeated appearance of REBOV in the Philippines. At the amino acid level, the GP genes of the five *Ebola virus* species differ between 27% and 41% (Sanchez et al. 1996; Towner et al. 2008).

Ebola virus genetic stability has been examined using viruses isolated from multiple patients during an outbreak (Rodriguez et al. 1999). Surprisingly, no nucleotide changes were observed over a 249-nt region of the GP gene sequence that had previously been shown to be variable. The observed lack of diversity contrasts with the notion that RNA viruses evolve rapidly because of a high error rate during RNA replication. Indeed, these data also contrast with a computational analysis that suggested filoviruses were subject to a similar mutation rate as other RNA viruses (Suzuki and Gojobori 1997). However, it was noted by Rodriguez et al. (1999) that there are examples of viruses that exhibit high genetic stability in nature despite the ability to evolve rapidly in cell culture (e.g., vesicular stomatitis virus). It will be interesting to see how these data compare to studies sequencing many virus isolates using the latest sequencing technologies that permit whole-genome sequencing and relatively easy identification of quasispecies.

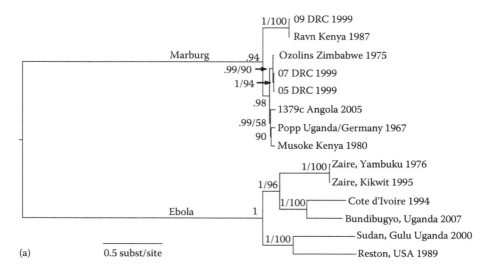

FIGURE 24.2 **(See color insert.)** (a) Phylogenetic tree comparing full-length genomes of *Ebola virus* and *Marburg virus* by Bayesian analysis. Posterior probabilities greater than 0.5 and maximum likelihood bootstrap values greater than 50 are indicated at the nodes. (Reproduced from Towner, J.S. et al., *PLOS Pathog.*, 4, e1000212, 2008. With permission.) (b) Viruses without borders: outbreaks of the African ebolaviruses occur close to the equator primarily across Gabon, RC, DRC, Southern Sudan, and Western Uganda. Note how the species appear to have clustered within tremendously diverse biomes with *Ivory Coast* western most, *Zaire* central, and *Sudan* eastern central with no intermingling, suggesting local ecological niches have been self-limiting to a degree so far. Insets show starting year of outbreak, village, country, fatalities/total cases, and mortality rate (occasionally, these numbers vary a little between reports). Outbreak data based on Leroy et al. (2009) and CDC/WHO Ebola websites (note that CDC is missing the Etoumbi 2005 outbreak at the time of writing) and locations were placed on a satellite image. The #7 *Zaire virus* outbreak is bracketed since a doctor who became infected in Gabon traveled to South Africa and infected a nurse who subsequently died. *Reston virus* has been imported in primate shipments to Reston, Virginia (1989, 1990), Philadelphia, Pennsylvania (1989), Alice, Texas (1989, 1990, and 1996), and Sienna (Italy 1992). *Reston virus* outbreaks in primate exporting facilities in the Philippines were detected in 1989 and 1996 and in swine in 2008. Laboratory-acquired infections have been reported to have occurred in Salisbury (UK, 1976), USAMRIID (2004), Vector (2004), and Hamburg (2009), which was immediately countered with a recombinant VSV-based vaccine (Tuffs 2009). All laboratorians except the vector case recovered. (The NASA satellite map of Africa used to create Figure 24.2b was obtained from http://www.zonu.com/detail-en/2009-11-08-10937/Africa-satellite-map.html.)

24.4 PROTEIN FUNCTIONS

24.4.1 NUCLEOPROTEIN (NP)

The ZEBOV NP is 739 amino acids in length, which makes it the largest nucleoprotein (NP) of the nonsegmented negative-stranded RNA viruses (Sanchez et al. 1989). Together with VP24 and VP35, NP is necessary and sufficient to form nucleocapsid structures that are indistinguishable from those seen in electron micrographs of infected cells (Huang et al. 2002).

24.4.2 VP35

The ZEBOV VP35 is 340 amino acids in length, and together with VP24 and NP, it forms the nucleo-capsid. Using a recombinant minigenome replication assay, VP35 was shown to be essential for replication and transcription (Mühlberger et al. 1999). The location in the viral genome, the importance for replication and transcription, and the interactions with NP and L have led to the proposal that

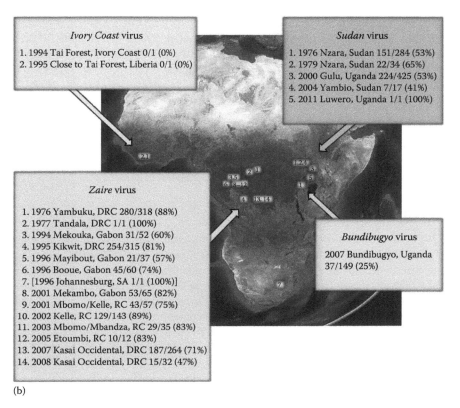

Ivory Coast virus

1. 1994 Tai Forest, Ivory Coast 0/1 (0%)
2. 1995 Close to Tai Forest, Liberia 0/1 (0%)

Sudan virus

1. 1976 Nzara, Sudan 151/284 (53%)
2. 1979 Nzara, Sudan 22/34 (65%)
3. 2000 Gulu, Uganda 224/425 (53%)
4. 2004 Yambio, Sudan 7/17 (41%)
5. 2011 Luwero, Uganda 1/1 (100%)

Zaire virus

1. 1976 Yambuku, DRC 280/318 (88%)
2. 1977 Tandala, DRC 1/1 (100%)
3. 1994 Mekouka, Gabon 31/52 (60%)
4. 1995 Kikwit, DRC 254/315 (81%)
5. 1996 Mayibout, Gabon 21/37 (57%)
6. 1996 Booue, Gabon 45/60 (74%)
7. [1996 Johannesburg, SA 1/1 (100%)]
8. 2001 Mekambo, Gabon 53/65 (82%)
9. 2001 Mbomo/Kelle, RC 43/57 (75%)
10. 2002 Kelle, RC 129/143 (89%)
11. 2003 Mbomo/Mbandza, RC 29/35 (83%)
12. 2005 Etoumbi, RC 10/12 (83%)
13. 2007 Kasai Occidental, DRC 187/264 (71%)
14. 2008 Kasai Occidental, DRC 15/32 (47%)

Bundibugyo virus

2007 Bundibugyo, Uganda 37/149 (25%)

(b)

FIGURE 24.2 (continued)   (See color insert.) (a) Phylogenetic tree comparing full-length genomes of *Ebola virus* and *Marburg virus* by Bayesian analysis. Posterior probabilities greater than 0.5 and maximum likelihood bootstrap values greater than 50 are indicated at the nodes. (Reproduced from Towner, J.S. et al., *PLOS Pathog.*, 4, e1000212, 2008. With permission.) (b) Viruses without borders: outbreaks of the African ebolaviruses occur close to the equator primarily across Gabon, RC, DRC, Southern Sudan, and Western Uganda. Note how the species appear to have clustered within tremendously diverse biomes with *Ivory Coast* western most, *Zaire* central, and *Sudan* eastern central with no intermingling, suggesting local ecological niches have been self-limiting to a degree so far. Insets show starting year of outbreak, village, country, fatalities/total cases, and mortality rate (occasionally, these numbers vary a little between reports). Outbreak data based on Leroy et al. (2009) and CDC/WHO Ebola websites (note that CDC is missing the Etoumbi 2005 outbreak at the time of writing) and locations were placed on a satellite image. The #7 *Zaire virus* outbreak is bracketed since a doctor who became infected in Gabon traveled to South Africa and infected a nurse who subsequently died. *Reston virus* has been imported in primate shipments to Reston, Virginia (1989, 1990), Philadelphia, Pennsylvania (1989), Alice, Texas (1989, 1990, and 1996), and Sienna (Italy 1992). *Reston virus* outbreaks in primate exporting facilities in the Philippines were detected in 1989 and 1996 and in swine in 2008. Laboratory-acquired infections have been reported to have occurred in Salisbury (UK, 1976), USAMRIID (2004), Vector (2004), and Hamburg (2009), which was immediately countered with a recombinant VSV-based vaccine (Tuffs 2009). All laboratorians except the vector case recovered. (The NASA satellite map of Africa used to create Figure 24.2b was obtained from http://www.zonu.com/detail-en/2009-11-08-10937/Africa-satellite-map.html.)

VP35 is orthologous to the P protein in other *Mononegavirales* (Mühlberger et al. 1999). VP35 is a multifunctional protein, and it has been shown to function as a type-I interferon (IFN) antagonist to prevent activation of the antiviral state, which can limit viral spread and affect adaptive immune responses (Basler et al. 2000). VP35 has a *C*-terminal domain that binds double-stranded RNA, which is likely important for inhibition of IFN activation (Basler et al. 2000, 2003; Cárdenas et al. 2006; Feng et al. 2007; Leung et al. 2009). The expression of VP35 prevents phosphorylation, dimerization, and nuclear translocation of IFN regulatory factor 3 (IRF3), which is induced by viral infection, thereby inhibiting IFN gene expression (Basler et al. 2003). Additionally, VP35 blocks the activation of the IFN-β promoter by some components of the RIG-I signaling pathway (Cardenas et al. 2006). A recent report has shown that VP35 binds to and is phosphorylated by both IKKε and TBK-1, which activate IRF-3 and IRF-7 upon RNA virus infection (Prins et al. 2009). These data suggest that VP35 blocks important interactions between these kinases and their normal interaction partners, to affect its IFN antagonist function. Another apparently independent function for VP35 in antagonizing the cellular antiviral response is to suppress RNA interference in a functionally equivalent manner to HIV-1 Tat protein (Haasnoot et al. 2007). Evidently, VP35 enters a complex with Dicer and its partners TRBP and PACT, although the only direct interactions are with TRBP and PACT, to antagonize Dicer activity (Fabozzi et al. 2011).

24.4.3   VP40

The ZEBOV VP40 is 326 amino acids in length and is a viral matrix protein and is the most abundant protein in viral particles (Elliott et al. 1985). Upon expression of VP40 in mammalian cells, the protein is released into the culture medium bound to membranes that are thought to be virosomes (Harty et al. 2000; Jasenosky et al. 2001; Timmins et al. 2001; Noda et al. 2002). Electron micrographs revealed that the virosomes look like spikeless viral particles. Thus, it appears that VP40 is the primary morphological determinant of filovirus particles and crucial to virus budding. There is a recent report showing that VP40 is a suppressor of RNA silencing (Fabozzi et al. 2011). Further work is required to determine the mechanism of action, but interestingly, VP40 was not observed to be associated with siRNA protein complexes. This provoked the speculation that VP40 may recruit components of RNAi that are associated with membranes.

24.4.4   GP, SGP, AND SSGP

The ZEBOV GP precursor is 676 amino acids in length, and it forms the peplomers that decorate the virion envelope. To date, three forms of GP have been shown to be generated from the GP0 precursor: GP1,2, soluble GP (sGP), and small sGP (ssGP). Maturation from the GP0 protein requires proteolytic cleavage by cellular furin (Volchkov et al. 1998). Multiple forms of the gene are generated following transcriptional editing of the *GP* transcript, and the relative abundance of the transcripts encoding these molecules is 70%, 25%, and 5%, respectively (Mehedi et al. 2011). The primary product of the gene is sGP, which is secreted and synthesized from non-edited transcripts (Volchkov et al. 1995; Sanchez et al. 1996). GP1,2 is generated as a consequence of the editing event, which results in a +1 shift in reading frame, eliminating the stop codon used in the synthesis of sGP. A third species (ssGP) has recently been identified and is the product of an editing event that yields a transcript with a +2 shift in reading frame (Mehedi et al. 2011). All GP proteins have identical *N*-terminal sequences but differ after amino acid 295.

Mature GP is a trimer of three disulfide-linked GP1,2 heterodimers that are generated following proteolytic cleavage of GP0 (Sanchez et al. 1998; Jeffers et al. 2002). GP1 appears to control virus–cell binding, and GP2 controls fusion of the virus and cellular membranes, and the arrangement of these proteins is analogous to other type-1 envelope proteins (Weissenhorn et al. 1998). In the endosome, cathepsins B and L are required to proteolyse GP1 in endosomes to trigger fusion.

sGP and ssGP lack a transmembrane domain and are therefore secreted from cells. sGP is a homodimer and so differs from the homotrimeric GP found on virions (Falzarano et al. 2006). It has been reported that sGP can bind to human neutrophils through an interaction with CD16b (Yang et al. 1998). However, this observation has proven to be controversial (Sui and Marasco 2002; Mehedi et al. 2011), and ssGP appears not to bind CD16b (Mehedi et al. 2011). sGP but not ssGP has been shown to rescue barrier function following treatment of endothelial cells with tumor necrosis factor alpha (Wahl-Jensen et al. 2005; Mehedi et al. 2011).

24.4.5 VP30

The ZEBOV VP30 is 288 nucleic acids in length and, together with VP24 and NP, forms the nucleocapsid. VP30 is an RNA-binding protein (John et al. 2007). Using the minigenome replication and transcription system, it was shown that VP30 was essential for replication; this contrasted with *Marburg virus*, where only NP, VP35, and L were required (Mühlberger et al. 1999). It was shown that VP30 was a positive regulator for transcription initiation acting as a transcription antitermination factor immediately following transcription initiation, while not affecting transcription elongation (Modrof et al. 2002; Weik et al. 2002; Modrof et al. 2003). Further work has shown that VP30 is important for transcriptional reinitiation during the stop–start process of *Ebola virus* transcription (Martinez et al. 2008). A recent report has suggested a role for VP30 in suppression of RNAi. It is postulated that VP30 may prevent loading of the RNA-induced silencing complex (RISC) and also prevent any RISC activity that requires PACT (Fabozzi et al. 2011). This activity was not dependent on the RNA-binding activity of the protein.

24.4.6 VP24

The ZEBOV VP24 is 251 amino acids in length and, together with VP35 and NP, it forms the nucleocapsid (Huang et al. 2002; Watanabe et al. 2006). VP24 is present in virions at relatively low levels compared to VP40 (Feldmann et al. 1999), and it localizes to the plasma membrane in infected cells (Han et al. 2003). The protein oligomerizes under physiological conditions, and this is thought to be important for promoting virion assembly (Han et al. 2003). Additionally, VP24 is thought to inhibit transcription and replication of the viral genome by direct association with the ribonucleoprotein complex (Watanabe et al. 2007). However, the biological significance of this effect is unclear. VP24 has a role in suppressing the cellular antiviral response. It inhibits IFN-α/β and IFN-γ-induced nuclear accumulation of tyrosine-phosphorylated STAT1 (PY-STAT1) and also inhibits IFN-α/β and IFN-γ-induced gene expression (Reid et al. 2006). This occurs by inhibiting the interaction of PY-STAT1 with karyopherin-α proteins (Reid et al. 2007).

24.4.7 Polymerase

The ZEBOV L protein is 2212 amino acids in length, making it the largest protein encoded by the virus. It is an RNA-dependent RNA polymerase, and together with VP35, L forms the polymerase complex, which serves to transcribe and replicate the viral genome (Mühlberger et al. 1999; Boehmann et al. 2005). The L protein is highly conserved among the *Ebola virus* species with approximately 75% amino acid identity (Boehmann et al. 2005).

24.5 ENTRY

24.5.1 Receptor Usage

Virus receptors on cells act to both bind viruses to cells and promote entry. Entry comprises both uptake through endocytosis and penetration into the cell cytoplasm through membrane fusion. For filoviruses, ZEBOV has been best studied. Five receptors have been identified on a variety of cell types and may reflect receptor use in different tissues. The folic acid receptor was first identified (Chan et al. 2001). While promoting infection once exogenously expressed in Jurkat cells, later work demonstrated that it was neither essential nor sufficient for infection of other cell types (Simmons et al. 2003b). Later, Axl, a member of the Tyro3 family, was identified (Shimojima et al. 2006, 2007). This tyrosine kinase is capable of triggering cell signaling through the PI3 kinase pathway. PI3 kinase activity was later shown to be important for EBOV infection of cells (Saeed et al. 2008), and so Axl is a potential candidate for mediating this signaling. Binding of EBOV to

Axl also promotes uptake through macropinocytic endocytosis (Hunt et al. 2011), which is a PI3 kinase–dependent process (discussed later). However, direct interaction of GP with Axl could not be directly demonstrated (Brindley et al. 2011), and so Axl may act by low-affinity interaction or indirect contact with virus. As for other viruses, like HIV and *West Nile virus*, DC-SIGN augments binding of EBOV to cells (Alvarez et al. 2002; Simmons et al. 2003a). DC-SIGN binds sugar modifications on the virus GP, tethering the virus to the cell surface and increasing dwell time for further interactions with entry receptors. More recently, TIM-1 was identified as important in promoting cell entry (Kondratowicz et al. 2011). Like Axl, TIM-1 is a tyrosine kinase, with a single transmembrane domain. Antibodies against TIM-1 block infection while its overexpression enhances infection. Unlike Axl, TIM-1 is predominantly expressed on airway epithelial cells and may play a key role in promoting infection by an aerosol. The most recent addition to the set of known *Ebola virus* receptors is a protein dysfunctional in Niemann–Pick disease, NPC1 (Carette et al. 2011; Cote et al. 2011). NPC1 plays important roles in endocytic trafficking of cholesterol and other proteins from the cell surface to endosomes and back to the surface (Ko et al. 2001; Lusa et al. 2001). The dysfunctional form of NPC1 results in the accumulation of cholesterol in late endosomes and lysosomes (Slotte et al. 1989). While NPC1 appears to bind virus GP, its complex role in cholesterol trafficking and controlling membrane lipid composition will complicate dissecting its role. The use of multiple receptors may explain why *Ebola virus* has such a broad tropism.

24.5.2 ENDOCYTOSIS

Ebola virus GP requires both cleavage by cathepsin proteases and a reduction in pH before membrane fusion is triggered (Chandran et al. 2005; Schornberg et al. 2006; Kaletsky et al. 2007; Hood et al. 2010; Martinez et al. 2010). In addition, cathepsins are pH dependent. This means that after binding to cells, EBOV must be taken up into an endocytic compartment that is both acidic and contains cathepsin. Integrins were shown important in uptake to cathepsin containing endosomes (Schornberg et al. 2009), and other work demonstrated a dependence on cholesterol content of cell membranes. This suggested a role for non-clathrin-related endocytic uptake mechanisms such as the cholesterol-rich caveolae (Bavari et al. 2002; Empig and Goldsmith 2002) but may suggest alternative roles for NPC1 aside from binding virus. Later, the work with pharmacological compounds suggested clathrin was important (Sanchez 2007; Bhattacharyya et al. 2010), but drug pleiotropy made results difficult to interpret. Recently, two independent groups, combining drugs, siRNA, and expression of dominant negative mutants of key endocytic proteins, demonstrated macropinocytosis was the principle uptake route (Nanbo et al. 2010; Saeed et al. 2010). Macropinocytosis gives uptake of large particles similar to filovirus particles (0.3–1.5 μm), and virus may have adapted to target this pathway. Interestingly, macropinocytic uptake was induced by virus binding to cells and is consistent with GP activation of PI3 kinase activity. Uptake may also occur through atypical macropinocytosis involving dynamin (Mulherkar et al. 2011). Interestingly, a role for exocytosis has been indicated (Miller et al. 2012) that proceeds uptake of virus. Exocytosis appears to bring acid sphingomyelinase to the cell surface, which EBOV depends upon for infection. From the macropinosome, EBOV particles are trafficked to the early endosome and the late endosome and have been seen associated with lysosomes (Murray et al. 2005; Saeed et al. 2010). It remains unclear from which compartment the capsid is released into the cell cytoplasm.

24.5.3 TRANSCRIPTION AND REPLICATION

Ebola virus transcription and replication occurs in the cytoplasm of an infected cell (Geisbert and Jahrling 1995). In broad terms, the replication of filoviruses is similar to replication of other viruses with nonsegmented negative-strand RNA virus (NNV) genomes (e.g., paramyxoviruses). Firstly, the genomic RNA from the infecting virion is transcribed into monocistronic polyadenylated mRNAs.

The 3′ and 5′ noncoding regions encode *cis*-acting elements that are important for control of replication and transcription (in addition to other functions) (Enterlein et al. 2009). Transcription follows a sequential stop–start mechanism where a downstream gene is dependent on the termination of synthesis of the upstream gene (Martinez et al. 2008).

As with other NNVs, the viral polymerase starts replication at the 3′ end of the genomic RNA, encounters a promoter region, and generates a complementary copy of the genome (known as an antigenome) (Kolakofsky et al. 2004). The antigenomic RNA is later used as a template for negative-stranded RNA.

24.5.4 ASSEMBLY AND BUDDING

Newly synthesized viral genomes associate with NP, VP30, VP35, and L proteins in the cell cytoplasm to form riboNP (RNP) complexes, which are eventually transported to budding sites, presumably lipid rafts on the plasma membrane (Geisbert and Jahrling 1995; Bavari et al. 2002; Panchal et al. 2003). There, RNPs associate with VP24 protein and the viral coat consisting of VP40 and GP complexes, and nascent virions are released. The mechanism of RNP recruitment to budding sites and the mechanism of RNP incorporation into virions are thought to depend on VP40–VP35 (Johnson et al. 2006b) and/or between VP40 and NP (Noda et al. 2006) interaction and most likely, host factors. VP40 is essential for virion assembly and release since it is sufficient to facilitate the formation and budding of *Ebola virus*–like particles (VLPs) from the cell surface (Harty et al. 2000; Jasenosky et al. 2001; Timmins et al. 2001; Noda et al. 2002). VP40-driven VLPs are indistinguishable from *Ebola virus* virions in density, length, and diameter, suggesting that VP40 dictates virus morphology (Johnson et al. 2006a; Noda et al. 2006). The co-expression of VP40 with either NP or GP results in enhanced release of VLPs (Licata et al. 2003, 2004; Liu and Harty 2010), suggesting cooperation between these proteins is necessary for efficient budding. Sec24C, a component of the COPII vesicle transport complex, is necessary for trafficking of VP40 to the plasma membrane and interacts with VP40 to mediate budding (Yamayoshi et al. 2008).

For many enveloped viruses, virion budding depends on "late domains" that comprise three sequence motifs, PT/SAP, PPxY, and YxxL (Freed 2002), which interact with cell proteins Tsg101, Nedd4, and ALIX/Aip1, respectively (Katzmann et al. 2001; Babst et al. 2002a,b; Demirov and Freed 2004). *Ebola virus* VP40 protein is unusual in that it contains two overlapping late domains in the sequence PTAPPEY (residues 7–13), both of which can function independently as late domains. The deletion of this motif severely impairs but does not prevent VLP formation (Harty et al. 2000; Jasenosky et al. 2001; Martin-Serrano et al. 2001; Licata et al. 2003). Roles for residues P (residue 53), KLR (212–214), and LPLGVA (96–101) in intracellular localization of VP40 and/or release of VLPs (McCarthy et al. 2007; Yamayoshi and Kawaoka 2007; Liu et al. 2010) have been reported. *Ebola virus* VP40 binds both Tsg101 and Nedd4 proteins and is required for virion release (Martin-Serrano et al. 2001; Licata et al. 2003; Timmins et al. 2003; Yasuda et al. 2003). *Marburg virus* NP also interacts with Tsg101 through a late domain-like motif (Dolnik et al. 2010). No interaction of *Ebola virus* NP with Tsg101 has been reported.

The cellular cytoskeleton plays important roles in *Ebola virus* particle assembly and/or release. Actin is incorporated into virions (Han and Harty 2005), and the alteration of cytosolic calcium levels, which affects actin polymerization, impacts particle assembly (Han and Harty 2007). VP40 interacts with microtubules to promote tubulin polymerization (Ruthel et al. 2005), and the disruption of microtubules inhibits VLP release (Noda et al. 2006).

In contrast to cell proteins important for *Ebola virus* budding, other cell proteins inhibit virion formation. IFN-stimulated gene 15 (ISG15) modification of Nedd4 interferes with VP40–Nedd4 interaction needed for virion budding (Okumura et al. 2008). Another IFN-stimulated protein, tetherin, acts to retain viruses on the cell surface to block their release (Neil et al. 2008; Jouvenet et al. 2009). Tetherin has been shown to also inhibit *Ebola virus* VP40–driven budding (Jouvenet et al. 2009).

During virion release, this inhibition might be antagonized by the GP through the direct binding to tetherin (Kaletsky et al. 2009), potentially explaining how GP enhances VP40-facilitated VLP release.

24.6   CLINICAL PRESENTATION

Human infection occurs as the result of contact with infected individuals or animals, with subsequent person-to-person spread. The incubation period of *Ebola virus* in humans ranges from 2 to 21 days (Bwaka et al. 1999). The onset of symptoms is quick with fever and nonspecific symptoms including, headache, malaise, and myalgia. Within several days after the onset of symptoms, the disease progresses with sore throat, nausea, and vomiting. Late stages of the disease are characterized by hemorrhagic manifestations including petechiae, both oral and conjunctival; ecchymosis; bleeding from gums; and bleed from site of venipuncture (Bwaka et al. 1999; Geisbert et al. 2003a). Maculopapular rash is seen in trunk and shoulders in 50% of cases (Bwaka et al. 1999). In fatal cases, death is usually preceded by shock and tachycardia. Death usually occurs from 6 to 16 days after onset of symptoms with fatality rates as high as 90% (Groseth et al. 2007).

High levels of viremia in human filovirus infections are associated with high mortality (Ksiazek et al. 1999a,b; Towner et al. 2004). A 2-log difference in titers is sufficient to predict clinical outcomes (Towner et al. 2004). In contrast, patients that survive *Ebola virus* infection show decreased viral titers and improved clinical appearance by 7–10 days after onset of symptoms (Mahanty and Bray 2004). Survivors also show an early appears of proinflammatory cytokines that is direct contrast to fatal cases that show spike in these cytokines late in the disease process (Wauquier et al. 2010). Furthermore, fatal outcomes are characterized by high levels of anti-inflammatory cytokines such as IL-10. The innate response in non-survivors causes a cytokine storm. This non-protective response results in disseminated intravascular coagulation, vascular dysfunction, and hypotension. In combination with the global lymphocyte apoptosis induced, these events lead to vascular collapse, multiple organ failure, and shock seen in fatal cases (Bray and Mahanty 2003).

24.7   PATHOLOGY AND PATHOGENESIS

Most of what is known about the pathogenesis of *Ebola virus* has been derived from experimental infection of nonhuman primates (NHPs). Mononuclear cells are the early site of virus replication (Stroher et al. 2001; Geisbert et al. 2003a,b). Virus-infected cells then migrate to regional lymph node where the virus subsequently disseminates to the lymph nodes, spleen, and other tissues (Geisbert et al. 2003b). At these distal sites, virus continues to amplify and affect hepatocytes and adrenal cortical cells resulting in focal necrosis. Infection with *Ebola virus* results in a number of hematological abnormalities. Human *Ebola virus* infection is marked by neutrophilia, lymphopenia, and thrombocytopenia (Bwaka et al. 1999). Indeed, abnormal platelet aggregation is a common feature of the disease. Biochemical finding includes elevated AST and ALT, whereas ALP and bilirubin levels are only moderately elevated (Geisbert et al. 2003a).

In similar fashion, biochemical analysis of blood from infected NHPs shows elevated levels of serum enzymes, primarily markers of liver function, including blood urea nitrogen (BUN), creatinine, gamma-glutamyl transferase (GGT), alkaline phosphatase (ALP), ALT, and AST. Hematology data from infected macaques reveal that white blood cell (WBC) and differential WBC counts show leukocytosis with increased neutrophilia and concomitant lymphopenia. At the late stage of disease, thromobocytopenia develops with increased level of d-dimers with a decrease in activated protein C. Coagulopathy is evident as prominent fibrin deposits in the liver and spleen. Indeed, widespread fibrin deposition is also common in human cases and appears to be a hallmark of coagulation abnormalities (Jaax et al. 1996; Geisbert et al. 2003b,c).

Gross examination of organs at necropsy reveals lymphadenopathy; liver characterized by enlargement, discoloration, and friability; enlarged tonsils; congestion of colon; and petechiae and hemorrhage in many tissue types including tonsils, gums, and stomach. Microscopic evaluation of tissue section reveals multifocal, random hepatocellular degeneration and necrosis, lymphoid apoptosis, and necrosis in the spleen and lymph nodes and fibrin deposits in the kidney, liver, and spleen (Geisbert et al. 2003a,c).

Filovirus infection of dendritic cells negatively impacts their function as modulators of both the innate and the adaptive responses. Dendritic cells are professional antigen presentation cells that function to activate T cells and stimulate the expansion and differentiation of B cells. Infected dendritic cells fail to mature or become activated and cannot upregulate MHC, directly affecting their ability to stimulate T cells. Dendritic cells also lose their ability to produce IFN, which are also impaired in macrophages and PBMCs. Two viral proteins, VP24 and VP35, have been shown to interfere with the IFN response. VP24 interrupts nuclear accumulation of tyrosine-phosphorylated STAT1 and STAT2 in infected cells, making them insensitive to IFN-a/b (Reid et al. 2006, 2007). Ebola VP35 functions as an IFN antagonist by binding to dsRNA or TANK-binding kinase 1, leading to the inhibition of IRF3 that induces the expression of antiviral genes and cytokines (Basler et al. 2000, 2003; Cárdenas et al. 2006; Hartman et al. 2006; Basler and Amarasinghe 2009; Prins et al. 2009).

Apoptosis of lymphocytes is a prominent feature in filovirus infection leading to immunosuppression of the adaptive immune response (Geisbert et al. 2000). Since the virus does not directly infect lymphocytes, apoptosis appears to be a bystander effect. In studies analyzing blood collecting during five *Ebola virus* outbreaks in Gabon and the Republic of Congo (RC), it was observed that the T CD4 and CD8 lymphocytes underwent massive apoptosis in ZEBOV fatalities whereas the level of lymphocyte apoptosis seen in the survivors was close to that found in the healthy controls (Wauquier et al. 2010). The mechanism of apoptosis appears to be mediated through extrinsic pathways (Bradfute et al. 2007). These findings are consistent with the marked bystander lymphocyte apoptosis associated with fatal ZEBOV infection in experimental animals (Geisbert et al. 2000, 2003a,b).

Recently, it has been suggested that *Ebola virus* may harbor superantigen (SAg) activity (Leroy et al. 2011). SAgs are proteins that bind MHCII molecules to the T-cell receptor Vb region causing amplification of T-cell Vb subsets that are deleted or become nonfunctional. It has been observed in both fatal and nonfatal human cases, *Ebola virus* downregulates transcripts associated with T-cell receptor Vb subsets. This SAg activity may contribute to the massive T-cell depletion seen in fatal cases (Leroy et al. 2011).

24.8 ECOLOGY AND EPIDEMIOLOGY

A constantly updated table of most of the *Ebola virus* outbreaks is maintained by CDC at http://www.cdc.gov/ncidod/dvrd/spb/mnpages/dispages/ebola/ebolatable.htm and serves to highlight their sporadic nature and is partially summarized in Figure 24.2b. Twenty-seven natural outbreaks involving approximately 2300 human cases and 1530 fatalities are noted since the first recorded emergence in 1976. The five species of *Ebola virus* have widely differing frequencies of occurrence and mortality rates with ZEBOV and SEBOV having caused the most outbreaks and *Bundibugyo ebolavirus* recently causing one in late 2007 and early 2008. CIEBOV has two known cases that both survived, and REBOV has so far produced asymptomatic infection of humans though it is highly lethal to nonhuman primates. REBOV heralds from the Philippines, while the others emerge in the tropical ecosystems of Central and Western equatorial Africa. *Ebola virus* is a zoonosis that is first transmitted to humans from encounters with a specific range of animals. The highly lethal hemorrhagic fever that ensues can serve to transmit the virus to other humans by direct contact with infectious bodily fluids yet also serves to limit the outbreak unlike many other zoonoses such as influenza A that have low lethality and high transmissibility. To appreciate the epidemiology and to have any chance of predicting or preventing future outbreaks, it is imperative to understand the ecology of the virus, and several excellent reviews exist on this topic including Gonzalez et al. (2005, 2007), Groseth et al. (2007), and Pourrut et al. (2005), and we aim to condense salient points here.

Though over 35 years since the first recorded emergences of *Ebola virus* simultaneously in Nzara in Southern Sudan and in Yambuku in Northern Zaire (now DRC), a potential reservoir host has only been tentatively identified relatively recently. In 2005, a small number of fruit bats *Hypsignathus monstrosus, Epomops franqueti,* and *Myonycteris torquata* sampled near the Gabon/RC border following the discovery of gorilla and chimpanzee carcasses were found to be seropositive for anti-*Ebola virus* IgG and contained ZEBOV RNA sequences (Leroy et al. 2005). In 2007, a serological survey extended the potential species range of *Zaire virus* to *Micropterus pusillus* and *Mops condylurus*, both insectivorous bats and another fruit bat *Rousettus aegyptiacus* (Pourrut et al. 2009). A proportion of *Rousettus amplexicaudatus* isolated from forests close to REBOV infections of monkeys (Miranda et al. 1999) and swine (Barrette et al. 2009) in the Philippines have also recently been shown to be seropositive to REBOV, yet no viral nucleotide sequences were detected (Taniguchi et al. 2011). The closely related *Marburg virus* has also been indicated in several bat species including *H. monstrous, E. franqueti,* and *R. aegyptiacus* by seropositivity (Pourrut et al. 2009), viral RNA detection (Swanepoel et al. 2007; Towner et al. 2007a, 2009), and even immune staining of *R. aegyptiacus* viral lesions (Towner et al. 2009). Strikingly, IgGs against both viruses in the same animal have also been found in *R. aegyptiacus* suggesting that this species may well be implicated in the maintenance and transmission of both viruses (Pourrut et al. 2009). At the time of writing, while *Marburg virus* has been cultured in vitro from *R. aegyptiacus* tissues with high viral RNA signatures (Towner et al. 2009), *Ebola virus* still has not been cultivated from any bat. The low PCR-positive load or high seroconversion, but not both simultaneously, might indicate replication of *Ebola virus* is more highly restricted by the onset of the bat immune response (Biek et al. 2006). Increased detection in pregnant bats may indicate the virus is taking advantage of transient changes in the immune system enabling growth of a partially allogeneic fetus. No information regarding potential bat hosts is currently available for SEBOV, CIEBOV, or *Bundibugyo ebolavirus*. However, it has been demonstrated that bats can have transient viremia when experimentally infected with *Ebola virus* (Swanepoel et al. 1996) and many other examples of viruses having bats as a host exist (Calisher et al. 2006; Wong et al. 2007).

The flight ranges of the bats studied encompass the known outbreak areas and beyond, ensuring a perennial threat of transmission, potentially even in areas untouched by recorded outbreaks. In 2010, one of the 262 migratory fruit bats, interestingly a pregnant female, *Eidolon helvum* sampled in Ghana was found to be seropositive for *Ebola virus* IgG though Ghana has not yet reported an *Ebola virus* outbreak (Hayman et al. 2010). A novel *Filovirus* named *Lloviu* virus was recently indicated in *Miniopterus schreibersii* in Northern Spain by a chip-based RNA detection and subsequent sequencing (Kuhn et al. 2010), though the pathogenicity of this agent to primates is currently unknown. A survey of the viromes of a handful of North American bats and bat guano, while not revealing *Ebola virus*–like sequences, did, however, discover a very rich variety of plant and animal viral nucleic acid signatures (Donaldson et al. 2010; Li et al. 2010), indicating that bats encounter a large segment of the virosphere. And on this note, filoviral ecologists are always careful to suggest the potential of other hosts being involved in the transmission pathway (Leroy et al. 2005) including parasites or insects seasonally active in the bat diet (Swanepoel et al. 2007). Although the filoviral-like particles detected by electron microscopy in leafhoppers collected in France (Lundsgaard 1997) were subsequently shown to be distinct from filoviruses yet still a member of Mononegavirales (Bock et al. 2004), it serves to indicate that studying the insect, arachnid, and plant life in or close to bat roosts and feeding areas might still be warranted using newer ultrasensitive detection techniques.

Filovirus-positive bats appear to be asymptomatic, and while the frequency of infection might be only a few percent, the large roosts of several tens of thousands of animals can increase the

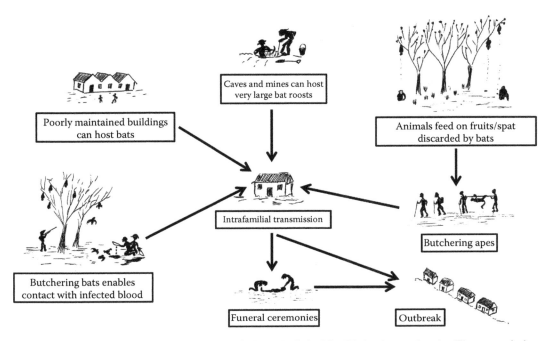

FIGURE 24.3 Epidemiology and ecology are intimately linked for *Ebola virus* outbreaks. The transmission paths are thought to originate from specific species of bats that have a low incidence of virus yet are present in sufficient numbers to drive species jumping directly to humans or via nonhuman primates, duikers, and perhaps other intermediate animals. Butchering bats and butchering bush meat are high-risk activities with fresh carcasses offering high titers of infectious virus. The index case returns to the home village to fall sick and is nursed by family and relatives that may succumb to infection by contact with bodily fluids and the outbreak takes hold.

likelihood of transmission to other hosts. Certainly, there was evidence of bats at a cotton factory in Nzara where index cases of both the 1976 and the 1979 Sudan outbreaks worked. Also, the 1994–1995 Gabon outbreak affected workers camps at forest gold mines, which are often roosts for bats. The hunting and butchering of bats for consumption has also been directly implicated in the 2007 DRC outbreak by detailed epidemiological investigation (Leroy et al. 2009) and is another route for emergence as shown in Figure 24.3. Gabon has a rich cave network within the forests, which has only recently started to be explored to reveal striking biodiversity including large roosts of bats (http://cavesintheheartofdarkness.blogspot.com/and http://www.abanda-expedition.org/).

Yet many human outbreak cases do not involve direct contact with bats and rather involve secondary hosts of duikers (forest antelopes) or nonhuman primates for which these viruses are typically extremely lethal and replicate to high titers (Leroy et al. 2004a). Indeed, large populations of western gorillas and common chimpanzees have been decimated by *Ebola virus* outbreaks (Walsh et al. 2003; Bermejo et al. 2006), and these events were often noted to precede human outbreaks (Leroy et al. 2004a; Rouquet et al. 2005; Walsh et al. 2005; Lahm et al. 2007) and might serve as future sentinel events. It is thought that the gorillas and chimpanzees come to feed at fruit tree sites and ingest partially consumed fruits and pulp dropped by the bats feeding in the canopy above, which are likely to be contaminated with saliva. The contamination of fruit sap by bats is known to be a transmission route to humans for *Nipah virus* (Luby et al. 2009a,b). Fluids from birthing have been also suggested as a further means of transmission, and though the mother usually consumes the placenta (http://www.conservationcentre.org/scase2.html), blood release could play a role. "The cost of sociality" enhances transmission between animals within the same groups as large social groups of apes succumb more frequently than lone males (Caillaud et al. 2006). Gorillas also rest by and closely inspect very recently dead or dying members of other groups and feed after one another in the same trees and even simultaneously with chimpanzees, thus potentially enhancing virus transmission even between other social groups (Walsh et al. 2007). Hunters may then come in to contact with infected animals, kill and butcher them, butcher fresh carcasses they may find on the forest floor [though their flesh biodegrades within days on the fertile rain-forest floor (Rouquet et al. 2005)], and portage them giving ample opportunity for infection (Allaranga et al. 2010). The 1994 CIEBOV case was of an ethnologist dissecting chimpanzee carcass in the Tai National Park (Le Guenno et al. 1995; Formenty et al. 1999), and a second CIEBOV case was of a soldier adjacent to the forest in Liberia (Le Guenno et al. 1999) perhaps consuming bush meat to survive. Almost all outbreaks apart from the bat-related transmissions mentioned earlier have been initially linked to cases of hunting, slaughtering, and butchering bush meat (Pourrut et al. 2005). The remaining cases that have not been traced to bats or bush meat may involve other means of transmission, perhaps even dogs that have been shown to have high incidences of anti-*Ebola virus* IgG from outbreak areas (Allela et al. 2005). The animals are kept by villagers though not formally fed and so may feed on infected nonhuman primate carcasses and may serve as intermediate hosts though not succumbing to lethal infection themselves.

Human-to-human transmission is a relatively inefficient process and primarily occurs through contact with infectious body fluids with fomites not thought to play a significant role (Bausch et al. 2007). Since standards of hygiene and education regarding infectious disease can be lacking in outbreak areas, conditions for spreading disease can be rife, and the main way an outbreak gets a foothold in the community is via intrafamilial transmission while caring for the sick and during funeral preparations where the dead are ceremonially washed (Allaranga et al. 2010). Transmission due to unsafe practices can also occur such as the use of an unsterilized blade used on patients by traditional healers (Georges et al. 1999; Allaranga et al. 2010). In clinics and hospitals, poor medical practice is often involved in amplifying spread with the absence of protective clothing and barrier nursing procedures, plus the improper sterilization of equipment including syringes and specula shown to be a contributing factor in several outbreaks (Peters 2005; Fisher-Hoch 2005). Nosocomial spread can be high under these conditions, and health workers often constitute a large portion of the infected within such an outbreak. Interagency teams sent to contain outbreaks have devised protocols that are continually revised, updated, and improved to implement good practice (World Health Organization 2008b) that when combined with the rapid tracing of contacts and subsequent patient quarantine can successfully contain *Ebola virus* outbreaks (Muyembe and Kipasa 1995; Khan et al. 1999). However, prior to the arrival of these teams, the danger of increasingly larger outbreaks, especially in urban settings where massive populations reside, is worrisome indeed with poverty, low education, and minimal services in existence. Consequently, the major factor in reducing this risk in the future will be to train and equip the health-care staff and to educate the public in these regions to demand higher standards to help sustain the improvements (Fisher-Hoch 2005).

Another layer in the interplay between epidemiology and ecology of *Ebola virus* has been revealed by retrospective analyses of sera from humans and nonhuman primates in epidemic and non-epidemic areas to reveal anti-*Ebola virus* antibodies. Here it is important to note that subclinical infections by normally pathogenic strains, recovery from disease, contact with noninfectious antigen, emergence of nonpathogenic strains and cross-reactivity with *Ebola virus*–like antigens from other viruses, and earlier assay techniques' specificity and sensitivity may all complicate this picture. Subclinical infections in humans have been demonstrated in several individuals in the 1996 Gabon outbreak who nursed family members infected with *Ebola virus* hemorrhagic fever. These individuals mounted immune responses and demonstrable viral RNA signatures for up to 3 weeks following exposure though showed no clinical signs of infection (Leroy et al. 2000b, 2001). A 15.3% seroprevalence for anti-Zaire IgG was recently found in 4349 individuals in 220 (10.7% of total) randomly chosen villages in Gabon, with the highest rates in forested areas (Becquart et al. 2010).

Prior to this, serum surveys of populations in several endemic areas had revealed low prevalence of antibodies reactive against one, other, or both *Ebola virus* and *Marburg virus*, often with no recollection of severe disease in both endemic and seemingly non-endemic areas [for a detailed summary, see Gonzalez et al. (2005)]. Nonhuman primates have also been surveyed and often found to have prevalence of *Ebola virus* antibodies in outbreak and non-outbreak regions suggesting they too might show the same wide range of response as humans following encounters with virus or viral antigens (Leroy et al. 2004b). Thus, it may well be that *Ebola virus* infections are more common than we might think and that only the most potent strains infecting individuals unable to counter pathogenesis go on to initiate an outbreak.

A contrasting theory based on phylogenetic analysis and reference to a "molecular clock" has ZEBOV isolates descended from the original Yambuku 1976 virus and spreading essentially westwards in a wavelike form with diversity increasing with distance (Walsh et al. 2005). Furthermore, it has been suggested that the association of ZEBOV with bats is a relatively recent event over the few years since the initial 1976 outbreak (Biek et al. 2006). Sequences identified in gorillas are also derived from the 1976 event with indications of recombination events generating a different lineage to account for outbreaks that cannot be linked epidemiologically (Wittmann et al. 2007). Performing fieldwork to constantly gather samples from wildlife for viral isolation and complete sequencing is not trivial to complete the ecology jigsaw puzzle starting to be assembled. Such work combined with climatic observations surrounding outbreaks (Pinzon et al. 2004) is critical to enhance our ability to predict an outbreak before it reaches a particular area. However, the unexpected *Bundibugyo ebolavirus* outbreak is a poignant reminder that a constant state of vigilance for hemorrhagic fevers needs to be maintained by health-care workers in the field, especially since laboratory confirmation can lag several months behind outbreak verification (Wamala et al. 2010).

ACKNOWLEDGMENTS

We apologize for not including so many more excellent references owing to space limitations. We are grateful to Maria Messenger for outstanding editorial assistance.

REFERENCES

Allaranga, Y., M. L. Kone, P. Formenty et al. 2010. Lessons learned during active epidemiological surveillance of Ebola and Marburg viral hemorrhagic fever epidemics in Africa. *East Afr J Public Health* 7:30–36.

Allela, L., O. Boury, R. Pouillot et al. 2005. Ebola virus antibody prevalence in dogs and human risk. *Emerg Infect Dis* 11:385–390.

Alvarez, C. P., F. Lasala, J. Carrillo, O. Muñiz, A. L. Corbí, and R. Delgado. 2002. C-type lectins DC-SIGN and L-SIGN mediate cellular entry by Ebola virus in cis and in trans. *J Virol* 76:6841–6844.

Babst, M., D. J. Katzmann, E. J. Estepa-Sabal, T. Meerloo, and S. D. Emr. 2002a. Escrt-III: an endosome-associated heterooligomeric protein complex required for mvb sorting. *Dev Cell* 3:271–282.

Babst, M., D. J. Katzmann, W. B. Synder, B. Wendland, and S. D. Emr. 2002b. Endosome-associated complex, ESCRT-II, recruits transport machinery for protein sorting at the multivesicular body. *Dev Cell* 3:283–289.

Barrette, R. W., S. A. Metwally, J. M. Rowland et al. 2009. Discovery of swine as a host for the Reston ebolavirus. *Science* 325:204–206.

Basler, C. F. and G. K. Amarasinghe. 2009. Evasion of interferon responses by Ebola and Marburg viruses. *J Interferon Cytokine Res* 29:511–520.

Basler, C. F., A. Mikulasova, L. Martinez-Sobrido et al. 2003. The Ebola virus VP35 protein inhibits activation of interferon regulatory factor 3. *J Virol* 77:7945–7956.

Basler, C. F., X. Wang, E. Mühlberger et al. 2000. The Ebola virus VP35 protein functions as a type I IFN antagonist. *Proc Natl Acad Sci USA* 97:12289–12294.

Bausch, D. G., J. S. Towner, S. F. Dowell et al. 2007. Assessment of the risk of Ebola virus transmission from bodily fluids and fomites. *J Infect Dis* 196(Suppl 2):S142–S147.

Bavari, S., C. M. Bosio, E. Wiegand et al. 2002. Lipid raft microdomains: a gateway for compartmentalized trafficking of Ebola and Marburg viruses. *J Exp Med* 195:593–602.

Becquart, P., N. Wauquier, T. Mahlakõiv et al. 2010. High prevalence of both humoral and cellular immunity to Zaire ebolavirus among rural populations in Gabon. *PLoS One* 5:e9126.

Bermejo, M., J. D. Rodriguez-Teijeiro, G. Illera et al. 2006. Ebola outbreak killed 5000 gorillas. *Science* 314:1564.

Bhattacharyya, S., K. L. Warfield, G. Ruthel, S. Bavari, M. J. Aman, and T. J. Hope. 2010. Ebola virus uses clathrin-mediated endocytosis as an entry pathway. *Virology* 401:18–28.

Biek, R., P. D. Walsh, E. M. Leroy, and L. A. Real. 2006. Recent common ancestry of Ebola Zaire virus found in a bat reservoir. *PLoS Pathog* 2:e90.

Bock, J. O., T. Lundsgaard, P. A. Pedersen, and L. S. Christensen. 2004. Identification and partial characterization of Taastrup virus: a newly identified member species of the Mononegavirales. *Virology* 319:49–59.

Boehmann, Y., S. Enterlein, A. Randolf, and E. Mühlberger. 2005. A reconstituted replication and transcription system for Ebola virus Reston and comparison with Ebola virus Zaire. *Virology* 332:406–417.

Bowen, E. T., G. Lloyd, W. J. Harris et al. 1977. Viral hemorrhagic fever in southern Sudan and northern Zaire. Preliminary studies on the aetiological agent. *Lancet* 1:571–573.

Bradfute, S. B., D. R. Braun, J. D. Shamblin et al. 2007. Lymphocyte death in a mouse model of Ebola virus infection. *J Infect Dis* 196(Suppl 2):S296–S304.

Bray, M. and S. Mahanty. 2003. Ebola hemorrhagic fever and septic shock. *J Infect Dis* 188:1613–1617.

Brindley, M. A., C. L. Hunt, A. S. Kondratowicz et al. 2011. Tyrosine kinase receptor Axl enhances entry of Zaire ebolavirus without direct interactions with the viral glycoprotein. *Virology* 415:83–94.

Bwaka, M. A., M. J. Bonnet, P. Calain et al. 1999. Ebola hemorrhagic fever in Kikwit, Democratic Republic of the Congo: clinical observations in 103 patients. *J Infect Dis* 179(Suppl 1):S1–S7.

Caillaud, D., F. Levréro, R. Cristescu et al. 2006. Gorilla susceptibility to Ebola virus: the cost of sociality. *Curr Biol* 16:R489–R491.

Calisher, C. H., J. E. Childs, H. E. Field, K. V. Holmes, and T. Schountz. 2006. Bats: important reservoir hosts of emerging viruses. *Clin Microbiol Rev* 19:531–545.

Cárdenas, W. B., Y. M. Loo, M. Gale Jr. et al. 2006. Ebola virus VP35 protein binds double-stranded RNA and inhibits alpha/beta interferon production induced by RIG-I signaling. *J Virol* 80:5168–5178.

Carette, J. E., M. Raaben, A. C. Wong et al. 2011. Ebola virus entry requires the cholesterol transporter Niemann-Pick C1. *Nature* 477:340–343.

Centers for Disease Control. 2008. Special pathogens branch. *Questions and Answers about Ebola Hemorrhagic Fever*. Atlanta, GA: Center for Disease Control. http://www.cdc.gov/ncidod/dvrd/spb/mnpages/dispages/Fact_Sheets/Ebola_Fact_Booklet.pdf (accessed October 3, 2011).

Chan, S. Y., C. J. Empig, F. J. Welte et al. 2001. Folate receptor-alpha is a cofactor for cellular entry by Marburg and Ebola viruses. *Cell* 106:117–126.

Chandran, K., N. J. Sullivan, U. Felbor, S. P. Whelan, and J. M. Cunningham. 2005. Endosomal proteolysis of the Ebola virus glycoprotein is necessary for infection. *Science* 308:1643–1645.

Côté, M., J. Misasi, T. Ren et al. 2011. Small molecule inhibitors reveal Niemann-Pick C1 is essential for Ebola virus infection. *Nature* 477:344–348.

Demirov, D. G. and E. O. Freed. 2004. Retrovirus budding. *Virus Res* 106:87–102.

Dolnik, O., L. Kolesnikova, L. Stevermann, and S. Becker. 2010. Tsg101 is recruited by a late domain of the nucleocapsid protein to support budding of Marburg virus-like particles. *J Virol* 84:7847–7856.

Donaldson, E. F., A. N. Haskew, J. E. Gates et al. 2010. Metagenomic analysis of the viromes of three North American bat species: viral diversity among different bat species that share a common habitat. *J Virol* 84:13004–13018.

Elliott, L. H., M. P. Kiley, and J. B. McCormick. 1985. Descriptive analysis of Ebola virus proteins. *Virology* 147:169–176.

Empig, C. J. and M. A. Goldsmith. 2002. Association of the caveola vesicular system with cellular entry by filoviruses. *J Virol* 76:5266–5270.

Enterlein, S., K. M. Schmidt, M. Schümann et al. 2009. The Marburg virus 3′ noncoding region structurally and functionally differs from that of Ebola virus. *J Virol* 83:4508–4519.

Fabozzi, G., C. S. Nabel, M. A. Dolan, and N. J. Sullivan. 2011. Ebolavirus proteins suppress the effects of small interfering RNA by direct interaction with the mammalian RNA interference pathway. *J Virol* 85:2512–2523.

Falzarano, D., O. Krokhin, V. Wahl-Jensen, J. Seebach, K. Wolf, H.-J. Schnittler, and H. Feldman. 2006. Structure-function analysis of the soluble glycoprotein, sGP, of *Ebola virus*. *ChemBioChem* 7:1605–1611.

Feldmann, H., E. Mühlberger, A. Randolf et al. 1992. Marburg virus, a filovirus: messenger RNAs, gene order, and regulatory elements of the replication cycle. *Virus Res* 24:1–19.

Feldmann, H., V. E. Volchkov, V. A. Volchkova, and H. D. Klenk. 1999. The glycoproteins of Marburg and Ebola virus and their potential roles in pathogenesis. *Arch Virol Suppl* 15:159–169.

Feng, Z., M. Cerveny, Z. Yan, and B. He. 2007. The VP35 protein of Ebola virus inhibits the antiviral effect mediated by double-stranded RNA-dependent protein kinase PKR. *J Virol* 81:182–192.

Fisher-Hoch, S. P. 2005. Lessons from nosocomial viral haemorrhagic fever outbreaks. *Br Med Bull* 73–74:123–137.

Formenty, P., C. Hatz, B. Le Guenno, A. Stoll, P. Rogenmoser, and A. Widmer. 1999. Human infection due to Ebola virus, subtype Cote d'Ivoire: clinical and biologic presentation. *J Infect Dis* 179(Suppl 1):S48–S53.

Freed, E. O. 2002. Viral late domains. *J Virol* 76:4679–4687.

Geisbert, T. W., L. E. Hensley, T. R. Gibb, K. E. Steele, N. K. Jaax, and P. B. Jahrling. 2000. Apoptosis induced in vitro and in vivo during infection by Ebola and Marburg viruses. *Lab Invest* 80:171–186.

Geisbert, T. W., L. E. Hensley, T. Larsen et al. 2003a. Pathogenesis of Ebola hemorrhagic fever in cynomolgus macaques: evidence that dendritic cells are early and sustained targets of infection. *Am J Pathol* 163:2347–2370.

Geisbert, T. W. and P. B. Jahrling. 1995. Differentiation of filoviruses by electron microscopy. *Virus Res* 39:129–150.

Geisbert, T. W., H. A. Young, P. B. Jahrling et al. 2003b. Pathogenesis of Ebola hemorrhagic fever in primate models: evidence that hemorrhage is not a direct effect of virus-induced cytolysis of endothelial cells. *Am J Pathol* 163:2371–2382.

Geisbert, T. W., H. A. Young, P. B. Jahrling, K. J. Davis, E. Kagan, and L. E. Hensley. 2003c. Mechanisms underlying coagulation abnormalities in Ebola hemorrhagic fever: overexpression of tissue factor in primate monocytes/macrophages is a key event. *J Infect Dis* 188:1618–1629.

Georges, A. J., E. M. Leroy, A. A. Renaut et al. 1999. Ebola hemorrhagic fever outbreaks in Gabon, 1994–1997: epidemiologic and health control issues. *J Infect Dis* 179(Suppl 1):S65–S75.

Georges-Courbot, M. C., A. Sanchez, C. Y. Lu et al. 1997. Isolation and phylogenetic characterization of Ebola viruses causing different outbreaks in Gabon. *Emerg Infect Dis* 3:59–62.

Gonzalez, J. P., V. Herbreteau, J. Morvan, and E. M. Leroy. 2005. Ebola virus circulation in Africa: a balance between clinical expression and epidemiological silence. *Bull Soc Pathol Exot* 98:210–217.

Gonzalez, J. P., X. Pourrut, and E. Leroy. 2007. Ebolavirus and other filoviruses. *Curr Top Microbiol Immunol* 315:363–387.

Groseth, A., H. Feldmann, and J. E. Strong. 2007. The ecology of Ebola virus. *Trends Microbiol* 15:408–416.

Groseth, A., U. Ströher, S. Theriault, and H. Feldmann. 2002. Molecular characterization of an isolate from the 1989/90 epizootic of Ebola virus Reston among macaques imported into the United States. *Virus Res* 87:155–163.

Haasnoot, J., W. de Vries, E. J. Geutjes, M. Prins, P. de Haan, and B. Berkhout. 2007. The Ebola virus VP35 protein is a suppressor of RNA silencing. *PLoS Pathog* 3:e86.

Han, Z., H. Boshra, J. O. Sunyer, S. H. Zwiers, J. Paragas, and R. N. Harty. 2003. Biochemical and functional characterization of the Ebola virus VP24 protein: implications for a role in virus assembly and budding. *J Virol* 77:1793–1800.

Han, Z. and R. N. Harty. 2005. Packaging of actin into Ebola virus VLPs. *Virol J* 2:92.

Han, Z. and R. N. Harty. 2007. Influence of calcium/calmodulin on budding of Ebola VLPs: implications for the involvement of the Ras/Raf/MEK/ERK pathway. *Virus Genes* 35:511–520.

Hartman, A. L., J. E. Dover, J. S. Towner, and S. T. Nichol. 2006. Reverse genetic generation of recombinant Zaire Ebola viruses containing disrupted IRF-3 inhibitory domains results in attenuated virus growth in vitro and higher levels of IRF-3 activation without inhibiting viral transcription or replication. *J Virol* 80:6430–6440.

Harty, R. N., M. E. Brown, G. Wang, J. Huibregtse, and F. P. Hayes. 2000. A PPxY motif within the VP40 protein of Ebola virus interacts physically and functionally with a ubiquitin ligase: implications for filovirus budding. *Proc Natl Acad Sci USA* 97:13871–13876.

Hayman, D. T., P. Emmerich, M. Yu, L. F. Wang et al. 2010. Long-term survival of an urban fruit bat seropositive for Ebola and Lagos bat viruses. *PLoS One* 5:e11978.

Hood, C. L., J. Abraham, J. C. Boyington, K. Leung, P. D. Kwong, and G. J. Nabel. 2010. Biochemical and structural characterization of cathepsin L-processed Ebola virus glycoprotein: implications for viral entry and immunogenicity. *J Virol* 84:2972–2982.

Huang, Y., L. Xu, Y. Sun, and G. J. Nabel. 2002. The assembly of Ebola virus nucleocapsid requires virion-associated proteins 35 and 24 and posttranslational modification of nucleoprotein. *Mol Cell* 10:307–316.

Hunt, C. L., A. A. Kolokoltsov, R. A. Davey, and W. Maury. 2011. The Tyro3 receptor kinase Axl enhances macropinocytosis of Zaire ebolavirus. *J Virol* 85:334–347.

Ikegami, T., A. B. Calaor, M. E. Miranda et al. 2001. Genome structure of Ebola virus subtype Reston: differences among Ebola subtypes. Brief report. *Arch Virol* 146:2021–2027.

International Committee on Taxonomy of Viruses (ICTV) Report. 2002. (August). http://www.ictvonline.org/codeOfVirusClassification_2002.asp (accessed October 3, 2011).

Isaacson, M., P. Sureau, G. Courteille, and S. R. Pattyn. 1977. Clinical aspects of Ebola virus disease at the Ngaliema Hospital, Kinshasa, Zaire, 1976. In: S. R. Pattyn (Ed.), *Ebola Virus Haemorrhagic Fever*, pp. 15–20.

Jaax, N. K., K. J. Davis, T. J. Geisbert et al. 1996. Lethal experimental infection of rhesus monkeys with Ebola-Zaire (Mayinga) virus by the oral and conjunctival route of exposure. *Arch Pathol Lab Med* 120:140–55.

Jasenosky, L. D., G. Neumann, I. Lukashevich, and Y. Kawaoka. 2001. Ebola virus VP40-induced particle formation and association with the lipid bilayer. *J Virol* 75:5205–5214.

Jeffers, S. A., D. A. Sanders, and A. Sanchez. 2002. Covalent modifications of the Ebola virus glycoprotein. *J Virol* 76:12463–12472.

John, S. P., T. Wang, S. Steffen, S. Longhi, C. S. Schmaljohn, and C. B. Jonsson. 2007. Ebola virus VP30 is an RNA binding protein. *J Virol* 81:8967–8976.

Johnson, K. M., J. V. Lange, P. A. Webb, and F. A. Murphy. 1977. Isolation and partial characterisation of a new virus causing acute hemorrhagic fever in Zaire. *Lancet* 1:569–571.

Johnson, R. F., P. Bell, and R. N. Harty. 2006a. Effect of Ebola virus proteins GP, NP and VP35 on VP40 VLP morphology. *Virol J* 3:31.

Johnson, R. F., S. E. McCarthy, P. J. Godlewski, and R. N. Harty. 2006b. Ebola virus VP35-VP40 interaction is sufficient for packaging 3E-5E minigenome RNA into virus-like particles. *J Virol* 80:5135–5144.

Jouvenet, N., S. J. Neil, M. Zhadina et al. 2009. Broad-spectrum inhibition of retroviral and filoviral particle release by tetherin. *J Virol* 83:1837–1844.

Kaletsky, R. L., J. R. Francica, C. Agrawal-Gamse, and P. Bates. 2009. Tetherin-mediated restriction of filovirus budding is antagonized by the Ebola glycoprotein. *Proc Natl Acad Sci USA* 106:2886–2891.

Kaletsky, R. L., G. Simmons, and P. Bates. 2007. Proteolysis of the Ebola virus glycoproteins enhances virus binding and infectivity. *J Virol* 81:13378–13384.

Katzmann, D. J., M. Babst, and S. D. Emr. 2001. Ubiquitin-dependent sorting into the multivesicular body pathway requires the function of a conserved endosomal protein sorting complex, ESCRT-I. *Cell* 106:145–155.

Khan, A. S., F. K. Tshioko, D. L. Heymann et al. 1999. The reemergence of Ebola hemorrhagic fever, Democratic Republic of the Congo, 1995. Commission de Lutte contre les Epidémies à Kikwit. *J Infect Dis* 179(Suppl 1):S76–S86.

Kissling, R. E., R. Q. Robinson, F. A. Murphy, and S. G. Whitfield. 1968. Agent of disease contracted from green monkeys. *Science* 160:888–890.

Ko, D. C., M. D. Gordon, J. Y. Jin, and M. P. Scott. 2001. Dynamic movements of organelles containing Niemann-Pick C1 protein: NPC1 involvement in late endocytic events. *Mol Biol Cell* 12:601–614.

Kolakofsky, D., P. Le Mercier, F. Iseni, and D. Garcin. 2004. Viral DNA polymerase scanning and the gymnastics of Sendai virus RNA synthesis. *Virology* 318:463–473.

Kondratowicz, A. S., N. J. Lennemann, P. L. Sinn et al. 2011. T-cell immunoglobulin and mucin domain 1 (TIM-1) is a receptor for Zaire Ebolavirus and Lake Victoria Marburgvirus. *Proc Natl Acad Sci USA* 108:8426–8431.

Ksiazek, T. G., P. E. Rollin, A. J. Williams et al. 1999a. Clinical virology of Ebola hemorrhagic fever (EHF): virus, virus antigen, and IgG and IgM antibody findings among EHF patients in Kikwit, Democratic Republic of the Congo, 1995. *J Infect Dis* 179(Suppl 1):S177–S187.

Ksiazek, T. G., C. P. West, P. E. Rollin, P. B. Jahrling, and C. J. Peters. 1999b. ELISA for the detection of antibodies to Ebola viruses. *J Infect Dis* 179(Suppl 1):S192–S198.

Kuhn, J. H., S. Becker, H. Ebihara et al. 2010. Proposal for a revised taxonomy of the family Filoviridae: classification, names of taxa and viruses, and virus abbreviations. *Arch Virol* 155:2083–2103.

Lahm, S. A., M. Kombila, R. Swanepoel, and R. F. Barnes. 2007. Morbidity and mortality of wild animals in relation to outbreaks of Ebola haemorrhagic fever in Gabon, 1994–2003. *Trans R Soc Trop Med Hyg* 101:64–78.

Le Guenno, B., P. Formenty, and C. Boesch. 1999. Ebola virus outbreaks in the Ivory Coast and Liberia, 1994–1995. *Curr Top Microbiol Immunol* 235:77–84.

Le Guenno, B., P. Formenty, M. Wyers, P. Gounon, F. Walker, and C. Boesch. 1995. Isolation and partial characterisation of a new strain of Ebola virus. *Lancet* 345:1271–1274.

Leroy, E. M., S. Baize, P. Debre, J. Lansoud-Soukate, and E. Mavoungou. 2001. Early immune responses accompanying human asymptomatic Ebola infections. *Clin Exp Immunol* 124:453–460.

Leroy, E. M., S. Baize, V. E. Volchkov et al. 2000b. Human asymptomatic Ebola infection and strong inflammatory response. *Lancet* 355:2210–2215.

Leroy, E. M., P. Becquart, N. Wauquier, and S. Baize. 2011. Evidence for Ebola virus superantigen activity. *J Virol* 85:4041–4042.

Leroy, E. M., A. Epelboin, V. Mondonge et al. 2009. Human Ebola outbreak resulting from direct exposure to fruit bats in Luebo, Democratic Republic of Congo, 2007. *Vector Borne Zoonotic Dis* 9:723–728.

Leroy, E. M., B. Kumulungui, X. Pourrut et al. 2005. Fruit bats as reservoirs of Ebola virus. *Nature* 438:575–576.

Leroy, E. M., P. Rouquet, P. Formenty et al. 2004a. Multiple Ebola virus transmission events and rapid decline of central African wildlife. *Science* 303:387–390.

Leroy, E. M., P. Telfer, B. Kumulungui et al. 2004b. A serological survey of Ebola virus infection in central African nonhuman primates. *J Infect Dis* 190:1895–1899.

Leung, D. W., N. D. Ginder, D. B. Fulton et al. 2009. Structure of the Ebola VP35 interferon inhibitory domain. *Proc Natl Acad Sci USA* 106:411–416.

Li, L., J. G. Victoria, C. Wang et al. 2010. Bat guano virome: predominance of dietary viruses from insects and plants plus novel mammalian viruses. *J Virol* 84:6955–6965.

Licata, J. M., R. F. Johnson, Z. Han, and R. N. Harty. 2004. Contribution of Ebola virus glycoprotein, nucleoprotein, and VP24 to budding of VP40 virus-like particles. *J Virol* 78:7344–7351.

Licata, J. M., M. Simpson-Holley, N. T. Wright, Z. Han, J. Paragas, and R. N. Harty. 2003. Overlapping motifs (PTAP and PPEY) within the Ebola virus VP40 protein function independently as late budding domains: involvement of host proteins TSG101 and VPS-4. *J Virol* 77:1812–1819.

Liu, Y., L. Cocka, A. Okumura, Y. A. Zhang, J. O. Sunyer, and R. N. Harty. 2010. Conserved motifs within Ebola and Marburg virus VP40 proteins are important for stability, localization, and subsequent budding of virus-like particles. *J Virol* 84:2294–2303.

Liu, Y. and R. N. Harty. 2010. Viral and host proteins that modulate filovirus budding. *Future Virol* 5:481–491.

Luby, S. P., E. S. Gurley, and M. J. Hossain. 2009a. Transmission of human infection with Nipah virus. *Clin Infect Dis* 49:1743–1748.

Luby, S. P., M. J. Hossain, E. S. Gurley et al. 2009b. Recurrent zoonotic transmission of Nipah virus into humans, Bangladesh, 2001–2007. *Emerg Infect Dis* 15:1229–1235.

Lundsgaard, T. 1997. Filovirus-like particles detected in the leafhopper *Psammotettix alienus*. *Virus Res* 48:35–40.

Lusa, S., T. S. Blom, E. L. Eskelinen et al. 2001. Depletion of rafts in late endocytic membranes is controlled by NPC1-dependent recycling of cholesterol to the plasma membrane. *J Cell Sci* 114:1893–1900.

Mahanty, S. and M. Bray. 2004. Pathogenesis of filoviral haemorrhagic fevers. *Lancet Infect Dis* 4:487–498.

Martinez, M. J., N. Biedenkopf, V. Volchkova et al. 2008. Role of Ebola virus VP30 in transcription reinitiation. *J Virol* 82:12569–12573.

Martinez, O., J. Johnson, B. Manicassamy et al. 2010. Zaire Ebola virus entry into human dendritic cells is insensitive to cathepsin L inhibition. *Cell Microbiol* 12:148–157.

Martin-Serrano, J., T. Zang, and P. D. Bieniasz. 2001. HIV-1 and Ebola virus encode small peptide motifs that recruit Tsg101 to sites of particle assembly to facilitate egress. *Nat Med* 7:1313–1319.

McCarthy, S. E., R. F. Johnson, Y. A. Zhang, J. O. Sunyer, and R. N. Harty. 2007. Role for amino acids 212KLR214 of Ebola virus VP40 in assembly and budding. *J Virol* 81:11452–11460.

Mehedi, M., D. Falzarano, J. Seebach et al. 2011. A new Ebola virus nonstructural glycoprotein expressed through RNA editing. *J Virol* 85:5406–5414.

Miller, M. E., S. Adhikary, A. A. Kolokoltsov, and R. A. Davey. 2012. Ebolavirus requires acid sphingomyelinase activity and plasma membrane sphingomyelin for infection. *J. Virol.* 86(14):7473–7483.

Miranda, M. E., T. G. Ksiazek, T. J. Retuya et al. 1999. Epidemiology of Ebola (subtype Reston) virus in the Philippines, 1996. *J Infect Dis* 179(Suppl 1):S115–S119.

Modrof, J., S. Becker, and E. Mühlberger. 2003. Ebola virus transcription activator VP30 is a zinc-binding protein. *J Virol* 77:3334–3338.

Modrof, J., E. Mühlberger, H. D. Klenk, and S. Becker. 2002. Phosphorylation of VP30 impairs Ebola virus transcription. *J Biol Chem* 277:33099–33104.

Mühlberger, E., M. Weik, V. E. Volchkov, H. D. Klenk, and S. Becker. 1999. Comparison of the transcription and replication strategies of Marburg virus and Ebola virus by using artificial replication systems. *J Virol* 73:2333–2342.

Mulherkar, N., M. Raaben, J. C. de la Torre, S. P. Whelan, and K. Chandran. 2011. The Ebola virus glycoprotein mediates entry via a non-classical dynamin-dependent macropinocytic pathway. *Virology* 419:72–83.

Murphy, F. A., G. van der Groen, S. G. Whitfield, and J. V. Lange. 1978. Ebola and Marburg virus morphology and taxonomy. In: F. A. Murphy (Ed.), *Ebola Virus Haemorrhagic Fever*, pp. 61–82. Amsterdam, the Netherlands: Elsevier/North-Holland.

Murray, J. L., M. Mavrakis, N. J. McDonald et al. 2005. Rab9 GTPase is required for replication of human immunodeficiency virus type 1, filoviruses, and measles virus. *J Virol* 79:11742–11751.

Muyembe, T. and M. Kipasa. 1995. Ebola haemorrhagic fever in Kikwit, Zaire. International scientific and technical committee and who collaborating centre for hemorrhagic fevers. *Lancet* 345:1448.

Nanbo, A., M. Imai, S. Watanabe et al. 2010. Ebolavirus is internalized into host cells via macropinocytosis in a viral glycoprotein-dependent manner. *PLoS Pathog* 6:e1001121.

Neil, S. J., T. Zang, and P. D. Bieniasz. 2008. Tetherin inhibits retrovirus release and is antagonized by HIV-1 Vpu. *Nature* 451:425–430.

Noda, T., H. Ebihara, Y. Muramoto et al. 2006. Assembly and budding of Ebolavirus. *PLoS Pathog* 2:e99.

Noda, T., H. Sagara, E. Suzuki, A. Takada, H. Kida, and Y. Kawaoka. 2002. Ebola virus VP40 drives the formation of virus-like filamentous particles along with GP. *J Virol* 76:4855–4865.

Okumura, A., P. M. Pitha, and R. N. Harty. 2008. ISG15 inhibits Ebola VP40 VLP budding in an L-domain-dependent manner by blocking Nedd4 ligase activity. *Proc Natl Acad Sci USA* 105:3974–3979.

Panchal, R. G., G. Ruthel, T. A. Kenny et al. 2003. In vivo oligomerization and raft localization of Ebola virus protein VP40 during vesicular budding. *Proc Natl Acad Sci USA* 100:15936–15941.

Pattyn, S., G. van der Groen, G. Courteille, W. Jacob, and P. Piot. 1977. Isolation of Marburg-like virus from a case of haemorrhagic fever in Zaire. *Lancet* 1:573–574.

Peters, C. J. 2005. Marburg and Ebola—arming ourselves against the deadly filoviruses. *N Engl J Med* 352:2571–2573.

Pinzon, J. E., J. M. Wilson, C. J. Tucker, R. Arthur, P. B. Jahrling, and P. Formenty. 2004. Trigger events: enviroclimatic coupling of Ebola hemorrhagic fever outbreaks. *Am J Trop Med Hyg* 71:664–674.

Pourrut, X., B. Kumulungui, T. Wittmann et al. 2005. The natural history of Ebola virus in Africa. *Microbes Infect* 7:1005–1014.

Pourrut, X., M. Souris, J. S. Towner et al. 2009. Large serological survey showing cocirculation of Ebola and Marburg viruses in Gabonese bat populations, and a high seroprevalence of both viruses in Rousettus aegyptiacus. *BMC Infect Dis* 9:159.

Prins, K. C., W. B. Cárdenas, and C. F. Basler. 2009. Ebola virus protein VP35 impairs the function of interferon regulatory factor-activating kinases IKKε and TBK-1. *J Virol* 83:3069–3077.

Regnery, R. L., K. M. Johnson, and M. P. Kiley. 1980. Virion nucleic acid of Ebola virus. *J Virol* 36:465–469.

Reid, S. P., L. W. Leung, A. L. Hartman et al. 2006. Ebola virus VP24 binds karyopherin alpha1 and blocks STAT1 nuclear accumulation. *J Virol* 80:5156–5167.

Reid, S. P., C. Valmas, O. Martinez, F. M. Sanchez, and C. F. Basler. 2007. Ebola virus VP24 proteins inhibit the interaction of NPI-1 subfamily karyopherin alpha proteins with activated STAT1. *J Virol* 81:13469–13477.

Rodriguez, L. L., A. De Roo, Y. Guimard et al. 1999. Persistence and genetic stability of Ebola virus during the outbreak in Kikwit, Democratic Republic of the Congo, 1995. *J Infect Dis* 179(Suppl 1):S170–S176.

Rouquet, P., J. M. Froment, M. Bermejo et al. 2005. Wild animal mortality monitoring and human Ebola outbreaks, Gabon and Republic of Congo, 2001–2003. *Emerg Infect Dis* 11:283–290.

Ruthel, G., G. L. Demmin, G. Kallstrom et al. 2005. Association of Ebola virus matrix protein VP40 with microtubules. *J Virol* 79:4709–4719.

Saeed, M. F., A. A. Kolokoltsov, T. Albrecht, and R. A. Davey. 2010. Cellular entry of Ebola virus involves uptake by a macropinocytosis-like mechanism and subsequent trafficking through early and late endosomes. *PLoS Pathog* 6:1001110.

Saeed, M. F., A. A. Kolokoltsov, A. N. Freiberg, M. R. Holbrook, and R. A. Davey. 2008. Phosphoinositide-3 kinase-Akt pathway controls cellular entry of Ebola virus. *PLoS Pathog* 4:e1000141.

Sanchez, A. 2007. Analysis of filovirus entry into vero e6 cells, using inhibitors of endocytosis, endosomal acidification, structural integrity, and cathepsin (B and L) activity. *J Infect Dis* 196(Suppl 2):S251–S258.

Sanchez, A., M. P. Kiley, B. P. Holloway, and D. D. Auperin. 1993. Sequence analysis of the Ebola virus genome: organization, genetic elements, and comparison with the genome of Marburg virus. *Virus Res* 29:215–240.

Sanchez, A., M. P. Kiley, B. P. Holloway, J. B. McCormick, and D. D. Auperin. 1989. The nucleoprotein gene of Ebola virus: cloning, sequencing, and in vitro expression. *Virology* 170:81–91.

Sanchez, A. and P. E. Rollin. 2005. Complete genome sequence of an Ebola virus (Sudan species) responsible for a 2000 outbreak of human disease in Uganda. *Virus Res* 113:16–25.

Sanchez, A., S. G. Trappier, B. W. Mahy, C. J. Peters, and S. T. Nichol. 1996. The virion glycoproteins of Ebola viruses are encoded in two reading frames and are expressed through transcriptional editing. *Proc Natl Acad Sci USA* 93:3602–3607.

Sanchez, A., Z. Y. Yang, L. Xu, G. J. Nabel, T. Crews, and C. J. Peters. 1998. Biochemical analysis of the secreted and virion glycoproteins of Ebola virus. *J Virol* 72:6442–6447.

Schornberg, K., S. Matsuyama, K. Kabsch, S. Delos, A. Bouton, and J. White. 2006. Role of endosomal cathepsins in entry mediated by the Ebola virus glycoprotein. *J Virol* 80:4174–4178.

Schornberg, K. L., C. J. Shoemaker, D. Dube et al. 2009. Alpha5beta1-integrin controls ebolavirus entry by regulating endosomal cathepsins. *Proc Natl Acad Sci USA* 106:8003–8008.

Shimojima, M., Y. Ikeda, and Y. Kawaoka. 2007. The mechanism of Axl-mediated Ebola virus infection. *J Infect Dis* 196(Suppl 2):S259–S263.

Shimojima, M., A. Takada, H. Ebihara et al. 2006. Tyro3 family-mediated cell entry of Ebola and Marburg viruses. *J Virol* 80:10109–10116.

Simmons, G., J. D. Reeves, C. C. Grogan et al. 2003a. DC-SIGN and DC-SIGNR bind Ebola glycoproteins and enhance infection of macrophages and endothelial cells. *Virology* 305:115–123.

Simmons, G., A. J. Rennekamp, N. Chai, L. H. Vandenberghe, J. L. Riley, and P. Bates. 2003b. Folate receptor alpha and caveolae are not required for Ebola virus glycoprotein-mediated viral infection. *J Virol* 77:13433–13438.

Slotte, J. P., G. Hedström, and E. L. Bierman. 1989. Intracellular transport of cholesterol in type C Niemann-Pick fibroblasts. *Biochim Biophys Acta* 1005:303–309.

Stroher, U., E. West, H. Bugany, H. D. Klenk, H. J. Schnittler, and H. Feldmann. 2001. Infection and activation of monocytes by Marburg and Ebola viruses. *J Virol* 75:11025–11033.

Sui, J. and W. A. Marasco. 2002. Evidence against Ebola virus sGP binding to human neutrophils by a specific receptor. *Virology* 303:9–14.

Suzuki, Y. and T. Gojobori. 1997. The origin and evolution of Ebola and Marburg viruses. *Mol Biol Evol* 14:800–806.

Swanepoel, R., P. A. Leman, F. J. Burt et al. 1996. Experimental inoculation of plants and animals with Ebola virus. *Emerg Infect Dis* 2:321–325.

Swanepoel, R., S. B. Smit, P. E. Rollin et al. 2007. Studies of reservoir hosts for Marburg virus. *Emerg Infect Dis* 13:1847–1851.

Taniguchi, S., S. Watanabe, J. S. Masangkay et al. 2011. Reston ebolavirus antibodies in bats, the Philippines. *Emerg Infect Dis* 17:1559–1560.

Timmins, J., G. Schoehn, S. Ricard-Blum et al. 2003. Ebola virus matrix protein VP40 interaction with human cellular factors Tsg101 and Nedd4. *J Mol Biol* 326:493–502.

Timmins, J., S. Scianimanico, G. Schoehn, and W. Weissenhorn. 2001. Vesicular release of Ebola virus matrix protein VP40. *Virology* 283:1–6.

Towner, J. S., B. R. Amman, T. K. Sealy et al. 2009. Isolation of genetically diverse Marburg viruses from Egyptian fruit bats. *PLoS Pathog* 5:e1000536.

Towner, J. S., X. Pourrut, C. G. Albariño et al. 2007a. Marburg virus infection detected in a common African bat. *PLoS One* 2:e764.

Towner, J. S., P. E. Rollin, D. G. Bausch et al. 2004. Rapid diagnosis of Ebola hemorrhagic fever by reverse transcription-PCR in an outbreak setting and assessment of patient viral load as a predictor of outcome. *J Virol* 78:4330–4341.

Towner, J. S., T. K. Sealy, M. L. Khristova et al. 2008. Newly discovered Ebola virus associated with hemorrhagic fever outbreak in Uganda. *PLoS Pathog* 4:e1000212.

Tuffs, A. 2009. Experimental vaccine may have saved Hamburg scientist from Ebola fever. *BMJ* 338:b1223.

Volchkov, V. E., S. Becker, V. A. Volchkova et al. 1995. GP mRNA of Ebola virus is edited by the Ebola virus polymerase and by T7 and vaccinia virus polymerases. *Virology* 214:421–430.

Volchkov, V. E., H. Feldmann, V. A. Volchkova, and H. D. Klenk. 1998. Processing of the Ebola virus glycoprotein by the proprotein convertase furin. *Proc Natl Acad Sci USA* 95:5762–5767.

Volchkov, V., V. Volchkova, C. Eckel et al. 1997. Emergence of subtype Zaire Ebola virus in Gabon. *Virology* 232:139–144.

Walsh, P. D., K. A. Abernethy, M. Bermejo et al. 2003. Catastrophic ape decline in western equatorial Africa. *Nature* 422:611–614.

Walsh, P. D., R. Biek, and L. A. Real. 2005. Wave-like spread of Ebola Zaire. *PLoS Biol* 3:e371.

Walsh, P. D., T. Breuer, C. Sanz, D. Morgan, and D. Doran-Sheehy. 2007. Potential for Ebola transmission between gorilla and chimpanzee social groups. *Am Nat* 169:684–689.

Wahl-Jensen, V., S. K. Kurz, P. R. Hazelton et al. 2005. Role of Ebola virus secreted glycoproteins and virus-like particles in activation of human macrophages. *J Virol* 79:2413–2419.

Wamala, J. F., L. Lukwago, M. Malimbo et al. 2010. Ebola hemorrhagic fever associated with novel virus strain, Uganda, 2007–2008. *Emerg Infect Dis* 16:1087–1092.

Watanabe, S., T. Noda, P. Halfmann, L. Jasenosky, and Y. Kawaoka. 2007. Ebola virus (EBOV) VP24 inhibits transcription and replication of the EBOV genome. *J Infect Dis* 196(Suppl 2):S284–S290.

Watanabe, S., T. Noda, and Y. Kawaoka. 2006. Functional mapping of the nucleoprotein of Ebola virus. *J Virol* 80:3743–3751.

Waterman, T. 1999. Ebola Cote d'Ivoire outbreaks. Honors thesis. Stanford University. http://virus.stanford.edu/filo/eboci.html (accessed October 3, 2011).

Wauquier, N., P. Becquart, C. Padilla, S. Baize, and E. M. Leroy. 2010. Human fatal Zaire Ebola virus infection is associated with an aberrant innate immunity and with massive lymphocyte apoptosis. *PLoS Negl Trop Dis* 4 pii:e837.

Weik, M., J. Modrof, H. D. Klenk, S. Becker, and E. Mühlberger. 2002. Ebola virus VP30-mediated transcription is regulated by RNA secondary structure formation. *J Virol* 76:8532–8539.

Weissenhorn, W., A. Carfi, K. H. Lee, J. J. Skehel, and D. C. Wiley. 1998. Crystal structure of the Ebola virus membrane fusion subunit, GP2, from the envelope glycoprotein ectodomain. *Molecular Cell* 2:605–616.

Wittmann, T. J., R. Biek, A. Hassanin et al. 2007. Isolates of Zaire ebolavirus from wild apes reveal genetic lineage and recombinants. *Proc Natl Acad Sci USA* 104:17123–17127.

Wong, S., S. Lau, P. Woo, and K. Y. Yuen. 2007. Bats as a continuing source of emerging infections in humans. *Rev Med Virol* 17:67–91.

World Health Organization. 2008a. End of Ebola outbreak in Uganda (February). http://www.who.int/csr/don/2007_02_20b/en/index.html (accessed October 3, 2011).

World Health Organization. 2008b. Interim infection control recommendation for care of patients with suspected or confirmed Filovirus (Ebola, Marburg) Hemorrhagic fever. Geneva (March 2008). http://www.who.int/csr/bioriskreduction/interim_recommendations_filovirus.pdf (accessed October 3, 2011).

Yamayoshi, S. and Y. Kawaoka. 2007. Mapping of a region of Ebola virus VP40 that is important in the production of virus-like particles. *J Infect Dis* 196(Suppl 2):S291–S295.

Yamayoshi, S., T. Noda, H. Ebihara et al. 2008. Ebola virus matrix protein VP40 uses the COPII transport system for its intracellular transport. *Cell Host Microbe* 3:168–177.

Yang, Z., R. Delgado, L. Xu et al. 1998. Distinct cellular interactions of secreted and transmembrane Ebola virus glycoproteins. *Science* 279:1034–1037.

Yasuda, J., M. Nakao, Y. Kawaoka, and H. Shida. 2003. Nedd4 regulates egress of Ebola virus-like particles from host cells. *J Virol* 77:9987–9992.

25 Marburg Virus Disease

Steven B. Bradfute, Sina Bavari,
Peter B. Jahrling, and Jens H. Kuhn

CONTENTS

25.1 INTRODUCTION

Marburgviruses (Marburg virus (MARV) and Ravn virus (RAVV)) cause a severe viral hemorrhagic fever, Marburg virus disease (MVD), in humans and are the founding members of the family *Filoviridae*. With an overall case fatality rate in humans of ≈82%, marburgviruses are more lethal than their more infamous cousins, the ebolaviruses (≈68%) (Kuhn et al., 2011b). Although they are endemic in Africa, cases of marburgvirus infections have been exported to Europe on several occasions and possibly to the United States by infected humans who traveled through Africa or nonhuman primates (NHPs) that were caught and exported from there (Kuhn, 2008). Marburgviruses are therefore a global threat.

Due to the sporadic nature of MVD outbreaks, as well as the requirement of maximum-containment laboratories for experiments, much is still unknown about marburgviruses. This chapter is meant to provide a survey of what is known about pathogenesis, animal models, and experimental treatments and vaccines for marburgvirus infections. A comprehensive source of marburgvirus (and other filovirus) information can be found in Kuhn (2008).

25.2 MARBURGVIRUS CLASSIFICATION, GENOME ORGANIZATION, AND REPLICATION

Marburgviruses, the members of the genus *Marburgvirus*, contain linear, single-stranded, negative-sense RNA genomes and form enveloped virions (Kuhn, 2008). Together with ebolaviruses (genus *Ebolavirus*) and "cuevaviruses" (tentative genus "*Cuevavirus*"), they constitute the family *Filoviridae* (Adams and Carstens, 2012; Kuhn et al., 2011a). The family is named for the filamentous shape of the virions (from the Latin "filo," meaning "thread"). Marburgvirions are ≈80 nm in width and ≈800 nm long and are pleomorphic, presenting rods, clubs, "shepherd's crooks," horseshoes, or ring shapes (Figure 25.1) (Geisbert and Jahrling, 1995).

The genus *Marburgvirus* contains a single species, *Marburg marburgvirus*, which has two members, MARV and RAVV (Adams and Carstens, 2012; Kuhn et al., 2010, 2011a). Several MARV and RAVV variants have been discovered, but animal studies suggest that immunization against one variant may confer protection against another variant (Daddario-DiCaprio et al., 2006a; Swenson et al., 2008b).

The marburgvirus genomic RNA is ≈19 kb in length, uncapped, and lacks a 3′ polyadenylation tail. It consists of seven monocistronic genes (in the order 3′-*NP/VP35/VP40/GP/VP30/VP24/L*-5′), each of which is flanked by conserved transcriptional termination and initiation sequences that are different from those of ebolaviruses and "cuevaviruses." These seven genes encode seven structural proteins (NP/VP35/VP40/GP$_{1,2}$/VP30/VP24/L). In contrast to ebolaviruses, marburgviruses do not seem to express nonstructural proteins (Feldmann et al., 1992).

25.3 MARBURGVIRUS PROTEINS

NP (94 kD) is the viral nucleoprotein (NP) and functions with VP35, VP30, and L to form the nucleocapsid complex (Becker et al., 1998); the addition of the viral genomic RNA forms the ribonucleocapsid complex. NP physically interacts with VP35 and VP30 (mediating their inclusion

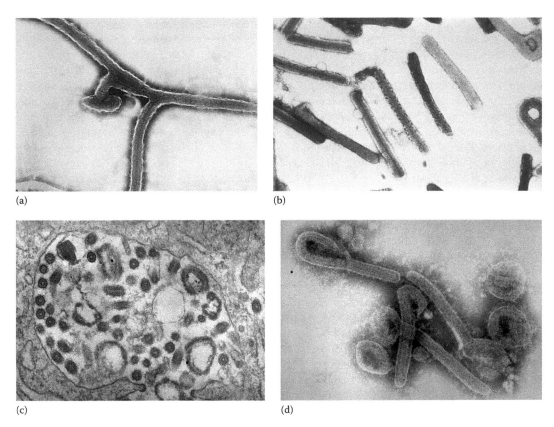

(a)

(b)

(c)

(d)

FIGURE 25.1 Electron micrographs of MARV virions. Note the pleomorphic nature of the virions. (a, b, and d): longitudinal view; (c): cross section. (Obtained from the CDC Public Health Image Library, http://phil. cdc.gov/phil/home.asp. Content providers: (a) CDC/Fred A. Murphy; (b) CDC/R. Regnery, E.L. Palmer, 1981; (c) CDC/Fred A. Murphy, Sylvia Whitfield, 1975; (d) CDC/Fred A. Murphy, J. Nakano, 1975.)

in the nucleocapsid) and encapsidates marburgviral genomes and antigenomes. It may also enhance the virion budding efficiency mediated by VP40 (Urata et al., 2007).

VP35 (32 kD) is a polymerase cofactor and is important in marburgvirus replication and transcription (Mühlberger et al., 1998). VP35 is required for the association of L with NP/genomic RNA. VP35 also inhibits type I interferon (IFN) production in human dendritic cells in vitro (Bosio et al., 2003).

VP40 (38 kD) is the major matrix protein of marburgviruses, residing in the virion between the envelope and the nucleocapsid (Bharat et al., 2011). It is responsible for budding of nascent virions from the cell (Kolesnikova et al., 2012). In fact, expression of VP40 alone is sufficient to form a self-assembling marburgvirus-like particle that is released from the cell (Kolesnikova et al., 2004; Swenson et al., 2004). Additionally, VP40 can inhibit type I IFN signaling of the host cell, which appears to be important for the virulence of marburgviruses (Valmas and Basler, 2011; Valmas et al., 2010). This function of VP40 is not apparent in the ebolavirus VP40 analogue. VP40 is also involved in recruiting nucleocapsids and $GP_{1,2}$ to the host cell plasma membrane, resulting in assembly and budding of mature virions.

The glycoprotein $GP_{1,2}$ (170–200 kD) is the single envelope protein required for virion attachment to host cells (Will et al., 1993). In contrast to ebolaviruses, which express two additional secreted nonstructural proteins from the *GP* gene (Sanchez et al., 1996; Volchkov et al., 1995), $GP_{1,2}$ is the only *GP* expression product. $GP_{1,2}$ is produced as a precursor (preGP) that is cleaved into two fragments, GP_1 and GP_2, by furin. The mature spike protein is composed of GP_1/GP_2 trimers (Volchkov et al., 2000). GP_1 comprises most of the ectodomain, which is responsible for binding to as yet unidentified cell-surface receptors (Kuhn et al., 2006). GP_2 is responsible for fusion of the viral membrane with that of the endosome (Miller and Chandran, 2012). Many receptor candidates have been identified, including C-type lectins (Becker et al., 1995; Gramberg et al., 2005; Marzi et al., 2004; Matsuno et al., 2010) and various receptor protein kinases (Kondratowicz et al., 2011). Cells without at least some of these receptors may still be infected with marburgviruses, however, suggesting that $GP_{1,2}$ can engage multiple receptors for surface attachment. Recent studies demonstrated that Niemann–Pick C1 (NPC1), an endosome-residing transporter of cholesterol, is an intracellular receptor for marburgviruses (as well as ebolaviruses) (Carette et al., 2011; Miller et al., 2012). NPC1 is required for exit of filoviruses from the endosome into the cytoplasm. Cells from patients with mutations in NPC1, or molecules that inhibit NPC1 function, also inhibit marburgvirus infection in vitro (Carette et al., 2011; Miller et al., 2012). $GP_{1,2}$ also serves as an antagonist to the cellular IFN-inducible antiviral protein tetherin (Jouvenet et al., 2009).

VP30 is a part of the nucleocapsid and binds to NP. It is important in amplification of viral transcription and replication (Fowler et al., 2005), although little else is known about this protein.

VP24 (24 kD) is a matrix protein that, with VP40, bridges the viral ribonucleocapsid and the envelope. Although it is not required for viral replication or transcription, knockdown of VP24 in marburgvirus-infected cells reduces output of infectious virions, suggesting a role in virion egress from the cell (Bamberg et al., 2005). Little is known about the function of VP24, in part because non-filoviruses, with the possible exception of "nyaviruses" (Herrel et al., 2012), do not have analogues of this protein.

L (220 kD) is the catalytic domain of the marburgviral RNA-dependent RNA polymerase (Mühlberger et al., 1992). It forms a replication complex with VP35, VP30, and NP (Mühlberger et al., 1998). L can generate positive-strand mRNA for viral genes, as well as negative-stranded RNA genomes or positive-strand RNA antigenomes during replication.

25.4 MARBURGVIRUS LIFE CYCLE

The glycoprotein ($GP_{1,2}$) is embedded into the virion envelope and coats the outside of the virion. It is responsible for attachment and entry of the virion into the host cell. Binding of the virion to a cell via $GP_{1,2}$ induces endocytosis, and fusion of the envelope with the endosomal membrane results in release of the ribonucleocapsid into the cytoplasm. Replication and transcription of the marburgvirus RNA is mediated by the ribonucleocapsid complex, consisting of the NP, polymerase cofactor (VP35), transcriptional activator (VP30), RNA-dependent RNA polymerase (L), and the viral RNA genome (Becker et al., 1998; Mühlberger et al., 1998). The ribonucleocapsid complex generates mRNA using the negative-stranded RNA genome as a template. Transcription occurs continuously, starting at the most 3′ gene (*NP*), and either stops after one of the genes or continues to the end of the genome. Therefore, genes located closer to the 3′ end of the genome are expressed in greater quantities than those closer to the 5′ end. After translation of viral mRNA, when NP reaches a critical concentration, the ribonucleocapsid complex begins to generate positive-stranded antigenomic RNA, which is then transcribed to form negative-stranded genomic progeny RNA (Becker et al., 1998; Mühlberger et al., 1998). The translated viral proteins and genomic RNA assemble near the cell membrane, and infectious virions are generated and released from the cell in a process that is not yet clearly understood. To achieve virion egress, MARV usurps two cellular vesicular transport pathways. The COPII pathway is used to transport the marburgvirus matrix protein VP40 to multivesicular bodies (Yamayoshi et al., 2008), from where marburgvirions bud (Kolesnikova et al., 2004). The ESCRT pathway facilitates the latter process (Dolnik et al., 2010; Kolesnikova et al., 2009).

25.5   DIFFERENCES BETWEEN MARBURGVIRUSES AND EBOLAVIRUSES

Although marburgviruses are similar in many ways to ebolaviruses, it is clear that there are many differences that are pertinent. From a genomic standpoint, marburgviruses do not express nonstructural proteins from the *GP* gene, while ebolaviruses and "cuevaviruses" express two (sGP and ssGP), as well as a secreted peptide called Δ-peptide (Basler et al., 2000; Sanchez et al., 1996; Volchkova et al., 1999). The functions of these products are an enigma. Marburgviruses utilize different methods to inhibit host cell type I IFN signaling (VP40) (Valmas and Basler, 2011; Valmas et al., 2010) than ebolaviruses (VP24 and VP35) (Basler et al., 2000, 2003; Reid et al., 2006, 2007). Marburgvirions also tend to be shorter than ebolavirions (Geisbert and Jahrling, 1995).

Pathogenesis in humans appears to be different between ebolavirus and marburgvirus infections. Most notably, cerebral edema and encephalopathy are often found to be severe effects of marburgvirus, but not ebolavirus, infections in humans (Bechtelsheimer et al., 1969; CDC, 2009; Gedigk et al., 1968; Stille et al., 1968; vanPaassen et al., 2012). However, this difference could be due to the more modern clinical setting and diagnostic methods available in European (German, Dutch) hospitals that had to deal with exported cases of MVD, in contrast to ebolavirus infections, which thus far have not been treated in modern hospitals.

25.6   MARBURG VIRUS DISEASE OUTBREAKS IN HUMANS

Symptoms of marburgvirus infections in humans include fever, fatigue, myalgia, vomiting, and diarrhea followed by the appearance of a rash (visible in light-skinned people), coagulopathy, and often confusion or cerebral edema. The outbreaks and the symptoms noted in each one are described as follows and in Table 25.1 and Figures 25.2 and 25.3.

25.6.1   Original Outbreak, West Germany and Yugoslavia, 1967

An outbreak of hemorrhagic fever occurred in Marburg and Frankfurt, West Germany (now Germany), and Belgrade, Yugoslavia (now Serbia), in workers handling blood and tissues from grivets (*Chlorocebus aethiops*) imported at roughly the same time from the same location in Uganda (Martini et al., 1968; Siegert et al., 1967; Stille et al., 1968). In all, 31 people were infected, and 7 died. The majority of patients (25) had direct contact with monkey tissues or cells, while the remainder contracted the disease from the infected patients. The incubation period of the disease ranged from 4 to 9 days, and disease began with an abrupt onset of fever, malaise, and severe headache. Nausea, vomiting, and diarrhea followed. Around 5–7 days after onset of illness, a characteristic maculopapular rash on the body (as well as enanthema) appeared. Laboratory tests revealed leukopenia (followed by leukocytosis) with left shift and appearance of activated lymphocytes. Another characteristic finding was platelet depletion, and a third of patients had accompanying bleeding from the nose, gums, and needle puncture sites. Aspartate aminotransferase (AST) and alanine aminotransferase (ALT) concentration in the serum increased, indicating liver damage. Kidney malfunction was noted in severe cases. Confusion and symptoms of nervous system alterations were also observed.

Antibiotics were given to patients, but did not have an effect on the outcome of disease. Treatments for the bleeding abnormalities were given, including platelets, clotting factors, blood, vitamin K, and fibrinogen. Electrolytes and, in some cases, albumin were given, as well as noramidopyrine methanesulfonate to reduce fever. Notably, four patients were given convalescent sera from recovered patients and had an overall mild disease course (Stille et al., 1968).

In fatal cases, autopsies revealed necroses in multiple organs, including liver, spleen, lymph nodes, kidneys, sex organs, adrenals, and pancreas. Lymphoid tissues did show evidence of proliferating

TABLE 25.1
History of MVD Outbreaks

Location	Year	# Human Cases	Case Fatality Rate (%)	Virus	Likely Source of Initial Infection	Person–Person Spread?	Treatment of Patients
Uganda → West Germany, Yugoslavia	1967	31	23	MARV	Imported grivets	Yes	Supportive care, antibiotics, clotting factors, blood, convalescent sera
Rhodesia	1975	3	33	MARV	?	Yes	Supportive care, heparin
Kenya	1980	2	50	MARV	Bats?	Yes	Fluids, antibiotics, antimalarials
Kenya	1987	1	100	RAVV	Bats?	No	Supportive care
USSR (laboratory accident)	1988	1	100	MARV	Laboratory accident	No	Not reported
USSR (laboratory accident)	1990	1	0	MARV	Laboratory accident	No	Plasma filtration
Democratic Republic of the Congo	1998–2000	154	83	MARV, RAVV	Bats?	Yes	Not reported
Angola	2004–2005	252	90	MARV	?	Yes	Antibiotics, antimalarials, antiemetics, and antipyretics
Uganda	2007	4	25	MARV, RAVV	Bats	Unknown	N/A
Uganda → Netherlands	2008	1	100	MARV	Bats?	No	N/A

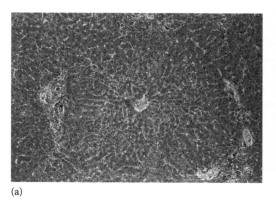

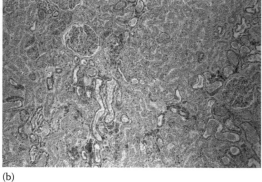

(a)   (b)

FIGURE 25.2 (**See color insert.**) Histopathology of liver and spleen. (a) Liver section from a patient that succumbed to MVD. Note bleeding into the sinusoids. (b) Kidney section from same patient. Glomerular fibrin thrombosis and tubular necrosis are present. (Obtained from the CDC Public Health Image Library, http://phil.cdc.gov/phil/home.asp. Content providers: CDC/J. Lyle Conrad, 1975.)

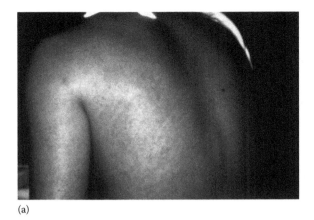

(a)

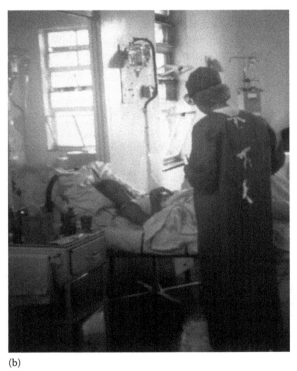

(b)

FIGURE 25.3 (**See color insert.**) (a) Characteristic rash found in MVD patient. This individual survived infection. (b) Same patient in isolation ward in hospital in South Africa. The doctor is wearing personal protective equipment to minimize the chance of infection. (Obtained from the CDC Public Health Image Library, http://phil.cdc.gov/phil/home.asp. Content providers: CDC/J. Lyle Conrad, 1975.)

lymphocytes indicative of an immune response. Cerebral edema and glial nodule encephalitis were evident (Bechtelsheimer et al., 1969; Gedigk et al., 1968), and renal damage and a "hemorrhagic state" were observed.

A number of known diseases (including yellow fever, leptospirosis, Q fever, syphilis, and rickettsioses) were ruled out by diagnostic testing (Siegert et al., 1968b), guinea pig infection tests, or ineffectiveness of antibiotics. Electron microscopy and animal experiments revealed that the infectious disease was caused by a rod-shaped virus subsequently named Marburg virus (Siegert et al., 1967).

25.6.2 Rhodesia, 1975

In 1975, an Australian man who had traveled through Rhodesia (now Zimbabwe) died from MVD after being hospitalized in South Africa (Gear et al., 1975). His traveling companion and a nurse treating him also contracted the disease after his death, but survived the infection. Notably, unlike the fatal case, these two individuals received supportive care, including prophylactic heparin to suppress intravascular coagulation. All three patients exhibited similar symptoms to those reported in the original outbreak in West Germany and Yugoslavia. The fatal case was due to liver failure and disseminated intravascular coagulation.

25.6.3 Kenya, 1980

A man working in Kenya was hospitalized with fever, myalgia, headache, and malaise, followed by vomiting and diarrhea two weeks after he had visited Kitum Cave in Mount Elgon National Park (Smith et al., 1982). He was treated with antibiotics and fluids, but his condition deteriorated, with hemorrhagic symptoms, including vomiting blood, diathesis, and bleeding from mouth and nose. He had increased AST and ALT concentrations and gastrointestinal hemorrhage. He succumbed to the illness, and necropsy revealed necrosis of the liver. Liver failure alongside hemorrhagic complications was determined to be the cause of death. MARV was detected postmortem by electron and fluorescent microscopic analysis of kidney tissue.

A physician attempted to resuscitate the patient by tracheal intubation and as a result was exposed to large quantities of the patient's blood on his hands. Nine days later, the physician became ill, with sore throat, headache, backache, and fever. He was treated with an antibiotic and an antimalarial. Four days later, he began vomiting and had severe diarrhea, followed by decreased kidney function and bloody stools. MARV was detected 17 days after the doctor's illness, but no virus was found in individuals who had contact with the patients. The physician recovered, but MARV was still detectable in his semen 2 months after the end of symptoms.

25.6.4 Kenya, 1987

A teenage Dutch boy, traveling with his family in Kenya, explored Kitum Cave in Mount Elgon National Park. Nine days later, he became sick and eventually died, exhibiting symptoms of marburgvirus infection (Johnson et al., 1996). The virus isolated from his tissues was found to be genetically distinct from MARV and is today referred to as RAVV.

25.6.5 Democratic Republic of the Congo, 1998–2000

A large, 2-year outbreak of marburgvirus infection in humans occurred in the Democratic Republic of the Congo from 1998 to 2000 (Bausch et al., 2006). In all, 154 individuals demonstrated disease, with an 83% case fatality rate. Fifty-two percent of the patients were gold miners, and many of these worked in proximity to bats, which are suspected to be the source of infection. Direct person-to-person spread was likely, as many of the other patients were family members of the miners. Sequencing of the obtained virus isolates suggested that nine or more distinct genetic marburgvirus (both MARV and RAVV) lineages were present over the course of the outbreak. Symptoms were characteristic, including fever, malaise, nausea, headache, anorexia, vomiting blood, and bleeding (Bausch et al., 2006). Interestingly, the rash observed during the original outbreak in Europe was not observed in many patients, most likely due to the dark skin tone of African patients. Of all symptoms noted, few were indicative of survival or death, with the odd exceptions of hiccups (found in 18% of survivors and 44% of fatalities) and red eyes (found in 14% of survivors and 42% of fatalities). It should be noted, though, that record keeping on the treatments and symptoms was poor (Colebunders et al., 2007).

25.6.6 Angola, 2004–2005

The MVD outbreak in Angola during 2004–2005 has been the largest to date, with 252 patients and a lethality of 90% (Towner et al., 2006). Little clinical data has been published from this outbreak. In contrast to the outbreak in the Democratic Republic of the Congo, genetic sequencing suggested that the outbreak was probably caused by a single virus introduction into the human population, followed by person-to-person transmission (Towner et al., 2006). Treatment at best consisted of antibiotics, antimalarials, antiemetics, and antipyretics (Jeffs et al., 2007).

25.6.7 Uganda, 2007

In 2007, four miners in Uganda were infected with marburgviruses (Adjemian et al., 2011). One died, but the other three survived. Sequencing suggested that two separate virus introductions (three cases of MARV and one of RAVV) occurred. Subsequently, studies were performed on bats found in this mine, and both viruses were isolated from these animals (Towner et al., 2009). Furthermore, sequencing suggested that the two viruses found in the miners were also found in the bats from that cave. This is the clearest evidence to date for bats (*Rousettus aegyptiacus*) being the natural carriers of marburgvirus and being responsible for transfer to humans or NHPs (but see Chapter 7 for a more thorough discussion on filovirus ecology).

25.6.8 The Netherlands, 2008

In 2008, a Dutch woman who had traveled in Uganda and visited a cave became ill after returning to the Netherlands. She had fever and chills, followed by liver and kidney failure, rash, hemorrhaging, conjunctivitis, and diarrhea (Timen et al., 2009; vanPaassen et al., 2012). The woman succumbed to disease and died due to complications from cerebral edema. Fellow travelers, hospital workers, and other contacts did not contract the disease.

25.6.9 Suspected Case in the United States, 2008

In 2008, a woman went on a 2-week safari in Uganda, including a visit to the same cave implicated in the Dutch case described previously. Upon her return to the United States, she developed chills, diarrhea, vomiting, headache, nausea, and a rash (CDC, 2009). A few days later, she complained of fatigue, continuing diarrhea, confusion, and weakness, and tests revealed elevated AST, ALT, and creatinine concentrations and leukopenia. Over the course of illness, she also demonstrated coagulopathy, pancreatitis, and encephalopathy (CDC, 2009). She was given fluids and antibiotics. Interestingly, testing for marburgviruses was negative. The patient eventually recovered, although confusion, pain, and fatigue persisted for over a year. Subsequent retrospective testing (performed after discovery of MARV infection in the Dutch patient) suggested that the patient had been infected with MARV, although no virus isolate was obtained. Interestingly, six of her traveling companions that also entered the cave tested negative for antibodies against marburgviruses. No hospital staff or other contacts appeared to have been infected.

25.6.10 Laboratory Infections

Several Russian laboratory research workers have become infected with marburgviruses (Alibek and Handelman, 1999; Kuhn, 2008). One was infected in 1988, when his coworker slipped during an animal injection and accidentally pricked him with a needle contaminated with MARV. The individual died several days later. A second case, associated with treatment of this case, may or may not have occurred, and may or may not have been lethal, depending on sources (Alibek and Handelman, 1999; Leitenberg et al., 2012). In 1990, another laboratory infection occurred, but the patient survived after treatment with plasma filtrations. A relapse occurred over 50 days later, but the patient survived this episode as well (Nikiforov et al., 1994).

25.7   MARBURGVIRUS HOSTS, RANGE, AND EPIDEMIOLOGY

Marburgvirus infections occur in arid woodlands in Equatorial Africa, including parts of Uganda, Zimbabwe, Kenya, the Democratic Republic of the Congo, and Angola (Kuhn, 2008). Individuals traveling through these endemic areas have carried marburgviruses back to their home countries where the viruses are not endemic. The original outbreaks in West Germany and Yugoslavia were caused by contact with infected grivets imported from Uganda, whereas people traveling through Uganda or Rhodesia have contracted the disease and become sick after returning to their home countries or touring other countries (the Netherlands, South Africa, and the United States).

One type of fruit bats, Egyptian rousettes (*Rousettus aegyptiacus*), has been proven to be a carrier of MARV and RAVV, as infectious viruses have been recovered from them (Towner et al., 2009). These bats do not appear to be harmed by marburgvirus infection. Many human MVD outbreaks can be linked to individuals who had been in mines or caves where bats, including Egyptian rousettes, reside. Egyptian rousettes have a wide geographic range, including most of Africa, the Middle East, and parts of India and Pakistan, suggesting that marburgviruses could be endemic to these areas as well. However, other factors, such as changes in weather patterns, ecosystems, and redistribution of people due to wars, may also have an impact on the distribution of marburgvirus infections (Kuhn, 2008).

NHPs are also known hosts of marburgviruses, but it appears that they usually succumb to infection. Other animals may very likely prove to be carriers or intermediate hosts for marburgviruses, but as yet none have been proven.

Primary transmission to humans appears to occur through close contact with infected animals (such as bats or NHPs) and their tissues or excretions. Human-to-human transmission can also occur through close contact, as family members and medical staff are often infected by relatives or patients. Although laboratory aerosol infection of NHPs results in disease, there is no evidence that this is a natural mode of transmission.

25.8   DETECTION

Several methods are available for the detection of marburgvirus infections. During the 1967 outbreak, researchers utilized convalescent sera from humans and guinea pigs to detect viral antigen in tissues (Siegert et al., 1968b; Slenczka et al., 1968), and later improvements on this technique provided ELISAs to test for the presence of MARV in blood [reviewed in Kuhn (2008)]. Polymerase chain reaction (PCR), a more sensitive assay to detect viral RNA, was developed later (Gibb et al., 2001). In fact, a mobile testing apparatus was used to screen blood and oral swabs from patients during the 2004–2005 Angola outbreak and could provide accurate results in less than 4 h (Grolla et al., 2011). Retrospective analysis of suspected cases can be performed by testing serum for antibodies against marburgviruses (2009).

25.9   ANIMAL MODELS

Animal models have been indispensable for the study of marburgviruses. To date, NHPs, guinea pigs, hamsters, and mice have been used to model disease (Table 25.2).

25.9.1   NHPs

Using blood and tissues obtained from infected human patients, investigators discovered that marburgviruses were lethal when injected into NHPs (Siegert et al., 1968c; Simpson et al., 1968).

TABLE 25.2
Lethality of Marburgvirus Infection in Animal Models

	Suckling Mice	Adult Mice	NHPs	Hamsters	Guinea Pig	Rodent-Adapted Variants?
MARV	Yes (Hofmann and Kunz 1970)	No (Bray 2001)	Yes (Bray 2001, Carrion et al., 2011, Simpson 1969, Warfield et al., 2007, Zlotnik and Simpson 1969)	Yes (Simpson, 1969, Zlotnik and Simpson 1969)	No (Siegert et al., 1968a)	Guinea pig (Hevey et al., 1997, Siegert et al., 1968c), mouse (Lofts et al., 2011)
RAVV	ND	No (Bray 2001)	Yes (Rhesus monkeys and crab-eating macaques) (Bray 2001, Johnson et al., 1996, Warfield et al., 2009)	ND	No (Hevey et al., 1997)	Guinea pig (Wang et al., 2006), mouse (Warfield et al., 2009)

Source: Modified from Bradfute, S.B. et al., *Hum. Vaccine*, 7, 701, 2011.

The animals perished ≈6–9 days after subcutaneous or intraperitoneal infection and had fever, weight loss, apathy, and similar histopathological characteristics as human patients. Other signs included mild rash, mild coagulation defects, and thrombocytopenia. Necrotic lesions in liver and depletion of lymphocytes and necrosis in spleen were also prominent.

Later studies described thrombocytopenia; rash; increased concentrations of AST, ALT, ASP, bilirubin, creatinine, blood urea nitrogen, and γ-glutamyltransferase; and lymphopenia followed by leukocytosis (Geisbert et al., 2007). Plasma protein C concentrations were found to decrease, while D-dimer concentrations and fibrin tissue deposition increased in lethal infection. A decrease in the percentage of blood natural killer (NK) cells, and an increase in B-cell percentage, was also found in lethal infection (Fritz et al., 2008).

MARV can also lethally infect NHPs by the aerosol route (Alves et al., 2010; Lub et al., 1995). This was accompanied by viral replication throughout the body, fever, and leukopenia; lymphocyte death via apoptosis was also found, along with characteristic histopathological lesions. The infection by the aerosol route is not known to occur naturally, but is important due to concern that marburgviruses could be misused as bioweapons (Borio et al., 2002).

25.9.2 GUINEA PIGS

MARV from human samples during the original outbreak was found to induce fever and weight loss when injected into guinea pigs (Siegert et al., 1968c). However, the animals survived and became resistant to reinfection. If the guinea pig blood was taken during the febrile period and injected into naïve guinea pigs (passaged), the disease became more severe. After 4–6 passages, the virus was lethal to guinea pigs, and induced coagulopathy, fever followed by a drop in temperature, and increased AST (Siegert et al., 1968c). Nearly 30 years later, a similar approach was taken using marburgviruses from various outbreaks (Hevey et al., 1997). Analysis of one of these strains showed only four amino acid changes in the guinea pig-adapted virus compared to the parent virus; one of these was in VP40, and three were in L (Lofts et al., 2007).

The disease produced in guinea pigs was found to be similar in many respects to that of NHPs and humans. Fever, high viral titer, and liver and kidney dysfunction were all present in infected guinea pigs. Platelet loss was also prominent, and fibrin deposition in tissues was present. The spleen was enlarged and contained lymphoblasts alongside debris from dying white blood cells. The liver contained necrotic lesions and degenerating hepatocytes.

The guinea pig model served a valuable role in the original generation of antisera for detection of MARV and as the animal model used to satisfy Koch's postulates and identify MARV as a novel pathogen (Siegert et al., 1968c; Slenczka et al., 1968). Additional research has utilized these animals for vaccine and therapeutic studies, as is described later. The main drawback of the guinea pig model is that there are few reagents available for detailed immunological studies and that transgenic or knockout animals are not available.

25.9.3 LABORATORY MICE

Marburgviruses do not induce disease in adult immunocompetent laboratory mice. They are lethal after intracerebral infection of newborn mice (Siegert et al., 1968c), as are many viruses, likely due to the immunodeficient state of these mice. However, adult mice lacking type I IFN pathways also succumb to marburgvirus infection. Additionally, mice lacking B and T cells (severe combined immunodeficiency, or SCID, mice) succumb to infection but only after ≈30 days. By passaging marburgviruses in SCID mice, variants that rapidly killed SCID mice (within 10 days) were generated (Warfield et al., 2007). These mice were used to test the ability of antibodies or antisense RNA molecules to prevent or delay death.

"SCIDapted" RAVV was then passaged in wild-type immunocompetent mice to generate a lethal mouse-adapted strain (Warfield et al., 2009). This mouse model of marburgvirus infection is similar in many ways to the NHP models or human disease: liver and kidney damage, lymphocyte death, lymphopenia, platelet loss, D-dimer concentration increases, and coagulopathy. However, these mice can only be lethally infected via the intraperitoneal route, and the hemorrhage does not seem as severe as in NHP models.

Genetic sequencing of the mouse-adapted marburgvirus revealed mutations leading to 19 amino acid changes compared to the parent virus (Lofts et al., 2011). Interestingly, the same VP40 mutation that occurs in guinea pig–adapted marburgvirus is present in the mouse virus.

25.9.4 HAMSTERS

Hamsters have been shown to succumb to infection by MARV that had been passaged nine times in guinea pigs, three times in NHPs, and nine times in hamsters (Simpson, 1969; Zlotnik and Simpson, 1969). Mortality was not complete, however, and intracranial infection was more apt to produce severe disease than intraperitoneal infection. Very little experimentation has been done in this model.

25.10   VACCINES

Several vaccines have been developed that protect NHPs from lethal marburgvirus disease (Bradfute et al., 2011). However, no vaccines have yet been approved for use in humans. Information on the most important marburgvirus vaccine candidates that have been successful in NHPs is presented in Table 25.3.

25.10.1   VRP

The first successful candidate vaccine reported to protect NHPs from MVD was a replication-defective Venezuelan equine encephalitis virus (VEEV) replicon particle (VRP) expressing MARV

TABLE 25.3

Summary of Marburgvirus Candidate Vaccines That Are Efficacious in NHPs

Platform	Viruses Tested	Antibody Titers[a]	Neutralizing Antibody Titers	T-cell Responses	References
Replication-competent VSIV	MARV, RAVV	Low to moderate	Undetectable to low	Not detected	Daddario-DiCaprio et al. (2006a), Geisbert et al. (2009), Jones et al. (2005)
Replication-incompetent VRP	MARV	Low to moderate	None	Not tested	Hevey et al. (1998)
Replication-incompetent AdV	MARV	Moderate to high	Undetectable to low	CD4+ and CD8+	Geisbert et al. (2010a), Swenson et al. (2008b)
DNA	MARV	Moderate	Undetectable to low	CD4+ and CD8+	Geisbert et al. (2010a), Riemenschneider et al. (2003)
DNA prime/ AdV boost	MARV	Moderate to high	Undetectable to low	CD4+ and CD8+	Geisbert et al. (2010a)
VLPs	MARV, RAVV	Moderate to high	Not tested	Not tested	Swenson et al. (2008b)

Source: Modified from Bradfute, S.B. et al., *Hum. Vaccine*, 7, 701, 2011.

[a]   Low titers are defined as endpoint dilution of 300–5,000; moderate as 5,000–50,000; and high as 50,000 or above.

GP$_{1,2}$ and/or NP (Hevey et al., 1998). Transgenes (in this case, MARV GP$_{1,2}$ or NP) replace structural proteins of VEEV, and the resultant vector undergoes a single cycle of replication while expressing high levels of the MARV transgene in infected cells. VRP vaccines expressing GP$_{1,2}$, NP, or VP35 protected guinea pigs from developing MVD, but VP30- or VP24-expressing VRPs were not efficacious. NHPs vaccinated with GP$_{1,2}$, NP, or both were also protected from lethal disease, but only GP$_{1,2}$ or NP-/GP$_{1,2}$-vaccinated animals were free from disease or MARV replication. The vaccine was protective without generating neutralizing antibodies, although low to moderate non-neutralizing antibodies were detected.

25.10.2 DNA and Adenovirus

Plasmid-based vaccines have been used alone or in combination with adenovirus (AdV)-based vaccines. Plasmid DNA can be injected into recipients and taken up into cells, where the DNA expresses the gene of interest. Modifications of the injection protocol have resulted in improved efficacy of DNA vaccines. Replication-defective AdV have been used in many human studies. Transgenes can be used to replace AdV genes that are needed for replication, resulting in a non-replicating vector that expresses high levels of the transgene.

Early studies showed that DNA-only vaccines expressing MARV GP$_{1,2}$ protected 4/6 NHPs from MVD (Riemenschneider et al., 2003). Vaccination induced low titers of antibodies in all animals.

Later studies using codon-optimized DNA plasmid vaccines with backbones that increased MARV GP$_{1,2}$ expression were found to protect 4/4 NHPs from disease, suggesting that these changes were more effective in providing protection (Geisbert et al., 2010a). This same study compared DNA-only vaccines, AdV-only vaccines, and DNA-prime/AdV-boost vaccines for antibody induction, T-cell responses, and overall survival. All vaccines expressed MARV GP$_{1,2}$. The DNA-only group was vaccinated four times, the AdV-only group 1 time, and the DNA-prime/AdV-boost group was vaccinated three times with DNA and once with AdV. Interestingly, all vaccinated NHPs survived MARV infection regardless of the regimen used. The only group that did not show viremia or any clinical signs of illness was the AdV-only group. In the DNA-only and DNA/AdV groups, there were some mild to moderate signs. In these groups, those animals that showed signs tended to have poorer CD8$^+$ T-cell responses than their asymptomatic cohorts. Neutralizing antibody generation was low to undetectable in all groups, and overall antibody titers were highest in the DNA/AdV groups.

A modified AdV vaccine that can harbor more than one transgene was used to test a multivalent vaccine targeting several filoviruses (Swenson et al., 2008a). Four different candidate vaccines were combined, expressing two MARV GP$_{1,2}$s, one RAVV GP$_{1,2}$, MARV NP, Ebola virus (EBOV) NP, and Sudan virus (SUDV) GP$_{1,2}$. This vaccine was used to vaccinate NHPs, which were then infected with various filoviruses. All animals survived, showing that a multivalent vaccine can generate protection against several different filoviruses.

25.10.3 VSIV

Live attenuated vesicular stomatitis Indiana virus (VSIV) vectors have been very effective against marburgviruses. In this vaccine platform, the glycoprotein of VSIV (G) is replaced with a marburgvirus GP$_{1,2}$. The resulting vaccine replicates in cells, but is attenuated. One injection of VSIV expressing MARV GP$_{1,2}$ protects NHPs from infection in the absence of any marburgvirus signs or viremia. One round of vaccination is sufficient for protection of NHPs against marburgvirus infection (Jones et al., 2005). Immunological analysis showed low or undetectable neutralizing antibody generation, low to moderate total antibody titers, and undetectable T-cell responses. Protection was also seen against aerosol MARV infection, an important result in the light of fears of a biological attack (Geisbert et al., 2008). VSIV/MARV vaccination protected against RAVV-induced disease, as well as against infection with heterotypic MARV variants (Daddario-DiCaprio et al., 2006a). A candidate vaccine comprised of four different VSIV vectors (expressing MARV $GP_{1,2}$, SUDV $GP_{1,2}$, Taï Forest virus (TAFV) $GP_{1,2}$, or EBOV $GP_{1,2}$) protected recipient NHPs from disease caused by these filoviruses (Geisbert et al., 2009).

The VSIV vaccine is also effective postexposure, likely due to its ability to replicate. When given within 30 min after MARV infection, VSIV expressing MARV $GP_{1,2}$ prevented the development of disease in NHPs (Daddario-DiCaprio et al., 2006b). These animals did not become viremic and did not show clinical signs. T-cell responses were not detected, and low antibody titers were generated beginning on day 6. Further experiments showed the vaccine is protective at later time points, with 83% survival of NHPs after 24 h, and 33% after 48 h (Geisbert et al., 2010b).

25.10.4 VLPs

Virus-like particles (VLPs) are generated by expression of VP40, resulting in particles that have the morphology of virions but cannot replicate (Bavari et al., 2002). Addition of $GP_{1,2}$ provides a target for neutralizing antibodies. VLPs (made of MARV VP40, $GP_{1,2}$, and NP) plus adjuvant protected NHPs from RAVV-induced disease and from disease caused by two different MARV variants (Swenson et al., 2008b). Animals did not have viremia, although one animal did have an increased ALT concentration, which resolved. High titers of antibodies were produced by VLP vaccination.

25.11 THERAPEUTICS

25.11.1 ANTICOAGULANTS

Due to the bleeding disorder present in MVD, treatment targeting coagulation pathways is an attractive therapy (Martini et al., 1968; Stille et al., 1968). Recombinant nematode anticoagulant protein C (rNAPc2), a protein that inhibits coagulation, provided only mild protection to NHPs. While 6/6 control NHPs died, 1/6 survived after rNAPc2 treatment, and 5/6 treated NHPs died with a slightly increased delay in time-to-death (Geisbert et al., 2007). Administration of platelets and other clotting factors, which was done during the 1967 outbreak (Stille et al., 1968), has not been tested in animal models to date.

25.11.2 CONVALESCENT SERA

Support for the idea that antibodies could serve as therapeutics for MVD was observed in the 1967 outbreak, when four patients received convalescent sera and developed only mild disease lacking AST/ALT concentration increases, kidney issues, or cerebral edema (Stille et al., 1968). However, no controlled studies were conducted, so it was impossible to prove that administration of convalescent sera was the reason for disease attenuation. However, guinea pigs receiving convalescent human sera could be at least partially protected from MVD (Siegert et al., 1968c). Sera from guinea pigs or horses vaccinated against MARV could provide protection to naïve guinea pigs (Borisevich et al., 2008; Hevey et al., 1997), whereas passive transfer of monoclonal antibodies to $GP_{1,2}$ provided partial to complete protection depending on the antibody used (Hevey et al., 2003; Razumov et al., 2001). Further studies in NHPs showed that convalescent IgG from crab-eating macaques could protect 100% of newly infected crab-eating macaques when given even as late as 48 h after infection (Dye et al., 2012). There are very little data on antibody production in lethal versus nonlethal marburgvirus infections in humans. Therefore, it is unknown if survivors generate a more effective antibody response that leads to survival. Antibody transfer is the most promising therapeutic treatment available to date, although it requires validation in studies in humans.

25.11.3 Supportive Treatment

The original outbreak in West Germany and Yugoslavia had a relatively low lethality (27%) compared to the larger outbreaks in Africa (≈85%). It should be noted that the patients in Africa were mostly treated with only intravenous fluids and antipyretic medications (Colebunders et al., 2007), whereas the medical treatment in West Germany and Yugoslavia also included administration of electrolytes and therapy for the bleeding tendency (including platelets, clotting factors, blood, and fibrinogen), while a few patients were treated with convalescent sera (Stille et al., 1968). The coagulation and antibody studies in animal models suggest that some of the original therapeutic treatments in 1967 were likely to have been effective. It is tempting to speculate that this supportive care drastically reduced the lethality of the original outbreak and that relatively simple measures available 45 years ago were actually quite effective for treatment of MVD.

25.11.4 Antisense Strategies

Phosphorodiamidate morpholino oligomers (PMO) are short, synthetic single-stranded RNA molecules that can inhibit gene expression by preventing translation of mRNA. PMOs that inhibit marburgviruses have been shown to protect 100% of infected NHPs when treatment is begun within 1 h of infection (Warren et al., 2010b).

25.11.5 Type I Interferon

The importance of type I IFN in controlling marburgvirus infection is clear from multiple experiments. Mice deficient in type I IFN signaling are susceptible to marburgviruses, whereas wild-type mice are not (Bray, 2001; Lever et al., 2012; Raymond et al., 2011). Furthermore, as discussed previously, marburgvirus VP40 inhibits type I IFN signaling (Valmas et al., 2010). Wild-type VP40 inhibits type I IFN signaling in humans cells, but not in mouse cells. However, VP40 from mouse-adapted marburgvirus inhibits type I IFN signaling in mouse cells (Valmas and Basler, 2011). It is possible that the mutations in VP40 confer IFN resistance of the virus to mice. This then suggests that inhibiting type I IFN is a major contributor to disease in humans. Treatment of MARV-infected NHPs with type I IFN (IFNα2) resulted in a significant delay in mean time-to-death (1.9–6.1 days, depending on the infectious dose) (Kolokol'tsov et al., 2001). Additional experiments, including generation of marburgviruses bearing only certain mutations from the mouse-adapted isolates, as well as measurement of IFN-induced genes in fatal versus nonfatal infections, need to be performed to further study this question.

25.11.6 Other Drugs

The small molecule FGI-103 has been shown to inhibit Marburgvirus growth in vitro (Warren et al., 2010a). This drug also protected mice from mouse-adapted MARV infection when given in a single dose 24 h after infection (Warren et al., 2010a).

25.12 CORRELATES OF IMMUNITY

Unfortunately, very little immunological data are available from human marburgvirus disease outbreaks. Much of what is known about correlates of immunity to marburgvirus infection is derived from animal studies [reviewed in Bradfute and Bavari (2011)], although significant gaps in knowledge still exist.

25.12.1 Antibody

Very few studies have measured antibody titers in nonfatal versus fatal human MVD cases. However, as discussed earlier, it is known that passive transfer of sera or purified polyclonal or monoclonal antibodies can protect against disease in animal models (Borisevich et al., 2008; Dye et al., 2012; Hevey et al.,

1997, 2003; Razumov et al., 2001; Siegert et al., 1968c). Four human patients receiving convalescent sera appeared to have less severe disease courses (Stille et al., 1968), although this is anecdotal.

The role of antibodies in protection varies according to the treatment regimen. Vaccine studies in guinea pigs showed that VLPs expressing MARV VP40 were not sufficient to protect animals; MARV $GP_{1,2}$ had to be expressed as well (Swenson et al., 2005). Marburgvirus VLPs generate high titers of antibodies; together, this suggests that antibodies against $GP_{1,2}$ may be important. However, many other vaccine platforms (VSIV, VRP) generate only low to moderate antibody titers and are still protective. Therefore, different vaccines may utilize different mechanisms of immunity to provide protection.

It is important to note that antibodies may function in many different ways to provide protection, including neutralization, complement fixation, and antibody-dependent cell-mediated cytotoxicity. Different vaccines or immune responses may induce multiple types of antibody responses. For example, total antibody titers range from low to high in successful marburgvirus vaccine studies, while neutralizing antibodies are usually either undetectable or low. However, the neutralizing antibody titers in passive transfer of convalescent IgG to NHPs that protects against disease are quite high (Dye et al., 2012). In addition, complement-fixing monoclonal antibodies against VP40 can protect guinea pigs from marburgvirus-induced disease (Razumov et al., 2001).

25.12.2   T Cells

Marburgvirus epitope-specific CD8+ T cells were generated in nonlethal, wild-type RAVV infection of mice (Kalina et al., 2009). Adoptive transfer of in vitro, specific peptide epitope-expanded, splenocytes from these mice protected naïve mice from mouse-adapted marburgvirus infection (Kalina et al., 2009). Vaccines that are effective in NHPs generate a wide range of CD4+ and CD8+ T-cell responses, ranging from undetectable to high. One study using DNA-only, AdV-only, or DNA-prime/AdV-boost vaccines found that these candidate vaccines all generated polyfunctional (i.e., multiple cytokine producing) T cells, but there was no correlation between clinical signs of survivors and magnitude of T-cell responses (Geisbert et al., 2010a). On the other hand, successful VSIV vaccination does not appear to generate detectable T-cell responses (Jones et al., 2005).

25.12.3   Cytokines

Many cytokines have been associated with lethal marburgvirus infections. IFN-γ, interleukin (IL)-5, MCP-1, MIG, IFN-α, and IL-12 concentrations all increased in fatal infection of mice with mouse-adapted RAVV relative to mice infected with nonfatal wild-type RAVV (Warfield et al., 2009). Tumor necrosis factor-α (TNF-α), another inflammatory cytokine, has been implicated in marburgvirus pathogenesis. Increased TNF-α is found in lethal NHP infection (Ignatyev et al., 1996), and antibody-mediated neutralization of TNF-α was found to increase guinea pig survival after marburgvirus infection (Ignat'ev et al., 1998).

As mentioned previously, type I IFN is also thought to be important in controlling marburgvirus pathogenesis. Type I IFN is inhibited by marburgvirus VP40 in a species-specific manner (Valmas and Basler, 2011; Valmas et al., 2010), and treatment with type I IFN delays time-to-death in marburgvirus-infected NHPs (Kolokol'tsov et al., 2001).

25.12.4   Dendritic Cells and Macrophages

Marburgviruses can replicate in monocytes and dendritic cells in vitro. Replication in monocytes in vitro is associated with increased proinflammatory cytokine production, including IL-6, IL-8, and TNF-α (Ströher et al., 2001). Replication in dendritic cells in vitro did not result in increased proinflammatory cytokine expression, but did inhibit IFN-α production after stimulation (Bosio et al., 2003). In vivo, marburgviruses have been shown to replicate in human macrophages (Geisbert and Jaax, 1998).

25.13   CONCLUSIONS AND FUTURE WORK

Much of what we know concerning MVD in humans stems from excellent record keeping, reporting, and analysis during the original outbreaks in 1967. Despite the lethal nature of the disease, experimental treatments (including passive transfer of antibody and possibly management of coagulopathy) and vaccines are in development that may alleviate the threat of marburgviruses. The development of NHP, guinea pig, and mouse animal models has greatly expanded our knowledge of marburgvirus pathogenesis. Still, there remain many aspects of the viruses and disease that are not known.

Very little immunological information, including cytokine, antibody, and T-cell responses, are known in human infections. In part, this is due to the rarity and the appearance of infections in isolated areas with limited accessibility and rudimentary health-care facilities. However, more studies in animal models with lethal and nonlethal infections could provide detailed understanding of successful versus unsuccessful immune responses. The advent of a reverse genetics system for MARV (Enterlein et al., 2006) provides an opportunity to study the mutations that are necessary for pathogenesis in various animal models, as well as the functions of marburgviral genes.

Further development of vaccines and therapeutics will no doubt be beneficial for individuals in endemic areas or military and research personnel. Relatively simple supportive care, including transfer of platelets, blood, and clotting factors, was used in the original outbreak. The lethality of this outbreak was only 23%, whereas case fatality rates approached 90% in other larger outbreaks during which this type of care was not available. Further research on the effectiveness of supportive care could provide an inexpensive and simple treatment regimen in resource-poor endemic areas.

REFERENCES

Adams, M.J., Carstens, E.B., 2012. Ratification vote on taxonomic proposals to the international committee on taxonomy of viruses (2012). *Arch Virol* 157, 1411–1422.

Adjemian, J., Farnon, E.C., Tschioko, F., Wamala, J.F., Byaruhanga, E., Bwire, G.S., Kansiime, E., Kagirita, A., Ahimbisibwe, S., Katunguka, F., Jeffs, B., Lutwama, J.J., Downing, R., Tappero, J.W., Formenty, P., Amman, B., Manning, C., Towner, J., Nichol, S.T., Rollin, P.E., 2011. Outbreak of Marburg hemorrhagic fever among miners in Kamwenge and Ibanda Districts, Uganda, 2007. *J Infect Dis* 204 Suppl 3, S796–S799.

Alibek, K., Handelman, S., 1999. *Biohazard. The Chilling True Story of the Largest Covert Biological Weapons Program in the World—Told From Inside by the Man Who Ran It*. Random House, New York.

Alves, D.A., Glynn, A.R., Steele, K.E., Lackemeyer, M.G., Garza, N.L., Buck, J.G., Mech, C., Reed, D.S., 2010. Aerosol exposure to the Angola strain of Marburg virus causes lethal viral hemorrhagic fever in cynomolgus macaques. *Vet Pathol* 47, 831–851.

Bamberg, S., Kolesnikova, L., Moller, P., Klenk, H.D., Becker, S., 2005. VP24 of Marburg virus influences formation of infectious particles. *J Virol* 79, 13421–13433.

Basler, C.F., Mikulasova, A., Martinez-Sobrido, L., Paragas, J., Mühlberger, E., Bray, M., Klenk, H.D., Palese, P., Garcia-Sastre, A., 2003. The Ebola virus VP35 protein inhibits activation of interferon regulatory factor 3. *J Virol* 77, 7945–7956.

Basler, C.F., Wang, X., Mühlberger, E., Volchkov, V., Paragas, J., Klenk, H.D., Garcia-Sastre, A., Palese, P., 2000. The Ebola virus VP35 protein functions as a type I IFN antagonist. *Proc Natl Acad Sci USA* 97, 12289–12294.

Bausch, D.G., Nichol, S.T., Muyembe-Tamfum, J.J., Borchert, M., Rollin, P.E., Sleurs, H., Campbell, P., Tshioko, F.K., Roth, C., Colebunders, R., Pirard, P., Mardel, S., Olinda, L.A., Zeller, H., Tshomba, A., Kulidri, A., Libande, M.L., Mulangu, S., Formenty, P., Grein, T., Leirs, H., Braack, L., Ksiazek, T., Zaki, S., Bowen, M.D., Smit, S.B., Leman, P.A., Burt, F.J., Kemp, A., Swanepoel, R., International, S., Technical Committee for Marburg Hemorrhagic Fever Control in the Democratic Republic of the, C., 2006. Marburg hemorrhagic fever associated with multiple genetic lineages of virus. *N Engl J Med* 355, 909–919.

Bavari, S., Bosio, C.M., Wiegand, E., Ruthel, G., Will, A.B., Geisbert, T.W., Hevey, M., Schmaljohn, C., Schmaljohn, A., Aman, M.J., 2002. Lipid raft microdomains: A gateway for compartmentalized trafficking of Ebola and Marburg viruses. *J Exp Med* 195, 593–602.

Bechtelsheimer, H., Jacob, H., Solcher, H., 1969. The neuropathology of an infectious disease transmitted by African green monkeys (Cercopithecus aethiops). *Ger Med Mon* 14, 10–12.

Becker, S., Rinne, C., Hofsass, U., Klenk, H.D., Mühlberger, E., 1998. Interactions of Marburg virus nucleocapsid proteins. *Virology* 249, 406–417.

Becker, S., Spiess, M., Klenk, H.D., 1995. The asialoglycoprotein receptor is a potential liver-specific receptor for Marburg virus. *J Gen Virol* 76 (Pt 2), 393–399.

Bharat, T.A., Riches, J.D., Kolesnikova, L., Welsch, S., Krahling, V., Davey, N., Parsy, M.L., Becker, S., Briggs, J.A., 2011. Cryo-electron tomography of Marburg virus particles and their morphogenesis within infected cells. *PLoS Biol* 9, e1001196.

Borio, L., Inglesby, T., Peters, C.J., Schmaljohn, A.L., Hughes, J.M., Jahrling, P.B., Ksiazek, T., Johnson, K.M., Meyerhoff, A., O'Toole, T., Ascher, M.S., Bartlett, J., Breman, J.G., Eitzen, E.M., Jr., Hamburg, M., Hauer, J., Henderson, D.A., Johnson, R.T., Kwik, G., Layton, M., Lillibridge, S., Nabel, G.J., Osterholm, M.T., Perl, T.M., Russell, P., Tonat, K., Working Group on Civilian, B., 2002. Hemorrhagic fever viruses as biological weapons: medical and public health management. *JAMA* 287, 2391–2405.

Borisevich, I.V., Potryvaeva, N.V., Mel'nikov, S.A., Evseev, A.A., Krasnianskii, V.P., Maksimov, V.A., 2008. Design of equine serum-based Marburg virus immunoglobulin. *Vopr Virusol* 53, 39–41.

Bosio, C.M., Aman, M.J., Grogan, C., Hogan, R., Ruthel, G., Negley, D., Mohamadzadeh, M., Bavari, S., Schmaljohn, A., 2003. Ebola and Marburg viruses replicate in monocyte-derived dendritic cells without inducing the production of cytokines and full maturation. *J Infect Dis* 188, 1630–1638.

Bradfute, S.B., Bavari, S., 2011. Correlates of immunity to filovirus infection. *Viruses* 3, 982–1000.

Bradfute, S.B., Dye, J.M., Jr., Bavari, S., 2011. Filovirus vaccines. *Hum Vaccine* 7, 701–711.

Bray, M., 2001. The role of the Type I interferon response in the resistance of mice to filovirus infection. *J Gen Virol* 82, 1365–1373.

Carette, J.E., Raaben, M., Wong, A.C., Herbert, A.S., Obernosterer, G., Mulherkar, N., Kuehne, A.I., Kranzusch, P.J., Griffin, A.M., Ruthel, G., Dal Cin, P., Dye, J.M., Whelan, S.P., Chandran, K., Brummelkamp, T.R., 2011. Ebola virus entry requires the cholesterol transporter Niemann-Pick C1. *Nature* 477, 340–343.

Carrion, R., Jr., Ro, Y., Hoosien, K., Ticer, A., Brasky, K., de la Garza, M., Mansfield, K., Patterson, J.L., 2011. A small nonhuman primate model for filovirus-induced disease. *Virology* 420, 117–124.

Centers for Disease Control and Prevention (CDC), 2009. Imported case of Marburg hemorrhagic fever—Colorado, 2008. *MMWR Morb Mortal Wkly Rep* 58, 1377–1381.

Colebunders, R., Tshomba, A., Van Kerkhove, M.D., Bausch, D.G., Campbell, P., Libande, M., Pirard, P., Tshioko, F., Mardel, S., Mulangu, S., Sleurs, H., Rollin, P.E., Muyembe-Tamfum, J.J., Jeffs, B., Borchert, M., International Scientific and Technical Committee "DRC Watsa/Durba 1999 Marburg Outbreak Investigation Group" 2007. Marburg hemorrhagic fever in Durba and Watsa, Democratic Republic of the Congo: clinical documentation, features of illness, and treatment. *J Infect Dis* 196 Suppl 2, S148–S153.

Daddario-DiCaprio, K.M., Geisbert, T.W., Geisbert, J.B., Ströher, U., Hensley, L.E., Grolla, A., Fritz, E.A., Feldmann, F., Feldmann, H., Jones, S.M., 2006a. Cross-protection against Marburg virus strains by using a live, attenuated recombinant vaccine. *J Virol* 80, 9659–9666.

Daddario-DiCaprio, K.M., Geisbert, T.W., Ströher, U., Geisbert, J.B., Grolla, A., Fritz, E.A., Fernando, L., Kagan, E., Jahrling, P.B., Hensley, L.E., Jones, S.M., Feldmann, H., 2006b. Postexposure protection against Marburg haemorrhagic fever with recombinant vesicular stomatitis virus vectors in non-human primates: an efficacy assessment. *Lancet* 367, 1399–1404.

Dolnik, O., Kolesnikova, L., Stevermann, L., Becker, S., 2010. Tsg101 is recruited by a late domain of the nucleocapsid protein to support budding of Marburg virus-like particles. *J Virol* 84, 7847–7856.

Dye, J.M., Herbert, A.S., Kuehne, A.I., Barth, J.F., Muhammad, M.A., Zak, S.E., Ortiz, R.A., Prugar, L.I., Pratt, W.D., 2012. Postexposure antibody prophylaxis protects nonhuman primates from filovirus disease. *Proc Natl Acad Sci USA* 109, 5034–5039.

Enterlein, S., Volchkov, V., Weik, M., Kolesnikova, L., Volchkova, V., Klenk, H.D., Mühlberger, E., 2006. Rescue of recombinant Marburg virus from cDNA is dependent on nucleocapsid protein VP30. *J Virol* 80, 1038–1043.

Feldmann, H., Mühlberger, E., Randolf, A., Will, C., Kiley, M.P., Sanchez, A., Klenk, H.D., 1992. Marburg virus, a filovirus: messenger RNAs, gene order, and regulatory elements of the replication cycle. *Virus Res* 24, 1–19.

Fowler, T., Bamberg, S., Moller, P., Klenk, H.D., Meyer, T.F., Becker, S., Rudel, T., 2005. Inhibition of Marburg virus protein expression and viral release by RNA interference. *J Gen Virol* 86, 1181–1188.

Fritz, E.A., Geisbert, J.B., Geisbert, T.W., Hensley, L.E., Reed, D.S., 2008. Cellular immune response to Marburg virus infection in cynomolgus macaques. *Viral Immunol* 21, 355–363.

Gear, J.S., Cassel, G.A., Gear, A.J., Trappler, B., Clausen, L., Meyers, A.M., Kew, M.C., Bothwell, T.H., Sher, R., Miller, G.B., Schneider, J., Koornhof, H.J., Gomperts, E.D., Isaacson, M., Gear, J.H., 1975. Outbreak of Marburg virus disease in Johannesburg. *Br Med J* 4, 489–493.

Gedigk, P., Bechtelsheimer, H., Korb, G., 1968. Pathological anatomy of the "Marburg virus" disease (the so-called "Marburg monkey disease"). *Dtsch Med Wochenschr* 93, 590–601.

Geisbert, T.W., Bailey, M., Geisbert, J.B., Asiedu, C., Roederer, M., Grazia-Pau, M., Custers, J., Jahrling, P., Goudsmit, J., Koup, R., Sullivan, N.J., 2010a. Vector choice determines immunogenicity and potency of genetic vaccines against Angola Marburg virus in nonhuman primates. *J Virol* 84, 10386–10394.

Geisbert, T.W., Daddario-Dicaprio, K.M., Geisbert, J.B., Reed, D.S., Feldmann, F., Grolla, A., Ströher, U., Fritz, E.A., Hensley, L.E., Jones, S.M., Feldmann, H., 2008. Vesicular stomatitis virus-based vaccines protect nonhuman primates against aerosol challenge with Ebola and Marburg viruses. *Vaccine* 26, 6894–6900.

Geisbert, T.W., Daddario-DiCaprio, K.M., Geisbert, J.B., Young, H.A., Formenty, P., Fritz, E.A., Larsen, T., Hensley, L.E., 2007. Marburg virus Angola infection of rhesus macaques: pathogenesis and treatment with recombinant nematode anticoagulant protein c2. *J Infect Dis* 196 Suppl 2, S372–S381.

Geisbert, T.W., Geisbert, J.B., Leung, A., Daddario-DiCaprio, K.M., Hensley, L.E., Grolla, A., Feldmann, H., 2009. Single-injection vaccine protects nonhuman primates against infection with Marburg virus and three species of Ebola virus. *J Virol* 83, 7296–7304.

Geisbert, T.W., Hensley, L.E., Geisbert, J.B., Leung, A., Johnson, J.C., Grolla, A., Feldmann, H., 2010b. Postexposure treatment of Marburg virus infection. *Emerg Infect Dis* 16, 1119–1122.

Geisbert, T.W., Jaax, N.K., 1998. Marburg hemorrhagic fever: report of a case studied by immunohistochemistry and electron microscopy. *Ultrastruct Pathol* 22, 3–17.

Geisbert, T.W., Jahrling, P.B., 1995. Differentiation of filoviruses by electron microscopy. *Virus Res* 39, 129–150.

Gibb, T.R., Norwood, D.A., Jr., Woollen, N., Henchal, E.A., 2001. Development and evaluation of a fluorogenic 5′-nuclease assay to identify Marburg virus. *Mol Cell Probes* 15, 259–266.

Gramberg, T., Hofmann, H., Moller, P., Lalor, P.F., Marzi, A., Geier, M., Krumbiegel, M., Winkler, T., Kirchhoff, F., Adams, D.H., Becker, S., Munch, J., Pohlmann, S., 2005. LSECtin interacts with filovirus glycoproteins and the spike protein of SARS coronavirus. *Virology* 340, 224–236.

Grolla, A., Jones, S.M., Fernando, L., Strong, J.E., Ströher, U., Moller, P., Paweska, J.T., Burt, F., Pablo, Palma, P., Sprecher, A., Formenty, P., Roth, C., Felmann, H., 2011. The use of a mobile laboratory unit in support of patient management and epidemiological surveillance during the 2005 Marburg Outbreak in Angola. *PLoS Negl Trop Dis* 5, e1183.

Herrel, M., Hoefs, N., Staeheli, P., Schneider, U., 2012. Tick-borne Nyamanini virus replicates in the nucleus and exhibits unusual genome and matrix protein properties. *J Virol* 86(19), 10739–10747.

Hevey, M., Negley, D., Geisbert, J., Jahrling, P., Schmaljohn, A., 1997. Antigenicity and vaccine potential of Marburg virus glycoprotein expressed by baculovirus recombinants. *Virology* 239, 206–216.

Hevey, M., Negley, D., Pushko, P., Smith, J., Schmaljohn, A., 1998. Marburg virus vaccines based upon alphavirus replicons protect guinea pigs and nonhuman primates. *Virology* 251, 28–37.

Hevey, M., Negley, D., Schmaljohn, A., 2003. Characterization of monoclonal antibodies to Marburg virus (strain Musoke) glycoprotein and identification of two protective epitopes. *Virology* 314, 350–357.

Hofmann, H., Kunz, C., 1970. A strain of "Marburg virus" (Rhabdovirus simiae) pathogenic to mice. *Arch Gesamte Virusforsch* 32, 244–248.

Ignat'ev, G.M., Strel'tsova, M.A., Kashentseva, E.A., Patrushev, N.A., 1998. Effects of tumor necrosis factor antiserum of the course of Marburg hemorrhagic fever. *Vestn Ross Akad Med Nauk* 3, 35–38.

Ignatyev, G.M., Agafonov, A.P., Streltsova, M.A., Kashentseva, E.A., 1996. Inactivated Marburg virus elicits a nonprotective immune response in Rhesus monkeys. *J Biotechnol* 44, 111–118.

Jeffs, B., Roddy, P., Weatherill, D., de la Rosa, O., Dorion, C., Iscla, M., Grovas, I., Palma, P.P., Villa, L., Bernal, O., Rodriguez-Martinez, J., Barcelo, B., Pou, D., Borchert, M., 2007. The Medecins Sans Frontieres intervention in the Marburg hemorrhagic fever epidemic, Uige, Angola, 2005. I. Lessons learned in the hospital. *J Infect Dis* 196 Suppl 2, S154–S161.

Johnson, E.D., Johnson, B.K., Silverstein, D., Tukei, P., Geisbert, T.W., Sanchez, A.N., Jahrling, P.B., 1996. Characterization of a new Marburg virus isolated from a 1987 fatal case in Kenya. *Arch Virol* Suppl 11, 101–114.

Jones, S.M., Feldmann, H., Ströher, U., Geisbert, J.B., Fernando, L., Grolla, A., Klenk, H.D., Sullivan, N.J., Volchkov, V.E., Fritz, E.A., Daddario, K.M., Hensley, L.E., Jahrling, P.B., Geisbert, T.W., 2005. Live attenuated recombinant vaccine protects nonhuman primates against Ebola and Marburg viruses. *Nat Med* 11, 786–790.

Jouvenet, N., Neil, S.J., Zhadina, M., Zang, T., Kratovac, Z., Lee, Y., McNatt, M., Hatziioannou, T., Bieniasz, P.D., 2009. Broad-spectrum inhibition of retroviral and filoviral particle release by tetherin. *J Virol* 83, 1837–1844.

Kalina, W.V., Warfield, K.L., Olinger, G.G., Bavari, S., 2009. Discovery of common marburgvirus protective epitopes in a BALB/c mouse model. *Virol J* 6, 132.

Kolesnikova, L., Berghofer, B., Bamberg, S., Becker, S., 2004. Multivesicular bodies as a platform for formation of the Marburg virus envelope. *J Virol* 78, 12277–12287.

Kolesnikova, L., Mittler, E., Schudt, G., Shams-Eldin, H., Becker, S., 2012. Phosphorylation of Marburg virus matrix protein VP40 triggers assembly of nucleocapsids with the viral envelope at the plasma membrane. *Cell Microbiol* 14, 182–197.

Kolesnikova, L., Strecker, T., Morita, E., Zielecki, F., Mittler, E., Crump, C., Becker, S., 2009. Vacuolar protein sorting pathway contributes to the release of Marburg virus. *J Virol* 83, 2327–2337.

Kolokol'tsov, A.A., Davidovich, I.A., Strel'tsova, M.A., Nesterov, A.E., Agafonova, O.A., Agafonov, A.P., 2001. The use of interferon for emergency prophylaxis of Marburg hemorrhagic fever in monkeys. *Bull Exp Biol Med* 132, 686–688.

Kondratowicz, A.S., Lennemann, N.J., Sinn, P.L., Davey, R.A., Hunt, C.L., Moller-Tank, S., Meyerholz, D.K., Rennert, P., Mullins, R.F., Brindley, M., Sandersfeld, L.M., Quinn, K., Weller, M., McCray, P.B., Jr., Chiorini, J., Maury, W., 2011. T-cell immunoglobulin and mucin domain 1 (TIM-1) is a receptor for Zaire Ebolavirus and Lake Victoria Marburgvirus. *Proc Natl Acad Sci USA* 108, 8426–8431.

Kuhn, J.H., 2008. Filoviruses. A compendium of 40 years of epidemiological, clinical, and laboratory studies. *Arch Virol* Suppl 20, 13–360.

Kuhn, J.H., Becker S., Ebihara H., Geisbert T.W., Jahrling P.B., Kawaoka Y., Netesov S. V., Nichol S. T., Peters C. J., Volchkov V. E., Ksiazek T. G., 2011a. Family *Filoviridae*, in: King A.M.Q., Adams M.J., Carstens E.B., Lefkowitz E.J. (Eds.), *Virus Taxonomy—Ninth Report of the International Committee on Taxonomy of Viruses*, London, U.K., pp. 665–671.

Kuhn, J.H., Becker, S., Ebihara, H., Geisbert, T.W., Johnson, K.M., Kawaoka, Y., Lipkin, W.I., Negredo, A.I., Netesov, S.V., Nichol, S.T., Palacios, G., Peters, C.J., Tenorio, A., Volchkov, V.E., Jahrling, P.B., 2010. Proposal for a revised taxonomy of the family *Filoviridae*: classification, names of taxa and viruses, and virus abbreviations. *Arch Virol* 155, 2083–2103.

Kuhn, J.H., Dodd, L.E., Wahl-Jensen, V., Radoshitzky, S.R., Bavari, S., Jahrling, P.B., 2011b. Evaluation of perceived threat differences posed by filovirus variants. *Biosecur Bioterror* 9, 361–371.

Kuhn, J.H., Radoshitzky, S.R., Guth, A.C., Warfield, K.L., Li, W., Vincent, M.J., Towner, J.S., Nichol, S.T., Bavari, S., Choe, H., Aman, M.J., Farzan, M., 2006. Conserved receptor-binding domains of Lake Victoria marburgvirus and Zaire ebolavirus bind a common receptor. *J Biol Chem* 281, 15951–15958.

Leitenberg, M., Zilinskas, R., Kuhn, J.H., 2012. *The Soviet Biological Weapons Program: A History*. Harvard University Press, Cambridge, MA.

Lever, M.S., Piercy, T.J., Steward, J.A., Eastaugh, L., Smither, S.J., Taylor, C., Salguero, F.J., Phillpotts, R.J., 2012. Lethality and pathogenesis of airborne infection with filoviruses in A129 alpha/beta -/- interferon receptor-deficient mice. *J Med Microbiol* 61, 8–15.

Lofts, L.L., Ibrahim, M.S., Negley, D.L., Hevey, M.C., Schmaljohn, A.L., 2007. Genomic differences between guinea pig lethal and nonlethal Marburg virus variants. *J Infect Dis* 196 Suppl 2, S305–S312.

Lofts, L.L., Wells, J.B., Bavari, S., Warfield, K.L., 2011. Key genomic changes necessary for an in vivo lethal mouse marburgvirus variant selection process. *J Virol* 85, 3905–3917.

Lub, M., Sergeev, A.N., P'iankov O, V., P'iankova O, G., Petrishchenko, V.A., Kotliarov, L.A., 1995. Certain pathogenetic characteristics of a disease in monkeys in infected with the Marburg virus by an airborne route. *Vopr Virusol* 40, 158–161.

Martini, G.A., Knauff, H.G., Schmidt, H.A., Mayer, G., Baltzer, G., 1968. On the hitherto unknown, in monkeys originating infectious disease: Marburg virus disease. *Dtsch Med Wochenschr* 93, 559–571.

Marzi, A., Gramberg, T., Simmons, G., Moller, P., Rennekamp, A.J., Krumbiegel, M., Geier, M., Eisemann, J., Turza, N., Saunier, B., Steinkasserer, A., Becker, S., Bates, P., Hofmann, H., Pohlmann, S., 2004. DC-SIGN and DC-SIGNR interact with the glycoprotein of Marburg virus and the S protein of severe acute respiratory syndrome coronavirus. *J Virol* 78, 12090–12095.

Matsuno, K., Kishida, N., Usami, K., Igarashi, M., Yoshida, R., Nakayama, E., Shimojima, M., Feldmann, H., Irimura, T., Kawaoka, Y., Takada, A., 2010. Different potential of C-type lectin-mediated entry between Marburg virus strains. *J Virol* 84, 5140–5147.

Miller, E.H., Chandran, K., 2012. Filovirus entry into cells–New insights. *Curr Opin Virol* 2, 206–214.

Miller, E.H., Obernosterer, G., Raaben, M., Herbert, A.S., Deffieu, M.S., Krishnan, A., Ndungo, E., Sandesara, R.G., Carette, J.E., Kuehne, A.I., Ruthel, G., Pfeffer, S.R., Dye, J.M., Whelan, S.P., Brummelkamp, T.R., Chandran, K., 2012. Ebola virus entry requires the host-programmed recognition of an intracellular receptor. *EMBO J* 31, 1947–1960.

Mühlberger, E., Lötfering, B., Klenk, H.D., Becker, S., 1998. Three of the four nucleocapsid proteins of Marburg virus, NP, VP35, and L, are sufficient to mediate replication and transcription of Marburg virus-specific monocistronic minigenomes. *J Virol* 72, 8756–8764.

Mühlberger, E., Sanchez, A., Randolf, A., Will, C., Kiley, M.P., Klenk, H.D., Feldmann, H., 1992. The nucleotide sequence of the L gene of Marburg virus, a filovirus: Homologies with paramyxoviruses and rhabdoviruses. *Virology* 187, 534–547.

Nikiforov, V.V., Turovskii Iu, I., Kalinin, P.P., Akinfeeva, L.A., Katkova, L.R., Barmin, V.S., Riabchikova, E.I., Popkova, N.I., Shestopalov, A.M., Nazarov, V.P. et al., 1994. A case of a laboratory infection with Marburg fever. *Zh Mikrobiol Epidemiol Immunobiol* May–Jun (3), 104–106.

Raymond, J., Bradfute, S., Bray, M., 2011. Filovirus infection of STAT-1 knockout mice. *J Infect Dis* 204 Suppl 3, S986–S990.

Razumov, I.A., Belanov, E.F., Bormotov, N.I., Kazachinskaia, E.I., 2001. Detection of antiviral activity of monoclonal antibodies, specific to Marburg virus proteins. *Vopr Virusol* 46, 33–37.

Reid, S.P., Leung, L.W., Hartman, A.L., Martinez, O., Shaw, M.L., Carbonnelle, C., Volchkov, V.E., Nichol, S.T., Basler, C.F., 2006. Ebola virus VP24 binds karyopherin alpha1 and blocks STAT1 nuclear accumulation. *J Virol* 80, 5156–5167.

Reid, S.P., Valmas, C., Martinez, O., Sanchez, F.M., Basler, C.F., 2007. Ebola virus VP24 proteins inhibit the interaction of NPI-1 subfamily karyopherin alpha proteins with activated STAT1. *J Virol* 81, 13469–13477.

Riemenschneider, J., Garrison, A., Geisbert, J., Jahrling, P., Hevey, M., Negley, D., Schmaljohn, A., Lee, J., Hart, M.K., Vanderzanden, L., Custer, D., Bray, M., Ruff, A., Ivins, B., Bassett, A., Rossi, C., Schmaljohn, C., 2003. Comparison of individual and combination DNA vaccines for B. anthracis, Ebola virus, Marburg virus and Venezuelan equine encephalitis virus. *Vaccine* 21, 4071–4080.

Sanchez, A., Trappier, S.G., Mahy, B.W., Peters, C.J., Nichol, S.T., 1996. The virion glycoproteins of Ebola viruses are encoded in two reading frames and are expressed through transcriptional editing. *Proc Natl Acad Sci USA* 93, 3602–3607.

Siegert, R., Shu, H.L., Slenczka, W. Peters, D., Muller, G., 1968a. The aetiology of an unknown human infection transmitted by monkeys (preliminary communication). *Ger Med Mon* 13, 1–2.

Siegert, R., Shu, H.L., Slenczka, W., 1968b. Detection of the "Marburg Virus" in patients. *Ger Med Mon* 13, 521–524.

Siegert, R., Shu, H.L., Slenczka, W., 1968c. Isolation and identification of the "Marburg virus". *Dtsch Med Wochenschr* 93, 604–612.

Siegert, R., Shu, H.L., Slenczka, W., Peters, D., Muller, G., 1967. On the etiology of an unknown human infection originating from monkeys. *Dtsch Med Wochenschr* 92, 2341–2343.

Simpson, D.I., 1969. Vervent monkey disease. Transmission to the hamster. *Br J Exp Pathol* 50, 389–392.

Simpson, D.I., Zlotnik, I., Rutter, D.A., 1968. Vervet monkey disease. Experiment infection of guinea pigs and monkeys with the causative agent. *Br J Exp Pathol* 49, 458–464.

Slenczka, W., Shu, H.L., Piepenberg, G., Siegert, R., 1968. Detection of the antigen of the "Marburg Virus" in the organs of infected guinea-pigs by immunofluorescence. *Ger Med Mon* 13, 524–529.

Smith, D.H., Johnson, B.K., Isaacson, M., Swanapoel, R., Johnson, K.M., Killey, M., Bagshawe, A., Siongok, T., Keruga, W.K., 1982. Marburg-virus disease in Kenya. *Lancet* 1, 816–820.

Stille, W., Bohle, E., Helm, E., van Rey, W., Siede, W., 1968. On an infectious disease transmitted by Cercopithecus aethiops. ("Green monkey disease"). *Dtsch Med Wochenschr* 93, 572–582.

Ströher, U., West, E., Bugany, H., Klenk, H.D., Schnittler, H.J., Feldmann, H., 2001. Infection and activation of monocytes by Marburg and Ebola viruses. *J Virol* 75, 11025–11033.

Swenson, D.L., Wang, D., Luo, M., Warfield, K.L., Woraratanadharm, J., Holman, D.H., Dong, J.Y., Pratt, W.D., 2008a. Vaccine to confer to nonhuman primates complete protection against multistrain Ebola and Marburg virus infections. *Clin Vaccine Immunol* 15, 460–467.

Swenson, D.L., Warfield, K.L., Kuehl, K., Larsen, T., Hevey, M.C., Schmaljohn, A., Bavari, S., Aman, M.J., 2004. Generation of Marburg virus-like particles by co-expression of glycoprotein and matrix protein. *FEMS Immunol Med Microbiol* 40, 27–31.

Swenson, D.L., Warfield, K.L., Larsen, T., Alves, D.A., Coberley, S.S., Bavari, S., 2008b. Monovalent virus-like particle vaccine protects guinea pigs and nonhuman primates against infection with multiple Marburg viruses. *Expert Rev Vaccines* 7, 417–429.

Swenson, D.L., Warfield, K.L., Negley, D.L., Schmaljohn, A., Aman, M.J., Bavari, S., 2005. Virus-like particles exhibit potential as a pan-filovirus vaccine for both Ebola and Marburg viral infections. *Vaccine* 23, 3033–3042.

Timen, A., Koopmans, M.P., Vossen, A.C., van Doornum, G.J., Gunther, S., van den Berkmortel, F., Verduin, K.M., Dittrich, S., Emmerich, P., Osterhaus, A.D., van Dissel, J.T., Coutinho, R.A., 2009. Response to imported case of Marburg hemorrhagic fever, the Netherland. *Emerg Infect Dis* 15, 1171–1175.

Towner, J.S., Amman, B.R., Sealy, T.K., Carroll, S.A., Comer, J.A., Kemp, A., Swanepoel, R., Paddock, C.D., Balinandi, S., Khristova, M.L., Formenty, P.B., Albarino, C.G., Miller, D.M., Reed, Z.D., Kayiwa, J.T., Mills, J.N., Cannon, D.L., Greer, P.W., Byaruhanga, E., Farnon, E.C., Atimnedi, P., Okware, S., Katongole-Mbidde, E., Downing, R., Tappero, J.W., Zaki, S.R., Ksiazek, T.G., Nichol, S.T., Rollin, P.E., 2009. Isolation of genetically diverse Marburg viruses from Egyptian fruit bats. *PLoS Pathog* 5, e1000536.

Towner, J.S., Khristova, M.L., Sealy, T.K., Vincent, M.J., Erickson, B.R., Bawiec, D.A., Hartman, A.L., Comer, J.A., Zaki, S.R., Ströher, U., Gomes da Silva, F., del Castillo, F., Rollin, P.E., Ksiazek, T.G., Nichol, S.T., 2006. Marburgvirus genomics and association with a large hemorrhagic fever outbreak in Angola. *J Virol* 80, 6497–6516.

Urata, S., Noda, T., Kawaoka, Y., Morikawa, S., Yokosawa, H., Yasuda, J., 2007. Interaction of Tsg101 with Marburg virus VP40 depends on the PPPY motif, but not the PT/SAP motif as in the case of Ebola virus, and Tsg101 plays a critical role in the budding of Marburg virus-like particles induced by VP40, NP, and GP. *J Virol* 81, 4895–4899.

Valmas, C., Basler, C.F., 2011. Marburg virus VP40 antagonizes interferon signaling in a species-specific manner. *J Virol* 85, 4309–4317.

Valmas, C., Grosch, M.N., Schumann, M., Olejnik, J., Martinez, O., Best, S.M., Krahling, V., Basler, C.F., Mühlberger, E., 2010. Marburg virus evades interferon responses by a mechanism distinct from Ebola virus. *PLoS Pathog* 6, e1000721.

van Paassen, J., Bauer, M.P., Arbous, M.S., Visser, L.G., Schmidt-Chanasit, J., Schilling, S., Olschlager, S., Rieger, T., Emmerich, P., Schmetz, C., van de Berkmortel, F., van Hoek, B., van Burgel, N.D., Osterhaus, A.D., Vossen, A.C., Gunther, S., van Dissel, J.T., 2012. Acute liver failure, multiorgan failure, cerebral oedema, and activation of proangiogenic and antiangiogenic factors in a case of Marburg haemorrhagic fever. *Lancet Infect Dis* 12, 635–642.

Volchkov, V.E., Becker, S., Volchkova, V.A., Ternovoj, V.A., Kotov, A.N., Netesov, S.V., Klenk, H.D., 1995. GP mRNA of Ebola virus is edited by the Ebola virus polymerase and by T7 and vaccinia virus polymerases. *Virology* 214, 421–430.

Volchkov, V.E., Volchkova, V.A., Ströher, U., Becker, S., Dolnik, O., Cieplik, M., Garten, W., Klenk, H.D., Feldmann, H., 2000. Proteolytic processing of Marburg virus glycoprotein. *Virology* 268, 1–6.

Volchkova, V.A., Klenk, H.D., Volchkov, V.E., 1999. Delta-peptide is the carboxy-terminal cleavage fragment of the nonstructural small glycoprotein sGP of Ebola virus. *Virology* 265, 164–171.

Wang, D., Hevey, M., Juompan, L.Y., Trubey, C.M., Raja, N.U., Deitz, S.B., Woraratanadharm, J., Luo, M., Yu, H., Swain, B.M., Moore, K.M., Dong, J.Y., 2006. Complex adenovirus-vectored vaccine protects guinea pigs from three strains of Marburg virus challenges. *Virology* 353, 324–332.

Warfield, K.L., Alves, D.A., Bradfute, S.B., Reed, D.K., VanTongeren, S., Kalina, W.V., Olinger, G.G., Bavari, S., 2007. Development of a model for marburgvirus based on severe-combined immunodeficiency mice. *Virol J* 4, 108.

Warfield, K.L., Bradfute, S.B., Wells, J., Lofts, L., Cooper, M.T., Alves, D.A., Reed, D.K., VanTongeren, S.A., Mech, C.A., Bavari, S., 2009. Development and characterization of a mouse model for Marburg hemorrhagic fever. *J Virol* 83, 6404–6415.

Warren, T.K., Warfield, K.L., Wells, J., Enterlein, S., Smith, M., Ruthel, G., Yunus, A.S., Kinch, M.S., Goldblatt, M., Aman, M.J., Bavari, S., 2010a. Antiviral activity of a small-molecule inhibitor of filovirus infection. *Antimicrob Agents Chemother* 54, 2152–2159.

Warren, T.K., Warfield, K.L., Wells, J., Swenson, D.L., Donner, K.S., Van Tongeren, S.A., Garza, N.L., Dong, L., Mourich, D.V., Crumley, S., Nichols, D.K., Iversen, P.L., Bavari, S., 2010b. Advanced antisense therapies for postexposure protection against lethal filovirus infections. *Nat Med* 16, 991–994.

Will, C., Mühlberger, E., Linder, D., Slenczka, W., Klenk, H.D., Feldmann, H., 1993. Marburg virus gene 4 encodes the virion membrane protein, a type I transmembrane glycoprotein. *J Virol* 67, 1203–1210.

Yamayoshi, S., Noda, T., Ebihara, H., Goto, H., Morikawa, Y., Lukashevich, I.S., Neumann, G., Feldmann, H., Kawaoka, Y., 2008. Ebola virus matrix protein VP40 uses the COPII transport system for its intracellular transport. *Cell Host Microbe* 3, 168–177.

Zlotnik, I., Simpson, D.I., 1969. The pathology of experimental vervet monkey disease in hamsters. *Br J Exp Pathol* 50, 393–399.

26 Dengue Hemorrhagic Fever

Chwan-Chuen King and Guey Chuen Perng

CONTENTS

26.1 INTRODUCTION

Dengue is an old disease known by numerous household names: break-borne fever, dandy fever, three-day fever, seven-day fever, or giraffe fever, to name a few (Simmons, 1931a). Historically, dengue epidemics have occurred in cycles, sporadically emerging and re-emerging throughout the past. Recently, dengue has become one of the fastest spreading human diseases in the world. It is now endemic in over 100 countries, though initially and mainly centralized in tropical and subtropical zones, and is continuing to spread, surfacing in new areas with sporadic outbreaks. Dynamic clinical presentations, ranging from asymptomatic, mild, undifferentiated fever to classical dengue fever (DF), and dengue hemorrhagic fever (DHF) associated with dengue shock syndrome (DSS), are well established. Some of these manifestations are similar to other common illnesses making dengue one of the most difficult diseases to diagnose and treat. With the availability of modern technologies and more adequate diagnostic tools, the global prevalence of this disease has come to light, prompting keen attention from the public health sectors in recent years.

The intimate relationship among human beings, mosquitoes, and dengue virus has been around for more than a thousand years. The earliest known official document registering a dengue-like illness was in a Chinese "encyclopedia of disease symptoms and remedies" edited in 992 during the Northern Song Dynasty (Nobuchi, 1979). Literary records have indicated that a dengue-like illness had spread widely throughout various geographic regions before the eighteenth century. For example, a suspected dengue-like illness was documented in the West Indies in 1635 and in Panama in 1699 (Howe, 1977; McSherry, 1982). The incidence of this disease seems not to be an isolated event but a series of occurrences implicating an expansion and continual trend. A much more descriptive clinical presentation was written in the literature from the first epidemic in the late eighteenth century by David Bylon in Java and by Benjamin Rush in Philadelphia (Rush, 1789). Since then, large epidemics occurred frequently in tropical and subtropical zones of all continents until early in the twentieth century (Hirsch, 1883; Howe, 1977; Pepper, 1941). DF has now been recognized as common and attributed to many epidemics in many parts of the world, especially in tropical Asia, Africa, and South America (WHO, 2010).

The involvement of *Aedes* spp. mosquitoes accounting for the spread of the disease was first suggested by Bancroft in 1906 (Bancroft, 1906). The observation was solidified and established with conclusive evidence supported by others (Cleland et al., 1916; Rosen et al., 1954; Siler et al., 1926; Simmons et al., 1931b). The etiological agent, isolated from tainted blood passed through a bacteria-tight filter, was first proved to be a virus by Ashburn and Craig in 1907 (Ashburn and Craig, 1907). The susceptibility of monkeys to dengue virus and the possibility they could serve as a natural reservoir were first demonstrated by Blanc et al. in 1929 (Blanc and Caminopetros, 1929), which was confirmed by Simmons et al. in 1931 (Simmons et al., 1931b). The etiological viral agent was isolated from dengue patients and characterized by Hotta in 1943 in Japan and by Sabin in 1944 in Hawaii (Kimura and Hotta, 1944; Sabin and Schlesinger, 1945).

Dengue immunity was first observed in humans by Siler et al. and Simmons et al., postulating that preexposure can induce protection (Siler et al., 1926; Simmons et al., 1931b). This has been the basis for the vaccine development against dengue virus infection; however, a protective vaccine appears to be far more challenging to design than initially thought. One of the main reasons for this difficulty is that four distinct serotypes are in circulation in endemic areas constantly (Hammon et al., 1960; Sabin, 1952b). Occasionally, concurrent infections with multiple serotypes in patients are observed (Gubler et al., 1985; Lorono-Pino et al., 1999; Noisakran et al., 2009b), which may result in interference of viral survival competitiveness and prone to rearrangement of epitope changes and perhaps immune escape as well (Anoop et al., 2010; Pepin et al., 2008). Preexposure to viruses of the same serotype is known to induce protective immunity to homotypic infections but often offers a limited protection to heterotypic strains. Another issue impeding the development of a useful vaccine is the absence of a correlate of protection or an immune response corresponding with reduced pathology that can be evaluated in clinical trials. Also there are likely multiple molecular mechanisms accounting for the many clinical presentations of disease seen by physicians. Without protection from multiple disease manifestations, a protective effect is not likely to be observed from one dengue vaccine. Hence, even up to now, an FDA-approved vaccine that can prevent dengue virus infection and disease is not yet available to the public.

To get a better understanding of the disease, appreciating the pathogenic parameters involved with the infected host is necessary. Since dengue is an acute illness with clinical features that change with time, it is not easy to obtain one sample or specimen that can comprehensively demonstrate the pathological features induced by the virus in human beings (Tsai et al., 2012). An attempt to understand the pathogenesis of dengue was made during the early and late 1950s, when a tsunami-level DHF epidemic occurred in the Philippines and Thailand, where thousands of patients succumbed to death. Autopsy and, in some cases, biopsy specimens were collected for morphological analysis, viral isolation, and histological staining for the presence of viral antigen in organs such as spleen, lymph node, liver, bone marrow, and skin (WHO, 1966). Although informative results were gathered, a definitive conclusion on the pathogenic mechanisms was unable to be drawn with these samples.

Attempting to reproduce the key features of the human dengue disease in animals was instituted during the early 1920s. All in all, more than 500 species were investigated, but the disease was impossible to be reproduced in the tested animals (Blanc and Caminopetros, 1929; Cleland et al., 1919; Rudnick and Lim, 1986; Sabin, 1952b; Simmons et al., 1931b). On a positive note, certain aspects of the clinical presentations were observed in monkeys, murine, and rodent animals (I.M.R., 1986; Kimura and Hotta, 1944; Sabin, 1952b). But on the downside, the pathogenic features of the virus derived from rodent animals were lost after only a few passages (Hotta, 1952; Sabin, 1952b). In summary, in spite of recent advancements in technology and knowledge on animal biology, finding a suitable model to investigate the pathogenic mechanisms remains a challenge (Cassetti et al., 2010). Consequently, the pathogenesis of dengue virus infection in human beings remains to be further investigated.

26.2 BIOLOGY OF THE DENGUE VIRUS

26.2.1 FAMILY OF DENGUE VIRUS

Classification of viruses is complicated even with today's available knowledge of virology. They can be categorized by either the classical or Baltimore classification systems (Flint et al., 2000). The classical system assigns virus lineage based on four structural characteristics: nature of the nuclei acid in the virion (DNA or RNA), symmetry of the protein shell (capsid), presence or absence of a lipid membrane (envelope), and dimensions of the virion and capsid. The Baltimore classification system is based upon the theme that all viruses must produce mRNA, the positive strand that contains immediately translatable information by cellular ribosomes; in other words, this system uses the nature and polarity of viral genomes to assort viruses into different classes (I–VII). Dengue virus is a class IV virus, which only goes through a negative-sense strand RNA intermediate to transcribe or replicate their genomes. These two classifications are not mutually exclusive but rather complement one another. According to the classification criteria, considering virion morphology and genome organization, dengue virus is placed under the genus of the *Flavivirus*, which is classified in the family of Flaviviridae, as do other genera, *Pestivirus* and *Hepacivirus* (Beasley, 2008). The prototype member of the *Flavivirus* genus is yellow fever virus. Due to their ecology, flaviviruses have been named arboviruses or arthropod-borne viruses because many are transmitted between vertebrate hosts by mosquitoes or ticks, although there are some members with no known arthropod vector. Under these criteria, there are several other families of viruses that are also named arboviruses, including members of the Togaviridae, Bunyaviridae, Reoviridae, Orthomyxoviridae, and Rhabdoviridae, even though they have a very different virion morphology and genome organization.

26.2.2 BIOLOGICAL PROPERTIES OF DENGUE VIRUS

A few methods have been developed and applied to characterize and establish relationships between flaviviruses (Guzman et al., 2010). Biological activities, such as hemagglutination inhibition and neutralization, are determined by serological assays (utilizing monoclonal or polyclonal antibodies). Complement-fixation associated with viral antigen (nonstructural protein 1) is an alternative identification method. Genetic relationships are established by nucleotide sequencing of certain regions of the genes within the viral genome of most flaviviruses (Holmes and Burch, 2000). Although cumulative data indicate that serological groupings closely resemble those formed based on genome homology, however compared to serological studies, the genetic analyses reveal much more detailed information about the *Flavivirus* genome (Vasilakis et al., 2010; Weaver and Barrett, 2004). Therefore, genetic groupings are the more favored in terms of determining classification. The availability of nucleotide sequences for portions of most flaviviruses in Genbank gives the notion that each flavivirus exists as a single serotype. In addition, evidence also suggests that the geographic origin of the virus strain can be determined by the genotypes. As a whole, four genetically and antigenically distinct serotypes have been accepted for dengue virus, termed dengue 1, dengue 2, dengue 3, and dengue 4, or DENV-1, DENV-2, DENV-3, and DENV-4, respectively (Hammon et al., 1960). The genome for dengue virus is a single positive-sense RNA strand with approximately 11,000 nucleotides. It has the same polarity as mRNA and does not require many unique host cell factors for conversion into protein; therefore, the genome by itself is infectious. A type I cap (m7GpppAmp) is at the 5′ terminus followed by the conserved AG dinucleotide, while the poly-A-tail at the 3′ terminus is absent (Boulton and Westaway, 1972; Cleaves and Dubin, 1979; Wengler and Gross, 1978). A single polyprotein is translated from one large open reading frame from the RNA genome. It is proteolytically cleaved by proteases derived from the host or from the viral encoded gene products into 10 known proteins: three structural proteins, capsid (C), membrane (prM/M), and envelope (E), and seven nonstructural (NS) proteins, NS1–NS5 (Smith and Wright, 1985; Stohlman et al., 1975). The order of the gene products is C–prM/M–E–NS1–NS2A–NS2B–NS3–NS4A–NS4B–NS5 (Gubler and Goro, 1997; Noisakran and Perng, 2008). The 5′ noncoding region (NCR), about 100 nucleotides long, is highly conserved in length among all flaviviruses, while the 3′ NCR is more variable, ranging from 400 to 600 nucleotides in size (Beasley, 2008).

26.2.3 Tissue Culture Infections

A general replication strategy similar to that of other viruses is observed in dengue virus such as attachment, entry, fusion, and uncoating (Beasley, 2008). However, evidence supporting the sequential events is unsubstantiated. Perceivably, a complete entry process involves the initial attachment, likely mediated by receptors, to the plasma membrane of the permissive cell followed by the involvement of pH-dependent fusion with the endocytic vesicle membrane. Although viral E protein appears to be a critical factor in entry, a consensus on the receptor(s) required for attachment and fusion to a permissive cell has not been established. Several surface-expressed proteins have been suggested to mediate this process, dependent upon the cell line. Dengue virus receptor candidates include but are not limited to heparin sulfate, heat-shock proteins (HSP90 and HSP70), CD14 molecule, DC-SIGN, mannose, and other unknown receptors (Guzman et al., 2010; Murphy and Whitehead, 2011; Rothman, 2011). However, other mechanisms of viral entry have been proposed as well. These include direct fusion to the plasma membrane and antibody-mediated via Fc or complement C3 receptors (Bargeron Clark et al., 2012; Murphy and Whitehead, 2011; Rothman, 2011). Dengue virus replicates dominantly in the cytoplasm of the permissive cell; hence, presumably the internal events are likely similar to other viruses (Gubler and Goro, 1997). Attempts to uncover the mechanisms detailing how dengue virus is transported via secretary pathways and released from the cell surface of infected cells remain an unresolved issue. The conserved sequences within the NCRs of flaviviruses have been implicated in having a role in viral RNA replication and translation (Edgil et al., 2006). It has been suggested that the stem-loop structures at the conserved regions of 5′ and 3′ NCR play a role in the initiation of the RNA replication and that NS3 and NS5 viral proteins are involved in the formation of the replication complex (Chiu et al., 2005; Clyde and Harris, 2006; Preugschat et al., 1990). However, in order to have efficient viral replication, circularization of the viral genome mediated by the complementary sequences encoded in the 5′ and 3′ NCRs of dengue virus is required as well as other host and viral proteins (Chambers et al., 1990; Weaver and Barrett, 2004).

In-depth descriptions of possible functions for each protein have been addressed elsewhere (Beasley, 2008); a brief summary on the key functions of these viral proteins is given. Two types of capsid (C) proteins have been found in infected cells: membrane-anchored (C_{anch}) and mature virion-associated (C_{vir}) forms. C protein has a role in virion structure and assembly, and to some extent, in the regulation of viral replication. The viral envelope is made of a lipid bilayer in which the E and M proteins are embedded. Although the protein composition of the viral envelope is orchestrated by the virus, two forms of M protein, premembrane (prM) and M, can be found, depending upon the maturity of the virus. prM has been implicated in acting as a chaperone for E protein folding in the ER and serving as a protective agent, preventing E protein from premature fusion during transport through acidic post-Golgi vesicles to the cell surface. M protein appears to participate in fusion activity in conjunction with the dominant viral envelope protein, E protein. E protein is also important in virus attachment to cells and has been implicated as the major target for neutralizing antibody in in vitro settings. The role of nonstructural protein NS1 in the context of viral replication is less clear. Although multiple forms with varying degrees of posttranslational modifications have been found in infected cells and in vivo, two major forms are mentioned frequently, membrane-bound NS1 and soluble NS1. As a result, several functions have been linked to NS1 protein, for example, roles in early replication events and in virion assembly/maturation. NS2, especially the NS2B, is a critical cofactor for the NS2–NS3 protease activity. In addition, several other functions have been demonstrated for NS3, for instance, a role in viral RNA replication through nucleotide triphosphatase (NTPase), RNA 5′ triphosphatase (RTPase), and helicase activities. A potential role of viroporins has been proposed for NS2A, NS2B, and NS4A since expression of these proteins in transfected mammalian cells led to altered membrane permeability and growth inhibition. The NS4B may function as an interferon antagonist via blocking of STAT1 activation. The last protein, NS5, is the largest in dengue viral genome. Two putative functional activities have been demonstrated: a viral RNA-dependent RNA polymerase (RdRp) encoded at the C-terminal portion and a potential methyltransferase at the N-terminal region.

26.2.4 IN VIVO INFECTIONS

Despite the dynamic range of cells that are permissive for dengue virus infection in vitro (Kurane et al., 1990), evidence that these cells become infected in vivo has not been substantiated. Dengue viral antigen, and rarely, viral RNA, has been shown in cells of numerous organs obtained from autopsy or from biopsy (Bhamarapravati and Boonyapaknavik, 1966; Jessie et al., 2004; Nisalak et al., 1970; Rosen et al., 1989; Yoksan and Bhamarparvathi, 1983). Owing to the morphological similarity of these cells to phagocytic cells, scientists have assumed that cells, such as macrophages and dendritic cells, are likely to be the target of the virus replication. However, isolation of virus from organs that are dominantly occupied by the phagocytic cells, such as liver, spleen, lymph node, and thymus, is an exceptional event. Surprisingly, isolation of virus from the bone marrow of dengue virus-infected patients appears to have the greatest success rate compared to organs from the rest of the body (Nelson et al., 1966; Nisalak et al., 1970). Nevertheless, the site of dengue virus replication and the source of viremia seen in peripheral blood of infected subjects remain unknown.

26.2.5 MORPHOLOGY OF DENGUE VIRAL PARTICLES

Several varieties of viral particles in dengue virus have been documented in the literature (Smith et al., 1970; Sriurairatna et al., 1973; Stevens and Schlesinger, 1965). Electron microscopy studies depict the presence of two types of virions: mature extracellular virus with M protein and immature intracellular particles containing prM, which becomes proteolytically cleaved during maturation to yield the mature form (Kuhn et al., 2002; Zhang et al., 2003). Both virions are infectious depending on the cell type (Rodenhuis-Zybert et al., 2011). Based upon buoyant density gradients, it appears as though two types of dengue virions form in different hosts (Smith et al., 1970; Stevens and Schlesinger, 1965); lipid-associated viral particles with a low buoyant density were demonstrated with purified virus from infected mouse brains, and classical/conventional viral forms can be observed in fibroblast, lymphoblast, or mosquito cell cultures (Sriurairatna et al., 1978). Two types of virions are observed in sucrose gradient preparation, and both have been demonstrated to be infectious (Junjhon et al., 2008). The main difference between the two depends on the protein content: the classical viral structures contain capsid, while the lipid-associated virions apparently do not have it (Junjhon et al., 2008). The significance of these viral morphologies relevant to disease severity has not been explored.

26.2.6 STAGES OF LIFE CYCLE

Interestingly, observations of classical viral particles in sera or plasmas obtained from acute dengue patients so far have not been achieved, in spite of several attempts covering more than six decades (Sabin, 1952b). Oddly, high viremia and infectious virus can be demonstrated by qRT-PCR, mosquito inoculation, or plaque assays. It has been speculated for more than six decades that the dengue virus may go through cycles as complicated as those of the plasmodia causing malaria (Simmons, 1931a). Evidence of two viral forms could support the hypothesis that there are at least two different life cycle stages of dengue virus that may exist; propagation through a classical dengue virus within a permissive cell and circulating as a lipid-associated virion, which buds or releases from the plasma membrane (Bargeron Clark et al., 2012). The strategy utilized by the virus may be a critical determinant of survival within the adverse host environment.

26.3 CLINICAL PRESENTATION

26.3.1 CLINICAL SYMPTOMS

There are two major categories of dengue clinical manifestations, asymptomatic and symptomatic, with the ratio of incidence between the two ranging from 1:1 to 20:1 dependent upon the geographic location, immune naivety of the population, and season (Balmaseda et al., 2010; Burke et al., 1988; Endy et al., 2011; Mammen et al., 2008; Porter et al., 2005; Wilder-Smith et al., 2009). For symptomatic dengue, two major clinical presentations are noticed: classical DF and DHF associated with or without DSS, a potential life-threatening illness characterized by plasma leakage resulting from an increase in vascular permeability. Classical DF is recognized as a self-limited illness, and patients normally recover within 2 weeks without any complications. However, a low percentage of the afflicted subjects progress to the severe form, DHF/DSS. Classical DF presentation bares similarity to other acute febrile illnesses but also initially resembles DHF, making the initial diagnosis of severe DHF very difficult and challenging to medical professionals. The unique feature that differentiates DHF from DF is the capillary/plasma leakage that is likely due to an increase in vascular permeability (WHO, 1997). In general, the onset of DHF and DSS are very abrupt, beginning with high fever. Although the common symptoms at the initial phase of the febrile stage are very similar between DF and DHF, such as headache, malaise, weakness, chills, aches and pains, sore eyes, nose bleeding, and gastrointestinal symptoms, their case definitions differ. Typical DHF is characterized by four major clinical manifestations: high fever, hemorrhagic phenomena, and often, hepatomegaly and circulatory failure. Upon physical examination, a number of conditions may be present: diffuse flushing or fleeting pinpoint eruptions may be observed on the face, neck and chest, lethargy, irritability (in young children), abdominal pain, hepatomegaly, petechial or maculopapular rashes, or bleeding manifestations (gastrointestinal bleeding, epistaxis, or gum bleeding). Atypical clinical presentations have been reported such as encephalopathy, severe hepatitis, respiratory failure, and myocarditis (WHO, 2009). DSS is one of the most serious complications in DHF. If not managed immediately, shock can lead to terminal death. The major symptoms resulting in the decease of patients suffering from shock are unconsciousness, tachycardia, decreased capillary refill, decreased body temperature, and hypertension. Hence, shock usually is the primary stage in DHF, while organ involvement and dysfunction are secondary, leading to DSS.

26.3.2 DIFFERENTIAL DIAGNOSIS

The abrupt nature of illness, the delay in time before patients seek help, and the diversity of clinical presentations make rapid and correct diagnosis a burden to physicians. Institution of the proper treatment needs to be done immediately to slow down or prevent the progression of the disease to the critical stage, DHF/DSS. During the initial febrile stage, clinical features in several common illnesses, such as bacteremia, influenza, acute respiratory illness, and measles, are very similar to that of DF, making the right differential diagnosis very challenging but is essential. Additionally, in many endemic areas, DF must be differentiated from chikungunya fever, another vector-borne viral disease that closely overlaps with DF in symptoms and epidemiology.

26.3.3 FACTORS AFFECTING THE OUTCOME OF SYMPTOMS

The wide spectrum of dengue presentations in infected patients may be the results of intrinsic and extrinsic factors. Intrinsic factors include age, ethnic background, and immune status of the individual, while extrinsic factors cover geographical locations, virus strains in circulation and serial sequence of infections, nutrition, and density, and competency of mosquitoes. As a whole, dengue virus can infect people of a wide range of ages, independent of other factors. Clinically, age has been attributed to differential presentations of the dengue features (WHO, 2009; Wilder-Smith and Schwartz, 2005). An undifferentiated febrile illness with maculopapular rash is likely to occur in young children (Burke et al., 1988), while classic dengue is more often displayed in older children, adolescents, and adults (Sharp et al., 1995). Although DF is generally benign, occasionally patients develop unusual hemorrhaging associated with severe muscle and joint pain (break-bone fever). Geographically differential clinical presentations have been documented as well. For example, DHF has been recognized as a disease primarily occurring in children under 15 years in hyperendemic areas, but in other regions, adults appear to be more likely to develop severe manifestations (Guilarde et al., 2008; Guzman and Kouri, 2003; Lin et al., 2010; Ooi et al., 2006). The reasons for the geographical variation in disease severity are currently unknown. Some advocate that different viral strains in circulation can explain the altered disease presentation (Guilarde et al., 2008; Messer et al., 2003; Rico-Hesse, 2003), while others attribute it to the genetic background (Guzman et al., 2010).

26.4 CLINICAL AND LABORATORY DIAGNOSIS

26.4.1 CLINICAL DIAGNOSIS

There are three critical stages for DHF/DSS (WHO, 1997). At the beginning of the febrile stage, clinical presentations of dengue make the disease one of the most difficult diseases to differentiate from other common febrile illnesses. DHF is characterized by a high-grade fever, hemorrhagic phenomena, hepatomegaly, and circulatory failure. The febrile stage is characterized by an acute onset of high fever, frequently as high as 39°C or 40°C, accompanied by vomiting and anorexia. The temperature usually declines to normal between the third and seventh day of illness. The patient's face is flushed, and the eyes become red and injected. When the disease progresses, most patients become critically ill on the third or fourth day after the onset of fever, accompanied with restlessness, depression, and apathy. Various forms of rash can be seen: morbilliform, scarlatiniform, petechial, purpuric, ecchymosis, or urticarial. The liver is usually enlarged and palpable early in the febrile phase, varying from just discernible to 2–4 cm below the right costal margin. These symptoms persist into the second stage with much more severe conditions. The toxic shock, hemorrhagic stage, begins with the fall in temperature, which occurs during the third to seventh day. The patients suffer from toxicity and restlessness rather than feel better. Shock associated with bleeding (melena, petechiae, and epistaxis) is one of the most dominant phenomena during this period. Shock lasts for a short period of time; the patient may die within 12–24 h or recover rapidly following volume-replacement therapy. Patients in shock are in danger of dying if they do not promptly receive the appropriate treatment. The blood pressure may fall to undetectable levels. If this occurs, the patient appears very sick with cold, clammy hands, but usually remains conscious. The pulse is aberrant, rapid, and then weak and imperceptible; death may follow after a short amount of time. Acute abdominal pain is a frequent complaint shortly before the onset of shock. Although liver size is not correlated with disease severity, hepatomegaly is frequently present in shock cases. Jaundice is infrequently observed, even in patients with an enlarged tender liver. Splenomegaly is also a rare event. Chest x-rays often reveal pleural effusion, mostly on the right side. The extent of pleural effusion is positively correlated with disease severity. At the convalescent stage, the positive tourniquet test (measurement of capillary fragility) disappears, usually on the 8th–10th day, and the patient improves rapidly; however, this depends on the severity of the symptoms and hemorrhaging. The first sign of improvement is often the return of appetite, which is followed by an increase in activity after a few days.

26.4.2 Laboratory Diagnosis

Laboratory investigations have shown that dengue patient blood hematocrit levels are as high as 50% at the time of the circulatory failure and return to normal by the convalescent stage (WHO, 1997). The leukocyte counts vary from mild leukopenia to slight leukocytosis. Day by day, whole blood counts depict a slight decrease to normal white blood cell levels on day 2 and a slight increase on days 3 and 4 of the illness. The leukocyte count stabilizes during the convalescent stage. A Schilling hematogram shifted to the left is a common finding. The number of band neutrophils increases, as does the number of monocytes. The appearance of atypical lymphocytes, ranging from 5% to 40% of the white blood cells, is a constant observation. A decrease in the number of platelets in the peripheral blood is a salient feature of DHF/DSS.

Thrombocytopenia is highly correlated with the severity of the disease and circulatory failure. The number of platelets increases very rapidly and reverts to normal within a few days to a week after the onset of shock. Studies reveal that bone marrow suppression occurs at the early stages of illness but rebounds to normal or hypercellular levels at later time points in disease. As such, dengue also has been referred to as break-bone fever. Considering all of the clinical details mentioned earlier, WHO has classified DHF/DSS into grades 1–4 according to the severity (WHO, 1997). Grade 1 includes fever accompanied by nonspecific symptoms with a suggestive hemorrhagic manifestation (confirmed by a positive tourniquet test). Grade 2 covers Grade 1 in conjunction with spontaneous bleeding, usually present in the skin. Grade 3 describes patients with circulatory failure manifested by rapid and weak pulse, narrowing of pulse pressure (20 mmHg or less), or hypotension, with the presence of cold clammy skin and restlessness. Grade 4 consists of profound shock with undetectable blood pressure and pulse. DHF grades 3 and 4 presenting with depletion of intravascular volume leading to shock are classified as DSS. Owing to abnormal or atypical clinical presentations, some of which are difficult to classify and may not fit the DHF definition, in 2009, another case classification system to improve guiding management is published by WHO (WHO, 2009). This new classification includes the dengue without warning signs, dengue with warning signs, and severe dengue (Barniol et al., 2011) and improves sensitivity but reduces specificity in detecting severe disease (Basuki et al., 2010). Therefore, this new classification was not widely adopted in many settings, and most of the surveillance systems still use the old classification of DF and DHF. Timing and accurate diagnosis are critical steps in initiating the proper treatment of dengue patients. With the complexity of clinical manifestations that are very similar to other common febrile viral infections, it is impossible to be 100% accurate in diagnosing dengue disease relying upon clinical criteria alone. Thus, a definite diagnosis is assisted by rapid test and confirmed by virus isolation and/or serology.

26.4.3 RAPID AND CONFIRMATORY DIAGNOSIS

Owing to the nature of dengue disease, with its similarity to other febrile illnesses, its ability to cause rapid deterioration, and the delayer in acquiring hospital care, it is imperative to make an accurate diagnosis at admission to avoid any further delay in treatment. There are many rapid diagnosis kits available for dengue (Guzman et al., 2010). The most commonly available commercial diagnostic kits are based upon antigen or antibody reactivity. Although these tests display high specificities, the sensitivities are usually unremarkable, ranging from very low to about 60%–70%. Therefore, laboratory confirmatory tests, such as viral isolation, dengue-specific ELISA and neutralization assays, and real-time RT-PCR, are implemented. Though these tests are time consuming, their specificities and sensitivities are usually high. One of the major drawbacks is the time required to obtain the results, which often occurs too late to dictate patient care. Thus, a better and highly sensitive rapid diagnostic tool is urgently needed for dengue diagnosis.

26.5 PATHOLOGY AND PATHOGENESIS

Dengue has been recognized as a unique entity among the flaviviruses for more than 200 years (Rush, 1789). Yet the pathogenic mechanisms remain largely unclear, due mainly to an unavailability of suitable animal models recapitulating the cardinal features of human dengue. In spite of the difficulty, some progress in understanding the pathogenesis of disease has been achieved in early pathological investigations with autopsies of deceased dengue patients. A lack of noticeable lesions in the major organs suggests that pathology cannot be considered pathognomonic; these patients normally recover very rapidly without sequelae (WHO, 1966). The occurrence of severe illness and yet very little structural injury lends credence to the hypothesis that biological mediators play a much more critical role and contribute to the systemic physiological dysfunction in patients (Gubler and Goro, 1997). Nonetheless, some of the salient features observed in cumulative autopsy specimens are worth mentioning.

26.5.1 PATHOLOGY

The acute nature of dengue disease implicates that the time of sample collection can be a critical factor in terms of pathologic findings in fatal cases. A few decades ago, when supportive or palliative care was not commonly practiced, dengue patients usually died within 24 h after admission to the hospital. In recent years, patient management has improved significantly, resulting in higher survival rates; some patients may live for a week or longer after shock, which may account for some of the unusual pathologic features documented in dengue patients (WHO, 2009). Furthermore, they may also be altered by the implementation of certain therapeutic and management procedures, such as administering high doses of steroids and excess fluid replacement, and the hemodynamic and anoxic effects of prolonged hypovolemic shock. Nevertheless, laboratory and histopathologic findings indicate that significant changes repeatedly occur in three major organs: the liver, the reticuloendothelial system, and the vascular system (WHO, 2009). The most pathognomonic changes are in the liver. Focal hyaline necrosis of liver cells and Kupffer cells and formation of Councilman bodies are a constant finding. Some of these features are reminiscent of early stages of yellow fever or other hemorrhagic fevers, such as Lassa fever and Ebola hemorrhagic fever (Rosen et al., 1989). Infrequently, fatty metamorphosis of liver cells or changes in circulatory disturbances such as focal hemorrhages, passive congestion, and centrilobular necrosis may occur (WHO, 1966). However, the pathologic profiles may be different with epidemics in different years and geographic locations (WHO, 2009). The reticuloendothelial system shows expansion en masse of mature, intermediate, and blastoid B lymphocytes. A marked increase in the activity of B lymphocytes, with active proliferation of plasma cells and lymphocytoid cells and the appearance of active germinal centers in lymphoid tissues, has been observed. There is an increase in the turnover of lymphocytes, including cytolysis and phagocytosis of lymphocytes. Dengue viral antigen has been observed in fixed reticuloendothelial cells in the spleen, liver, lungs, and thymus gland (WHO, 2009). However, due to the phagocytic nature of cells in these organs, the significance of localizing viral antigen in these tissues is not clear. Whether these cells serve as a reservoir for dengue virus remains in question. The levels of viral antigen in these organs appear to be correlated with age; higher levels are found in children <1 year of age than in older children. The underlying difference is poorly understood, but it has been suggested that primary dengue virus infection is more dominant in the former group, while secondary infections in the latter. There are two major categories of pathologic abnormalities in the vascular system: leakage of high-protein exudates (>4 g of protein/dL) into extravascular and serosal spaces and hemorrhages associated with diapedesis but without detectable signs of inflammation or necrosis. Considering the minimal degree of tissue death and structural injury and those DHF/DSS survivors always recover very rapidly, this suggests that the physiological dysfunction is secondary to the action of biologic mediators.

Complement components C3a and C5a, to name a few, have been implicated to be the front-runners among possible mediators, mainly owing to the enhanced activation of the complement system in DHF/DSS. Pathologic conditions from other organs have been documented even though it is generally perceived that these features are abnormal. Thickening of interstitial septa with an increase in the numbers of mononuclear cells, alveolar macrophages, and megakaryocytes in the lungs has been observed. Although the brain and spinal cord may exhibit perivascular edema and hemorrhaging, there is no histopathologic evidence of meningitis, encephalitis, neuronal necrosis, or proliferation of glial cells. In the kidneys, no evidence of acute tubular or bilateral cortical necrosis has been found even though increases in the cellularity of the glomerular capillary tuffs and anoxic tubular changes do transpire. Also the heart typically shows flame-shaped subendocardial hemorrhages in the left ventricular while there is no evidence of myocarditis; however, edema and focal perivascular hemorrhages have been documented in this tissue.

26.5.2 PATHOGENESIS

There is a limited understanding of dengue virus pathogenesis-induced hemorrhages. Variation in the degree of hemorrhaging has been associated with the age of patients (WHO, 2009). In infants and children, hypovolemia appears to be due mainly to the reduction in plasma volume rather than to the formation of hemorrhages. In adolescents and young adults or in children with prolonged shock, hemorrhaging appears to play a more prominent role. In addition, disseminated intravascular coagulation (DIC) can only be detected in adolescents and adults and not in young children with severe DHF/DSS. As a whole, hemorrhaging in DHF/DSS is a combination of multiple factors, including thrombocytopenia, coagulopathy, and undefined vasculopathy.

26.5.2.1 Thrombocytopenia

Thrombocytopenia is one of the major hallmarks of DHF/DSS. Importantly, the degree of thrombocytopenia corresponds with disease severity and is one of the diagnosis criteria for dengue patients. Several testable hypotheses have been proposed to address the cause of the low platelet counts in dengue patients. These include decreased production, increased consumption, viral tropism and antibody-mediated lysis resulting in dysfunction, aggregation with leukocytes, and engulfment by phagocytic cells (Funahara et al., 1987a; Mitrakul et al., 1977; Noisakran et al., 2009a; Srichaikul and Nimmannitya, 2000; Tsai et al., 2011; Wang et al., 1995). A series of studies of peripheral blood and bone marrow in patients with DHF suggest that dengue virus exerts a depressive effect on bone marrow elements, including megakaryocytes, in early course of disease (Bierman and Nelson,

1965; Nakao et al., 1989; Noisakran et al., 2012; Phanichyakarn et al., 1977; Rothwell et al., 1996; Srichaikul et al., 1989; WHO, 1966). Platelet dysfunction and shortened survival times have been demonstrated by ^{125}I-labeled isologous platelets in patients with DHF (Mitrakul et al., 1977). The labeled platelets were sequestered primarily in the liver, with a shift to the spleen as the patient recovered. Immune-mediated platelet lysis is supported by the demonstration of immune complexes containing dengue viral antigen via Fc receptors on the platelet surfaces (Boonpucknavig et al., 1979b). Direct viral tropism was subsequently shown by the observation of dengue viral particles in platelets isolated from DHF patients (Noisakran et al., 2009b). Furthermore, evidence of increased consumption was validated when dengue virus-containing platelets were engulfed by monocytes in infected rhesus monkeys (Onlamoon et al., 2010) and activated platelets aggregated with monocytes during defervescent stage in humans (Tsai et al., 2011). Similarly performed in vitro experiments incorporating antibodies also support these observations, in which antibody-coated dengue virus can bind platelets (Wang et al., 1995). Thus, viral tropism, binding, and engulfment by phagocytic cells are likely associated with the decrease in platelet counts.

26.5.2.2 Coagulopathy

Coagulation defects in patients with DHF/DSS have been investigated extensively and well documented (Srichaikul and Nimmannitya, 2000). Generally speaking, although DIC is not a required component in the pathogenesis of DHF/DSS, relevant elements, such as prolonged partial thromboplastin time, prothrombin time, and thrombin time, and variable decreases in the activity of factors II, V, VII, VIII, IX, and X are common findings (Funahara et al., 1987b; Isarangkura et al., 1987; Srichaikul et al., 1977; Suvatte et al., 1973; Weiss and Halstead, 1965).

26.5.2.3 Vasculopathy

Vasculopathy remains a poorly understood factor underlying bleeding and increased vascular permeability in DHF/DSS. Cumulative evidence indicates that hemorrhages in the form of erythrocyte diapedesis are probably localized to capillaries, whereas leakage of plasma occurs from venules (Bhamarapravati, 1989). Skin biopsy specimens from individuals with petechial rashes contain lesions in the microvasculature of dermal papillae and lymphocyte and mononuclear phagocyte infiltrates in the vessel walls (Boonpucknavig et al., 1979a). Some of these cells contain dengue viral antigen, deposits of IgM, complement, and fibrinogen (Boonpucknavig et al., 1979a). However, electron micrograms of dermal vessels revealed only nonspecific changes associated with increased numbers of transport vesicles in the cytoplasm of endothelial cells (Bhamarapravati et al., 1967). Importantly, dengue virus has not been detected in any cells of the skin by electron microscopy (Bhamarapravati, 1989).

26.6 EPIDEMIOLOGY

Dengue is a mosquito-borne viral disease and recently has become a significant global disease burden. The latest number indicates that more than 100 countries worldwide are affected by dengue, and the regions most vulnerable to epidemics are in Southeast Asia, South America, and the West Pacific. The WHO reports that about 2.5 billion people or about two-fifth of the world's population are at risk of infection and 50–100 million cases of dengue occur annually (WHO, 2009).

26.6.1 HISTORY OF OUTBREAK

The very first description of a possible dengue outbreak was documented by Benjamin Rush in 1789 (Rush, 1789). The disease appeared during an unusually warm season with a high density of mosquitoes. The literature also reveals that sporadic outbreaks took place on different continents during the eighteenth century, potentially attributable to an increase in trading and commercial shipping. These activities were likely practiced in the torrid zones, consequently establishing dengue disease widely throughout the tropics and subtropics.

26.6.2 Re-Emerging Dengue Outbreak

Dengue was prevalent in Southeast Asia throughout WWII; with the deployment of the soldiers and personnel back and forth to war zones, rapid spread of the disease to other parts of the world incited a pandemic (Sabin, 1952a). Research on the causative agent of dengue was not hindered by the war. The virus was isolated almost simultaneously by researchers from Japan, Dr. Susumu Hotta in 1943 (Kimura and Hotta, 1944), and from the United States, Dr. Albert Sabin in 1944 (Sabin and Schlesinger, 1945). With the available tools, methods, such as the complement-fixation and neutralization assays, were developed to detect the virus (Sabin, 1952b). Subsequently, different dengue serotypes were discovered. A major outbreak of DHF in Philippines and Thailand in late 1950 led to more serotypes being isolated and verified (Hammon et al., 1960). These different serotypes became persistently established and co-circulated in the tropical and subtropical regions causing endemics annually thereafter.

With factors such as climate change, increasing travel, and unplanned urbanization, the serotypes continued to spread globally, eventually reaching areas of South and Central America, Cuba, Puerto Rico, and recently the United States (WHO, 2009). The incidence of dengue, particularly DHF, may continue to escalate with the establishment of more dengue serotypes in new parts of the world.

26.6.3 Risk Factors for Outbreak

In dengue-endemic regions, the majority of people have antibody to dengue virus. But interestingly, only a small percentage of infected subjects develop clinical symptoms and severe DHF/DSS. The reasons for this are unclear, despite significant interest and research emphasis for several decades. Several potential risk factors are involved, including age, individual's immunity, genetic background, nutritional status, underlying illness, and viral strains (WHO, 1997).

26.7 PROGNOSIS AND TREATMENT

26.7.1 Prognosis

Although patients infected with dengue display a wide array of symptoms, the majority of subjects are asymptomatic (WHO, 1997). For those experiencing a primary infection, the disease is normally mild, associated with malaise. The prognosis for this group is excellent, most patients recovering without any sequelae. For those who have an underlying illness or take immunosuppressive medicines, accompanying complications may occur with dengue, and the prognosis is fair to good. For those living in dengue-endemic zones, despite having antibody to one or more dengue serotypes, reinfections by similar or other dengue serotypes are still prominent. Thus, people in high-risk regions are likely to develop much more complicated illnesses, which results in a less optimal prognosis. Subjects who suffer DHF/DSS can vary in clinical outcomes from good to very poor, depending on the patient's underlying medical conditions and how quickly supportive measures are given. The return of appetite is a good prognostic sign for this group. The recovery time for this group may be slow, with associated weakness and mental depression. In some patients, continued pain in the bones and joints is felt during the recovery phase. The mortality rate is higher in high-risk patients who are more prone to serious complications (WHO, 2009). The rate can reach 15% in patients who receive improper management and 100% in those with DSS and massive uncontrolled bleeding. However, because of the awareness and experience of the attending physicians and progress made in comprehensive care techniques, the mortality rate among patients with DHF/DSS has progressively declined from 20% to below 0.1% in recent years in dengue-endemic regions. As a whole, the fatality rate averages at 1% for all dengue virus hospitalizations, though in some endemic countries, the rate can be as high as 5%, death predominantly occurring in children under age of 5 (WHO, 2009).

26.7.2 Treatment

Since there is no specific treatment or antiviral therapy for DHF/DSS, timing palliative care is critical in patient survival. Survival correlates with early hospitalization and aggressive supportive treatment. Patients who do not receive the proper treatment usually die within 24 h after shock. Aspirin and other nonsteroidal anti-inflammatory drugs, such as ibuprofen, are not recommended. Hence, vigilance and constant clinical and laboratory monitoring by attending physicians and nurses are the most important aspect in managing patients with DHF/DSS.

26.7.3 Vaccines and Antiviral Drugs

The best strategy to reduce dengue disease would be administering a vaccine preventing dengue virus infection. The development of a dengue vaccine has been attempted for more than six decades (WHO, 2010). Various forms of formula have been tried, including inactivated and attenuated—live, subunits, chimera-backbone, and DNA-based vaccines (Gusman, 2011; Murphy and Whitehead, 2011; Perng et al., 2011; Rothman, 2011; Schmitz et al., 2011; Swaminathan et al., 2010; Thomas and Endy, 2011; Webster et al., 2009). Most of the vaccines under development show good immunogenicity and reactogenicity, but there are no approved vaccines available yet for public use. However, within the next few years, one of the vaccines from the pipelines may become available. The major challenge in dengue vaccine development so far is to create and guarantee a formula that can elicit equal immunogenicity among all four serotypes. The unique scenario of dengue virus infection in infected patients is that viremia is often at the peak or downward at the time that afflicted patients seek help. Thus, technically speaking, though it is possible, it is a very difficult and highly challenging task to develop a therapeutic antiviral drug that can have a significant impact on the viral load in patients within a short time frame.

26.8 CONCLUSIONS

Dengue, a vector-borne human disease, is a re-emerging public health threat. The disease is characterized by an acute onset of fever accompanied with a dynamic spectrum of other symptoms; these clinical manifestations are highly similar and confused with other common febrile illnesses. Making the differential diagnosis early in the febrile stage and administering the proper treatment in a timely manner remain a challenging task to medical professionals. Although, the majority of affected patients experience DF, a self-limited illness, some subjects progress to the life-threatening condition, DHF/DSS, potentially resulting in death. The clinical presentations of DF and DHF/DSS are very similar, but the distinct features in DHF/DSS are hemoconcentration and plasma leakage, likely resulting from increased vascular permeability. Despite intensive efforts made to understand the factors that may contribute to the wide clinical spectrum of disease, the pathogenic mechanisms remain poorly understood. Without a suitable animal model recapitulating the cardinal features of human dengue, gaps in our understanding on the aspects of pathogenesis, such as viral tropisms and immune-mediated thrombocytopenia, will persist. Several control measures can be applied; these include developing better mosquito control programs and designing vaccines and antiviral drugs. Mosquito control may be easy to practice, but the efficacy is uncertain. Currently, there is no vaccine or antiviral therapeutic drug available to prevent or treat dengue. Supportive care is the only approved practice to mitigate the progression of the illness. Considering the viremic nature of dengue virus infections and that the majority of infected individuals are asymptomatic, dengue virus contamination is a likely incidence and should be treated as a major threat to the blood supply in dengue-endemic countries (Wilder-Smith et al., 2009).

REFERENCES

Anoop, M., Issac, A., Mathew, T., Philip, S., Kareem, N.A., Unnikrishnan, R., Sreekumar, E., 2010. Genetic characterization of dengue virus serotypes causing concurrent infection in an outbreak in Ernakulam, Kerala, South India. *Indian J Exp Biol* 48, 849–857.

Ashburn, P.M., Craig., C.F., 1907. Experimental investigations regarding the etiology of dengue. *J Infect Dis* 4, 440–475.

Balmaseda, A., Standish, K., Mercado, J.C., Matute, J.C., Tellez, Y., Saborio, S., Hammond, S.N. et al., 2010. Trends in patterns of dengue transmission over 4 years in a pediatric cohort study in Nicaragua. *J Infect Dis* 201, 5–14.

Bancroft, T.L., 1906. On the etiology of dengue fever. *Australas Med Gaz* 25, 17.

Bargeron Clark, K., Hsiao, H.M., Noisakran, S., Tsai, J.J., Perng, G.C., 2012. Role of microparticles in dengue virus infection and its impact on medical intervention strategies. *Yale J Biol Med* 85, 3–18.

Barniol, J., Gaczkowski, R., Barbato, E.V., da Cunha, R.V., Salgado, D., Martinez, E., Segarra, C.S. et al., 2011. Usefulness and applicability of the revised dengue case classification by disease: Multi-centre study in 18 countries. *BMC Infect Dis* 11, 106.

Basuki, P.S., Budiyanto, Puspitasari, D., Husada, D., Darmowandowo, W., Ismoedijanto, Soegijanto, S., Yamanaka, A., 2010. Application of revised dengue classification criteria as a severity marker of dengue viral infection in Indonesia. *Southeast Asian J Trop Med Public Health* 41, 1088–1094.

Beasley, D.W.C., Barret, A.D., 2008. The infectious agent, in: Halstead, S.B. (Ed.), *Dengue*. Imperial College Press, London, U.K., pp. 29–73.

Bhamarapravati, N., 1989. Hemostatic defects in dengue hemorrhagic fever. *Rev Infect Dis* 11(Suppl II), s826–s829.

Bhamarapravati, N., Boonyapaknavik, V., 1966. Pathogenetic studies on Thai haemorrhagic fever: immuno-fluorescent localization of dengue virus in human tissue. *Bull World Health Organ* 35, 50–51.

Bhamarapravati, N., Tuchinda, P., Boonyapaknavik, V., 1967. Pathology of Thailand haemorrhagic fever: A study of 100 autopsy cases. *Ann Trop Med Parasitol* 61, 500–510.

Bierman, H.R., Nelson, E.R., 1965. Hematodepressive virus diseases of Thailand. *Ann Intern Med* 62, 867–884.

Blanc, G., Caminopetros, J., 1929. Contributions to the study of vaccination against dengue. *Bull Acad Med* 102, 40–47.

Boonpucknavig, S., Boonpucknavig, V., Bhamarapravati, N., Nimmannitya, S., 1979a. Immunofluorescence study of skin rash in patients with dengue hemorrhagic fever. *Arch Pathol Lab Med* 103, 463–466.

Boonpucknavig, S., Vuttiviroj, O., Bunnag, C., Bhamarapravati, N., Nimmanitya, S., 1979b. Demonstration of dengue antibody complexes on the surface of platelets from patients with dengue hemorrhagic fever. *Am J Trop Med Hyg* 28, 881–884.

Boulton, R.W., Westaway, E.G., 1972. Comparisons of togaviruses: Sindbis virus (group A) and Kunjin virus (group B). *Virology* 49, 283–289.

Burke, D.S., Nisalak, A., Johnson, D.E., Scott, R.M., 1988. A prospective study of dengue infections in Bangkok. *Am J Trop Med Hyg* 38, 172–180.

Cassetti, M.C., Durbin, A., Harris, E., Rico-Hesse, R., Roehrig, J., Rothman, A., Whitehead, S., Natarajan, R., Laughlin, C., 2010. Report of an NIAID workshop on dengue animal models. *Vaccine* 28, 4229–4234.

Chambers, T.J., Hahn, C.S., Galler, R., Rice, C.M., 1990. Flavivirus genome organization, expression, and replication. *Annu Rev Microbiol* 44, 649–688.

Chiu, W.W., Kinney, R.M., Dreher, T.W., 2005. Control of translation by the 5′- and 3′-terminal regions of the dengue virus genome. *J Virol* 79, 8303–8315.

Cleaves, G.R., Dubin, D.T., 1979. Methylation status of intracellular dengue type 2 40 S RNA. *Virology* 96, 159–165.

Cleland, J.B., Bradley, B., McDonald, W., 1916. On the transmission of Australian Dengue by the Mosquito Stegomyia Fasciata. *Med J Australia* 2, 179–184.

Cleland, J.B., Bradley, B., MacDonald, W., 1919. Further experiments in the etiology of dengue fever. *J Hyg (Lond)* 18, 217–254.

Clyde, K., Harris, E., 2006. RNA secondary structure in the coding region of dengue virus type 2 directs translation start codon selection and is required for viral replication. *J Virol* 80, 2170–2182.

Edgil, D., Polacek, C., Harris, E., 2006. Dengue virus utilizes a novel strategy for translation initiation when cap-dependent translation is inhibited. *J Virol* 80, 2976–2986.

Endy, T.P., Anderson, K.B., Nisalak, A., Yoon, I.K., Green, S., Rothman, A.L., Thomas, S.J., Jarman, R.G., Libraty, D.H., Gibbons, R.V., 2011. Determinants of inapparent and symptomatic dengue infection in a prospective study of primary school children in Kamphaeng Phet, Thailand. *PLoS Negl Trop Dis* 5, e975.

Flint, S.J., Enquist, L.W., Racaniello, V.R., Skalka, A.M., 2000. *Principles of Virology,* 2nd edn. ASM Press, Washington, DC.

Funahara, Y., Ogawa, K., Fujita, N., Okuno, Y., 1987a. Three possible triggers to induce thrombocytopenia in dengue virus infection. *Southeast Asian J Trop Med Public Health* 18, 351–355.

Funahara, Y., Sumarmo, Shirahata, A., Setiabudy-Dharma, R., 1987b. DHF characterized by acute type DIC with increased vascular permeability. *Southeast Asian J Trop Med Public Health* 18, 346–350.

Gubler, D.J., Goro, K., 1997. *Dengue and Dengue Hemorrhagic Fever*, 1st edn. CABI, Wallingford, U.K.

Gubler, D.J., Kuno, G., Sather, G.E., Waterman, S.H., 1985. A case of natural concurrent human infection with two dengue viruses. *Am J Trop Med Hyg* 34, 170–173.

Guilarde, A.O., Turchi, M.D., Siqueira, J.B., Jr., Feres, V.C., Rocha, B., Levi, J.E., Souza, V.A., Boas, L.S., Pannuti, C.S., Martelli, C.M., 2008. Dengue and dengue hemorrhagic fever among adults: clinical outcomes related to viremia, serotypes, and antibody response. *J Infect Dis* 197, 817–824.

Gusman, M.G., 2011. Dengue vaccines: new developments. *Drugs Future* 36, 45–62.

Guzman, M.G., Halstead, S.B., Artsob, H., Buchy, P., Farrar, J., Gubler, D.J., Hunsperger, E. et al., 2010. Dengue: a continuing global threat. *Nat Rev Microbiol* 8, S7–S16.

Guzman, M.G., Kouri, G., 2003. Dengue and dengue hemorrhagic fever in the Americas: lessons and challenges. *J Clin Virol* 27, 1–13.

Hammon, W.M., Rudnick, A., Sather, G.E., 1960. Viruses associated with epidemic hemorrhagic fevers of the Philippines and Thailand. *Science* 131, 1102–1103.

Hirsch, A., 1883. *Dengue, a Comparatively New Disease: Its Symptoms, Handbook of Geographical and Historical Pathology*. Sydenham Society, London, U.K., pp. 55–81.

Holmes, E.C., Burch, S.S., 2000. The causes and consequences of genetic variation in dengue virus. *Trends Microbiol* 8, 74–77.

Hotta, S., 1952. Experimental studies on dengue I isolation, identification and modification of the virus. *J Infect Dis* 90, 1–9.

Howe, G.M., 1977. *A World Geography of Human Diseases*. Academic Press, New York.

Isarangkura, P.B., Pongpanich, B., Pintadit, P., Phanichyakarn, P., Valyasevi, A., 1987. Hemostatic derangement in dengue haemorrhagic fever. *Southeast Asian J Trop Med Public Health* 18, 331–339.

Jessie, K., Fong, M.Y., Devi, S., Lam, S.K., Wong, K.T., 2004. Localization of dengue virus in naturally infected human tissues, by immunohistochemistry and in situ hybridization. *J Infect Dis* 189, 1411–1418.

Junjhon, J., Lausumpao, M., Supasa, S., Noisakran, S., Songjaeng, A., Saraithong, P., Chaichoun, K. et al., 2008. Differential modulation of prM cleavage, extracellular particle distribution, and virus infectivity by conserved residues at nonfurin consensus positions of the dengue virus pr-M junction. *J Virol* 82, 10776–10791.

Kimura, R., Hotta, S., 1944. Studies on dengue fever (VI). On the inoculation of dengue virus into mice (in Japanese). *Nippom Igaku*, 3379, 629–633.

Kuhn, R.J., Zhang, W., Rossmann, M.G., Pletnev, S.V., Corver, J., Lenches, E., Jones, C.T. et al., 2002. Structure of dengue virus: implications for flavivirus organization, maturation, and fusion. *Cell* 108, 717–725.

Kurane, I., Kontny, U., Janus, J., Ennis, F.A., 1990. Dengue-2 virus infection of human mononuclear cell lines and establishment of persistent infections. *Arch Virol* 110, 91–101.

Lin, C.C., Huang, Y.H., Shu, P.Y., Wu, H.S., Lin, Y.S., Yeh, T.M., Liu, H.S., Liu, C.C., Lei, H.Y., 2010. Characteristic of dengue disease in Taiwan: 2002–2007. *Am J Trop Med Hyg* 82, 731–739.

Lorono-Pino, M.A., Cropp, C.B., Farfan, J.A., Vorndam, A.V., Rodriguez-Angulo, E.M., Rosado-Paredes, E.P., Flores-Flores, L.F., Beaty, B.J., Gubler, D.J., 1999. Common occurrence of concurrent infections by multiple dengue virus serotypes. *Am J Trop Med Hyg* 61, 725–730.

Mammen, M.P., Pimgate, C., Koenraadt, C.J., Rothman, A.L., Aldstadt, J., Nisalak, A., Jarman, R.G. et al., 2008. Spatial and temporal clustering of dengue virus transmission in Thai villages. *PLoS Med* 5, e205.

McSherry, J.A., 1982. Some medical aspects of the Darien scheme: was it dengue. *Scott Med J* 27, 183–184.

Messer, W.B., Gubler, D.J., Harris, E., Sivananthan, K., de Silva, A.M., 2003. Emergence and global spread of a dengue serotype 3, subtype III virus. *Emerg Infect Dis* 9, 800–809.

Mitrakul, C., Poshyachinda, M., Futrakul, P., Sangkawibha, N., Ahandrik, S., 1977. Hemostatic and platelet kinetic studies in dengue hemorrhagic fever. *Am J Trop Med Hyg* 26, 975–984.

Murphy, B.R., Whitehead, S.S., 2011. Immune response to dengue virus and prospects for a vaccine. *Annu Rev Immunol* 29, 587–619.

Nakao, S., Lai, C.J., Young, N.S., 1989. Dengue virus, a flavivirus, propagates in human bone marrow progenitors and hematopoietic cell lines. *Blood* 74, 1235–1240.

Nelson, E.R., Bierman, H.R., Chulajata, R., 1966. Hematologic phagocytosis in postmortem bone marrows of dengue hemorrhagic fever. (Hematologic phagocytosis in Thai hemorrhagic fever). *Am J Med Sci* 252, 68–74.

Nisalak, A., Halstead, S.B., Singharaj, P., Udomsakdi, S., Nye, S.W., Vinijchaikul, K., 1970. Observations related to pathogenesis of dengue hemorrhagic fever. 3. Virologic studies of fatal disease. *Yale J Biol Med* 42, 293–310.

Nobuchi, H., 1979. The symptoms of a dengue-like illness recorded in a Chinese medical encyclopedia (in Japanese). *Kanp Rinsho* 26, 422–425.

Noisakran, S., Chokephaibulkit, K., Songprakhon, P., Onlamoon, N., Hsiao, H.M., Villinger, F., Ansari, A., Perng, G.C., 2009a. A re-evaluation of the mechanisms leading to dengue hemorrhagic fever. *Ann NY Acad Sci* 1171(Suppl 1), E24–E35.

Noisakran, S., Gibbons, R.V., Songprakhon, P., Jairungsri, A., Ajariyakhajorn, C., Nisalak, A., Jarman, R.G., Malasit, P., Chokephaibulkit, K., Perng, G.C., 2009b. Detection of dengue virus in platelets isolated from dengue patients. *Southeast Asian J Trop Med Public Health* 40, 253–262.

Noisakran, S., Onlamoon, N., Hsiao, H.M., Clark, K.B., Villinger, F., Ansari, A.A., Perng, G.C., 2012. Infection of bone marrow cells by dengue virus in vivo. *Exp Hematol* 40, 250–259 e254.

Noisakran, S., Perng, G.C., 2008. Alternate hypothesis on the pathogenesis of Dengue Hemorrhagic Fever (DHF)/Dengue Shock Syndrome (DSS) in dengue virus infection. *Exp Biol Med (Maywood)* 233, 401–408.

Onlamoon, N., Noisakran, S., Hsiao, H.M., Duncan, A., Villinger, F., Ansari, A.A., Perng, G.C., 2010. Dengue virus-induced hemorrhage in a nonhuman primate model. *Blood* 115, 1823–1834.

Ooi, E.E., Goh, K.T., Gubler, D.J., 2006. Dengue prevention and 35 years of vector control in Singapore. *Emerg Infect Dis* 12, 887–893.

Pepin, K.M., Lambeth, K., Hanley, K.A., 2008. Asymmetric competitive suppression between strains of dengue virus. *BMC Microbiol* 8, 28.

Pepper, O.H.P., 1941. A note on David Bylon and Dengue. *Ann Med Hist* 3, 363–368.

Perng, G.C., Lei,H.Y., lin, Y.S., Choekphaibulkit, K., 2011. Dengue vaccines: Challenge and confrontation. *World J Vac* 1, 109–130.

Phanichyakarn, P., Pongpanich, B., Israngkura, P.B., Dhanamitta, S., Valyasevi, A., 1977. Studies on Dengue hemorrhagic fever. III. Serum complement (C3) and platelet studies. *J Med Assoc Thai* 60, 301–306.

Porter, K.R., Beckett, C.G., Kosasih, H., Tan, R.I., Alisjahbana, B., Rudiman, P.I., Widjaja, S. et al., 2005. Epidemiology of dengue and dengue hemorrhagic fever in a cohort of adults living in Bandung, West Java, Indonesia. *Am J Trop Med Hyg* 72, 60–66.

Preugschat, F., Yao, C.W., Strauss, J.H., 1990. In vitro processing of dengue virus type 2 nonstructural proteins NS2A, NS2B, and NS3. *J Virol* 64, 4364–4374.

Rico-Hesse, R., 2003. Microevolution and virulence of dengue viruses. *Adv Virus Res* 59, 315–341.

Rodenhuis-Zybert, I.A., Moesker, B., da Silva Voorham, J.M., van der Ende-Metselaar, H., Diamond, M.S., Wilschut, J., Smit, J.M., 2011. A fusion-loop antibody enhances the infectious properties of immature flavivirus particles. *J Virol* 85, 11800–11808.

Rosen, L., Khin, M.M., U, T., 1989. Recovery of virus from the liver of children with fatal dengue: reflections on the pathogenesis of the disease and its possible analogy with that of yellow fever. *Res Virol* 140, 351–360.

Rosen, L., Rozeboom, L.E., Sweet, B.H., Sabin, A.B., 1954. The transmission of dengue by Aedes polynesiensis Marks. *Am J Trop Med Hyg* 3, 878–882.

Rothman, A.L., 2011. Immunity to dengue virus: a tale of original antigenic sin and tropical cytokine storms. *Nat Rev Immunol* 11, 532–543.

Rothwell, S.W., Putnak, R., La Russa, V.F., 1996. Dengue-2 virus infection of human bone marrow: Characterization of dengue-2 antigen-positive stromal cells. *Am J Trop Med Hyg* 54, 503–510.

Rudnick, A., Lim, T.W. 1986. I.M.R. *Dengue Fever Studies in Malaysia*. The Institute for Medical Research, Kuala Lumpur, Malaysia.

Rush, A.B., 1789. *An Account of the Bilious Remitting Fever, as It Appeared in Philadelphia in the Summer and Autumn of the Year, Medical Inquiries and Observation.* Prichard $ Hall, Philadelphia, PA, pp. 104–117.

Sabin, A.B., 1952a. *Dengue*, 2nd edn. J.B. Lippincott Co., Philadelphia, PA.

Sabin, A.B., 1952b. Research on dengue during World War II. *Am J Trop Med Hyg* 1, 30–50.

Sabin, A.B., Schlesinger, R.W., 1945. Production of immunity to dengue with virus modified by propagation in mice. *Science* 101, 640–642.

Schmitz, J., Roehrig, J., Barrett, A., Hombach, J., 2011. Next generation dengue vaccines: A review of candidates in preclinical development. *Vaccine* 29(42), 7276–7294. WHO 210. Published on 2010, Reference number SEA-DEN-9. New Delhi, India.

Sharp, T.W., Wallace, M.R., Hayes, C.G., Sanchez, J.L., DeFraites, R.F., Arthur, R.R., Thornton, S.A. et al., 1995. Dengue fever in U.S. troops during Operation Restore Hope, Somalia, 1992–1993. *Am J Trop Med Hyg* 53, 89–94.

Siler, J.F., Hall, M.W., Hitchins, A.P., 1926. Dengue: Its history, epidemiology, mechanism of transmission, etiology, clinical manifestations, immunity and prevention. *Philippine J Sci* 29, 1–340.

Simmons, J.S., 1931a. Dengue fever. *Am J Trop Med* 11, 77–102.

Simmons, J.S., St. John, J.H., Reynolds, F.H.K., 1931b. Experimental studies of dengue. *Philippine J Sci* 44, 1–251.

Smith, G.W., Wright, P.J., 1985. Synthesis of proteins and glycoproteins in dengue type 2 virus-infected vero and Aedes albopictus cells. *J Gen Virol* 66 (Pt 3), 559–571.

Smith, T.J., Brandt, W.E., Swanson, J.L., McCown, J.M., Buescher, E.L., 1970. Physical and biological properties of dengue-2 virus and associated antigens. *J Virol* 5, 524–532.

Srichaikul, T., Nimmannitya, S., 2000. Haematology in dengue and dengue haemorrhagic fever. *Baillieres Best Pract Res Clin Haematol* 13, 261–276.

Srichaikul, T., Nimmanitaya, S., Artchararit, N., Siriasawakul, T., Sungpeuk, P., 1977. Fibrinogen metabolism and disseminated intravascular coagulation in dengue hemorrhagic fever. *Am J Trop Med Hyg* 26, 525–532.

Srichaikul, T., Nimmannitya, S., Sripaisarn, T., Kamolsilpa, M., Pulgate, C., 1989. Platelet function during the acute phase of dengue hemorrhagic fever. *Southeast Asian J Trop Med Public Health* 20, 19–25.

Sriurairatna, S., Bhamarapravati, N., Diwan, A.R., Halstead, S.B., 1978. Ultrastructural studies on dengue virus infection of human lymphoblasts. *Infect Immun* 20, 173–179.

Sriurairatna, S., Bhamarapravati, N., Phalavadhtana, O., 1973. Dengue virus infection of mice: morphology and morphogenesis of dengue type-2 virus in suckling mouse neurones. *Infect Immun* 8, 1017–1028.

Stevens, T.M., Schlesinger, R.W., 1965. Studies on the nature of dengue viruses. I. Correlation of particle density, infectivity, and RNA content of type 2 virus. *Virology* 27, 103–112.

Stohlman, S.A., Wisseman, C.L., Jr., Eylar, O.R., Silverman, D.J., 1975. Dengue virus-induced modifications of host cell membranes. *J Virol* 16, 1017–1026.

Suvatte, V., Pongpipat, D., Tuchinda, S., Ratanawongs, A., Tuchinda, P., Bukkavesa, S., 1973. Studies on serum complement C3 and fibrin degradation products in Thai hemorrhagic fever. *J Med Assoc Thai* 56, 24–32.

Swaminathan, S., Batra, G., Khanna, N., 2010. Dengue vaccines: state of the art. *Expert Opin Ther Pat* 20, 819–835.

Thomas, S.J., Endy, T.P., 2011. Vaccines for the prevention of dengue: Development update. *Hum Vac* 7, 674–684.

Tsai, J.J., Jen, Y.H., Chang, J.S., Hsiao, H.M., Noisakran, S., Perng, G.C., 2011. Frequency alterations in key innate immune cell components in the peripheral blood of dengue patients detected by FACS analysis. *J Innate Immun* 3, 530–540.

Tsai, J.-J., Liu, L.-T., Chang, K., Wang, S.-H., Hsiao, H.-M., Clark, K.B., Perng, G.C., 2012. The importance of hematopoietic progenitor cells in dengue. *Ther Adv Hematol* 3, 59–71.

Vasilakis, N., Cardosa, J., Diallo, M., Sall, A.A., Holmes, E.C., Hanley, K.A., Weaver, S.C., Mota, J., Rico-Hesse, R., 2010. Sylvatic dengue viruses share the pathogenic potential of urban/endemic dengue viruses. *J Virol* 84, 3726–3727; author reply 3727–3728.

Wang, S., He, R., Patarapotikul, J., Innis, B.L., Anderson, R., 1995. Antibody-enhanced binding of dengue-2 virus to human platelets. *Virology* 213, 254–257.

Weaver, S.C., Barrett, A.D., 2004. Transmission cycles, host range, evolution and emergence of arboviral disease. *Nat Rev Microbiol* 2, 789–801.

Webster, D.P., Farrar, J., Rowland-Jones, S., 2009. Progress towards a dengue vaccine. *Lancet Infect Dis* 9, 678–687.

Weiss, H.J., Halstead, S.B., 1965. Studies of hemostasis in Thai hemorrhagic fever. *J Pediatr* 66, 918–926.

Wengler, G., Gross, H.J., 1978. Studies on virus-specific nucleic acids synthesized in vertebrate and mosquito cells infected with flaviviruses. *Virology* 89, 423–437.

WHO, 1966. Summaries of papers presented at the WHO inter-regional seminar on mosquito-borne haemorrhagic fevers in the South-East Asia and Western Pacific regions. Bulletin 35, 1–95.

WHO, 1997. *Dengue Haemorrhagic Fever: Diagnosis, Treatment, Prevention and Control*, 2nd edn. World Health Organization, Geneva, Switzerland.

WHO, 2009. *Dengue Guidelines for Diagnosis, Treatment, Prevention and Control*, World Health Organization, Geneva, Switzerland.

WHO, 2010. Dengue vaccine development: the role of the WHO South-East Asia Regional Office. Published on 2010, Reference number SEA-DEN-9. New Delhi, India.

Wilder-Smith, A., Chen, L.H., Massad, E., Wilson, M.E., 2009. Threat of dengue to blood safety in dengue-endemic countries. *Emerg Infect Dis* 15, 8–11.

Wilder-Smith, A., Schwartz, E., 2005. Dengue in travelers. *N Engl J Med* 353, 924–932.

Yoksan, S., Bhamarapravati, N., 1983. Localization of dengue antigens in tissue specimens from fatal cases of dengue hemorrhagic fever, in: Pang, T., and Pathmanathan, R. (Ed.), *Proceedings of the International Conference on Dengue/Dengue Hemorrhagic Fever*, University of Malaya, Kuala Lumpur, Malaysia, pp. 406–410.

Zhang, Y., Corver, J., Chipman, P.R., Zhang, W., Pletnev, S.V., Sedlak, D., Baker, T.S., Strauss, J.H., Kuhn, R.J., Rossmann, M.G., 2003. Structures of immature flavivirus particles. *EMBO J* 22, 2604–2613.

27 Yellow Fever Virus Infection

Kate D. Ryman and William B. Klimstra

CONTENTS

27.1 INTRODUCTION

In the eighteenth and nineteenth centuries, a disease known as "yellow fever" (abbreviated as YF) because of its proclivity to cause severe jaundice in the afflicted was one of the most feared diseases in the United States, particularly in port cities along the Atlantic Coast and Gulf Coast. Epidemics struck with extreme ferocity and indiscriminately of age, gender, race, or wealth. Doctors were uncertain of the cause and incapable of treating or even alleviating the suffering of the victims. During this period, the YF death toll in the United States was in the *hundreds of thousands*, with major epidemics occurring in the southeastern states and as far north as New York City, Boston, and Philadelphia (Patterson, 1992). It is now believed that YF was imported into the Americas and Europe from Africa as a consequence of the slave trade and intercontinental shipping (Barrett and Monath, 2003); one of the earliest examples of intercontinental travelers spreading disease. Retrospective analyses have revealed that the first recorded epidemic of YF-like disease in the Western Hemisphere occurred in 1648 on the Yucatan peninsula in Mexico (Staples and Monath, 2008).

FIGURE 27.1   Major Walter Reed, MD (1851–1902). Dr. Reed was a U.S. Army physician who led the team in 1900 that confirmed the theory of Dr. Carlos Finlay that YF is transmitted from human to human by mosquitoes, rather than by direct contact.

Early on, although scientists understood that YF was not contagious, its source was incorrectly attributed to environmental miasmas. A major breakthrough came toward the end of the nineteenth century, when Cuban scientist Carlos Finlay postulated that the disease was disseminated by mosquitoes, although he was never able to prove his theory. Walter Reed (Figure 27.1) and his colleagues were dispatched to Cuba by the U.S. government to determine the cause of YF after the disease decimated troops during the Spanish–American War. Indeed, YF was responsible for the deaths of vastly more soldiers than the war itself. Dr. Reed's team confirmed that disease transmission to humans was primarily effected by the *Aedes aegypti* mosquito (often called the "yellow fever" mosquito, Figure 27.2), and, in a historic series of experiments, they demonstrated that the disease was caused by a filterable agent in patients' blood (Staples and Monath, 2008). Led by a U.S. army surgeon named William Gorgas, campaigns to eradicate *A. aegypti* mosquitoes from Havana, Cuba, and subsequently from Panama during the construction of the Panama Canal proved very effective in eliminating YF epidemics. Unfortunately, it also became clear that complete eradication

FIGURE 27.2   *A. aegypti,* the "yellow fever mosquito." *A. aegypti* is the vector for urban YF epidemics. (Photograph by James Gathany (2006), Courtesy of the Centers for Disease Control and Prevention (CDC), Atlanta, GA.)

of YF would never be possible because the virus is maintained in a zoonotic cycle high in the rain-forest canopies, transmitted between monkeys by sylvatic (forest or jungle) mosquito species such as *Haemagogus* spp. with occasional spillover into rural-dwelling human populations. This remains a largely insurmountable problem today in both Latin American and sub-Saharan Africa.

The causative agent of YF disease, YF virus (YFV), was first isolated in Ghana, West Africa, in 1927 (Staples and Monath, 2008). The Asibi strain of YFV, named for the patient from whom it was isolated, is still widely used in modern laboratories as a "wild-type" virus (McElroy et al., 2005; Meier et al., 2009). In the 1930s, Max Theiler and colleagues at the Rockefeller Institute produced a live attenuated vaccine strain from the Asibi virus, designated 17D, which was attenuated for viscerotropic disease in monkeys and humans but remained immunogenic (Theiler and Smith, 1937a,b, 2000). The live attenuated YF vaccines used today derive from the original 17D strain, and Dr. Theiler (Figure 27.3) was awarded the Nobel Prize for his life-saving research in 1951. However, vaccine manufacturing processes are antiquated, and the stockpile of vaccine doses is insufficient in years with several outbreaks worldwide. Almost concurrently, a second live attenuated vaccine was developed from the French viscerotropic virus (FVV) YFV strain isolated in Senegal in 1927. The French neurotropic vaccine (FNV) was widely used from the 1940s to the 1960s in French-speaking African countries, virtually eradicating human disease in these areas. However, its use was discontinued in 1980 due to an unacceptably high rate of severe adverse event (SAE) incidence in vaccinees, and YF has reemerged in these countries in the last few years.

In the modern world, outbreaks of YF are largely restricted to a zone in which the virus is endemic including tropical and subtropical regions of Africa and the Americas (Figure 27.4). While devastating epidemics of this potentially fatal, viral hemorrhagic fever (VHF) still occur, mass vaccinations have, for the most part, effectively controlled these outbreaks. However, in recent years, the risk of major epidemics, especially in densely populated, poor urban settings, has once again

FIGURE 27.3 Max Theiler, MD (1899–1972). Dr. Theiler won the 1951 Nobel Prize in physiology or medicine for his development of the live attenuated YFV vaccine, 17D, from which the modern 17D-204 and 17DD vaccine substrains originate.

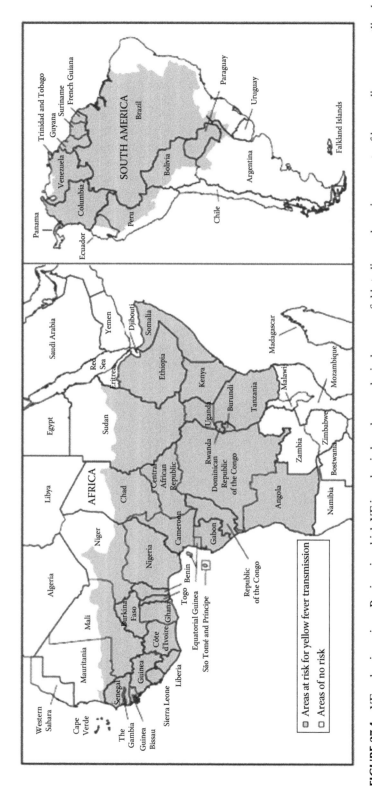

FIGURE 27.4 YF endemic regions. Regions in which YF is endemic based on serological surveys, field studies, and previous reports of human disease are outlined. Regions of the world outside the YF endemic zone infested with *A. aegypti* and thus receptive to the introduction and spread of the disease include coastal areas of South America, Central America, the Caribbean, the southern United States, South Africa, India, Southeast Asia, Australia (Queensland), southern China, Taiwan, and the Pacific Islands.

risen due to reinvasion of urban settings by *A. aegypti*, rapid urbanization in the endemic zone (particularly in parts of Africa where populations are shifting at a tremendous rate from rural to predominantly urban), and by waning immunization coverage. Consequently, YF is a greater threat today than it has been in decades.

27.2   BIOLOGICAL PROPERTIES OF YELLOW FEVER VIRUS

YFV was the first identified member of the genus *Flavivirus*, family *Flaviviridae*. The name derives from the Latin "flavus" meaning "yellow," referring to the jaundiced appearance of YF patients. The *Flaviviridae* family contains a large number of important human and veterinary pathogens (Westaway, 1987), divided into three genera: the *Flavivirus* genus, which includes YFV, dengue (DENV), Japanese encephalitis (JEV), and West Nile (WNV) viruses, among others; the *Pestivirus* genus, which includes bovine viral diarrhea (BVDV) and classical swine fever (CSFV) viruses; and the *Hepacivirus* genus, which contains only hepatitis C virus (HCV), although the unassigned GB viruses are closely related to HCV. Flaviviruses are epidemiologically classified as *ar*thropod-*bo*rne (arbo) viruses (Figure 27.5), primarily vectored by mosquitoes or ticks, whereas pestiviruses and HCV are not.

27.2.1   STRUCTURE OF THE YFV VIRION

A mature YFV virion has an icosahedral nucleocapsid built from capsid (C) protein subunits, enveloped in a lipid bilayer, which is acquired from host cell membranes during egress (Figure 27.6). The virion is approximately 40 nm in diameter, with surface projections of 5–10 nm. The outer envelope of the mature virus particle is studded with heterodimeric pairs of envelope (E) glycoprotein and membrane (M) protein (Li et al., 2008). The E glycoprotein is the major component of the

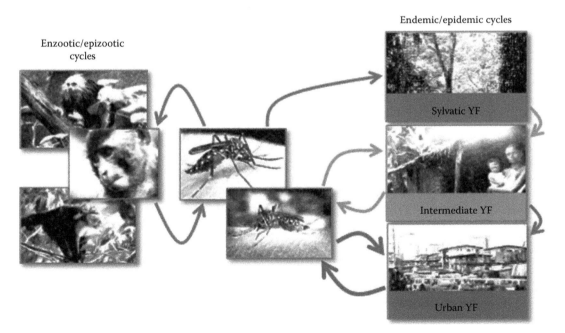

FIGURE 27.5   YF transmission cycles. The YFV is transmitted between human and NHP hosts by mosquitoes in three cycles: the sylvatic (jungle) cycle in which mosquitoes of the forest canopy transmit virus to monkeys and secondarily to humans entering the jungle; the intermediate cycle (or zone of emergence) in which virus enters rural towns and villages bordering jungle areas; and the urban cycle in which humans serve as the viremic host and virus is transmitted from human to human by the domesticated *A. aegypti* mosquito.

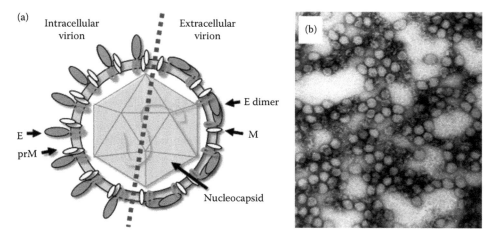

FIGURE 27.6 The YFV virion. The virus particle is small, icosahedral, and enveloped. (a) Diagrammatic representation of the immature (intracellular) and mature, infectious (extracellular) virion. The single-stranded, infectious RNA genome is packaged in an icosahedral nucleoside with a lipid envelope and viral spike proteins, prM/M and E. The prM protein is processed to M by furin-mediated cleavage immediately prior to egress. (b) Photomicrograph showing multiple YFV virions (×234,000 magnification). (Image from the Public Health Image Library, CDC.)

virion surface (Chambers et al., 1990a) and provides most of the biological activity, including cell-surface receptor binding, fusion activity at low pH, virion assembly and egress, and immunogenicity (Nasidi et al., 1989).

The virion-packaged genome is a single-stranded RNA molecule, approximately 11 kb in length (Rice et al., 1985) (Figure 27.7). In 1985, the entire genomic sequence of the 17D-204 vaccine substrain of YFV was determined by Rice et al. (1985). The virus genome was shown to encode only 10 proteins, from genes in a single open reading frame (ORF) (Chambers et al., 1990a; Rice et al., 1985). The structural proteins (C, prM, and E) that form the virion are encoded in the 5′-quarter of the genome, while the nonstructural (NS) proteins that perform enzymatic functions in viral replication (NS1, NS2A, NS2B, NS3, NS4A, 2K, NS4B, and NS5) are encoded in the remaining three-quarters (Chambers et al., 1990a; Rice et al., 1985). The viral RNA genome is of "positive polarity" meaning that it resembles a cellular messenger RNA (mRNA) molecule that is translated into protein by the host cell's ribosomes. A single long polyprotein is produced that is sequentially cleaved into the individual viral proteins by host and viral proteases. Like the host cell's mRNA, the genomic RNA has a 5′ terminal cap structure methylated at the N-7 and 2′-O positions of the 5′ guanosine cap (7mGpppNm). While N-7 methylation is essential for genome translation and stability (Liu et al., 2010), 2′-O methylation appears to shield the flavivirus genome from detection by some of the host cell's innate antiviral responses (Daffis et al., 2010). However, unlike most host

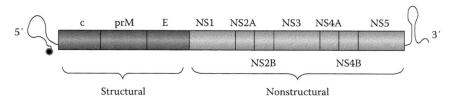

FIGURE 27.7 The YFV genome. The genome is a single-stranded RNA molecule of positive polarity (i.e., can be translated), with highly structured 5′ and 3′ NTR, a 5′ terminal cap, and a single ORF encoding the 10 viral proteins, three structural and seven NS.

mRNAs, the genomes of YFV and most other flaviviruses are not polyadenylated at the 3′ terminus (Rice et al., 1985); instead, a stable, highly conserved stem-loop structure forms in the 3′ non-translated region (NTR), which protects the genome from degradation and provides signals for circularization, initiation of translation, and RNA synthesis (Brinton et al., 1986).

27.2.2 Replication of YFV

The interaction between the virus and the host has proven to be extremely complex and efficient at the molecular, cellular, and organismal level, in spite of the fact that YFV and the other flaviviruses appear to be molecularly quite unsophisticated, encoding only 10 proteins on a single RNA strand (Figure 27.7). The virus has evolved very successfully to exploit the host cell's machinery for macromolecular synthesis for its own amplification and to antagonize or circumvent the cell's antiviral detection and defense systems (Fernandez-Garcia et al., 2009), in the equivalent of an arms race. Detection of the invading virus by cellular receptors that recognize pathogen-associated molecular patterns (PAMPs) is modulated by the virus, stress, and antiviral responses that are compromised or rerouted, and membranous structures within cells are commandeered as replication and assembly sites, all with the goal of promoting crucial steps in the virus replication and propagation cycle (Fernandez-Garcia et al., 2009). Today, research groups are seeking to characterize the YFV–host cell interaction at all levels, to identify points of vulnerability in the virus life cycle that can be targeted with antiviral drugs or utilized in the design of vaccine candidates (Schoggins et al., 2011; Zhou et al., 2011).

The YFV virion binds to receptors on the surface of cells, which have yet to be definitively identified, and is internalized by receptor-mediated endocytosis. As the pH environment of the endosome lowers, the heterodimeric M/E protein spikes on the virion surface rearrange to expose fusion domains, which triggers the mixing of the virus particle's lipid envelope with the endosomal membrane and the release of the nucleocapsid into the cell's cytoplasm (Bressanelli et al., 2004; Modis et al., 2004). The nucleocapsid disassembles making the viral RNA genome available for translation on host ribosomes to synthesize the polyprotein. It appears that the genome circularizes through base-pairing of two short, conserved sequence elements, CS1 and CS2, located in the 5′ and 3′ putative secondary structure sequences, respectively, probably compensating for the absence of 3′ polyadenylation. Circularization facilitates genome translation, replication, and/or packaging (Brinton and Dispoto, 1988). Cap-dependent translation of the long ORF initiates at an AUG codon near the 5′ end of the RNA genome (Chambers et al., 1990a), borrowing host eukaryotic translation initiation factors (eIFs) including the eIF4F complex, membrane-bound ribosomes, and a variety of other proteins (Figure 27.8, Step 1). It has also been suggested that, under certain conditions, flaviviruses can employ a novel, cap-independent translation mechanism to synthesize the polyprotein (Edgil et al., 2006), perhaps to better compete with cap-dependent translation of host mRNAs. Co- and posttranslational processing of the polyprotein into individual mature proteins involves temporally regulated, sequential cleavages mediated by both host and viral proteases (Figure 27.8, Step 2) (Chambers et al., 1990a; Fernandez-Garcia et al., 2009; Lindenbach and Rice, 2003). Interestingly, mutations in the genome that alter (decrease or increase) the efficiency of these cleavage steps typically are highly detrimental to virus replication, suggesting that virus evolution has arrived at optimal cleavage efficiency.

Newly translated and processed viral NS proteins associate in infected cells to form the viral replication complex (RC). The small, hydrophobic NS2A, NS2B, NS4A, and NS4B proteins contain multiple predicted membrane-spanning domains that are believed to anchor the large NS3 and NS5 proteins to cell membranes via protein–protein interactions, leading to assembly of the RC (Lindenbach and Rice, 2003; Miller et al., 2007; Paul et al., 2011; Roosendaal et al., 2006) via association with host proteins (Nagy and Pogany, 2012; Urcuqui-Inchima et al., 2010; Yi et al., 2012). Interestingly, a dramatic proliferation of spherical invaginations known as vesicle packets (VPs) in the perinuclear region of the endoplasmic reticulum (ER) is observed in flavivirus-infected cells

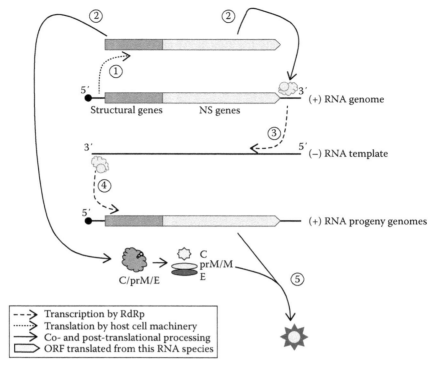

FIGURE 27.8 (See color insert.) Diagrammatic representation of YFV replication in a permissive cell. The replication cycle is depicted as a series of temporally regulated steps: (1) translation of the polyprotein; (2) co- and posttranslational processing to produce the structural proteins (C, prM, and E) and the NS proteins (NS1, NS2A, NS2B, NS3, NS4A, NS4B, and NS5); (3) synthesis of complementary, negative-sense RNA by the RdRp NS5 and other viral replicase components; (4) synthesis of progeny genomes by transcription of the negative strand; and (5) packaging of progeny genomes into nucleocapsids and bud intracellularly to acquire envelope.

(Roosendaal et al., 2006; Uchil and Satchidanandam, 2003; Welsch et al., 2009; Westaway et al., 1997a, 1999), at least in part triggered by the NS4A protein (Miller et al., 2007; Roosendaal et al., 2006). The membrane-associated viral RC appears to be sequestered within these VP structures (Nisbet et al., 2005; Westaway et al., 1997b), likely serving to concentrate virus and host components required for replication and perhaps also transiently concealing viral PAMPs from the cell's antiviral surveillance systems (Hoenen et al., 2007).

The assembled viral RC recognizes and binds to conserved secondary structure in the 3′ terminus of the incoming genomic RNA. Through its C-terminal RNA-dependent RNA polymerase (RdRp) activity, the NS5 protein initiates the synthesis of full-length negative-sense RNA copies from this template (Chu and Westaway, 1987; Grun and Brinton, 1986; Guyatt et al., 2001; Figure 27.8, Step 3) that are then rapidly transcribed to produce progeny positive-sense RNA genomes (Figure 27.8, Step 4). The replication of YFV RNA is typically detectable within 3–6 h after infection. Flavivirus RNA replication is asymmetric with negative-strand synthesis considerably less efficient than positive-strand synthesis, probably due to differential ability of the RC to recognize and initiate transcription from the stem-loop structures in the 3′-termini of the positive- and negative-strand RNAs, respectively (Brinton and Dispoto, 1988). The NS3 carboxy-terminal NTPase/helicase facilitates RNA replication by separating the RNA strands in an ATP-dependent mechanism (Arias et al., 1993; Benarroch et al., 2004; Chambers et al., 1990c, 1991; Falgout et al., 1991; Li et al., 1999; Nestorowicz et al., 1994; Warrener et al., 1993; Wu et al., 2005). Nascent positive-sense RNA progeny genomes are capped by s-adenosyl methyltransferase activity located in the N-terminus of the NS5 protein (Egloff et al., 2002; Issur et al., 2009; Ray et al., 2006).

As replication progresses, progeny viral genomes complex with C protein subunits that become enveloped in an ER-derived lipid bilayer with embedded heterodimers of the M protein precursor (prM) and E proteins when they bud intracellularly into Golgi apparatus vesicles (Lorenz et al., 2003; Mackenzie and Westaway, 2001). Noninfectious, immature virion-containing vesicles are transported to the plasma membrane where exocytotic events release the vesicle contents including virions into the extracellular environment (Ishak et al., 1988). During assembly and transport, prM protects E proteins from undergoing irreversible conformational changes in acidic compartments of the secretory pathway (Guirakhoo et al., 1991). The final steps of virion maturation occur in the trans-Golgi network with a furin protease–mediated cleavage of prM to form M (Kiehn et al., 2008; Li et al., 2008), which triggers the rearrangement of E/M proteins on the virion surface and render the virion infectious (Chambers et al., 1990b) immediately prior to their release into the extracellular medium (Figure 27.8, Step 5). The first progeny virions are typically released within 12 h of initial infection events.

The third large, highly conserved NS protein, NS1, is an atypical glycosylated protein, which is critical to replication, virulence, and pathogenesis. This multifunctional protein can be cell-associated (Westaway and Goodman, 1987), on the cell surface (Somnuke et al., 2011; Westaway and Goodman, 1987) or extracellularly (Lee et al., 1989; Mason, 1989; Winkler et al., 1988). In some circumstances, extracellular NS1 exacerbates disease by inhibiting complement cascade activity (Chung et al., 2006). However, protective antibodies are elicited against NS1 (Bray et al., 1989; Zhang et al., 1988) via Fc receptor-dependent and receptor-independent mechanisms (Chung et al., 2006, 2007), ameliorating disease.

27.3 CLINICAL PRESENTATION OF YELLOW FEVER INFECTION

YF is the original VHF (Figure 27.9), sharing clinical features with other, more recently recognized, VHFs such as DEN hemorrhagic fever/DEN shock syndrome (DHF/DSS), Lassa fever

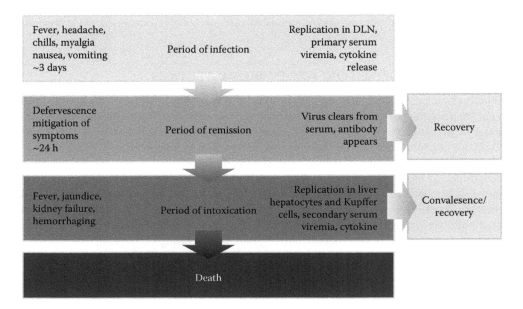

FIGURE 27.9 The phases of YF.

and Crimean–Congo hemorrhagic fever, although the hepatic involvement in YF is considered a distinguishing feature (Nasidi et al., 1989). After a 3–6-day incubation period following mosquito bite, symptoms appear suddenly with patients experiencing fever, chills, malaise, headache, lower back pain, generalized myalgia, nausea, and dizziness and often manifesting Faget's sign (increasing temperature with decreasing pulse rate). During these first few days, levels of virus in the bloodstream (viremia) are sufficiently high for transmission back to biting mosquitoes, completing the transmission cycle and facilitating disease outbreaks. For a few hours, symptoms may then remit with rapid abatement of fever. In 15%–20% of patients, illness reappears in more severe form with high fever, vomiting, epigastric pain, prostration, and dehydration. Severe liver damage causes patients to have a jaundiced appearance, and hepatic-induced coagulopathy commonly leads to hemorrhagic manifestations including petechiae, ecchymoses, epistaxis (bleeding of the gums), and hematemesis (gastrointestinal hemorrhage). Hematemesis is responsible for the "black vomit," characteristic of severe YF and used as a descriptor for this disease in the 1800s. Extreme albuminuria, acute renal failure, and late central nervous system (CNS) manifestations, such as confusion, seizure, and coma, often presage death, which occurs in roughly half of the severe cases and generally within 10 days of symptom onset. However, if the patient survives the disease, convalescence is often remarkably rapid.

Although we tend to think of YF in its severe, potentially fatal form, in reality the majority of YFV infections are completely asymptomatic or relatively mild and self-limiting, and only recognized retrospectively via serology studies. In endemic areas where the virus is active, up to 50% of adults may be seropositive (Monath, 2001); however, it is difficult to distinguish undiagnosed infection from childhood immunization, and thus overall infection rates are rarely enumerated. In mild YFV infections, symptoms of infection are typically nonspecific and "flu-like," manifest as fever, headache, and constitutional problems from which patients recover in a few days with no lasting *sequelae*. In essence, these individuals experience the earliest symptoms of YF, but the infection resolves without further symptoms, the virus is cleared from the circulation coincident with the appearance of anti-YFV antibodies, and lifelong protection is acquired (Monath, 2007). The factors that determine differential susceptibility of individuals to the infection are unknown at this time, although there is little to no evidence of virus strain-dependent virulence differences. It is tempting to speculate that predisposition of some individuals to severe disease might be controlled by host genetics, immune responses, and/or comorbid disease, but this remains to be systematically investigated.

27.4 DIAGNOSIS OF YELLOW FEVER VIRUS INFECTION

Epidemiological surveillance has proven invaluable for monitoring the incidence of YF and allowing the prediction and early detection of potential outbreak foci. Consequently, the monitoring of control measures and case reporting is required by the International Health Regulations, and the World Health Organization (WHO) recommends that every at-risk country has at least one national laboratory where basic YF blood tests can be performed. In spite of this, it is feared that YF cases are underreported by an estimated 250-fold; in other words, less than 0.5% of YF cases may be recognized and confirmed as YF worldwide; the rest remain as cases of undiagnosed febrile or hemorrhagic illness. Today, *one confirmed YF case* in an unvaccinated population is considered a potential outbreak, and a confirmed case in any context demands a full investigation. Investigative teams are required to assess the outbreak and respond by generating both emergency measures for outbreak control and longer-term immunization. Rapid response to YF detection through emergency mass vaccination campaigns is essential for epidemic control and has proven very successful in recent years if implemented correctly. Note that little emphasis is placed on the treatment of infected individuals for outbreak control, although they are transported to an intensive care unit (ICU) and/or quarantined wherever possible.

27.4.1 CLINICAL DIAGNOSIS OF YELLOW FEVER

Presentation of a patient with a sudden fever, relative bradycardia and signs of jaundice within the YFV endemic zone (or with a history of recent travel to an endemic area) should lead physicians to suspect YF and obtain CBC, urinalysis, liver function tests, coagulation tests, viral blood culture, and serologic tests. Common in YF patients are leukopenia with a relative neutropenia, thrombocytopenia, prolonged clotting, and increased prothrombin time. Bilirubin and aminotransferase levels may be elevated acutely and for several months after the onset of symptoms, while albuminuria indicating kidney dysfunction occurs in 90% of patients and helps with the differential diagnosis of YF from other causes of hepatitis. Obviously, clinical diagnosis of YF in the field, particularly for an isolated case, is tremendously difficult. Case-by-case differences in severity, and in the particular subset of observed symptoms, make this a difficult disease to recognize. Although classical YF cases should be fairly easily spotted, jaundice is not observed even in a majority of patients, and YF may not be included in the differential diagnosis of patients presenting with symptoms of headache, nausea, backache, and fever, especially during the early stages of illness or in mild cases. YF is easily confused with many other infectious diseases including but not limited to DHF/DSS, Lassa fever, Ebola, malaria, typhoid, hepatitis, and other diseases, with which it overlaps geographically, as well as poisoning. All too often, mild disease likely escapes diagnosis entirely, and severe or fatal disease is diagnosed after the fact, by serology or at postmortem, respectively.

27.4.2 LABORATORY DIAGNOSIS OF YELLOW FEVER

Case definitions state that YF should be suspected in a patient presenting with acute onset of fever followed by jaundice within 2 weeks of the onset of the first symptoms (bearing in mind that less than half of YF patients exhibit jaundice). However, a confirmed YF case additionally requires laboratory confirmation or an epidemiological link to a laboratory-confirmed case or outbreak. Either of the following is considered a criterion for laboratory diagnosis: (1) detection of YFV-specific IgM or a greater than fourfold rise in IgG levels between acute and convalescent sera in the absence of recent vaccination or (2) the detection of YFV virions by amplification in cell culture, YFV genomic RNA in blood or tissues by reverse transcriptase polymerase chain reaction (RT-PCR), or YFV antigen by immunohistochemical staining in tissue sections (often postmortem). Unfortunately, laboratory confirmation of YF requires trained laboratory staff with access to specialized equipment and materials, often scarce or unavailable in YF endemic areas. If a diagnosis is made at all, it is frequently not until the patient has either recovered or succumbed and thus is useful only for surveillance and outbreak control, not for patient care.

27.5 PATHOLOGY AND PATHOGENESIS OF YELLOW FEVER VIRUS INFECTION

For the most part, our knowledge of pathologies in human YF comes from postmortem samples from the most severe of cases, obviously taken at the terminal stage of disease. The acquisition of acute samples is virtually impossible and even postmortem samples are difficult to procure primarily because of the lack of advanced medical facilities in areas with active YF, the contraindication for biopsy or phlebotomy in patients with VHF, and the frequent failure to diagnose until the infection has resolved for better or worse.

During acute YF, hemorrhagic symptoms and fatal outcome are strongly correlated with highly elevated pro- and anti-inflammatory cytokines (ter Meulen et al., 2004), suggesting a contribution to disease. Although the source of the cytokines is not known, hepatocytes (Quaresma et al., 2006), endothelial cells (Khaiboullina et al., 2005), and/or activated Küpffer cells (Woodson et al., 2011) and other macrophages (Monath, 2008; Monath and Barrett, 2003) may contribute. The characteristic hemorrhagic manifestations of YF most likely result primarily from disseminated intravascular coagulation (DIC): a pathological activation of coagulation mechanisms that can be triggered by a variety of diseases. Deregulation of the coagulation and fibrinolytic pathways causes the widespread formation of small blood clots, which consumes coagulation proteins and platelets, disrupts coagulation homeostasis, and causes abnormal bleeding from the skin (*petechiae*), the nose and gums (*epistaxis*), the gastrointestinal tract (*hematemesis* or "black vomit"), and other organs, contributing to multiple organ dysfunction (MOD) and death. After vascular damage occurs in DIC, perhaps resulting from elevated cytokines such as interleukin (IL)-1 and tumor necrosis factor-alpha (TNF-α) or directly due to virus replication in endothelial cells, the tissue factor (TF) transmembrane glycoprotein is released into the circulation, becomes activated, and binds with coagulation factors triggering the extrinsic and subsequently the intrinsic coagulation pathways. As a further complication, all but one of the clotting factors and several coagulation regulators are normally synthesized in the liver, which in the YF patient is being attacked and damaged by the virus directly.

Gross pathology of fatal human YF reveals that the liver, the characteristic organ of YF infection, is normal or slightly enlarged in size, but icteric in appearance with lobular markings obliterated (Quaresma et al., 2005). Micropathologic changes in the liver include swelling and necrosis of hepatocytes in the midzone of the lobes, while cells in the portal area and surrounding the central veins are often spared. The presence of apoptotic hepatocytes (known for decades as councilman bodies) coincident with the disarray of the midzonal hepatocyte plate and microvesicular steatosis (lipid accumulation) is the hallmark of fatal human YF infection (DeBrito et al., 1992; Nasidi et al., 1989). Viral antigen and RNA can be detected by immunocytochemistry and nucleic acid hybridization in cells undergoing these pathologic changes, and cytopathology appears to be mediated by direct viral injury (DeBrito et al., 1992; Nasidi et al., 1989) via apoptosis (Quaresma et al., 2006). Furthermore, immune clearance attempts may exacerbate viral pathogenesis via viral hepatitis (Monath, 2008), but surprisingly, inflammatory changes are often absent or minimal, and patients with hepatitis who recover do not develop residual scarring or cirrhosis. The kidneys and heart are often overtly enlarged, congested, and edematous. Acute tubular necrosis in the kidneys is most likely due to small DIC-related clots disrupting normal blood flow to the organs (including the kidneys), which may malfunction as a result, not necessarily to direct virus-mediated damage. Focal degeneration of cardiac muscle cells may be seen in the heart (DeBrito et al., 1992) and B-cell follicles in the spleen and lymph nodes are involuted. The brain may be edematous with evidence of petechial hemorrhage, but true viral encephalitis is a rare event. Alternatively, late CNS manifestations may be due to acute liver failure and MODS.

27.6 TRANSMISSION CYCLES OF YELLOW FEVER VIRUS

The epidemiology of YFV is driven by the absolute requirement for mosquito-borne transmission between vertebrate hosts, like many other viruses in the *Flavivirus* genus. *Flavivirus* infections are not contagious (in other words, transmitted from host to host by physical contact), despite the presence of virus in the blood and bodily secretions during acute infection. Consequently, available "reservoir hosts" for infectious virus (human and nonhuman primates (NHPs) in the case of YFV) and high levels of mosquito vector populations are prerequisites for epidemic outbreaks. In fact, it might not be outrageous to think of YFV as being primate borne between mosquito reservoir hosts, particularly because vertical (or transovarial) transmission of virus in mosquitoes is critical to the maintenance and overwintering of YFV in endemic areas (Rosen, 1981, 1987a,b).

YFV infects humans and NHPs in tropical sub-Saharan Africa and Latin America, transmitted between primates in the saliva of infected mosquitoes (Monath, 2001). The natural epidemiology of YF on both continents is a cycling of the virus between forest canopy mosquitoes and wild NHPs. Secondary accidental transmission to humans occurs in three cycles of escalating importance (Figure 27.5). The *sylvatic (jungle) YF transmission cycle* occurs in the tropical rainforests of Africa and South America when the virus is "accidentally" transmitted to humans who enter these areas, causing sporadic cases, usually in male forestry workers. The *intermediate YF transmission cycle* results in small-scale epidemics in rural villages of the African savannah when infected mosquitoes of semidomestic species indiscriminately feed on both monkeys and human hosts. This has been the most common type of African outbreak in recent decades. The major cause for concern, however, comes from the *urban YF transmission cycle*, initiated when virus is introduced into areas with high human population density and mosquitoes transmit YFV from human to human, resulting in explosive, large-scale epidemics. In the urban cycle, humans are the primary host and the urban-dwelling mosquito species, *A. aegypti,* is the vector. Urban YF occurs indiscriminately among naïve humans, spread either by the movement of viremic human hosts or by the accidental transportation of infected mosquitoes (e.g., in used tires). Any outbreak in which *A. aegypti* is identified as the vector is classified as urban YF by the WHO, regardless of the area, while outbreaks involving other mosquito species are classified as jungle YF (Bres, 1986). On the edge of the jungle, the cycles intermingle, and infected workers returning from jungle areas often form the focus of an urban outbreak.

27.7 EPIDEMIOLOGY OF YELLOW FEVER

During the seventeenth, eighteenth, and nineteenth centuries, devastating epidemics of urban YF were recorded not only in Africa and South and Central America but also in the port cities of Europe and along the Eastern Seaboard and Gulf Coast of North America (Monath, 2001). With no sylvatic cycle, improved sanitation and mosquito abatement programs in Europe and the United States eliminated epidemic urban YF, with the last outbreak occurring in New Orleans, 1905 (Staples and Monath, 2008). However, in tropical regions of Africa and South and Central America, although urban YF was reduced, it is still estimated that YF afflicts over 200,000 persons annually, with at least 30,000 fatalities (Monath, 2001). Forty-four African and Latin American countries (Figure 27.4) are within the modern YF endemic zone, putting almost 900 million people at risk of infection, an estimated 508 million people in 32 African countries and the remainder in 12 South and Central American countries (Srisuphanunt et al., 2007). Today, the majority of YF cases occur in sub-Saharan Africa, including periodic, unpredictable urban outbreaks. However, in recent years, an alarming increase in virus activity and expansion of the endemic zones have been detected in Africa (Onyango et al., 2004; Robertson et al., 1996) and South America (Bryant et al., 2007), raising the possibility of an explosion of urban epidemics for which we are poorly prepared.

27.7.1 YELLOW FEVER IN AFRICA

In Africa, YFV infects many NHP species, causing a subclinical, but viremic, infection indicating coevolution of the host and the virus. In the jungle, YFV most frequently is vectored by *Aedes africanus* mosquitoes, which are night feeders preferentially taking blood meals from monkeys and rarely biting humans. Consequently, cases of true jungle YF in humans are uncommon in Africa. Much more commonly, humans are infected by the bite of semidomesticated *Aedes simpsoni* mosquitoes, known as "bridging vectors" because they feed indiscriminately on NHPs and humans. An important epidemiological ramification of this seemingly minor distinction is that exposure to infection results from the proximity of human dwellings to the forest and consequently all household members tend to be equally at risk.

Urban YF was quite effectively controlled in Africa by mass vaccination campaigns in the 1940s and efforts to eradicate *A. aegypti* breeding sites, particularly in French-speaking West African countries where vaccination with the FNV was compulsory (Monath, 1991). Unfortunately, by the 1960s, vaccine coverage had waned or was absent in large areas, and the use of FNV was completely discontinued in the 1980s due to the risk of postvaccinal encephalitis. Current vaccination coverage with the 17D vaccine in African countries is variable but generally insufficient. Exacerbating this situation, an extremely rapid urbanization is occurring in Africa, with cities growing in size and number. The shift from rural to urban population is occurring at an annual rate of 4%, crowding unvaccinated, naïve populations into areas with poor housing and inadequate sanitation. With little access to running water, the common practice of storing drinking water in large open containers near dwellings provides perfect breeding sites for *A. aegypti* mosquitoes. As these conditions worsen, the risk of large-scale urban YF epidemics in West African cities becomes greater and the number of countries reporting YF activity to the WHO is increasing. This reveals a disconcerting increase in the circulation of the virus in nonimmune human populations and reemergence in geographical areas that have long been free of YF. Several large cities have already experienced urban YF outbreaks in the past few years: Abidjan, Côte d'Ivoire (2001); Dakar, Senegal (2002); Touba, Senegal (2002); Conakry, Guinea (2002); and Bobo Dioulasso, Burkina Faso (2004). The outbreaks were rapidly controlled by emergency vaccination campaigns with the administration of millions of vaccine doses to at-risk populations, and the number of YF cases has been restricted. For example, in the 2001 outbreak in Abidjan, 2.61 million people were vaccinated within 2 weeks after urban YF was detected. Disturbingly, however, if multiple outbreaks occur simultaneously in different locations, the response capacity of affected countries and support capabilities of the international community are greatly stressed, while vaccine stockpiles become severely depleted.

27.7.2 YELLOW FEVER IN THE AMERICAS

As in Africa, urban YF was remarkably well controlled by the 1940s through improved sanitation, mosquito abatement programs, and mass vaccination campaigns. However, it was clearly impossible to eradicate the jungle *Haemagogus* spp. mosquito vectors or the NHP reservoir hosts and thus jungle YF continued to occur, primarily afflicting young male forestry and agricultural workers in the Orinoco and Amazon River basins. Unlike the African situation and most likely due to the more recent introduction of YFV into the Americas, several NHP species that act as viremic hosts are highly susceptible to disease, and local YF activity can first be indicated by large die-offs of these animals in the forests, aiding surveillance efforts (Barrett and Higgs, 2007).

Recently, the Pan American Health Organization (PAHO) has reported greatly increased jungle YF incidence in the Americas, from Argentina, Paraguay, and Brazil in the southern part of the continent to Colombia and Venezuela in the Andean region and Trinidad and Tobago in the Caribbean. *A. aegypti* mosquitoes have reinfested most major urban centers in Central and South America and in the southeastern United States including cities that were historically centers of urban YF, now inhabited by large nonimmune populations. Immunization coverage has dropped dramatically, putting these areas at greater risk of urban epidemics today than at any time in the past 50 years. The last documented urban YF epidemic in the Americas occurred in 1928 in Rio de Janeiro, Brazil (Bryant et al., 2007). Although there have been reports of sporadic cases in residents of urban areas in Brazil (1942), Trinidad (1954), and Bolivia (1999), verification of a true urban YF cycle in which humans serve as the primary host and virus is transmitted by *A. aegypti* is controversial. After an absence of over three decades, YF returned to Paraguay in 2008, causing a cluster of urban YF cases in Asunción (Johansson et al., 2012). Two million vaccine doses were urgently requested from the global stockpile (which had been greatly depleted by several African outbreaks in the same year), and, although there was public panic, the outbreak was successfully contained by mass vaccination. However, in computational simulations of the Asunción outbreak, local outbreaks were predicted in 12.8% of simulations and international spread in 2.0% (Johansson et al., 2012), and replenishment of vaccine lots took many months.

27.7.3 YELLOW FEVER IN TRAVELERS

YF is considered a significant threat to those that travel to countries in the YF endemic zone, and demonstration of vaccination is often required for entry (Monath, 2001). The traveler's risk of YFV infection is dependent upon immunization status, travel destination, season, length of visit, and activities. The ease of international travel also makes the introduction and spread of YF into new areas infested with competent *A. aegypti* vectors possible, theoretically placing parts of

Asia, Australia, Europe, and North America at risk for local transmission. During recent decades, there have been several documented cases of the human importation of YFV to non-endemic areas. Since 1964, a total of nine YF cases have been documented in European and North American tourists after returning home from visits to West Africa and South America, but secondary transmission has not yet occurred. Why YFV has not spread to other regions infested with *A. aegypti*, particularly East African and Asian countries, is unclear, but clearly the traditional geographic barriers to YF are gradually breaking down. Although it is probable that YF cases in developed countries will be quickly identified, index patient(s) quarantined, and the outbreak controlled by mass vaccination, even in the best-case scenario, the potential for public panic and the stress on the medical system would be significant.

27.7.3.1 How Do We Cope with the Threat of Future Urban Yellow Fever Epidemics?

This is likely to be an important question in the coming years. The modern strategies for YF control are as follows: (1) routine immunization of infants aged 6 months or above, (2) early detection of potential urban YF cases, (3) rapid mass vaccination campaigns to prevent epidemics, and (4) control/abatement of *A. aegypti* in urban centers. In 1988, the WHO recommended that vaccination against YF be included in routine infant immunization programs (Monath and Nasidi, 1993). In 2006, the Yellow Fever Initiative was launched to procure the resources to confront the challenge presented by YF worldwide that, as discussed earlier, if not addressed, might result in large-scale urban epidemics, affecting millions of people. The initiative assesses risk of YF outbreaks at the district level based on location in an ecological risk area, notification of suspected or confirmed cases since 1960 in this or an adjacent district, the number of these cases and the number of years since 1960 in which they have been reported, and the susceptibility of the population (i.e., the proportion not covered by vaccination). During the first stage of the initiative in 2007, 12 African countries with large nonimmune populations were considered to be at high risk, and immunization efforts were intensified. As of 2009, 22 African countries and 14 South American countries have added the YF vaccine to routine infant immunization programs, and overall vaccine coverage rates are increasing in South American and many African countries.

27.8 PROGNOSIS AND TREATMENT FOR YELLOW FEVER VIRUS INFECTION

There is no specific licensed antiviral drug available to treat YF, but a number of compounds with in vitro antiviral activity against flaviviruses have been described in the scientific literature. Trials with several of these antivirals have yielded promising results in newly developed small animal models (Botting and Kuhn, 2012). Interferon gamma treatment of monkeys resulted in delayed onset of viremia and illness, but had no effect on survival (Arroyo et al., 1988).

Supportive care is critical for YF patients as there is no specific treatment. Many patients afflicted by YF live in rural or squalid conditions and do not have access to state-of-the-art medical care (Monath and Nasidi, 1993), but, under ideal circumstances, a severely ill patient is admitted to the ICU and provided with vasoactive medications, fluid resuscitation, and ventilator support. DIC, hemorrhage, renal and hepatic dysfunction, and possible secondary infection are treated symptomatically, although the use of salicylates is contraindicated in YF cases because of the bleeding risk. Although YFV is not contagious, viremic patients are isolated with mosquito netting in areas with potential vector transmission and should be quarantined until differential diagnoses for VHF and/or hepatitis are rejected. Beyond this, there is little that can be done, and there is not good evidence that these symptomatic treatments significantly ameliorate disease.

27.8.1 Vaccination against Yellow Fever Virus

Two contemporary YF vaccines were developed in the 1920s and 1930s, but attenuations of virulence were achieved by distinctly different methods. Both the FNV and 17D strains lost the ability to cause visceral YF in NHPs or humans and lost their mosquito transmission competence (an important consideration for a live attenuated vaccine to prevent reversion to virulence and vector transmission), but retained their immunogenicity.

27.8.1.1 Developing Live Attenuated Yellow Fever Vaccines

The FNV live attenuated vaccine strain was isolated after 128 sequential passes in the brains of mice beginning with the wild-type FVV, which was isolated in Senegal in 1927 (Staples and Monath, 2008). During the 1940s and the early 1950s, nearly 40 million doses of FNV were administered in a mandatory vaccination campaign in French-speaking countries of West Africa, typically by scarification of the skin along with the smallpox vaccine (Staples and Monath, 2008). This resulted in a dramatic decline in YF cases in these countries, but the administration of FNV was associated with a high incidence of encephalitis in children. Eventually, it was decided that the numbers of these FNV vaccination–associated SAEs in children under the age of 10 were unacceptably high, the vaccine was abandoned, and manufacture was discontinued in the 1980s. The mechanisms underlying attenuation and immunogenicity of FNV remain unknown, although the 77 nucleotide/45 amino acid mutations accumulated in the FVV genome during attenuation have been identified (Jennings et al., 1993; Wang et al., 1995).

The 17D vaccine was developed by Theiler and Smith in 1937 (Theiler and Smith, 2000). A wild-type isolate called Asibi after the man from whom it was isolated was empirically attenuated by 176 sequential passes in cultured murine and chick embryo tissues, resulting in loss of viscerotropism in monkeys and loss of mosquito competence (Theiler and Smith, 2000). A virus seed-lot system was established by the WHO in 1945 after early difficulties with over- or under-attenuation of different vaccine lots, which requires that no vaccine shall be manufactured that is more than one passage level from a seed lot that has passed all safety tests. The 17D vaccine strain is generally regarded as one of the safest and most effective live attenuated viral vaccines ever developed (Barrett, 1997; Monath, 2005).

Today, two substrains of 17D (17D-204 and 17DD) are in use as vaccines, with a total of over 500 million doses administered since licensure. The vaccines are manufactured at four sites worldwide (Institut Pasteur in Dakar, Senegal; Bio-Manguinhos/Fiocruz, Rio de Janeiro, Brazil; and Sanofi Pasteur in the United States and France), which produce a combined total of 20–25 million doses annually. As described previously, protection in areas at high risk of YF transmission, the WHO's dual strategy for prevention of YF epidemics, relies on preventative mass immunization campaigns followed by infant routine immunization. One vaccine dose, containing approximately 10^5 plaque-forming units of virus, provides protective immunity for at least 10 years after which the WHO recommends revaccination. Approximately 90% of vaccines seroconvert by 10 days and 99% of vaccinees seroconvert by 30 days post-immunization without any booster shots (Monath, 2005; Monath et al., 2005; Reinhardt et al., 1998). YFV-specific neutralizing antibodies are detectable for over 35 years (Niedrig et al., 1999; Poland et al., 1981). Given these statistics, the 17D substrains are truly remarkable vaccines.

The nucleotide sequences of wild-type YFV isolate genomes and their attenuated derivatives have provided clues as to the molecular basis of virulence/attenuation (Hahn et al., 1987; Wang et al., 1995; Xie et al., 1998), but although the number of accumulated mutations are relatively low, their distribution across the entire genome has complicated interpretation of these comparative data. Comparison of the complete genomic sequences of 17D-204 and two other 17D substrains, 17DD and 17D-213 (the original 17D virus is not available) (dos Santos et al., 1995; Galler et al., 1998; Rice et al., 1985), with those of wild-type YFV isolates including the parental Asibi strain (Hahn et al., 1987) identified 20 coding and 28 noncoding nucleotide mutations, scattered throughout the genome. As these mutations are common to all 17D substrains (dos Santos et al., 1995), it is most likely that they arose prior to divergence of the 17D-204 and 17DD substrains from the common attenuated lineage at passage 176, and therefore these changes are considered to be primary attenuating candidates. Five mutations from Asibi clustered in the E protein occur at sites conserved in natural YFV isolates from Africa and South America and consequently are particularly likely to be implicated in the attenuation process, and their location relative to functional domains in the E protein support this hypothesis. Two (E-52 and E-200) lie in the fusion domain (Guirakhoo et al., 1991), while mutations at E-173 and E-305 map to neutralization-escape epitopes (Ryman et al., 1997, 1998). The loss of a 17D-204 substrain-specific epitope from the 17D virus, selected for neutralization escape, resulted in dramatic changes in neurovirulence ranging from avirulence to increased virulence in suckling mice (Ryman et al., 1998). These E protein mutations have been implicated in attenuation resulting from cell culture adaptive heparan sulfate binding (Lee and Lobigs, 2008), known to be a mechanism of attenuation in arboviruses (Klimstra et al., 1998). Although, less well studied, mutations in other locations, particularly in the nonstructural proteins, are also being considered as determinants of YFV virulence (Dunster et al., 1999; McArthur et al., 2003; Xie et al., 1998).

27.8.1.2 Immune Response to Vaccination

In cohorts of human vaccinees, during the first few days after immunization with the live attenuated 17D vaccine, low levels of viremia are detectable (<100 pfu/mL) (Reinhardt et al., 1998). The live attenuated 17D vaccine strain infects human dendritic cells in vitro and stimulates their activation and maturation via multiple toll-like receptors (Barba-Spaeth et al., 2005; Querec and Pulendran, 2007). Proinflammatory cytokines are released into the circulation (Hacker et al., 1998; Kohler et al. 2012; Reinhardt et al., 1998), and markers of the type I interferon (IFN-α/β) response are expressed by peripheral blood cells (Bonnevie-Nielsen et al., 1995; Roers et al., 1994). Humoral immune responses, involving CD4+ T-cell and B-cell activation, occur between days 4 and 15 with rapid elicitation of YFV-specific IgM and, subsequently, neutralizing IgG antibodies. The later have long been believed to play a major role in host defense against YFV infection (Bonnevie-Nielsen et al., 1995) and are used by the WHO and many scientists as the primary correlate of protection. However, the potential importance of the early innate immune response in shaping immune memory (Kasturi et al., 2011; Pulendran, 2009; Querec et al., 2009; Querec and Pulendran, 2007) and cell-mediated immunity, especially class I–restricted CD8+ T cells, in controlling primary YFV infections should not be underestimated (Barrett and Teuwen, 2009; Co et al., 2002; vanderMost et al., 2002; Reinhardt et al., 1998). Activation of YFV-17D-specific CD8+ T cells in PBMC populations is detectable as early as day 14, with maximal expansion around day 30 (Miller et al., 2008). Antigen-specific cells initially displayed an effector cell phenotype and gradually developed into highly functional YFV-17D-specific memory. Overall, 17D vaccination induces a specific, protective adaptive immune response, with cytotoxic T cells, a balanced Th1/Th2 T-cell profile, and potent neutralizing antibodies. Systems biology studies revealed a strong positive correlation between induction of the B-cell growth factor TNFRS17 and level of neutralizing antibody titers (Querec et al., 2009). Attempts are now being made to elucidate in detail the early cellular events leading to such specific activation of all three branches of the adaptive immune response (Kohler et al., 2012). Recent improvements made to murine models of 17D immunization and YF disease will facilitate mechanistic studies in this area (Meier et al., 2009).

27.8.1.3 Vaccine-Associated Severe Adverse Events

Usually only mild side effects are reported by 17D vaccinees, appearing in ~25% of people and including injection site pain or redness, headache, malaise, and myalgia. However, over the last few years, possibly due to improved reporting, some vaccine-associated problems have arisen (Barrett et al., 2007). Vaccine production technologies are antiquated, not having significantly changed since 1945 when the seed-lot system was introduced. Tremendously limiting production potential, the 17D virus is propagated in embryonated eggs yielding around 400 vaccine doses per egg. The eggs must be free of adventitious agents such as avian retrovirus. As mentioned earlier, the inability to rapidly replace vaccine stockpiles has critical implications for availability in years with greater YF activity.

Two types of SAEs associated with vaccination have now been officially recognized: (1) vaccine-associated neurotropic disease (YEL-AND) (Merlo et al., 1993; Schoub et al., 1990) and (2) vaccine-associated viscerotropic disease (YEL-AVD) (Belsher et al., 2007; Chan et al., 2001; Gerasimon and Lowry, 2005; Martin et al., 2001; Vasconcelos et al., 2001). YEL-AND was first described in the 1940s as "postvaccinal encephalitis," most often in young children. Consequently, in 1945, vaccination was contraindicated for infants younger than 6 months and only recommended for those over 9 months, which complicates inclusion in routine infant immunization programs and greatly increases the expense of providing the vaccine. In addition, a monkey neurovirulence test was employed to evaluate the attenuation of vaccine seeds. Together, these measures reduced incidence of YEL-AND, but the development of the vaccine adverse event reporting system (VAERS) in the United States revealed that YEL-AND was not so rare a phenomenon, occurring at a frequency of 4–5 per million doses and with case fatality rates of <5%.

The second category of SAE, YEL-AVD, resembles wild-type YF with pansystemic inflammatory disease and high virus titers in many organs, especially the liver (Belsher et al., 2007; Struchiner et al., 2004; Vasconcelos et al., 2001). YEL-AVD was first described in 1999 in Brazil (Engel et al., 2006). The introduction of routine monitoring by VAERS revealed that YEL-AVD occurs with a frequency of 3–4 per million doses administered but has a 60% case fatality rate. It has been proposed that host genetics may to be at least in part responsible for YEL-AVD (Hayes, 2007) or there may be issues with certain vaccine lots/reversion to virulence (Barrett et al., 2007). No one apparent causal factor explains all cases. However, a cluster of five cases of YEL-AVD in Peru (four of them fatal) in October 2007 (Whittembury et al., 2009) following administration of 42,742 doses from one vaccine lot (i.e., one in 10,000 vaccinees died following administration of this vaccine lot) has caused great concern. However, the process from discovery to licensure for a new vaccine takes at least 15 years and risk–benefit evaluations of the current 17D vaccine clearly indicate that it is highly efficacious and that its withdrawal would leave an enormous number of people unprotected in disease-endemic areas. Furthermore, considerable investment has been made in the development of chimeric vaccines for other flaviviruses that the 17D vaccine as a "backbone." The "ChimeriVax" platform is being used by Sanofi Pasteur and Acambis to develop candidate DENV, WNV, and JEV vaccines (Pugachev et al., 2005). Currently, the U.S. Advisory Committee on Immunization Practices and the WHO Global Advisory Committee on Vaccine Safety are reviewing the vaccine and contraindications for the vaccine, and the WHO Expert Committee on Biological Standardization met in late 2008 to discuss the manufacturing and quality control processes for the vaccine (2012). This committee last reviewed the vaccine in 1998.

27.9 CONCLUSIONS AND FUTURE PERSPECTIVES

This chapter reveals that there are as yet many unanswered questions about YF disease and the YFV agent, and disturbingly the endemic zone is expanding, and the threat of urban outbreaks increase annually. At the same time, we still have little clue as to how to manage the treatment of patients with this disease or with vaccine-associated SAEs. Treatment of YF by supportive care is virtually ineffective, and even admission to the ICU does not seem to improve the prognosis or change the mortality rate. There is a desperate need for the development of specific antiviral drugs and improved rapid diagnostic tests for this and other flaviviral diseases. Finally, the 17D vaccine, which has been touted for years as one of the safest and most efficacious vaccines in use and upon which outbreak control entirely rests, has been recently associated with an increased rate of fatal adverse events. Although this may well be due to increased monitoring for and reporting of such events, improvements to the vaccine's safety will almost certainly be required. Our ability to address all of the aforementioned will come only by performing studies to understand the complex interactions between the virus and host cell factors that control replication and innate and adaptive immune responses, an area that has been somewhat neglected and overshadowed by other flaviviral diseases in recent years.

REFERENCES

(1998). WHO Expert Committee on Biological Standardization. Forty-sixth Report. *World Health Organ Tech Rep Ser* **872**(i–vii), 1–90.

(2012). WHO Expert Committee on Biological Standardization. *World Health Organ Tech Rep Ser* (964), 1–228, back cover.

Arias, C. F., Preugschat, F., and Strauss, J. H. (1993). Dengue 2 virus NS2B and NS3 form a stable complex that can cleave NS3 within the helicase domain. *Virology* **193**(2), 888–899.

Arroyo, J. I., Apperson, S. A., Cropp, C. B., Marafino, B. J., Jr., Monath, T. P., Tesh, R. B., Shope, R. E., and Garcia-Blanco, M. A. (1988). Effect of human gamma interferon on yellow fever virus infection. *Am J Trop Med Hyg* **38**(3), 647–650.

Barba-Spaeth, G., Longman, R. S., Albert, M. L., and Rice, C. M. (2005). Live attenuated yellow fever 17D infects human DCs and allows for presentation of endogenous and recombinant T cell epitopes. *J Exp Med* **202**(9), 1179–1184.

Barrett, A. D. (1997). Yellow fever vaccines. *Biologicals* **25**(1), 17–25.

Barrett, A. D. and Higgs, S. (2007). Yellow fever: A disease that has yet to be conquered. *Annu Rev Entomol* **52**, 209–229.

Barrett, A. D. and Monath, T. P. (2003). Epidemiology and ecology of yellow fever virus. *Adv Virus Res* **61**, 291–315.

Barrett, A. D., Monath, T. P., Barban, V., Niedrig, M., and Teuwen, D. E. (2007). 17D yellow fever vaccines: New insights. A report of a workshop held during the World Congress on medicine and health in the tropics, Marseille, France, Monday 12 September 2005. *Vaccine* **25**(15), 2758–2765.

Barrett, A. D. and Teuwen, D. E. (2009). Yellow fever vaccine—How does it work and why do rare cases of serious adverse events take place? *Curr Opin Immunol* **21**(3), 308–313.

Belsher, J. L., Gay, P., Brinton, M., DellaValla, J., Ridenour, R., Lanciotti, R., Perelygin, A., Zaki, S., Paddock, C., Querec, T., Zhu, T., Pulendran, B., Eidex, R. B., and Hayes, E. (2007). Fatal multiorgan failure due to yellow fever vaccine-associated viscerotropic disease. *Vaccine* **25**(50), 8480–8485.

Benarroch, D., Selisko, B., Locatelli, G. A., Maga, G., Romette, J. L., and Canard, B. (2004). The RNA helicase, nucleotide 5′-triphosphatase, and RNA 5′-triphosphatase activities of Dengue virus protein NS3 are Mg2+-dependent and require a functional Walker B motif in the helicase catalytic core. *Virology* **328**(2), 208–218.

Bonnevie-Nielsen, V., Heron, I., Monath, T. P., and Calisher, C. H. (1995). Lymphocytic 2′,5′-oligoadenylate synthetase activity increases prior to the appearance of neutralizing antibodies and immunoglobulin M and immunoglobulin G antibodies after primary and secondary immunization with yellow fever vaccine. *Clin Diagn Lab Immunol* **2**(3), 302–306.

Botting, C. and Kuhn, R. J. (2012). Novel approaches to flavivirus drug discovery. *Expert Opin Drug Discov* **7**(5), 417–428.

Bray, M., Zhao, B. T., Markoff, L., Eckels, K. H., Chanock, R. M., and Lai, C. J. (1989). Mice immunized with recombinant vaccinia virus expressing dengue 4 virus structural proteins with or without nonstructural protein NS1 are protected against fatal dengue virus encephalitis. *J Virol* **63**(6), 2853–2856.

Bres, P. L. (1986). A century of progress in combating yellow fever. *Bull World Health Organ* **64**(6), 775–786.

Bressanelli, S., Stiasny, K., Allison, S. L., Stura, E. A., Duquerroy, S., Lescar, J., Heinz, F. X., and Rey, F. A. (2004). Structure of a flavivirus envelope glycoprotein in its low-pH-induced membrane fusion conformation. *EMBO J* **23**(4), 728–738.

Brinton, M. A. and Dispoto, J. H. (1988). Sequence and secondary structure analysis of the 5′-terminal region of flavivirus genome RNA. *Virology* **162**(2), 290–299.

Brinton, M. A., Fernandez, A. V., and Dispoto, J. H. (1986). The 3′-nucleotides of flavivirus genomic RNA form a conserved secondary structure. *Virology* **153**(1), 113–121.

Bryant, J. E., Holmes, E. C., and Barrett, A. D. (2007). Out of Africa: A molecular perspective on the introduction of yellow fever virus into the Americas. *PLoS Pathog* **3**(5), e75.

Chambers, T. J., Grakoui, A., and Rice, C. M. (1991). Processing of the yellow fever virus nonstructural polyprotein: A catalytically active NS3 proteinase domain and NS2B are required for cleavages at dibasic sites. *J Virol* **65**(11), 6042–6050.

Chambers, T. J., Hahn, C. S., Galler, R., and Rice, C. M. (1990a). Flavivirus genome organization, expression, and replication. *Annu Rev Microbiol* **44**, 649–688.

Chambers, T. J., McCourt, D. W., and Rice, C. M. (1990b). Production of yellow fever virus proteins in infected cells: Identification of discrete polyprotein species and analysis of cleavage kinetics using region-specific polyclonal antisera. *Virology* **177**(1), 159–174.

Chambers, T. J., Weir, R. C., Grakoui, A., McCourt, D. W., Bazan, J. F., Fletterick, R. J., and Rice, C. M. (1990c). Evidence that the N-terminal domain of nonstructural protein NS3 from yellow fever virus is a serine protease responsible for site-specific cleavages in the viral polyprotein. *Proc Natl Acad Sci USA* **87**(22), 8898–8902.

Chan, R. C., Penney, D. J., Little, D., Carter, I. W., Roberts, J. A., and Rawlinson, W. D. (2001). Hepatitis and death following vaccination with 17D-204 yellow fever vaccine. *Lancet* **358**(9276), 121–122.

Chu, P. W. and Westaway, E. G. (1987). Characterization of Kunjin virus RNA-dependent RNA polymerase: Reinitiation of synthesis in vitro. *Virology* **157**(2), 330–337.

Chung, K. M., Liszewski, M. K., Nybakken, G., Davis, A. E., Townsend, R. R., Fremont, D. H., Atkinson, J. P., and Diamond, M. S. (2006). West Nile virus nonstructural protein NS1 inhibits complement activation by binding the regulatory protein factor H. *Proc Natl Acad Sci USA* **103**(50), 19111–19116.

Chung, K. M., Thompson, B. S., Fremont, D. H., and Diamond, M. S. (2007). Antibody recognition of cell surface-associated NS1 triggers Fc-gamma receptor-mediated phagocytosis and clearance of West Nile Virus-infected cells. *J Virol* **81**(17), 9551–9555.

Co, M. D., Terajima, M., Cruz, J., Ennis, F. A., and Rothman, A. L. (2002). Human cytotoxic T lymphocyte responses to live attenuated 17D yellow fever vaccine: Identification of HLA-B35-restricted CTL epitopes on nonstructural proteins NS1, NS2b, NS3, and the structural protein E. *Virology* **293**(1), 151–163.

Daffis, S., Szretter, K. J., Schriewer, J., Li, J., Youn, S., Errett, J., Lin, T. Y., Schneller, S., Zust, R., Dong, H., Thiel, V., Sen, G. C., Fensterl, V., Klimstra, W. B., Pierson, T. C., Buller, R. M., Gale, M., Jr., Shi, P. Y., and Diamond, M. S. (2010). 2′-O methylation of the viral mRNA cap evades host restriction by IFIT family members. *Nature* **468**(7322), 452–456.

De Brito, T., Siqueira, S. A., Santos, R. T., Nassar, E. S., Coimbra, T. L., and Alves, V. A. (1992). Human fatal yellow fever. Immunohistochemical detection of viral antigens in the liver, kidney and heart. *Pathol Res Pract* **188**(1–2), 177–181.

Dunster, L. M., Wang, H., Ryman, K. D., Miller, B. R., Watowich, S. J., Minor, P. D., and Barrett, A. D. (1999). Molecular and biological changes associated with HeLa cell attenuation of wild-type yellow fever virus. *Virology* **261**(2), 309–318.

Edgil, D., Polacek, C., and Harris, E. (2006). Dengue virus utilizes a novel strategy for translation initiation when cap-dependent translation is inhibited. *J Virol* **80**(6), 2976–2986.

Egloff, M. P., Benarroch, D., Selisko, B., Romette, J. L., and Canard, B. (2002). An RNA cap (nucleoside-2′-O-)-methyltransferase in the flavivirus RNA polymerase NS5: Crystal structure and functional characterization. *EMBO J* **21**(11), 2757–2768.

Engel, A. R., Vasconcelos, P. F., McArthur, M. A., and Barrett, A. D. (2006). Characterization of a viscerotropic yellow fever vaccine variant from a patient in Brazil. *Vaccine* **24**(15), 2803–2909.

Falgout, B., Pethel, M., Zhang, Y. M., and Lai, C. J. (1991). Both nonstructural proteins NS2B and NS3 are required for the proteolytic processing of dengue virus nonstructural proteins. *J Virol* **65**(5), 2467–2475.

Fernandez-Garcia, M. D., Mazzon, M., Jacobs, M., and Amara, A. (2009). Pathogenesis of flavivirus infections: Using and abusing the host cell. *Cell Host Microbe* **5**(4), 318–328.

Galler, R., Post, P. R., Santos, C. N., and Ferreira, II (1998). Genetic variability among yellow fever virus 17D substrains. *Vaccine* **16**(9–10), 1024–1028.

Gerasimon, G. and Lowry, K. (2005). Rare case of fatal yellow fever vaccine-associated viscerotropic disease. *South Med J* **98**(6), 653–656.

Grun, J. B. and Brinton, M. A. (1986). Characterization of West Nile virus RNA-dependent RNA polymerase and cellular terminal adenylyl and uridylyl transferases in cell-free extracts. *J Virol* **60**(3), 1113–1124.

Guirakhoo, F., Heinz, F. X., Mandl, C. W., Holzmann, H., and Kunz, C. (1991). Fusion activity of flaviviruses: Comparison of mature and immature (prM-containing) tick-borne encephalitis virions. *J Gen Virol* **72**(Pt 6), 1323–1329.

Guyatt, K. J., Westaway, E. G., and Khromykh, A. A. (2001). Expression and purification of enzymatically active recombinant RNA-dependent RNA polymerase (NS5) of the flavivirus Kunjin. *J Virol Methods* **92**(1), 37–44.

Hacker, U. T., Jelinek, T., Erhardt, S., Eigler, A., Hartmann, G., Nothdurft, H. D., and Endres, S. (1998). In vivo synthesis of tumor necrosis factor-alpha in healthy humans after live yellow fever vaccination. *J Infect Dis* **177**(3), 774–778.

Hahn, C. S., Dalrymple, J. M., Strauss, J. H., and Rice, C. M. (1987). Comparison of the virulent Asibi strain of yellow fever virus with the 17D vaccine strain derived from it. *Proc Natl Acad Sci USA* **84**(7), 2019–2023.

Hayes, E. B. (2007). Acute viscerotropic disease following vaccination against yellow fever. *Trans R Soc Trop Med Hyg* **101**(10), 967–971.

Hoenen, A., Liu, W., Kochs, G., Khromykh, A. A., and Mackenzie, J. M. (2007). West Nile virus-induced cytoplasmic membrane structures provide partial protection against the interferon-induced antiviral MxA protein. *J Gen Virol* **88**(Pt 11), 3013–3017.

Ishak, R., Tovey, D. G., and Howard, C. R. (1988). Morphogenesis of yellow fever virus 17D in infected cell cultures. *J Gen Virol* **69**(Pt 2), 325–335.

Issur, M., Geiss, B. J., Bougie, I., Picard-Jean, F., Despins, S., Mayette, J., Hobdey, S. E., and Bisaillon, M. (2009). The flavivirus NS5 protein is a true RNA guanylyltransferase that catalyzes a two-step reaction to form the RNA cap structure. *RNA* **15**(12), 2340–2350.

Jennings, A. D., Whitby, J. E., Minor, P. D., and Barrett, A. D. (1993). Comparison of the nucleotide and deduced amino acid sequences of the envelope protein genes of the wild-type French viscerotropic strain of yellow fever virus and the live vaccine strain, French neurotropic vaccine, derived from it. *Virology* **192**(2), 692–695.

Johansson, M. A., Arana-Vizcarrondo, N., Biggerstaff, B. J., Gallagher, N., Marano, N., and Staples, J. E. (2012). Assessing the risk of international spread of yellow fever virus: A mathematical analysis of an urban outbreak in Asuncion, 2008. *Am J Trop Med Hyg* **86**(2), 349–358.

Kasturi, S. P., Skountzou, I., Albrecht, R. A., Koutsonanos, D., Hua, T., Nakaya, H. I., Ravindran, R., Stewart, S., Alam, M., Kwissa, M., Villinger, F., Murthy, N., Steel, J., Jacob, J., Hogan, R. J., Garcia-Sastre, A., Compans, R., and Pulendran, B. (2011). Programming the magnitude and persistence of antibody responses with innate immunity. *Nature* **470**(7335), 543–547.

Khaiboullina, S. F., Rizvanov, A. A., Holbrook, M. R., and St Jeor, S. (2005). Yellow fever virus strains Asibi and 17D-204 infect human umbilical cord endothelial cells and induce novel changes in gene expression. *Virology* **342**(2), 167–176.

Kiehn, L., Murphy, K. E., Yudin, M. H., and Loeb, M. (2008). Self-reported protective behaviour against West Nile Virus among pregnant women in Toronto. *J Obstet Gynaecol Can* **30**(12), 1103–1109.

Klimstra, W. B., Ryman, K. D., and Johnston, R. E. (1998). Adaptation of Sindbis virus to BHK cells selects for use of heparan sulfate as an attachment receptor. *J Virol* **72**(9), 7357–7366.

Kohler, S., Bethke, N., Bothe, M., Sommerick, S., Frentsch, M., Romagnani, C., Niedrig, M., and Thiel, A. (2012). The early cellular signatures of protective immunity induced by live viral vaccination. *Eur J Immunol* **42**(9), 2363–2373.

Lee, E. and Lobigs, M. (2008). E protein domain III determinants of yellow fever virus 17D vaccine strain enhance binding to glycosaminoglycans, impede virus spread, and attenuate virulence. *J Virol* **82**(12), 6024–6033.

Lee, J. M., Crooks, A. J., and Stephenson, J. R. (1989). The synthesis and maturation of a non-structural extracellular antigen from tick-borne encephalitis virus and its relationship to the intracellular NS1 protein. *J Gen Virol* **70**(Pt 2), 335–343.

Li, H., Clum, S., You, S., Ebner, K. E., and Padmanabhan, R. (1999). The serine protease and RNA-stimulated nucleoside triphosphatase and RNA helicase functional domains of dengue virus type 2 NS3 converge within a region of 20 amino acids. *J Virol* **73**(4), 3108–3116.

Li, L., Lok, S. M., Yu, I. M., Zhang, Y., Kuhn, R. J., Chen, J., and Rossmann, M. G. (2008). The flavivirus precursor membrane-envelope protein complex: Structure and maturation. *Science* **319**(5871), 1830–1834.

Lindenbach, B. D. and Rice, C. M. (2003). Molecular biology of flaviviruses. *Adv Virus Res* **59**, 23–61.

Liu, L., Dong, H., Chen, H., Zhang, J., Ling, H., Li, Z., Shi, P. Y., and Li, H. (2010). Flavivirus RNA cap methyltransferase: Structure, function, and inhibition. *Front Biol* **5**(4), 286–303.

Lorenz, I. C., Kartenbeck, J., Mezzacasa, A., Allison, S. L., Heinz, F. X., and Helenius, A. (2003). Intracellular assembly and secretion of recombinant subviral particles from tick-borne encephalitis virus. *J Virol* **77**(7), 4370–4382.

Mackenzie, J. M. and Westaway, E. G. (2001). Assembly and maturation of the flavivirus Kunjin virus appear to occur in the rough endoplasmic reticulum and along the secretory pathway, respectively. *J Virol* **75**(22), 10787–10799.

Martin, M., Tsai, T. F., Cropp, B., Chang, G. J., Holmes, D. A., Tseng, J., Shieh, W., Zaki, S. R., Al-Sanouri, I., Cutrona, A. F., Ray, G., Weld, L. H., and Cetron, M. S. (2001). Fever and multisystem organ failure associated with 17D-204 yellow fever vaccination: A report of four cases. *Lancet* **358**(9276), 98–104.

Mason, P. W. (1989). Maturation of Japanese encephalitis virus glycoproteins produced by infected mammalian and mosquito cells. *Virology* **169**(2), 354–364.

McArthur, M. A., Suderman, M. T., Mutebi, J. P., Xiao, S. Y., and Barrett, A. D. (2003). Molecular characterization of a hamster viscerotropic strain of yellow fever virus. *J Virol* **77**(2), 1462–1468.

McElroy, K. L., Tsetsarkin, K. A., Vanlandingham, D. L., and Higgs, S. (2005). Characterization of an infectious clone of the wild-type yellow fever virus Asibi strain that is able to infect and disseminate in mosquitoes. *J Gen Virol* **86**(Pt 6), 1747–1751.

Meier, K. C., Gardner, C. L., Khoretonenko, M. V., Klimstra, W. B., and Ryman, K. D. (2009). A mouse model for studying viscerotropic disease caused by yellow fever virus infection. *PLoS Pathog* **5**(10), e1000614.

Merlo, C., Steffen, R., Landis, T., Tsai, T., and Karabatsos, N. (1993). Possible association of encephalitis and 17D yellow fever vaccination in a 29-year-old traveller. *Vaccine* **11**(6), 691.

ter Meulen, J., Sakho, M., Koulemou, K., Magassouba, N., Bah, A., Preiser, W., Daffis, S., Klewitz, C., Bae, H. G., Niedrig, M., Zeller, H., Heinzel-Gutenbrunner, M., Koivogui, L., and Kaufmann, A. (2004). Activation of the cytokine network and unfavorable outcome in patients with yellow fever. *J Infect Dis* **190**(10), 1821–1827.

Miller, S., Kastner, S., Krijnse-Locker, J., Buhler, S., and Bartenschlager, R. (2007). The non-structural protein 4A of dengue virus is an integral membrane protein inducing membrane alterations in a 2K-regulated manner. *J Biol Chem* **282**(12), 8873–8882.

Miller, J. D., van der Most, R. G., Akondy, R. S., Glidewell, J. T., Albott, S., Masopust, D., Murali-Krishna, K., Mahar, P. L., Edupuganti, S., Lalor, S., Germon, S., Del Rio, C., Mulligan, M. J., Staprans, S. I., Altman, J. D., Feinberg, M. B., and Ahmed, R. (2008). Human effector and memory CD8+ T cell responses to smallpox and yellow fever vaccines. *Immunity* **28**(5), 710–722.

Modis, Y., Ogata, S., Clements, D., and Harrison, S. C. (2004). Structure of the dengue virus envelope protein after membrane fusion. *Nature* **427**(6972), 313–319.

Monath, T. P. (1991). Yellow fever: Victor, Victoria? Conqueror, conquest? Epidemics and research in the last forty years and prospects for the future. *Am J Trop Med Hyg* **45**(1), 1–43.

Monath, T. P. (2001). Yellow fever: An update. *Lancet Infect Dis* **1**(1), 11–20.

Monath, T. P. (2005). Yellow fever vaccine. *Expert Rev Vaccines* **4**(4), 553–574.

Monath, T. P. (2007). Dengue and yellow fever—Challenges for the development and use of vaccines. *N Engl J Med* **357**(22), 2222–2225.

Monath, T. P. (2008). Treatment of yellow fever. *Antiviral Res* **78**(1), 116–124.

Monath, T. P. and Barrett, A. D. (2003). Pathogenesis and pathophysiology of yellow fever. *Adv Virus Res* **60**, 343–395.

Monath, T. P., Myers, G. A., Beck, R. A., Knauber, M., Scappaticci, K., Pullano, T., Archambault, W. T., Catalan, J., Miller, C., Zhang, Z. X., Shin, S., Pugachev, K., Draper, K., Levenbook, I. S., and Guirakhoo, F. (2005). Safety testing for neurovirulence of novel live, attenuated flavivirus vaccines: Infant mice provide an accurate surrogate for the test in monkeys. *Biologicals* **33**(3), 131–144.

Monath, T. P. and Nasidi, A. (1993). Should yellow fever vaccine be included in the expanded program of immunization in Africa? A cost-effectiveness analysis for Nigeria. *Am J Trop Med Hyg* **48**(2), 274–299.

van der Most, R. G., Harrington, L. E., Giuggio, V., Mahar, P. L., and Ahmed, R. (2002). Yellow fever virus 17D envelope and NS3 proteins are major targets of the antiviral T cell response in mice. *Virology* **296**(1), 117–124.

Nagy, P. D. and Pogany, J. (2012). The dependence of viral RNA replication on co-opted host factors. *Nat Rev Microbiol* **10**(2), 137–149.

Nasidi, A., Monath, T. P., DeCock, K., Tomori, O., Cordellier, R., Olaleye, O. D., Harry, T. O., Adeniyi, J. A., Sorungbe, A. O., Ajose-Coker, A. O., and et al. (1989). Urban yellow fever epidemic in western Nigeria, 1987. *Trans R Soc Trop Med Hyg* **83**(3), 401–406.

Nestorowicz, A., Chambers, T. J., and Rice, C. M. (1994). Mutagenesis of the yellow fever virus NS2A/2B cleavage site: Effects on proteolytic processing, viral replication, and evidence for alternative processing of the NS2A protein. *Virology* **199**(1), 114–123.

Niedrig, M., Lademann, M., Emmerich, P., and Lafrenz, M. (1999). Assessment of IgG antibodies against yellow fever virus after vaccination with 17D by different assays: Neutralization test, haemagglutination inhibition test, immunofluorescence assay and ELISA. *Trop Med Int Health* **4**(12), 867–871.

Nisbet, D. J., Lee, K. J., van den Hurk, A. F., Johansen, C. A., Kuno, G., Chang, G. J., Mackenzie, J. S., Ritchie, S. A., and Hall, R. A. (2005). Identification of new flaviviruses in the Kokobera virus complex. *J Gen Virol* **86**(Pt 1), 121–124.

Onyango, C. O., Grobbelaar, A. A., Gibson, G. V., Sang, R. C., Sow, A., Swaneopel, R., and Burt, F. J. (2004). Yellow fever outbreak, southern Sudan, 2003. *Emerg Infect Dis* **10**(9), 1668–1670.

Patterson, K. D. (1992). Yellow fever epidemics and mortality in the United States, 1693–1905. *Soc Sci Med* **34**(8), 855–865.

Paul, D., Romero-Brey, I., Gouttenoire, J., Stoitsova, S., Krijnse-Locker, J., Moradpour, D., and Bartenschlager, R. (2011). NS4B self-interaction through conserved C-terminal elements is required for the establishment of functional hepatitis C virus replication complexes. *J Virol* **85**(14), 6963–6976.

Poland, J. D., Calisher, C. H., Monath, T. P., Downs, W. G., and Murphy, K. (1981). Persistence of neutralizing antibody 30–35 years after immunization with 17D yellow fever vaccine. *Bull World Health Organ* **59**(6), 895–900.

Pugachev, K. V., Guirakhoo, F., and Monath, T. P. (2005). New developments in flavivirus vaccines with special attention to yellow fever. *Curr Opin Infect Dis* **18**(5), 387–394.

Pulendran, B. (2009). Learning immunology from the yellow fever vaccine: Innate immunity to systems vaccinology. *Nat Rev Immunol* **9**(10), 741–747.

Quaresma, J. A., Barros, V. L., Fernandes, E. R., Pagliari, C., Takakura, C., da Costa Vasconcelos, P. F., de Andrade, H. F., Jr., and Duarte, M. I. (2005). Reconsideration of histopathology and ultrastructural aspects of the human liver in yellow fever. *Acta Trop* **94**(2), 116–127.

Quaresma, J. A., Barros, V. L., Pagliari, C., Fernandes, E. R., Guedes, F., Takakura, C. F., Andrade, H. F., Jr., Vasconcelos, P. F., and Duarte, M. I. (2006). Revisiting the liver in human yellow fever: Virus-induced apoptosis in hepatocytes associated with TGF-beta, TNF-alpha and NK cells activity. *Virology* **345**(1), 22–30.

Querec, T. D., Akondy, R. S., Lee, E. K., Cao, W., Nakaya, H. I., Teuwen, D., Pirani, A., Gernert, K., Deng, J., Marzolf, B., Kennedy, K., Wu, H., Bennouna, S., Oluoch, H., Miller, J., Vencio, R. Z., Mulligan, M., Aderem, A., Ahmed, R., and Pulendran, B. (2009). Systems biology approach predicts immunogenicity of the yellow fever vaccine in humans. *Nat Immunol* **10**(1), 116–125.

Querec, T. D. and Pulendran, B. (2007). Understanding the role of innate immunity in the mechanism of action of the live attenuated Yellow Fever Vaccine 17D. *Adv Exp Med Biol* **590**, 43–53.

Ray, D., Shah, A., Tilgner, M., Guo, Y., Zhao, Y., Dong, H., Deas, T. S., Zhou, Y., Li, H., and Shi, P. Y. (2006). West Nile virus 5′-cap structure is formed by sequential guanine N-7 and ribose 2′-O methylations by nonstructural protein 5. *J Virol* **80**(17), 8362–8370.

Reinhardt, B., Jaspert, R., Niedrig, M., Kostner, C., and L'Age-Stehr, J. (1998). Development of viremia and humoral and cellular parameters of immune activation after vaccination with yellow fever virus strain 17D: A model of human flavivirus infection. *J Med Virol* **56**(2), 159–167.

Rice, C. M., Lenches, E. M., Eddy, S. R., Shin, S. J., Sheets, R. L., and Strauss, J. H. (1985). Nucleotide sequence of yellow fever virus: Implications for flavivirus gene expression and evolution. *Science* **229**(4715), 726–733.

Robertson, S. E., Hull, B. P., Tomori, O., Bele, O., LeDuc, J. W., and Esteves, K. (1996). Yellow fever: A decade of reemergence. *JAMA* **276**(14), 1157–1162.

Roers, A., Hochkeppel, H. K., Horisberger, M. A., Hovanessian, A., and Haller, O. (1994). MxA gene expression after live virus vaccination: A sensitive marker for endogenous type I interferon. *J Infect Dis* **169**(4), 807–813.

Roosendaal, J., Westaway, E. G., Khromykh, A., and Mackenzie, J. M. (2006). Regulated cleavages at the West Nile virus NS4A-2K-NS4B junctions play a major role in rearranging cytoplasmic membranes and Golgi trafficking of the NS4A protein. *J Virol* **80**(9), 4623–4632.

Rosen, L. (1981). Transovarial transmission of arboviruses by mosquitoes (author's transl). *Med Trop (Mars)* **41**(1), 23–29.

Rosen, L. (1987a). Mechanism of vertical transmission of the dengue virus in mosquitoes. *C R Acad Sci III* **304**(13), 347–350.

Rosen, L. (1987b). Overwintering mechanisms of mosquito-borne arboviruses in temperate climates. *Am J Trop Med Hyg* **37**(3 Suppl), 69S–76S.

Ryman, K. D., Ledger, T. N., Campbell, G. A., Watowich, S. J., and Barrett, A. D. (1998). Mutation in a 17D-204 vaccine substrain-specific envelope protein epitope alters the pathogenesis of yellow fever virus in mice. *Virology* **244**(1), 59–65.

Ryman, K. D., Xie, H., Ledger, T. N., Campbell, G. A., and Barrett, A. D. (1997). Antigenic variants of yellow fever virus with an altered neurovirulence phenotype in mice. *Virology* **230**(2), 376–380.

dos Santos, C. N., Post, P. R., Carvalho, R., Ferreira, II, Rice, C. M., and Galler, R. (1995). Complete nucleotide sequence of yellow fever virus vaccine strains 17DD and 17D-213. *Virus Res* **35**(1), 35–41.

Schoggins, J. W., Wilson, S. J., Panis, M., Murphy, M. Y., Jones, C. T., Bieniasz, P., and Rice, C. M. (2011). A diverse range of gene products are effectors of the type I interferon antiviral response. *Nature* **472**(7344), 481–485.

Schoub, B. D., Dommann, C. J., Johnson, S., Downie, C., and Patel, P. L. (1990). Encephalitis in a 13-year-old boy following 17D yellow fever vaccine. *J Infect* **21**(1), 105–106.

Somnuke, P., Hauhart, R. E., Atkinson, J. P., Diamond, M. S., and Avirutnan, P. (2011). N-linked glycosylation of dengue virus NS1 protein modulates secretion, cell-surface expression, hexamer stability, and interactions with human complement. *Virology* **413**(2), 253–264.

Srisuphanunt, M., Sithiprasasna, R., Patpoparn, S., Attatippaholkun, W., and Wiwanitkit, V. (2007). ELISA as an alternative tool for epidemiological surveillance for dengue in mosquitoes: A report from Thailand. *J Vector Borne Dis* **44**(4), 272–276.

Staples, J. E. and Monath, T. P. (2008). Yellow fever: 100 years of discovery. *JAMA* **300**(8), 960–962.

Struchiner, C. J., Luz, P. M., Dourado, I., Sato, H. K., Aguiar, S. G., Ribeiro, J. G., Soares, R. C., and Codeco, C. T. (2004). Risk of fatal adverse events associated with 17DD yellow fever vaccine. *Epidemiol Infect* **132**(5), 939–946.

Theiler, M. and Smith, H. H. (1937a). The effect of prolonged cultivation in vitro upon the pathogenicity of yellow fever virus. *J Exp Med* **65**(6), 767–786.

Theiler, M. and Smith, H. H. (1937b). The Use of Yellow Fever Virus Modified by in vitro Cultivation for Human Immunization. *J Exp Med* **65**(6) 787–800.

Theiler, M. and Smith, H. H. (2000). The use of yellow fever virus modified by in vitro cultivation for human immunization. *J Exp Med* **65**, 787–800 (1937). *Rev Med Virol* **10**(1), 6–16; discussion 3–5.

Uchil, P. D. and Satchidanandam, V. (2003). Architecture of the flaviviral replication complex. Protease, nuclease, and detergents reveal encasement within double-layered membrane compartments. *J Biol Chem* **278**(27), 24388–24398.

Urcuqui-Inchima, S., Patino, C., Torres, S., Haenni, A. L., and Diaz, F. J. (2010). Recent developments in understanding dengue virus replication. *Adv Virus Res* **77**, 1–39.

Vasconcelos, P. F., Luna, E. J., Galler, R., Silva, L. J., Coimbra, T. L., Barros, V. L., Monath, T. P., Rodigues, S. G., Laval, C., Costa, Z. G., Vilela, M. F., Santos, C. L., Papaiordanou, P. M., Alves, V. A., Andrade, L. D., Sato, H. K., Rosa, E. S., Froguas, G. B., Lacava, E., Almeida, L. M., Cruz, A. C., Rocco, I. M., Santos, R. T., and Oliva, O. F. (2001). Serious adverse events associated with yellow fever 17DD vaccine in Brazil: A report of two cases. *Lancet* **358**(9276), 91–97.

Wang, E., Ryman, K. D., Jennings, A. D., Wood, D. J., Taffs, F., Minor, P. D., Sanders, P. G., and Barrett, A. D. (1995). Comparison of the genomes of the wild-type French viscerotropic strain of yellow fever virus with its vaccine derivative French neurotropic vaccine. *J Gen Virol* **76**(Pt 11), 2749–2755.

Warrener, P., Tamura, J. K., and Collett, M. S. (1993). RNA-stimulated NTPase activity associated with yellow fever virus NS3 protein expressed in bacteria. *J Virol* **67**(2), 989–996.

Welsch, S., Miller, S., Romero-Brey, I., Merz, A., Bleck, C. K., Walther, P., Fuller, S. D., Antony, C., Krijnse-Locker, J., and Bartenschlager, R. (2009). Composition and three-dimensional architecture of the dengue virus replication and assembly sites. *Cell Host Microbe* **5**(4), 365–375.

Westaway, E. G. (1987). Flavivirus replication strategy. *Adv Virus Res* **33**, 45–90.

Westaway, E. G. and Goodman, M. R. (1987). Variation in distribution of the three flavivirus-specified glycoproteins detected by immunofluorescence in infected Vero cells. *Arch Virol* **94**(3–4), 215–228.

Westaway, E. G., Khromykh, A. A., Kenney, M. T., Mackenzie, J. M., and Jones, M. K. (1997a). Proteins C and NS4B of the flavivirus Kunjin translocate independently into the nucleus. *Virology* **234**(1), 31–41.

Westaway, E. G., Khromykh, A. A., and Mackenzie, J. M. (1999). Nascent flavivirus RNA colocalized in situ with double-stranded RNA in stable replication complexes. *Virology* **258**(1), 108–117.

Westaway, E. G., Mackenzie, J. M., Kenney, M. T., Jones, M. K., and Khromykh, A. A. (1997b). Ultrastructure of Kunjin virus-infected cells: Colocalization of NS1 and NS3 with double-stranded RNA, and of NS2B with NS3, in virus-induced membrane structures. *J Virol* **71**(9), 6650–6661.

Whittembury, A., Ramirez, G., Hernandez, H., Ropero, A. M., Waterman, S., Ticona, M., Brinton, M., Uchuya, J., Gershman, M., Toledo, W., Staples, E., Campos, C., Martinez, M., Chang, G. J., Cabezas, C., Lanciotti, R., Zaki, S., Montgomery, J. M., Monath, T., and Hayes, E. (2009). Viscerotropic disease following yellow fever vaccination in Peru. *Vaccine* **27**(43), 5974–5981.

Winkler, G., Randolph, V. B., Cleaves, G. R., Ryan, T. E., and Stollar, V. (1988). Evidence that the mature form of the flavivirus nonstructural protein NS1 is a dimer. *Virology* **162**(1), 187–196.

Woodson, S. E., Freiberg, A. N., and Holbrook, M. R. (2011). Differential cytokine responses from primary human Kupffer cells following infection with wild-type or vaccine strain yellow fever virus. *Virology* **412**(1), 188–195.

Wu, J., Bera, A. K., Kuhn, R. J., and Smith, J. L. (2005). Structure of the Flavivirus helicase: Implications for catalytic activity, protein interactions, and proteolytic processing. *J Virol* **79**(16), 10268–10277.

Xie, H., Ryman, K. D., Campbell, G. A., and Barrett, A. D. (1998). Mutation in NS5 protein attenuates mouse neurovirulence of yellow fever 17D vaccine virus. *J Gen Virol* **79**(Pt 8), 1895–1899.

Yi, Z., Yuan, Z., Rice, C. M., and Macdonald, M. R. (2012). Flavivirus replication complex assembly revealed by DNAJC14 functional mapping. *J Virol* **86**(21), 11815–11832.

Zhang, Y. M., Hayes, E. P., McCarty, T. C., Dubois, D. R., Summers, P. L., Eckels, K. H., Chanock, R. M., and Lai, C. J. (1988). Immunization of mice with dengue structural proteins and nonstructural protein NS1 expressed by baculovirus recombinant induces resistance to dengue virus encephalitis. *J Virol* **62**(8), 3027–3031.

Zhou, Z., Wang, N., Woodson, S. E., Dong, Q., Wang, J., Liang, Y., Rijnbrand, R., Wei, L., Nichols, J. E., Guo, J. T., Holbrook, M. R., Lemon, S. M., and Li, K. (2011). Antiviral activities of ISG20 in positive-strand RNA virus infections. *Virology* **409**(2), 175–188.

28 Kyasanur Forest Disease

Gerhard Dobler and Frank Hufert

CONTENTS

28.1 INTRODUCTION

Kyasanur Forest disease virus (KFDV) was detected after a new febrile and sometimes fatal disease was described in Shimoga district of Karnataka State in India. Human suspected cases of typhoid occurred at least since 1956 in this area (Rau, 1957). However, serological tests to detect typhoid fever, like Widal or Weil–Felix, turned out to be nonreactive, so the suspected diagnosis of typhoid and typhus seemed unlikely. A field excursion of the Indian Virus Research Center in March and April 1957 finally discovered a filterable agent from a dead langur (*Presbytis entellus*) and two dead bonnets (*Macaca radiata*), which were found near a street along the Kyasanur State Forest (Work and Trapido 1957a). At the same time as human cases of this new severe illness were reported, a large number of sick or dead monkeys had been noticed by the local population. This observation at that time also caused severe concerns of an outbreak of sylvatic yellow fever in India. Later, the filterable agent could be isolated from patients' blood and organs and the transmission cycle could be explored after a similar virus was isolated from ticks of the genus *Haemaphysalis* (Trapido et al., 1959). The agent was named Kyasanur forest disease virus after the location of the original discovery. Serological tests revealed that KFDV was a member of the tick-borne virus serocomplex of the group B arboviruses, now the family Flaviviridae.

Over the next more than 20 years, the KFDV spread at least into three new districts of the Karnataka State. In 1989, a KFDV strain possibly was isolated from a patient in Yunnan, China. In 1994, a genetically related, but distinct virus, Alkhurma hemorrhagic fever virus (AHFV), was isolated from patients with severe hemorrhagic fever near Medina in Saudi Arabia.

28.2 VIROLOGY

28.2.1 STRUCTURE

KFDV is a member of the family Flaviviridae. There, it is classified into the "Mammalian tick-borne virus group." Flaviviruses are about 50 nm in diameter and have a spherical shape. They are enveloped. The genome consists of a single-stranded RNA of positive polarity. The type I cap at the 5′-end distinguishes the genus *Flavivirus* from all other genera in the family Flaviviridae. Although more than 100 partial sequences of KFDV strains are deposited at the NCBI data bank (December 2011), only one single whole genome, the complete viral genome of a KFDV, strain KFDV P9605, is available. This strain's genome consists of a total of 10,774 nucleotides. No poly-A-tail can be detected at the 3′-end. The genome has the typical structure as other flaviviruses, with the structural proteins at the 5′-end and the nonstructural proteins found at the 3′-end of virus genome.

As with other flaviviruses, the genome of the KFDV codes for three structural proteins: C protein (11 kDa), E protein (50 kDa), and a prM–M protein (26–28 kDa). The structural proteins form the virion particle with prM in the immature virus particle and M in the mature virion. Replication needs another seven viral nonstructural proteins: NS1 protein (46 kDa), NS2A protein (22 kDa), NS2B protein (14 kDa), NS3 protein (70 kDa), NS4A protein (16 kDa), NS4B protein (27 kDa), and NS5 protein (103 kDa). The E (envelope) protein is thought to be a main determinant for host range and pathogenicity through its receptor function. The virus particle contains lipids that derive from the host cell membranes and also carbohydrates in the prM, E, and NS1 proteins.

28.2.2 ORIGIN OF KYASANUR FOREST VIRUS

The origin of KFDV is not known. Due to the high genetic homology to Omsk hemorrhagic fever virus and to Langat virus, it might be speculated that KFDV could have originated from OHFV some time ago and then had been introduced by migrating birds from Siberia to India. New whole-genome analysis of KFDV and AHFV shows that these two viruses already diverged from each other about 500–700 years ago (Dodd et al., 2011). Earlier studies based on genetic analyses of fragments of the E genes and NS 5 genes only, however, suggested that AHFV was introduced from India to the Saudi Arabian Peninsula only between 66 and 177 years ago (Mehla et al., 2009). These authors speculated that camel ticks might have spread the virus along the Silk Road or by ship transport of camels from India to Saudi Arabia. There is some phylogenetic evidence that all viruses of the mammalian tick-borne virus group of Flaviviridae evolved some 2500 years ago in Asia and started a cline over the Eurasian continent. Several of the tick-borne flaviviruses only partially fit into this model, among them KFDV. One explanation could be that KFDV spread was influenced by human activities or by unusual bird migration. According to Work, no villager out of hundreds who had been interviewed ever realized any dead monkey in the Kyasanur Forest area before 1955 (Work, 1958). However, there is the possibility that KFDV circulated in Shimoga district for a long time and was not noticed. Possibly due to changes in ecology (e.g., increased farm animals in forests, clearing of forests, increased entering of forests for wood cutting) or due to changes in the pathogenicity of the virus, the clinical disease in monkeys and humans became evident in the mid-1950s. Although the knowledge of the origin of these viruses might be useful for understanding the spread of this virus and tick-borne viruses in general, due to the few data available, so far, the available data have to be interpreted with caution only.

28.2.3 Strain Variation

So far, only one subtype of KFDV has been identified. Intensive genetic studies including 48 virus strains that had been isolated over a period of 50 years from different Indian locations and different sources (humans, ticks, rodents, monkeys) show a genetic stability of KFDV with a diversity of only 1.2% at nucleotide level and 0.5% at amino acid level in a 720 nt fragment of the E gene and a 620 nt fragment of the NS5 gene (Mehla et al., 2009). Further genetic analyses show that the closest genetically related virus is AHFV, which was isolated from Saudi Arabian Peninsula in 1994. The genetic divergence between AHFV and KFDV is about 8.4% at the nucleotide level and 3.0% at the amino acid level (Dodd et al., 2011). Diversity to other members of the Mammalian tick-borne flavivirus group ranges from 35% to 45% at the nucleotide level (Wang et al., 2009).

There are only few studies available on the biological variation of KFDV isolates. The prototype strain showed reduced virulence in mice after 169 serial passages in primary kidney cells of *Macaca radiata*. However, after a further 23 passages in primary chick kidney cells, an increased virulence in mice was noticed again. Differences in plaque size and in growth kinetics were noticed, which may indicate different virulence properties.

28.2.4 Antigenic Variation

Antigenically, KFDV strains from different isolation sources and locations cannot be distinguished. However, KFDV can be distinguished antigenically from other members of the tick-borne flavivirus group. In a serologic study using cross-neutralization using polyclonal hyperimmune sera, KFDV showed the closest antigenic relationship to tick-borne encephalitis virus (Far Eastern/Siberian subtype) and only lower antigenic relationships to Omsk hemorrhagic fever virus and Langat virus (Table 28.1; Calisher et al., 1989).

TABLE 28.1
Cross-Neutralization of Tick-Borne Serocomplex Flaviviruses

Virus	Reciprocal Neutralizing Antibody Titer										
	TBE-FE	OHF	LI	KFD	LGT	NEG	POW	KSI	RF	CI	PPB
TBE-FE	**1,280**	320	20	320	320	<20	<20	<20	<20	<20	<20
OHF	20,000	**20,000**	1280	320	5120	1280	20	20	<20	<20	<20
TBE-W	20,000	1,280	5120	1280	1280	1280	<20	<20	<20	<20	<20
LI	1,280	320	**1280**	320	320	5120	80	<20	<20	<20	<20
KFD	1,280	80	20	**5120**	80	<20	<20	<20	<20	<20	<20
LGT	1,280	1,280	80	1280	**5120**	1280	20	<20	<20	80	<20
NEG	80	40	20	40	40	**2560**	<20	<20	<20	<20	<20
POW	80	40	20	40	40	80	**2560**	80	<20	<20	<20
KSI	80	<20	20	<20	<20	20	80	**320**	<20	<20	<20
RF	20	<20	<20	20	20	<20	<20	<20	**320**	<20	<20
CI	<20	<20	<20	<20	<20	<20	<20	<20	<20	**160**	<20
PPB	<20	<20	<20	<20	<20	<20	<20	<20	<20	40	**640**

Source: Modified after Calisher et al., *J. Gen. Virol.*, 70, 37, 1989.

Note: Homologous fibers are shown in bold.

TBE-FE, Tick-borne encephalitis virus, Far Eastern subtype; OHF, Omsk hemorrhagic fever virus; TBR-W, Tick-borne encephalitis virus, Western subtype; LI, Louping ill virus; KFD, Kyasanur Forest disease virus; LGT, Langat virus; NEG, Negishi virus; POW, Powassan virus; KSI, Karshi virus; RF, Royal Farm virus; CI, Carey Island virus; PPB, Phnom Penh Bat virus.

28.2.5 GEOGRAPHIC DISTRIBUTION

KFDV was originally isolated in Kyasanur Forest of Shimoga district of Karnataka State in southwestern India (Work and Trapido, 1957b). The known geographic range of KFDV was restricted to this original location until 1972, when a new focus in the Uttara Kannada district of Karnataka State was detected (Mehla et al., 2009). The virus spread further and, in 1972, established a third focus some kilometers south of the original focus in the district of Shimoga. A forth focus was detected in 1973 in Uttara Kannada near the village of Kodani, more than 50 km northwestern of the original focus. In 1975, the virus spread further east and established a focus around the village of Mandagadde, around 20 km east of the original focus. The so far final new focus which established was detected in 1982 in Patrame in the district of Dakshina Kannada, 80 km south of the original focus in Shimoga district. The KFDV now occurs in the four districts of Shimoga, Chikmagalur, Uttara Kannada, and Dakshina Kannada of the Indian State of Karnataka (Figure 28.1).

In 1989, a virus that was preliminarily named Nanjianyin was isolated from a febrile patient in Nanjian County in Yunnan State of southern China, bordering Burma (Wang et al., 2009). This isolate constitutes the only known KFDV isolate from outside India so far. There is some speculation that this virus, however, might be a laboratory contamination, as it turned out to be almost identical to the prototype strain, P9605. This prototype virus strain was used as antigen at that time for serological testing and was distributed to many arbovirus laboratories all over the world.

Serological results imply that KFDV or a related flavivirus might be prevalent in the other parts of India. Early serological studies detected antibodies against KFDV in the population in the districts of Kutch and Saurashtra in the Indian State of Gujarat, some 1200 km north of the original endemic area in Karnataka (Banerjee, 1989). The antibody reactivity against tick-borne virus in these sera that had been found in the early 1950s, before KFDV had been discovered, turned out to be specific antibodies against KFDV some years later. Other KFDV seropositive sera had been detected in scattered locations of the Indian State of West Bengal (Sarkar and Chatterjee, 1962). Although these data have to be interpreted with caution due to ample cross-reactivities among flaviviruses, these old data may imply a far wider geographical distribution on the Indian mainland as supposed today. A more recent study detected high seroprevalence rates of antibodies against KFDV in populations of the Andaman and Nicobar islands, two island groups that politically belong to India and are located in the Gulf of Bengal, some 1300 km southeast of the Indian mainland (Padbidri et al., 2002). So far, neither any clinical human disease nor any virus isolations have been reported from these islands. The serological results await further confirmation by isolation of KFDV from outside the known endemic focus area in Karnataka.

28.2.6 TRANSMISSION CYCLE AND ECOLOGY

KFDV belongs to the ecologically defined group of arboviruses. KFDV in nature circulates in a transmission cycle between ticks as vectors and reservoir and of vertebrates as hosts. This type of circulation in nature and transmission cycle became obvious already shortly after the first occurrence of KFDV in 1956 (Work, 1958). It is well known that the life cycle of arthropods and also of vertebrates will be influenced by many ecological factors, like geography, season, temperature, spatial activity, temporal activity, or host-seeking behavior. Any of these factors may interact with one another. Therefore, minor changes in the ecozone of an arbovirus may increase or decrease its activity and therefore cause its distinction or epidemic occurrence.

The known area of distribution of KFDV is located mainly at the crest of the Western Ghats in the Indian State of Karnataka in southwestern India (Figure 28.1). This area constitutes a unique form of ecotope as it is composed of different diverse biotopes as forests, grassland, and cultivated valleys. The forests are formed by a mixture of different vegetation types, namely, tropical evergreen, semi-evergreen, deciduous forest, and bamboo brakes. The ground usually is covered by

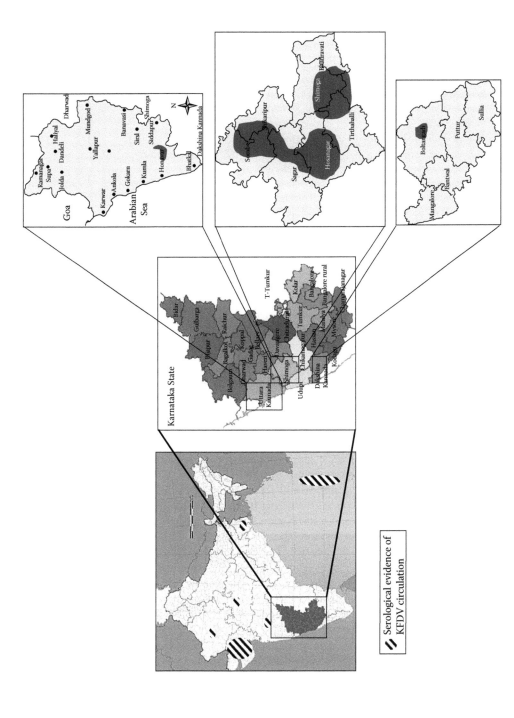

FIGURE 28.1 Known geographic distribution of KFDV in the districts of Uttara Kannada, Shimoga, and Mangalore, and serological evidence of KFDV distribution in India.

TABLE 28.2

Tick Species Potential to Transmit KFDV and Proportion of Single Tick Species in Naturally Infected Ticks

Tick Species	Infection Type	Proportion in Naturally Infected Ticks (913 Positive Ticks) (%)
Haemaphysalis spinigera	Natural infection	57
Haemaphysalis turturis	Natural infection	32
Haemaphysalis kinneari	Natural infection	5
Haemaphysalis minuta	Natural infection	3
Haemaphysalis cuspidata	Natural infection	0.9
Haemaphysalis kyasanurensis	Natural infection	
Haemaphysalis bispinosa	Natural infection	<0.5
Haemaphysalis wellingtoni	Natural infection	
Haemaphysalis aculeata	Natural infection	
Haemaphysalis obesa	Experimental infection	
Rhipicephalus haemaphysaloides	Natural infection	
Hyalomma marginatum isaaci	Natural infection	
Ornithodoros crossi	Experimental infection	
Ornithodoros savignyi	Experimental infection	
Argas persicus	Experimental infection	
Argas arboreus	Experimental infection	
Ixodes petauristae	Natural infection	
Dermacentor auratus	Natural infection	<0.5
Ixodes ceylonensis	Natural infection	
Non-specified	Natural infection	1.8

Sources: Sreenivasan et al., *Prog. Acarol.*, 1, 37, 1975; Banerjee, K., Kyasanur Forest disease, in Monath, T., (Ed.) *The Arboviruses: Epidemiology and Ecology*, Vol. III, CRC Press, Boca Raton, pp. 93–116, 1989; Pattnaik, P., *Rev. Med. Virol.*, 16, 151, 2006.

a layer of residual scrub of different types. This ecotope forms an ideal biotope for ticks as the vectors and small mammals as major vertebrate hosts of the KFDV.

Comprehensive studies in the original natural focus in Shimoga district detected KFDV in at least 9 out of the more than 30 tick species occurring in the area (Sreenivasan et al., 1975). Further experimental work on the infection and transmission of ticks of different species finally identified at least 19 species that will or may be able to replicate and transmit KFDV (Table 28.2). The rate of infected ticks was tested mainly in *Haemaphysalis spinigera*. Depending on the year and the location tested, the rate of infected ticks ranges from 0.1% to 10% (Banerjee, 1989).

These early studies (Table 28.2) show that the highest proportions of naturally infected ticks are among the hard tick species *Haemaphysalis spinigera*, *Haemaphysalis turturis*, and *Haemaphysalis kinneari*. These three species constitute about 95% of all KFDV-positive ticks found in the Kyasanur Forest and therefore are thought to have a major role as vectors of KFDV. According to the climatic conditions in India, *Haemaphysalis* ticks may complete a whole life cycle from eggs to adult stages within 1 year as a continuous process. Nevertheless, there is a clear seasonal synchronicity of biological processes regarding tick maturation, which may have a deep impact on the occurrence of KFD during the year and also on the KFD epidemics and epizootics (Figure 28.2). Adult tick activity seems to be synchronized with the time of monsoon as these months seem to provide the necessary high humidity for the development of tick eggs. At the end of the monsoon months, the developed larvae may migrate from the soil litter, the place of development, and may search for

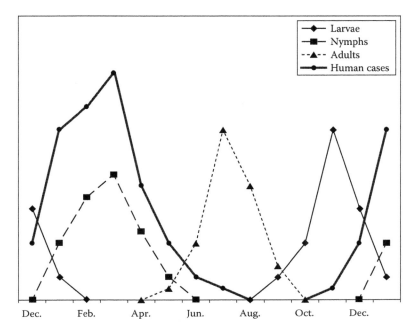

FIGURE 28.2 Seasonal activity of different developmental stages of *Haemaphysalis spinigera* and temporal coincidence with human cases.

the blood meals in small ground-living mammals. The development from larvae to nymphs takes place during the post-monsoon months from October to December. During this time, the humidity in the litter zone seems still quite high as to support the life of larvae and also the development from larvae to nymphs. Nymphs are active during the months of December to June. Adults will be dormant in the litter until they become active when the first rain of the monsoon arrives. The adults may be active for 2–3 months when they gradually migrate horizontally and vertically. Four weeks after starting their activity, they reach an activity maximum, which will decrease during the following months until November.

The different stages of ticks need different hosts and sources for blood. The larvae are mainly searching for small, ground-living rodents. Among these rodents are rats, shrews, and squirrels that turned out to support efficiently the growth of KFDV. Nymphs are mainly searching for rodents and medium-sized forest animals. The nymphs are thought to be the main tick stage responsible for the transmission of KFDV to monkeys and to human beings. Adults are rarely found on humans. They are searching cattle, goats, or larger game animals like porcupines, buffalos, samba deers, and spotted deers. KFDV may be transmitted transovarially from females to eggs in some tick species. While *Haemaphysalis spinigera* does support transovarial transmission, this way of transmission was detected to be an effective mechanism in *Ixodes petauristae*. All tick species that support virus replication of KFDV are able to pass the virus from one stage to the next (transstadial transmission). Therefore, ticks form the vectors but also seem to be the most important reservoir of KFDV. When a tick takes up KFDV at any stage, it will stay infectious for the rest of their lives (Table 28.3).

Virus isolation and serological data from wild animals show that a number of rodents are infected with KFDV and may also be able to develop viremias that are high enough to transmit the KFDV to other blood-sucking ticks. Natural and experimental studies show the special importance of several rodent species for the transmission of the KFDV. The Blanford's rat, the common house shrew, and the jungle palm squirrel develop high viremias (>10^7 LD^{50}), which support the transmission to new blood-sucking ticks. Viremias last for 1–2 weeks with a

TABLE 28.3

Non-Primate Vertebrates (Mammals, Birds) with Evidence of Kyasanur Forest Disease Virus Natural or Experimental Infection

Vertebrate	Name	KFD Infection
Rattus blanfordi	White-tailed wood rat	Natural infection, experimental transmission
Suncus murinus	Asian house shrew, Asian musk shrew	Natural infection, experimental transmission
Funambulus tristriatus	Jungle palm squirrel	Experimental infection, seropositive
Rattus rattus wroughtoni	Field rat	Seropositive, natural infection
Rattus rattus rufescens	House rat	Seropositive
Golunda ellioti	Indian bush rat	Seropositive
Mus booduga	Little Indian field mouse	Seropositive
Vandeleuria oleracea	Asiatic long-tailed climbing mouse	Natural infection, experimental infection
Funambulus pennanti	Northern palm squirrel	Seropositive
Tatera indica	South Indian gerbile	Seropositive
Petaurista petaurista	Red giant flying squirrel	Experimental transmission
Rousettus leschenaulti	Leschenaulti's Roussette fruit bat	Seropositive
Eonycteris spelaea	Cave nectar bat, dawn bat	Seropositive
Cynopterus sphinx	Greater short-nosed fruit bat	Experimental transmission; seropositive
Rhinolophus rouxi	Rufous horseshoe bat	Seropositive
Hipposideros speoris	Schneider's leaf-nosed bat	Seropositive
Hipposideros lankadiva	Indian leaf-nosed bat, Indian roundleaf bat	Seropositive
Miniopterus schreibersii	Long-winged bat	Seropositive
Mus plathytrix	Brown spiny mouse, flat-haired mouse	Experimental transmission; seropositive
Lepus nigricollis	Black-naped hare	Experimental transmission
Tephrodornis gularis	Large woodshrike	Seropositive
Galloperdix spadicea	Red spur fowl	Natural infection, experimental infection
Gallus sonneratii	Jungle fowl	Experimental infection
Megalaima zeylanica	Brown-headed barbet, large green barbet	Seropositive
Chalcophaps indica	Common Emerald dove	Seropositive
Treron pompadora	Pompadour green pigeon	Seropositive
Rhopocichla atriceps	Dark-fronted babbler	Seropositive
Hystrix indica	Indian crested porcupine	Natural infection; experimental infection

Notes: Seropositive: antibodies detected in wild animals; natural infection: KFDV isolation; experimental infection: experimental infection.

peak on the fourth to fifth day (Boshell et al., 1968). Also the giant flying squirrel, a common squirrel in the Kyasanur Forest, develops high viremias and is thought to support the transmission cycle of KFDV. Members of the genus *Mus* (little Indian field mouse, brown spiny mouse) and *Vandeleuria* (Asiatic long-tailed climbing mouse) develop low viremic titers that probably may not or may only randomly support virus transmission to new ticks. Also the black-naped hare, a very common inhabitant of the forests, does not develop viremias high enough to support the virus transmission cycle.

The role of bats for the transmission cycle of KFDV has not been finally clarified. The Leschenaulti's Roussette fruit bat and the greater short-nosed fruit bat develop viremias high enough to support virus transmission to new ticks. Also the experimentally infected insectivorous Rufous horseshoe bat developed viremias high enough to propagate KFDV to other blood-sucking ticks. However, bat ticks usually show a highly specialized host-seeking behavior with using only bats as a blood source. Therefore, it is unclear whether bats really play a role in sustaining the transmission

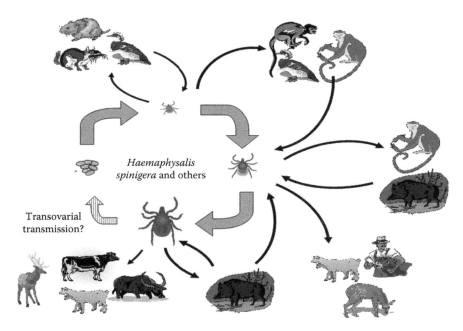

FIGURE 28.3 Proposed transmission cycle of KFD virus.

cycle or whether KFDV only circulates by chance in these animals. While experimentally infected Leschenaulti's Roussette fruit bats did not show any clinical signs after infection and developed neutralizing antibodies, the greater short-nosed fruit bat showed clinical disease and death after infection with KFDV.

Larger game and house animals seem not to support the virus replication and form viremias high enough to transmit the virus to other ticks. They, however, seem to be important to sustain the adult tick populations. Adult ticks mainly parasitize on the larger game and farm animals. One exception seems to be the Indian crested porcupine. This animal is very abundant in the KFDV endemic forest zones. Experimentally infected porcupines develop high viremias. All stages of *Haemaphysalis turturis* and *Haemaphysalis kinnearis* and immature stages of *Haemaphysalis spinigera* are found on the porcupines. The developed viremia titer seems high enough and long enough to support the effective transmission to noninfected blood-sucking ticks. Therefore, porcupines may act as a kind of amplifying hosts to distribute and increase the number of infected ticks.

The role of birds was studied in several early experimental studies. Out of 29 bird species tested, two abundant bird species, the Red spur fowl and the Jungle fowl, were found to develop viremias high enough to play a role for the virus transmission cycle (Figure 28.3).

28.2.7 EPIDEMIOLOGY

There are only incomplete and sometimes inconsistent data available on the epidemiology of KFDV in India. The number of human cases ranges from several dozens to more than 1500 cases per year (Table 28.4).

The epidemic season tends to start in October with a maximum of human cases from February to April. Only few or no cases are reported from June to September. The time of the increased number of human cases corresponds with the drier season of the year, while during the monsoon season from June to October, the number decreases or even halts. A similar pattern of seasonal distribution is seen in monkey epizootics. One main reason for the occurrence in the drier season may be the

TABLE 28.4

Reported Annual Human Cases of KFD in India

Year	KFD Cases Reported
1957	466
1958	181
1981	>550
1982/1983	1555
1984	>3000[a]
1999	10–85[b]
2000	>130[b]
2001	>435[b]
2002	98–625[b]
2003	253–920[b]
2004	27–606[b]
2005	>21[c]
2006	>16[c]
2011	>11[c]

[a] *Source:* According to Nichter (1987).
[b] Inconsistent data.
[c] According to ProMEDmail.

main time of activity of tick nymphs that are thought to be of major importance for the transmission of the disease to monkeys and humans, while during the rainy seasons, mainly larvae are active, which do not frequently attack humans or primates. There is some evidence that the total number of nymphs and the relative rate of infected ticks correlate well with the number of human cases and with the intensity of epidemics and epizootics.

The main risk factor for the infection with KFDV in humans is a history of exposure in the forest. The age group of 10–40 years is mainly affected, and males are 2.6 times more frequently affected than females (Banerjee, 1989). This age and sex distribution reflects the occupational exposure in forested areas.

The effect of vaccination on the epidemiology of KFD is unclear. Every year, between 50,000 and 100,000 doses are administered. In total, up to 1 million inhabitants may live in or near the KFDV natural endemic areas of the four affected districts of Karnataka State. Assuming an attack rate of 870/100,000 as observed in the vaccine trial of 1992–1994 in the non-vaccinated study arm, a total number of up to 9000 infections may occur per year (Dandawate et al., 1994). However, no information on the manifestation rate of KFD is available. As with other tick-borne flaviviruses, perhaps only a minor part of infections may represent the full clinical picture of severe hemorrhagic fever. Therefore, a major part of infections might go unrecognized. The annual vaccination campaigns with between 50,000 and 100,000 administered doses may modify the epidemiology of KFD in the affected human population. In a disease with a similar epidemiology, tick-borne encephalitis, the example of Austria shows that the rate of vaccinated population correlates indirectly to the number of cases reported in humans. Seroprevalence studies may be only of limited value as vaccine-induced antibodies may interfere with the conventional serological tests used unless non-structural proteins (NS1) may be used as antigen as with other flaviviruses (e.g., Japanese encephalitis virus).

28.3 CLINICAL PICTURE AND PATHOGENESIS

28.3.1 HUMANS

The incubation period of KFD ranges between 2 and 8 days. In laboratory exposures, the incubation period could be determined to range from 4 to 6 days. The clinical symptoms start suddenly with chills and severe frontal headache. The onset is so sudden that many patients can give an exact time of onset. Fever is following within hours and will rise to more than 40°C. Temperature is continuously high for 1–2 weeks. Patients may complain of generalized myalgia, which may show a severity similar to the myalgias reported with dengue fever. Often muscle pains are described as cramp-like and are especially severe at the neck and lumbar region and in the calf muscle. From the beginning of symptoms, the patients show general signs of infection like severe prostration, conjunctival suffusions, and photophobia. The cervical and axillary nymph nodes are enlarged and are palpable. In almost all human cases, papulovesicular lesions are detected in the soft palate. On the third or fourth day of the disease, gastrointestinal symptoms including vomiting and diarrhea are described.

A relative bradycardia is reported in many patients with a heartbeat rate of 50–65 during the febrile period. This bradycardia may be prolonged up to 3 weeks. There may be a fall in blood pressure with the fall of temperature in the second week of disease. A splenomegaly may occur in up to 20% of patients while hepatomegaly is uncommon.

Hemorrhagic symptoms may only develop in a minor part of all cases. In a study on ocular manifestations in patients with KFD, 17/98 (17.3%) patients developed any symptoms of hemorrhage. Hemorrhages may present as early as day 3 of clinical symptoms, but may also develop after the fever disappeared, up to 12–14 days after symptoms started. The main hemorrhagic symptoms are gastrointestinal bleeding with hematemesis and/or blood in the stool. Also bleeding from gums and from nose may appear.

Neurological symptoms may appear during the course of disease. Mental confusion, disorientation, and drowsiness may be found. Rarely behavioral changes are described.

Often KFD shows a biphasic course. After a first febrile phase of 1–2 weeks, an afebrile phase may appear that can last for 1–2 weeks, followed by a new increase in temperature between the third and fourth weeks. Besides fever, this second phase of symptoms usually is dominated by neurological symptoms resembling Omsk hemorrhagic fever or tick-borne encephalitis.

The convalescence may be prolonged and lasts one to several months until the patient gains his full activity again. During this time, an extreme weakness of the muscles and tremors after only slight effort are noticed (Figure 28.4).

Death occurs mainly in the second week of the disease. Persistent high fever, slurred speech, tremors, and dyspnea may proceed a fatal outcome for 2–12 h. In few fatal human cases that were examined, the cause of death was found to be due to pulmonary edema. The fatality rate of KFD ranges between 2% and 10% (Work and Trapido 1957a). While during the original epidemic in 1957/1958 a fatality rate of 2%–4% had been reported, during the period of high virus activity from 1982 to 1984, up to 10% of the hospitalized patients died. It is unclear whether this higher fatality rate was biased with mainly severe cases hospitalized or whether there were differences in the virulence of KFDV. In surviving patients, antibody response can be detected at the end of the second week of illness.

Regarding hematological parameters, leukopenia is a constant feature of KFD. Leukopenia is due to neutropenia and lymphopenia and in most cases will be less than 2000 cells/μL. Also an eosinopenia is found during the first week of illness. Another hematological feature is a prominent thrombocytopenia. Coagulation was only partially impaired with a prolonged bleeding time and mean coagulation time, while prothrombin time and fibrinogen content were not found abnormal. A low serum albumin, increased gammaglobulin, alkaline phosphatase, liver enzymes, and bilirubin were found commonly in patients with KFD.

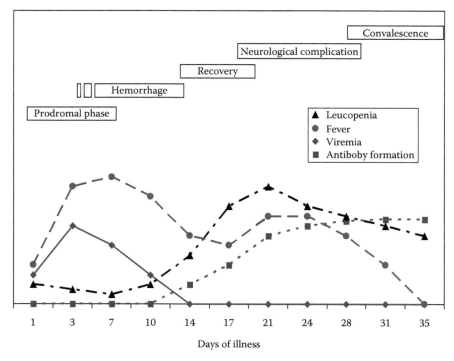

FIGURE 28.4   Clinical course of Kyasanur Forest disease.

The pathophysiology of KFD is far from known. The very few conducted studies showed elevated levels of circulating interferon-α (Pavri, 1989). Interferon-α may interact with IgE and activate macrophages. According to one hypothesis, the hemorrhagic symptoms of KFD may be caused by mediators, like prostaglandins and platelet-activating factor, which may be released by activated macrophages (Pavri, 1989).

28.3.2   Domestic Animals

There is no evidence that the infection with KFDV in domestic animals will cause any clinical disease. In cattle, which are exposed intensively to ticks and to KFDV, no evidence of viremia could be detected. No clinical symptoms could be detected in sheep, goat, horses, donkeys, dogs, or cats (Banerjee, 1989).

28.3.3   Laboratory Animals

In the original attempts to isolate KFDV, baby mice and 3-week-old mice were inoculated intracerebrally with original serum and heart muscle, skeletal muscle, lung liver, spleen, kidney, and brain of the dead langur and the bonnet monkeys. Most suckling mice showed symptoms on the third day post inoculation and died on the fourth day after infection. The sub-adult mice died little later between the fourth and sixth days post inoculation. Sub-adult mice could be also infected by the subcutaneous and intraperitoneal route. By these inoculation routes, the prodromal phase was prolonged, and virus titers in the brains of the animals were up to 3 logs lower than that in suckling mice (Work, 1958).

During the original isolation attempts, also suckling hamsters and cotton rats were inoculated intracerebrally and intraperitoneally with serum from the first dead langur. The hamsters became ill and died of the infection while the cotton rats did not exhibit any clinical symptoms.

The intraperitoneal inoculation of brain, liver, spleen, and heart muscle of the same monkey into adult hamsters and guinea pigs did not cause any clinical symptoms in the experimentally inoculated animals. The subsequent serological testing showed that hamsters and cotton rats did develop neutralizing antibodies against KFDV, while no immune response could be detected in inoculated guinea pigs (Work, 1958).

28.3.4 WILD ANIMALS

KFDV was originally isolated from dead langurs (*Presbytis entellus*) and bonnet monkeys (*Macaca radiata*). There is evidence that the greater short-nosed fruit bat (*Cynopterus sphinx*) sometimes develops clinical symptoms with fatal outcome after experimental infection. So far, the two monkey species and the bat species mentioned are the only wild animals known to become sick from KFDV infection. Infection in the two monkey species resembles human infection. The incubation period in experimentally infected rhesus monkeys was 4–6 days depending on the route of inoculation. The viremia lasted from day 4 to day 12 for a minimum of 7 days.

28.4 DIAGNOSTIC PROCEDURE

A first suspicion of KFD should become evident if a patient with a generalized febrile illness will have spent time in a known or suspected endemic region of the KFDV. Any febrile illness after an anamnesis of a stay in an endemic area should force a diagnostics to confirm or exclude KFD. Diagnostic procedure during the first 8–10 days (up to day 13) after the beginning of clinical symptoms (the viremic phase) is the detection of virus. The highest level of viremia can be detected between the third and sixth days. The detection of KFDV may be done using molecular techniques (reverse transcriptase polymerase chain reaction) or by virus isolation. Virus isolation can be conducted using intracerebral inoculation of animals like baby mice or cell culture. KFDV grows in various commonly used cell cultures (Vero, BHK-21). For inoculation serum, plasma or homogenized organ material can be used. So far, no virus was detected either in milk or in urine. Virus isolation must be conducted in a BSL-3 laboratory or in a BSL-4 laboratory. The KFDV is highly infectious, and before introduction of laboratories of BSL-3 or BSL-4, numerous laboratory infections were reported.

In fatal cases, KFDV can be detected in organs like liver, heart, lung, spleen, or kidney. In liver cells, a brownish pigment may be detected with focal hepatic necrosis. Kupffer cells are prominent and sometimes multinucleated as evidence for erythrophagocytosis. In the lungs, a marked exudation of plasma, erythrocytes, and leukocytes in the alveoli is found. In the kidneys, variable degrees of degenerative changes in the cortical convoluted tubules are typical. KFDV antigen may be detected by immunochemistry in the affected organs.

After day 10 of illness, besides virus detection, also the detection of specific antibodies against KFDV should be conducted. IgM and/or IgG detection may be conducted using ELISA or immunofluorescence. There are no commercial tests available, and therefore antibody detection should be done only by experienced laboratories. The detected antibody titer should be confirmed by a second blood sample 10–14 days later. If a significant increase in IgG and/or IgM is detected, the KFD is confirmed if any other flavivirus infection can be excluded. There is no information available about the cross-reaction of tick-borne encephalitis or Japanese encephalitis or yellow fever vaccinated people in serological tests using KFDV as antigen. There is also no information available whether and to what extend cross-reactions between KFDV antibody positive sera and other flavivirus antigens (e.g., dengue, yellow fever, West Nile, tick-borne encephalitis virus) exist and may disturb serological reactions against KFDV. In general, RT-PCR is preferred as acute diagnostic testing, while virus isolation may last days to weeks until confirmation of the virus. Antibody detection may help in retrospective diagnosis and in post-acute diagnosis of KFD.

28.4.1 Differential Diagnosis

The differential diagnosis of KFD includes all febrile infections with or without hemorrhage. Besides malaria and influenza, also typhoid and epidemic typhus and murine typhus may be excluded. Leptospirosis and spotted fever group rickettsioses as well as Q fever, Tsutsugamushi fever, and other arbovirus infections like dengue fever and chikungunya fever have to be included in the differential diagnosis.

28.5 TREATMENT

There is no specific treatment for KFD available. Therefore, all therapeutic efforts have to concentrate on the stabilization of physiologic functions. In severe cases, patients will have to be monitored under intensive care. Dehydration may be treated by using intravenous infusion therapy. In cases of hemorrhage, blood and plasma transfusions may be applied. Furthermore, supportive and vitamin-enriched diet may be beneficial for the patients.

28.6 PREVENTION AND CONTROL

Shortly after the detection of KFDV, an inactivated mouse-brain vaccine of tick-borne encephalitis virus Far Eastern subtype was applied to more than 4000 people living in the endemic region. However, the TBE vaccine proved to be noneffective in preventing or reducing the risk of KFD.

Since the late 1960s, an inactivated formalinized cell culture-grown vaccine for KFDV is available in India. The vaccine strain used is P9605. The virus is grown in primary chick embryo cells and purified, concentrated, and inactivated by formalinization. No adjuvants are mentioned to be added. No data on protein content, formalin residues, stabilizers or residues of cells or antibiotics, etc., are available. In a clinical trial, the vaccine was administered to 1985 inhabitants of the endemic areas. One single dose was received by 1405 participants, and 480 participants received two doses of vaccine with the second shot 4 weeks after the first (Dandawate et al., 1980). The serological testing of 214 paired sera exhibited a seroconversion rate with neutralizing antibodies of 59%. It was discussed that pre-existing antibodies against other flaviviruses (e.g., West Nile virus) might have interfered with the formation of antibodies. Using the same vaccine in a small group of highly exposed laboratory personnel showed a serological response rate of 72.5% (Banerjee et al., 1969). The difference between the two studies was explained by the different conditions in the laboratory and in the field.

In a large field study in 1990–1992 in 72 of the affected villages in the endemic area in three affected districts of Karnataka State, the vaccine efficacy in the field was tested by following more than 30,000 vaccinated inhabitants of the study area. In the unvaccinated population, 325 infections were found out of 37,374 study participants (attack rate 87/10,000). Among 9072 participants receiving one dose of vaccine, a total of 14 infections were detected (attack rate 15/10,000). In 21,083 participants receiving two doses of vaccine, 10 infections with KFDV were detected (attack rate 5/10,000). The vaccine efficacies were calculated at 82.4% after one single dose and at 94.8% for two doses of vaccine (Dandawate et al., 1994). No vaccine-related severe adverse events were reported during the field trial. However, detailed information on the safety of the vaccine is not available. The vaccine is produced by the State Institute of Animal Health and Veterinary Biologicals. Annually, between 100,000 and 150,000 doses of KFD vaccine are purchased by the Karnataka State and distributed to the afflicted populations.

Besides immunological prevention, measures for tick control and tick exposure were tested and used. In a comparative study, six different repellents were tested for their effectiveness against eight different Indian ticks. The repellents DEET and pyrethrum exhibited the highest effectiveness against ticks of the genus *Haemaphysalis* (Kulkarni and Naik 1985). Therefore, the use of impregnated clothes will be another possibility in reducing the potential risk of exposition to ticks when entering the endemic forest areas. It was shown that the risk of becoming infected for humans is higher in the vicinity of dead monkeys. Therefore, the use of insecticides within a radius of 50 m around a dead monkey was recommended to reduce the risk of exposition to nymphs and larvae that were seeking for new hosts after leaving the dead monkey (Banerjee, 1989). This procedure may be applied around houses to reduce the risk of infection for inhabitants. Furthermore, treatment of cattle with acaricides may also reduce the tick populations around homes and therefore diminish the risk of people living at the forest edges and working at or in the forests.

REFERENCES

Banerjee, K., 1989. Kyasanur Forest disease. In: Monath, T. (Ed.) *The Arboviruses: Epidemiology and Ecology*, Vol. III, CRC Press, Boca Raton, FL, pp. 93–116.

Banerjee, K., Dandawate, C.N., Bhatt, P.N., Ramachandra Rao, T., 1969. Serological response in humans to a formolized Kyasanur Forest disease vaccine. *Indian J. Med. Res.* 57, 969–974.

Boshell, M.J., Goverdhan, M.K., Rajagopalan, P.K., 1968. Preliminary studies on the susceptibility of wild rodents and shrews to KFS virus. *Indian J. Med. Res.* 56 (Suppl), 614–627.

Calisher, C.H., Karabatsos, N., Dalrymple, I.M., Shope, R.E., Porterfield, J.S., Westaway, E.G., Brandt, W.E., 1989. Antigenic relationships between flaviviruses as determined by cross-neutralization test with polyclonal antisera. *J. Gen. Virol.* 70, 37–43.

Dandawate, C.N., Desai, G.B., Achar, T.R., Banerjee, K., 1994. Field evaluation of formalin inactivated Kyasanur Forest disease tissue culture vaccine in three districts of Karnataka state. *Indian J. Med. Res.* 99, 152–158.

Dandawate, C.N., Upadhyaya, S., Banerjee, K., 1980. Serological response to formolized Kyasanur Forest disease virus vaccine in humans at Sagar and Sorab Talukas of Shimoga district. *J. Biol. Stand.* 8, 1–6.

Dodd, K.A., Bird, B.H., Khristova, M.L., Albarino, C.G., Carroll, S.A., Comer, J.A., Erickson, B.R., Rollin, P.E., Nichol, S.T., 2011. Ancient ancestry of KFDV and AHFV revealed by complete genome analyses of viruses isolated from ticks and mammalian hosts. *PloS Negl. Dis.* 5, e1352.

Kulkarni, S.M., Naik, V.M., 1985. Laboratory evaluation of six repellents against some Indian ticks. *Indian J. Med. Res.* 82, 14–18.

Mehla, R., Kumar, S.R.P., Yadav, P., Barde, P.V., Yergolkar, P.N., Erickson, B.R., Carroll, S.A., Mishra, A.C., Nichol, S.T., Mourya, D.T., 2009. Recent ancestry of Kyasanur Forest disease virus. *Emerg. Infect. Dis.* 15, 1431–1437.

Nichter, M., 1978. Kyasanur Forest disease: An ethnography of a disease of development. *Med. Anthropol. Rev.* 1, 406–423.

Padbidri, V.S., Wairagkar, N.S., Joshi, G.D., Umarani, U.B., Risbud, A.R., Gaikwad, D.L., Bedekar, S.S., Divekar, A.D., Rodrigues, F.M., 2002. A serological survey of arboviral diseases among the human population of the Andaman and Nicobar islands, India. *Southeast Asian J. Trop. Med. Public Health* 33, 794–800.

Pattnaik, P., 2006. Kyasanur Forest disease: An epidemiological view in India. *Rev. Med. Virol.* 16, 151–165.

Pavri, K., 1989. Clinical, clinicopathologic, and hematologic features of Kyasanur Forest disease. *Rev. Infect. Dis.* 11 (Suppl. 4), S854–S859.

Rau, S., 1957. Preliminary report on epidemic of continuous fever in human beings in some villages of Sorab taluk, Shimoga district (Malnad area) in Mysore. *Indian J. Public Health* 45, 195–196.

Sarkar, J.K., Chatterjee, S.N., 1962. Survey of antibodies against arthropod-borne viruses in the human sera collected from Calcutta and other areas of West Bengal. *Indian J. Med. Res.* 50, 833–841.

Sreenivasan, M.A., Rajagopalan, P.K., Bhat, H.R., 1975. Isolation of Kyasanur Forest disease virus from Ixodid ticks collected between 1965 and 1972. *Prog. Acarol.* 1, 37–44.

Trapido, H., Rajagopalan, P.K., Work, T.H., Varma, M.G.R., 1959. Kyasanur Forest disease: VIII. Isolation of Kyasanur Forest disease virus from naturally infected ticks of the genus *Haemaphysalis*. *Indian J. Med. Res.* 47, 133–138.

Wang, J., Zhang, H., Fu, S., Wang, H., Ni, D., Nasci, R., Tang, Q., Liang, G., 2009. Isolation of Kyasanur Forest disease virus from febrile patient, Yunnan, China. *Emerg. Infect. Dis.* 15, 326–328.

Work, T.H., 1958. Russian spring-summer virus in India. *Prog. Med. Virol.* 1, 248–277.

Work, T.H., Trapido, H., 1957a. Kyasanur Forest disease, a new virus disease in India. *Indian J. Med. Sci.* 11, 341–345.

Work, T.H., Trapido, H., 1957b. Summary of preliminary report of investigations of the Virus Research Centre on an epidemic disease affecting forest villagers and wild monkeys of Shimoga District, Mysore. *Indian J. Med. Sci.* 11, 340–341.

29 Alkhurma Virus Hemorrhagic Fever

Ziad A. Memish, Shamsudeen Fagbo, and Ali Zaki

CONTENTS

29.1 INTRODUCTION

Recent research has shown that most (75%) of the recently emerging pathogens affecting man are zoonotic with viruses substantially contributing to this observed emergent phenomenon (Taylor et al., 2001). One of these viruses that emerged in Saudi Arabia before the end of the twentieth century is the Alkhurma hemorrhagic fever virus (AHFV). It is a mammalian tick-borne flavivirus genetically related to the Kyasanur Forest disease virus (KFDV). Though AHFV may have existed unrecognized for years in the Arabian Peninsula, the first recorded tick-borne mammalian flavivirus in Saudi Arabia was Kadam virus: it was isolated from Hyalomma ticks collected from a dead camel in Thumaamah, about 70 km from the capital, Riyadh (Wood et al., 1982).

The first reported case of acute AHFV infection was that of a 32 year old Makkah-based butcher in 1995. He had slaughtered a sheep that originated from the small town known as Alkhurma in the Taif governorate, one of the several governorates constituting the Makkah province. Alkhurma is about 200 km from Taif city, which has to be transited before reaching Makkah. Initially, the butcher's hemorrhagic fever signs prompted suspicion of infection with Congo hemorrhagic fever virus (CCHFV), a Bunyavirus endemic in the Makkah province (el-Azazy and Scrimgeour, 1997). Early testing for CCHF using fluorescent antibody testing was positive. However, as this result could not be confirmed in the lab of Dr. Ali Zaki using a CCHFV specific IgM capture enzyme-linked immunosorbent assay (ELISA), an acute CCHF infection was ruled out. The patient's serum was then inoculated into adult and suckling mice: the adults were inoculated intraperitoneally and the suckling mice, intracerebrally and intraperitoneally. Within 5–7 days post inoculation, the suckling mice developed symptoms and died. The time to death—about 3–5 days—was shorter after passage in another litter. The mice were convulsive, irritable with tremors and encephalatic prior to death. Death occurred in adult mice after 6–8 days (Zak, 1997).

Simultaneously, attempts were made to isolate the virus in mammalian (Vero) and mosquito (C6/36) cells directly from the patient's serum. There was no evidence of cytopathic effect (CPE) in C6/36 cells directly from the patient's blood or after passage from infected mice brain suspension. On the contrary, primary isolation in Vero cells inoculated with the patient's blood. Nevertheless, this was characterized by poor and slow CPE development. However, when these cells were inoculated with mice brain suspension from infected suckling mice, CPE indicative of viral growth was observed as early as 5 days in a part of the monolayer. The affected cells became rounded, shrank, and detached from the monolayer. To determine the virus responsible for the observed CPE, the following viruses were tested: West Nile fever virus (WNV), Rift Valley fever virus (RVFV), CCHFV, dengue virus (DENV), and yellow fever virus (YFV). An indirect fluorescent assay (IFA) utilizing monoclonal/polyclonal antibodies (DENV) and mouse ascitic fluid (CCHFV, RVFV and YFV) specific for these viruses was performed. Additionally, a panflavivirus monoclonal antibody, 4G2, was also used. Positive results were only obtained with the 4G2 antibody. The isolates were sent to the Special Pathogens Branch of the CDC in Atlanta, Georgia, where they were further characterized by sequencing the NS5 region of flavivirus genome and determined to be closely related to the KFDV (Zak, 1997), a tick-borne flavivirus known to be endemic only in India (Padbidri et al., 2002).

29.2 BIOLOGICAL PROPERTIES

29.2.1 VIRUS

The virus is a positive-sense single-stranded RNA virus and a genetic variant of KFDV, genus *Flavivirus*, family Flaviviridae. A large number of these viruses are pathogenic in humans (Gould and Solomon, 2008). Flaviviruses are either tick borne, mosquito borne, or have no known arthropod vectors. The KDFV and AHFV are tick borne. Species of tick-borne flavivirus fall under two groups: the mammalian tick-borne group and the seabird tick-borne group mammalian flavivirus. The KDFV and its genetic variant, AHFV, belong to the former. Other viruses in this group are the Gadgets Gully, Langat, Louping ill, Omsk hemorrhagic fever, Powassan, Royal Farm, and tick-borne encephalitis viruses (TBEVs). Their genomes are almost 11 kb long. The polyprotein encoded is cleaved into three structural proteins (C, E, and M) and seven nonstructural proteins—NS1, NS2a, NS2b, NS3, NS4a, NS4b, and NS5 (Thiel et al., 2005). The E protein is the most important viral surface glycoprotein. It plays a major role in receptor binding, and membrane has three domains, DI, DII, and DIII. DII consists of two loops protruding from DI, the tip of which is highly conserved and interacts with the membranes of the target cell during fusion (Modis et al. 2004). Based on modeling studies using TBEV and Dengue-2 virus E proteins as templates, sequence and structure homologies of the fusogenic motif (aa 98–111 aa) were detected between KFDV and TBEV. This could indicate a possible common mechanism of envelope fusion for these two tick-borne flaviviruses (Pattnaik, 2006).

Much of our understanding of AHFV and KFDV structural and functional properties is gleaned from studies on other flaviviruses as well as comparative analyses. Recently, a computer-based comparison of the E protein of AHFV, KFDV, and TBEV suggests an almost equal evolutionary distance of E protein for these three mammalian tick-borne viruses (Mohabatkar, 2011). In another study, the conserved region of the prM protein of the TBEV was shown to be a molecular determinant of flaviviral assembly (Yoshii et al., 2012). During replication, the M protein is found as a precursor protein, prM, in intracellular immature virions (Thiel et al., 2005). However, French researchers (Pastorino et al., 2006) have carried out AHFV-specific work aimed at elucidating viral protein functions with potential therapeutic endpoints. Unlike other Flaviviridae viruses, flaviviruses lack an NS2 autoprotease. However, this segment of the flavivirus genome has two hydrophobic proteins: NS2A and NS2B. These French workers focused on this part of the AHFV genome and have shown that it encodes a serine protease (NS2B–NS3) that is critical for intracellular AHFV replication. Two residues on the NS2 protein, V88 and Q77, essential for AHFV NS3 activation, were also identified (Pastorino et al., 2006). Earlier work (Bessaud et al., 2005) identified the AHFV protease-encoding sequences: on the AHFV polyprotein, NS2B was positioned between 1361 and 1491 and NS3 between 1492 and 2112.

It has been previously established that flaviviruses suppress type 1 interferon response: NS proteins are thought to play this suppressive role (Robertson et al., 2009). Recently, Cook et al. (2012) demonstrated that the KFDV NS5 protein was mainly responsible for interferon response suppression. When compared to other NS proteins, NS4B and NS4B-2k, the NS5 demonstrated greater interferon suppression.

It is not known whether the effects of AHFV infections in humans are dependent on virus-dependent variable pathogenicity. It has been suggested that the lower diversity of AHFV human isolates could be due to a decreased pathogenicity of these isolates (Dodd et al., 2011).

Earlier work demonstrating the relatedness of AHFV with KDFV reported a very strong homology between the two (Charrel et al., 2001; Mehla et al., 2009). Recently, the most comprehensive molecular study on AHFV isolates was published by Dodd et al. (2011). They carried out complete sequencing of the genomes of 16 AHFV isolates from ticks as well as human cases. Additionally, the full genome of three archived KDFV isolates were sequenced: It was shown that the AHFV and KFDV differed by only one nucleotide. The observed genomic sizes for the viruses were 10,775 and 10,774, respectively.

29.2.2 CLINICAL, EPIDEMIOLOGICAL, AND LABORATORY CHARACTERISTICS OF ALKHURMA VIRUS INFECTION

The incubation period in laboratory-confirmed cases of AHFV infection is usually between 4 and 6 days after exposure to source of infection. Clinical presentations in AHFV-infected patients may either be severe with possible fatal outcomes, mild or asymptomatic. Initial observations did tend to suggest a high case-fatality rate approaching 30% for AHFV-infected persons (Charrel et al., 2005). However, recent data suggest that mild or even asymptomatic AHFV infections do occur (Alzahrani et al., 2010). For example, in 2009, 59 cases were documented nationwide with no associated fatalities. In 2010, there were 81 cases with two fatalities (Table 29.1, Figure 29.1). The increase in the following years (Figure 29.2) may be due to increased awareness, enhanced surveillance, and testing of suspected patients. It is expected that this increasing pattern could persist for some time till perhaps a plateau is achieved. The Saudi Ministry of Health (MOH) held an international meeting to better understand the challenges posed by AHFV locally and internationally at the beginning of 2010 (Memish et al., 2009), and heightened efforts of the MOH may have contributed to the

TABLE 29.1
Confirmed AHFV Cases—Saudi Arabia (2009/2010)*—Fatalities

Region	2009	2010
Makkah	0	5
Jeddah	4	14
Najran	55	61
Jizan	0	1
Total	**59(0)***	**81(2)***

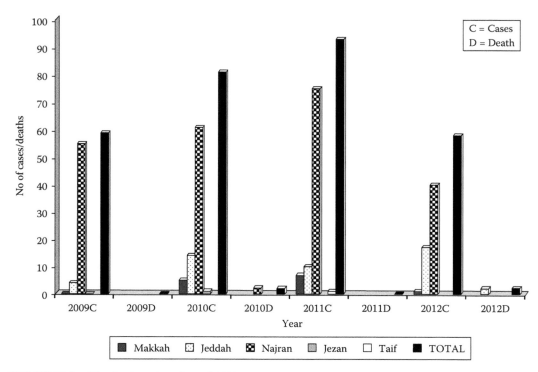

FIGURE 29.1 Distribution of confirmed Alkhurma cases, by region—2009/2010.

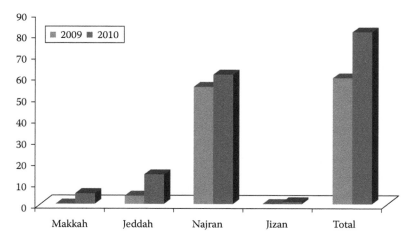

FIGURE 29.2 Graph depicting increase in Alkhurma cases.

observed increases in reported cases. Nonetheless, it is not unlikely that cases are being under-reported from large parts of the Kingdom sharing similar livestock and tick ecology with areas presently known to be AHFV endemic. This is supported by a recent seroprevalence study of Saudi military personnel posted to the South Western province of Jizan (Memish et al., 2011). However, a limitation of this study was that positive samples were not confirmed by serum neutralization to rule possible cross-reactions with other flaviviruses (WNV, Usutu virus, Kadam virus, etc.) that may be endemic in the country albeit undetected. Based on available data, the worldwide distribution of AHFV is depicted in Figure 29.3.

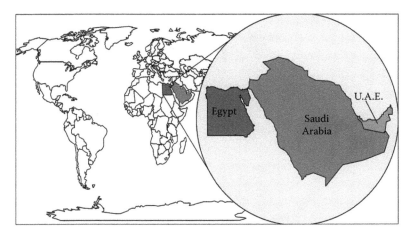

FIGURE 29.3 Worldwide distribution of AHFV.

Symptoms observed in patients include fever, generalized body aches, headache, anorexia, vomiting, and myalgia. CNS involvement manifests as confusion, drowsiness, convulsion, and coma. Hemorrhagia is characterized by observations including epistaxis and hematemesis. Other symptoms include skin rash, diarrhea, and cough with respiratory symptoms. Clinical laboratory data often characterize patients as being leukopenic and thrombocytopenic. They also have elevated enzymes including liver transaminases and creatinine phosphokinase.

Illustratively, three recent cases are briefly described. The first case, diagnosed in 2010, is that of a 49 year old male who developed symptoms 4 days after partaking in the slaughtering of sheep. During the slaughtering process, the patient reported excessive exposure to blood from slaughtered sheep. His fever evolved into an encephalitic episode and eventually he became comatose. The second case is that of a 58 year old woman who developed fever and respiratory symptoms 5 days after packing fresh sheep meat for distribution during the Hajj (pilgrimage) days in Jeddah. Along with her respiratory symptoms, she became confused. Both of these cases were diagnosed by RT-PCR, IgM capture ELISA and IgG ELISA. Another recent case of AHFV infection was initially diagnosed as community-acquired pneumonia.

Diseases to be considered as differential diagnoses for AHFV infection include dengue, RVF, brucellosis, Q fever, malaria, and leptospirosis.

29.3   RISK FACTORS

29.3.1   Work-Related Risk

A large number of confirmed AHFV infections have occurred in persons occupationally exposed to infected livestock blood while performing slaughter and post-slaughter-related tasks. This risk factor has been documented in multiple studies (Alzahrani et al., 2010; Charrel et al., 2005; Madani, 2005; Zaki, 1997). It can be concluded that direct contact with infected blood appears to be associated with a higher risk of AHFV infection. Shepherds are also at risk for AHFV infection. Lastly, housewives and maids have developed clinical illness due to handling as a result of infected meat in the kitchen.

29.3.2   Drinking Raw Unpasteurized Milk

Some AHFV-infected patients have reported consumption of raw camel milk as the only risk factor prior to disease onset (Charrel et al., 2005; Fagbo, personal data) (Figure 29.4). Consumption

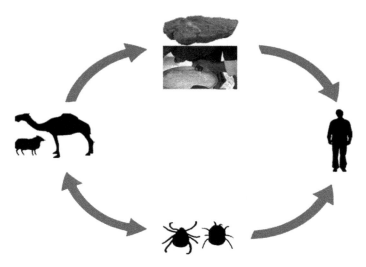

FIGURE 29.4 A simplified AHFV cycle.

of raw milk has been associated with the transmission of TBEV to humans (Gritsun et al., 2003). Presently, it remains unknown whether the co-endemicity of other milk-borne infections in Saudi Arabia adversely affects the clinical recognition of AHFV infection transmitted by the ingestion of raw unpasteurized milk. However, our observations suggest that such a position cannot be ruled out. A classical example is brucellosis, which remains hyperendemic in most regions of the country. In particular, it remains endemic in the Najran province, where confirmed AHFV cases are the highest. In another province, Qasseem, several cases of laboratory-negative clinically suspected Brucellosis cases remain etiologically unresolved. These cases share the same occupational and food borne risks for AHFV transmission.

29.3.3 TICK-BORNE TRANSMISSION

Tick-associated transmission of AHFV has been documented (Alzahrani et al., 2010; Charrel et al., 2005; Zaki, 1997). Additionally one of the two Italian cases exposed to AHFV in Egypt reported being bitten by a tick (Carletti et al., 2010).

The AHFV has been isolated and viral RNA detected in soft (*Ornithodoros savignyi*) and hard (*Hyalomma dromedarii*) ticks in parts of Saudi Arabia (Jeddah and Najran) known to be endemic for AHFV (Charrel et al., 2007; Grard et al., 2007; Mahdi et al., 2011). The first detection was in an *Ornithodoros savignyi* tick collected from camel yard: the camels were being kept awaiting slaughter for human consumption.

29.3.4 MOSQUITO-BORNE TRANSMISSION

Strong scientific evidence describing tick-associated transmission of AHFV in clinically ill patients has been documented (Alzahrani et al., 2010; Carletti et al., 2010; Charrel et al., 2005; Zaki, 1997). Additionally, there is strong molecular, virological, and entomological evidence affirming AHFV to be a tick-borne mammalian flavivirus (Charrel et al., 2005, 2007; Dodd et al., 2011; Mahdi et al., 2011; Zaki, 1997). However, a mosquito-borne transmission of AHFV has been suggested (Madani, 2005; Madani et al. 2011, 2012). This proposition is not only confounding but remains scientifically unsubstantiated (Charrel et al., 2006). In the first paper (Madani, 2005), eight (40%) of the 20 purportedly laboratory confirmed acute cases of AHFV were only IgG positive. Notably, the sera of these IgG only positive patients were collected in the acute phase of their illnesses. This testing was based on sera taken in the acute phase of the illness. None of the acutely ill 20 patients had positive RT-PCR results. Confirmatory neutralization tests that could have ruled out cross-reaction with other flaviviruses were not carried out. In addition, the author failed to report the results of TBEV tests said to have been carried out. Finally, an understanding of the comparative epidemiology of mosquito- and tick-borne pathogens is completely lacking. If AHFV were truly mosquito borne, more cases would have been recorded in AHFV-endemic Jeddah and Makkah, where the persistent dengue endemicity is indicative of active mosquito transmission of a similar virus. Classical examples abound elsewhere: In India, where mosquito populations abound and mosquito-borne flaviviruses such as dengue Japanese encephalitis virus (JEV), the yearly incidence of JEV surpasses that of KDFV infections. In the United States, where tick-borne flaviviruses such as Powassan have been endemic prior to the introduction of the WNV in 1999, we have seen the incidence of WNV infections rise remarkably leaving far behind these tick-borne viruses as well as other tick-borne diseases like Lyme disease.

29.4 DIAGNOSIS

In Saudi Arabia and elsewhere, AHFV infection should be considered in cases of acute dengue-like illness or influenza-like episodes.

29.4.1 VIRUS ISOLATION

The most effective method of virus isolation is by inoculation into suckling mice by the intracerebral route. Blood samples from acutely ill patients can be collected in heparinized or plain tubes and serum separated and kept at −80°C until tested. The mice are observed for 5 days or less during which they develop symptoms. These manifest in the form of tremors, irritability, weakness, and death. The virus can be recovered from the brains of dead mice, and the presence of AHFV RNA directly confirmed using RT-PCR. Additionally, brain suspensions may also be inoculated into Vero cells, and AHFV presence confirmed using an IFA with monoclonal or polyclonal antibodies. Initial attempts to culture AHFV C6/36 mosquito cells, routinely used in Ali Zaki's lab for virus isolation, proved unsuccessful.

However, a recent study seems to indicate that AHFV could grow in these same cells (Madani et al., 2012). A downside of this study was that the authors did not sequence their virus isolates: such data would have allowed additional confirmation of their results by comparison with AHFV sequences available in public databases. Additionally, this study would have had greater value if simultaneous attempts were made to isolate AHFV from the same clinical samples using established tick cell lines. While reproducing the results in this work by other independent workers could give it much needed credibility, early virological work on AHFV strongly suggests a cautious approach to interpreting such results. Interestingly, the first study on AHFV included archived samples from dengue-suspected patients that failed to grow on C6/36 mosquito cells as well as having negative serology (IgM ELISA and IgG immunoflourescent assay) results (Zak, 1997). These samples were drawn from patients during a 1994 dengue fever epidemic in Jeddah. Some of these patients have severe or fatal outcomes. These samples—archived at −80°C— yielded AHFV when inoculated in suckling mice as described earlier.

29.4.2 SEROLOGY

Detection of IgM and IgG antibodies in acute and convalescent samples is done using IgM capture ELISA and IgG ELISA. An IgM capture ELISA was developed that is based on a generic flavivirus monoclonal antibody peroxidase labeled as indicator system and was used to diagnose acute case in conjunction with virus isolation. Briefly, ELISA plates are coated overnight with goat-prepared antihuman IgM u chain. Plates are washed and blocked using 5% fetal bovine serum in PBS for 1 h. Serum samples, diluted at 1/100, are added in duplicates and incubated for 1 h. The positive and negative antigen prepared in suckling mice are added to plates after wash and incubated overnight at 4°C. Plates are then washed, and the flavivirus generic peroxidase-labeled conjugate added in 100 μL volumes at 1/6000 dilution. After incubation for 1 h, the plates are washed and substrate added. Plates are read at 450 nm using an ELISA reader. Cutoff is calculated based on three standard deviations plus the mean of a group of known negative sera.

For the IgG ELISA, plates are coated with monoclonal or polyclonal antibodies to AHFV overnight at 4°C. Next day, plates are washed, blocked, and incubated for 1 h at 37°C and then rewashed. Antigen positive and negative are added to adjacent columns of the ELISA plates and incubated overnight at 4°C. On the following morning, the plates are washed and serum diluted 1/300 added in duplicates to adjacent wells of ELISA plates incubated at 37°C and then washed. Antihuman IgG peroxidase-labeled 100 μL per well is added, and plates incubated at 37°C and then washed. Substrate is added and plates are read at 450 nm after adding a stopping solution. Cutoff is calculated as for the IgM assay.

To exclude cross-reactions due to dengue or any other flaviviral infection, it is suggested that serological tests for such viruses be carried out in parallel to the AHFV IgM and IgG ELISA. Due to this, the MOH makes it mandatory to screen all suspected hemorrhagic fever patients for AHFV and DENV as well as RVFV and CCHFV using RT-PCR and serology. Historically, AHFV was considered a differential diagnosis for the first human cases of RVF reported in Saudi Arabia (Shoemaker et al., 2002). Although the IFA is a valuable serological tool in AHFV as was shown in its early history (Zaki, 1997), the ELISA is more preferred for convenience. Serum neutralization or western blotting techniques could be employed to resolve cases of suspected cross-reactions.

29.4.3 MOLECULAR TEST

It is possible to detect AHFV RNA in clinical samples using PCR methods. Detection of AHFV RNA can be achieved using reverse-transcriptase-PCR (RT-PCR) or quantitative real-time RT-PCR. Serum separated from blood collected in plain or EDTA-containing tubes are used for RT-PCR. A semi-nested RT-PCR is used for direct detection of the virus using primers in the envelope region of the virus. Primers are ALK-ES1 (CTAATGAGTCCCACAGCAATC), and ELK-ER (TTCCAAGCAAACTTTGATCCC) for first round RT-PCR, and primers ALK-ES2 (ACACGAGGGCGCCCATGAA) and ALK-ER for nested PCR. The first round yields a 487 bp amplicon product and the nested 261 bp amplicon (Charrel et al., 2008). Recently, other workers have used a set of primers (GTG AGT GGC GCT TTG TTTG TA and CCC CCT TTC CTT TAA GGA CG) for use in a one-step real-time RT-PCR system combining superscript reverse transcriptase with platinum Taq-polymerase (Madani et al. 2011). Using a similar system, Dodd et al. (2011) used a different set of primers—see Table 29.2—to yield PCR products approximately 4 kb. These primers amplified both AHFV and KFDV viruses.

A commercial RT-PCR system for detection of AHFV nucleic acid has been developed and optimized for use with Roche kits. The system amplifies a 110 bp fragment of the AHFV poly protein gene (TIB MOLBIOL 2009). It is labeled for research use and is yet to be clinically validated.

29.4.4 PROGNOSIS AND TREATMENT

Presently, as with almost all flaviviral infections, there are no antivirals to treat persons acutely infected with AHFV. Patients are symptomatically managed and undergo supportive therapy. Monoclonal antibody therapy has been suggested for the clinical management of flaviviral infections, and much work has been done in this regard (Rodenhuis-Zybert et al., 2011). Specifically, studies have focused on the E protein of flaviviruses (Crill and Chang, 2004; Crill and Roehrig, 2001; Deng et al., 2011). One recent study focused on the neutralizing effects of monoclonal

TABLE 29.2
List of Primers Used in AHFV Detection

PCR Format	Sequences	Reference
1. Semi-nested	CTAATGAGTCCCACAGCAATC; TTCCAAGCAAACTTTGATCCC; ACACGAGGGCGCCCATGAA	Charrel et al. (2008)
2. One-step RT-PCR	GTG AGTGGCG CTTTGTTTGTA; CCC CCTTTCCTTTAAGGACG	Madani et al. (2011)
3. One-step RT-PCR	GTTTCAGACAACGTGAGTGG; GAAGCGTTAACGTGTTGAGG; CAGTTACGACTAGGCCAAG; AACAGGGTGGTCTGGTGAG; AGGGAGTGAACAGAGAGACG; GGAAACAGCGAAGTAGCG; AGCGGATGTTTTTTCCGAAAC	Dodd et al. (2011)

antibodies on the NS1 protein of mosquito-borne flaviviruses (WNV and JEV). While the results are promising, the effects on tick-borne flaviviruses such as AHFV, KDFV, and TBEV remain to be elucidated (Lee et al., 2012). Another promising work (Deng et al., 2011) included the TBEV in the test panels. These workers determined a novel epitope on the highly conserved fusion loop peptide. Structural analysis, using the DENV2 E protein as a template, revealed that the three amino acid sequences therein are highly conserved in the four serotypes of DENV as well as WNV, JEV, YFV, and TBEV. Though the antiviral ribavirin is known to have some anti-flaviviral activity in vitro, its effect on flaviviruses has been shown to have lesser susceptibility than other viruses (Leyssen et al., 2005).

29.5 CONCLUSIONS

A review of the literature shows that translational research into AHFV and KDFV has not been as forthcoming as would be expected for viruses that are not only highly pathogenic and zoonotic, but cause hemorrhagic and encephalitic episodes. For example, the complete sequencing of KDFV, discovered in 1957, did not occur until recently (Cook et al., 2012; Dodd et al., 2011). Comparatively, the complete sequencing of the AHFV, discovered almost 40 years after it, was achieved more than a decade earlier.

There is a need to better understand the ecological basis of AHFV transmission and evolution. More work is needed to characterize how the biology of the tick fauna endemic in Saudi Arabia (Hoogstraal et al., 1981, 1983) contributes to AHFV transmission and genomic evolution as well as other co-endemic tick pathogens (Keung et al., 1995). Are bird-associated ticks involved in AHFV transmission, as have been reported for KDFV? The existence of such birds and their expected migratory patterns may help understand the geographical spread of AHFV and the existence of AHFV/KDFV-like viruses as suggested recently (Dodd et al., 2011). It has been suggested that such birds may serve as evolutionary bridge in viral genetic diversity (Grard et al., 2007).

There are opportunities to use a proteomic approach to further develop tools for AHFV diagnosis as well as templates for AHFV immunoprophylaxis. Specifically, recombinant protein technology can be used to design peptides targeting epitopic regions of the envelope protein of the AHFV that exhibit variability. These peptides can be biotinylated and used to coat ELISA plates. There will be the need for validation and comparison with existing ELISAs for AHFV as well as other mosquito- or tick-borne flaviviruses.

These recombinant proteins can also provide a basis for a Western blot system that can achieve flaviviral differential diagnosis outside high containment facilities (Oceguera et al., 2007). Additionally, these proteins can be used to generate monoclonal antibodies that may be used to immunotherapeutically manage clinical cases of AHFV infection.

Vaccine development work may build on the work done on similar viruses, namely, the TBEV and the KFDV (Demicheli et al., 2009; Pattnaik, 2006). In Europe where TBEV is mainly endemic, vaccines have been produced and are available commercially. It may be possible to adapt these vaccines to local efforts to prevent AHFV infection.

REFERENCES

Alzahrani AG et al. (2010). Alkhurma hemorrhagic fever in humans, Najran, Saudi Arabia. *Emerg Infect Dis*; **16**(12):1882–1888.

el-Azazy OM and Scrimgeour EM. (1997) May–June. Crimean-Congo haemorrhagic fever virus infection in the western province of Saudi Arabia. *Trans R Soc Trop Med Hyg*; **91**(3):275–278.

Bessaud M et al. (2005) January. Identification and enzymatic characterization of NS2B-NS3 protease of Alkhurma virus, a class-4 flavivirus. *Virus Res*; **107**(1):57–62.

Carletti F et al. (2010) December. Alkhurma hemorrhagic fever in travelers returning from Egypt, 2010. *Emerg Infect Dis*; **16**(12):1979–1982.

Charrel RN et al. (2001) September. Complete coding sequence of the Alkhurma virus, a tick-borne flavivirus causing severe hemorrhagic fever in humans in Saudi Arabia. *Biochem Biophys Res Commun*; **287**(2):455–461.

Charrel RN et al. (2005). Low diversity of Alkhurma hemorrhagic fever virus, Saudi Arabia, 1994–1999. *Emerg Infect Dis*; **11**(5):683–688.

Charrel RN et al. (2006) June. Alkhurma hemorrhagic fever virus is an emerging tick-borne flavivirus. *J Infect*; **52**(6):463–464. Epub 2005 October 10.

Charrel RN et al. (2007) January. Alkhurma hemorrhagic fever virus in Ornithodoros savignyi ticks. *Emerg Infect Dis*; **13**(1):153–155.

Charrel RN et al. (2008). Human cases of hemorrhagic fever in Saudi Arabia due to a newly discovered Flavivirus, Alkhurma Hemorrhagic Fever Virus. In: Lu Y, Essex M, Roberts B (Eds.). *Emerging Infections in Asia*. New York: Springer; pp. 179–192.

Cook BW et al. (2012) February. The generation of a reverse genetics system for Kyasanur Forest Disease Virus and the ability to antagonize the induction of the antiviral state in vitro. *Virus Res*; **163**(2):431–438. (Epub 2011 November).

Crill WD and Chang GJ. (2004) December. Localization and characterization of flavivirus envelope glycoprotein cross-reactive epitopes. *J Virol*; **78**(24):13975–13986.

Crill WD and Roehrig JT. (2001) August. Monoclonal antibodies that bind to domain III of dengue virus E glycoprotein are the most efficient blockers of virus adsorption to Vero cells. *J Virol*; **75**(16):7769–7773.

Demicheli V et al. (2009) Cochrane Database Syst Rev; 1:CD000977.

Deng YQ et al. (2011) January. A broadly flavivirus cross-neutralizing monoclonal antibody that recognizes a novel epitope within the fusion loop of E protein. *PLoS One*. **11**; 6(1):e16059.

Dodd KA et al. (2011) October. Ancient ancestry of KFDV and AHFV revealed by complete genome analyses of viruses isolated from ticks and mammalian hosts. *PLoS Negl Trop Dis*; **5**(10):e1352.

Gould EA and Solomon T. (2008) February. Pathogenic flaviviruses. *Lancet 9*; **371**(9611):500–509. Review. PubMed.

Grard G et al. (2007) April. Genetic characterization of tick-borne flaviviruses: New insights into evolution, pathogenetic determinants and taxonomy. *Virology 25*; **361**(1):80–92.

Gritsun TS et al. (2003). Tick-borne flaviviruses. *Adv Virus Res*; **61**:317–371.

Hoogstraal H et al. (1981). Ticks (Acarina) of Saudi Arabia Family Argasidae, Ixodidae. In: Wittmer W and Buttiker W (Eds.). *Fauna of Saudi Arabia*, Vol. 3, Basel: Ciba Geigy Ltd; pp. 25–110.

Hoogstraal et al. (1983). Hyalomma (Hyalommina) arabica (Fam. Ixodidae), a parasite of goats and sheep in Saudi Arabia. Fauna of Saudi Arabia 3: In: Wittmer W and Buttiker W (Eds.) Fauna of Saudi Arabia, Vol. 3, Basel, Switzerland: Ciba Geigy Ltd; pp. 117–120.

Keung YK et al. (1995) August. Borreliosis as a cause of fever in a woman who recently returned from Saudi Arabia. *Clin Infect Dis*; **21**(2):447–448.

Lee TH et al. (2012) January. A cross-protective mAb recognizes a novel epitope within the flavivirus NS1 protein. *J Gen Virol*; **93**(Pt 1):20–26.

Leyssen P et al. (2005) February. The predominant mechanism by which ribavirin exerts its antiviral activity in vitro against flaviviruses and paramyxoviruses is mediated by inhibition of IMP dehydrogenase. *J Virol*; **79**(3):1943–1947.

Madani TA. (2005) August. Alkhumra virus infection, a new viral hemorrhagic fever in Saudi Arabia. *J Infect*; **51**(2):91–97.

Madani TA et al. (2011) January. Alkhumra (Alkhurma) virus outbreak in Najran, Saudi Arabia: Epidemiological, clinical, and laboratory characteristics. *J Infect*; **62**(1):67–76.

Madani TA et al. (2012) March. Successful propagation of Alkhumra (misnamed as Alkhurma) virus in C6/36 mosquito cells. *Trans R Soc Trop Med Hyg*; **106**(3):180–185.

Mahdi M et al. (2011) May. Kyasanur forest disease virus Alkhurma subtype in ticks, Najran province, Saudi Arabia. *Emerg Infect Dis*; **17**(5):945–947.

Mehla R et al. (2009) September. Recent ancestry of Kyasanur Forest disease virus. *Emerg Infect Dis*; **15**(9):1431–1437.

Memish ZA et al. (2009). Alkhurma viral hemorrhagic fever virus: Guidelines for detection, prevention and control TIB MOLBIOL. http://www.roche-as.es/logs/LightMix_40-0581-16_Alkhurma_Virus_V_090701.pdf (accessed 23.01.2012).

Memish ZA et al. (2011) December. Seroprevalence of Alkhurma and other hemorrhagic fever viruses, Saudi Arabia. *Emerg Infect Dis*; **17**(12):2316–2318. doi: 10.3201/eid1712.110658.

Modis Y et al. (2004) January. Structure of the dengue virus envelope protein after membrane fusion. *Nature* 22; **427**(6972):313–319.

Mohabatkar H (2011) June. Computer-based comparison of structural features of envelope protein of Alkhurma hemorrhagic fever virus with the homologous proteins of two closest viruses. *Protein Pept Lett*; **18**(6):559–567.

Oceguera LF 3rd et al. (2007) July. Flavivirus serology by Western blot analysis. *Am J Trop Med Hyg*; **77**(1):159–163.

Padbidri VS et al. (2002) December. A serological survey of arboviral diseases among the human population of the Andaman and Nicobar Islands, India. *Southeast Asian J Trop Med Public Health*; **33**(4):794–800.

Pastorino BA et al. (2006) November. Mutagenesis analysis of the NS2B determinants of the Alkhurma virus NS2B-NS3 protease activation. *J Gen Virol*; **87**(Pt 11):3279–3283.

Pattnaik P (2006) May–June. Kyasanur forest disease: an epidemiological view in India. *Rev Med Virol*; **16**(3):151–165.

Robertson SJ et al. (2009). Tick-borne flaviviruses: Dissecting host immune responses and virus countermeasures. *Immunol Res*; **43**(1–3):172–186.

Rodenhuis-Zybert IA et al. (2011) November. A fusion-loop antibody enhances the infectious properties of immature flavivirus particles. *J Virol*; **85**(22):11800–11808.

Shoemaker T et al. (2002). Genetic analysis of viruses associated with emergence of Rift Valley fever in Saudi Arabia and Yemen, 2000–2001. *Emerg Infect Dis*; **8**(12):1415–1420.

Taylor LH et al. (2001) July 29. Risk factors for human disease emergence. *Philos Trans R Soc Lond B Biol Sci*; **356**(1411):983–989.

Thiel H et al. (2005). Family Flaviviridae. In: Fauquet CM, Mayo MA, Maniloff J, Desselberger U, Ball LA (Eds.) *Virus Taxonomy. Eighth Report of the International Committee on Taxonomy of Viruses*. London, U.K.: Elsevier; pp. 981–998.

Wood OL et al. (1982) March. Kadam virus (Togaviridae, Flavivirus) infecting camel-parasitizing Hyalomma dromedarii ticks (Acari: Ixodidae) in Saudi Arabia. *J Med Entomol*; **19**(2):207–208.

Yoshii K et al. (2012) January. A conserved region in the prM protein is a critical determinant in the assembly of flavivirus particles. *J Gen Virol*; **93**(Pt 1):27–38.

Zaki AM. (1997) March–April. Isolation of a flavivirus related to the tick-borne encephalitis complex from human cases in Saudi Arabia. *Trans R Soc Trop Med Hyg*; **91**(2):179–181.

30 Omsk Hemorrhagic Fever

Daniel Růžek, Valeriy V. Yakimenko, Lyudmila S. Karan, and Sergey E. Tkachev

CONTENTS

30.1 INTRODUCTION

The first epidemic outbreaks of Omsk hemorrhagic fever (OHF) were recorded in a number of rural regions of Omsk region, Russia, in 1943–1945. OHF was firstly described as a new disease in the years 1945–1946, when physicians in the northern-lake steppe and forest-steppe areas of Omsk region recorded the sporadic cases of acute febrile disease with abundant hemorrhagic signs (bleeding from the nose, mouth, uterus and skin, and hemorrhagic rash) and leukopenia (Chumakov, 1948; Mazbich and Netsky, 1952). The expedition of Russian scientists under the supervision of M. P. Chumakov took place in Omsk region with the purpose to investigate this new disease, identify its causative agent, and reveal the mode of its transmission. During this expedition, a new virus, Omsk hemorrhagic fever virus (OHFV), was isolated in 1947 from a human patient and later from *Dermacentor reticulatus* ticks. *D. reticulatus* ticks were then identified as the vector of OHFV (Chumakov, 1948; Mazich and Netsky, 1952). Later, however, a predominant pattern of contact with muskrats (*Ondatra zibethicus*) was seen among new cases: the illness occurred predominantly in muskrat hunters and their family members, who participated in muskrat skinning and preparing skins (Kharitonova and Leonov, 1985). Muskrats are an alien animal species in Russia that were introduced to Siberia from Canada for industrial fur production purposes (Neronov et al., 2008). Breeding of muskrats, however, did not reach its economic potential in Siberia because of fatal epizootics. It is supposed that OHFV existed silently in Siberia before muskrat's release, but the introduction of these highly susceptible animals greatly amplified infection rates in other animals including human beings (Korsh, 1971; Růžek et al., 2010). In 1946–1970, 76 different epizootics of OHF were recorded in Tyumen, Kurgan, Omsk, and Novosibirsk regions, and this was followed by human cases of OHF in Omsk and Novosibirsk regions. Epizootic activity of OHFV was then recorded starting from the late 1980s in Tyumen (1987), Omsk (1988, 1999–2007), Novosibirsk (1989–present), and Kurgan (1992) regions. Between 1946 and 1958, a total of 972 human cases of OHF were officially recorded. In 1960–1970, the OHF incidence decreased remarkably. In the last 20 years, the disease was reported only in Novosibirsk region with the highest numbers of cases in 1990 (29 cases) and 1991 (38 cases) (Růžek et al., 2010, 2011).

30.2 VIRUS

OHFV is a member of tick-borne group of family *Flaviviridae*, genus *Flavivirus*. OHFV is closely phylogenetically related to tick-borne encephalitis virus (TBEV). Using classical methods, like neutralization test or complement-fixation test, it was very difficult to differentiate between OHFV and TBEV, which suggested the close relationship of these two viruses. The OHFV morphology, structural features, and mode of replication are the same as those of TBEV (Shestopalova et al., 1965). The genetic factors that determine virus association with hemorrhagic manifestation rather than encephalitis are unknown, but are under intensive investigation.

Virions of flaviviruses, including OHFV, are spherical particles, approximately 50–60 nm in diameter with a nucleocapsid composed of (+)ssRNA genome enclosed in capsid (C) protein and surrounded by a host cell–derived lipid bilayer. Membrane (M) and envelope (E) proteins are integrated in the bilayer. The M protein, produced during the maturation of nascent virus particles within the secretory pathway, is a small proteolytic fragment of the precursor prM protein. Glycoprotein E, the main component of the virus surface, mediates receptor binding and fusion activity after uptake by a receptor-mediated endocytosis. This protein is the main target of neutralizing antibodies and induces protective immunity in the infected organisms. For its functional importance, it is believed that the E protein is also an important determinant of virulence (Lindenbach and Rice, 2003).

OHFV has a genome length of 10,787 bases with an open reading frame (ORF) of 10,242 nucleotides encoding 3414 amino acids. ORF is flanked by 5' and 3' untranslated regions (UTRs). The 5' UTR of OHFV contains a 5'-cap. The character of 5' UTR is considerably different from other TBE complex viruses through an approximately 30 nucleotide stretch, while the remainder of the 5' UTR is highly homologous. The 3' UTR lacks 3'-poly(A) tail like other flaviviruses and is slightly shorter in comparison to TBEV, but otherwise is similar to the 3' UTR observed in Far-Eastern and Siberian strains of TBEV (Lin et al., 2003). ORF encodes a large polyprotein that is co- and posttranslationally cleaved by cellular and viral proteases into three structural proteins (C, prM, E) and seven nonstructural proteins (NS1, NS2A, NS2B, NS3, NS4A, NS4B, and NS5). Viral nonstructural proteins have several functions during the virus replication in the host cells. For example, they form the RNA-dependent RNA polymerase replication complex and provide a serine protease needed to cleave the polyprotein.

The strains of OHFV isolated from various sources (ixodid ticks, muskrats, blood and organs of sick people) were differentiated into two groups based on biological properties (Clarke, 1964; Kornilova et al., 1970). Based on the comparison of nucleotide sequence of E gene, it was demonstrated that OHFV is most closely related to European subtype of TBEV, rather than to Far-Eastern and Siberian subtypes. Phylogenetic analysis of E and NS5 genes of various OHFV strains revealed two genotypes (Li et al., 2004). The first genotype is formed predominantly by the strains from Novosibirsk and most of the strains from Omsk regions. The second genotype includes the strains from Kurgan region and two strains from the western part of Omsk region. Within the first genotype, the level of E gene homology is 96.7%; the second genotype has a homology of 98.1%–100%. The homology between genotypes is 87.2%–89.0%. Homology with the most closely related TBEV is 81.7%–83.4% (genotype 1), and 81.9%–82.1% (genotype 2), respectively. In the case of NS5 gene, the homology between strains inside the genotype is 96.4%–99.8% and 88.6%–89.5% between the two genotypes. The NS5 gene homology between OHFV and the most closely related TBEV is 83.2%–83.9% (Yakimenko, 2011).

30.3 ECOLOGY OF OHFV

Human beings can be infected with OHFV by transmissive (i.e., via infected tick's bite) or non-transmissive (respiratory or probably alimentary) routes.

At the time of OHF discovery, most attention of the researchers was given to investigations of transmissive infections because most of the initial cases had transmissive features (i.e., via vector).

The only exceptions were the cases in a village Gornostayevka in 1945 (Sargat district, Omsk region), where potential non-transmissive OHF cases were reported. In the active OHFV foci, high abundance of *Dermacentor reticulatus* (*pictus*) ticks was observed, suggesting a connection between these ticks and the new disease. Later, this connection was subsequently confirmed. Narrow-skulled vole (*Microtus gregalis*) was considered to be the main natural host of OHFV and *D. reticulatus* ticks as vector as well as the main reservoir of the virus. Low levels of transovarial transmission of OHFV in *D. reticulatus* ticks suggested that this mode does not play a crucial role in virus maintenance in the natural focus. *D. reticulatus* imago feed on wild ungulates and humans, whereas immature forms feed mainly on water voles (*Microtus gregalis*) in forest-steppe habitats. Vole populations are cyclic, and expansion of the virus-infected tick population coincides with the increases in vole populations (Hoogstraal, 1985; Estrada-Peña and Jongejan, 1999). In the steppe regions of Southern and Western Siberia, the virus is transmitted mainly by *D. marginatus* ticks. Gamasid mites and the taiga tick (*Ixodes persulcatus*) are believed to be involved in the sylvatic cycle of OHFV (Kondrashova, 1970). Although OHFV was also isolated from several species of mosquitoes (*Ochlerotatus excrucians, Coquilletidia richiardii,* and *Ochlerotatus flavescens*), their role in the circulation of OHFV in nature is unclear.

As early as in 1946–1948, OHF cases followed the contacts of laboratory workers with muskrats were reported. Subsequently, it was demonstrated that OHFV-infected muskrats are really infectious for humans. In 1954, OHFV was isolated from the brain of dead muskrat for the first time. It was experimentally demonstrated that muskrats are highly sensitive to OHFV infection. During 1946–1970, 76 OHF epizootic outbreaks in muskrats were reported. The OHF epizootic outbreaks occurred in four regions of Russia (Tyumen, Kurgan, Omsk, and Novosibirsk), mostly in the areas of boundaries of steppe and forest-steppe zones. The development of the outbreak is usually slow, not occurring simultaneously in the whole lake. In the beginning, there is usually only a local focus of OHFV, and the development of local outbreak takes from 1 to 3 years. During 1971–1989, there were no registered OHF outbreaks in muskrats (Figure 30.1). This might be caused by changes in agricultural activities, leading to a decrease in the numbers of *D. reticulatus* and narrow-skulled voles as the main host of the preimaginal stages of these ticks. At present, there is virtually no OHFV detected in ixodid ticks in Siberia. As a consequence, the numbers of transmissive human OHF cases is very low (less than 7.4% of all OHF cases) (Busygin et al., 1996).

The key role of muskrats in the dissemination of OHFV in the endemic regions is well accepted. The introduction of muskrats to Siberia amplified the infection and caused the emergence of this disease. Urine and other excretes of the infected muskrats contain high titers of the virus. In water, the virus remains stable for more than 2 weeks in summer and for 3.5 months in winter (Kharitonova and Leonov, 1978). Water animals or those ones living near water may get infected via lake water contaminated from the muskrat corpses or feces. Latent infection can be observed in tundra voles (*Microtus oeconomus*), water voles (*Arvicola terrestris*), and several other rodent species that live close to the water reservoirs. These animals disseminate the virus with their urine and excrements. The virus transmission can occur following direct contact of the animals, but also via aerosol or water contaminated with infected urine or feces. Non-transmissive infection with OHFV occurs mainly after close contact with infected muskrats; that is, the patients are generally rural residents, agricultural workers, and people involved in hunting and skinning these animals.

Development and establishment of natural foci for OHFV is favored by a combination of landscape, climatic, and biotic factors. OHFV foci can be found in forested areas, to some degree, and steppe with multiple marshes and lakes.

Seroepidemiological studies suggest that many animal species are in contact with OHFV including rodents, insectivores, birds, ungulates, and domestic animals. Some wild hosts develop latent chronic infections, and the others develop the acute (root vole [*Microtus oeconomus*], narrow-skulled vole [*M. gregalis*], red-cheeked suslik [*Citellus erythrogenys*], hedgehog, etc.) and, in some cases, fatal infections. Multiple hosts are, therefore, involved in the OHFV natural foci, in particular water voles (*Arvicola amphibius*) and narrow-skulled voles (*M. gregalis*) (Avakyan et al., 1955; Korsh, 1971).

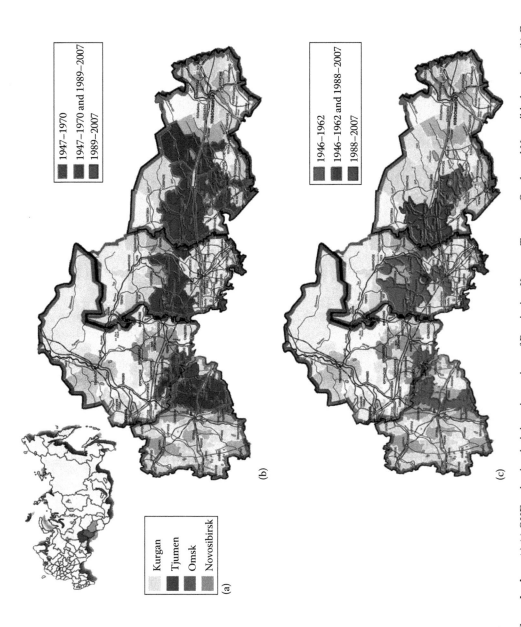

FIGURE 30.1 **(See color insert.)** (a) OHF endemic administrative regions of Russia, i.e., Kurgan, Tyumen, Omsk, and Novosibirsk regions. (b) Geographical distribution of epizootics of OHF in the muskrat population. (c) Geographical distribution of areas of OHF human morbidity since 1946.

30.4 EPIDEMIOLOGY

OHF natural foci are situated in Omsk, Tyumen, Novosibirsk, and Kurgan regions of Russia; that is, in regions with forests, steppe with multiple marshes, lakes, and reed thickets (Figure 30.1). Seasonal morbidity of OHF in humans has two peaks. The first peak with its height in May and June correlates with the activity of *D. marginatus* (in the southern and western areas of Siberia) or *D. reticulatus* (in the northern forest-steppe regions of Siberia). The second peak of OHF morbidity occurs between August and September. This correlates with the muskrat hunting season, and most of the patients were infected following contact with the muskrats harboring OHFV.

Between the years 1946 and 1958, 972 OHF cases were officially recorded; however, it is assumed that the number of cases was much higher because mild cases were not reported. In 1960–1970, the incidence decreased remarkably: only single cases were recorded. This probably correlated with the decrease in OHFV natural foci, decrease in the numbers of *D. reticulatus* and narrow-skulled voles as the main hosts of the preimaginal stages of these ticks. During the past 20 years, OHF cases were reported only in Novosibirsk region. The last largest outbreaks of OHF were in 1990 (29 cases) and 1991 (38 cases). Most of the recent OHF cases represent muskrat hunter and poachers. Only about 10% of cases are associated with tick bites. In 1998, seven OHF cases were reported, one was fatal and three were severe. However, exact numbers of OHF cases are not available, since this disease is frequently misdiagnosed or mild cases are not reported (Netesov and Conrad, 2001). Humans are susceptible to the infection at any age, but 40- to 50-year-old patients predominate.

30.5 CLINICAL SIGNS AND SYMPTOMS

The incubation period of OHF is 3–7 days, usually without prodromal signs, but sometimes malaise, aches, and pains are recorded. There is sudden onset of headache and hemorrhagic signs (bleeding from the nose, mouth, and uterus). Cough, muscle pain, gastrointestinal symptoms, and dehydration are also frequent symptoms. Fever is high (39°C–40°C), lasting 8–15 days, and frequently accompanied by chill. Hypotension and bradycardia; hyperemia of the face, neck, and breast; acute scleral injection; bright colorization and light edema of the tunica mucosa in the mouth and throat; unusual dryness of mucous membranes, especially on the tongue; putrid odor from the mouth; and most predominantly, an enlarged liver are seen. Face becomes slightly puffy and labial fissures and crusts appear (Akhrem-Akhremovich, 1948, 1959; Lebedev et al., 1975; Belov et al., 1995; Růžek et al., 2010). On the third and fourth day of the clinical disease, the signs described earlier are further progressing; for example, face hyperemia and sclera injection are pronounced more intensely, the face becomes puffy, and pharynx hyperemia is intensified (looks like "flaming"). The dryness of mucous coats increases and labial fissure and crusts appear. Also the permanent gingivitis without pronounced stomatorrhagia, tonsils, and soft palate hyperemia and uvula edema (without inflammatory changes) are observed. In rare cases, the surface necroses in the pharynx (which usually disappear rapidly) may be observed. Poignant skin hyperesthesia and muscle pain may be recorded. The skin is very sensitive. The patients exhibit arterial hypotension and bradycardia. The frequency of hemorrhagic signs increases, and this is an important diagnostic marker. Also the blood spitting, uterine bleeding, skin hemorrhagia, and gastrointestinal and pulmonary bleeding (in serious cases) may be frequently observed. Petechial rash was seen in up to 22% of cases during the last outbreaks. Typhoid maculopapular rash on the skin of abdomen and upper and lower extremities is found more rarely (Akhrem-Akhremovich, 1948, 1959; Lebedev et al., 1975; Růžek et al., 2010).

After 1–2 weeks of symptomatic disease, some patients recover without any complications, but in 30%–50% of cases, the second phase occurs. The second phase is characterized with fever and signs of (meningo)encephalitis. The duration of the relapse is about 5, 10, or 14 days. The patients complain of permanent headache, meningism, nausea, chill and present reddening of the face and sclerae, nasal and gingival bleeding, hematuria, and uterine bleeding. Sometimes, petechial rash may appear with bruises at the site of pressure or injections. Laboratory analysis of blood shows leukopenia, thrombocytopenia, and plasmacytosis, and some patients have pathology in internal organs (pneumonia, nephrosis) (Akhrem-Akhremovich, 1948, 1959; Lebedev et al., 1975; Růžek et al., 2010).

The disease usually ends with complete recovery after quite a long period of asthenia. Rarely, there are permanent complications after the recovery; these include weakness, hearing loss, hair loss, and behavioral, psychological, or psychiatric difficulties associated with neurological conditions (poor memory, reduced ability to concentrate, reduced ability to work) (Jelínková-Skalová et al., 1974). Pareses are not seen in patients with OHF (Akhrem-Akhremovich, 1948, 1959; Lebedev et al., 1975; Růžek et al., 2010).

There are no pathomorphological changes in the internal organs typical for OHF. In brain, focal diffuse degenerative necrotic and inflammatory changes, proliferation of microglial cells, and perivascular cell infiltrates can be seen. Blood–brain barrier is usually severely affected (Novitskiy, 1948).

The prognosis is usually favorable. The case fatality rate is quite low, reaching 0.4%–2.5%. In fatal cases, the patients die either in the period of rapid increase in hemorrhagic signs (gastric and intestinal bleeding) or in later period of disease as a result of septic condition (suppurative bilateral parotititis, empyema). The patients are highly sensitive to secondary infections (Lebedev et al., 1975; Růžek et al., 2010).

Apart from classical form of OHF, asymptomatic or influenza-like courses without hemorrhagic signs of OHF are also reported. In literature, the subfebrile cases with equivocal clinical signs, but with characteristic blood picture were described. Such an atypical clinical course of OHF is seen more frequently in the patients infected after the contact with ill muskrat and dominates during more recent outbreaks. The typical cases with clear hemorrhagic signs were seen in less than 20% of the patients. Mortality was about 1% (Lebedev et al., 1975; Belov et al., 1995; Busygin, 2000; Růžek et al., 2010).

30.6 PATHOGENESIS

The pathogenesis of OHF has been investigated mostly using laboratory mice (Holbrook et al., 2005) and nonhuman primates (Kenyon et al., 1992). The laboratory mice are sensitive to the infection with OHFV and develop fatal neuroinfection. The clinical signs of OHFV-infected mice include spasms, paresis, and paralysis. The animals are weakened, lose mobility and appetite, lay in a corner, and become hyperpneic. Animals usually die within a few hours, up to 1 day, after the onset of the disease. The most of the virus is accumulated in the cerebellum and brain hemispheres. However, similar virus titers were also found in lungs, kidney, blood, and feces. A lower titer was observed in the spleen and the least in the liver (Shestopalova et al., 1972). Early and prominent induction of IL-1α, TNF-α, IL-12p10, MCP-1, MIP-1α, and MIP-1β in the spleen of infected mice is observed (Tigabu et al., 2010).

Experimentally infected macaques (*Macaca radiata*) do not develop any signs of clinical disease, and no virus can be isolated from tissues or blood. However, the animals seroconvert, and elevated levels of serum aminotranferase can be detected (Kenyon et al., 1992).

30.7 DIAGNOSIS

The diagnosis of OHF is usually based on clinical and epidemiological observations. Laboratory diagnosis is based on serological methods, which represent the gold standard in OHF diagnostics. The antibodies to OHFV can be detected by ELISA, and seroconversion with paired sera is investigated by hemagglutination-inhibition, complement-fixation, and neutralization tests. The neutralization test is considered to be the most specific for laboratory diagnosis of OHF. Neutralizing antibodies can be detected within the first week of the onset of the disease and persist for years and possibly lifelong (Sizemova, 1957; Casals, 1961). However, the antibodies against the other tick-borne flaviviruses, in particular to TBEV, also have the ability to cross-neutralize OHFV (Clarke, 1964; Calisher et al., 1989).

PCR-based methods are not suitable for the laboratory diagnosis of OHF. The virus seems to be detectable in the blood during the first phase of the illness, but not during the second. Since most patients seek medical advice during the second phase of the illness, the molecular detection of the virus is likely to be unsuccessful. So the molecular methods are suitable for screening of ticks and animals for the presence of OHFV or for postmortem investigations. Postmortem samples can be also used for virus isolation or can be investigated by immunofluorescence or electron microscopy. Virus isolation from blood can be done only within the first days of the disease and is performed by inoculation of the sample into cell cultures or into brains of suckling mice. During the second phase of the illness, the virus isolation has never been successful (Jelínková-Skalová et al., 1974). OHFV is a BSL-3 or BSL-4 agent in several countries, and therefore, all work with the viable virus or with samples potentially containing viable virus must be done in appropriate facilities (Von Lubitz, 2004).

30.8 THERAPY AND PREVENTION

No specific therapy against OHFV is available. The disease is generally self-limiting, but the patients must keep strict bed rest. The main focus of therapy is to control the symptoms. Abundant intake of the liquids and nutritious diet is recommended. Administration of chloride potassium, glucose, and vitamins K and C is beneficial for the patients. The long convalescent period is necessary to reduce any permanent complications. Any complications that might occur during OHF like pneumonia, heart insufficiency, bacterial infections, etc., should be treated accordingly. Transfusions are indicated in cases with severe blood loss. The hemostatic drugs that strengthen vascular walls can be used (Gaidamovich, 1995).

Several compounds exhibit anti-OHFV activity during the experiments in cell culture or in laboratory animals. Recombinant human α-2b interferon completely inhibits OHFV reproduction in cell culture. Ribavirin and interferon inducers larifan and rifastin exhibit moderate virus growth inhibition in cell culture; larifan has the highest antiviral activity when tested in laboratory animals infected with OHFV. However, no data on the effectiveness of OHF patients' treatment with these compounds are available (Loginova et al., 2002).

A vaccine against OHFV was developed as early as in 1948. This vaccine was highly effective, but its production was discontinued owing to the adverse reactions to the mouse-brain components of the vaccine and because the disease incidence has dropped (Stephenson, 1988). Vaccines against TBEV can be used to prevent the infection with OHFV, since there is high antigenic similarity between these two viruses (Clarke, 1964). These vaccines were used as a preventive measure against OHFV infection during the 1991 outbreak. Recently, a high protective effect of the European TBEV vaccines against OHFV was demonstrated (Orlinger et al., 2011; Ngulube et al., 2012).

ACKNOWLEDGMENTS

The authors acknowledge financial support by the Czech Science Foundation project No. P302/10/P438 and No. P502/11/2116, and the AdmireVet project No. CZ.1.05./2.1.00/01.006.

REFERENCES

Akhrem-Akhremovich, R.M., 1948. Spring-autumn fever in Omsk region. *Proc. OmGMI* 13, 3–43.
Akhrem-Akhremovich, R.M., 1959. Problems of hemorrhagic fevers. *Proc. OmGMI* 25, 107–116.
Avakyan, A.A., Lebedov, A.D., Ravdonikas, O.V., Chumakov, M.P., 1955. On the question of importance of mammals in forming natural reservoirs of Omsk hemorrhagic fever. *Zool. Zh.* 34, 605–609.
Belov, G.F., Tofanyuk, E.V., Kurzhukov, G.P., Kuznetsova, V.G., 1995. Clinico-epidemiologial characterization of Omsk hemorrhagic fever at the period of 1988–1992. *Zh. Mikrobiol.* 4, 88–91.

Busygin, G.G., 2000. Omks hemorrhagic fever—current status of the problem. *Vopr. Virusol.* 45, 4–9.

Calisher, C.H., Karabatsos, N., Dalrymple, J.M., Shope, R.E., Porterfield, J.S., Westaway, E.G., and Brandt, W.E., 1989. Antigenic relationships between flaviviruses as determined by cross-neutralization tests with polyclonal antisera. *J. Gen. Virol.* 70 (Pt 1), 37–43.

Casals, J., 1961. Procedures for identification of arthropod-borne viruses. *Bull. World Health Organ.* 24, 723–734.

Chumakov, M.P., 1948. Results of a study made of Omsk hemorrhagic fever (OL) by an expedition of the Institute of Neurology. *Vestnik Acad. Med. Nauk SSSR* 2, 19–26.

Clarke, D.H., 1964. Further studies on antigenic relationships among the viruses of the group B tick-borne complex. *Bull. World Health Organ.* 31, 45–56.

Estrada-Peña, A. and Jongejan, F., 1999. Ticks feeding on humans: A review of records on human-biting Ixodoidea with special reference to pathogen transmission. *Exp. Appl. Acarol.* 23(9), 685–715.

Gaidamovich, S.Y., 1995. Tick-borne flavivirus infections. In: Porterfield, J.S. (Ed.), *Exotic Viral Infections.* London, U.K., Chapman and Hall, 203–221.

Holbrook, M.R., Aronson, J.F., Campbell, G.A., Jones, S., Feldmann, H., and Barrett, A.D., 2005. An animal model for the tickborne flavivirus—Omsk hemorrhagic fever virus. *J. Infect. Dis.* 191(1), 100–108.

Hoogstraal, H., 1985. Argasid and Nuttalliellid ticks as parasites and vectors. *Adv. Parasitol.* 24, 135–238.

Jelínková-Skalová, E., Tesařová, J., Burešová, D., Kouba, K., and Hronovský, V., 1974. Laboratory infection with virus of Omsk hemorrhagic fever with neurological and psychiatric symptomatology. *Čs. Epidemiol. Mikrobiol. Immunol.* 23, 290–293.

Kenyon, R.H., Rippy, M.K., McKee, K.T., Zack, P.M., and Peters, C.J., 1992. Infection of Macaca radiata with viruses of the tick-borne encephalitis group. *Microb. Pathogenesis* 13, 399–409.

Kharitonova, N.N. and Leonov, Yu.A., 1985. Omsk hemorrhagic fever. *Ecology of the Agent and Epizootology.* Amerid, New Delhi, 230 pp.

Kondrashova, Z.N., 1970. Study of Omsk hemorrhagic fever virus preservation in Ixodes persulcatus ticks during mass dosated infection. *Med. Parasitol. Parasit. Dis.* 39, 274–277.

Kornilova, E.A., Gagarina, A.V., and Chumakov, M.P., 1970. Comparison of the strains of Omsk hemorrhagic fever virus isolated from different objects of a natural focus. *Vopr. Virus.* 15, 232–236.

Korsh, P.V., 1971. Epizootic characteristics of Omsk hemorrhagic fever natural foci at the south of Western Siberia. The problems of infectious pathology. *Omsk Russ. Acad. Sci.,* 70–75.

Lebedev, E.P., Sizemova, G.A., and Busygin, F.F., 1975. Clinical and epidemiological characteristics of Omsk hemorrhagic fever. *Zh. Mikrobiol.* 11, 132–133.

Li, L., Rollin, P.E., Nichol, S.T., Shope, R.E., Barrett, A.D., and Holbrook, M.R., 2004. Molecular determinants of antigenicity of two subtypes of the tick-borne flavivirus Omsk haemorrhagic fever virus. *J. Gen. Virol.* 85(Pt 6), 1619–1624.

Lin, D., Li, L., Dick, D., Shope, R.E., Feldmann, H., Barrett, A.D., and Holbrook, M.R., 2003. Analysis of the complete genome of the tick-borne flavivirus Omsk hemorrhagic fever virus. *Virology* 313(1), 81–90.

Lindenbach, B.D. and Rice, C.M., 2003. Molecular biology of flaviviruses. *Adv. Virus Res.* 59, 23–61.

Loginova, S.Ia., Efanova, T.N., Koval'chuk, A.V., Faldina, V.N., Androshchuk, I.A., Pistsov, M.N., Borisevich, S.V., Kopylova, N.K., Pashchenko, Iu.I., Khamitov, P.A., Maksimov, V.A., and Vasil'ev, N.T., 2002. Effectiveness of virazol, realdiron and interferon inductors in experimental Omsk hemorrhagic fever. *Vopr. Virusol.* 47(6), 27–30.

Mazbich, I.B. and Netsky, G.I., 1952. Three years of study of Omsk hemorrhagic fever (1946–1948). *Trud. Omsk Inst. Epidemiol. Microbiol. Gigien* 1, 51–67.

Neronov, V.M., Khlyap, L.A., Bobrov, V.V., and Warshavsky, A.A., 2008. Alien species of mammals and their impact on natural ecosystems in the biosphere reserves of Russia. *Integr. Zool.* 3, 83–94.

Netesov, S.V. and Conrad, J.L., 1991. Emerging infectious diseases in Russia, 1990–1999. *Emerg. Infect. Dis.* 7(1), 1–5.

Ngulube, C.N., Yoshii, K., and Kariwa, H., 2012. Evaluation of antigenic cross-reactivity of tick-borne encephalitis virus and Omsk hemorrhagic fever virus. Abstract No. P1–7. *The 4th International Young Researcher Seminar for Zoonosis Control.* Graduate School of Veterinary Medicine, Hokkaido University, Sapporo, Japan, September 19–20.

Novitskiy, V.S., 1948. Pathologic anatomy of spring-summer fever in Omsk region. *Proc. OmGMI* 13, 97–134.

Orlinger, K.K., Hofmeister, Y., Fritz, R., Holzer, G.W., Falkner, F.G., Unger, B., Loew-Baselli, A., Poellabauer, E.M., Ehrlich, H.J., Barrett, P.N., and Kreil, T.R., 2011. A tick-borne encephalitis virus vaccine based on the European prototype strain induces broadly reactive cross-neutralizing antibodies in humans. *J. Infect. Dis.* 203(11), 1556–1564.

Růžek, D., Yakimenko, V.V., Karan, L.S., and Tkachev, S.E., 2010. Omsk hemorrhagic fever. *The Lancet* 376(9758), 2104–2113.

Růžek, D., Yakimenko, V.V., Karan, L.S., and Tkachev, S.E., 2011. Omsk hemorrhagic fever virus. In: Liu D. (Ed.) *Molecular Detection of Human Viral Pathogens*. CRC Press, Boca Raton, FL, London, New York, 231–239.

Shestopalova, N.M., Reingold, V.N., Gagarina, A.V., Kornilova, E.A., Popov, G.V., and Chumakov, M.P., 1972. Electron microscopic study of the central nervous system in mice infected by Omsk hemorrhagic fever (OHF) virus. *J. Ultrastruct. Res.* 40, 458–469.

Shestopalova, N.M., Reyngold, V.N., Gavrilovskaya, I.N., Belyayeva, A.P., and Chumakov, M.P., 1965. Electron microscope studies into the morphology and localization of Omsk hemorrhagic fever virus in infected tissue culture cells. *Vopr. Virus.* 4, 425–430.

Sizemova, G.A., 1957. Diagnostics of Omsk hemorrhagic fever. *Proc. OmGMI* 21, 256–260.

Stephenson, J.R., 1988. Flavivirus vaccines. *Vaccine* 6, 471–480.

Tigabu, B., Juelich, T., and Holbrook, M.R., 2010. Comparative analysis of immune responses to Russian spring-summer encephalitis and Omsk hemorrhagic fever viruses in mouse models. *Virology* 408(1), 57–63.

Von Lubitz, D.K.J.E., 2004. *Bioterrorism. Field Guide to Disease Identification and Initial Patient Management.* CRC Press, Washington, DC, 175 pp.

Yakimenko, V.V., 2011. Omsk hemorrhagic fever virus: Epidemiological and clinical aspects. Tick-borne infections in Siberia region. *Siberian Branch of Russ. Acad. Sci.*, 279–295.

Index